Carbohydrate Chemistry

Volume 29

A Specialist Periodical Report

Carbohydrate Chemistry
Monosaccharides, Disaccharides
and Specific Oligosaccharides
Volume 29

A Review of the Literature Published during 1995

Senior Reporter
R.J. Ferrier, *Victoria University of Wellington, New Zealand*

Reporters
R. Blattner, *Industrial Research Limited, Lower Hutt, New Zealand*
K. Clinch, *Industrial Research Limited, Lower Hutt, New Zealand*
R.H. Furneaux, *Industrial Research Limited, Lower Hutt, New Zealand*
J.M. Gardiner, *UMIST, Manchester, UK*
P.C. Tyler, *Industrial Research Limited, Lower Hutt, New Zealand*
R.H. Wightman, *Heriot Watt University, Edinburgh, UK*

THE ROYAL
SOCIETY OF
CHEMISTRY
Information
Services

ISBN 0-85404-213-X
ISSN 0951-8428

Published by The Royal Society of Chemistry,
Thomas Graham House, Science Park, Milton Road, Cambridge CB4 4WF, UK

Typeset by Computape (Pickering) Ltd, Pickering, North Yorkshire, UK
Printed and bound by Athenaeum Press Ltd, Gateshead, Tyne and Wear, UK

Preface

The old order has indeed changed, and beginning with this volume, each member of the abstracting/writing team has been contracted to produce chapters of agreed maximum lengths. In consequence, it has been necessary to compress some abstracting and reporting yet further, so that in many instances the main objective has been to call attention to original work and indicate its source rather than to provide a meaningful summary from which the interested reader could work. The reporters are acutely aware that brief notes do scant justice to developing literature of ever-increasing complexity.

More and more carbohydrate research is being developed and stimulated by chemists who are not specialized in the field – and the subject is being greatly enhanced in consequence – and more and more it is merging with organic chemistry generally. Increasingly, important new work is appearing in non-specialized journals, the first 3 issues of *Journal of Organic Chemistry* for 1996 for example each containing about 10 papers to be cited in *Specialist Periodical Reports, Carbohydrate Chemistry*.

This year's volume has been produced by a depleted team and the members are due major thanks for taking on additional work under particularly frustrating circumstances.

Mrs Janet Freshwater and Mr Alan Cubitt, The Royal Society of Chemistry, are thanked most warmly for the cooperation and help they have provided.

R.J. Ferrier
May, 1997

Contents

Abbreviations

The following abbreviations have been used:

Ac	acetyl
Ade	adenin-9-yl
AIBN	2,2 -azobisisobutyronitrile
All	allyl
Ar	aryl
Ara	arabinose
Asp	Aspartic acid
BBN	9-borabicyclo[3.3.1]nonane
Bn	benzyl
Boc	*t*-butoxycarbonyl
Bu	butyl
Bz	benzoyl
CAN	ceric ammonium nitrate
Cbz	benzyloxycarbonyl
CD	circular dichroism
Cer	ceramide
c.i.	chemical ionization
Cp	cyclopentadienyl
Cyt	cytosin-1-yl
Dahp	3-deoxy- D-*arabino*-2-heptulosonic acid 7-phosphate
DAST	diethylaminosulfur trifluoride
DBU	1,5-diazabicyclo[5.4.0]undec-5-ene
DCC	dicyclohexylcarbodi-imide
DDQ	2,3-dichloro-5,6-dicyano-1,4-benzoquinone
DEAD	diethyl azodicarboxylate
DIBAL	di-isobutylaluminium hydride
DMAD	dimethylacetylene dicarboxylate
DMAP	4-dimethylaminopyridine
DMF	*N,N*-dimethylformamide
DMSO	dimethyl sulfoxide
Dmtr	dimethoxytrityl
e.e.	enantiomeric excess
Ee	1-ethoxyethyl
ESR	electron spin resonance
Et	ethyl
FAB	fast-atom bombardment
Fmoc	9-fluorenylmethylcarbonyl
Fru	fructose
Fuc	fucose
Gal	galactose

GalNAc	2-acetamido-2-deoxy-D-glactose
GLC	gas liquid chromatography
Glc	glucose
GlcNAc	2-acetamido-2-deoxy-D-glucose
Gly	glycine
Gua	guanin-9-yl
Hep	L-*glycero*-D-*manno*-heptose
HMPA	hexmethylphosphoric triamide
HMPT	hexamethylphosphorous triamide
Ido	idose
Im	imidazolyl
IR	infrared
Kdo	3-deoxy-D-*manno*-2-octulosonic acid
LAH	lithium aluminium hydride
LDA	lithium di-isopropylamide
Leu	leucine
LTBH	lithium triethylborohydride
Lyx	lyxose
Man	mannose
MCPBA	*m*-chloropherbenzoic acid
Me	methyl
Mem	methoxyethoxymethyl
Mmtr	monomethoxytrityl
Mom	methoxymethyl
MS	mass spectrometry
Ms	methanesulfonyl (mesyl)
NMR	nuclear magnetic resonance
NAD	nicotinamide adenine dinucleotide
NBS	*N*-bromosuccinimide
NeuNAc	*N*-acetylneuraminic acid
NIS	*N*-iodosuccinimide
NMSO	*N*-methylmorpholine-*N*-oxide
ORD	optical rotatory dispersion
PCC	pyridinium chlorochromate
PDC	pyridinium dichromate
Ph	phenyl
Phe	phenylalanine
Piv	pivaloyl
Pmb	*p*-methoxybenzyl
Pr	propyl
Pro	proline
p.t.c.	phase transfer catalysis
Py	pyridine
Rha	rhamnose
Rib	ribose
Ser	serine

SIMS	secondary-ion mass spectrometry
TASF	tris(dimethylamino)sulfonium(trimethylsilyl)difluoride
Tbdms	*t*-butyldimethylsilyl
Tbdps	*t*-butyldiphenylsilyl
Tf	trifluoromethanesulfonyl(trifyl)
Tfa	trifluoroacetyl
TFA	trifluoroacetic acid
THF	tetrahydrofuran
Thp	tetrahydropyranyl
Thr	threonine
Thy	thymin-1-yl
Tips	1,1,3,3-tetraisopropyldisilox-1,3-diyl
Tms	trimethylsilyl
TPP	triphenylphosphine
Tps	tri-isopropylbenzenesulfonyl
Tr	triphenylmethyl(trityl)
Ts	toluene-*p*-sulfonyl(tosyl)
Ura	uracil-1-yl
UDP	uridine diphosphate
UDPG	uridine diphosphate glucose
UV	ultraviolet
Xyl	xylose

1
Introduction and General Aspects

An appreciation has been written of the seminal, major contributions of H.S. Isbell to carbohydrate chemistry.[1]

Several carbohydrate applications have been included in a review of microwave-assisted organic reactions,[2] and a survey of chemical reagents in photoaffinity labelling included the role of various base-labelled azido- and thionucleosides, sugar azide derivatives and p-benzoylbenzoate esters of nucleosides.[3] A further review intitled 'Reverse Anomeric Effect: Fact or Fiction?' refers to early evidence based on studies of glycosyl pyridinium and imidazolium salts. It concludes that its origin as an electronic effect is not supported by theory or experimental results.[4]

A review in 'Advances in Carbohydrate Chemistry and Biochemistry' has covered $^{13}C-^{1}H$ coupling in sugar derivatives and included theoretical aspects, experimental techniques and conformational dependencies.[5] A related survey was produced on ^{13}C nuclear magnetic relaxation and motional behaviour of carbohydrates in solution. Theoretical and observed data were included.[6]

Part I of a review on the uses of enzymes in carbohydrate chemistry has appeared, this part dealing with biological recognition and syntheses of monosaccharides and analogues including amino-sugars.[7] A further essay on applications of enzymes deals with the use of glucansucrase in the synthesis of oligosaccharides and polysaccharides.[8] The chemotherapy of HIV infection has been surveyed with attention being given to the various applications used and targets identified. Nucleoside analogues and glucosidase inhibitors were discussed amongst the drugs used.[9]

An in-depth treatment of the organic chemistry of the monosaccharides also briefly covers their role in natural products, for example glycosides, oligosaccharides, polysaccharides and glycoproteins. Such topics as the chemical synthesis of oligosaccharides and the use of sugars in the synthesis of enantiomerically pure non-sugar products are dealt with after an extensive mechanistic treatment of the reactions of the monosaccharides. Containing 1000 references, the book provides a convenient means of access to the primary literature.[10]

References

1   H.S. El Khadam, *Adv. Carbohydr. Chem. Biochem.*, 1995, **51**, 1.
2   S. Caddick, *Tetrahedron*, 1995, **51**, 10403.

3   S.A. Fleming, *Tetrahedron*, 1995, **51**, 12479.
4   C.L. Perrin, *Tetrahedron*, 1995, **51**, 11901.
5   I. Tvaroska and F.R. Taravel, *Adv. Carbohydr. Chem. Biochem.*, 1995, **51**, 15.
6   P. Dais, *Adv. Carbohydr. Chem. Biochem.*, 1995, **51**, 63.
7   C.-H. Wong, R.L .Halcomb, Y. Ichikawa and T. Kajimoto, *Angew. Chem. Int. Ed. Engl.*, 1995, **34**, 412.
8   J.F. Robyt, *Adv. Carbohydr. Chem. Biochem.*, 1995, **51**, 133.
9   E. de Clercq, *J. Med. Chem.*, 1995, **38**, 2491.
10   P.M. Collins and R.J. Ferrier, *Monosaccharides*, John Wiley and Sons, Chichester, 1995, p. 574.

2
Free Sugars

1 Theoretical Aspects

A sytematic search has been made for all low energy structures of α- and β-D-glucopyranose, α- and β-D-galactopyranose, β-D-talopyranose and β-D-allopyranose in the solid state, the energies being minimized with respect to nine lattice and rigid-body parameters and six intramolecular dihedral angles. In four cases, the observed crystal structures corresponded to one of the lowest energy structures; in the other two cases, the observed structures were of more than 20 kJ mol^{-1} higher energy than the calculated minima.[1] An extension of the GROMOS force field allowed an improved crystal structure determination of α-D-galactopyranose.[2]

The hydroxyl acidities of sucrose, assessed through semiempirical calculations of the deprotonation enthalpies, followed the order OH-2g >> OH-3g > OH-3f > OH-1f = OH-4g > OH-4f >> OH-6g > OH-6f.[3] The molecular electrostatic potential profile of sucrose in polar, aprotic solvents indicated likewise that in the main conformation OH-2g is the most electropositive hydroxyl group; the preparations of selectively benzylated and acetylated sucrose derivatives on the basis of these findings are referred to in Chapters 5 and 7, respectively.[3a]

The hydration of α-maltose has been investigated using molecular modelling and thermodynamic methods. Due to crystallinity, the observed non-freezing water content was lower than that calculated.[4]

Conformational studies on several free sugars are covered in Chapter 21.

2 Synthesis

A review article on enzymatic oxidoreduction in organic synthesis included the following carbohydrate-related topics: i) oxidation of polyols to polyhydroxy aldehydes by use of galactose oxidase (yields 10-15%); ii) the reduction of 2-acetylfuran by alcohol dehydrogenases from two different sources to furnish enantiospecifically either the (R)-alcohol **1** or its (S)-isomer **2**, which serve as potential building blocks for the synthesis of D- and L-sugars, respectively.[5]

Aldolase- and transketolase-catalysed synthesis of pentoses, hexoses and higher sugars continues to receive considerable attention, especially by C.-H. Wong and co-workers who have covered these processes in Part 1 of a review

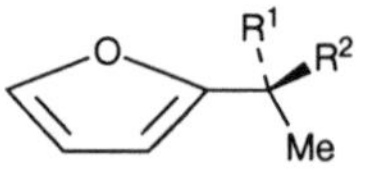

1 $R^1 = OH$, $R^2 = H$
2 $R^1 = H$, $R^2 = OH$

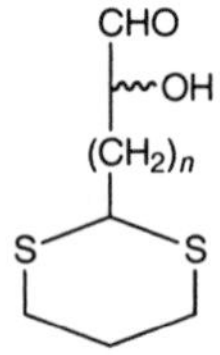

3 $n = 0$
4 $n = 1$

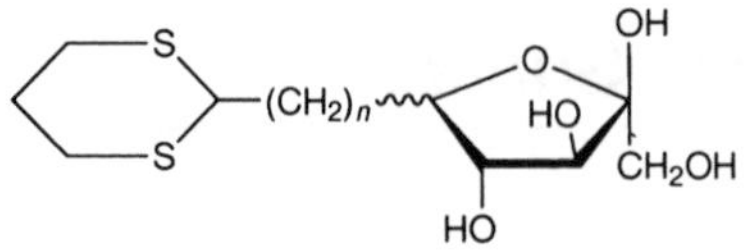

5 $n = 0$
6 $n = 1$

article (60 refs.) on the use of enzymes in carbohydrate chemistry.[6] DERA (deoxyribose 5-phosphate aldolase, EC 4.1.2.4) has been prepared on a large scale by recombinant techniques.[7] It was employed, either by itself or in combination with RAMA (rabbit muscle aldolase, a fructose-1,6-diphosphate aldolase, EC 4.1.2.13) or NeuAc aldolase (EC 4.1.3.3) and, if necessary, phosphatase in sequential condensations to afford deoxyaldoses from -pentoses to -nonoses from small substrates such as acetaldehyde, glyceraldehyde and dihydroxyacetone phosphate.[8,9] Condensations catalysed by L-fuculose 1-phosphate aldolase (EC 4.1.2.17)[10] and L-rhamnulose 1-phosphate aldolase (EC 4.1.2.19),[11] in combination with phosphatase and the appropriate isomerases, resulted in efficient syntheses of L-fucose and L-glucose, respectively. An alternative strategy to prepare aldoses by this process which initially produces uloses, is illustrated in Scheme 1.[12] Similar condensations involving the masked hydroxydialdehydes **3** and **4** were used in the synthesis of aldoketose derivatives **5** and **6**, respectively, required as precursors for the microbial preparation of deoxynojirimycin analogues.[13] The condensation of pyruvate with simple hydroxyaldehydes, *e.g.*, D-glyceraldehyde or D-erythrose, under catalysis by an aldolase from *Aspergillus terreus* gave 3-deoxy-2-ulosonic acids with (*R*)-configuration at the new chiral centre (C-4).[14]

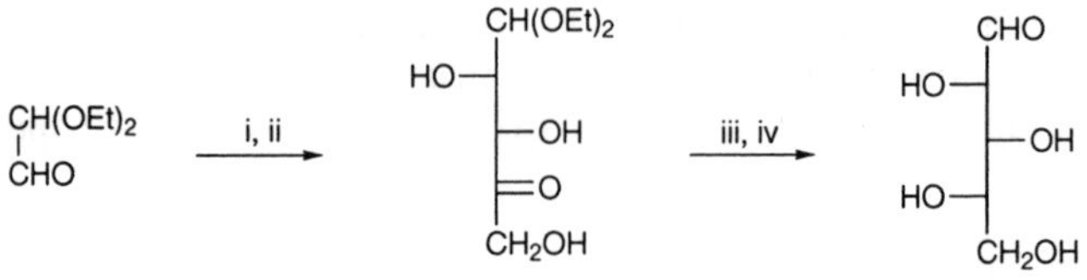

Reagents: i, DHAP, fructose 1,6-diphosphate aldolase; ii, phosphatase; iii, L-iditol dehydrogenase; iv, H_3O^+

Scheme 1

2.1 Tetroses and Pentoses. – The application of the Strecker synthesis to the preparation of 2-amino-2-deoxytetrose derivatives is covered in Chapter 9 and a lipase-mediated route to 4-carbon diols and triols is referred to in Chapter 18.

The CaCl$_2$/KOH-promoted aldol condensation of dihydroxyacetone with formaldehyde has been optimized for the production of D,L-*threo*-3-pentulose.[15] L-Ribofuranose derivatives have been obtained in good yields from 1,2-*O*-isopropylidene-5-*O*-trityl-α-L-xylofuranose by oxidation-reduction at C-3.[16] The efficient preparation of the D-ribose derivative **8** from the corresponding lactone **7** was one of twelve examples in a paper describing the catalytic reduction of 1,4- and 1,5-lactones to the corresponding lactols by use of the reagents indicated in Scheme 2.[17] A new synthesis of 1-deoxy-D-*threo*-pentulose in eight steps from (−)-tartaric acid is referred to in Chapter 12 and the preparation of D-lyxose and D-arabinose in high yields by oxidative degradation of D-galactose and sodium D-gluconate, respectively, is covered in Section 5 of this Chapter.

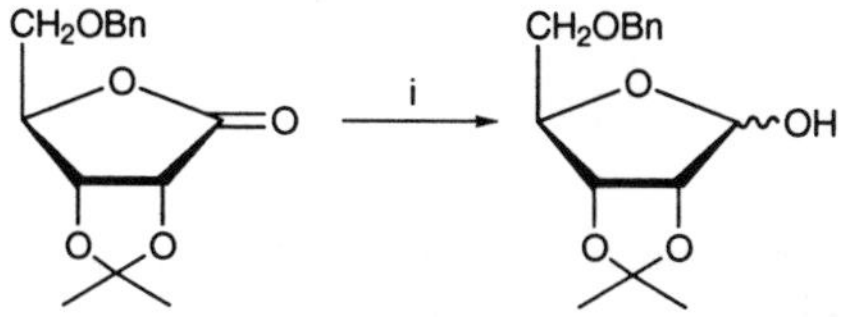

Reagents: i, Cp-2Ti(OC6H4-*p*Cl)2, Bu-4NF on Al2O3, polymethylhydroxysilane

Scheme 2

2.2 Hexoses. – Isopropylidene-D-glyceraldehyde (**9**) has been converted to a six-carbon sugar precursor with introduction of one new chiral centre in a lengthy reaction sequence involving Wittig-Horner elongation by two carbons and Sharpless asymmetric epoxidation, followed by one-carbon extension *via* diazoketone **10**, which gave **11** on photolysis (Scheme 3).[18]

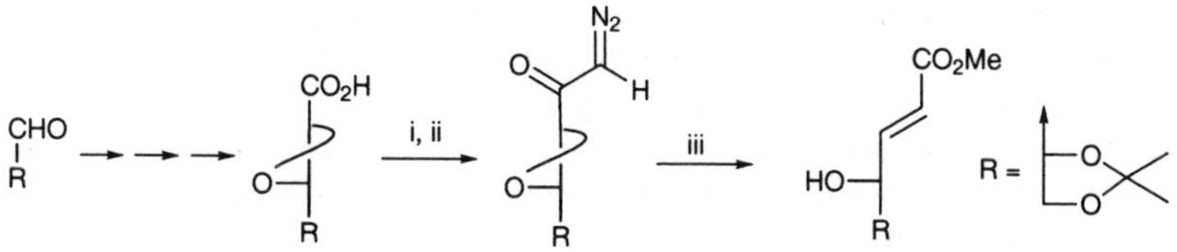

Reagents: i, Bu^{i}O2CCl, Et3N; ii, CH2N2; iii, *hv*, MeOH

Scheme 3

Efficient procedures for the conversion of levoglucosenone to rare sugars, *e.g.*, D-altrose, D-allose, 4-deoxy-D-mannose, have been devised.[19] Hydrolytic opening of the L-sorbopyranose-derived epoxide **12** with nucleophilic attack at C-4 was the key-step in the transformation of L-sorbose to L-fructose, which requires inversion of configuration at C-3 and C-4.[20] The formation of D-tagatose from D-galactose involving oxidation at C-2 and reduction at C-1 was facilitated by the spontaneous cyclization of the osulose intermediate **13** to the bicyclic hemiketal **14**. This was readily *O*-methylated to **15**, thus offering convenient protection for the carbonyl group during the subsequent reduction step.[21]

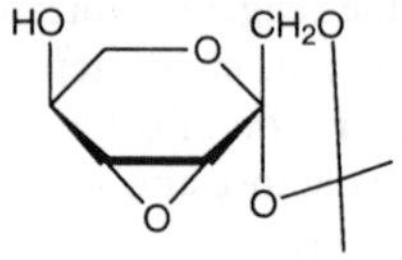

12 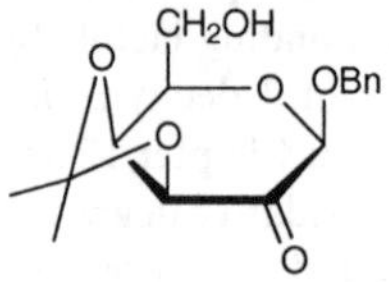   **13** 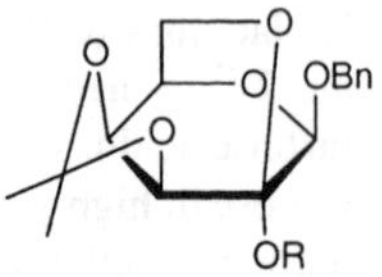   **14** R = H
                                                                        15 R = Me

D-Tagatose 3-epimerase (see Vol. 28, Chapter 2, Ref. 64) immobilized on Chitopearl beads, was employed for converting D-tagatose to D-sorbose on a multigram scale.[22] By use of the same enzyme preparation, D-psicose was available from D-fructose or, in the simultaneous presence of D-xylose isomerase, from D-glucose.[23] The preparation of L-tagatose from 1,5-anhydro-D-galactitol via 'L-2-sorbal' is mentioned in Chapter 13 and a synthesis of L-fucose from D-galactose is covered in Chapter 12.

2.3   Chain-extended Sugars. – A review on two-directional synthesis involving desymmetrization of chain-extension products contained several examples of higher alditols, aldoses and aldonic acids.[24] A section on the synthesis of higher carbon sugars has been included in a review dealing with the application of furan- and pyrrole-based siloxydienes.[25] The cycloaddition product formed from penta-2,4-dienol (**16**) and sodium glyoxalate (**17**) at 100 °C in aqueous medium was converted to the racemic methyl 2,3-dideoxyheptulosonate triacetates **18**, as shown in Scheme 4. These were further processed to furnish 3-deoxy-2-heptulosonic acid derivatives **19**.[26] By use of tetrose-based dienes **20**, this approach has also been adapted to the synthesis of 3-deoxy-2-nonulosonic acid derivatives.[27] Many chain-extended monosaccharides have been prepared by enzymatic aldol condensations (see Refs. 7–14 above).

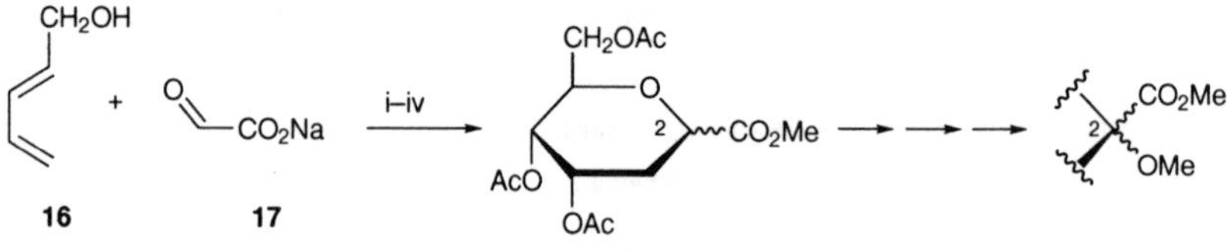

16   **17**   **18**   **19**

Reagents: i, H$_2$O, 100 °C; ii, MeI, DMF; iii, OsO$_4$, NMO; iv, Ac$_2$O, Py

Scheme 4

2.3.1 Chain-extension at the 'Non-reducing' End. – As usual, protected dialdoses have been used extensively for this purpose. Stereoselective hydroxymethylation of **21** to give the methyl L-*glycero*-D-*manno*-heptopyranoside derivative **23** was achieved by treatment with [dimethyl(thiophenylmethyl)silyl]methylmagnesium chloride followed by oxidative desilylation of the addition product **22**.[28] The addition of methyl nitroacetate to **24** furnished an inseparable mixture **25** in mediocre yield; reduction of the nitro groups was inefficient but the resulting amines **26** could be separated.[29] The indium-mediated allylation of **27** (see Vol 28, Chapter 2, Ref. 47) showed high *anti*-diastereofacial selectivity in the presence of Lewis acids, especially Y(OTf)$_3$, thus affording mainly the 4,5-*erythro* product **28**.[30]

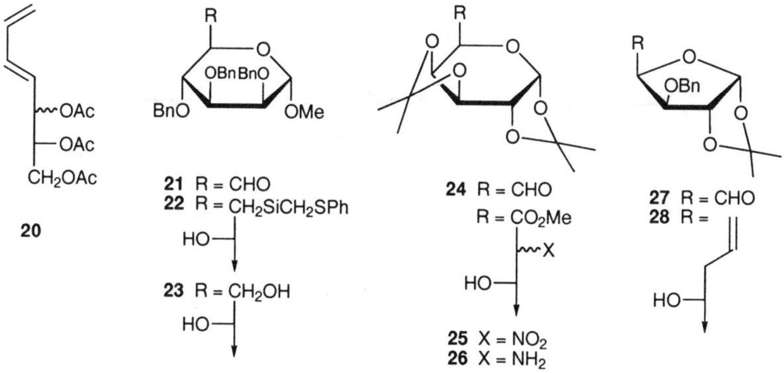

erythro-Addition was also favoured in the condensations of sugar-derived aldehydes with acetyliron anions (see Vol. 28, Chapter 2, Ref. 28) when counterions other than Li[30] (*e.g.*, Sn^{2+}, Zr^{4+}, Et$_2$Al$^+$) were used; as an example, the preparation of 6-deoxy-β-D-*allo*-heptopyranose pentaacetate (**30**) from **29** is given in Scheme 5.[31]

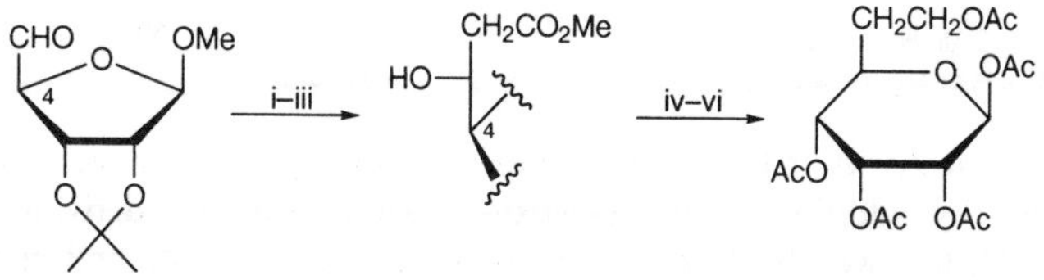

Reagents: i, FeCp(CO)PPh$_3$Ac, BuLi; ii, SnCl$_2$; iii, NBS; iv, DIBAL; v, aq. HOAc; vi, Ac$_2$O, Py

Scheme 5

Addition of alkylmanganese reagents, prepared *in situ* from the corresponding alkyl lithium compounds and MnI$_2$ in ether, to substrates **24** and **27** proceeded diastereoselectively to give 4,5- and 5,6-*threo* products, respectively.[32] Spiroketals

31 have been synthesized from methyl 2,3,4-tri-*O*-methyl-α-D-glucopyranoside by a radical method similar to that shown in Scheme 7 below.[33] Reformatzki reaction of **32** with ethyl 2-bromomethylacrylate and activated zinc gave the α,β-unsaturated lactones **33** with an exocyclic double bond in mediocre yield. An analogue with an endocyclic double bond, the fungicidal sugar butenolide **35**, was obtained by opening of epoxide **34** with the dilithium salt of phenylselenoacetic acid and subsequent oxidative elimination.[34] The selectivities of the intramolecular nitrone cycloaddition reactions of 3-*O*-allyl-D-glucofuranose and 3-*O*-allyl-D-allofuranose derivatives **36** and **37**, respectively, to give oxepans **38** and/or pyrans **39** have been investigated (see Vol. 27, Chapter 18, Scheme 1).[35]

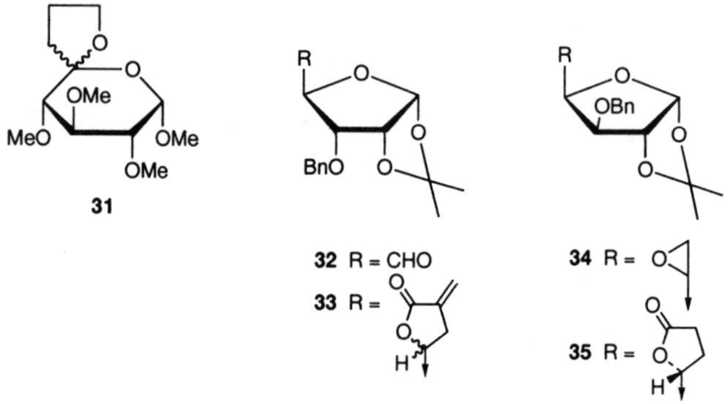

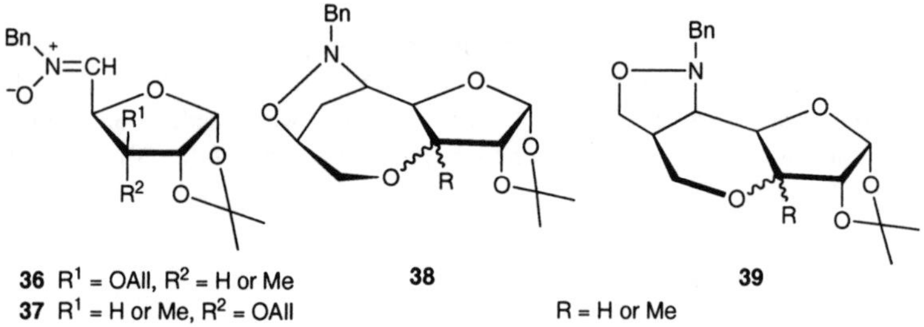

The *C*-disaccharide **43** (2-deaminotunicamycin) and its 7-epimer have been synthesized by addition of the 6-deoxy-6-diazo-D-galactose derivative **40** to aldehyde **29**, followed by LiBHEt₃-reduction of the ketone **41** and epoxide **42** thus obtained as a separable mixture (Scheme 6).[36] The preparation of the first *pseudo-C*-disaccharide by use of the alkenyllithium compound derived from dibromoalkene **44** is covered in Chapter 18.

2.3.2 Chain-extension at the Reducing End. – Two-carbon Wittig elongation of 3,4:5,6-di-*O*-isopropylidene-aldehydo-D-glucose gave adduct **45**, which on Mitsunobu inversion at C-4, hydrogenation of the double bond and lactonization furnished the 4,5-dideoxy-D-*manno*-oct-1,4-lactone derivative **46**, a synthetic KDO precursor.[37] Anomeric spiroketals have been constructed by alkoxy radical promoted intramolecular hydrogen abstraction/cyclization, as depicted in Scheme 7. Similar procedures were used to prepare hexose-derived dioxaspiro[5.5]undecanes and pentose-derived dioxaspiro[4.4]nonanes.[33]

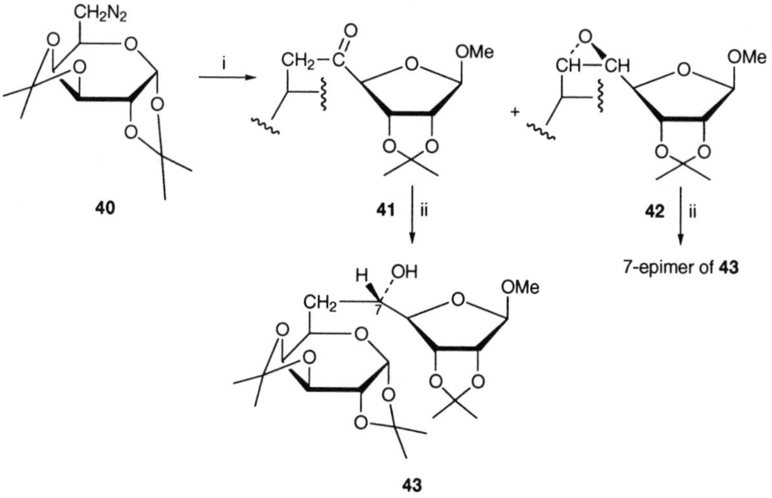

Reagents: i, compound **29**; ii, LiBHEt$_3$

Scheme 6

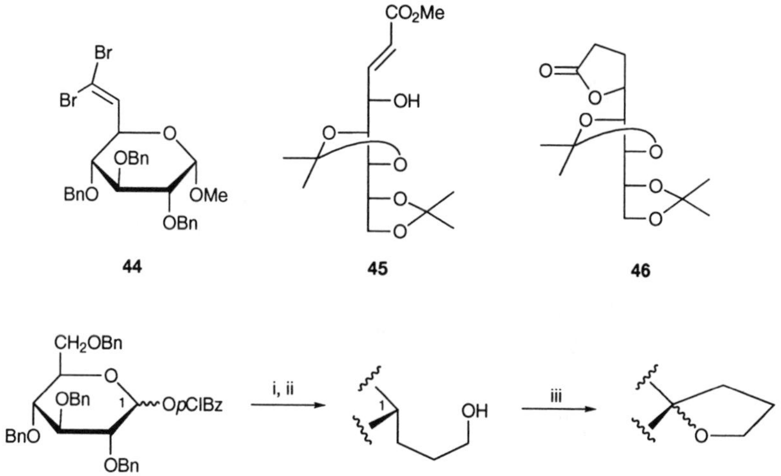

Reagents: i, AllTms, BF$_3$·Et$_2$O; ii, BH$_3$, then H$_2$O$_2$, OH$^-$; iii, (diacetoxyiodo)benzene, I$_2$, C$_6$H$_{12}$, *hv*

Scheme 7

The α-*C*-furanosylglycine derivative **47** was prepared in 37% overall yield by use of *N*-Boc-*t*-butyldimethylsilyloxypyrole as a masked glycine anion, as shown in Scheme 8.[38] Condensation of 4,6-*O*-benzylidene-D-glucose with nitroethane in the presence of DBU and 2-hydoxypyridine as 1,3 proton transfer catalyst gave, after reduction, the (α-aminoethyl) β-D-*C*-glucoside derivative **48**.[39] Further *C*-disaccharides are referred to in Chapter 3.

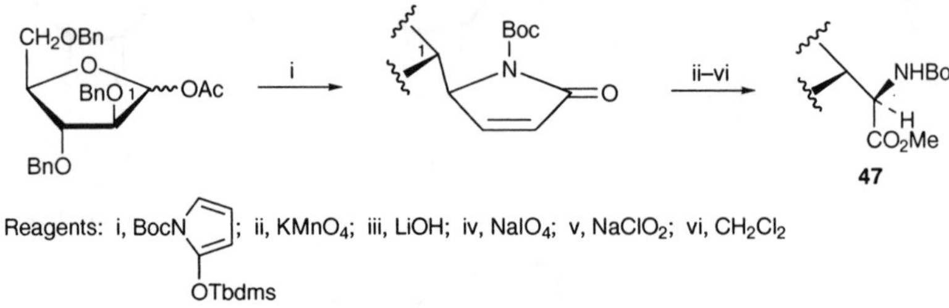

Reagents: i, BocN⟨⟩; ii, KMnO₄; iii, LiOH; iv, NaIO₄; v, NaClO₂; vi, CH₂Cl₂
OTbdms

Scheme 8

Application of the intramolecular nitrone-olefin cycloaddition to 3-*O*-allyl-D-glucose derivatives **49** and **50** and their D-*allo* isomers **51** and **52** furnished pyranoisoxazolidines **53** with varying degrees of diastereoselectivity. The main product obtained from **50** gave **54** on trimming of the side chain.[40]

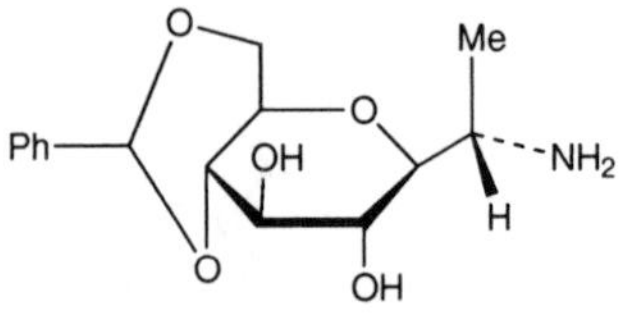

48

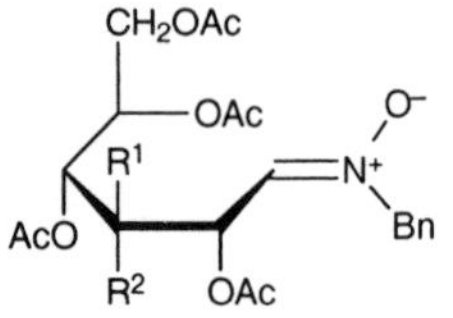

49 R¹ = OAll, R² = H or Me
50 R¹ = OAll, R² = Me
51 R¹ = H, R² = H
52 R¹ = H, R² = Me

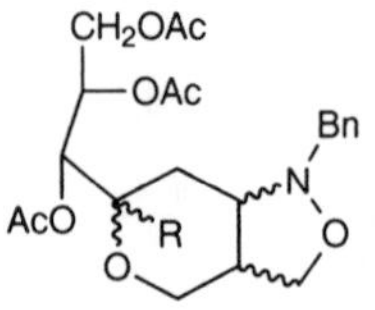

53 R = H or Me 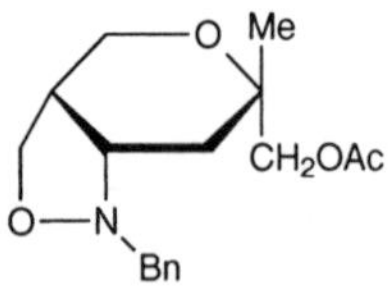   **54**

3   Physical Measurements

The nature of the relaxation process in supercooled, glassy carbohydrates has been examined by differential scanning calorimetry and dielectric relaxation measurements at 77-400 K over the frequency range 10^{-6}-10^{-3}Hz.[41] A gravimetric method during desorption of water has been used to determine the mean diffusion coefficients for maltose-water mixtures close to the glass transition temperature T_g.[42] The effects of water structure enhancers (ethanol, tetramethylammonium chloride) and water structure breakers (urea, guanidine hydrochloride) on the stability of concentrated sucrose solutions have been studied by polarimetry and ion chromatography,[43] and the attractive interaction between saccharides and monolignols has been estimated by measuring the solubilities of *p*-coumaryl-, coniferyl- and sinapyl-alcohol in aqueous solutions of D-glucose, D-galactose, D-mannose and D-xylose. The apparent association constant K_{app} for coniferyl alcohol, for example, followed the order Xyl > Man > Gal > Glc.[44]

4   Isomerization

Examples of enzymic epimerization are given in Refs. 22 and 23 above.

5   Oxidation

The mechanism of the oxidation of D-glucose in alkaline solution on single crystal platinum electrodes has been investigated.[45] A process for the preparation of D-arabinose from sodium D-gluconate in an electrochemical reactor with a fluidized bed electrode has been developed.[46]

The oxidation of D-glucose to D-gluconic acid by molecular oxygen has been performed with bismuth-containing palladium-on-charcoal in water,[47] with palladium-on-alumina in weakly basic media,[48] and with sodium nitrite as catalyst in strongly acidic solution.[49] The low temperature oxidation of non-reducing sugars (methyl D-glucosides, sucrose) and alditols (glucitol, mannitol) by oxygen in alkaline media has been studied in the presence of various catalysts, such as Cu(II) salts.[50] The global kinetics of the 'classical' and of the anthraquinone-2-sulfonate catalysed alkaline oxidative degradation of lactose and related carbohydrates have been described.[51] A reliable liquid chromatographic method for monitoring the oxidation of saccharides by oxygen in alkaline solution is referred to in Chapter 23.

The degradation of aldoses by aqueous alkaline hydrogen peroxide has been made much more selective by the addition of borate ions and/or EDTA; thus, galactose, lactose, maltose and cellobiose were each converted in one high yielding step to the next lower aldoses and formic acid.[52] A kinetic study of the ozonation of 2-amino-2-deoxy-D-glucose and its *N*-acetate indicated that the former reacts relatively fast through ozone attack on the amino group, whereas the latter is more resistant, resembling D-glucose in reactivity.[53] It has been demonstrated that the oxidation of 2-deoxy D-ribose with excess of periodate depends on pH, type of buffer, temperature and reaction time.[54]

Kinetic studies have been reported for the oxidation of 2-deoxy-D-glucose and 2-acetamido-2-deoxy-D-glucose by Cr(VI) in perchloric acid solution,[55,56] D-galactose by Cu(II) in acetate buffer at pH 4.0-4.75 at 110 °C,[57] maltose, lactose, melibiose and cellobiose by Cu(II) in the presence of ammonium hydroxide,[58] D-fructose[59] and dextrose[60] by Ce(IV) in the presence of bromide ions in aqueous sulfuric acid, and of D-arabinose, D-ribose, D-xylose, D-glucose, D-galactose, D-fructose, 2-deoxy-D-glucose and methyl α-D-glucopyranoside by permanganate in aqueous alkaline media.[61] The reducing abilities of several hexoses and pentoses towards V(V) and Mo(VI) in aqueous HCl have been compared with those of ascorbic acid and cysteine.[62]

An investigation of the cleavage patterns in DNA and RNA used the oxidation of D-ribose, 2-deoxy-D-ribose and several nucleotides by Ru(2,2'-bypyridine)(2,2',2''-tripyridine)O^{2+} as model reactions.[63] The Ru(III)-catalysed oxidation of D-glucitol by N-bromoacetamide is covered in Chapter 16.

Aerobic incubation of D-glucose with a crude fungal enzyme extract (white rot fungus) resulted in the transient accumulation of D-*arabino*-hexos-2-ulose which was then converted to D-*erythro*-hexos-2,3-diulose. A single enzyme, pyranose-2-oxidase (EC 1.1.3.10) is believed to be responsible for both oxidation steps.[64] A review on enzymatic oxidoreduction in organic synthesis which includes applications to carbohydrate-related topics is referred to in Section 2 above.

6   Other Aspects

3-Deoxy-D-hexonic acids **55**, detected in small quantities among the hydrothermolysis products of glucose, fructose, cellobiose and β-cyclodextrin, provided evidence that D-*erythro*-hexos-2-ulose plays a role in carbohydrate degradation not only under alkaline and acidic, but also under neutral conditions.[65] General acid catalysis has been established for the dehydration of fructose to 5-(hydroxymethyl)furfural.[66,67] The formation of 2-furfuraldehyde and formic acid from pentoses in slightly acidic D$_2$O has been monitored by ^{1}H-NMR spectroscopy.[68] During the acid-catalysed melt thermolysis of sucrose, fructosyl cation **56** and D-glucose were produced. Cation **56** may react with hydroxyl groups on sucrose and glucose and other materials present in the melt.[69] The formation of levulinic acid from glucose in aqueous solution was promoted by MgCl$_2$ and other alkaline and alkaline earth metal salts, especially in slightly acidic media. The reaction was completely inhibited by NO$_3{}^-$.[70]

Five-fold deprotonated D-mannose acted as a ligand in homoleptic dinuclear metalates of trivalent Al, Cr, Fe, and V. The polyolate was derived from the β-furanose form, the only one to have all hydroxyl groups on one face of the ring.[71] Complexes of D-glucose with boronic or diboronic acid derivatives of p-toluene, biphenyl, anthracene, ferrocene, calix[4]arenes, and other macrocyclic species are covered in Chapter 17.

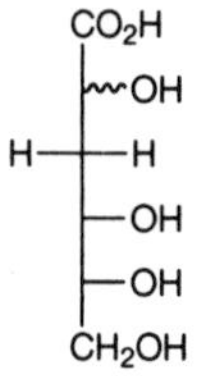

55 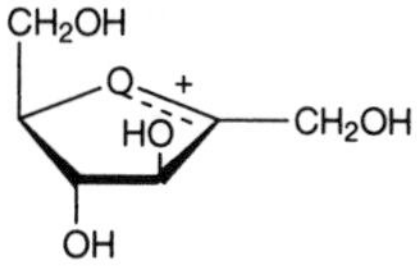   56 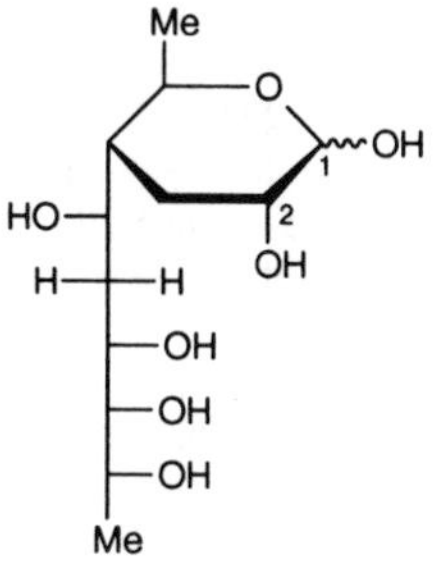   57

A new higher sugar named caryophyllose, isolated from the lipopolisaccharide fraction of *Pseudomonas caryophylli*[72] has been identified as the branched tetradeoxydodecose **57**.[73]

References

1   B.P. van Eijck, W.T.M. Mooij and J. Kroon, *Acta Crystallogr.*, 1995, **B51**, 99.
2   M.L.C.E. Kouwijzer, B.P. van Eijck, H. Kooijman and J. Kroon, *Acta Crystallogr.*, 1995, **B51**, 209.
3   S. Houdier and S. Pérez, *J. Carbohydr. Chem.*, 1995, **14**, 1117.
3a   F.W. Lichtenthaler, S. Immel and P. Pokinskyi, *Liebigs Ann. Chem.*, 1995, 1939.
4   C. Fringant, I. Tvaroska, K. Mazeau, M. Rinaudo and J. Debrieres, *Carbohydr. Res.*, 1995, **278**, 27.
5   J.-M. Fang, C.-H. Lin, C.W. Bradshaw and C.-H. Wong, *J. Chem. Soc., Perkin Trans. 1*, 1995, 967.
6   C.-H. Wong, R.L. Halcomb, Y. Ichikawa and T. Kajimoto, *Angew. Chem., Int. Ed. Engl.*, 1995, **34**, 412.
7   C.-H. Wong, E. García-Junceda, L. Chen, O. Blanco, H.J.M. Gissen and D.H. Steensma, *J. Am. Chem. Soc.*, 1995, **117**, 3333.
8   H.J.M. Gissen and C.-H. Wong, *J. Am. Chem. Soc.*, 1995, **117**, 7585.
9   H.J.M. Gissen and C.-H. Wong, *J. Am. Chem. Soc.*, 1995, **117**, 2947.
10   C.-H. Wong, R. Alajarn, F. Mors-Varas, O. Blanco and E. García-Junceda, *J. Org. Chem.*, 1995, **60**, 7360.
11   R. Alajarín, E. García-Junceda and C.-H. Wong, *J. Org. Chem.*, 1995, **60**, 4294.
12   R. Duncan and D.G. Drueckhammer, *J. Org. Chem.*, 1995, **60**, 7394.
13   M. Lemaire, M.L. Valentin, L. Hecquet, C. Demuynck and J. Bolte, *Tetrahedron: Asymm.*, 1995, **6**, 67.
14   C. Augé and V. Delest, *Tetrahedron: Asymm.*, 1995, **6**, 863.
15   H. Saimoto, S. Yatani, H. Sashiwa and Y. Shigemasa, *Tetrahedron Lett.*, 1995, **36**, 937.
16   E. Chelain, O. Floch and S. Czernecki, *J. Carbohydr. Chem.*, 1995, **14**, 1251.
17   X. Verdaguer, S.C. Berk and S.L. Buchwald, *J. Am. Chem. Soc.*, 1995, **117**, 12641.
18   M.P.M. van Aar, L. Thijs and B. Zwanenburg, *Tetrahedron.*, 1995, **51**, 9699.
19   T. Ebata, H. Kawakami, K. Matsumoto and H. Mashushita, *Front. Biomed. Biotechnol.*, 1994, **2**, 73 (*Chem. Abstr.*, 1995, **122**, 240 307).

20   Y.Gizaw and J.N. BeMiller, *Carbohydr. Res.*, 1995, **266**, 81.

21   P.L. Barili, G. Berti, G. Catelani, F. D'Andrea and L. Miarelli, *Carbohydr. Res.*, 1995, **274**, 197.

22   H. Itoh, T. Sato, T. Takeuchi, A.R. Khan and K. Izumori, *J. Ferment. Bioeng.*, 1995, **79**, 184 (*Chem. Abstr.*, 1995, **122**, 291 341).

23   H. Itoh, T. Sato and K. Izumori, *J. Ferment. Bioeng.*, 1995, **80**, 101 (*Chem. Abstr.*, 1995, **123**, 314 276).

24   S.R. Magnuson, *Tetrahedron.*, 1995, **51**, 2167.

25   S. Casiraghi and G. Rassu, *Synthesis*, 1995, 607.

26   A. Lubineau and Y. Queneau, *J. Carbohydr. Chem.*, 1995, **14**, 1295.

27   A. Lubineau, H. Arcostanzo and Y. Queneau, *J. Carbohydr. Chem.*, 1995, **14**, 1307.

28   F.L. van Delft, G.A. van der Marel and J.H. van Boon, *Synlett*, 1995, 1069.

29   P. Borracher, M.J. Diánez, M.D. Estrada, M. Goméz-Guillén, A. Goméz-Sánchez, A. López-Castro and S. Pérez-Garrido, *Carbohydr. Res.*, 1995, **271**, 79.

30   R. Wang, C.-M. Lim, C.-H. Tan, B.-K. Lim, K.-Y. Sim and T.-P. Loh, *Tetrahedron: Asymm.*, 1995, **6**, 1825.

31   Z. Pakulski and A. Zanojski, *Tetrahedron.*, 1995, **51**, 871.

32   A.N. Kasatkin, I.P. Podlipchuk, R.K. Biktimirov and G.A. Tolstikov, *Izv. Akad. Nauk, Ser Khim.*, 1993, 1122 (*Chem. Abstr.*, 1995, **122**, 187 886).

33   A. Martn, J.A. Salazar and E. Suárez, *Tetrahedron Lett.*, 1995, **36**, 4489.

34   A.P. Rauter, M.J. Ferreira, J. Font, A. Virgili, M. Figueiredo, J.A. Figueiredo, M.I. Ismael and T.L. Canda, *J. Carbohydr. Chem.*, 1995, **14**, 929.

35   A. Bhattacharjee, A. Bhattacharjya and A. Patra, *Tetrahedron Lett.*, 1995, **36**, 4677.

36   F. Sarabia-García, F.J. Lopez-Herrera and M.S. Pino Gonzales, *Tetrahedron*, 1995, **51**, 5491.

37   S. Jarosz, E. Kozlowska and Z. Ciunik, *J. Polish Chem.*, 1994, **68**, 2209 (*Chem. Abstr.*, 1995, **122**, 265 872).

38   G. Rassu, F. Zanardi, L. Battistin and G. Casiraghi, *Tetrahedron: Asymm.*, 1995, **6**, 371.

39   X. Wang and C.H. Gross, *Liebigs Ann. Chem.*, 1995, 1367.

40   R. Mukhapadhay, A.P. Kundu and A. Bhattacharjya, *Tetrahedron Lett.*, 1995, **36**, 7729

41   Gangasharan and S.S.N. Murthy, *J. Phys. Chem.*, 1995, **99**, 12349 (*Chem. Abstr.*, 1995, **123**, 83 856).

42   R. Parker and S.G. Ring, *Carbohydr. Res.*, 1995, **273**, 147.

43   G. Eggleston, G.A. Vercellotti, L. Edye and M.A. Clarke, *J. Carbohydr. Chem.*, 1995, **14**, 1035.

44   M. Shigematsu, A. Goto, S. Shinichi, M. Tanahashi and Y. Shinoda, *Mokuzai Gakkaishi*, 1994, **40**, 321 (*Chem. Abstr.*, 1995, **122**, 81 758).

45   K.D. Popovic, A.V. Tripkovic and R.R. Adzic, *J. Serb. Chem. Soc.*, 1994, **59**, 677 (*Chem. Abstr.*, 1995, **122**, 10 381).

46   V. Jiricny and V. Stanek, *Collect. Czech. Chem. Commun.*, 1995, **60**, 863 (*Chem. Abstr.*, 1995, **123**, 286 408).

47   M. Besson, F.Lahmer, P. Gallezot, P. Fuertes and G. Fleche, *J. Catal.*, 1995, **152**, 116 (*Chem. Abstr.*, 1995, **122**, 291 340).

48   I. Nikov and K. Paev, *Catal. Today*, 1995, **24**, 41 (*Chem. Abstr.*, 1995, **123**, 228 665).

49   I. Grigoryeva, T. Stelmah, S. Trusov and S. Chornaya, *Latv. Kim. Z.*, 1993, 707 (*Chem. Abstr.*, 1995, **122**, 161 065).

50   A.M. Sakharov and I.P.Skibida, *Stud. Surf. Sci. Catal.*, 1994, **82**, 629 (*Chem. Abstr.*, 1995, **123**, 314 318).

51   H.E.J. Hendriks, B.F.M. Kuster and G.B. Marin, *J. Mol. Catal.*, 1994, **93**, 317 (*Chem. Abstr.*, 1995, **122**, 31 792).

52   R. van den Berg, J.A. Peters and H. van Bekkum, *Carbohydr. Res.*, 1995, **267**, 65.

53   R.P. Rey, A.N. Selles, C. Baluja and M.L. Otero, *Ozone: Sci. Eng.*, 1995, **17**, 463 (*Chem. Abstr.*, 1995, **123**, 340 660).

54   S.S. Razic and Z.S. Dusic, *J. Serb. Chem. Soc.*, 1995, **60**, 675 (*Chem. Abstr.*, 1995, **123**, 340 574).

55   M. Rizotto, S. Signorella, M.I. Frascaroli, V. Daier and L.F. Sala, *J. Carbohydr. Chem.*, 1995, **14**, 45.

56   L.F. Sala, C. Palopoli and S. Signorella, *Polyhedron*, 1995, **14**, 1725 (*Chem. Abstr.*, 1995, **123**, 257 176).

57   L.F. Sala, L. Ciullo, R. Lafarga and S. Signorella, *Polyhedron*, 1995, **14**, 1207 (*Chem. Abstr.*, 1995, **123**, 144 459).

58   R.S. Singh, *Acta Cienc. Indica, Chem.*, 1993, **19**, 54 (*Chem. Abstr.*, 1995, **123**, 199 242).

59   M.P. Sah, *J. Indian Chem. Soc.*, 1995, **72**, 173 (*Chem. Abstr.*, 1995, **123**, 340 571).

60   J. Sharma and M.P. Sah, *J. Indian Chem. Soc.*, 1994, **71**, 613 (*Chem. Abstr.*, 1995, **123**, 340 573).

61   K.V. Rao, M.T. Rao and M. Adinarayana, *Int. J. Chem. Kinet.*, 1995, **27**, 555 (*Chem. Abstr.*, 1995, **123**, 144 391).

62   R.P. Bandwar and C.P. Rao, *Carbohydr. Res.*, 1995, **277**, 197.

63   G.A. Neyhart, C.-C. Cheng and H.H. Thorp, *J. Am. Chem. Soc.*, 1995, **117**, 1463.

64   J. Volc, P. Sedmera, V. Havlicek, V. Prikrylova and G. Daniel, *Carbohydr. Res.*, 1995, **278**, 59.

65   G.C.A. Luijkx, F. van Ratwijk, H. van Bekkum and M.J. Antal, Jr., *Carbohydr. Res.*, 1995, **272**, 191.

66   S.A. Grin, S.R. Tsimbaliev and S.Yu. Gel'fand, *Zh. Prikl. Khim.*, 1993, **66**, 2267 (*Chem. Abstr.*, 1995, **122**, 31 769).

67   S.A. Grin, S.R. Tsimbaliev and S.Yu. Gel'fand, *Khim. Fiz.*, 1994, **13**, 113 (*Chem. Abstr.*, 1995, **122**, 81 763).

68   T. Ahmad, L. Kenne, K. Olsson and O. Theander, *Carbohydr. Res.*, 1995, **276**, 309.

69   M. Manley-Harris and G.N. Richards, *Carbohydr. Res.*, 1995, **278**, 363.

70   S.K. Tyrlik, D. Szerszen, B. Kurzak and K. Bal, *Starch/Staerke*, 1995, **47**, 171 (*Chem. Abstr.*, 1995, **123**, 144 458).

71   J. Burger, C. Gack and P. Klüfers, *Angew. Chem., Int. Ed. Engl.*, 1995, **34**, 2647.

72   M. Adinolfi, M.M. Corsaro, C. De Castro, R. Lanzetta, M. Parrilli, A. Evidente and P. Lavermicocca, *Carbohydr. Res.*, 1995, **267**, 307.

73   M. Adinolfi, M.M. Corsaro, C. De Castro, A. Evidente, R. Lanzetta, L. Mangoni and M. Parrilli, *Carbohydr. Res.*, 1995, **274**, 223.

3
Glycosides and Disaccharides

1   O-Glycosides

1.1   Synthesis of Monosaccharide Glycosides. – Attention will first be given to
the methods which are used in glycoside synthesis and then the types of
glycosides produced will be considered.

1.1.1   Methods of synthesis of glycosides. – The 2-(dimethyloctylsilyl)ethyl group
can be used at a hydrophobic aglycon which allows the use of reversed phase
adsorbants for easy product isolation following enzymic glycosylation.[1]

Several papers have reported the preparation of glycosides directly from free
sugars. Hexoses treated in THF with octanol and iron(III) chloride in the
presence of barium or calcium chloride afforded glycosides as follows: glucose,
the furanosides in 65% yield, α/β ratio 1:3.4; mannose, the furanosides in 58%
yield, α/β ratio 19:1; galactose, the furanosides in 56% yield, α/β ratio 1:4.9 and
fructose, the β-pyranoside isolated in 23% yield, it being formed almost exclusively.[2] A paper reporting analogous work with D-fructose describes methods of
preparing the glycosides without degradation to 5-(hydroxymethyl)furfural.[3]
Two reports have been made of the reactions of fluoro-aromatics containing
strongly electron-withdrawing substituents with unprotected sugars in the presence of mild base. One describes the use of fluoro-2,4-dinitrobenzene and
sodium bicarbonate in aqueous ethanol as giving 1,2-*trans*-related 2,4-dinitro-
phenyl glycosides,[4] while the second is more extensive and describes analogous
reaction of glucose and its 2,3,4,6-tetra-O-acetyl or O-benzyl derivatives with, for
example, pentafluoropyridine in the presence of sodium hydride and a crown
ether in dichloromethane as giving 90% of the 3-(glycosyloxy)tetrafluoropyridine.
Analogous results were obtained by use of phenylsulfonyl 2,4-dinitrobenzene.[5]

A range of glycosyl esters continue to be useful in glycoside syntheses. For
example, glycosyl acetates are commonly used and a particular report has
appeared on the formation from them of long-chain alkyl glycosides by the use of
the required alcohols in the presence of trimethylsilyl triflate. As expected, high
selectivity in favour of 1,2-*trans*-related products was observed. Otherwise, for
example, 2,3,4,6-tetra-O-benzyl-D-glucose treated with these alcohols in the
presence of thrimethylsilyl bromide and cobalt(II) bromide and tetrabutyl
ammonium bromide can be used.[6] The 1-trichloroacetate group has been
employed as leaving group with tetra-O-benzyl-D-galactose and boron trifluoride
or trimethylsilyl triflate as catalysts. A wide range of alcohols, phenols, acids and

thiols were coupled, giving products with α,β-ratios ranging from approximately 2 to 4, but with phenol and thiophenol only the α-configurated products were observed.[7] The same group of Chinese workers carried out comparable studies using tetra-*O*-benzyl-α-D-galactopyranosyl trifluoroacetate.[8] Further work has been reported on the use of 1-methoxyacetyl esters activated with ytterbium(III) perchlorate. (see Vol 27, Chap. 3, Refs 15, 16). α-Glucopyranosides and β-ribofuranosides were reported as highly stereoselectively favoured products when ether was used as solvent.[9] α-Mannopyranosides, including disaccharides, are the highly favoured products of reaction of tetra-*O*-benzyl α-D-mannopyranosyl 3,5-dinitrobenzoate used in dichloromethane at −20 °C with trimethylsilyl triflate as catalyst.[10] The *O*-acetylated 1,2-orthoester method applied in the D-glucose and -maltose series has given access to the glycosides of echinochrome and related hydroxynaphthazarines.[11] Different trityl ethers have been assessed as acceptors in glycosylations induced using 1,2-*O*-cyanoethylidene derivatives.[12]

Tetra-*O*-benzyl-D-glucopyranosyl ethyl phosphite with trimethylsilyl triflate gives in 80% yield the β-glycosides derived using either primary or secondary sugar alcohols. α,β-Ratios were 1:20 and 1:3 for the primary and secondary cases respectively.[13] The corresponding dibenzyl glycosyl phosphite of the fully acetylated neuraminic acid methyl ester has been used as a glycosylating agent,[14] and in related work glycosyl diphenyl phosphates have been used; for example the β-mannosyl compound **1** and a closely related L-guloside were prepared in this way and elaborated into analogues of bleomycin A2.[15]

Unsaturated compounds continue to be of value in glycoside synthesis - see some examples given in Chapter 13.

Thioglycosides are now of considerable use in disaccharide and higher saccharide syntheses, and many of the products described in later sections of this Chapter and in Chapter 4 are produced following use of these compounds. The new activating agent *O*-mesitylenesulfonylhydroxylamine allows good yields of glycosides from methyl tetra-*O*-benzyl-1-thio-β-D-glucopyranoside.[16] The analogous *S*-phenyl glycoside carrying either a *p*-methyl or 2,4,6-trimethoxy groups affords access to glycosides by electrochemical means (cf. *Tetrahedron Lett.*, **1990**, *31*, 5761). In the case of the *p*-tolyl compound, products were obtained in 90% yield; α,β-ratio 3:7. With the analogous thioglycoside based on 3,4,6-tri-*O*-acetyl-*N*-phthaloyl glucosamine yields were similarly high, but only the β-compounds were obtained. Simple primary and secondary alcohols were used in the work.[17] *S*-Ethyl thioglycosides of heptuloses may be used in glycoside or disaccharide syntheses with various activating agents.[18] A further example of the phenylthioglycoside sulfoxide procedure (*J. Am. Chem. Soc.*, **1993**, *115*, 1580) led to a fucosylated 1-adamantanol in 53% yield. The β-sulfoxide activated with triflic anhydride reacted with retention of configuration mainly.[19]

As always, glycosyl halides are of value in glycoside synthesis and fluorides are now commonly used. Tetra-*O*-benzyl-β-D-glucosyl fluoride, for example, gives about 70% mainly of β-products when treated with alcohols in the presence of ytterbium triflate in acetonitrile. Change of solvent to ether affords 90% yields of products having α,β-ratio of 4:1, it being considered that an α-acetonitrilium ion is involved as a reaction intermediate. Disaccharides linked through primary or secondary positions were obtained.[20] The same glycosyl fluoride in the presence of silver perchlorate gave good α-selectivity when treated with ethyl 5-alkoxy-4-hydroxypent-2-enoate in which the substituents were extended chain alkyl groups. Ozonolysis then led to 1-alkylated-2-glycosylated glycerols.[21] (See also Chapter 19). Tri-*O*-benzyl-L-fucopyranosyl fluoride or trichloroacetimidate, activated with lithium perchlorate in dichloromethane, give approximately quantitative yields of disaccharides even when secondary sugar alcohols are involved. Very high α-selectivity is observed.[22] A novel way of making *O*-protected glycosyl fluorides is from compounds with free hydroxy groups at C-1 which give glycosyl tetrazoles with 5-chloro-1-phenyltetrazole. These with HF in pyridine, give the glycosyl fluorides generally in greater than 80% yield.[23]

Normal Koenigs-Knorr reactions have been applied to the synthesis of the β-galactopyranosides of several simple ols and diols to provide substrates for a range of β-galactosidases. One enzyme released butane-1,3-diol, butan-2-ol and 1,2-isopropylideneglycerol with e.e.s in the range 60-75%.[24] Acetobromoglycose reactions with the sodium enolate derived from malondialdehyde afforded the β-enol glycosides in the glucose, galactose and lactose series and the α-glycoside in the fucose series. The α-mannoside was made by use of the tosyl enolate. Methylenation then led to the 1-glycosyloxybutadienes.[25] The glycosyl bromide derived from **2** was then condensed with the alcohol **3** in studies related to the synthesis of concanamycin A.[26]

1   2   3

Anhydro-compounds are still of use in the synthesis of glycosides; for example 1,2-anhydro-3,4-di-*O*-benzyl-D-fucose normally reacts readily to give only β-glycosides, whereas on treatment with alcohols in the presence of silver triflate α-compounds are obtained preferentially. With serine methyl ester the α-anomer was produced exclusively, and with 1,2:3,4-di-*O*-isopropylidene-D-galactose the α-product was again favoured, but the α,β-ratio was 2:1.[27] In related work, serine methyl esters carrying different *N*-protecting groups were condensed with the 1,2-anhydrides of 3,4-di-*O*-benzyl-α-D-xylose, 3,4,6-tri-benzyl-α-D-galactose and -β-D-mannose in the presence of Lewis acid catalysts. High yields and high selectivities of 1,2-*trans*-related compounds were produced.[28]

Hanessian has filed a patent based on his use of *O*-glycosides carrying heterocyclic aglycons as glycosyl donors. Pyridine and pyrimidine rings containing a range of alkoxy substituents were claimed with glucose and 2-amino-2-deoxy-glucose derivatives. A wide range of *O*-substituents, activators and heterocycles were employed, and conditions were found to permit good α-selectivity in the glucoside series and β-selectivity with 2-acetamido-2-deoxy compounds. Disaccharides and trisaccharides were produced in the course of this work.[29]

Modification of relatively readily available glycosides still remains a potent means of producing more complex compounds. Allyl glycosides have been reported in four papers to give access to modified derivatives by refunctionalization of the double bonds. Hydroxylation affords 1-glycosyl glycerols;[30] bromination affords dibromides from which optically active 2,3-dibromo-1-propanol and epibromohydrin are obtainable;[31] hydroboration yields 3-glycosyloxypropanols and hence, in the D-xylose series, a compound which, following phosphorylation at positions 3 and 4 in the sugar and in the side chain, gave a triester claimed to be a mimic of inositol 1,4,5-trisphosphate; ozonolysis of the allyl xyloside allowed access to 2-hydroxylethyl 2-*O*-benzyl-β-D-xylopyranoside and hence an analogous trisphosphate.[32] A related paper reported a similar preparation manner of 2-hydroxyethyl α-D-glucopyranoside bearing phosphate groups in the aglycon and at O-3 and O-4, again as a mimic of the secondary messenger.[33]

The use of enzymes in glycoside synthetic work continues to develop. Allyl and benzyl β-D-glucopyranosides have been made from glucose by use of almond β-glucosidase,[34] and in related work the preparation of *n*-alkyl β-glucosides was described, the alkyl groups having six to twelve carbon atoms, and a biphasic system being employed.[35] A β-glucosidase of a thermophilic organism was examined using immobilized cells, crude homogenates, native enzyme and recombinant enzyme, and various β-D-glucopyranosides derived from a range of alcohols, diols and aromatic hydroxy compounds were prepared.[36] Allyl β-D-galactopyranoside has been prepared,[34] as has the analogous glycoside of kojic acid.[37] Selective enzymic acetylations and deacetylations have afforded means of obtaining 1-*O*- and 3-*O*-β-D-glucopyranosyl-*sn*-glycerols.[38] Syntheses of α-galactosides, including corresponding oligosaccharides, have been effected using an appropriate enzyme from a *candida* species.[39] An *N*-acetylglucosamine-based insoluble polysaccharide on hydrolysis with sulfuric acid as catalyst gave a set of oligosaccharides which were then converted into simple alkyl β-glycosides by enzymic methods.[40] A very different use of enzymes involves the preparation of *p*-hydroxyphenyl β-D-glucopyranoside by use of a glucuronidase. Peroxidase then coupled the aromatic rings of the product and gave a polymer, following hydrolysis with acid, having structure **4**.[41]

1.1.2 Classes of glycosides. – Using the *N*-tetrachlorophthaloyl-protected derivative of *O*-acetylated β-glucosamine trichloroacetimidate Schmidt has reported a good yield of β-glycosides following the use of tin triflate in acetonitrile,[42] and the same types of products were obtained by Mukaiyama's group by use of the glycosyl acetates of glucosamine and galactosamine carrying the trichloroethoxycarbonyl and allyloxycarbonyl *N*-protecting groups with tin triflate or ytterbium triflate as catalysts.[43] Schmidt's group has also employed the *N,N*-diacetylamino group in the glucosamine and galactosamine series together with methyl thioglycosides, this approach also leading to β-configurated products.[44] In the galactosamine series *N*-acetyl protection together with α-trichloroacetimidate activation at the anomeric centre allows good access to α-derivatives.[45] Access to *N*-acetylmannosamine glycosides has been provided by an azide/triflate exchange at C-2 of a glucoside, and related reactions afforded β-mannosides having a free hydroxyl group or a deoxy group at C-2.[46]

Direct glycoside formation from glucuronic acid and galacturonic acid using long chain alcohols and boron trifluoride etherate as catalyst give, with the former acid, the furanoid uronosides of the 3,6-lactone, while with D-galacturonic acid, the pyranosides were produced with high α,β-selectivities. On the other hand, with iron(III) chloride as catalyst, together with calcium chloride, the galacturonofuranosides were formed with α,β-ratios of approximately 1:10.[47] Specific pyranoid β-D-glucuronides to have been produced are those from the anti-mitotic agent combretastatin A4 (**5**),[48] impiramine (**6**),[49] several *O*-alkyl-5-fluorouracils[50] and compound **7** which is a hydroxylated metabolite of the antimalarial artemesinin.[51]

4   5   6   7

Much activity has been reported in the area of aromatic glycosides. A novel synthesis of 2-deoxyglycosides follows from the finding that 1-thiono-1,2-naphthaquinone reacts with *O*-substituted glucals to give cyclic adducts **8** which, on reductive desulfurization, afford the 2-deoxy compounds.[52] A more orthodox route to such deoxy compounds involves addition of phenylselenyl chloride or phenylsulfinyl chloride to glycals, followed by hydrolysis of the glycosyl chloride to the free hydroxy compounds, and coupling with phenols using Mitsunobu technology. Compounds such as those found in aureolic acids have been made in this way.[53] *N*-Tosylglycono-1,4-lactone hydrazones give access to aryl glycosides on treatment with phenols under light irradiation. (See also Chapter 10).[54] Further work has been published on glycosyloxyporphyrins made from the *O*-protected *p*-glycosyloxybenzaldehyde and pyrrole (cf. *J. Org. Chem.*, 1993, *58*, 2774).[55]

A set of substituted phenyl β-glucosides carrying a carboxyl and related groups in the aromatic rings have been produced to investigate the mechanism of *agrobacterium* β-glycosidase activity.[56] *p*-Aminophenyl α-D-galactopyranoside and the *p*-isothiocyanatophenyl β-D-glucopyranoside have been used as means of bonding aryl glycosides to bovine serum albumin,[57] and four *p*-aminophenyl 2-acetamido-2-deoxy-β-D-glucopyranosides have been prepared as substrates for *N*-acetyl-β-D-glucosaminidase.[58] (4-Ethoxycarbonyl-2,6-dimethoxy)phenyl β-D-glucopyranoside, an intermediate for the production of clemochinenoside A, has been made using ethyl syringate and trichloroacetimidate technology.[59] A set of biphenyl α-D-mannopyranosides, containing carboxylic acid groups on substituents and also other substituents on the aromatic rings, have been assessed as siaLe^x mimics.[60]

Substituted aryl glycosides can be made by manipulations of preformed aryl glycosides. Thus bromination of *O*-acetylated β-D-glucopyranosides using bromine in dichloromethane at $-20\,°C$ has resulted in substitutions which show that hydroxy and methoxy groups on aromatic rings are more strongly activating towards electrophiles than is the tetra-*O*-acetyl-β-D-glucopyranosyloxy group.[61] Halogenation of various tolyl α-D-glucopyranosides has resulted in the production of halomethyl analogues which were tested as enzyme-activated irreversible inhibitors of yeast α-glucosidase. This work is being carried out in connection with assessment of HIV activity.[62] *p*-Bromophenyl β-D-glucopyranoside tetraacetate has been coupled with a substituted phenylboronic acid in the presence of a palladium catalyst to give biphenyl glycosides which showed liquid crystal properties.[63]

A full account of the use of 2,6-anhydro-2-thio sugars for the stereo-controlled synthesis of 2,6-dideoxy-α- and β-glycosides has appeared, and the method has been applied to the synthesis of erythromycin A and olivomycin A trisaccharide.[64]

1.2 Synthesis of Glycosylated Natural Products. – In the area of glycosylated amino acids and peptides the 1,2-cyanoethylidene/trityl ether method has been used to couple glucose to serine and serine-containing peptides,[65] and related work involving use of the β-peracetate has given rise to β-galactosides of serine and threonine and related compounds.[66] β-Galactosides have also been made of serine amide-bonded to long chain alkylamines; these and other related compounds were required as amphiphiles for liposomal drug carrying systems. Analogues containing fluorine in the alkyl chains were produced during this work.[67]

Considerable interest is being taken in the coupling of 2-amino-2-deoxy sugars with amino acids and peptides. *N*-Acetylglucosamine has been linked to serine and threonine derivatives for solid phase peptide syntheses by use of trichloroacetimidate technology.[68] 2-Azido-2-deoxy-α-D-galactopyranose has been *O*-linked to threonine, and a disaccharide comprising β-D-galactose linked 1,3- to 2-azido-2-deoxygalactose has been condensed with serine and threonine for solid phase peptide syntheses and for the preparation of specifically and multiply glycosylated peptides.[69] Imino substitution of the amino group affords serine and threonine derivatives which are much better glycosyl acceptors than are the normal Fmoc derivatives, and this observation has been used in the preparation of several 2-acetamido-2-deoxy-α- and β-galactopyranosyl derivatives of these amino acids.[70] Solid phase peptide synthetic methods using *O*-protected *N*-acetylglucosamine and 2-azido-2-deoxygalactose linked to serine and asparagine have been applied to the synthesis of four major histo-compatibility complexes.[71] Glucosylation of hydroxyproline derivatives has been effected using both the paracetylated sugar with boron trifluoride as catalyst and acetobromoglucose with silver triflate, but better yields were obtained using the perbenzoylated bromide.[72]

Solid phase technology has now come into its own for the production of glycosylated peptides. Compounds to have been produced by this procedure include a vasopressin compound having a β-D-galactopyranosyl group linked to a serine component,[73] the peptide SIKVAV having α-mannose, α-galactosamine, β-lactose or β-lactosamine O-linked to serine,[74] a fragment of RNA polymerase II carrying a β-D-N-acetylglucosamine unit[75] and a fucosylated glycopeptide of human factor IX with α-L-fucose O-bonded to serine.[76]

An α-D-galactosylceramide isolated from a sponge and having anti-tumour activity (Vol. 27, p. 21) has acted as a model for several analogues prepared using, for example, tetra-O-benzyl-D-galactosyl fluoride as glycosylating agent.[77] Related syntheses of glucosides and galactosides led to closely related marine sponge products.[78] The α-galactosylceramides AGL 506, 512, 514, 517, 525 and 564 were shown to be effective radio-protectants for mice irradiated with X-rays.[79] The *penicillium* glucosylcerebroside 9, which is the fruiting inducer against a *Schizophyllum* commune, has been synthesized,[80] and likewise the two glycosy-

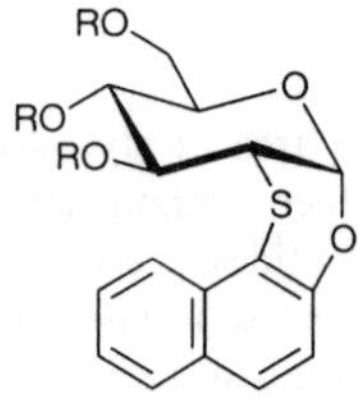

8

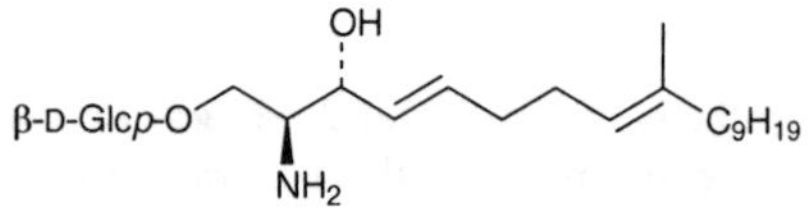

9

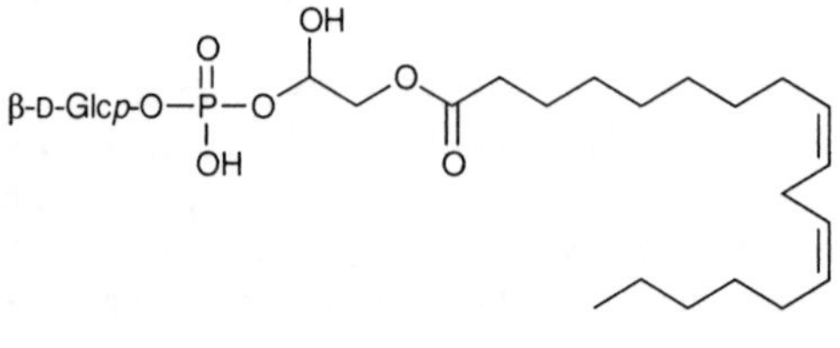

10

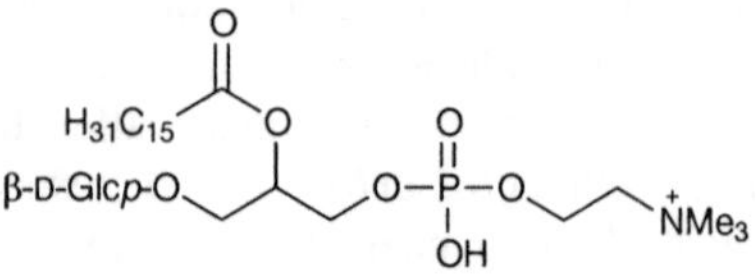

11

lated phosphatidylcholines **10** and **11**, which belong to the same series as a new antifungal agent, have been prepared.[81] A closely related but new type of glucosyl phospholipid to have been made by chemical methods is based on glycerol carrying a long chain fatty ester group at O-1, α-D-glucose at O-2 and a choline phosphate group at O-3.[82]

The synthetic cluster galactoside **12** and a related shorter chain compound have high affinity for the hepatic asialoglycoprotein receptor,[83] and an *N*-acetylglucosaminide having a long amide-containing aglycon with an attached fluorescein substituent is referred to in Chapter 10.

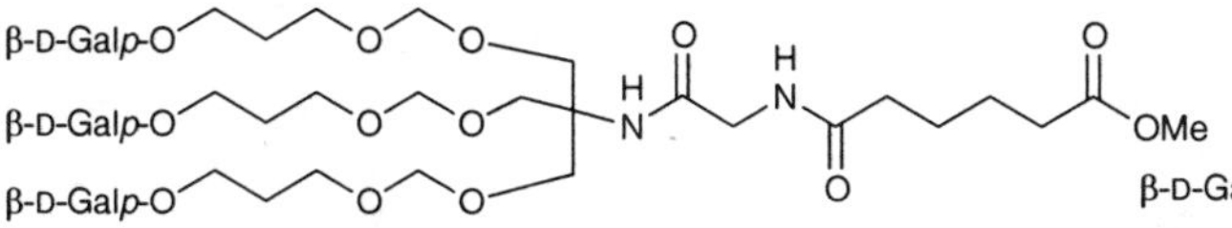

12

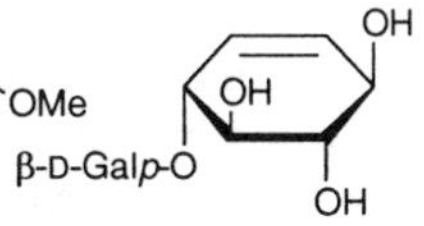

13

In the area of glycosylated inositol derivaties (±) condurital B, on enzymic galactosylation, afforded compound **13** and (+) conduritol B, the reaction representing a means of resolving the conduritol.[84] The *myo*-inositol derivative **14**, which is of potential value for the synthesis of some glycosylphosphatidyl inositols, has been made by standard procedures,[85] as has the α-2-linked mannosyl-*myo*-inositol **15** carrying a phosphate linked ceramide at O-1 of the inositol.[86] The glucuronic acid derivative **16** was prepared by glycosylation of a suitably protected deoxyinosidiamine, itself made by carbocyclization of methyl 2-acetamido-2-deoxy-α-D-glucopyranoside.[87] Further reference to glycosyl inositols produced by chemical synthesis of the carbocyclic ring is made in Chapter 18.

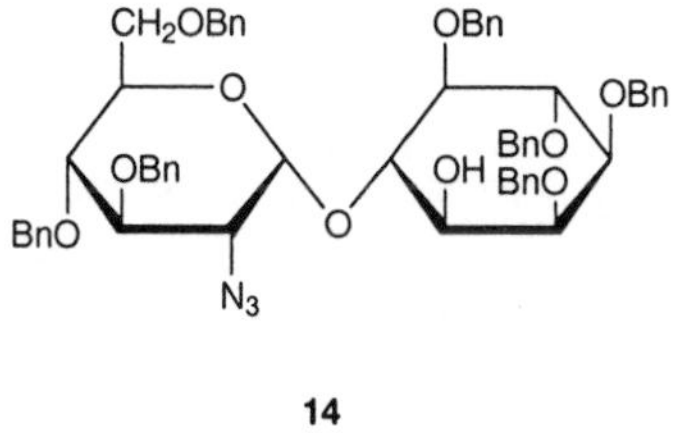

14

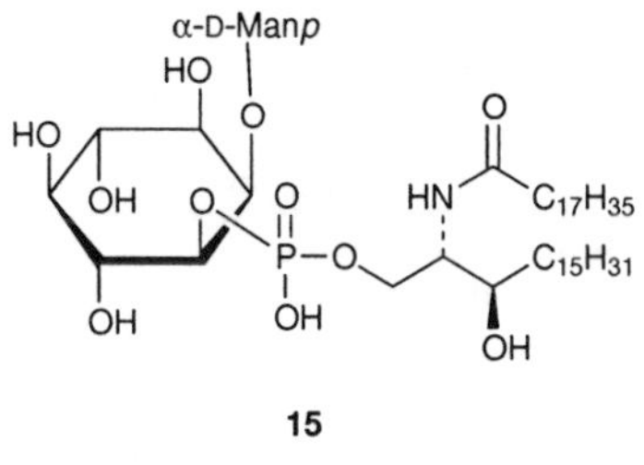

15

In connection with the finding that perilloside A **17** inhibits rat lens aldose reductase, an extensive range of monoterpene glucosides and their tetraacetates have been synthesized by standard Koenigs-Knorr methods,[88] and the same approach has been used to prepare the 3- and 25-monoglucuronides of 25-

hydroxyvitamin D. The substitutions were carried out on provitamin D and the final products were obtained by photochemical and thermal isomerization.[89] Reports have appeared on the preparation of Δ^0, Δ^5 and Δ^7 β-steryl glycosides,[90] and glucosylation and galactosylation of luteolin[91] and 5,7,3′,4′-tetra-*O*-benzoyl-quercitin,[92] respectively, have been reported.

Several reports of glycosylation of compounds containing *N*-heterocyclic rings have appeared. Morphine and simple derivatives have been notable examples. An improved synthesis of the 6-glucuronide, involving use of the glycosyl bromide has been reported,[93] and a related report indicates this compound could not be made by enzymic coupling and went on to describe its preparation by the selective enzymic hydrolysis of the 3,6-diglucuronide.[94] A standard chemical method was used to prepare the 6-glucuronide of *O*-ethylmorphine,[95] and a further analogue to have been produced is the 6-glucuronide of morphine carrying a 4-aminobutyl substituent on nitrogen which was made as a hapten for use in the radioimmuno assay of morphine 6-glucuronide.[96] Conventional methods were used to prepare the 6-*O*-α- and 6-*O*-β-D-glucopyranosyl derivatives of morphine and codeine.[97] The natural products **18** and **19** have been prepared using the trichloroacetimi-date method,[98] and mono- and di-ribosylation of biopterins afforded compounds **20** and **21**.[99] (2-Indol-3′-yl) ethyl β-D-xylopyranoside,[100] and the corresponding β-D-galactopyranoside[101] were synthesized, and the tetraacetate of the latter was subjected to X-ray diffraction analysis. β-L-Xylopyranosides of known and new hydroxypyrrolidines have been studied as glycoside bond cleavage transition state analogues.[102]

16   **17**

18 R¹ = H, R² = OMe
19 R¹ = OH, R² = H

Miroside **22**, an anti-fungal glucoside obtained from the leaves of a New Zealand tree, was prepared by use of an *O*-silylated-1,2-anhydro-D-glucose which gave the β- and α-anomers in the ratio 2:1. In the course of the work the isomer **23** was prepared by use of acetobromoglucose and the corresponding sodium enolate.[103] Norisoprenoid glucosides and related compounds, for example the alkyne **24**, were prepared by conventional methods.[104] Sucrose phosphorylase was used in the transglycosylation of α-D-glucopyranose from sucrose to (−)-epigallocatechin and the 4′-α-D-glucoside and the 4′,4″-diglucoside were obtained.[105]

Glycosylation was effected during the preparation of the deoxytetrahydrodes-mycosin 1, a macrocyclic antibiotic containing mycinose and desosamine.[106]

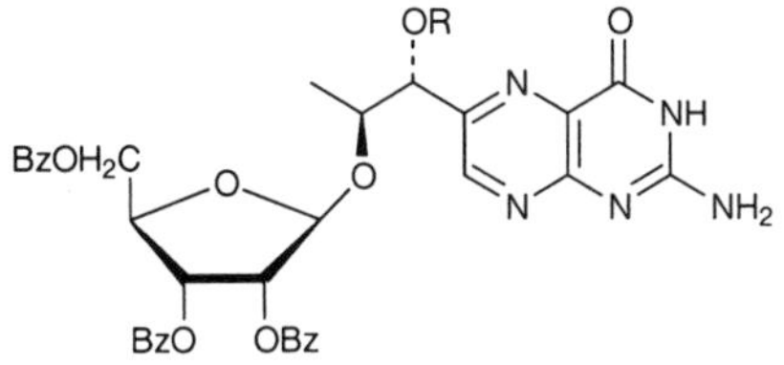

20 R = H
21 R = tri-*O*-benzoyl-D-ribosyl

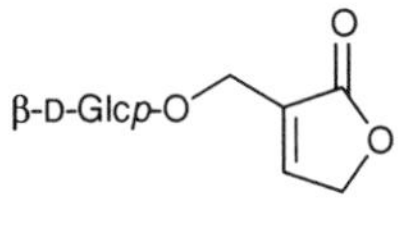

22 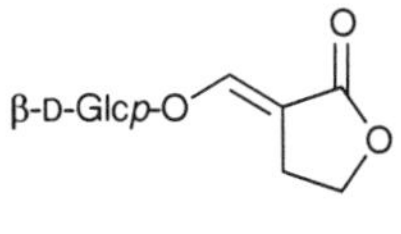   **23**

1.3   *O*-Glycosides Isolated from Natural Products. – As usual, only compounds showing special features, usually within the sugar moieties, are dealt with in this section. A comprehensive review has been written on steroidal glycosides, which includes a description of their biosynthesis and digitalis binding site structures,[107] and a further has appeared in Rodd's Chemistry of Carbon Compounds series on glycosides, saponins and sapogenins.[108]

A polyhydroxylated steroidal saponin from the plant *Nobina recurvata*, unusually, contains a β-D-fructofuranosyl residue, this being the first report of such a unit in a saponin.[109] A green sulfur bacterium has yielded a biopterin glycoside of *N*-acetylglucosamine.[110] The composition of cyanogenic glycosides in a new cassava to be investigated was unusual in that some commonly found cyclopentenyl derivatives were absent. Two compounds that were found were derived by glucosylation or gentiobiosylation of the cyanohydrins of acetone and butanone.[111]

The absolute configuration of compounds **25**, **26** was determined by the bichromophoric exciton chirality method,[112] and the interesting observation has been made that irradiation of Scots pine seedlings with ultraviolet B radiation induces the production of sunscreen pigments **27** and **28**.[113]

1.4   Synthesis of Disaccharides and their Derivatives. – T. Ogawa, C.H. Wong and colleagues have produced a short review on methods, including enzymic procedures, used for coupling sugars[114] and a 44 page compendium of reactions leading to the formation of glycosidically linked compounds contains about 700 entries, all of them relating to 1994 work.[115]

In the field of non-reducing disaccharides trehalose has been converted into the hexabenzoate having free hydroxyl groups at both 4-positions and thereby, by

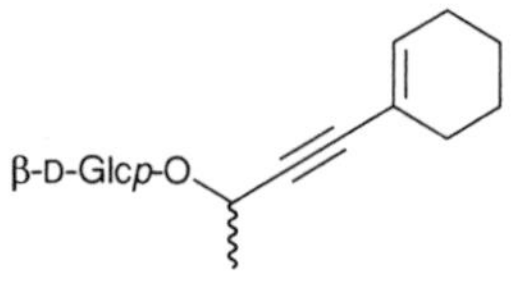

24

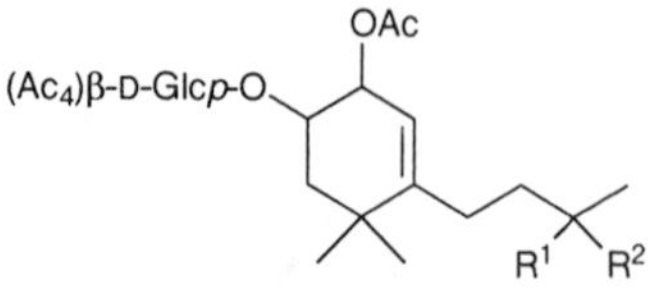

25 R^1, R^2 = O
26 R^1 = H, R^2 = OAc

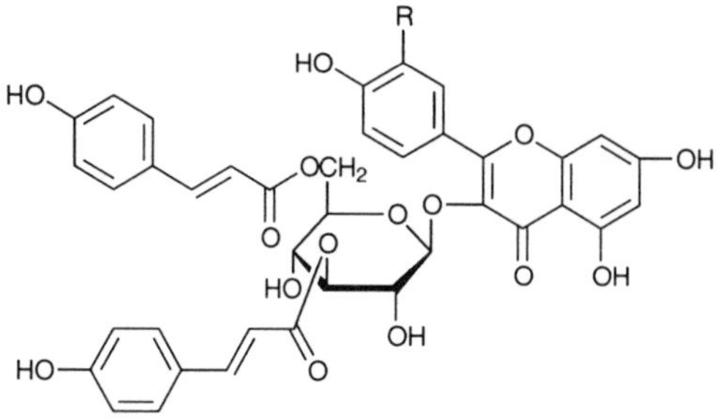

27 R = OH
28 R = H

triflate displacement, into α-D-galactopyranosyl α-D-galactopyranoside.[116] Benzo-bromoglucose has been coupled with 2,3,4,6-tetra-*O*-benzyl-D-glucose, -mannose and -galactose to give mainly α,β-, α,α- and α,α- products respectively; in the last two cases the main products were isolated crystalline in 40% yield.[117] A novel intramolecular glucosyltransferase from *Primelobacter* converts maltose into trehalose,[118] and α-D-galactopyranosyl α-D-galactopyranoside was obtained during work in which a *Candida* α-galactosidase was used in reverse mode.[119] Coupling of UDP-glucose and D-lyxose in the presence of sucrose synthase has led to the isolation of α-D-glucopyranosyl α-D-lyxopyranoside and α-D-glucopyr-anosyl β-D-*threo*-pentulofuranoside.[120] The α,α-linked dimer of 2,6-dideoxy-D-*ribo*-hexopyranose is bitter in taste, and a stereochemical basis for this observa-tion has been proposed. The tetrabenzoate of the compound was subjected to X-ray crystallographic analysis.[121]

Two molecules of α,α-trehalose have been linked into a macrocycle by way of two thiourea bridges involving the C-6 positions.[122]

Much attention continues to be paid to inter-unit linking methods for the production of reducing disaccharides and many methods referred to at the beginning of this Chapter are relevant in this context. New intra-molecular procedures utilize dicarboxylic acids as linking agents prior to the coupling of the glycosyl donor and acceptor. The use of the succinyl bridging is illustrated in Scheme 1[123] and analogous chemistry has led with good efficiency to the synthesis of a derivative of 4-*O*-α-L-rhamnosyl-D-glucose.[124] The succinic acid based bridge was found to be more effective than malonic or phthalic acid bridges, and best results were obtained when linkages were between the O-3 of the acceptor and O-

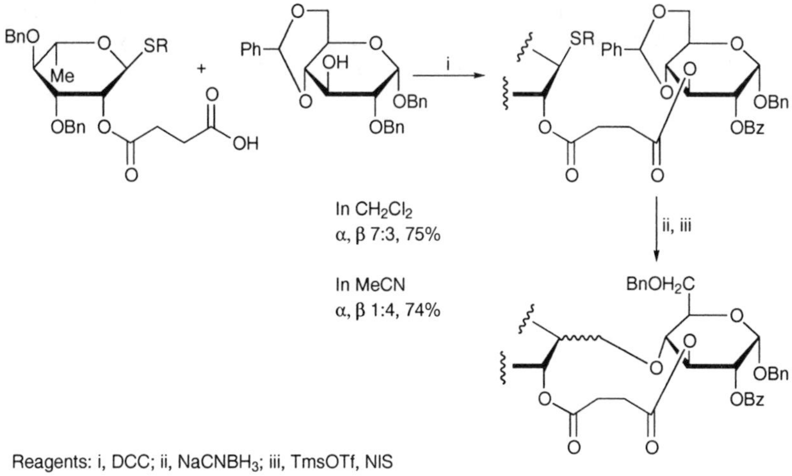

Reagents: i, DCC; ii, NaCNBH₃; iii, TmsOTf, NIS

Scheme 1

2 of the donor.[123] Other workers however have used the phthalic acid linkage and from diester **29** they prepared β-3-linked glucobiosides achieving 65-80% efficiencies for the cyclizations.[125]

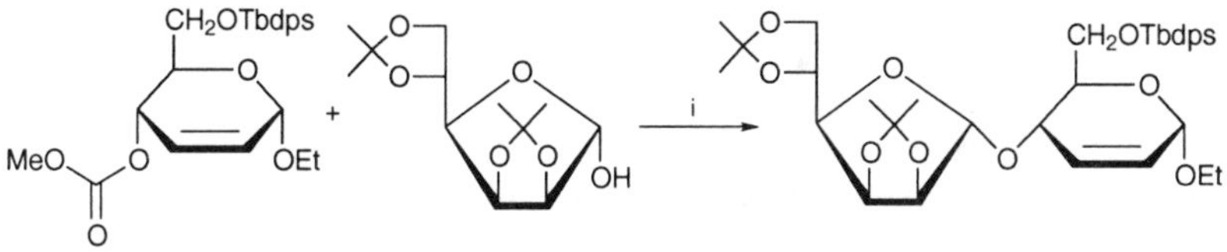

Reagents: i, Pd(dba)₃, Ph₂P⌣⌣PPh₂

Scheme 2

A novel disaccharide synthesis is illustrated in Scheme 2 and involves a palladium-catalysed allylic displacement reaction.[126] Ketene acetal **30** has been adapted as a glycosylating agent, and with 1,2;3,4-di-*O*-isopropylidene-D-galactose in the presence of camphorsulfonic acid, undergoes addition at the double bond to give an orthoester derivative. Treatment with trimethylsilyl triflate then gives access to 6-*O*-β-D-glucopyranosyl-D-galactose produced in 61% overall yield from **30**.[127]

Hashimoto and co-workers have shown that glycopyranosyl diethyl phosphite, with such participating groups as benzoyloxy or phthalimido at C-2 and trimethylsilyl triflate as promoter, gives high yields of 1,2-*trans*-related disacchar-ides,[128] and more surprisingly they also find significant β-selectivity when they applied the method with several 2-deoxy sugar phosphites.[129]

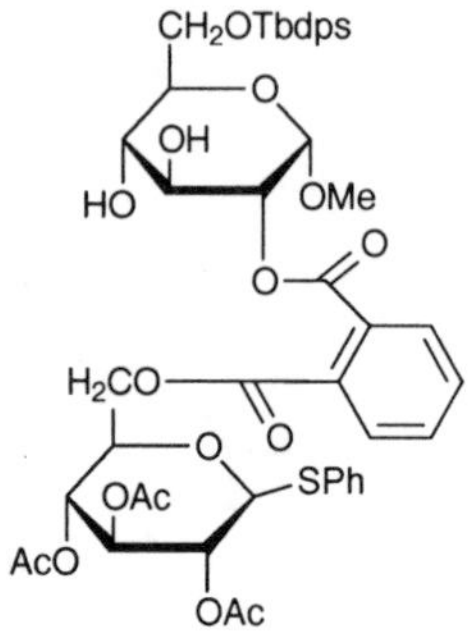

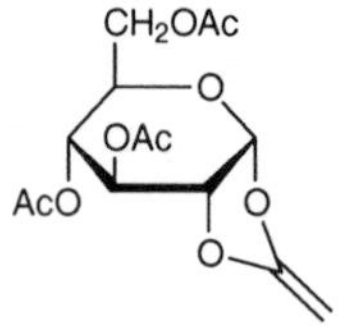

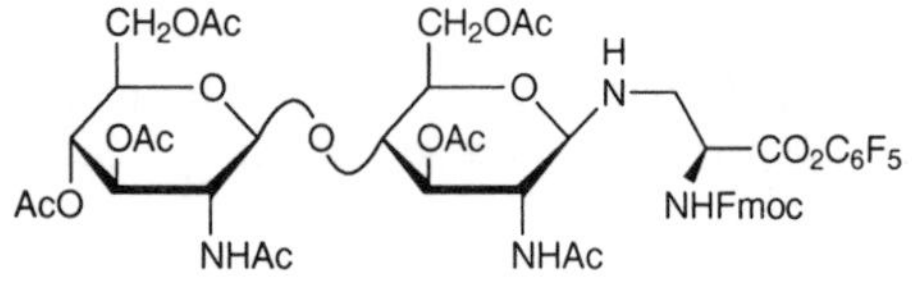

Pyrimidin-2-yl 1-thio-β-D-glycopyranosides, which are stable odourless com-pounds, appear suited for the production of 1,2-*cis*-related disaccharides. For example, the 2,3,4,6-tetrabenzyl-β-D-galactopyranoside condensed with methyl 2,4,6-tri-*O*-benzyl-β-D-mannopyranoside gave the α-linked disaccharide in 93% yield under these conditions.[130] Electrochemical glycosylation involving the use, for example, of *p*-tolyl 2,3,4,6-tetra-*O*-benzyl-1-thio-β-D-glucopyranoside gives very good coupling conversion with methyl 2,3,4-tri-*O*-benzyl-α-D-glucopyrano-side with the α,β-ratio being 1:3. Good yields and selectivities with secondary sugar alcohols were also claimed.[131] Glycosyl fluorides couple well with acceptor trimethylsilyl ethers in the presence of lanthanum perchlorate as catalyst, for example tetrabenzyl-β-D-glucosyl fluoride reacts with methyl tribenzyl-6-*O*-trimethylsilyl-α-D-glucopyranoside to give high yields of the 1,6-linked products in acetonitrile, the α,β-ratio being 1:10, while when the reaction was conducted in ether the ratio was 75:25.[132] A rather unusual application of the Koenigs-Knorr reaction involves coupling of acetobromogalactose with 2-(trimethylsilyl)ethyl β-D-galactopyranoside. Good selectivity for coupling at the primary position was reported with, however, 5-10% of other products being formed, with this bromide and also with acetobromolactose and the analogous compound from 2-deoxy-2-phthalimidoglucose.[133] *N*-Protection with the novel dithiasuccinoyl protecting group and trichloroacetimidate methods has led to the formation of β-linked glucosamine disaccharides and derivatives, in particular the *N*-linked chitobiosyl-amino compound **31**.[134]

Reducing disaccharides will now be dealt with according to the nature of the residues at the non-reducing termini. Yields ranging from 60-80% were obtained following glucosylation, galactosylation and mannosylation with glycosyl bromides or thioglycosides of methyl hexopyranosides with unsubstituted hydroxyl groups which were first stannylated. This had the effect of selective activation of the primary hydroxyl groups and the method, therefore, looks a rather straightforward means of obtaining 1,6-linked compounds.[135] Compound **32**, on the other hand, on treatment with acetobromoglucose, was selectively substituted at O-3, giving 44% of the product and only traces of other isomers. In the case of the L-rhamnosylamine analogue, however, 30% of O-3-linked product and 24% of O-4-linked product were isolated.[136] When D-glucosamine and 2-deoxyglucose were separately taken with equimolar proportions of α-D-glucose-1-phosphate on the 100 mM scale and in the presence of cellobiose phosphorylase, the β-(1→4)-linked disaccharides were obtained in 55 and 50% respectively.[137] Several methyl maltoside analogues having sulfur as the ring atom and various hetero atoms in the inter-unit position are referred to in Chapter 11.

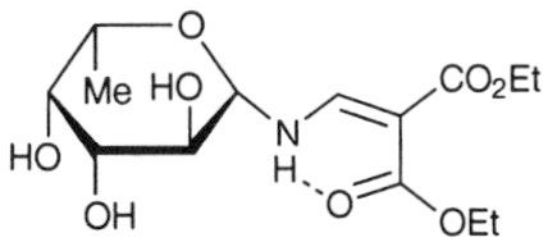

32

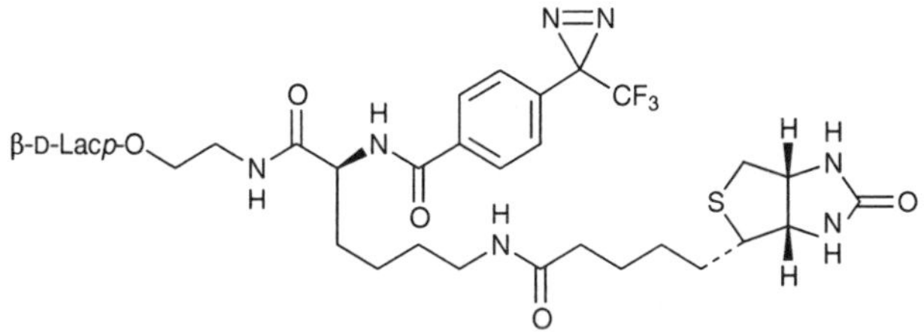

33

Considerable interest continues in the chemistry of disaccharides terminating in D-galactose. 4,6-O-Benzylidene-1-O-Tbdms lactose, on treatment with benzyl bromide in the presence of sodium hydride, underwent silyl migration from O-1 to O-2 and subsequent O-benzylation. Desilylation and pivaloylation then afforded a very suitable building block for lactose derivatives having glycosyl substitution at O-2.[138] The lactosylceramide analogue **33** has been produced as a novel photoreactive substrate for GM$_3$ synthase,[139] and a lactose-containing peptide **34** was made from an *O*-acetylated glycosyl azide which was reduced and coupled initially with glycine prior to peptide development.[140] Compound **35**, a bifluorescence-labelled substrate for ceramide glucanase, was prepared and its usefulness demonstrated; in particular it showed sensitivity and allowed rapid analysis consistent with a continuous assay.[141] The 1,2-; 1,3- and 1,6-β-D-galactosyl-D-glucoses have been produced using β-galactosidase.[142]

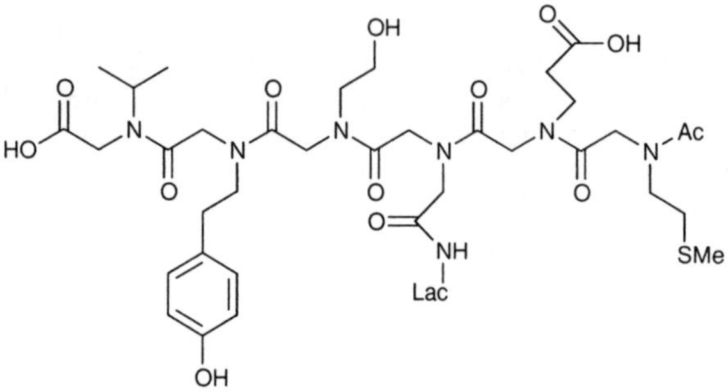

34

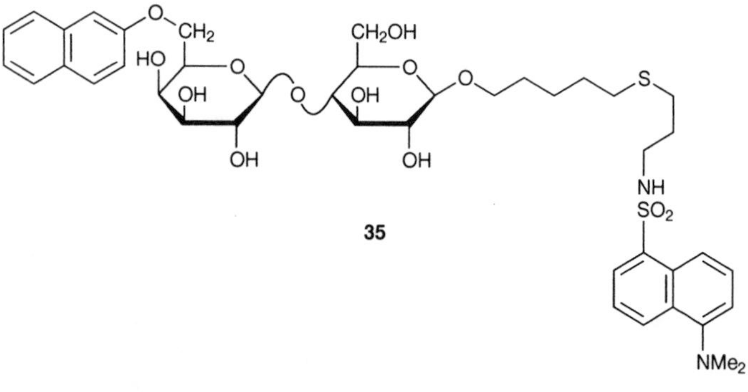

35

A set of dendromers, of which compounds **36-38** are the most complex, were made with the intention of using compounds of the set **37** to make sialyl Le^x analogues by enzymic sialylation and fucosylation.[143] *N*-Acetyllactosamine has also been linked to the three hydroxyl groups of compound **38a**.[144] Several other papers have also been concerned with *N*-acetyllactosamine and its glycosides. Methods have been developed for efficient conversion of the known oxime derivative **39** to *p*-methoxybenzyl glycosides of α- and β-*N*-acetyllactosamine,[145] and the pyruvyl acetal derivative **40** was prepared in studies related to the aggregation factor of a marine sponge.[146] All six monosulfates of the β-*N*-acetyllactosaminyl 8-methoxycarbonyloctanol have been synthesized and characterized by high resolution NMR spectroscopy.[147] A complex enzymic system

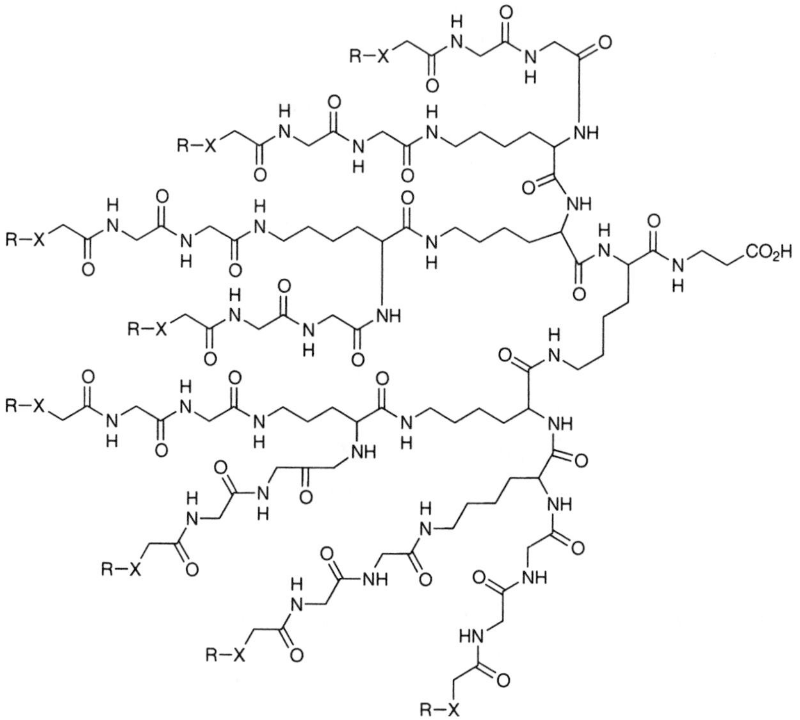

38a

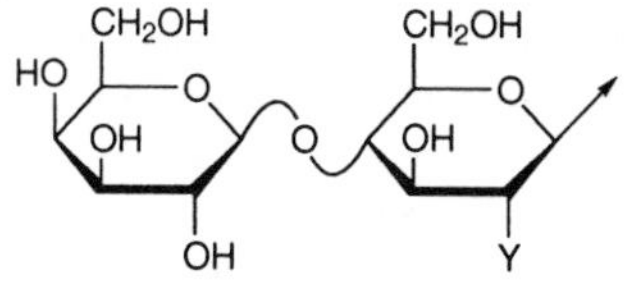

	Y	X
36	OH	no spacer group
37	NHAc	NHCOCH$_2$CH$_2$
38	OH	NHCOCH$_2$CH$_2$

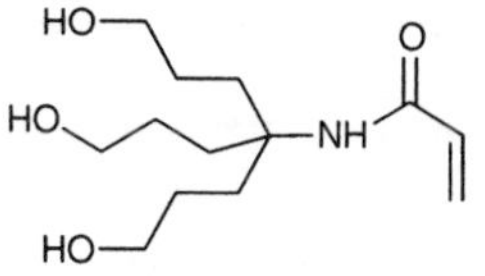

involving several catalysts has been used in an impressive 4-β-galactosylation of glycosides of *N*-acetylglucosamine, the ultimate source of the required glucose-1-phosphate was inorganic phosphate and sucrose, therefore avoiding the need to obtain the ester commercially.[148] Enzymic methods were also involved in the preparation of a polymer derived from *p*-(acrylylamino)phenyl *N*-acetyllactosaminide.[149] Formal chemical methods were used to obtain the disaccharide based on α-D-galactose 1,3-linked to methyl α-D-*N*-acetylglucosaminide.[150]

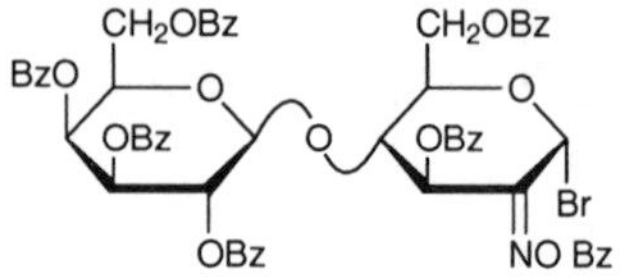

39

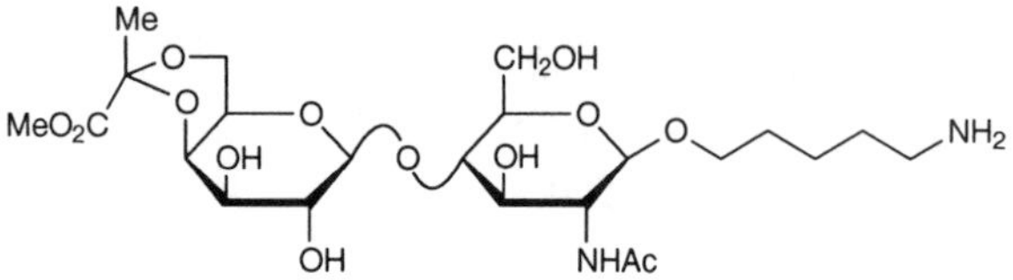

40

Reports have appeared in the proceedings of a symposium on the synthesis of a new ceramide α-1,4-linked digalactoside,[151] The 1,1-; 1,2-; 1,3- and 1,6-α-D-galactosyl-D-galactoses have been produced using enzymic methods.[119] β-D-Gal-(1→3)-β-D-Gal-*O*-2-naphthyl, the disaccharide of which is related to sub-units of heparin and chondriotin sulfates, was synthesized together with related glycosides to test their ability to serve as primers of oligosaccharide synthesis in animal cells.[152] A successful synthesis of 2-azido-2-deoxy-3-*O*-β-D-galactopyranosyl-D-galactose and its *O*-linking to serine and L-threonine derivatives, and subsequent development of the peptide chains has been reported,[153] and a related study led to the development of means of linking this disaccharide by α- or β-bonds to various alcohols.[154] A further synthetic effort led to the description of several specific monosulfates and monomethyl ethers of the allyl glycoside of this disaccharide in the 2-acetamido-2-deoxy form,[155] and the 3',6-dimethyl ether and the 3'-methyl ether were prepared of the benzyl α-glycoside.[156]

Enzymic procedures have been used to introduce a β-D-galactopyranosyl unit at O-5' of β-D-arabinofuranosylcytosine,[157] and at the same position of several ribo- and 2-deoxyribonucleosides.[158] An α-galactosidase from *Candida* transfers the sugar to several other sugars, including pentoses and 6-deoxyhexoses.[159] β-D-Gal*p*-(1→4)-β-D-Xyl*p*-*O*-2-naphthyl was found to act as a primer for the biosynthesis of glycosaminoglycan chains,[152] and the hexaacetate of this disaccharide glycoside was more active as a chain-primer suggesting that cells took up the substituted compound easily prior to causing deacetylation and glycosylation. Acid-catalysed reaction of D-galactose with 5-hydroxypent-1-ene led to a large proportion of furanosides. The derived acetates afforded a means of making the β-D-galactofuranoside selectively, as did the corresponding benzyl ether. In the course of the work a β-1,2-galactofuranosyl mannopyranoside was made.[160]

Mannose-terminating disaccharides have proved very popular targets of synthesis, the α-1,2-linked mannobiose having been prepared as its 6-(trifluoroacetyl-amino)hexyl glycoside,[161] and also as its specifically 6-deoxygenated methyl α-D-pyranoside.[162] An α-mannosidase acting on concentrated aqueous solutions of mannose led to the formation of this same 1,2-linked compound, as well as to the 1,3- and 1,6-isomers and also to trisaccharides.[163] The α-1,2-linked compound has also been produced using mannose glycosidically linked to poly(ethyleneglycol)monomethyl ether.[164] Chemical methods have been employed to prepare the α-1,2- α-1,3- and α-1,6-linked mannobioses which were condensed to partially protected hydroxy-containing amino acids as starting materials for the synthesis of glycopeptides.[165] α-1,4-And α-1,6-linked dimers have been made and linked to spacer arms.[166] A metal complex derived on the β-face of a mannose derivative fully O-substituted except at O-1, on condensation with 1,6-anhydro-2,3-di-O-benzyl-4-O-triflyl-D-galactose, gave 61% of the α-1,4-linked mannosylglucose derivative,[167] and a mannofuranosyl α-1,6-linked galactose was formed by way of 1,2-anhydro-3,5,6-tri-O-benzyl-β-D-mannose.[168] A new route has been reported to 2-O-α-(3-O-carbamoyl-α-D-mannopyranosyl)-L-gulose which utilized a mannose derivative chain extended by a Wittig methylenation and configurationally inverted at C-6 of this acyclic compound to give access to L-gulose.[169] The same disaccharide, in the form of peracetate glycosidically linked to hydroxyhistamine, has been elaborated to an analogue of bleomycin A2.[170]

Disaccharides ending in pentose units include a fully substituted 2-O-α-D-ribofuranosyl-L-rhamnose derivative produced from a glucosyl acetate or glycosyl trichloroacetimidate.[171] 2,3,5-Tri-O-benzoyl-α-L-arabinofuranosyl chloride coupled with specifically di-O-substituted methyl α-L-arabinofuranosides led to 1,2- 1,3- and 1,5-linked disaccharides.[172] 1,3-Anhydro-2,4-di-O-benzyl-L-arabi-nose with 1,2:3,4-di-O-isopropylidene-D-galactose in the presence of zinc chloride gave 80% of the 6-O-arabinosylgalactoses with α,β-ratio 3:1.[173] Various 1,4-linked β-D-xylobiosides in the form of nitrophenyl glycosides have been used as inhibitors of xylanases.[174] Coupling of 1,2-anhydro-3,4-di-O-benzyl-D-lyxose and L-ribose with 1,2:3,4-di-O-isopropylidene-α-D-galactose, in the presence of molecular sieves, gave the α-D-lyxoside and β-L-riboside in high yield and with good stereoselectivity.[175]

The heptobiose α-D-Hepp-(1→3)-α-D-Hepp has been synthesized as its 4-phosphate and 4-diphosphate.[176]

Amino sugars commonly feature as non-reducing termini of disaccharides. Coupling of N-acetylglucosamine and its β-p-nitrophenyl glycoside in the presence of an appropriate β-transferase gave rise to the β-1,4- and the β-1,6-linked dimers, and in the course of the same work analogous transfers of N-acetylgalactosamine were effected.[177] Chitobiose α-1-phosphate has been extended to corresponding diphosphate diesters carrying long chain alkyl ester groups.[178] The disaccharide comprising N-acetylglucosamine β-(1→4)-linked to 2-deoxy-3-O-muramoyl-2-N-palmitoylamino-D-glucose has been prepared as a key substance for the synthesis of new lipid analogues of muramoyl dipeptide.[179] The synthesis of an analogue of the biosynthesis precursor of Lipid A with (*S*) chirality in 3-hydroxytetradecanoic acid side chains has been reported,[180] as has an improved route to the phosphono-oxyethyl analogue of lipid A.[181] β-D-Glc*p*NAc(1→4)-β-D-Man*p*OH and its *p*-nitrophenyl β-glycoside have been made by lysozyme-catalysed transfer of *N*-acetylglucosamine. The regioselectivity of the reaction is dependent on the anomeric configuration of the acceptor.[182] The α- and β-inter-unit isomers of D-Glc*N*Ac(1→4)-β-D-GlcA have been tested for their binding affinity to fibroblast growth factor as have α-L-IdoA-(1→4)-α-D-GlcNac and β-D-GlcA(1→4)-α-D-GlcNSO₃, and trisaccharides related to these compounds, this work being carried out in connection with the affinity of heparin and its components and the growth factor.[183]

A peracetate of an α-1,3-linked dimer of 2-azido-2-deoxy-α-D-galactose glyco-sidically bonded to a derivative of L-threonine has been prepared for use in solid phase synthesis of glycopeptides,[184] and *N*-acetylglucosamine carrying a sulfate at O-4 and β-1,4-linked to D-glucuronic acid has been made; it is a component of chondroitin sulfate.[185] β-D-Man*N*Ac*p*-(1→X)-α-D-Rha*p* where X = 2, 3 or 4 have been made in connection with the study of bacterial lipopolysaccharide repeating units.[186]

Considerable work has continued in the area of deoxy-sugar disaccharides. Methyl α-isomaltoside and methyl α-isomaltotrioside analogues specifically de-oxygenated at C-2 in one or more of the glucose moieties have been made by use of a 1-Tbdms ether of 2-deoxyglucose as glycosyl donor with TmsOTf as activator.[187] A parallel report on isomers bearing deoxy groups at C-4 has been published.[188] Enzymic condensation of 2-deoxy-D-glucose afforded α-linked products from disaccharides to pentamers. Different enzymes produced different ratios of the 1,3-; 1,4- and 1,6-linked products. *Aspergillus niger* enzyme produced mainly the last of these.[189] α-L-Rha*p*-(1→2)-β-D-Glc*p* has been made as a triterpenoid glycoside as part of a study of the structure of natural saponins,[190] and osladin, a sweet principle of a fern, which contains α-L-Rha*p*-(1→2)-D-Glc*p* as part of its steroidal structure, has been discussed in a review of intensely sweet glycosides. One of the compounds reported is from a Chinese drug, and it contains β-D-Xyl*p*-(1→2)-D-Glc*p* and is 250 times sweeter than sucrose.[191] 1,2-Anhydro-3,4-di-*O*-benzyl-α-D-fucose is reported to act with serine to give the α-glycoside rather than the anticipated β-compound and with 1,2:3,4-di-*O*-isopro-pylidene α-D-galactose to give the mixed anomers of the 1,6-linked fucosylgalac-tose.[192] Two reports have appeared on the synthesis of the marine toxin polycavernoside A disaccharide which has the structure 2,3-di-*O*-methyl-α-L-fucopyranosyl-(1→3)-2,4-di-*O*-methyl-D-xylose.[193,194]   2,3-di-*O*-Benzoyl-4,6-dideoxy-α-L-*lyxo*-hexopyranosyl chloride has been made from methyl α-L-rham-nopyranoside and condensed to form an α-(1→2)-linked α-D-galactopyrano-side.[195] Selective methods have been developed to isolate the β-1,4-linked dimer of 2,6-dideoxy-D-*ribo*-hexose from digitalis glycoside.[196] Chapter 19 contains reference to other dideoxy sugar glycosides.

Sialic acids continue to gain prominence, and the chemical synthesis of sialic acid dimers and related matters has been treated in a short Japanese review.[197] α-D-Neuraminic acid 2→6-linked to methyl β-D-galactoside has been reported,[198] and also to 2-*N*-acetylgalactosamine through the same type of linkage, the product being glycosidically linked to threonine for use in solid phase peptide synthesis.[199] Neuraminic acid linked to the 6-position of *N*-methyldeoxynojirimycin is described in Chapter 18.

Lepidimoide (**41**), which is a plant growth promotion factor, is exuded by seeds of all plant species, but with larger amounts coming from sunflower and buckwheat.[200]

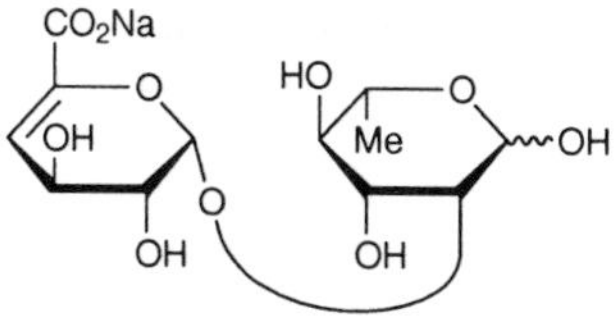

41

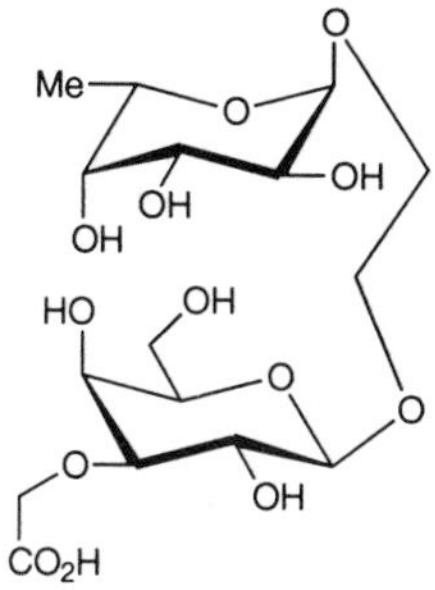

42

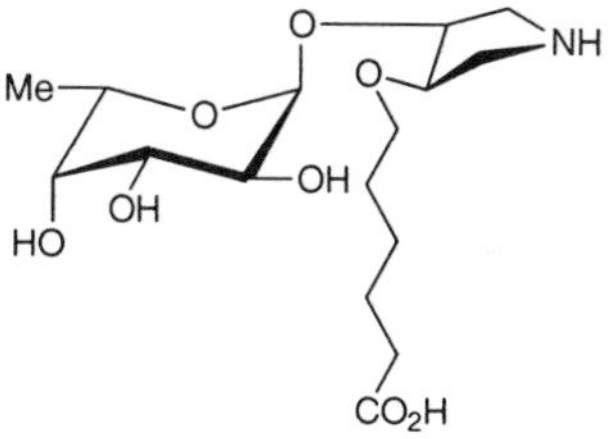

43

1.5 Disaccharides with Anomalous Linking. – Compounds **42** and **43** which mimic parts of the sialyl Le^x trisaccharide have been reported,[201] and a set of compounds represented by **44** have been produced in the course of new studies on intramolecular glycosidation.[202] Work on this topic is covered earlier in this section (cf. Scheme 1). Diglycoside **45** was treated with di-*O*-tosyldiethyleneglycol in the presence of sodium hydride to give crown ether **46**.[203] Bolaamphiphiles are α,ω-surfactants which consist of hydrophobic central and hydrophilic end groups; compounds of set **47** have been made by glycosylation of the corresponding α,ω-diols.[204] Compound **47a**, containing the unusual amidine component, has been prepared and is a good inhibitor of α-mannosidase.[205]

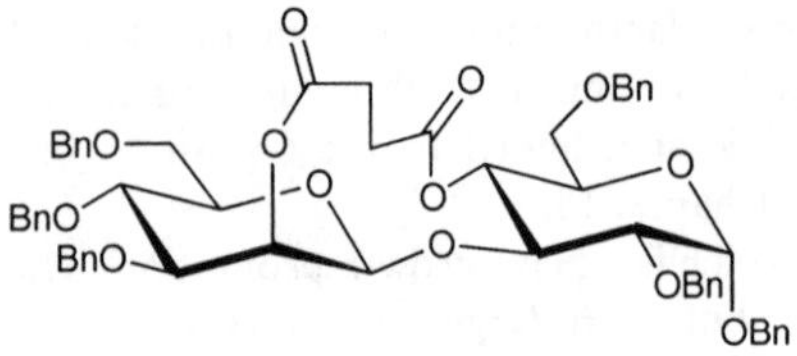

44

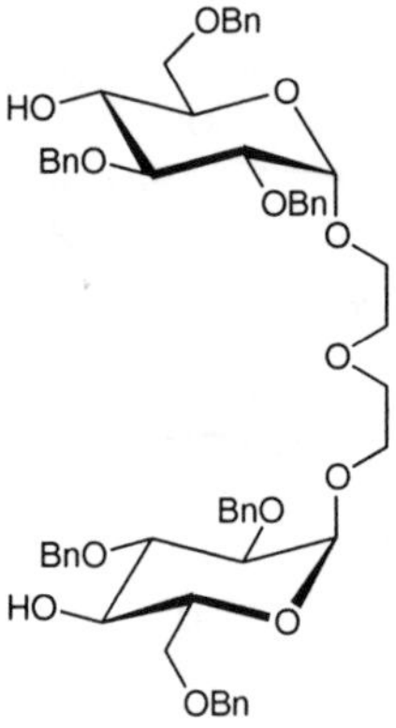

45

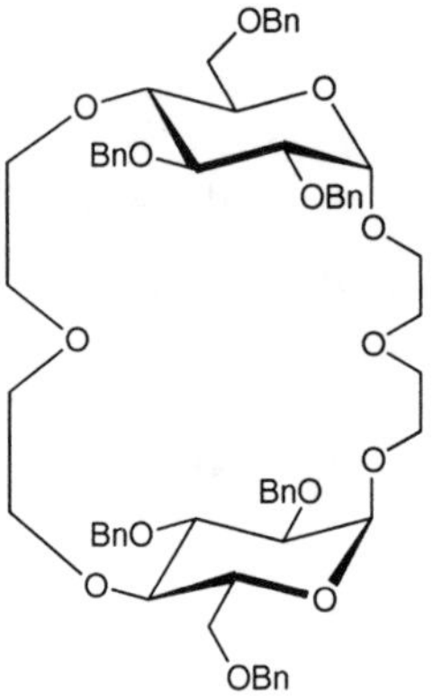

46

RO—(CH₂)ₙOR

47 n = 6–12,16

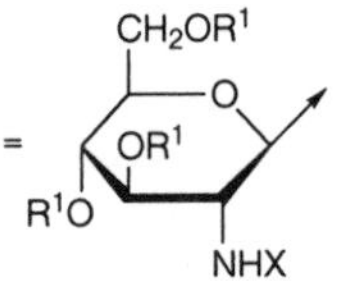

$R^1 = Ac, X = COAll$

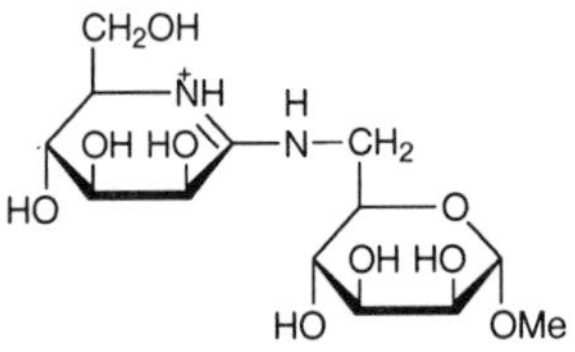

47a

1.6 Reactions, Intermolecular Complexation and other Features of Glycosides. –
A theoretical modelling of the mechanism of hydrolysis of methyl D-
glucopyranoside has been reported,[206] and anomeric pairs of *n*-pentenyl
glycosides have been concurrently hydrolysed, following activation with *N*-
bromosuccinimide, to give insight into the relative rates of anomer reactions in
the absence of strong acids. *O*-Protecting groups, for example cyclic acetals, were

shown to exert a profound effect on the rates of reactions.[207] The kinetics of the acetolysis and concurrent anomerizations of ethyl tetra-*O*-acetyl-α- and β-D-glucopyranoside induced by use of acetic anhydride, acetic acid and sulfuric acid have been reported. Reactions were followed by gas chromatography of the products, and their mechanisms were discussed.[208] Boron trifluoride etherate in the presence of borane.methyl sulfide or borane.amine complexes cause reductive cleavage of glycosidic linkages of fully methylated monosaccharides or polysaccharides, and reductive cleavage analysis of the latter is shown to be particularly straightforward by this method.[209] *p*-Methoxyphenyl glycosides may be deprotected by anodic oxidation, and the reaction has been studied in the cases of several α-mannosides, -glucosides and -galactosides including di- and tri-saccharide derivatives; yields were in the range 74-100%.[210] Enzymic hydrolysis of an *N*-acetylneuraminic acid α-glycosides with bacterial sialidase occurs with retention of anomeric configuration.[211]

The 2-deoxygalactosides **48** in dichloromethane on treatment with three equivalents of boron trifluoride etherate resulted in the formation of the tricyclic compound **49** in 65% yield. The reaction, repeated in the presence of five equivalents of the Lewis acid, however, gave the more expected β-*C*-glycoside **50** (Scheme 3).[212]

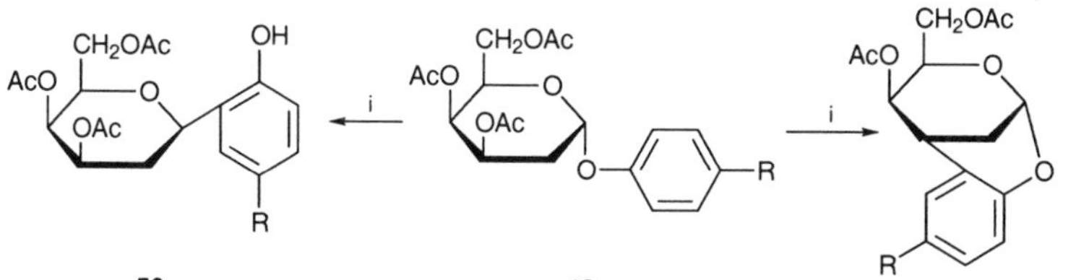

50

48

49 R = Me, Et, OMe

Reagents: BF$_3$·Et$_2$O

Scheme 3

It is increasingly being recognized that glycosides can be complexed by various macrocyclic species, and several references to this are given in the boronic acid section of Chapter 17. A related paper has described a complex receptor comprising a benzo-crown ether and an aryl boronic acid group which binds *p*-nitrophenyl β-D-glucopyranoside and transports it through liquid organic membranes at pH 11. The efficiency of transport was higher than with a mixture of the appropriate benzcrown ether and phenylboronic acid.[213] Long chain alkyl β-D-glucopyranosides and maltopyranosides with C$_8$, C$_{10}$ and C$_{12}$ alkyl groups can be converted into micelles which have been studied alone and as complexes with biomolecules, for example porphyrins, bilirubins and glycolipids.[214] The use of zinc porphorin as a synthetic receptor of long chain alkyl glucosides, mannosides and galactosides has been reported,[215] and a bis-8-quinolyl porphy-rin.zinc complex exhibits strong affinity for octyl β-D-glucopyranoside.[216] In parallel work Diederich and co-workers have examined the complexation of octyl glycosides by compound **51**[217] and also a set of chiral binaphthylcyclophanes, for example **52**.[218]

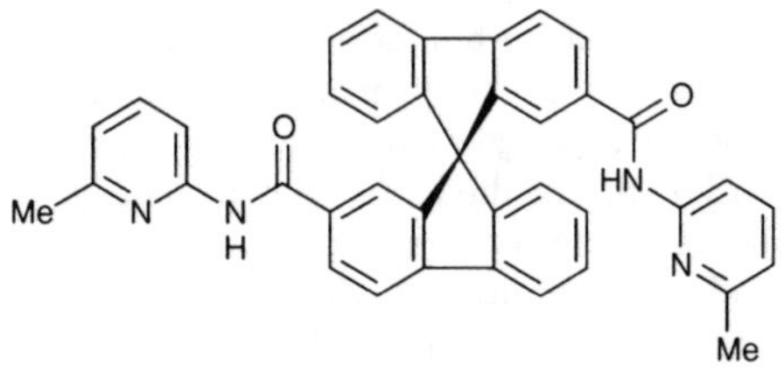

51

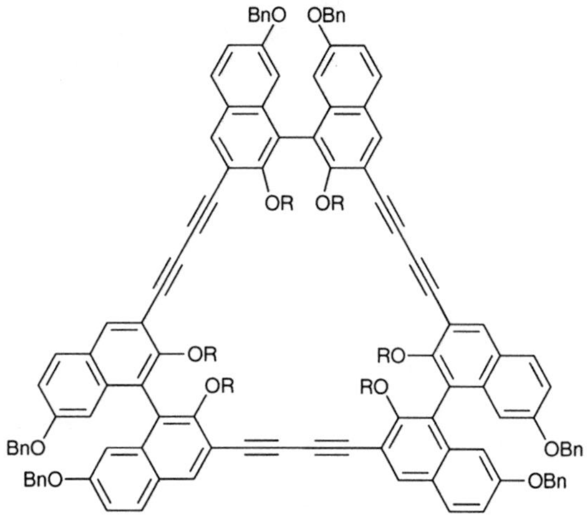

52

2   S- and *Se*-Glycosides

A novel route to *S*-alkyl and *S*-aryl thioglycosides involves treatment of free sugars protected at all positions except *O*-1 with corresponding disulfides in the presence of triphenylphosphine and boron triflouride etherate; the anomeric configurations of the products were widely variable.[219] Treatment of tri-*O*-acetyl-α-L-fucopyranosyl chloride with ethanethiol and its sodium salt afforded three ethyl α-thioglycosides carrying two ester groups, each in approximately 25% yield, which were used as starting materials for preparing the 1,2- 1,3- and 1,4-linked difucosyl disaccharides.[220] Compound **53** has been made as a spin labelled glycolipid analogue, together with several *O*-linked and *C*-linked glycosidic analogues,[221] and compound **54** together with its α-anomer has been prepared using the *S*-tributylstannyl thioglycerol derivative.[222] Compound **55** was made from the *p*-(aminobenzyl)thioglycoside for affinity chromatography purposes,[223] and the heterocyclic glycosides **56** and **57** were prepared from the acetohalogen sugars and the 2-thiols.[224]

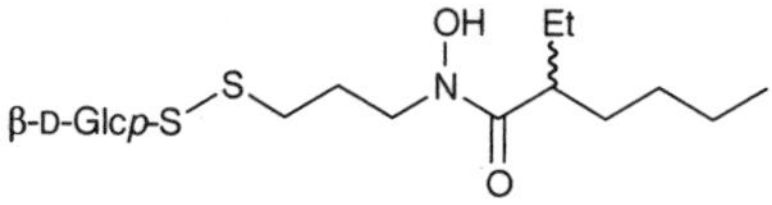

53

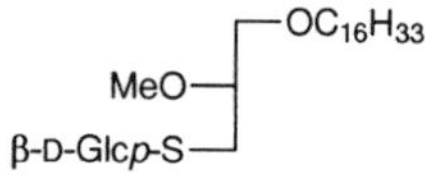

54

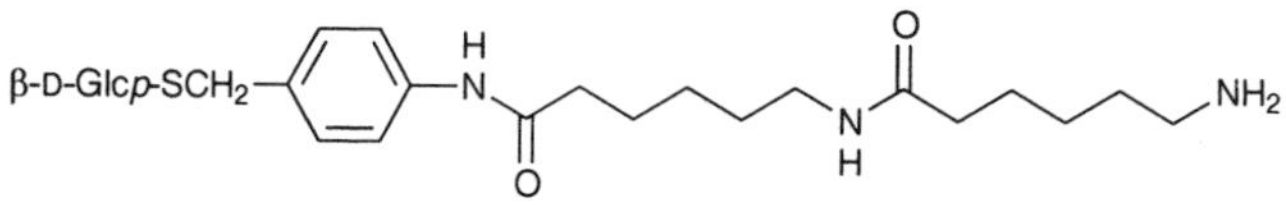

55

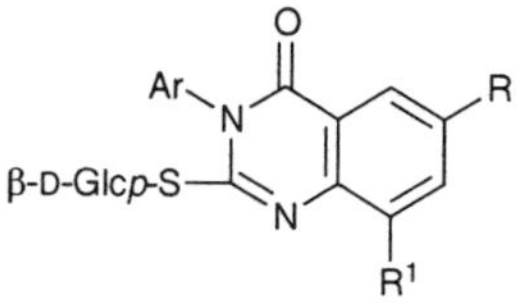

56

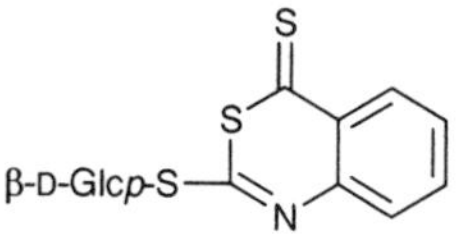

57

In the field of sialic acids compound **58** has been derived by appropriate additions to the 2,3-ene and appears to be a potentially important sialylating agent giving almost 100% α-selectivity; several appropriate disaccharides and trisaccharides were made.[225] From the 3-deoxy-2-acetylthio derivative several *S*-alkyl thioglycosides have been made by treatment with the alkyl bromides in alkaline conditions.[226] Synthesis of the thiocoumaryl α-glycosides **59** was effected to afford potential fluorogenic substrates for sialidases.[227]

The number of disaccharide derivatives involving sulfur linkage between the sugars is increasing, and compounds **60** were made following the addition of a 1-thiosugar to levoglucosenone and tested as inhibitors of α-L-fucosidase.[228] The ganglioside analogue **61** has been described[229] as have the lactosyl ceramide analogue **62** (together with several derivatives),[230] and the bis-saccharide compounds **63** containing the *O*-protected thiohydroximate linkage.[231]

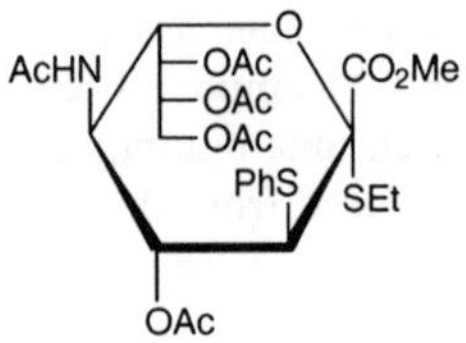

58

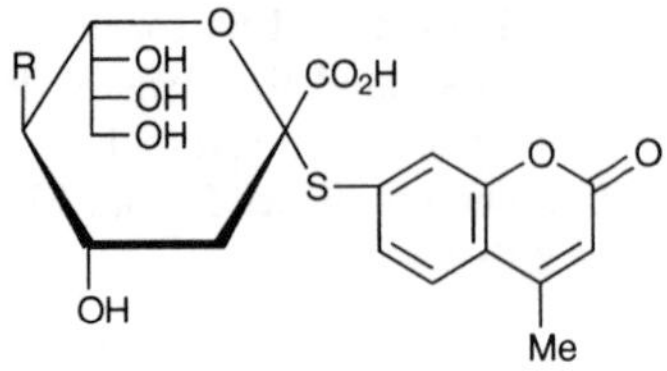

59 R = NHAc, NHCOCH$_2$OH, OH

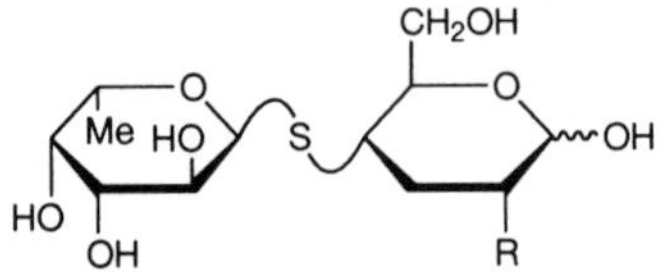

60 R = OH, NHAc

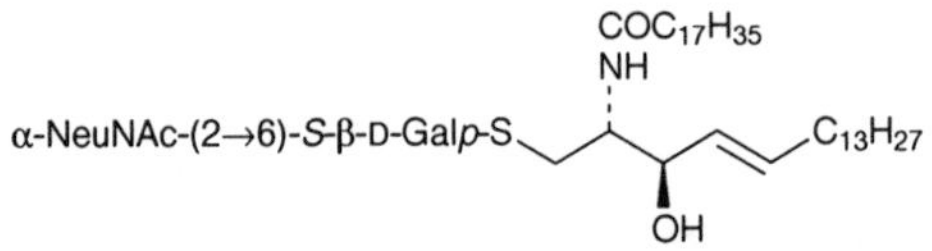

α-NeuNAc-(2→6)-S-β-D-Galp-S

61

β-D-Galp-(1→4)-S-β-D-Glcp-O

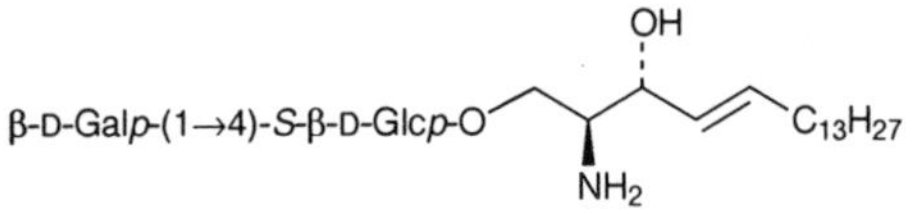

62

The phenylthio groups at anomeric positions of mono- and oligo-saccharide compounds have been cleaved hydrolytically using N-bromosuccinimide as activator.[232] Such activation can be dependent on the size of the anomeric leaving groups as is evidenced by the selective activation of the ethylthio group by iodonium dicholidine perchlorate in the presence of an analogous thioglycoside carrying the (dicyclohexyl)methyl S-substituent.[233] Thioglycosides may be activated as glycosyl donors by N-bromosuccinimide used with salts of strong acids such as t-butylammonium triflate or perchlorate in acetonitrile, which lead it to afford β-selectivity with either 2-O-acyl or 2-O-benzyl groups present in the donors. On the other hand, lithium perchlorate or lithium nitrate afford α-products preferentially when the donors have 2-O-benzyl substituents.[234] In the thiofuranoside series glycosylation of S-phenyl 3,5-di-O-benzyl-1-thio-β-D-ribofuranoside with *Se*-phenyl 2-*O*-benzoyl-3,5-di-*O*-benzyl-1-seleno-β-D-ribofurano-side occurs with good chemoselectivity showing that the selenoglycoside may be activated preferentially in the presence of the phenyl thioglycoside notwith-standing its having a disarming group at C-2.[235]

Certain thioxylosides are good substrates for a glycosyl transferase *in vivo*, and on oral administration to rats elicit a large increase in plasma concentrations of glycosaminoglycans. The same compounds have important anti-thrombotic activity in rats.[236]

Partial deacetylation of thioglycoside peracetates during synthesis has been reported above and many disaccharides and oligosaccharides mentioned in Chapters 3 and 4 have been made by use of such compounds.

3   *C*-Glycosides

3.1   Pyranoid Compounds. – From the relatively simple compound **64** an extensive range of *C*-glycosides with variations within the sugar and aglycon were produced.[237] The related compound **65** was made by treatment of 4,6-*O*-benzylidene-D-glucose with nitroethane in the presence of DBU and 2-hydroxypyridine used as a 1,3-proton transfer catalyst. Nitropropane led to the analogous compound **66** together with its epimer.[238] Reaction of the corresponding glycosyl chloride with 2-methyl-1-(trimethylsilyloxy)propene gave good yields of compounds **67** with αβ ratio 1:10. An extensive study was made of this reaction which opens a good way of making 2-deoxy-β-*C*-glucopyranosides.[239] The coupling reaction illustrated in Scheme 4 is subject to

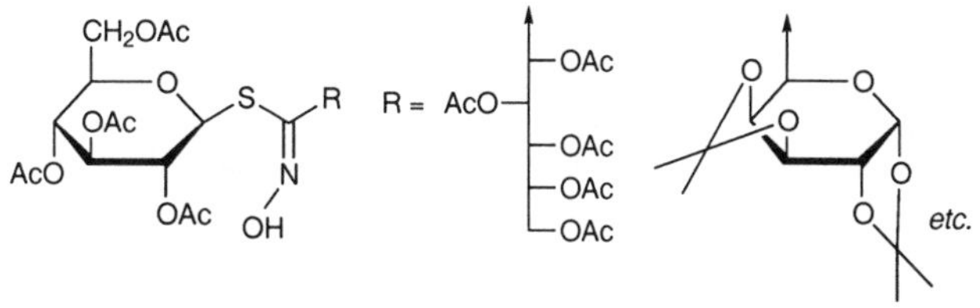

63

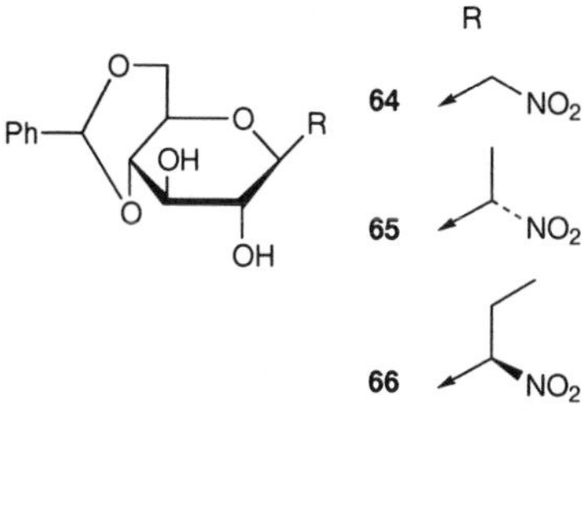

the influence of chiral ligands, dipyridyl compounds giving appreciably increased stereoselectivity with high proportions of the illustrated diastereomer.[240]

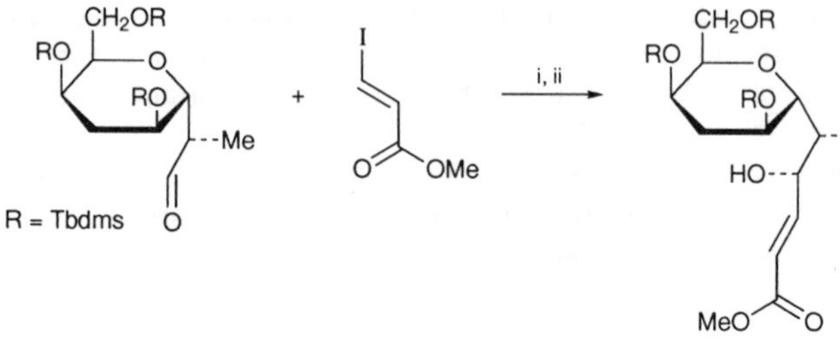

Reagents: i, NiCl$_2$, CrCl$_2$; ii, chiral ligands

Scheme 4

A photo-irradiation study has been carried out on phenacyl tetra-*O*-benzyl-glycosides in the α- and β-D-glucopyranose and α-D-mannopyranose series. Reactions led in the main to lactones together with spiro-oxetanes e.g. **70** (Scheme 5); somewhat surprisingly the α- and β-glucose derivatives reacted at similar rates.[241] The glycosylated α- and β-homomannojirimycins **68** have been prepared as potential inhibitors of *endo*-mannosidases,[242] and the esters **69** were produced by palladium, copper-mediated cross-coupling of glycosyl tributylstan-

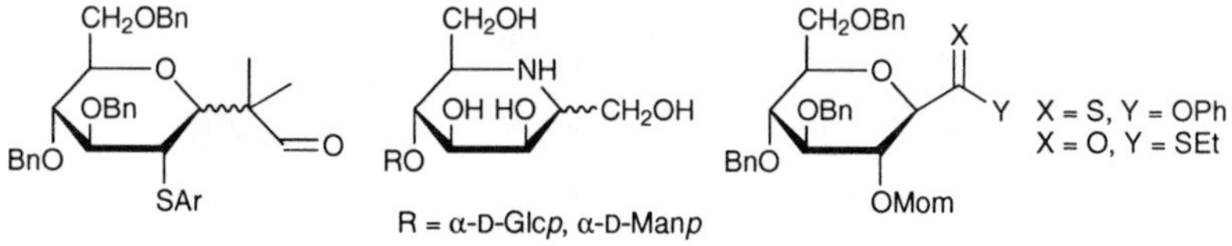

R = α-D-Glc*p*, α-D-Man*p*

X = S, Y = OPh
X = O, Y = SEt

67   **68**   **69**

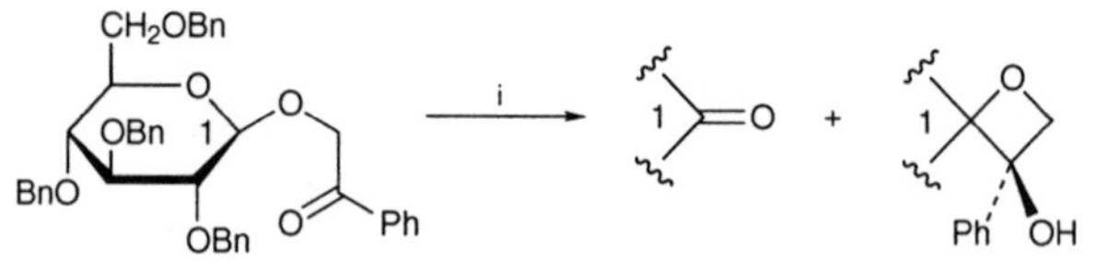

Reagent: *hv*   **70**

Scheme 5

nanes with thiono- and thiol-chloroformates.[243] Elaborate *ab initio* methods were used to produce the thioglycoside **71** and hence the cyclized *C*-glycoside **72** *en route* to 2,3-dideoxy-D-*manno*-2-octulopyranosonic acid.[244] A series of *C*-glycosidic compounds based on 1,2-cyclopropanated sugars have been produced by Simmons-Smith additions to glycal derivatives. A range of compounds related to **73** were described.[245]

The exoalkene **74**, on epoxidation, gave isomeric spiro-compounds which could be directed with nucleophiles to open in two directions leading to *C*-glycosides with a range of substituents on aglycon carbon atom and also at the anomeric centre.[246] In related work arylsulfenyl chlorides were added to the double bond of **74** to give access to a range of *C*-glycosides with various substituents (OH, OMe, CN, Ph, All etc.) at the anomeric centre.[247] Detailed

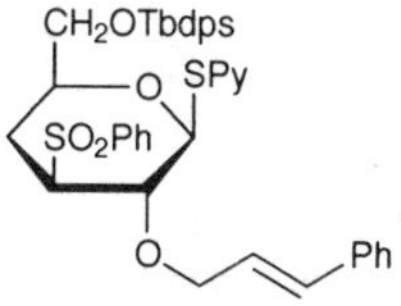

71 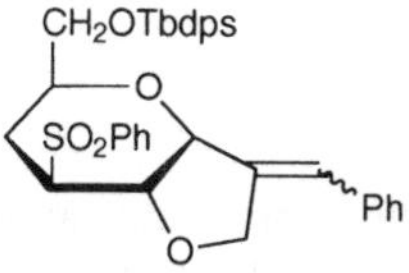   **72** 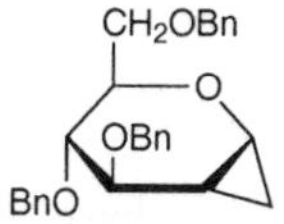   **73** 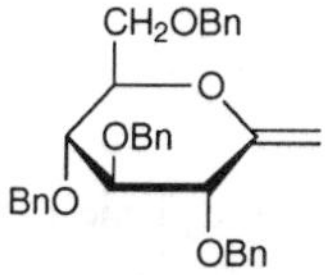   **74**

studies have been carried out on the conjugate addition of nucleophiles to various *C*-glycosides having unsaturated aglycons which occurs with treatment with methyl lithium and lithium bromide followed by fluoride. Compound **75** was converted into **76** with almost complete specificity. Several nucleophilic additions were examined and the modes of addition were switchable according to whether α- or β-chelation control was used.[248] Radical cyclization of compound **77** in the presence of acrylonitrile gave the α-*C*-glycoside **78** in 75% yield; appreciable related work was reported.[249] Epoxidation of allyl *C*-α-D-glucopyranoside gave compounds which were inhibitors of a bacterial glucosyltransferase.[250] *C*-Allyl-β-D-galactopyranoside, made by an allylmagnesium

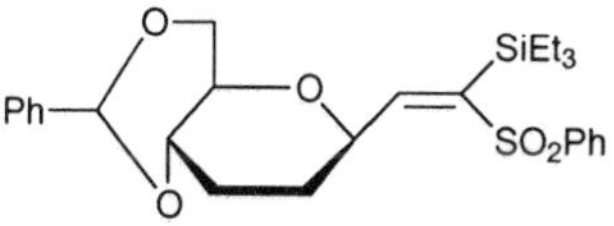

75 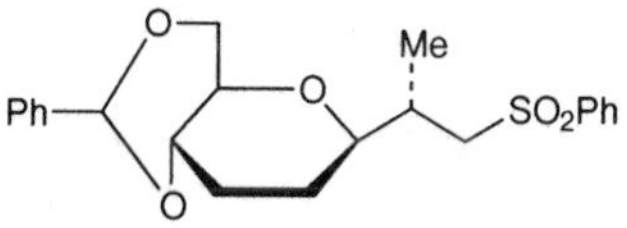   **76**

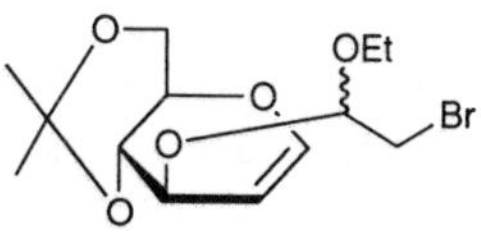

77

bromide treatment of acetobromogalactose, was used to make the anomalously linked disaccharide compound **79**, a Wittig reaction being used for the key bond-forming reaction.[251]

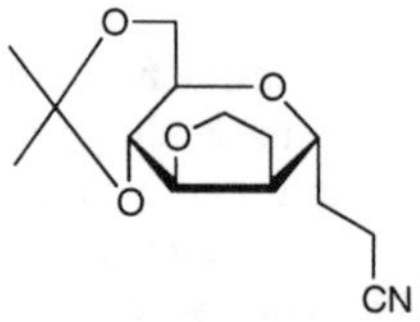

Acetylenic *C*-glycosides have attracted appreciable attention. Reaction of the appropriate (tributylstannyl)acetylene with the corresponding glycosyl chloride afforded compound **80** which was readily converted into the hexacarbonyldicobalt complex **81** and thence the β-anomer of **80**. The *C*-glycosides gave access to a range of related compounds.[252] The 2,3-Wittig rearrangement conducted with butyllithium on compound **82** afforded **83** with greater than 99% diastereoisomeric selectivity, and hence alkynes of general type **84** were produced.[253] Vasella and his co-workers have conducted very elegant work in this area leading to acetylenic-linked oligosaccharide analogues, the key step being outlined in Scheme 6.[254] Subsequent work leading to oligomeric compounds of this type was based on further couplings of related kinds.[255, 256]

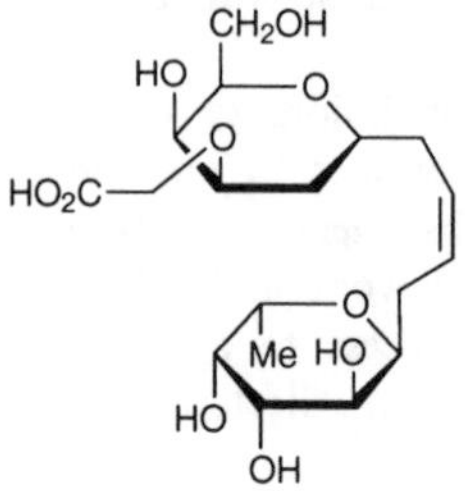

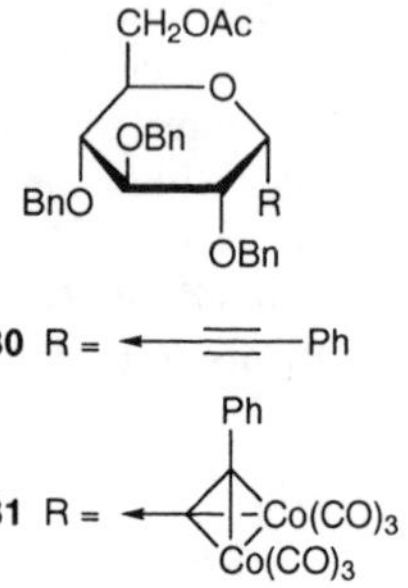

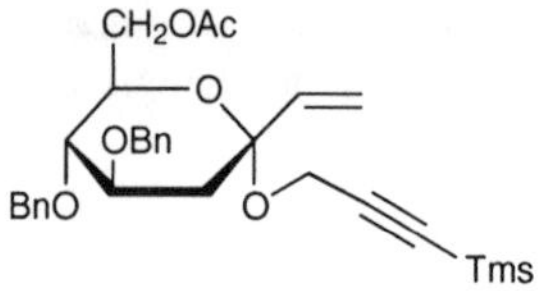

C-Aryl compounds continue to attract attention and a short review has appeared on the preparation of natural products with this structural feature.[257] A synthesis of an aryl *C*-glycoside which occurred by glycosyl migration from *O* to *C* is illustrated in Scheme 7.[258] See Scheme 3 for a related example. A more novel route involving the aromatization of a quinol ketal glycoside is shown in Scheme 8 which also illustrates a rearrangement reaction.[259] Attempted free radical bromination of tetra-*O*-acetyl-β-D-glucopyranosylbenzene with *N*-bromosuccin-

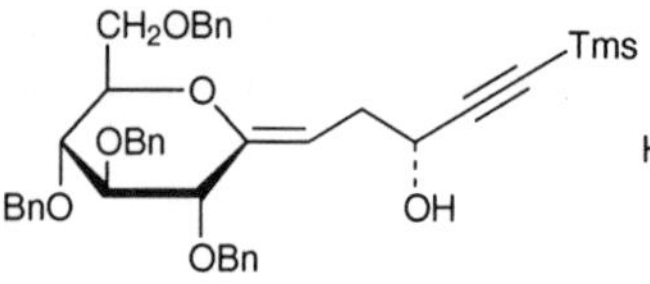

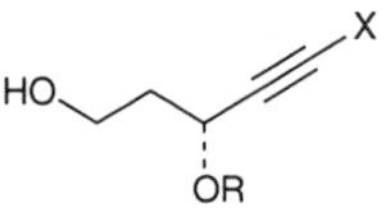

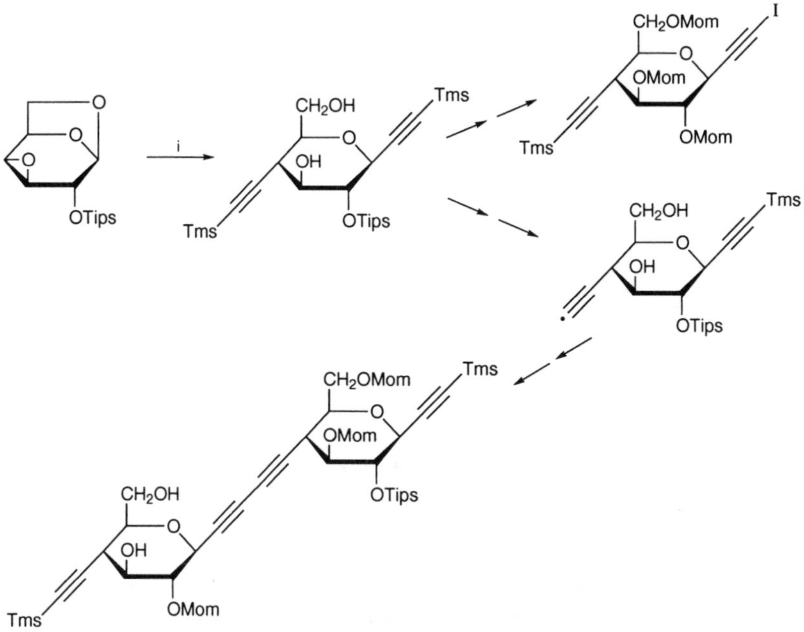

Reagents: i, TmsC≡CLi, Et₂AlCl

Scheme 6

imide gave a low yield of mixed products. However, when the reaction was repeated using bromine under light and in moist carbon tetrachloride, the ketose derivative **85** was produced very efficiently. Analogous reactions occurred in the α-glucose and β-mannose series.[260]

Reaction of tri-*O*-acetyl-D-glucal with phenyl alkyl ethers in the presence of tin(IV) chloride gave predominantly 2,3-unsaturated glycosides linked through

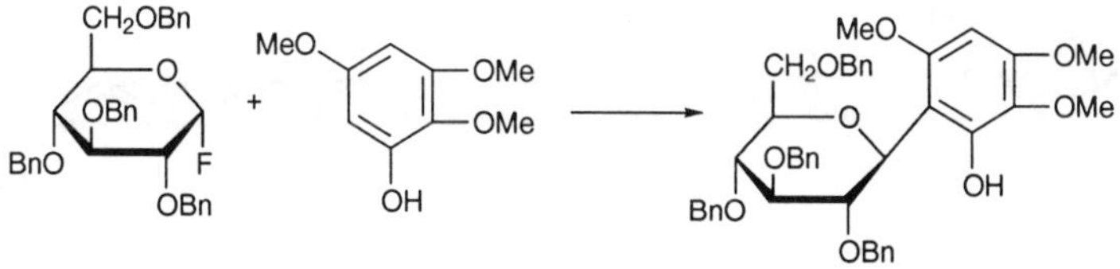

Reagents: BF₃.Et₂O

Scheme 7

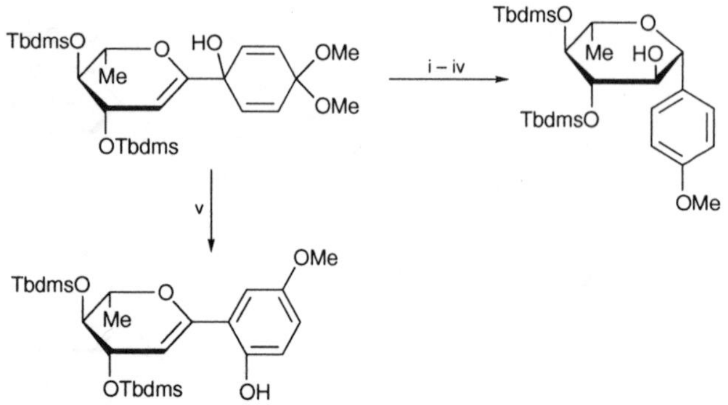

Reagents: i, DIBAL; ii, POCl₃, Py; iii, BH₃; iv, H₂O₂, NaOH; v, ZnCl₂, Et₂O

Scheme 8

the *para*-positions. Several derivatives were prepared and liquid crystals were encountered in this set of compounds when long chain alkyl substituents were incorporated.[261] Reaction of 4,6-di-*O*-benzyl 2,3-unsaturated *p-tert*-butylphenyl glycosides with a range of Grignard reagents afforded 2,3-unsaturated aryl *C*-glycosides. With palladium dichloride as catalyst the α-anomers were produced exclusively, whereas with nickel dichloride the β-products were the only products.[262] Base-catalysed addition of an imidazole derivative to tetra-*O*-benzyl-D-glucono-δ-lactone gave a ketose derivative, which on reduction, gave access to the *C*-imidazolyl glycoside **86** which inhibits an almond β-glucosidase, and the α-anomer, also produced during the work, inhibits yeast α-glucosidase.[263]

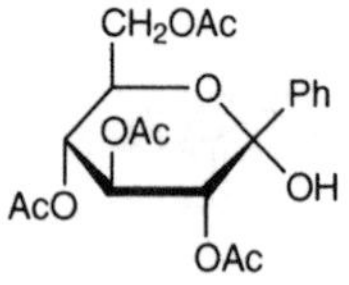

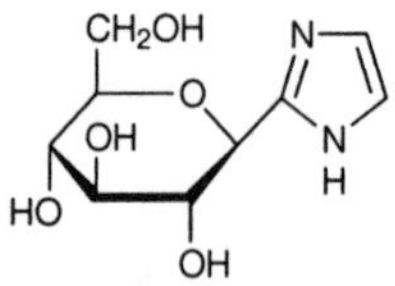

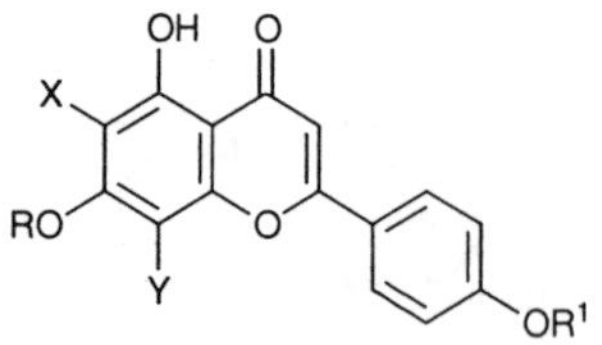

	X	Y	R	R¹
87	H	β-D-Glc*p*	H	H
88	β-D-Glc*p*	H	H	H
89	H	β-D-Glc*p*	Me	H

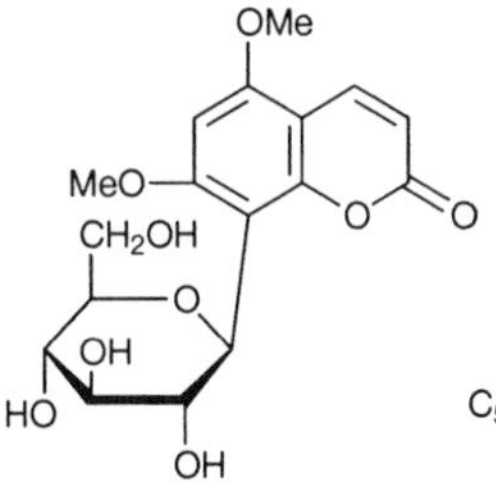

90

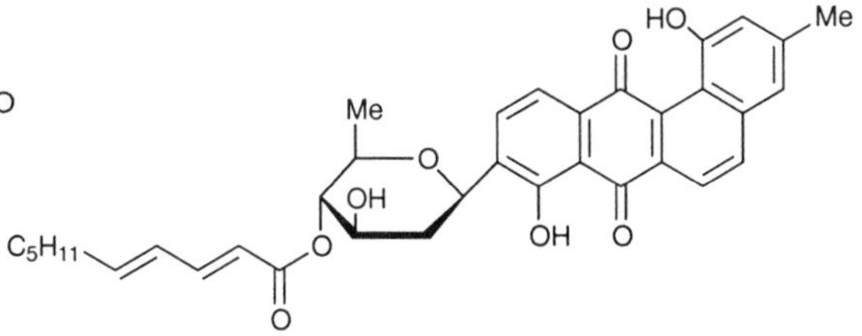

91

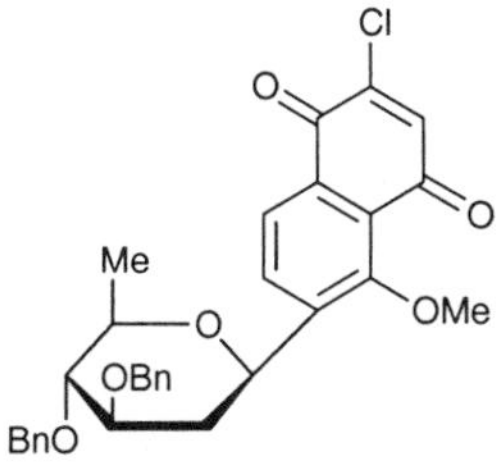

92

In the area of naturally occurring *C*-glycosides a new flavone compound chrysin 8-*C*-β-D-glucopyranoside has been characterized.[264] The trichloroacetimidate procedure, used with appropriate phenols in the presence of trimethylsilyl triflate, afforded *C*-glycosides by way of initial *O*-linked products which were thereby converted into vitexin, isovitexin and isoembegenin 87-89[265] and the coumarin analogue 90.[266] The total synthesis of angucycline antibiotic C104, (91) utilized the appropriate glycosyl acetate, the corresponding rearrangement of the *O*-glycoside giving access to the required product,[267] and the same workers made compound 92 en route to *C*-glycosyl juglones.[268] The preparation of the antifungal papulacandin D (93) in outlined in Scheme 9.[269] The component molecule 94 of the same natural product has been made as illustrated in Scheme 10 with the aromatic ring having been synthesized from acetylenic components, and in the course of the same work the vineomycinone B molecule 95 was made by a similar approach involving use of a 1,2-unsaturated acetylenic *C*-glycoside.[270]

In the field of *C*-glycosyl amino acid derivatives *C*-glycosyl radical addition to compound 96 was employed to give the diastereoisomer 97 in 88% yield exclusively. Hydrogenation gave one isomer of α-*C*-glucosyl alanine.[271] The peptide analogue 98 and its β-anomer were made efficiently from the corresponding glycosylcarboxylic acid by amine coupling.[272] Related phosphonates 99,[273] and 100[274] and the related phosphate 101[274] have been made from corresponding *C*-glycosides having simple substituted methyl aglycons.

C-Glycosidically linked disaccharides have now become of considerable importance and a review has covered the preparation of such compounds as well

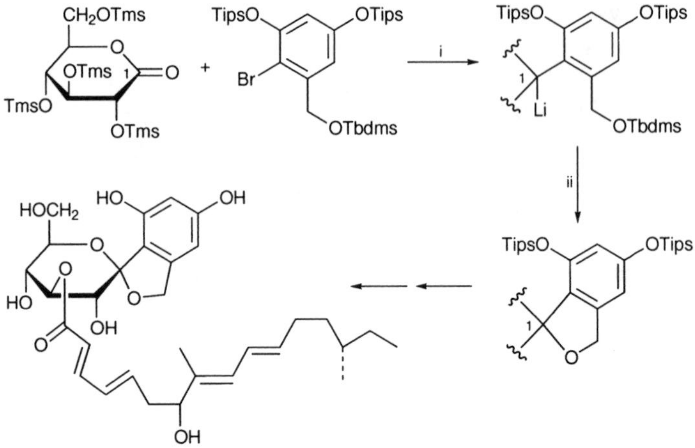

93

Reagents: i, BuLi; ii, Amberlite IR120

Scheme 9

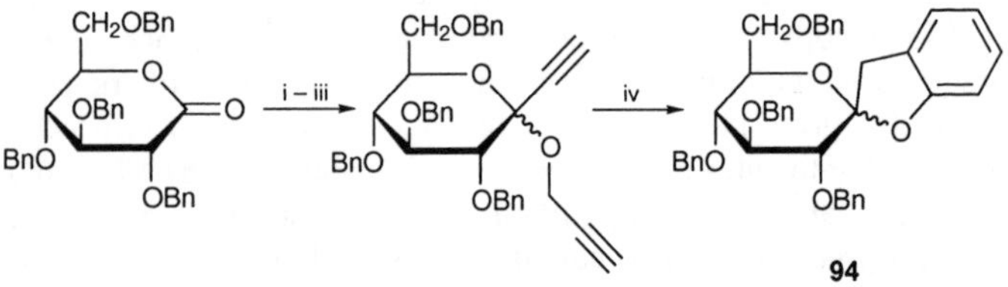

94

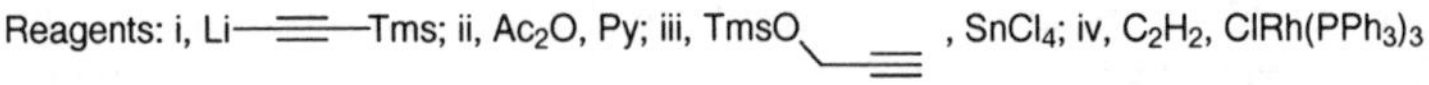

Reagents: i, Li——≡——Tms; ii, Ac$_2$O, Py; iii, TmsO⌣⌣≡ , SnCl$_4$; iv, C$_2$H$_2$, ClRh(PPh$_3$)$_3$

Scheme 10

as *C*-glycosylflavanoids.[275] The directly linked compounds **102** and **103** are examples of enzymic amplification of dihydroxy dialdehydes produced following double bond cleavage of specific cyclohexene diols.[276] In the main, disaccharide *C*-glycosidic mimics have the sugars linked by one carbon unit and, commonly, are made by reaction of glycosyl nucleophiles acting on compounds carrying *C*-formyl groups. Compound **106** reacts very readily with samarium(II) iodide and carbonyl compounds to give *C*-glycosides; thus when taken with the aldehyde **107** it afforded *C*-linked disaccharide **108** (Scheme 11) in 75% yield.[277] In related work the sulfoxide **104** was converted to the 1-lithio derivative and coupled with the appropriate 4-*C*-formyl compound to give access to the *N*-acetyllactosamine

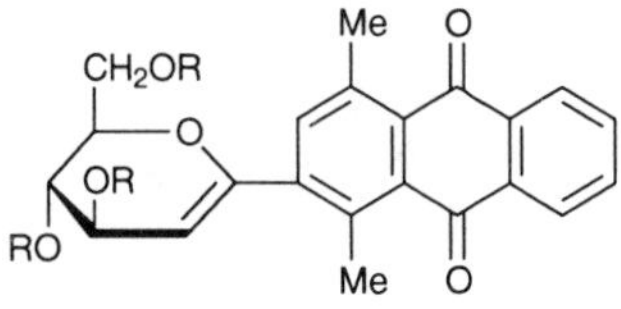

95 R = Tbdps

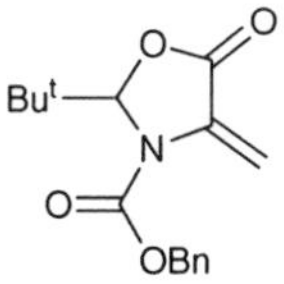

96

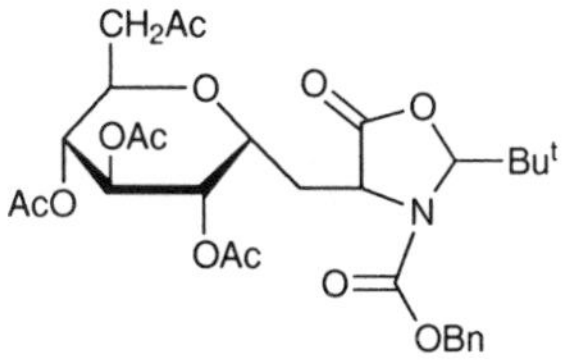

97

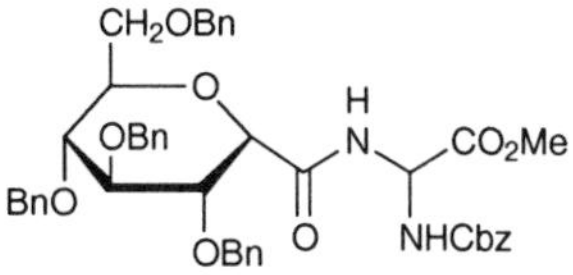

98

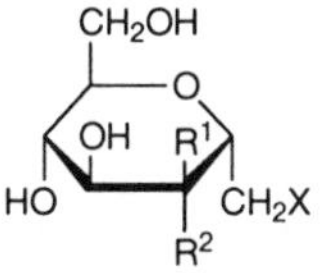

	R^1	R^2	X
99	NHAc	H	PO_3Et_3
100	H	OH	$P(O)(OH)OR^3$
101	H	OH	$CH_2OP(O)(OH)OR^3$

$R^3 = C_{16}H_{33}$

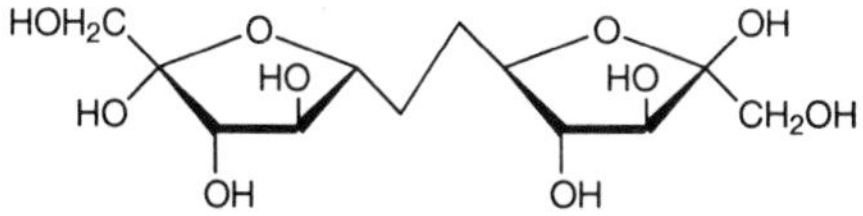

102

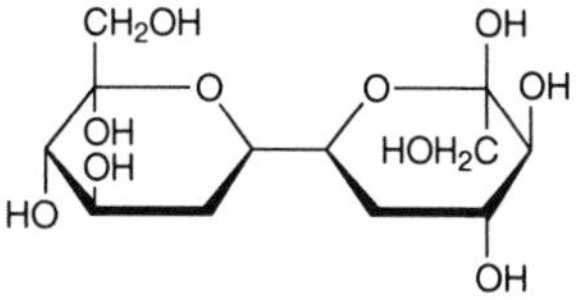

103

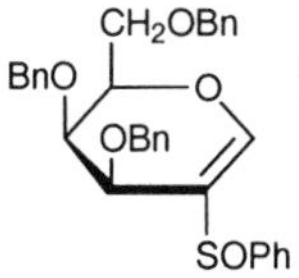

104

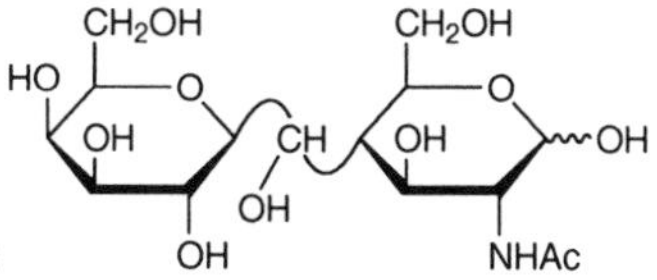

105

analogue **105**,[278] and parallel work has afforded the undecose backbone of the herbicidins in the form of the *C*-linked disaccharide **109**.[279] Two ulosyl bromides have been used in related reactions and, for example, compound 110, and a set of analogues, have been prepared.[280] Conversely compound **111** was made by addition of a carbanion, derived from an iodomethyl substituent at C-4 of a glucopyranoside derivative, at C-1 of an aldehydo sugar. From the product compounds **112** and **113** were produced and from the latter, the novel tricyclic ketoside **114** was obtained.[281]

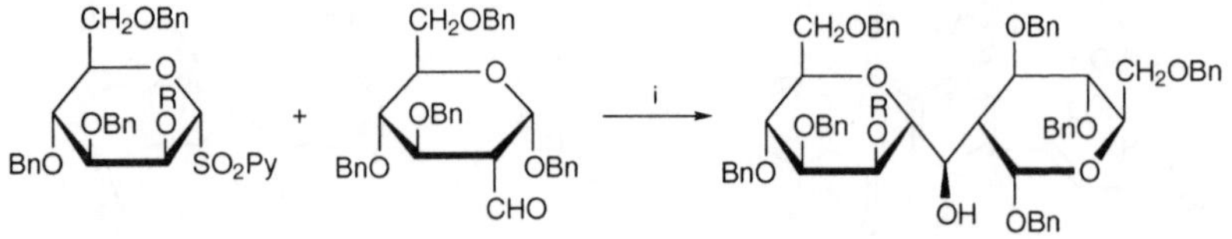

106 R = Tms   **107**   **108**

Reagents: i, SmI₂

Scheme 11

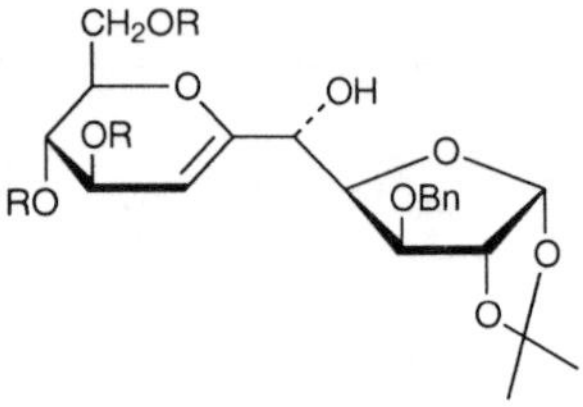

109 R = Tips

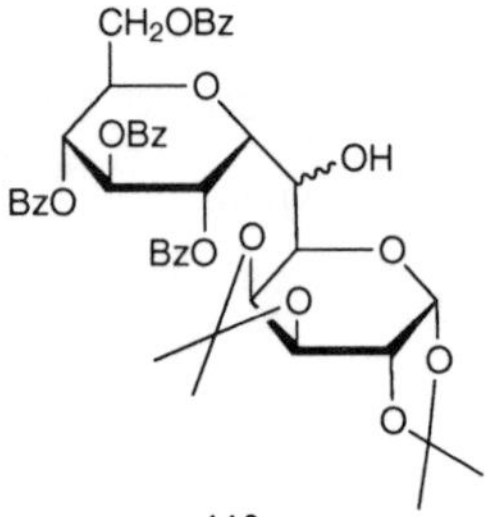

110

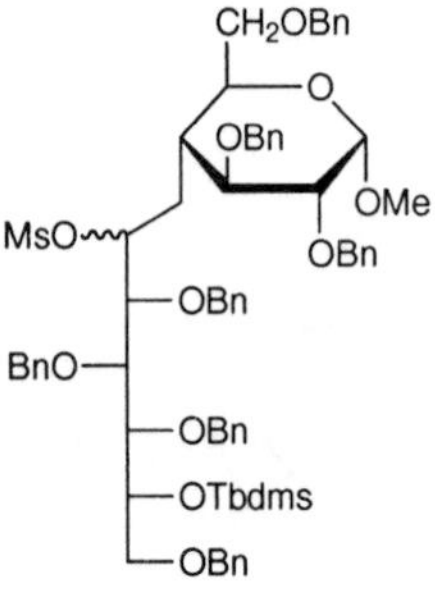

111

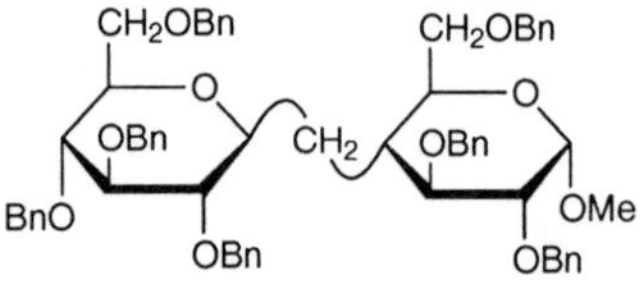

112

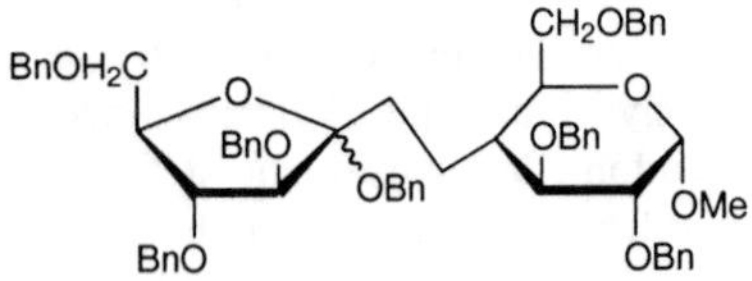

113 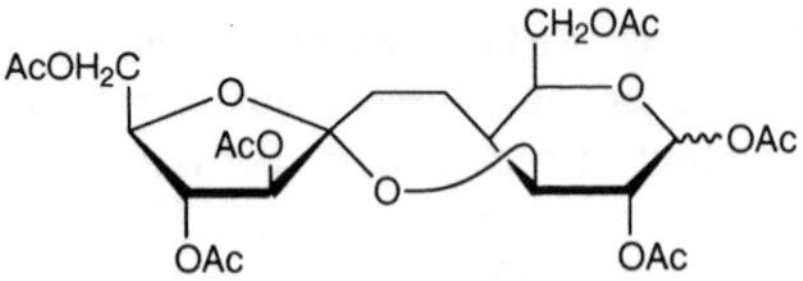   **114**

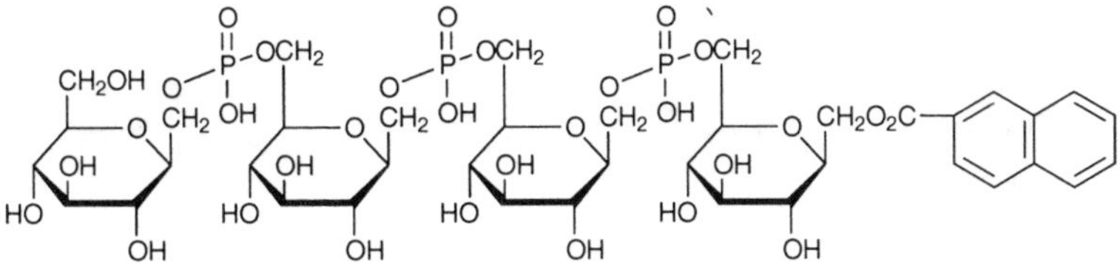

115

To indicate how the field is expanding, the doubly *C*-linked analogue of trisaccharide α-L-Fuc*p*-(1→2)-β-D-Gal*p*-(1→4)-D-GlcNAc, which is the human blood group determinant H-II, has been made with a methylene group replacing oxygen at both inter-unit positions. Several analogues were also made and NMR conformational analyses were carried out.[282] So-called carbonucleotoids having phosphate linkages between *C*-glycoside components have been investigated, and compounds such as **115** have been made for use in pharmaceutical work.[283] Continuation by Vogel of his work on carbocyclic analogues has led to the *C*-linked α-D-galactopyranosides of carbapentapyranose derivatives (Scheme 12).[284] A parallel publication has described the α-*C*-glucopyranosyl analogues.[285]

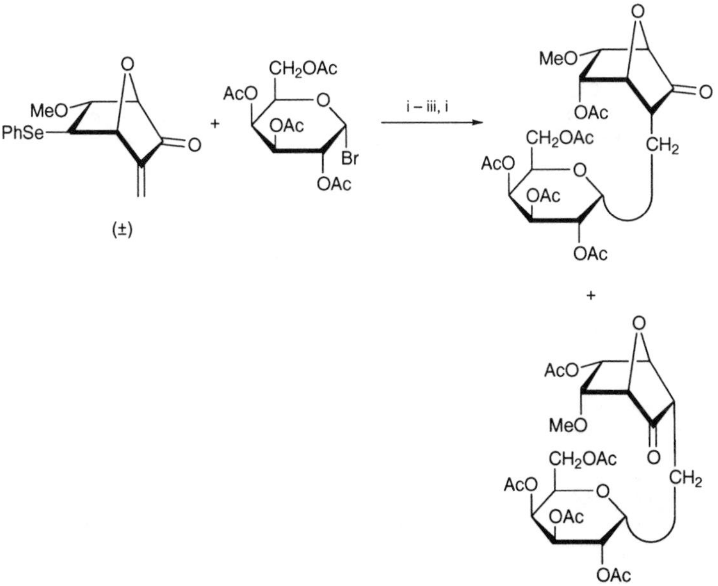

Reagents: i, Bu₃SnH, AIBN; ii, *m*-CPBA; iii, Ac₂O, NaOAc

Scheme 12

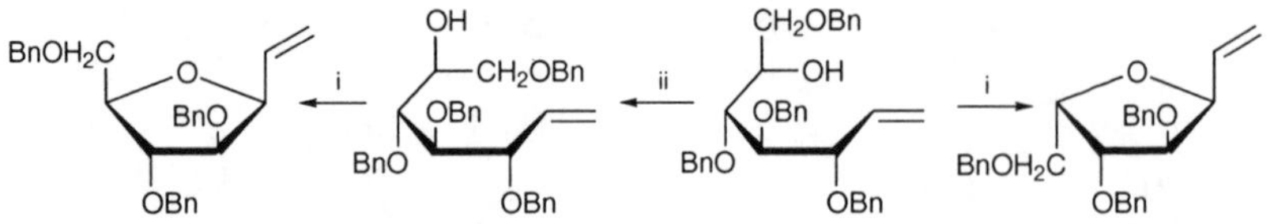

Reagents: i, Tf₂O, Py; ii, Ph₃P, DEAD

Scheme 13

3.2 Furanoid Compounds. – A review has appeared on the synthesis of gilvocarcin *C*-glycosides, for example compound **116**.[286] The synthesis of a simple furanoid aryl *C*-glycoside has been achieved by reaction of a glycosyl chloride with diaryl cadmiums,[287] and a further report has appeared on the intramolecular formation of aryl *C*-glycosides by, for example, the cyclization of methyl 2-*O*-benzyl-3,5,6-tri-*O*-methyl-D-glucofuranoside which, in the presence of HF, gives compound **117**.[288] A potentially very useful way of making vinyl furanosyl *C*-glycosides is illustrated in Scheme 13; the reaction seems to be applicable over a range of appropriate compounds and gives yields upwards of 80%.[289, 290] It appears to occur, unexpectedly, by the attack of the allylic oxygen atom on the

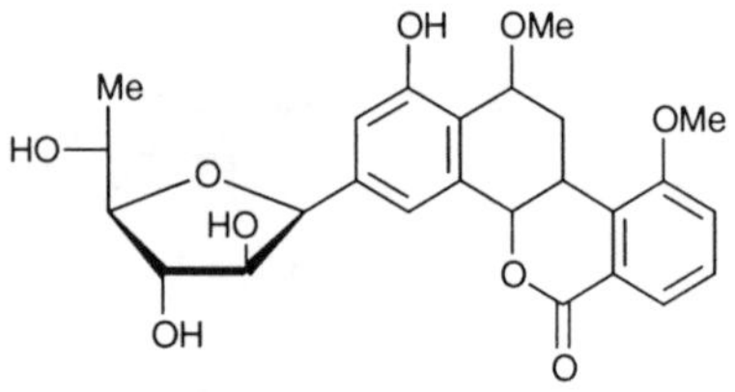

116 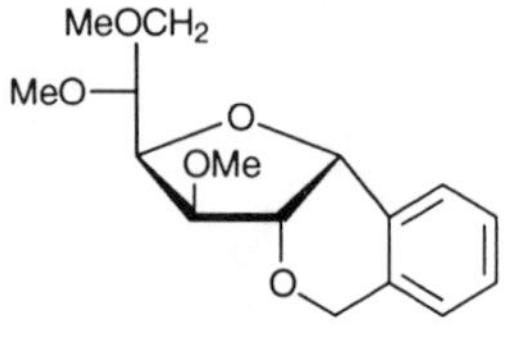   **117**

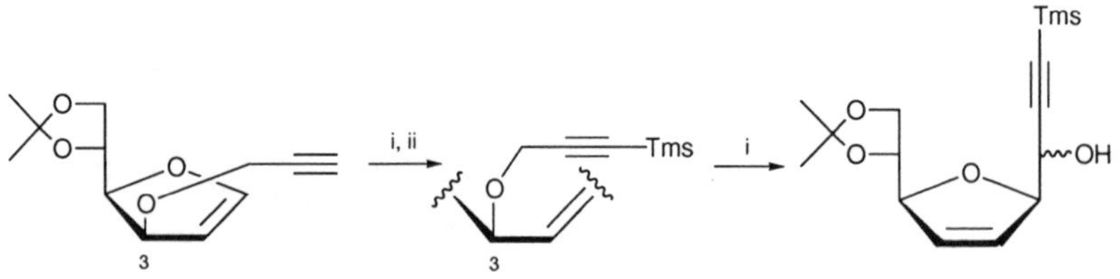

Reagents: i, BuLi; ii, TmsCl

Scheme 14

centre activated by esterification of the hydroxyl group. An appropriate allyl *C*-furanoside was used as starting material to obtain the artificial nucleotide-like compound **118** which forms a triple complex with the dinucleotide Cyt, Gua base pair.[291] Reaction of tri-*O*-benzyl-D-arabino-γ-lactone with appropriate organolithiums, followed by triethylsilane reduction, gives α-*C*-glycofuranosides including methyl and ethynyl compounds.[292]

The *C*-glycosylglycine compound **119** was made by use of the glycosyl acetate and the pyrrole **120**, the intermediate glycosyl pyrrolidone being degraded oxidatively.[293] The 2,3-Wittig rearrangement illustrated in Scheme 14 offers potential starting materials for some acetogenins.[294] Other *C*-glycosidic compounds of the furanoid type are mentioned in Chapters 19 and 20.

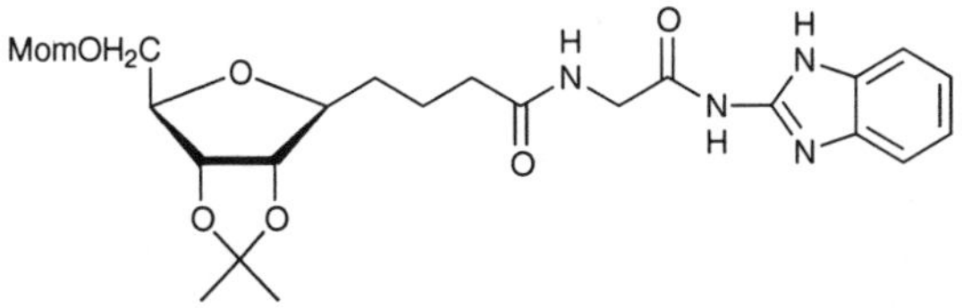

118

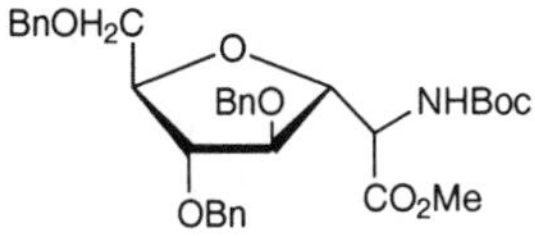

119

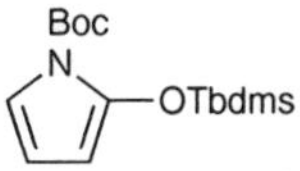

120

References

1   P. Stangier, M.M. Palcic and D.R. Bundle, *Carbohydr. Res.*, 1995, **267**, 153.

2   V. Ferrières, J.-N. Bertho and D. Plusquellec, *Tetrahedron Lett.*, 1995, **36**, 2749.

3   A.T.J.W. de Goede, F. van Rantwijk and H. van Bekkum, *Starch/Staerke*, 1995, **47**, 233 (*Chem. Abstr.*, 1995, **123**, 314 272).

4   S.K. Sharma, G. Corrales and S. Penadés, *Tetrahedron Lett.*, 1995, **36**, 5627.

5   V. Huchel, C. Schmidt and R.R. Schmidt, *Tetrahedron Lett.*, 1995, **36**, 9457.

6   F. Wilhelm, S.K. Chatterjee, B. Rattay, P. Nuhn, R. Benecke and J. Ortwein, *Liebigs Ann. Chem.*, 1995, 1673.

7   J. Mao, H. Chen, J. Zhang and M. Cai, *Synth. Commun.*, 1995, **25**, 1563.

8   J.M. Mao, H.M. Chen, J. Zhang and M.S. Cai, *Chin. Chem. Lett.*, 1994, **5**, 465 (*Chem. Abstr.*, 1995, **122**, 56 324).

9   J. Inanaga, Y. Yokoyama and T. Hanamoto, *Mem. Fac. Sci., Kyushu Univ., Ser. C*, 1994, **19**, 125 (*Chem. Abstr.*, 1995, **122**, 56 336).

10   Z.J. Li, H.Q. Huang and M.S. Cai, *Chin. Chem. Lett.*, 1994, **5**, 477 (*Chem. Abstr.*, 1995, **122**, 214 357).

11   S.G. Polonik, A.M. Tolkach and N.I. Uvarova, *Zh. Org. Khim.*, 1994, **30**, 248
      (*Chem. Abstr.*, 1995, **122**, 187 894).

12   P.I. Kitov, Y.E. Tsvetkov, L.V. Backinowsky and N.K. Kochetkov, *Izv. Akad.
      Nauk, Ser. Khim.*, 1993, 1992 (*Chem. Abstr.*, 1995, **123**, 33 532).

13   S.-i. Hashimoto, K. Umeo, A. Sano, N. Watanabe, M. Nakajima and S. Ikegami,
      Tetrahedron Lett., 1995, **36**, 2251.

14   S. Aoki, H. Kondo and C.-H. Wong, *Methods Enzymol.*, 1994, **247**, 193 (*Chem.
      Abstr.*, 1995, **123**, 83 893).

15   D.L. Boger, S. Teramoto and J. Zhou, *J. Am. Chem. Soc.*, 1995, **117**, 7344.

16   Y. Gama, Y. Kawabata and I. Kusakabe, *Yukagaku*, 1994, **43**, 520 (*Chem. Abstr.*,
      1995, **122**, 187 907).

17   G. Balavoine, S. Berteina, A. Gref, J. Fischer and A. Lubineau, *J. Carbohydr.
      Chem.*, 1995, **14**, 1217.

18   B.M. Heskamp, G.H. Veeneman, G.A. van der Marel, C.A.A. van Boeckel and J.H.
      van Boom, *Tetrahedron*, 1995, **51**, 5657.

19   L. Yan and D. Kahne, *Synlett.*, 1995, 523.

20   S. Hosono, W.-S. Kim, H. Sasai and M. Shibasaki, *J. Org. Chem.*, 1995, **60**, 4.

21   M. Mickeleit, T. Weider, K. Buchner, C. Geilen, J. Mulzer and W. Reuth, *Angew.
      Chem. Int. Ed. Engl.*, 1995, **34**, 2667.

22   G. Bohm and H. Waldmann, *Tetrahedron Lett.*, 1995, **36**, 3843.

23   M. Palme and A. Vasella, *Helv. Chim. Acta*, 1995, **78**, 959.

24   B. Werschkun, W.A. König, V. Křen and J. Thiem, *J. Chem. Soc., Perkin Trans. 1*,
      1995, 2459.

25   A. Lubineau, H. Bienayame and Y. Queneau, *Carbohydr. Res.*, 1995, **270**, 163.

26   I. Paterson and M.D. McLeod, *Tetrahedron Lett.*, 1995, **36**, 9065.

27   Y. Du and F. Kong, *J. Carbohydr. Chem.*, 1995, **14**, 413.

28   Y. Du and F. Kong, *J. Carbohydr. Chem.*, 1995, **14**, 341.

29   S. Hanessian, Can. Pat. Appl. CA 2,100,821, 1995 (*Chem. Abstr.*, 1995, **123**, 340
      752).

30   A.V. Lyubeshkin, M.V. Anikin, A.V. Gurina and Y.L. Sebyakin, *Zh. Org. Khim.*,
      1994, **30**, 567 (*Chem. Abstr.*, 1995, **122**, 187 905).

31   G. Belluci, C. Chiappa and F. D'Andrea, *Tetrahedron: Asymm.*, 1995, **6**, 221.

32   N. Moitessier, F. Chrétien, Y. Chapleur and C. Humeau, *Tetrahedron Lett.*, 1995,
      36, 8023.

33   D.J. Jenkins and B.V.L. Potter, *J. Chem. Soc., Chem. Commun.*, 1995, 116.

34   G. Vic and D.H.G. Crout, *Carbohydr. Res.*, 1995, **279**, 315.

35   C. Panintrarux, S. Adachi, Y. Araki, Y. Kimura and R. Matsuno, *Enzyme Microb.
      Technol.*, 1995, **17**, 32 (*Chem. Abstr.*, 1995, **122**, 187 928).

36   A. Trincone, R. Improta, R. Nucci, M. Rossi and A. Gambacorta, *Bicatalysis*, 1994,
      10, 195 (*Chem. Abstr.*, 1995, **122**, 265 810).

37   M.A. Hassan, F. Ismail, S. Yamamoto, H. Yamada and K. Nakanishi, *Biosci.
      Biotech. Biochem*, 1995, **59**, 543 (*Chem. Abstr.*, 1995, **123**, 83 860).

38   D. Colombo, F. Ronchetti, A. Scala, I.M. Taino, F.M. Albini and L. Toma,
      Tetrahedron Lett., 1995, **36**, 4865.

39   H. Hashimoto, C. Katayama and K. Ikura, *Trends Glycosci. Glycotechnol.*, 1995, **7**,
      149 (*Chem. Abstr.*, 1995, **123**, 314 287).

40   V.I. Maksimov and D.V. Zashikhina, *Biotekhnologiya*, 1994, **2**, 26 (*Chem. Abstr.*,
      1995, **122**, 31 799).

41   P. Wang, B.D. Martin, S. Parida, D.G. Rethwisch and J.S. Dordick, *J. Am. Chem.
      Soc.*, 1995, **117**, 12885.

42   J.C. Castro-Palomino and R.R. Schmidt, *Tetrahedron Lett.*, 1995, **36**, 5343.

43   K. Matsubara and T. Mukaiyama, *Pol.J. Chem.*, 1994, **68**, 2365 (*Chem. Abstr.*, 1995, **122**, 265 850).

44   J.C. Castro-Palomino and R.R. Schmidt, *Tetrahedron Lett.*, 1995, **36**, 6871.

45   J.E. Yule, T.C. Wong, S.S. Gandhi, D. Qiu, M.A. Riopel and R.R. Koganty, *Tetrahedron Lett.*, 1995, **36**, 6839.

46   K. Sato and A. Yoshitomo, *Chem. Lett.*, 1995, 39.

47   J.-N. Bertho, V. Ferrières and D. Plusquellec, *J. Chem. Soc., Chem. Commun.*, 1995, 1391.

48   R.T. Brown, B.W. Fox, J.A. Hadfield, A.T. McGown, S.P. Mayalarp, G.R. Pettit and J.A. Woods, *J. Chem. Soc., Perkin Trans. 1*, 1995, 577.

49   M. Adamczyk, J.R. Fishpaugh, J.C. Gebler and K.J. Heuser, *Org. Prep. Proced. Int.*, 1994, **26**, 706 (*Chem. Abstr.*, 1995, **122**, 161 180).

50   C.-J. Sun, Y.-G. Wang, W.-F. Hu, Z.-C. Chen, P. Xue, J.-G. Zhang, B.-L. Xu, Y.R. Zhao and M.-L. Wang, *Hecheng Huaxue.*, 1994, **2**, 246 (*Chem. Abstr.*, 1995, **122**, 314 953).

51   K. Ramu and J.K. Baker, *J. Med. Chem.*, 1995, **38**, 1911.

52   G. Capozzi, C. Falciani, S. Menichetti, C. Nativi and R.W. Franck, *Tetrahedron Lett.*, 1995, **36**, 6755.

53   W.R. Roush and X.-F. Lin, *J. Am. Chem. Soc.*, 1995, **117**, 2236.

54   S.E. Mangholz and A. Vasella, *Helv. Chim. Acta*, 1995, **78**, 1020.

55   D. Oulmi, P. Maillard, J.-L. Guerquin-Kern, C. Huel and M. Momenteau, *J. Org. Chem.*, 1995, **60**, 1554.

56   Q. Wang and S.G. Withers, *J. Am. Chem. Soc.*, 1995, **117**, 10137.

57   C.M. Reichert, C.E. Hayes and I.J. Goldstein, *Methods Enzymol.*, 1994, **242**, 108 (*Chem. Abstr.*, 1995, **122**, 265 848).

58   K. Kasai, K. Okada and N. Yamaji, *Chem. Pharm. Bull.*, 1995, **43**, 266.

59   J.M. Mao, J. Zhang, H.M. Chen and M.S. Cai, *Chin. Chem. Lett.*, 1994, **5**, 1003 (*Chem. Abstr.*, 1995, **122**, 214 352).

60   T.P. Kogan, B. Dupré, K.M. Keller, I.L. Scott, H. Bui, R.V. Market, P.J. Beck, J.A. Voytus, B.M. Revelle and D. Scott, *J. Med. Chem.*, 1995, **38**, 4976.

61   S.M. Mabic and J.-P. Lepoittevin, *Tetrahedron Lett.*, 1995, **36**, 1705.

62   J.C. Briggs, A.H. Haines and R.J.K. Taylor, *J. Chem. Soc., Perkin Trans. 1*, 1995, 27.

63   H. Müller and C. Tschierske, *J. Chem. Soc., Chem. Commun.*, 1995, 645.

64   K. Toshima, Y. Nozaki, S. Mukaiyama, T. Tamai, M. Nakata, K. Tatsuta and M. Kinoshita, *J. Am. Chem. Soc.*, 1995, **117**, 3717.

65   A. Rajca and M. Wiessler, *Carbohydr. Res.*, 1995, **274**, 123.

66   L.A. Salvador, M. Elofsson and J. Kihlberg, *Tetrahedron*, 1995, **51**, 5643.

67   L. Clary, J. Greiner, C. Santaella and P. Vierling, *Tetrahedron Lett.*, 1995, **36**, 539.

68   E. Meinjohanns, M. Meldal and K. Bock, *Tetrahedron Lett.*, 1995, **36**, 9205.

69   H. Paulsen, S. Peters, T. Biefeldt, M. Meldal and K. Bock, *Carbohydr. Res.*, 1995, **268**, 17.

70   L. Szabó, J. Ramza, C. Langdon and R. Polt, *Carbohydr. Res.*, 1995, **274**, 11.

71   G. Arsequell, J.S. Haurum, T. Elliott, R.A. Dwek and A.C. Lellouch, *J. Chem. Soc., Perkin Trans. 1*, 1995, 1739.

72   G. Arsequell, N. Sàrries and G. Valencia, *Tetrahedron Lett.*, 1995, **36**, 7323.

73   J. Kihlberg, J. Ahman, B. Walse, T. Drakenberg, A. Nilsson, C. Söderberg-Ahlm, B. Bengtsson and H. Olsson, *J. Med. Chem.*, 1995, **38**, 161.

74   J.J. Barchi Jr., P. Russ, B. Johnson, A. Otaka, M. Nomizu and Y. Yamada, *Bioorg. Med. Chem. Lett.*, 1995, **5**, 711.

75   E. Meinjohanns, A. Vargas-Berenguel, M. Meldal, H. Paulsen and K. Bock, *J. Chem. Soc., Perkin Trans. 1*, 1995, 2165.

76   S. Peters, T.L. Lowary, O. Hindsgaul, M. Meldal and K. Bock, *J. Chem. Soc., Perkin Trans. 1*, 1995, 3017.

77   M. Morita, K. Motoki, K. Akimoto, T. Natori, T. Sakai, E. Sawa, K. Yamaji, Y. Koezuka, E. Kobayashi and H. Fukushima, *J. Med. Chem.*, 1995, **38**, 2176.

78   M. Morita, T. Natori, K. Akimoto, T. Osawa, H. Fukushima and Y. Koezuka, *Bioorg. Med. Chem. Lett.*, 1995, **5**, 699.

79   K. Motoki, E. Kobayashi, M. Morita, T. Uchida, K. Akimoto, H. Fukushima and Y. Koezuka, *Bioorg. Med. Chem. Lett.*, 1995, **5**, 2413.

80   T. Abe and K. Mori, *Biosci. Biotech. Biochem*, 1994, **58**, 1671 (*Chem. Abstr.*, 1995, **122**, 81 775).

81   D.E. Bierer, L.B. Dubenko, J. Litvak, R.E. Gerber, J. Chu, D.L. Thai, M.S. Tempesta and T.V. Truong, *J. Org. Chem.*, 1995, **60**, 7646.

82   M. Mickeleit, T. Weider, K. Buchner, C. Geilen, J. Mulzer and W. Reutter, *Angew. Chem. Int. Ed. Engl.*, 1995, **34**, 2667.

83   E.A.L. Biessen, D.M. Beuting, H.C.P.F. Roelen, G.A. van de Marel, J.H. van Boom and T.J.C. van Berkel, *J. Med. Chem.*, 1995, **38**, 1538.

84   L. Yu, R. Cabrera, J. Ramirez, V.A. Malinovsku, K. Brew and P.G. Wang, *Tetrahedron Lett.*, 1995, **36**, 2897.

85   S. Cottaz, J.S. Brimacombe and M.A.J. Ferguson, *Carbohydr. Res.*, 1995, **270**, 85.

86   A.Y. Zamyatina and V.I. Shvets, *Chem. Phys. Lipids*, 1995, **76**, 225 (*Chem. Abstr.*, 1995, **123**, 340 640).

87   L.-X. Wang, N. Sakairi and H. Kuzuhara, *Carbohydr. Res.*, 1995, **275**, 33.

88   T. Fujita, K. Ohira, K. Miyatake, Y. Nakano and M. Nakayama, *Chem. Pharm. Bull.*, 1995, **43**, 920.

89   K. Shimada, K. Sugaya, H. Kaji, I. Nakatani, K. Mitamura and N. Tsutsumi, *Chem. Pharm. Bull.*, 1995, **43**, 1379.

90   D. Todorova, A. Ivanova and T. Milkova, *Dokl. Bulg. Akad. Nauk.*, 1994, **47**, 45 (*Chem. Abstr.*, 1995, **123**, 314 285).

91   Y. Xing, S. Sun, W. Hao and X. Han, *Zhongguo Yiyao Gongye Zazhi.*, 1994, **25**, 484 (*Chem. Abstr.*, 1995, **123**, 9780).

92   Z.L. Jiang, Z.Y. Zhu, Y.H. Wu, Z.Q. Wang and W.S. Zhou, *Yaoxue Xuebao.*, 1994, **29**, 874 (*Chem. Abstr.*, 1995, **122**, 133 558).

93   C. Lacy and M. Sainsbury, *Tetrahedron Lett.*, 1995, **36**, 3949.

94   R.T. Brown, N.E. Carter, F. Scheinmann and N.J. Turner, *Tetrahedron Lett.*, 1995, **36**, 1117.

95   A. Bugge, T. Aasmundstad, A.J. Aasen, A.S. Christophersen, S. Morgenlie and J. Mørland, *Acta Chem. Scand.*, 1995, **49**, 380.

96   R.T. Brown, N.E. Carter, K.W. Lumbard and F. Scheinmann, *Tetrahedron Lett.*, 1995, **36**, 8661.

97   P. Kováβ and K.C. Rice, *Heterocycles*, 1995, **41**, 697.

98   H. Hartenstein, C. Vogt, I. Fortsch and D. Sicker, *Phytochemistry* 1995, **38**, 1233.

99   H. Yamamoto, T. Hanaya, K. Torigoe and W. Pfleiderer, *Adv. Exp. Med. Biol.*, 1993, **338**, 21 (*Chem. Abstr.*, 1995, **122**, 133 549).

100   S. Tomic, B.P. van Eijck, B. Kojić-Prodić, J. Kroon, V. Magnus, B. Nigovic, G. Lacan, N. Ilic, H. Duddeck and M. Hiegemann, *Carbohydr. Res.*, 1995, **270**, 11.

101 S. Tomić, B. Kojić-Prodić, V. Magnus, G. Laćan, H. Duddeck and M. Hiegemann, *Carbohydr. Res.*, 1995, **279**, 1.
102 G. Mikkelsen, T.V. Christensen, M. Bols, I. Lundt and M.R. Stierks, *Tetrahedron Lett.*, 1995, **36**, 6541.
103 S.D. Lorimer, S.D. Mawson, N.B. Perry and R.T. Weavers, *Tetrahedron*, 1995, **51**, 7287.
104 G.K. Skouroumounis, R.A. Massy-Westropp, M.A. Sefton and P.J. Williams, *J. Agric. Food Chem.*, 1995, **43**, 974 (*Chem. Abstr.*, 1995, **122**, 265 800).
105 S. Kitao, T. Matsudo, M. Saitoh, T. Horiuchi and H. Sekine, *Biosci. Biotech. Biochem*, 1995, **59**, 2167.
106 A. Narandja, K. Kelneric, L. Kolacny-Babic and S. Djokic, *J. Antibiot.*, 1995, **48**, 248 (*Chem. Abstr.*, 1995, **123**, 56 449).
107 K.R.H. Repke, R. Megges, J. Weiland and R. Schön, *Angew. Chem. Int. Ed. Engl.*, 1995, **34**, 282.
108 S.B. Mahato, *Rodd's Chem. Carbon Compd. (2nd Suppl. B, C, D)* 1994, **2**, 509.
109 Y. Takaashi, Y. Mimaki, M. Kuroda, Y. Sashida, T. Nikaido and T. Ohmoto, *Tetrahedron*, 1995, **51**, 2281.
110 K.W. Cha, W. Pfleiderer and J.J. Yim, *Helv. Chim. Acta*, 1995, **78**, 600.
111 J. Lykkesfeldt and B.L. Møller, *Acta Chem. Scand.*, 1995, **49**, 540.
112 H.-U. Humpf, N. Zhao, N. Berova, K. Nakanishi and P. Schreier, *J. Nat. Prod.*, 1994, **57**, 1761 (*Chem. Abstr.*, 1995, **122**, 265 797).
113 T.P. Jungblut, J.-P. Schnitzler, W. Heller, N. Hertkorn, J.W. Metzger, W. Szymczak and H. Sandermann, Jr., *Angew. Chem. Int. Ed. Engl.*, 1995, **34**, 312.
114 Y. Nakahara, T. Ogawa, T. Kajimoto and C.-H. Wong, *Gurikobaioroji Shirizu*, 1994, **5**, 140 (*Chem. Abstr.*, 1995, **123**, 314 269).
115 F. Barresi and O. Hindsgaul, *J. Carbohydr. Chem.*, 1995, **14**, 1043.
116 R.H. Youssef, R.W. Bassily, A.N. Asaad, R.I. El-Sokkary and M.A. Nashed, *Carbohydr. Res.*, 1995, **277**, 347.
117 T.E.C.L. Rønnow, M. Meldal and K. Bock, *J. Carbohydr. Chem.*, 1995, **14**, 197.
118 T. Nishimoto, M. Nakano, S. Ikegami, H. Chaen, S. Fukuda, T. Sugimoto, M. Kurimoto and Y. Tsujisaka, *Biosci. Biotech. Biochem*, 1995, **59**, 2189.
119 H. Hashimoto, C. Katayama, M. Goto, T. Okinaga and S. Kitahata, *Biosci. Biotech. Biochem*, 1995, **59**, 179.
120 M. Grothus, A. Steigel, M.-R. Kula and L. Elling, *Carbohydr. Lett.*, 1994, **1**, 83.
121 C.K. Lee and A. Linden, *J. Carbohydr. Chem.*, 1995, **14**, 9.
122 J.M. García Fernández, J.L. Jiménez Blanco, C. Orbiz Mellet and J. Fuentes, *J. Chem. Soc., Chem. Commun.*, 1995, 57.
123 R. Lau, G. Schüle, U. Schwaneberg and T. Ziegler, *Liebigs Ann. Chem.*, 1995, 1745.
124 T. Ziegler and R. Lau, *Tetrahedron Lett.*, 1995, **36**, 1417.
125 S. Valverde, A.M. Gómez, A. Hernández, B. Herradón and J.C. Lopéz, *J. Chem. Soc., Chem. Commun.*, 1995, 2005.
126 D. Sinou, I. Frappa, P. Lhoste, S. Porwanski and B. Kryczka, *Tetrahedron Lett.*, 1995, **36**, 1251.
127 M.L. Sznaidman, S.C. Johnson, C. Crasto and S.M. Hecht, *J. Org. Chem.*, 1995, **60**, 3942.
128 S.-i. Hashimoto, A. Sano, K. Umeo, M. Nakajima and S. Ikegami, *Chem. Pharm. Bull.*, 1995, **43**, 2267.
129 S.-i. Hashimoto, A. Sano, H. Sakamoto, M. Nakajima and S. Ikegami, *Synlett.*, 1995, 1271.
130 Q. Chen and F. Kong, *Carbohydr. Res.*, 1995, **272**, 149.

131 G. Balavoine, S. Berteina, A. Gref, J. Fischer and A. Lubineau, *J. Carbohydr. Chem.*, 1995, **14**, 1237.

132 W.-S. Kim, S. Hosono, H. Sasai and M. Shibasaki, *Tetrahedron Lett.*, 1995, **36**, 4443.

133 R.K.P. Kartha, M. Kiso, A. Hasegawa and H.J. Jennings, *J. Chem. Soc., Perkin Trans. 1*, 1995, 3023.

134 E. Meinjohanns, M. Meldal, H. Paulsen and K. Bock, *J. Chem. Soc., Perkin Trans. 1*, 1995, 405.

135 P.J. Garegg, J.-L. Maloisel and S. Oscarson, *Synthesis*, 1995, 409.

136 J. Fuentes, E. González-Eulate, E. Lopez-Barba and I. Robina, *J. Carbohydr. Chem.*, 1995, **14**, 79.

137 M.A. Tariq, K. Hayashi, K. Tokuyasu and T. Nagata, *Carbohydr. Res.*, 1995, **275**, 67.

138 J.M. Lassaletta and R.R. Schmidt, *Synlett.*, 1995, 925.

139 Y. Hatanaka, M. Hashimoto, K. I.-P. Jwa Hidari, Y. Sanai, Y. Nagai and Y. Kanaoka, *Bioorg. Med. Chem. Lett.*, 1995, **5**, 2859.

140 U.K. Saha and R. Roy, *J. Chem. Soc., Chem. Commun.*, 1995, 2571.

141 K. Matsuoka, S. Ichiro Nishimura, Y.C. Lee, *Carbohydr. Res.*, 1995, **276**, 31.

142 S. Yanahira, T. Kobayashi, T. Suguri, M. Nakakoshi, S. Miura, H. Ishikawa and I. Nakajima, *Biosci. Biotech. Biochem*, 1995, **59**, 1021.

143 D. Zanini, W.K.C. Park and R. Roy, *Tetrahedron Lett.*, 1995, **36**, 7383.

144 T. Furuike, N. Nishi, S. Tokura and S.I. Nishimura, *Chem. Lett.*, 1995, 823.

145 E. Kaji and F.W. Lichtenthaler, *J. Carbohydr. Chem.*, 1995, **14**, 791.

146 T. Ziegler, *Liebigs Ann. Chem.*, 1995, 949.

147 R.A. Field, A. Otter, W. Fu and O. Hindsgaul, *Carbohydr. Res.*, 1995, **276**, 347.

148 M. Ichikawa, R.L. Schnaar and Y. Ichikawa, *Tetrahedron Lett.*, 1995, **36**, 8731.

149 K. Kobayashi, T. Akaike and T. Usui, *Methods Enzymol.*, 1994, **242**, 226 (*Chem. Abstr.*, 1995, **123**, 33 544).

150 V. Pozsgay and B. Coxon, *Carbohydr. Res.*, 1995, **277**, 171.

151 H. Nakashima, T. Hiraki, Y. Yamagiwa and T. Kamikawa, *Tennen Yuki Kagobutsu Toronkai Koen Yoshishu*, 1994, **36**, 752 (*Chem. Abstr.*, 1995, **123**, 144 400).

152 A.K. Sarkar and J.D. Esko, *Carbohydr. Res.*, 1995, **279**, 161.

153 J. Rademann and R.R. Schmidt, *Carbohydr. Res.*, 1995, **269**, 217.

154 M. Wilstermann and G. Magnusson, *Carbohydr. Res.*, 1995, **272**, 1.

155 R.K. Jain, C.F. Piskorz and K.L. Matta, *Carbohydr. Res.*, 1995, **275**, 231.

156 R.K. Jain, C.F. Piskorz, E.V. Chandrasekaran and K.L. Matta, *Carbohydr. Res.*, 1995, **271**, 247.

157 J.J. Krepinsky, D.M. Whitfield, S.P. Douglas, N. Lupescu, D. Pulleyblank and F.L. Moolten, *Methods Enzymol.*, 1994, **247**, 144 (*Chem. Abstr.*, 1995, **122**, 315 019).

158 W.H. Binder, H. Kählig and W. Schmid, *Tetrahedron: Asymm.*, 1995, **6**, 1703.

159 H. Hashimoto, C. Katayama, M. Goto, T. Okinaga and S. Kitahata, *Biosci. Biotech. Biochem*, 1995, **59**, 619.

160 A. Arasappan and B. Fraser-Reid, *Tetrahedron Lett.*, 1995, **36**, 7967.

161 R.T. Lee and Y.C. Lee, *Carbohydr. Res.*, 1995, **271**, 131.

162 S. Oscarson and U. Tedebark, *Carbohydr. Res.*, 1995, **278**, 271.

163 K. Ajisaka, I. Matsuo, M. Isomura, H. Fujimoto, M. Shirakabe and M. Okawa, *Carbohydr. Res.*, 1995, **270**, 123.

164 J.J. Krepinsky, S.P. Douglas and D.M. Whitfield, *Methods Enzymol.*, 1994, **242**, 280 (*Chem. Abstr.*, 1995, **122**, 265 814).

165 H. Franzyk, M. Meldal, H. Paulsen and K. Bock, *J. Chem. Soc., Perkin Trans. 1*, 1995, 2883.

166 S. Akhtar, A. Routledge, R. Patel and J.M. Gardiner, *Tetrahedron Lett.*, 1995, **36**, 7333.
167 J. Tamura and R.R. Schmidt, *J. Carbohydr. Chem.*, 1995, **14**, 895.
168 Y. Du and F. Kong, *Tetrahedron Lett.*, 1995, **36**, 427.
169 T. Oshitari and S. Kobayashi, *Tetrahedron Lett.*, 1995, **36**, 1089.
170 D.L. Boger, S. Teramoto, T. Honda and J. Zhou, *J. Am. Chem. Soc.*, 1995, **117**, 7338.
171 I. Chiu-Machado, J.C. Castro-Palomino, O. Madrazo-Alonso, C. Lopetegui-Palacios and V. Verez-Bencomo, *J. Carbohydr. Chem.*, 1995, **14**, 551.
172 Y. Kawabata, S. Kaneko, I. Kusakabe and Y. Gama, *Carbohydr. Res.*, 1995, **267**, 39.
173 Y. Du and F. Kong, *Carbohydr. Res.*, 1995, **275**, 259.
174 L. Ziser, I. Setyawati and S.G. Withers, *Carbohydr. Res.*, 1995, **274**, 137.
175 G. Yang and F. Kong, *Carbohydr. Lett.*, 1995, **1**, 137.
176 K. Ekelöf and S. Oscarson, *J. Carbohydr. Chem.*, 1995, **14**, 299.
177 S. Singh, D.H.G. Crout and J. Packwood, *Carbohydr. Res.*, 1995, **279**, 321.
178 X. Fang, B.S. Gibbs and J.K. Coward, *Bioorg. Med. Chem. Lett.*, 1995, **5**, 2701.
179 S.S. Pertel, A.L. Kadun and V.Y. Chirva, *Bioorg. Khim.*, 1995, **21**, 226 (*Chem. Abstr.*, 1994, **123**, 340 564).
180 K. Fukase, W.-C. Liu, Y. Suda, M. Oikawa, A. Wada, S. Mori, A.J. Ulmer, E.T. Rietschel and S. Kusumoto, *Tetrahedron Lett.*, 1995, **36**, 7455.
181 K. Fukase, Y. Aoki, I. Kinoshita, Y. Suda, M. Kurosawa, V. Zähringer, E.T. Rietschel and S. Kusumoto, *Tetrahedron Lett.*, 1995, **36**, 8645.
182 Y. Matahira, K. Ohno, M. Kawaguchi, H. Kawagishi and T. Usui, *J. Carbohydr. Chem.*, 1995, **14**, 213.
183 J. Westman, M. Nilsson, D.M. Ornitz and C.-M. Svahn, *J. Carbohydr. Chem.*, 1995, **14**, 95.
184 S. Rio-Anneheim, H. Paulsen, M. Meldal and K. Bock, *J. Chem. Soc., Perkin Trans. 1*, 1995, 1071.
185 J.-I. Tamura, K.W. Neumann and T. Ogawa, *Bioorg. Med. Chem. Lett.*, 1995, **5**, 1351.
186 E. Kaji, N. Anabuki and S. Zen, *Chem. Pharm. Bull.*, 1995, **43**, 1441.
187 E. Petrakova and C.P.J. Glaudemans, *Carbohydr. Res.*, 1995, **268**, 35.
188 E. Petrakova and C.P.J. Glaudemans, *Carbohydr. Res.*, 1995, **279**, 133.
189 H. Nakano, K-i Hamayasu, K. Fujita, K. Hara, M. Ohi, H. Yushizumi and S. Kitahata, *Biosci. Biotech. Biochem*, 1995, **59**, 1732.
190 H. Yamada and M. Nishizawa, *J. Org. Chem.*, 1995, **60**, 386.
191 M. Nishizawa and H. Yamada, *Synlett.*, 1995, 785.
192 Y. Du and F. Kong, *Carbohydr. Res.*, 1995, **275**, 413.
193 J.N. Johnston and L.A. Paquette, *Tetrahedron Lett.*, 1995, **36**, 4341.
194 K. Fujiwara, S. Amano and A. Murai, *Chem. Lett.*, 1995, 191.
195 L.A. Mulard and C.P.J. Glaudemans, *Carbohydr. Res.*, 1995, **274**, 209.
196 M. Adamczyk and J. Grote, *Tetrahedron Lett.*, 1995, **36**, 63.
197 H. Ishida, M. Kiso and A. Hasegawa, *Kagaku (Kyoto)*, 1995, **50**, 518 (*Chem. Abstr.*, 1995, **123**, 169 979).
198 T. Mukaiyama, T. Sasaki, E. Iwashita and K. Matsubara, *Chem. Lett.*, 1995, 455.
199 M. Elofsson and J. Kihlberg, *Tetrahedron Lett.*, 1995, **36**, 7499.
200 K. Yamada, T. Anai and K. Hasegawa, *Phytochemistry*, 1995, **39**, 1031.
201 H. Huang and C.-H. Wong, *J. Org. Chem.*, 1995, **60**, 3100.
202 T. Ziegler, G. Lemanski and A. Rakoczy, *Tetrahedron Lett.*, 1995, **36**, 8973.

203   P.P. Kanakamma, N.S. Mani and V. Nair, *Synth. Commun.*, 1995, **25**, 3777.
204   D. Lafont, P. Boullanger and Y. Chevalier, *J. Carbohydr. Chem.*, 1995, **14**, 533.
205   Y. Blériot, T. Dintinger, N. Guillo and C. Tellier, *Tetrahedron Lett.*, 1995, **36**, 5175.
206   J.A. Barnes and I.H. Williams, *Spec. Publ. - R. Soc. Chem.*, 1995, **148**, 437 (*Chem. Abstr.*, 1995, **123**, 228 736).
207   B.G. Wilson and B. Fraser-Reid, *J. Org. Chem.*, 1995, **60**, 317.
208   J. Kaczmarek, M. Preyss, H. Lönnberg and J. Szafranek, *Carbohydr. Res.*, 1995, **279**, 107.
209   I.H. Oh and G.R. Gray, *Carbohydr. Res.*, 1995, **278**, 329.
210   S. Iacobucci, N. Filippova and M. d'Alarcao, *Carbohydr. Res.*, 1995, **277**, 321.
211   J.C. Wilson, D.I. Angus and M. von Itzstein, *J. Am. Chem. Soc.*, 1995, **117**, 4214.
212   C. Booma and K.K. Balasubramanian, *Tetrahedron Lett.*, 1995, **36**, 5807.
213   J.T. Bien, M. Shang and B.D. Smith, *J. Org. Chem.*, 1995, **60**, 2147.
214   T. Kano and T. Ishimura, *J. Chem. Soc., Perkin Trans. 2*, 1995, 1655.
215   R.P. Bonar-Law and J.K.M. Sanders, *J. Am. Chem. Soc.*, 1995, **117**, 259.
216   T. Mizutani, T. Murakami, N. Matsumi, T. Kurahashi and H. Ogoshi, *J. Chem. Soc., Chem. Commun.*, 1995, 1257.
217   J. Cuntze, L. Owens, V. Alcázar, P. Seiler and F. Diederich, *Helv. Chim. Acta*, 1995, **78**, 367.
218   S. Anderson, U. Neidlein, V. Gramlich and F. Diederich, *Angew. Chem. Int. Ed. Engl.*, 1995, **34**, 1597.
219   P. Li, L. Sun, D.W. Landry and K. Zhao, *Carbohydr. Res.*, 1995, **275**, 179.
220   G. Dekany, P. Ward and I. Toth, *J. Carbohydr. Chem.*, 1995, **14**, 227.
221   J.M.J. Tronchet, M. Zsély and M. Geoffroy, *Carbohydr. Res.*, 1995, **275**, 245.
222   H.-S. Byun and R. Bittman, *Tetrahedron Lett.*, 1995, **36**, 5143.
223   R.M. Kessler, Z.I. Glebova and Y.A. Zhdanov, *Zh. Obshch. Khim.*, 1994, **64**, 848 *Chem. Abstr.*, 1995, **122**, 187 881).
224   M.F. Abdel-Mageed, M.A. Saleh, Y.L. Aly and I.M. Abdo, *Nucleosides Nucleotides*, 1995, **14**, 1985.
225   T. Ercégovic and G. Magnusson, *J. Org. Chem.*, 1995, **60**, 3378.
226   D.I. Angus and M. von Itzstein, *Carbohydr. Res.*, 1995, **274**, 279.
227   M. Tanaka, T. Kai, X.-L. Sun, H. Takayanagi, Y. Uda and K. Furuhata , *Chem. Pharm. Bull.*, 1995, **43**, 1844.
228   Z.J. Witczak, J. Sun and R. Mielguj, *Bioorg. Med. Chem. Lett.*, 1995, **5**, 2169.
229   H. Ishida, M. Kiso and A. Hasegawa, *Methods Enzymol.*, 1994, **242**, 183 (*Chem. Abstr.*, 1995, **122**, 291 389).
230   B. Albrecht, U. Pütz and G. Schwarzmann, *Carbohydr. Res.*, 1995, **276**, 289.
231   B. Joseph and P. Rollin, *Carbohydr. Res.*, 1995, **266**, 321.
232   M.S. Motavia, J. Marcussen and B.L. Møller, *J. Carbohydr. Chem.*, 1995, **14**, 1279.
233   G.-J. Boons, R. Geurtsen and D. Holmes, *Tetrahedron Lett.*, 1995, **36**, 6325.
234   K. Fukase, A. Hasuoka, I. Kinoshita, Y. Aoki and S. Kusumoto, *Tetrahedron*, 1995, **51**, 4923.
235   L.A.J.M. Sliedregt, H.J.G. Broxterman, G.A. Van Der Marel and J.H. Van Boom, *Carbohydr. Lett.*, 1994, **1**, 61.
236   F. Bellamy, V. Barberousse, N. Martin, P. Masson, J. Millet, S. Samreth, C. Sepulchre, J. Thevaniaux and D. Horton, *Eur.J. Med. Chem.*, 1995, **30**, 101 (*Chem. Abstr.*, 1995, **123**, 257 145).
237   X. Wang and P.H. Gross, *J. Org. Chem.*, 1995, **60**, 1201.
238   X. Wang and P.H. Gross, *Liebigs Ann. Chem.*, 1995, 1367.

239 I.P. Smoliakova, R. Caple, D. Gregory, W.A. Smit, A.S. Shashkov and O.S. Chizhov, *J. Org. Chem.*, 1995, **60**, 1221.
240 C. Chen, K. Tagami and Y. Kishi, *J. Org. Chem.*, 1995, **60**, 5386.
241 J. Brunckova and D. Crich, *Tetrahedron*, 1995, **51**, 11945.
242 Y. Suhara and K. Achiwa, *Chem. Pharm. Bull.*, 1995, **43**, 414.
243 Y.Y. Belosludtsev, R.K. Bhatt and J.R. Falck, *Tetrahedron Lett.*, 1995, **36**, 5881.
244 D. Craig, M.W. Pennington and P. Warner, *Tetrahedron Lett.*, 1995, **36**, 5815.
245 R. Murali, C.V. Ramana and M. Nagarajan, *J. Chem. Soc., Chem. Commun.*, 1995, 217.
246 L. Lay, F. Nicotra, L. Panza and G. Russo, *Synlett.*, 1995, 167.
247 I.P. Smoliakova, Y.H. Kim, M.J. Barnes, R. Caple, W.A. Smit and A.S. Shashkov, *Mendeleev Commun.*, 1995, **1**, 15 (*Chem. Abstr.*, 1995, **122**, 265 796).
248 M. Isobe and Y. Jiang, *Tetrahedron Lett.*, 1995, **36**, 567.
249 J.C. López, A.M. Gómez and B. Fraser-Reid, *J. Org. Chem.*, 1995, **60**, 3871.
250 S. Bombard, M. Maillet and M.-L. Capmau, *Carbohydr. Res.*, 1995, **275**, 433.
251 T. Uchiyama, V.P. Vassilev, T. Kajimoto, W. Wong, H. Huang, C.-C. Lin and C.-H. Wong, *J. Am. Chem. Soc.*, 1995, **117**, 5395.
252 J. Desire and A. Veyrieres, *Carbohydr. Res.*, 1995, **268**, 177.
253 K. Tomooka, Y. Nakamura and T. Nakai, *Synlett.*, 1995, 321.
254 J. Alzeer and A. Vasella, *Helv. Chim. Acta*, 1995, **78**, 177.
255 C. Cai and A. Vasella, *Helv. Chim. Acta*, 1995, **78**, 732.
256 J. Alzeer and A. Vasella, *Helv. Chim. Acta*, 1995, **78**, 1219.
257 K.A. Parker, *Pure Appl. Chem.*, 1994, **66**, 2135.
258 T. Kumazawa, K. Ohki, M. Ishida, S. Sato, J.-i. Onodera and S. Matsuba, *Bull. Chem. Soc. Jpn.*, 1995, **68**, 1379.
259 K.A. Parker, C.A. Coburn and Y.-h. Koh, *J. Org. Chem.*, 1995, **60**, 2938.
260 P. Cettour, G. Descotes and J.-P. Praly, *J. Carbohydr. Chem.*, 1995, **14**, 451.
261 V. Vill and H.-W. Tunger, *Liebigs Ann. Chem.*, 1995, 1055.
262 C. Moineau, V. Bolitt and D. Sinou, *J. Chem. Soc., Chem. Commun.*, 1995, 1103.
263 T. Granier and A. Vasella, *Helv. Chim. Acta*, 1995, **78**, 1738.
264 Y.Y. Zhang, X. Li, Y.Z. Gou, Y. Harigaya, M. Onda, K. Hashimoto, Y. Ikeya, M. Okada and M. Maruno, *Chin. Chem. Lett.*, 1994, **5**, 849 (*Chem. Abstr.*, 1995, **122**, 133 555).
265 J.A. Mahling, K.-H. Jung and R.R. Schmidt, *Liebigs Ann. Chem.*, 1995, 461.
266 J.A. Mahling and R.R. Schmidt, *Liebigs Ann. Chem.*, 1995, 467.
267 T. Matsumoto, T. Sohma, H. Yamaguchi, S. Kurata and K. Suzuki, *Tetrahedron*, 1995, **51**, 7347.
268 T. Matsumoto, T. Sohma, H. Yamaguchi and K. Suzuki, *Chem. Lett.*, 1995, 677.
269 A.G.M. Barrett, M. Peña and J.A. Willardsen, *J. Chem. Soc., Chem. Commun.*, 1995, 1147.
270 F.E. McDonald, H.Y.H. Zhu and C.R. Holmquist, *J. Am. Chem. Soc.*, 1995, **117**, 6605.
271 J.R. Axon and A.L. Beckwith, *J. Chem. Soc., Chem. Commun.*, 1995, 549.
272 O. Frey, M. Hoffmann and H. Kessler, *Angew. Chem. Int. Ed. Engl.*, 1995, **34**, 2026.
273 L. Cipolla, L. Lay, F. Nicotra, L. Panza and G. Russo, *J. Chem. Soc., Chem. Commun.*, 1995, 1993.
274 G. Brooks, P.D. Edwards, J.D.I. Hatto, T.C. Smale and R. Southgate, *Tetrahedron*, 1995, **51**, 7999.
275 Z.J. Witczak, *Pure Appl. Chem.*, 1994, **66**, 2189.
276 O. Eyrisch and W.-D. Fessner, *Angew. Chem. Int. Ed. Engl.*, 1995, **34**, 1639.

277   D. Mazéas, T. Skrydstrup and J.-M. Beau, *Angew. Chem. Int. Ed. Engl.*, 1995, **34**, 909.

278   T. Eisele, H. Ishida, G. Hummel and R.R. Schmidt, *Liebigs Ann. Chem.*, 1995, 2113.

279   J.R. Bearder, M.L. Dewis and D.A. Whiting, *J. Chem. Soc., Perkin Trans. 1*, 1995, 227.

280   H.M. Birich, A.M. Griffin, S. Schwidetzky, M.V.J. Ramsay, T. Gallagher and F.W. Lichtenthaler, *J. Chem. Soc., Chem. Commun.*, 1995, 967.

281   A.T. Khan, P. Sharma and R.R. Schmidt, *J. Carbohydr. Chem.*, 1995, **14**, 1353.

282   A. Wei, A. Haudrechy, C. Audin, H.-S. Jun, N. Haudrechy-Bretel and Y. Kishi, *J. Org. Chem.*, 1995, **60**, 2160.

283   K.C. Nicolaou, H. Flörke, M.G. Egan, T. Barth and V.A. Estevez, *Tetrahedron Lett.*, 1995, **36**, 1775.

284   J. Cossy, J.-L. Ranaivosata, V. Bellosta, J. Ancerewicz, R. Ferritto and P. Vogel, *J. Org. Chem.*, 1995, **60**, 8351.

285   R. Ferritto and P. Vogel, *Tetrahedron Lett.*, 1995, **36**, 3517.

286   D.H. Hua and S. Saha, *Recl. Trav. Chim. Pays-Bas*, 1995, **114**, 341.

287   N.C. Chaudhuri and E.T. Kool, *Tetrahedron Lett.*, 1995, **36**, 1795.

288   R. Miethchen and T. Gabriel, *J. Fluorine Chem.*, 1994, **67**, 11 (*Chem. Abstr.*, 1995, **122**, 10 376).

289   O.R. Martin, F. Yang and F. Xie, *Tetrahedron Lett.*, 1995, **36**, 47.

290   B.-H. Yang, J.-Q. Jiang, K. Ma and H.-M. Wu, *Tetrahedron Lett.*, 1995, **36**, 2831.

291   S. Sagaki, S. Nakashima, F. Nagatsugi, Y. Tanaka, M. Hisatome and M. Maeda, *Tetrahedron Lett.*, 1995, **36**, 9521.

292   E. Calzada, C.A. Clarke, C. Roussin-Bouchard and R.H. Wightman, *J. Chem. Soc., Perkin Trans. 1*, 1995, 517.

293   G. Rassu, F. Zanardi, L. Battistin and G. Casiraghi, *Tetrahedron: Asymm.*, 1995, **6**, 371.

294   P. Bertrand, J.-P. Gesson, B. Renoux and I. Tranoy, *Tetrahedron Lett.*, 1995, **36**, 4073.

4
Oligosaccharides

1 General

As previously, this Chapter deals with specific tri- and higher oligosaccharides; most references relate to their chemical or enzymic or chemoenzymic syntheses, and an increasing number of compounds which can be considered to be analogues of oligosaccharides have been reported. Likewise the cyclodextrins continue to attract increasing activity. Their chemistry is dealt with separately (Section 10).

The number of possible structural isomers of hexasaccharides comprising D-hexoses alone has been shown to be greater than 10^{12} which illustrates the great difficulty in developing a microchemical method for structural analysis comparable with the Edman protein and Sanger DNA procedures used for sequencing in these fields. It also illustrates well the difficulty involved in the synthesis of compounds of this degree of complexity.[1] Nevertheless, sophisticated mass spectrometric techniques involving laser desorption methods[2] or electrospray-ionization tandem procedures[3] have been put to good use.

A major compilation has appeared on the 700 or so glycosidation reactions that were reported in the 1994 literature.[4] Hindsgaul's group has also reviewed the syntheses of oligosaccharides to be used as acceptors for glycosyl transferases.[5] A further review on large scale and efficient production of complex human oligosaccharides with pharmaceutical properties has been produced.[6] Lehmann's group has reviewed the photolabile, spacer-modified oligosaccharides that can be used for regioselective probing of receptor binding sites,[7] and the methods available for putting together difficult sequences in oligosaccharides by using chemical, enzymic and mixed methods have also been reviewed.[8] The well-known method involving 1,2-O-(1-cyano)ethylidene derivatives as glycosyl donors and sugar trityl ethers as acceptors, and its application to the synthesis of regular homo- and hetero-oligo- and poly-saccharides, has also been surveyed.[9]

Further reviews appeared on the following topics: the elongation of oligosaccharide chains, synthetic studies on oligosaccharins, total synthesis of glycolipids, of cycloglycans, glycoproteins, glycosaminoglycans and proteoglycans;[10] the synthesis of glycosphingolipids;[11] enzymic glycosylations.[12] This last report is part 2 of a review on the use of enzymes in carbohydrate chemistry.

The potentially important topic of the solid-phase syntheses of oligosaccharides has been reviewed (in Chinese).[13] Enzymic solid-phase methods for production of oligosaccharides and of glycopeptides are becoming efficient, and the promise of this approach is illustrated in a further review.[14] The base-labile anchoring unit **1**, together with the *S*-phenyl thioglycoside sulfoxide glycosylating method, have been developed for the polymer supported synthesis of oligosaccharides,[15] and an ACS symposium was held on the subject of oligosaccharide-polyacrylamide conjugates of immunological interest. Their preparations and applications were covered,[16] and the solid phase approach was applied to produce the oligosaccharide mimic **2**.[17]

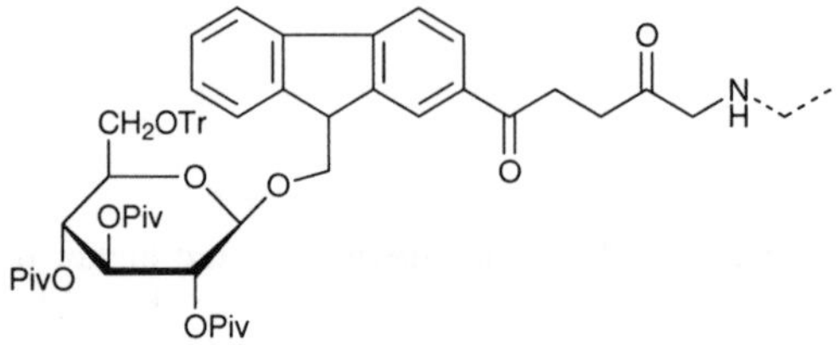

1

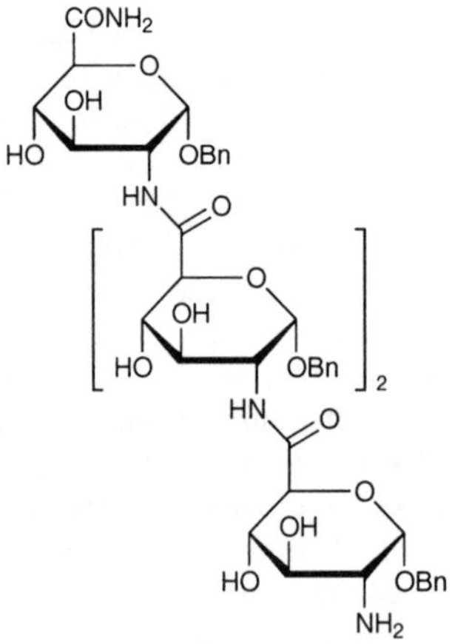

2

Considerable attention has been given to oligosaccharides containing sialic acid units. The synthesis of sialooligosaccharides and their ceramide conjugates, used for elucidating biological functions of gangliosides, have been reviewed,[18] as has the chemical synthesis of sialylated glycoconjugates.[19] An efficient method for the synthesis of sialyl Le^x ganglioside and analogues has been reported in *Methods in Enzymology*.[20] An impressive example of the power of electrospray-ionization tandem mass spectrometry was provided in the structural determination of some underivatized polysialogangliosides containing as many as eight sugar units.[3] Compound **58** of Chapter 3 is a new, potent sialylating agent.

Several reports have described the preparation of chitin oligosaccharides. The polymer itself, on treatment with HCl under ultrasound, has yielded a set of oligomers up to the heptamer with the yields increasing on the use of ultrasound, with time, and with concentration of catalyst. Decreases were observed on increase of the concentration of the chitin.[21] The fluorolysis and fluorohydrolysis of chitosan, which result in deacetylated oligomers, have been briefly reviewed by David.[22]

The reverse approach, *i.e.* oligomerization of simple compounds, has also been successfully attempted. Treatment of short chitose oligosaccharides with an appropriate enzyme from an *Aspergillus* led to the formation mainly of higher saccharides with chain lengths extending by usually two or three sugar units.[23] Lysozyme, on the other hand, applied with tri-*N*-chloroacetyl chitotriose and tri-*N*-acetyl chitotriose led to oligomers with a degree of polymerization 4-12.[24] In related work, UDP-*N*-acetyl-D-glucosamine and UDP-glucuronic acid, in the presence of hyaluronic acid synthase, led to a hyaluronic acid-like polymer,[25] and semi-synthetic heparin sulfate-like polymers have been prepared from an *E. coli* polysaccharide by controlled chemical deacetylation, *N*-sulfation, partial enzymic C-5-epimerization and *O*-sulfation. These represent major specific changes effected at the polymer level.[26]

Compound **3** has been used as a key material in the preparation of β-D-xylo-oligosaccharides of DP 4-10,[27] and specific oligosaccharides containing 10-20 sugar units have been made by application of the following specific enzymes: *N*-acetyl-β-D-glucosaminyl (1→3)transferase and (1→6)transferase, β-D-galactosyl (1→4)transferase and α-D-galactosyl (1→3)transferase.[28]

A new method of coupling oligosaccharides with proteins involves initial reductive amination with *p*(*N*-trifluoroacetylamino)aniline followed by conversion of the trifluoroacetyl amino group to the isothiocyanato group which affords a receptor for protein amino functions.[29]

The fluorescent labelling compound 2-amino-6-amidobiotinylpyridine can be used to follow oligosaccharides during purification and complex formation; improved methodology has been descsribed.[30]

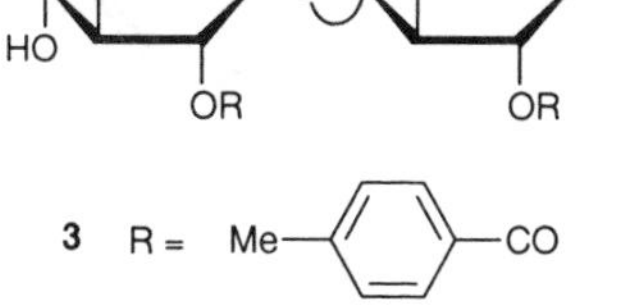

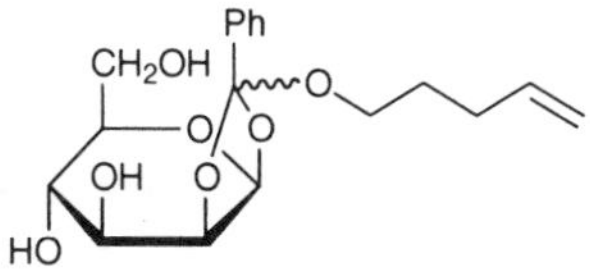

3   R =   Me—⟨ ⟩—CO

2   Trisaccharides

With the advent of combinatorial chemistry new problems are raised for writers of review material which refers to specific compounds. Thus, the random glycosylation of an *N*-acetyllactosaminide with an L-fucosylating agent affords a range of trisaccharides some of which are linear while others are branched.[31] It may well be that a new section will be required shortly in surveys of this series to deal with this type of development.

Compounds in sections 2.1 - 2.3 are now categorized according to their non-reducing end sugars.

2.1 Linear Homotrisaccharides. – The nephritogenoside trisaccharide unit α-D-Glc-(1→6)-β-D-Glc-(1→6)-α-D-Glc has been prepared by application of the glycosyl sulfoxide procedure.[32,33] Cellotriose can be made from cellobiose by use of a sesame seed transferase which also transfers from other disaccharide substrates.[34] Treatment of highly concentrated aqueous solutions of D-mannose with an α-mannosidase from *Aspergillus niger* has resulted in the isolation of the trimers α-D-Man-(1→2)-α-D-Man-(1→6)-D-Man and α-D-Man-(1→2)-α-D-Man-(1→2)-D-Man,[35] and this latter trisaccharide has been formally synthesized by Fraser-Reid and colleagues in studies of the mannan components of phosphatidylinositol membrane anchors. In the course of the work they used the orthoesters **4** as versatile synthetic intermediates.[36,37]

N-Acetylchitobiose has been formed enzymically starting from *N*-acetylgluco-samine and its *p*-nitrophenyl β-glycoside, and subsequently chitotriose from the biose.[38] Also in the field of amino-sugars the α-1,2-linked trimer of 4-amino-4,6-dideoxy-D-mannose (perosamine) has been synthesized with each of the amino groups carrying 2,4-dihydroxybutanoyl substituents, the trimer representing the terminal unit of the *O*-antigen of a *Vibrio cholerae* bacterium.[39] Deacetylation of lantanoside C (**5**) and partial hydrolysis of the product by use of barley seed β-D-glucosidase afforded the clinically useful digoxin compound **6**.[40] The use of 2,6-anhydro-2-thio sugars for stereocontrolled synthesis of 2,6-dideoxy-α- and β-glycosides has described the preparation of the olivomycin A trisaccharide. (See Chapters 3 and 11).[41] The Kdo trimer β-Kdo-(2→8)-α-Kdo-(2→4)-Kdo has been synthesized in conjunction with studies of chlamydial and enterobacterial lipo-polysaccharides.[42]

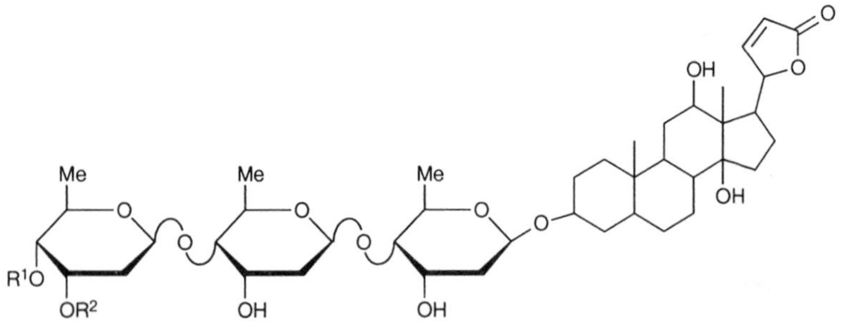

5 R¹ = β-D-Glc, R² = Ac
6 R¹ = R² = H

2.2 Linear Heterotrisaccharides. – An α-glucosidase from *Aspergillus*, acting on leucrose, afforded α-D-Glc-(1→6)-α-D-Glc-(1→5)-D-Fru and the tetramer with a further 1,6-linked α-D-glucose moiety at the non-reducing end.[43] A lactosaminyl fluoride derivative has been shown to be a good lactosamine donor, and by its use β-D-Gal-(1→4)-β-D-GlcNAc-(1→3)-D-Gal was produced.[44] The Le[a] determinant β-D-Gal-(1→3)-α-L-Fuc-(1→4)-D-Glc was prepared as its β-allyl glycoside with sulfate ester groups separately on O-3 and O-6 of the galactose unit and together at O-3 and O-6 of the same unit.[45] Enzymic transglycosylation of α-D-galactose from melibiose to itself resulted in α-D-Gal-(1→6)-α-D-Gal-(1→6)-D-Glc, and lactose, maltose and sucrose were acceptors at the primary centres of the non-reducing moieties (O-6 of the glucosyl moiety of sucrose).[46] Treatment of lactose with a *B. circulans* β-galactosidase gave a mixture containing 1,2-; 1,3- and 1,6-linked isomers of the starting material and eight trisaccharides made up of two galactose and one glucose units.[47] α-D-Gal-(1→4)-β-D-Gal-(1→4)-β-D-Glc, having deoxy groups at O-2, O-3 or O-6 of the central galactose moeity, were linked to the sulfone $HOCH_2-CH(CH_2SO_2C_{19}H_{39})_2$ by use of the trichloroacetimidate method to give bis-sulfone neoglycolipids containing deoxygloboltrioses.[48]

A β-D-fructofuranosidase of an *Aspergillus* caused transfer of the fructosyl unit from sucrose to the 6-position of trehalose and then elongation of the fructose chain to give ultimately oligofructosyl trehaloses.[49] In the field of amino sugar trisaccharides, various phenyl thioglycoside derivatives of glucosamine were developed as glycosylating agents by use of which the chondroitin 4-sulfate trisaccharide was produced, the glucosamine being converted to *N*-acetylglucos-amine units following the glycosylation step.[50] In the course of studies of fucosylation of branched blood group I-type oligo-(*N*-acetyllactosamino)glycans by human milk transferases, it was shown that reaction was restricted to distyl *N*-acetyllactosamine units. In the course of the work trisaccharide D-GlcNAc-(1→3)-β-D-Gal-(1→4)-GlcNAc was found to be a good acceptor of an α-*N*-fucose unit at the terminal 3-position. The β-1,6-β-1,4-linked isomer was a much poorer acceptor.[51] The mannosamine-based trimer β-D-Man-NH$_2$-(1→4)-α-D-Glc-(1→2)-L-Rha, which is the repeating trisaccharide of the capsular polysac-charide of a *Streptoccucus*, was made using a 2-oximo glycosyl bromide,[52] and in related work the α-1,4-α-1,2-isomer was produced.[53]

Details of the interaction of the H-type 2 human blood group determinant α-L-Fuc-(1→2)-β-D-Gal-(1→4)-β-D-GlcNAc-OMe with various lectins have been reviewed,[54] and Kishi and co-workers have reported detailed studies on the conformations and protein binding characteristics of the same trisaccharide and analogues having methylene instead of oxygen inter-unit links. The compounds studied were 8-carboxyoctyl glycosides and the *C*-glycosyl analogues.[55] Five trisaccharides represented by α-L-Hex-(1→3)-β-D-GlcNAc-(1→3)-D-Gal and ter-minating in different 6-deoxy-L-hexoses have been made in connection with studies of sialyl Le[x] ganglioside analogues. In the course of the same work several relevant pentasaccharides were also produced (see section 4).[56] The 6-deoxy-hexose trimer α-L-Fuc-(1→3)-α-L-Rha-(1→3)-L-Rha with *O*-methyl groups at all positions of the fucose moiety and O-2 of the terminal rhamnose was made, this compound being a component of the glycolipid fraction of a *Mycobacterium* lipopolysaccharide.[57]

The glucuronic acid terminating compound β-D-GlcA-(1→3)-α-D-Gal-(1→3)-α-D-Man-OMe, has been made,[58] and compounds α-L-IdoA-(1→4)-α-D-GlcNSO₃-(1→4)-β-D-GlcA and β-D-GlcA-(1→4)-α-D-GlcNAc-(1→4)-β-D-GlcA, which are trisaccharides related to heparin and heparan sulfate, have been synthesized and tested for binding of fibroblastic growth factor.[59]

Enzymic transfer of sialic acid from colominic acid to lactose has resulted in the production of small amounts of α-D-NeuAc-(2→3)-β-D-Gal-(1→4)-D-Glc and the (2→6),(1→4)-linked isomer.[60] The ganglioside analogue of G_{M3}, α-D-NeuAc-(2→3)-β-D-Gal-(1→4)-β-D-Glc-OR, containing a ceramide mimic group as aglycon, was synthesized and found to bind to influenza virus A as well as the natural ganglioside,[61] and in related work the same natural ganglioside was produced as well as various analogues containing a variety of modified sialic acids and ceramides.[62] In an ingenious enzymic one-pot synthesis the sialyl T-antigen α-D-NeuNAc-(2→3)-β-D-Gal-(1→3)-GalNAc was prepared, the procedure involving several enzymes,[63] and the closely related compound α-NeuNAc-(2→6)-β-D-Gal-(1→4)-β-D-GlcNAc-O(CH₂)₆NH₂ was produced using solid-phase technology and a peptidase-sensitive linkage. The glycosylation techniques were again enzymic in character.[64]

NeuNH₂-(2→3)-β-D-Gal-(1→4)-β-D-Glc-1-Cer, the de-*N*-acetylated form of ganglioside G_{M3}, has been converted to the *N*-thioacetyl derivative.[65]

Chemical syntheses have been reported of the xylose-terminating trimer α-D-Xyl-(1→3)-α-D-Gal-(1→3)-L-Fuc and the (1→3)-, (1→4)-linked isomer, each compound carrying a specific sulfate ester grouping and being required to elucidate the structure of a starfish-derived trisaccharide sulfate.[66]

2.3   Branched Homotrisaccharides. – Vacuum and solution force field calculations were used to determine the favoured conformation of β-D-Glc-(1→3)-[β-D-Glc-(1→6)]-α-D-Glc-OMe which is essential to the elicitor activity of a β-D-glucohexapyranosyl-D-glucitol.[67] A variety of specific deoxy derivatives and *O*-methyl ethers of α-D-Man-(1→3)-[α-D-Man-(1→6)]-β-D-Man have been described,[68] and Paulsen and colleagues have also modified the octyl glycoside of this compound by substitution at OH-4 or OH-6 of the (1→3)-linked residue by pentyl groups carrying reactive substituents at the 5-position, for testing as potential inhibitors of *N*-acetylglucosaminyl transferase I.[69]

Methyl 3,5-di-*O*-α-L-arabinofuranosyl-α-L-arabinofuranoside, which is the core unit of naturally occurring arabinans, has been described.[70]

2.4   Branched Heterotrisaccharides. – Compounds in this section are categorized according to their reducing end sugars.

As always, considerable interest has been shown in branched oligosaccharides having *N*-acetylhexosamines at the reducing termini. Enzymic fucosylation has been used to produce the Le^a antigenic compound β-D-Gal-(1→3)-[α-L-Fuc-(1→4)]-D-GlcNAc carrying a sulfate ester at position 6 of the amino-sugar unit,[71] and a related trisaccharide sulfated at O-3 of the galactose unit was produced by selective sulfation procedures.[72] The closely related β-D-Xyl-(1→3)-[α-L-Fuc-(1→4)]-D-GlcNH$_2$ has been produced as a phenyl thioglycoside thereby affording a compound which could be transferred to other carbohydrate acceptors.[73] Sialylated trimers β-D-Gal-(1→3)-[α,β-D-NeuAc-(2→6)]-α-GalNAc-o-C$_6$H$_4$NO$_2$, carrying a sulfate ester group at position 6 of the galactose unit, have been reported in connection with studies of adhesion of proteins to carbohydrates.[74] Detailed ^{13}C NMR studies applied to the trimer β-D-GlcNAc-(1→2)-[β-D-GlcNAc-(1→3)]-L-Rha have suggested that the N-acetylamino groups are strongly involved in interunit interactions.[75]

2.5 Analogues of Trisaccharides. – A thesis has been presented on the total synthesis of the carbohydrate part of calicheamicin which contains a trisaccharide component having two units joined by a hydroxamino linkage,[76] and the final steps in total synthesis of calicheamicin γ$_1$^{I} have been described.[77,78]

Several analogues of trisaccharides containing atoms other than oxygen in ring positions have been reported by S. Ogawa and his group as follows: β-D-GlcNAc-(1→2)-α-D-Man-(1→6)-β-D-Glc-O-(CH$_2$)$_7$Me, with the mannose unit replaced by the carba-analogue, *i.e.* it contains a methylene rather than oxygen atom in the ring position;[79] α-D-Man-(1→6)-[α-D-Man-(1→3)]-D-Man with the two 'non-reducing' units again being carbamannose;[80] and α-D-Man-(1→6)-[α-D-Man-(1→3)]-D-Man wherein either the glucose is carbaglucose or the two mannose units are carbomannoses and the inter-unit linkages are NH.[81] In the course of their work on calicheamycin, Danishefsky and his group have encountered the unusual nitrogen in the ring analogue **8** produced during the desilylation of compound **7**.[78] Trisaccharide analogue **9** is an acaricide obtained from *streptomyces*, the synthesis of which has now been reported.[82]

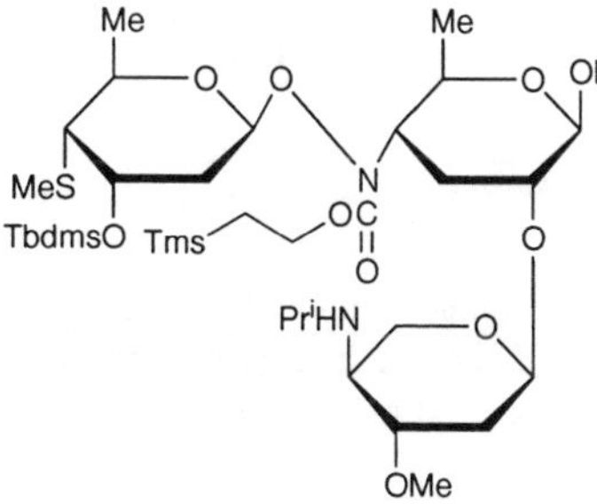

7   8

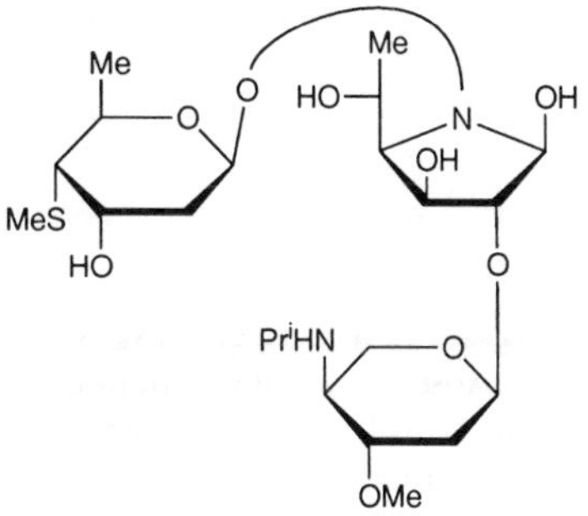

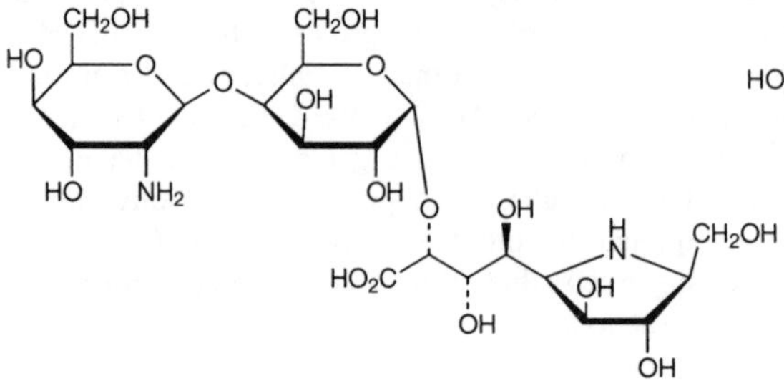

9

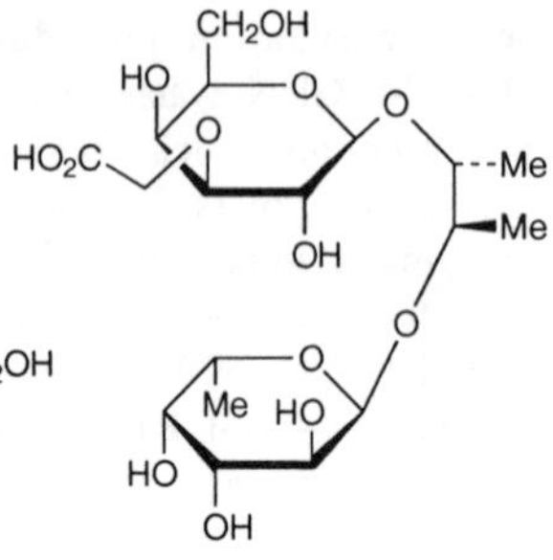

10

Four papers have dealt with the preparation of compounds which have been prepared as analogues of sialyl Le^x. Their structures are illustrated: 10,[83] 11,[84] 12,[85] and 13.[86] The *m*- as well as the *p*-linked isomer of the last of these were also prepared.

The chitobiosyl analogue 14 of the D-allosamine-containing compound allosamidin has been shown to be an inhibitor of two chitinases.[87]

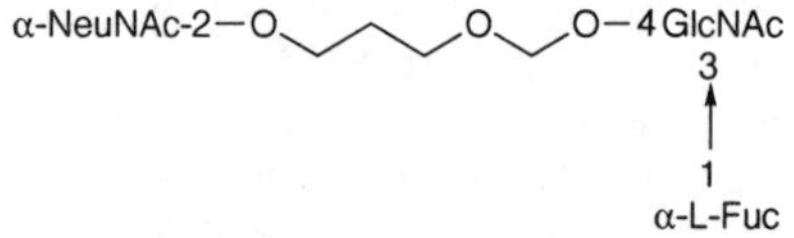

α-NeuNAc-2—O⟍⟍⟍O⟍⟍O—4 GlcNAc
                                  3
                                  ↑
                                  1
                             α-L-Fuc

11

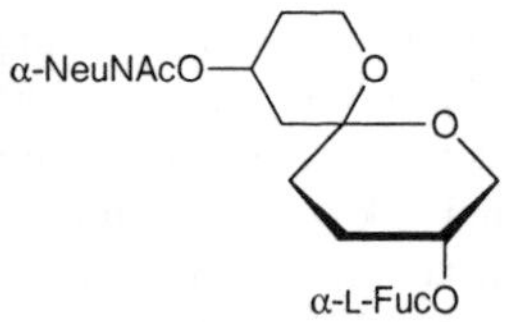

α-NeuNAcO

α-L-FucO

12

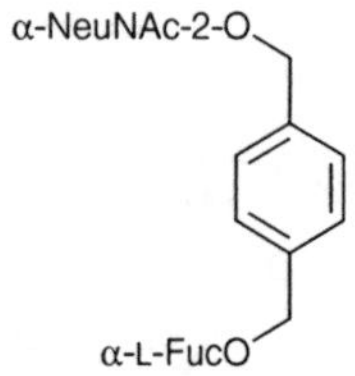

α-NeuNAc-2-O⟍

α-L-FucO

13

β-D-GlcNAc-(1⟶ 4)-β-D-GlcNAcO--

CH$_2$OH

NHMe

OH

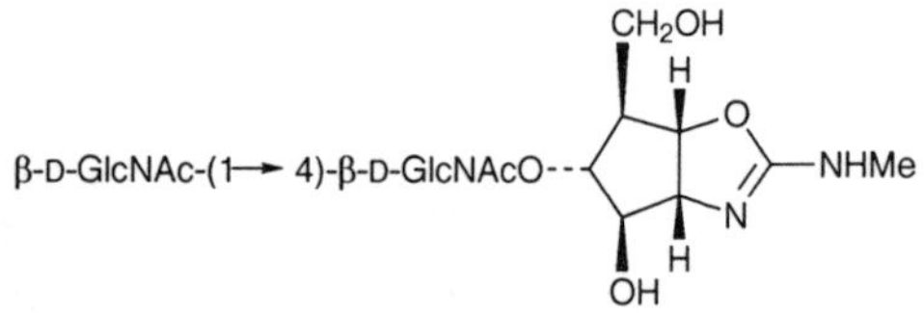

14

Digoxin is a steroidal glycoside of the β-1,4-linked trimer of 2,6-dideoxy-D-*ribo*-hexose. Periodate oxidation and reaction of the product with ω-aminocarboxylic acids led to compounds 15 which were assessed in enzymic and immunological studies.[88]

3   Tetrasaccharides

Compounds of this set are classified according to whether they have linear or branched structures and then by the nature of the sugars at the reducing termini.

3.1 Linear Homotetrasaccharides. – A set of tetrasaccharides of this series have been produced by coupling glucobiosyl bromides with suitably protected α,α-trehaloses. Cellobiose, isomaltose, melibiose, gentiobiose and sophorose were used, and also lactose and rutinose to give heterotetrasaccharides. The products were sulfated and tested for their antiproliferative activity on smooth muscle cells.[89] Sucrose and trehalose having α-isomaltose units substituted at O-6 of the glucose moieties were produced by enzymic methods.[90] D-Galactal 3,4-carbonate, linked by way of a silyl ether at O-6 to a polymer, was epoxidized to give the 1,2-anhydro-α-galactose compound and then treated with the initial carbonate to give a β-D-galactose configured dimer. Repetition of this cycle gave access to the β-1→6-linked galactotetraose. Danishefsky's approach has been extended to produce Le[x] blood group oligosaccharides, and is particularly effective for creating branched oligosaccharides with branching through the 2-position.[91] In an impressive one-pot procedure Ley and colleagues have produced compound **16** which is a derivative of an α-(1→2)-α-(1→2)-α-(1→3)-linked mannatetraose.

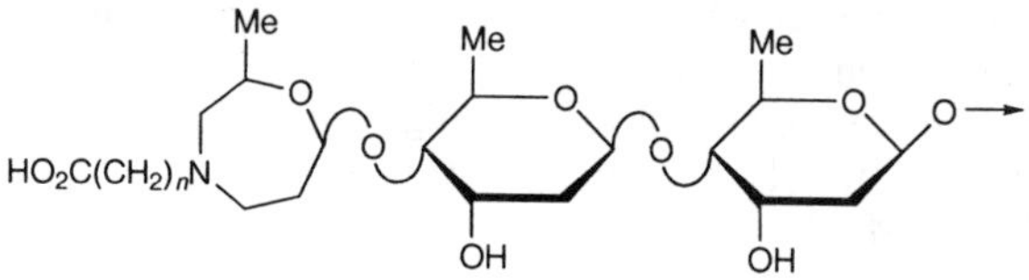

15

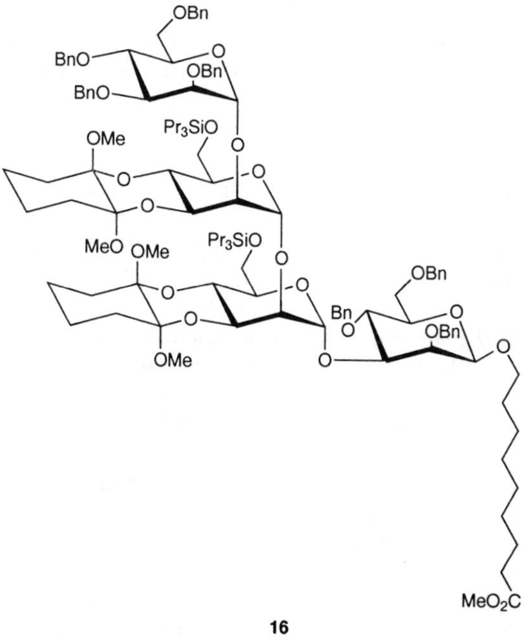

16

The non-reducing biose was attached by use of a phenylselenyl glycoside to an ethylthio mannoside having a free hydroxyl group at C-2. The resulting trisaccharide was then linked to the terminal mannoside having a free hydroxyl at O-3.[92]

3.2   Linear Heterotetrasaccharides. – The disialyl lactose α-NeuNAc-(2→8)-α-NeuNAc-(2→3)-β-D-Gal-(1→4)-D-Glc has been isolated from buffalo colostrum.[93]   Tetrasaccharide   α-L-Fuc-(1→2)-β-D-Gal-(1→3)-β-D-GalNAc-(1→3)-D-Gal, which is related to a hexamer overexpressed in breast cancer cells, has been made and tested as an inhibitor of antibodies to tumour cells.[94] The same workers have shown that the molecule has high flexibility within the disaccharide unit at the reducing end.[95] With maltose and glucuronic acid as starter materials, tetramer α-D-Glc-(1→4)-β-D-GlcA-(1→4)-α-D-Glc-(1→4)-β-D-GlcA-(1→4)-O-Me has been prepared as a simple analogue of heparin oligosaccharides.[96]   Compound   β-D-GalNAc-(1→4)-β-D-GlcA-(1→3)-β-D-GalNAc-(1→4)-β-D-GlcA carrying sulfate groups at position 4 of both N-acetylgalactosamine units, which has been synthesized, represents a closer model of a repeating unit of chondroitin sulfate.[97] The fucose-terminating tetrasaccharide repeating unit of the K antigen of *Klebsiella* type 16 has been produced as its methyl glycoside: β-D-Gal-(1→4)-α-D-Glc-(1→4)-β-D-GlcA-(1→4)-α-L-Fuc-O-Me.[98]

3.3   Branched Homotetrasaccharides. – Using the Danishefsky glycal approach Van Boom and co-workers have prepared β-D-Glc-(1→6)-[β-D-Glc-(1→3)]-β-D-Glc-(1→6)-α-D-Glc-O-Me, the branching glucose unit having been introduced initially as the glycal component.[99] The synthesis of the galactotetraose α-D-Gal-(1→2)-α-D-Gal-(1→6)-[α-D-Gal-(1→2)]-D-Gal, which is a component of the glycosyl phosphatidyl inositol anchor of the glycoprotein of *Trypanosoma brucei*, has been described.[100]

3.4   Branched Heterotetrasaccharides. – This group represents much the most common type of tetrasaccharide to have been prepared by synthesis. Compound β-D-Glc-(1→2)-[β-D-Xyl-(1→3)]-β-D-Glc-(1→4)-β-D-Gal-O-R has been prepared by Danishefsky and colleagues by use of glycal technology, all units except the galactose moiety having started as glycals. The R group is a triterpenoid unit and the product is a *Digitalis* saponin.[101] Compound α-NeuNAc-(2→3)-β-D-Gal-(1→4)-[α-L-Fuc-(1→3)]-β-D-Glc-O-Cer has been prepared as an analogue of sialyl Le[x] having glucose in place of N-acetylglucosamine.[102] The tetramer α-L-Ara-(1→4)-[β-D-Glc-(1→3)]-α-L-Rha-(1→2)-β-D-Glc has been isolated from a saponin by ester bond cleavage,[103] and β-D-Glc-(1→3)-α-L-Rha-(1→2)-[β-D-Xyl-(1→3)]-D-Glc is likewise a saponin component, the natural product inhibiting phospholipid synthesis in human cells without being cytotoxic,[104] and β-D-Xyl-(1→2)-β-D-Fuc-(1→6)-[β-D-Glc-(1→2)]-D-Glc is a tetrasaccharide of terpene glycosides from plant stem bark.[105]

Several tetrasaccharides terminating with D-mannose at the reducing end have been reported. Compound β-D-GlcA-(1→3)-α-D-Gal-(1→3)-[β-D-Gal-(1→2)]-D-

Man, which is the repeating unit of the antigen from *Klebsiella* type 20, has been prepared,[106] as has β-D-Glc-(1→3)-β-D-Glc-(1→4)-[α-D-GlcA-(1→3)]-D-Man, which is the repeating component of the *Klebsiella* type 2 antigen.[107] The glycosyl phosphate β-D-Gal-(1→4)-α-D-Man-(1→6)-[β-D-Gal-(1→4)]-α-D-Man-*O*-PO$_3$H$_2$ is a fragment of the phosphoglycan part of *Leishmania donovani* lipophosphoglycan. Several related phosphates were synthesized in the course of this work.[108] Two analogues of tetramer β-D-GlcNAc-(1→2)-α-D-Man-(1→3)-[α-D-Man-(1→6)]-D-Man have been made as inhibitors of *N*-acetylglucosaminyl transferase: the 2-*O*-methyl- and the 2-deoxy-compounds with these modifications in the branching mannose unit.[109]

Further reports have appeared on the synthesis of sialyl Le^x and sialyl Le^a tetrasaccharides,[110] analogues of the former having GalNAc replacing NeuNAc and glucose replacing GlcNAc,[111] and sialyl Le^x bonded through the anomeric centre by nitrogen to a peptide[112] has been reported. Sialyl Le^x has been elaborated into the 'trivalent' glycosides **17** and **18** and the glycosyl amide **19**,[113] and a sulfate ester of sialyl Le^a with the ester group at O-6 of the galactose unit was described.[114] In related studies T. Ogawa and colleagues have reported on the preparation of the disialyl tetrasaccharide α-NeuNAc-(2→3)-β-D-Gal-(1→3)-[α-NeuNAc-(2→6)]-α-D-GalNAc-*O*-serine and -threonine.[115] The allyl glycoside of tetrasaccharide β-D-GlcA-(1→3)-β-D-GalNAc-(1→6)-[β-D-GlcA-(1→3)]-β-D-GalNAc, which is a part of the *O*-linked polysaccharide chain of the circulating anodic antigen of *Schistosoma mansoni*, has been prepared,[116] as has the blood group A substance α-D-GalNAc-(1→3)-[α-L-Rha-(1→2)]-β-D-Gal-(1→3)-α-D-GalNH$_2$-*O*-serine, linked to an α-amino acid.[117]

[Sia Le^x–O(CH$_2$)$_6$NCO(CH$_2$)$_2$]$_3$CNO$_2$ [Sia Le^x–O(CH$_2$)$_6$NHCO(CH$_2$)$_5$NHCO(CH$_2$)$_2$]$_3$CNO$_2$

17   **18**

[Sia Le^x–NHCO(CH$_2$)$_2$]$_3$CNO$_2$

19

In the field of tetrasaccharides terminating at the reducing end with deoxy-sugars, α-D-GlcA-(1→3)-α-D-Gal-(1→3)-[β-D-Glc-(1→4)]-α-L-Rha, which is the repeating unit of the antigen from *Klebsiella* type 55, has been described.[118] The plant growth regulator calonyctin, which contains the tetrasaccharide β-D-Qui-(1→3)-[α-L-Rha-(1→2)]-β-D-Qui-(1→2)-β-D-Qui, bound glycosidically to a long-chain hydroxy fatty acid, which is lactone-closed to *O*-2 of the non-reducing quinovose unit, has been synthesized and thereby the structure of the natural product confirmed.[119]

3.5   Tetrasaccharide Analogues. – Several analogues of sialyl Le^x antigen determinant have been reported, variations on the structure of the *N*-acetylglucosamine reducing moiety having been concentrated on. Thus, the following compounds have been described: the *O*-substituted tetrasaccharide glycal and congeners derived from it,[120] analogues produced by reduction of the glycal,[121] and the compound having a 1-deoxy-glucosamine unit.[122] A more extensive modification of the *N*-acetylglucosamine unit has involved preparation of the analogue having cyclohexane 1,2-*trans*-diol replacing this hexosamine.[123] The preparation of ganglioside lactams, corresponding to analogues of G_{M1}, G_{M2}, G_{M3}, and G_{M4}, and represented by β-D-GalNAc-(1→4)-[α-NeuNAc]-β-D-Gal-(1→4)-D-Glc with the sialic acid amide linked to the amino function of galactosamine has been reported.[124]

Compound **20**, a *C*-linked analogue of a known sophorosyl trehalose, has been made following coupling of the known *C*-linked compound **21** and a trehalose acceptor in the presence of trimethylsilyl triflate.[125] Coupling of disaccharide analogues afforded the novel tetramer **22**, and subsequent coupling of compounds derived from it gave an octamer.[126] A saccharide-peptide hybrid comprising four molecules of benzyl 2-amino-2-deoxy-α-D-glucopyranoside linked by *O*-3 glycolyl ethers amide bonded to the amino functions has been reported.[127]

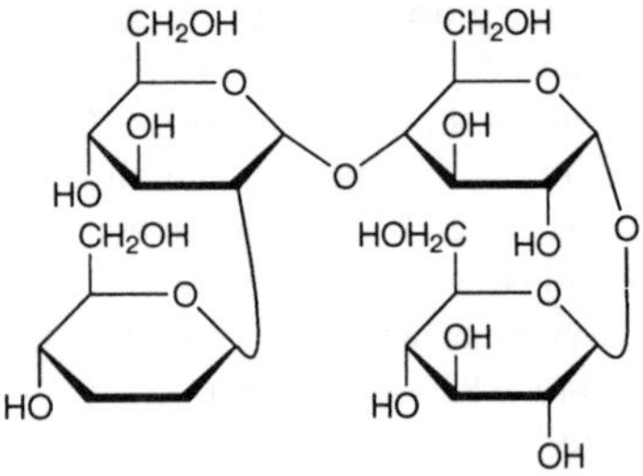

20 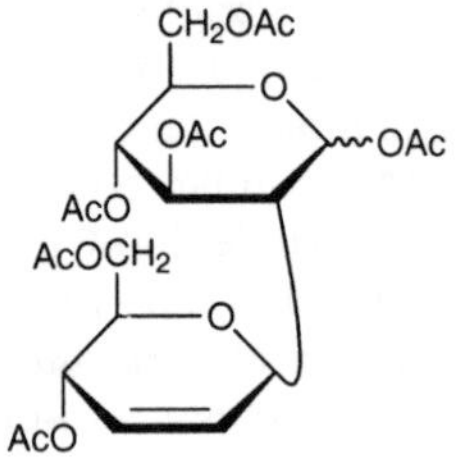   21

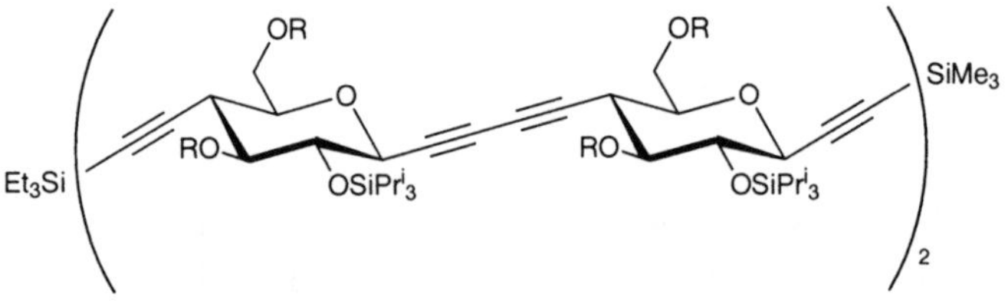

22 R = Morn

4   Pentasaccharides

4.1   Linear Pentasaccharides. – Several glucopentaoses have been reported. The α-1,2-linked compound which is a pentamer fragment of polysaccharide II of *Mycobacterium tuberculosis*, has been synthesized,[128] and the β-1,3-linked isomer has been obtained by acetolysis, followed by hydrolysis, of curdlan and converted to the dodecyl glycoside which was then partially sulfated to give a product showing some anti-HIV activity.[129] Chemo-enzymic methods have been used in the preparation of the pentamer made up of α-1,4-linked glucotetraose having an α-D-glucosylthio-substituent at C-6 of the the non-reducing glucose unit.[130] A 4-methylumbelifferyl glycoside of laminarobiose has been used as substrate for chemical glycosylation, and several oligosaccharides have been appended at O-4 of the non-reducing glucose unit. Consequently the trimer, tetramer and pentamer were produced as substrates for a β-D-glucanase assay.[131] The non-reducing pentamer α-D-Glc-(1→4)-α-D-Glc-(1→4)-β-D-Glc-(1→4)-α-D-Glc-(1→1)-α-D-Glc has been synthesized for conversion to sulfate esters[132] and another maltotriosyl trehalose was produced from maltopentaose by intramolecular enzymic transfer.[133]

The hetero-pentamer β-D-Gal-(1→3)-β-D-GalNAc-(1→3)-α-D-Gal-(1→4)-β-D-Gal-(1→4)-β-D-Glc has been made as its 2-(trimethylsilyl)ethyl glycoside prior to coupling to bovine serum albumin and sepharose,[134] and as a ceramide glycoside for use as an embryogenesis marker.[135] The analogues α-KDN-(2→6)-β-D-Gal-(1→3 or 4)-β-D-GlcNAc-(1→3)-β-D-Gal-(1→4)-β-D-Glc-(1→1)-Cer of the sialyl lactotetraosyl- and neolactotetraosylceramides with KDN instead of sialic acid have been described.[136] To act as substrates for N-linked glycoprotein-processing enzymes the pentamer α-D-Glc-(1→3)-α-D-Man-(1→2)-α-D-Man-(1→2)-α-D-Man-(1→3)-α-D-Man-O-Me and related compounds have been prepared.[137]

4.2 Branched Pentasaccharides. – The synthesis of the pentamer based on maltotriose having a maltosyl unit at O-6 of the central glucose moiety has been effected.[138] The branched compound β-D-Gal-(1→3)-β-D-GalNAc-(1→4)-[α-NeuNAc-(2→3)]-β-D-Gal-(1→4)-D-Glc[139] and the analogue β-D-Gal-(1→3)-[α-L-Fuc-(1→4)]-β-D-GlcNAc-(1→3)-β-D-Gal-(1→4)-D-Glc have been prepared, respectively, as the G_{M1} ganglioside pentasaccharide and the Le[a] pentasaccharide ceramide glycoside carrying a sulfate group at O-3 of the terminal galactose unit.[140] The former has been isolated from calf brain and converted to the analogue having an N-thioacetyl group in place of the N-acetyl on the sialic acid unit.[65] Enzymic methods were used to produce the pentamer comprising two units of β-D-GlcNAc-(1→2)-α-D-Man bonded to O-3 and O-6 of octyl β-D-mannopyranoside.[141] The sialyl Le[x] pentamer NeuNAc-(2→3)-β-D-Gal-(1→4)-[α-L-Fuc-(1→3)]-β-D-GlcNAc-(1→3)-D-Gal has been prepared largely by enzymic procedures having a sulfate ester group at O-6 of the N-acetylglucosamine unit[142] and an analogue having a glycolyl group in place of the acetyl group in the terminal unit has been made.[143] In related work the unsulfated pentamer was attached to both oxygen atoms of pentane 1,5-diol to produce a 'bivalent' analogue as a potential inhibitor of E-selectin-mediated cell adhesion.[144] The preparation of the further galactose-terminating pentamer α-D-3,6-dideoxy-D-xylo-hexose-(1→3)-[α-D-Gal-(1→2)]-α-D-Man-(1→4)-α-L-Rha-(1→3)-D-Gal has been reported.[145]

A pentamer comprising β-1,4-linked glucosamine units and a fucose moiety linked to O-6 of the reducing terminal unit, the pentamer being glycosidically bound to glycerol, is an Nod factor produced by a *rhizobium* species. Several stereo- and regio-isomers were synthesized.[146] Enzymic glycosylation of the tetramer derived by β-1,3-linking of two units of N-acetyllactosamine has afforded β-D-Gal-(1→4)-[α-L-Fuc-(1→3)]-β-D-GlcNAc-(1→3)-β-D-Gal-(1→4)-D-GlcNAc, which is related to Le^x compounds, and an analogue having *N*-acetylglucosamine β-1,6-linked to the non-terminal galactose unit.[147] Synthesis of the pentasaccharide core of asparagine-linked glycoproteins α-D-Man-(1→6)-[α-D-Man-(1→3)]-β-D-Man-(1→4)-β-D-GlcNAc-(1→4)-β-D-GlcNAc has been made with the aid of Ogawa's β-mannosylation procedure,[148,149] The pentamer β-D-Gal-(1→4)-β-D-Xyl-(1→4)-[β-D-Api*f*-(1→3)]-α-L-Rha-(1→2)-D-Fuc, occurring with a *p*-methoxycinnamoyl ester group at O-4 of the fucosyl unit, is the carbohydrate component of the senegasasaponins found in the root of *Polygala senega* as potent inhibitors of alcohol absorption and hypoglycemia.[150] The doubly branched 5-*O*-(3,4,6-tri-*O*-β-D-glucopyranosyl-α-D-glucopyranosyl)-α-D-Kdo, which corresponds to a structural feature of a lipopolysaccharide of *Moraxella catarrhalis*, has been synthesized as its 2-(*p*-aminophenyl)ethyl glycoside.[151]

4.3   Pentasaccharide Analogues. – A pentamer analogue which is the inositol-based fragment of *Mycobacterium tuberculosis* polysaccharide has been made using 2-pyridyl-1-thio-mannosides as donors. It consists of myo-inositol carrying an α-mannosyl unit and also an α-1,6-linked mannotriosyl unit.[152] The core pentasaccharide α-D-Man-(1→2)-α-D-Man-(1→6)-α-D-Man-(1→4)-α-D-GlcNH₂-1→inositol of the glycophosphatidylinositol anchors of membrane-bound proteins and some analogues have been assembled using the *n*-pentenyl glycosylation procedure in a key step.[153]

5   Hexasaccharides

As has become customary in these volumes, an abbreviated method is now used for representing higher saccharides. Sugars will be numbered as follows, and linkages will be indicated in the usual way:

1 D-Glc*p*	2 D-Man*p*	3 D-Gal*p*
4 D-Glc*p*NAc	5 D-Gal*p*NAc	6 Neu*p*Ac
7 L-Rha*p*	8 L-Fuc*p*	9 D-Xyl*p*
10 D-Glc*p*NH₂	11 D-Glc*p*A	12 D-Qui (6-deoxy-D-glucose)
13 L-*Glycero*-D-*manno*-heptose	14 L-Ara*f*	15 Kdn

5.1   Linear Hexasaccharides. – A glucohexaose comprising alternating α-1,4- and β-1,4-linkages has been made as a model for heparin.[154] Danishefsky's group largely using their glycal technology have synthesized compound **23**, which is the human breast tumour associated antigen.[155] Several analogues of the compound were then made and comparisons of binding with the key receptor were conducted, it being found that the tetrasaccharide at the non-reducing end is critical for binding.[156] Lassaletta and Schmidt have described the synthesis of the analogue of compound **23**, having a sialyl group 1,3-linked in the terminal position instead of the 1,2-linked L-fucose group.[157]

(α8) 1→2 (β3) 1→3 (β5) 1→3 (α3) 1→4 (β3) 1→4 (β1)-*O*-Cer

23

(α2) 1→2 (α2) 1→3 (β4) 1→2 (β12) 1→3 (α5) 1→4 (α11)

24

Hexasaccharide **24**, bearing a 3-deoxy-3-formamido group in the quinovose unit has been identified as the repeating unit of the O-specific polysaccharide of *Hafnia alvei*.[158] An unusual hexasaccharide found in *Vitrio cholerae* and comprising six units of α-1,2-linked of 4-amino-4,6-dideoxy-D-mannose, with the amino groups carrying 2,4-dihydroxybutanoyl groups, has been synthesized.[159]

5.2 Branched Hexasaccharides. – During a symposium on elicitor-active compounds a report was made on the one-pot glycosylation of the glucose hexamer comprising four β-1,6-linked units with β-1,3-substituents on glucose 1 and glucose 3.[160] The related compounds **25**,[161] **26**, [162] and **27**[163] have been prepared in the course of work aimed at the study of gangliosides and related compounds. The structure of a disialyl ganglioside of human erythrocytes has been revised to be **28**.[164]

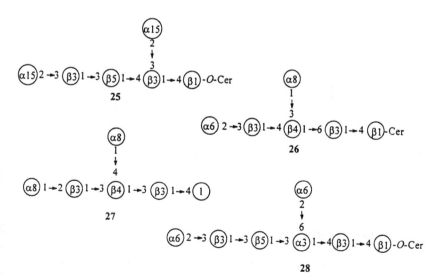

6   Heptasaccharides

Enzymic fucosylation of an *N*-acetyllactosamine-derived hexamer afforded two branched heptasaccharides and a difucosylated product. Substitutions occurred at O-3 of the two glucosamine units other than that at the reducing end.[165] A heptasaccharide anchor compound based on the inositol pentamer compound noted in reference 146, and carrying a GalNAc unit β-(1→4)-linked on the central mannose moiety and a further α-(1→2)-linked D-mannose unit on the 'non-reducing' terminal mannose molecule, has been synthesized in a phosphorylated form.[153,166]

7   Octasaccharides

Chemical oligomerisation of β-D-Glc-(1→4)-D-GlcNH$_2$ afforded means of making an octasaccharide with alternating β-(1→4)-linked glucose and glucosamine units.[167] Interest in ganglioside science has led to the synthesis of the following octasaccharides: **29**,[168] **30**,[169] and **31**.[170]

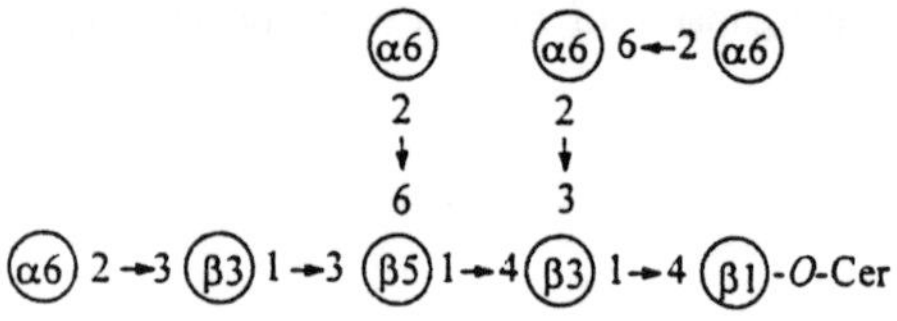

29

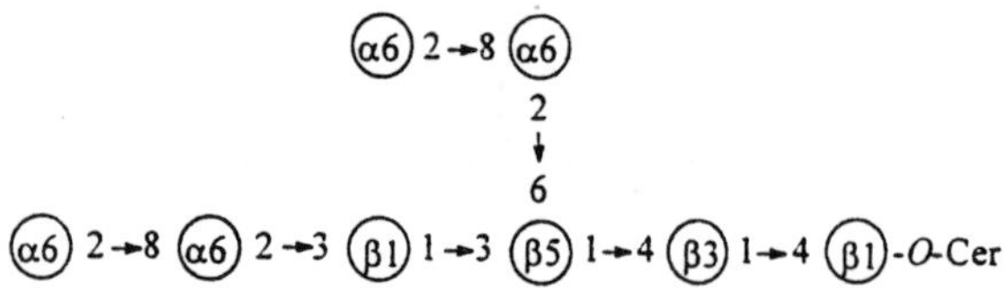

30

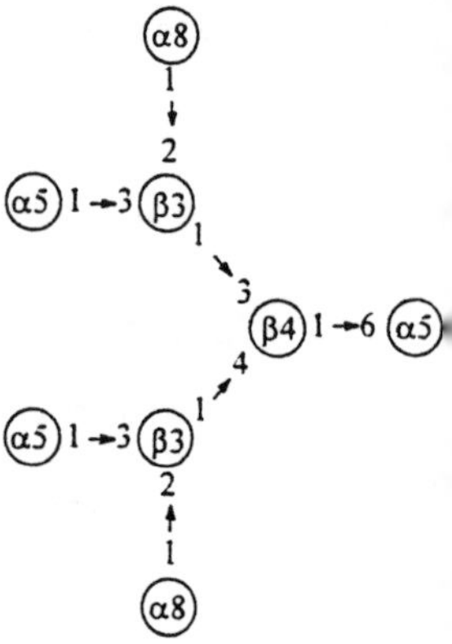

31

8   Nonasaccharides

The synthesis of the tumor-associated glycolipid antigen sialyl Le^x dimer **32** has been reported,[171,172] and glycolipid compound **33** has also been prepared.[173]

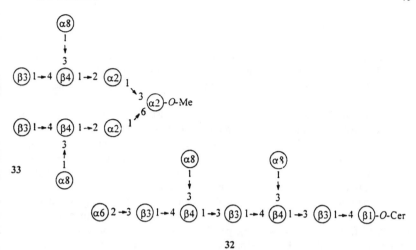

33

32

9 Higher Saccharides

The octamer comprised of α-D-glucopyranose linked in sequence through positions 6, 4, 4, 4, 6, 4 and 4, was dimerized by chemical methods to give a glucohexadecasaccharide in connection with studies of glycogen storage disease.[174] The further hexadecamer fragment of the O-polysaccharide of *Shigella dysenteriae* type I has been made from the tetrasaccharide repeating unit α-L-Rha-(1→2)-α-D-Gal-(1→3)-α-D-GlcNAc-(1→3)-α-L-Rha. Four of these moieties joined by the terminal rhamnose units with α-1→3-linkages, were used in the preparation.[175] Illustrating well the power of enzymological methods, the synthesis of the octadecamer **34** has been achieved by stepwise addition of appropriate sugar units starting from **35** as the initial substrate.[176]

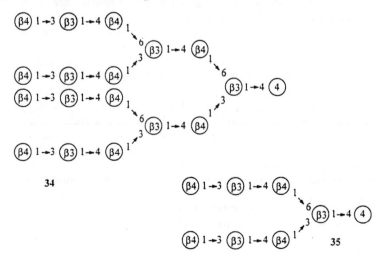

34

35

10   Cyclodextrins

This report concentrates almost exclusively on the chemistry of the cyclodextrins themselves and disregards the large amount of material that has been reported on the properties of these compounds as binding hosts and the properties of the complexes they form.

A symposium report on the problem of selective derivatization of the secondary hydroxyl groups of β-cyclodextrin has appeared and included a description of a novel procedure employing *N*-tosylaziridine as electrophile. Extension of this procedure for further functionalization at secondary positions was discussed.[177] Procedures for the preparation of α-, β- and γ-cyclodextrins have appeared in *Methods in Carbohydrate Chemistry*.[178]

Cyclomaltododecaose has been isolated in 0.01% yield from commercial cyclodextrins.[179] The β-cyclodextrin based on α-D-altrose has been produced in 73% yield from the corresponding per-2,3-anhydro-*manno*-compound.[180] Condensation of 1, 2 ,2′, 3, 3′, 4′, 6-hepta-*O*-acetyl-β-maltose and 6-*O*-trityl maltose heptaacetate in the presence of tin(IV) chloride led to cyclodextrins comprising anhydroglucoses linked β-1,6- and α-1,4- alternately. The products with 6-, 8- and 10-anhydroglucoses were obtained in 30%, 40% and 20% yield respectively.[181]

Acid catalysed acetolysis of fully benzoylated cyclodextrins causes ring-opening and the production of acyclic perbenzoylated compounds with acetyl groups at O-1 of the reducing moiety and O-4 at the non-reducing end. The α-cyclodextrin-derived product was converted to the phenylthio β-glycoside.[182] In analogous manner, acetolysis of peracetylated α-cyclodextrin gave the fully acetylated acyclic hexamer which likewise was converted to the phenylthio β-glycoside and, following various conversions which leave O-4 unsubstituted on the non-reducing moiety, was glycosylated and a route to 2-amino-2-deoxy-cyclomaltoheptaose was thus opened.[183]

Methylation of cyclodextrins under phase-transfer catalysis has been examined and conditions were found to produce 60-70% of heptakis(2,6-di-*O*-methyl)-β-cyclodextrin together with 10-15% of the 2,3,6-trimethyl analogue.[184] Permethylated β-cyclodextrin allows distinction between enantiomers of trisubstituted allenes.[185] Per 2,6-di-*O*-Tbdms γ-cyclodextrin isomerizes in strongly basic conditions to the 3,6-disubstituted ether from which the octa 2-benzyl ether can be made. Desilylation, *O*-methylation and finally debenzylation gives access to the per 3,6-dimethyl ether. The per 3,6-anhydride was made by way of the per 6-deoxy-6-iodo-3-methyl compound.[186] Various 3,6-anhydro derivatives of α-cyclodextrin, notably the pentakis and several tetrakis isomers, were produced from the 6-sulfonates.[187] Per 2,6-di-*O*-methyl-3-pentyl-γ-cyclodextrin and the 2,6-dimethyl-3,(4-oxopentyl) analogue were made, together with related compounds, to act as stationary phases for GLC separation.[188] Selective substitution at O-2 of cyclodextrins can be effected using, for example, allyl bromide together with sodium and lithium iodide in DMSO, the higher kinetic reactivity of O-2 being exploited. Several related ethers were produced, and in the same paper several mono-, di- and tri-6-deoxy-6-iodo derivatives of α-, β-, and γ-cyclodextrins were described.[189]

New β-cyclodextrin ethers having the carbohydrate linked through O-2 to calix[4]arene by way of o- and p-(aminomethyl)benzyl alcohol bridges have been reported, and some of the complexing characteristics described.[190] Complexes formed between metals and a related benzylically-linked otherwise fully methylated β-cyclodextrin bonded to 2,2′-bipyridyl were reported in a 1992 symposium.[191] β-Cyclodextrins carrying alkyl substituents with terminal amino functions substituted with terephthaloyl acyl groups were examined. Intramolecular inclusion was observed, and in the case of the compound with a pentyl link this novel type of association was complete.[192]

Various peracylated β-cyclodextrins with acyl substituents having carbon chains up to C-12 in length were made as potential sustained-release carriers for water-soluble drugs.[193] Bifunctional bridging compounds have been used to link pairs of β-cyclodextrin molecules as bifunctional binding agents.[194] The per-(phenylcarbamate) ester of β-cyclodextrin shows selective binding for the enantiomers of the β-blocker atenolol.[195] The 'capped' cyclodextrin having a biphenyl 4,4′-disulfonate group spanning O-6, O-6 of the A,D rings was converted by DMSO oxidation to the dialdehyde and dicarboxylic acid thereby selectively oxidised at these sites.[196] α-Cyclodextrin has been converted to a sulfonate having a 1-naphthalenesulfonyl ester group at O-2 and a mesitylenesulfonate at one of the other positions. All six isomers were produced.[197] Acyl chlorides in the presence of aqueous alkali and acetonitrile have been used to introduce benzoyl groups and α- and β-naphthoyl groups at the secondary positions of β-cyclodextrins.[198] 2-(Hydroxymethyl)bipyridyl has been linked by way of a succinic acid bridge to β-cyclodextrin and the product fully O-methylated prior to the examination of some metal complexing. Further studies on the analogous ether-linked compound mentioned in reference 191 were described.[199] β-Cyclodextrin has been ester-linked to provide an HPLC stationary-phase column and used in the separation of galacturonic acid oligomers up to the heptamer.[200]

Benzoyl and naphthoyl esters derived from cyclodextrin primary hydroxyls have also been described (compare reference 198), and in the course of this work 6-O-tosyl-β-cyclodextrin was treated with 2-aminoethanol and ethanediamine to give 6-amino-compounds carrying hydroxyethyl or aminoethyl N-substituents. These were then functionalized at the new extended chain positions.[201] 6-Amino-6-deoxy-β- and γ-cyclodextrins have been condensed with N-protected amino acids, and with some of the products self-inclusion complexes were noted.[202] In related work 6-deoxy-6-iodo-β-cyclodextrin, on treatment separately with the enantiomers of α-methylbenzylamine, showed considerable selectivity for the S-enantiomer.[203] Coupling of 6-amino-6-deoxy-β-cyclodextrin with 5-carboxypentyl β-N-acetylglucosaminide gave the corresponding amide which was then enzymically galactosylated using UDPG as starting material and a 4-epimerase, followed by a galactosyl transferase, to give the N-acetyllactosamine derivative **36**.[204] 6-[2-(9-Anthrylmethylamino)ethylamino]-6-deoxy-β-cyclodextrin has been produced as an allosterically switched DNA intercalator.[205] 6-Deoxy-6-hydrazino and -N-hydroxylamino derivatives have been produced, and gave access to various adducts at these new functional groups. Cyclodextrin 2,3-epoxide, on treatment with hydroxylamine, has been converted to the 2-deoxy-2-N-hydroxylamino derivative and hence, by oxidation, to the oximo compound.[206] 3-Deoxy-3-imidazolyl β-cyclodextrin, with the D-*altro* configuration at the newly functionalized sugar unit, and 6-deoxy-6-imidazolyl β-cyclodextrin have been produced from the corresponding D-*manno* epoxide and 6-tosylate respectively. The latter product showed enantioselectivity in its catalysis of the hydrolysis of a protected amino-acid ester.[207]

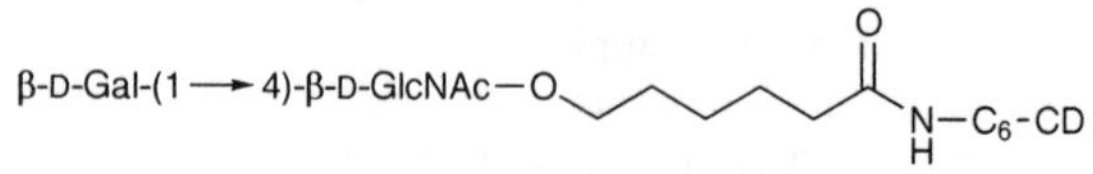

β-D-Gal-(1 ⟶ 4)-β-D-GlcNAc—O

36

Hexakis(6-deoxy-6-iodo-2,3-di-*O*-methyl) α-cyclodextrin, on treatment with alkoxides and aryloxides, gave products derived by nucleophilic displacements of the iodide at five residues and elimination of hydrogen iodide to give the 6-deoxy-5-ene function at the sixth residue.[208] β-Cyclodextrin was converted to the per-3-thio-D-*altro* analogue and then treated with the tetra-substituted benzene **37** to give the compound **38** which contains four cyclodextrin units and forms a 1:1 complex with tetraarylporphyrins.[209] Disulfides, produced by linking monothio-cyclodextrin, have been described and include the 6,6-dimer, which is known, and the new 3,3- and 3,6-linked dimers.[210]

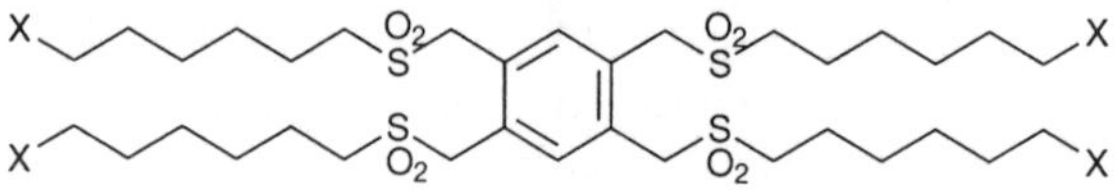

37 X = I

38 X = S-β-CD

Considerable interest continues in cyclodextrins carrying branching sugar units. The mono- and hepta-derivatives of thio-β-cyclodextrin carrying one or seven glucose units at thiol substituents positions at C-6 have been reported.[211] Maltosyl- and mannosyl- and galactosyl-substituted cyclodextrins, formed by enzymic methods, have been described,[212] and the same authors produced mono- and di-α-D-galactosylated products in extensions of the same work.[213] They have then described the $6^1,6^2$-, $6^1,6^3$-, $6^1,6^4$-, and $6^1,6^5$-di-*O*-glucopyranosyl γ-cyclodextrins, produced by way of ditrityl- and di-Tbdms-derivatives by chemical methods.[214] β-D-Glucose, β-D-galactose, α-D-mannose, β-L- and β-D-fucose have been linked by way of a C-9 spacer chain to β-cyclodextrin.[215]

The solubility and primary nucleation characteristics of β-cyclodextrin in water between 15 and 85 °C have been examined.[216]

References

1   R.A. Laine, *Glycobiology*, 1994, **4**, 759 (*Chem. Abstr.*, 1995, **122**, 214 367).
2   B. Spengler, D. Kirsch, R. Kaufmann and J. Lemoine, *Org. Mass. Spectrom.*, 1994, **29**, 782 (*Chem. Abstr.*, 1995, **122**, 240 218).
3   T. Ii, Y. Ohashi and Y. Nagai, *Carbohydr. Res.*, 1995, **273**, 27.
4   F. Barresi and O. Hindsgaul, *J. Carbohydr. Chem.*, 1995, **14**, 1043.
5   S.H. Khan and O. Hindsgaul, *Mol. Glycobiol.*, 1994, 206 (*Chem. Abstr.*, 1995, **123**, 144 373).
6   S. Roth and D. Zopf, *Biol. Approaches Ration. Drug Des.*, 1995, 131 (*Chem. Abstr.*, 1995, **123**, 112 522).
7   J. Lehmann, S. Petry and M. Schmidt-Schuchardt, *ACS Symp. Ser.*, 1994, **560**, 198 (*Chem. Abstr.*, 1995, **122**, 31 754).
8   Y. Ito, *Riken Rev.*, 1995, **8**, 7 (*Chem. Abstr.*, 1995, **123**, 144 374).
9   L.V. Backinowsky, *ACS Symp. Ser.*, 1994, **560**, 36 (*Chem. Abstr.*, 1995, **122**, 56 338).
10   T. Ogawa, *Nippon Nogei Kagaku Kaishi*, 1995, **69**, 999 (*Chem. Abstr.*, 1995, **123**, 169 980).
11   S. Nunomura, Y. Matsuzaki, Y. Ito and T. Ogawa, *Pure Appl. Chem.*, 1994, **66**, 2123 (*Chem. Abstr.*, 1995, **122**, 31 762).
12   C.-H. Wong, R.L. Halcomb, Y. Ichikawa and T. Kajimoto, *Angew. Chem. Int. Ed. Engl.*, 1995, **34**, 521.
13   W.-S. Liao and D.-P. Lu, *Youji Huaxue*, 1994, **14**, 571 (*Chem. Abstr.*, 1995, **122**, 133 537).
14   M. Meldal, *Curr. Opin. Struct. Biol.*, 1994, **4**, 710 (*Chem. Abstr.*, 1995, **122**, 81 754).
15   Y. Wang, H. Zhang and W. Voelter, *Chem. Lett.*, 1995, 273.
16   A.Y. Chernyak, *ACS Symp. Ser.*, 1994, **560**, 133 (*Chem. Abstr.*, 1995, **122**, 31 752).
17   C. Müller, E. Kitas and H.P. Wessel, *J. Chem. Soc., Chem. Commun.*, 1995, 2425.
18   A. Hasegawa, *ACS Symp. Ser.*, 1994, **560**, 184 (*Chem. Abstr.*, 1995, **122**, 31 753).
19   R.R. Schmidt, *ACS Symp. Ser.*, 1994, **560**, 276 (*Chem. Abstr.*, 1995, **122**, 31 756).
20   A. Hasegawa and M. Kiso, *Methods Enzymol.*, 1994, **242**, 158 (*Chem. Abstr.*, 1995, **123**, 199 268).
21   Y. Takahashi, F. Miki and K. Nagase, *Bull. Chem. Soc. Jpn.*, 1995, **68**, 1851.
22   S. David, *Chemtracts: Org. Chem.*, 1994, **7**, 388 (*Chem. Abstr.*, 1995, **123**, 9765).
23   S. Singh, R. Gallagher, P.J. Derrick and D.H.G. Crout, *Tetrahedron: Asymm.*, 1995, **6**, 2803.
24   K. Akiyama, K. Kawazu and A. Kobayashi, *Carbohydr. Res.*, 1995, **279**, 151.
25   C. De Luca, M. Lansing, I. Martini, F. Crescenzi, G.-J. Shen, M. O'Regan and C. H. Wong, *J. Am. Chem. Soc.*, 1995, **117**, 5869.
26   B. Casu, G. Grazioli, H.H. Hannesson, B. Jann, K. Jann, U. Lindahl, A. Naggi, P. Oreste, N. Razi, G. Torri, F. Tursi and G. Zoppetti, *Carbohydr. Lett.*, 1995, **1**, 107.
27   K. Takeo, Y. Ohguchi, R. Hasegawa and S. Kitamura, *Carbohydr. Res.*, 1995, **278**, 301.
28   J. Helin, H. Maaheimo, A. Seppo, A. Keane and O. Renkonen, *Carbohydr. Res.*, 1995, **266**, 191.
29   E. Kallin, *Methods Enzymol.*, 1994, **242**, 119 (*Chem. Abstr.*, 1995, **122**, 265 849).
30   D.K. Toomre and A. Varki, *Glycobiology*, 1994, **4**, 653 (*Chem. Abstr.*, 1995, **122**, 106 312).

31   O. Kanie, F. Barresi, Y. Ding, J. Labbe, A. Otter, L.S. Forsberg, B. Ernst and
      O. Hindsgaul, *Angew. Chem. Int. Ed. Engl.*, 1995, **34**, 2720.

32   H. Zhang, Y. Wang and W. Voelter, *Tetrahedron Lett.*, 1995, **36**, 1243.

33   Y. Wang, H. Zhang and W. Voelter, Z. *Naturforsch., B: Chem. Sci.*, 1995, **50**, 661
      (*Chem. Abstr.*, 1995, **123**, 170 008).

34   K-i. Kuriyama, K-y. Tsuchiya and T. Murui, *Biosci. Biotech. Biochem*, 1995, **59**,
      1142.

35   K. Ajisaka, I. Matsuo, M. Isomura, H. Fujimoto, M. Shirakabe and M. Okawa,
      Carbohydr. Res., 1995, **270**, 123.

36   C. Roberts, R. Madsen and B. Fraser-Reid, *J. Am. Chem. Soc.*, 1995, **117**, 1546.

37   C. Roberts, C.L. May and B. Fraser-Reid, *Carbohydr. Lett.*, 1994, **1**, 89.

38   S. Singh, J. Packwood, C.J. Samuel, P. Critchley and D.H.G. Crout, *Carbohydr.
      Res.*, 1995, **279**, 293.

39   P. Lei, Y. Ogawa and P. Kováč, *Carbohydr. Res.*, 1995, **279**, 117.

40   Z. Lepojević and B. Pekić, *Carbohydr. Res.*, 1995, **271**, 119.

41   K. Toshima, Y. Nozaki, S. Mukaiyama, T. Tamai, M. Nakata, K. Tatsuta and M.
      Kinoshita, *J. Am. Chem. Soc.*, 1995, **117**, 3717.

42   P. Kosma, F.W. D'Souza and H. Brade, *J. Endotoxin Res.*, 1995, **2**, 63 (*Chem.
      Abstr.*, 1995, **123**, 228 705).

43   T. Anindyawati, H. Yamaguchi, K. Furuichi, M. Iizuka and N. Minamiura, *Biosci.
      Biotech. Biochem*, 1995, **59**, 2146.

44   E. Kaji, F.W. Lichtenthaler, Y. Osa and S. Zen, *Bull. Chem. Soc. Jpn.*, 1995, **68**,
      1172.

45   D.D. Manning, C.R. Bertozzi, N.L. Pohl, S.D. Rosen and L.L. Kiessling, *J. Org.
      Chem.*, 1995, **60**, 6254.

46   H. Hashimoto, C. Katayama, M. Goto, T. Okinaga and S. Kitahata, *Biosci. Biotech.
      Biochem*, 1995, **59**, 619.

47   S. Yanahira, T. Kobayashi, T. Suguri, M. Nakakoshi, S. Miura, H. Ishikawa and
      I. Nakajima, *Biosci. Biotech. Biochem*, 1995, **59**, 1021.

48   Z. Zhang and G. Magnusson, *J. Org. Chem.*, 1995, **60**, 7304.

49   M. Muramatsu and T. Nakakuki, *Biosci. Biotechnol. Biochem.*, 1995, **59**, 208 (*Chem.
      Abstr.*, 1995, **122**, 314 967).

50   C. Coutant and J.-C. Jacquinet, *J. Chem. Soc., Perkin Trans. 1*, 1995, 1573.

51   R. Niemelae, J. Natunen, E. Brotherus, A. Saarikangas and O. Renkonen,
      Glycoconjugate J., 1995, **12**, 36 (*Chem. Abstr.*, 1995, **123**, 228 717).

52   E. Kaji, F.W. Lichtenthaler, Y. Osa, K. Takahashi and S. Zen, *Bull. Chem. Soc.
      Jpn.*, 1995, **68**, 2401.

53   E. Kaji, E. Matsui, M. Kobayashi and S. Zen, *Bull. Chem. Soc. Jpn.*, 1995, **68**, 1449.

54   M.-H. Du, U. Spohr and R.U. Lemieux, *Glycoconjugate J.*, 1994, **11**, 443 (*Chem.
      Abstr.*, 1995, **122**, 265 846).

55   A. Wei, K.M. Boy and Y. Kishi, *J. Am. Chem. Soc.*, 1995, **117**, 9432.

56   A. Hasegawa, M. Kato, T. Ando, H. Ishida and M. Kiso, *Carbohydr. Res.*, 1995,
      274, 155; 165.

57   M.K. Gurjar, G. Vishwanadham and L. Ghosh, *J. Indian Inst. Sci.*, 1994, **74**, 287
      (*Chem. Abstr.*, 1995, **123**, 9782).

58   S.K. Das and N. Roy, *Synth. Commun.*, 1995, **25**, 1699.

59   J. Westman, M. Nilsson, D.M. Ornitz and C.-M. Svahn, *J. Carbohydr. Chem.*, 1995,
      14, 95.

60   H. Tanaka, F. Ito and T. Iwasaki, *Biosci. Biotech. Biochem*, 1995, **59**, 638.

61   Y. Nagao, T. Nekado, K. Ikeda and K. Achiwa, *Chem. Pharm. Bull.*, 1995, **43**, 1536.

62   M. Kiso and A. Hasegawa, *Methods Enzymol.*, 1994, **242**, 173 (*Chem. Abstr.*, 1995, **122**, 291 388).

63   V. Křen and J. Thiem, *Angew. Chem. Int. Ed. Engl.*, 1995, **34**, 892.

64   K. Yamada and S.-I. Nishimura, *Tetrahedron Lett.*, 1995, **36**, 9493.

65   R. Isecke, R. Brossmer and S. Sonino, *Bioorg. Med. Chem. Lett.*, 1995, **5**, 2805.

66   K. Hiruma and H. Hashimoto, *J. Carbohydr. Chem.*, 1995, **14**, 879.

67   H. Yamada, T. Harada and T. Takahashi, *Tetrahedron Lett.*, 1995, **36**, 3185.

68   H. Paulsen, M. Springer, F. Reck, E. Meinjohanns, I. Brockhausen and H. Schachter, *Liebigs Ann. Chem.*, 1995, 53.

69   H. Paulsen, M. Springer, F. Reck, I. Brockhausen and H. Schachter, *Liebigs Ann. Chem.*, 1995, 67.

70   S. Kaneko, Y. Kawabata, T. Ishii, Y. Gama and I. Kusakabe, *Carbohydr. Res.*, 1995, **268**, 307.

71   R. Vig, R.K. Jain and K.L. Matta, *Carbohydr. Res.*, 1995, **266**, 279.

72   T.V. Zemlyanukhina, N.E. Nifant'ev, L.O. Kononov, A.S. Shashkov and N.V. Bovin, *Bioorg. Khim.*, 1994, **20**, 556 (*Chem. Abstr.*, 1995, **122**, 187 916).

73   L. Yan and D. Kahne, *Synlett.*, 1995, 523.

74   R.K. Jain, C.F. Piskorz and K.L. Matta, *Bioorg. Med. Chem. Lett.*, 1995, **5**, 1389.

75   N.E. Nifant'ev, A.S. Shashkov, E.A. Khatuntseva, Y.E. Tsvetkov, A.A. Sherman and N.K. Kochetkov, *Bioorg. Khim.*, 1994, **20**, 1001 (*Chem. Abstr.*, 1995, **122**, 187 926).

76   S.H. Boyer, *Diss. Abstr. Int. B*, 1995, **55**, 3307 (*Chem. Abstr.*, 1995, **122**, 291 345).

77   S.A. Hitchcock, M.Y. Chu-Moyer, S.H. Boyer, S.H. Olson and S.J. Danishefsky, *J. Am. Chem. Soc.*, 1995, **117**, 5750.

78   R.L. Halcomb, S.H. Boyer, M.D. Wittman, S.H. Olson, D.J. Denhart, K.K.C. Liu and S.J. Danishefsky, *J. Am. Chem. Soc.*, 1995, **117**, 5720.

79   S. Ogawa, T. Furuya, H. Tsunoda, O. Hindsgaul, K. Stangier and M.M. Palcic, *Carbohydr. Res.*, 1995, **271**, 197.

80   S. Ogawa, S. Saki and H. Tsunoda, *Carbohydr. Res.*, 1995, **274**, 183.

81   S. Ogawa, *Methods Enzymol.*, 1994, **247**, 128 (*Chem. Abstr.*, 1995, **122**, 314 993).

82   K. Tatsuta and M. Kitagawa, *Tetrahedron Lett.*, 1995, **36**, 6717.

83   J.C. Prodger, M.J. Bamford, P.M. Gore, D.S. Holmes, V. Saez and P. Ward, *Tetrahedron Lett.*, 1995, **36**, 2339.

84   B.M. Heskamp, G.H. Veenaman, G.A. van der Marel, C.A.A. van Boeckel and J.H. van Boom, *Recl. Trans. Chim. Pays-Bas*, 1995, **114**, 398.

85   A.A. Birkbeck, S.V. Ley and J.C. Prodger, *Bioorg. Med. Chem. Lett.*, 1995, **5**, 2637.

86   N. Kaila, H.-A. Yu and Y. Xiang, *Tetrahedron Lett.*, 1995, **36**, 5503.

87   S. Takahashi, H. Terayama, H. Kuzuhara, S. Sakuda and Y. Yamada, *Biosci. Biotechnol. Biochem.*, 1994, **58**, 2301 (*Chem. Abstr.*, 1995, **122**, 214 388).

88   M. Adamczyk, J.C. Gebler and J. Grote, *J. Org. Chem.*, 1995, **60**, 3557.

89   H.P. Wessel, E. Vieira, M. Trumtel, T.B. Tschopp and N. Iberg, *Bioorg. Med. Chem. Lett.*, 1995, **5**, 437.

90   Y-K. Kim, Y. Tsumuraya and Y. Sakano, *Biosci. Biotech. Biochem*, 1995, **59**, 1367,

91   J.T. Randolph, K.F. McClure and S.J. Danishefsky, *J. Am. Chem. Soc.*, 19°°, 5712.   , *Synlett.*,

92   P. Grice, S.V. Ley, J. Pietruszka, A.W.M. Priepke and E.P.E. V 1995, 781.

93   H.S. Aparna and P.V. Salimath, *Carbohydr. Res.*, 1995, **268**

86 Carbohydrate Chemistry

94 L. Lay, L. Panza, G. Russo, D. Colombo, F. Ronchetti, E. Adobati and S. Canevari, *Helv. Chim. Acta*, 1995, **78**, 533.

95 L. Toma, D. Colombo, F. Ronchetti, L. Panza and G. Russo, *Helv. Chim. Acta*, 1995, **78**, 636.

96 J. Westman and M. Nilsson, *J. Carbohydr. Chem.*, 1995, **14**, 949.

97 J.-I. Tamura, K.W. Neumann and T. Ogawa, *Bioorg. Med. Chem. Lett.*, 1995, **5**, 1351.

98 A.K. Choudhury, A.K. Ray and N. Roy, *J. Carbohydr. Chem.*, 1995, **14**, 1153.

99 C.M. Timmers, G.A. van der Marel and J.H. van Boom, *Chem. Eur. J.*, 1995, **1**, 161.

100 N. Khiar and M.M. Lomas, *J. Org. Chem.*, 1995, **60**, 7017.

101 J.T. Randolph and S.J. Danishefsky, *J. Am. Chem. Soc.*, 1995, **117**, 5693.

102 A. Hasegawa, K. Ito, H. Ishida and M. Kiso, *J. Carbohydr. Chem.*, 1995, **14**, 353.

103 T. Ikeda, T. Kajimoto, T. Nohara, J.-e. Kinjo and C.-H. Wong, *Tetrahedron Lett.*, 1995, **36**, 1509.

104 Y. Miyaki, O. Nakamura, Y. Sashida, K. Koike, T. Nikaido, T. Ohmoto, A. Nishino, Y. Satomi and H. Nishino, *Chem. Pharm. Bull.*, 1995, **43**, 971.

105 T. Ikeda, S. Fujiwara, J. Kinzo, T. Nohara, Y. Ida, J. Shoji, T. Shingu, R. Isobe and T. Kajimoto, *Bull. Chem. Soc. Jpn.*, 1995, **68**, 3483.

106 S.K. Das and N. Roy, *Carbohydr. Res.*, 1995, **271**, 177.

107 A.K. Misra and N. Roy, *Carbohydr. Res.*, 1995, **278**, 103.

108 A.V. Nikolaev, T.J. Rutherford, M.A.J. Ferguson and J.S. Brimacombe, *J. Chem. Soc., Perkin Trans. 1*, 1995, 1977.

109 H. Paulsen, M. Springer, F. Reck, I. Brockhausen and H. Schachter, *Carbohydr. Res.*, 1995, **275**, 403.

110 N.E. Nifant'ev, Y.E. Tsvetkov, A.S. Shashkov, A.B. Tuzikov, I.V. Maslennikov, I.S. Popova and N.V. Bovin, *Bioorg. Khim.*, 1994, **20**, 551 (*Chem. Abstr.*, 1995, **122**, 214 361).

111 J.M. Coterón, K. Singh, J.L. Asensio, M. Dominguez-Dalda, A. Fernández-Mayoralas, J. Jiménez-Barbero and M. Martín-Lomas, *J. Org. Chem.*, 1995, **60**, 1502.

112 U. Sprengard, G. Kretzschmar, E. Bartnik, C. Hüls and H. Kunz, *Angew. Chem. Int. Ed. Engl.*, 1995, **34**, 990.

113 G. Kretzschmar, U. Sprengard, H. Kunz, E. Bartnik, W. Schmidt, A. Toepfer, B. Horsch, M. Krausse and D. Seiffge, *Tetrahedron*, 1995, **51**, 13015.

114 R. Vig, R.K. Jain, C.F. Piskorz and K.L. Matta, *J. Chem. Soc., Chem. Commun.*, 1995, 2073.

115 Y. Nakahara, H. Iijima and T. Ogawa, *Carbohydr. Lett.*, 1995, **1**, 102.

116 K.M. Halkes, T.M. Slaghek, H.J. Vermeer, J.P. Kamerling and J.F.G. Vliegenthart, *Tetrahedron Lett.*, 1995, **36**, 6137.

117 W.M. Macindoe, H. Ijima, Y. Nakahara and T. Ogawa, *Carbohydr. Res.*, 1995, **269**, 227.

118 S.K. Das and N. Roy, *J. Carbohydr. Chem.*, 1995, **14**, 417.

119 Z.-H. Jiang, A. Geyer and R.R. Schmidt, *Angew. Chem. Int. Ed. Engl.*, 1995, **34**, 2520.

120 S.J. Danishefsky, J. Gervay, J.M. Peterson, F.E. McDonald, K. Koseki, D.A. Griffith, T. Oriyama and S.P. Marsden, *J. Am. Chem. Soc.*, 1995, **117**, 1940.

'1 Maeda, K. Ito, H. Ishida, M. Kiso and A. Hasegawa, *J. Carbohydr. Chem.*, 1995, ^7.

H. Ishida, M. Kiso and A. Hasegawa, *J. Carbohydr. Chem.*, 1995, **14**,

123   A. Toepfer, G. Kretzchmar and E. Bartnik, *Tetrahedron Lett.*, 1995, **36**, 9161.
124   M. Wilstermann, L.O. Kononov, U. Nilsson, A.K. Ray and G. Magnusson, *J. Am. Chem. Soc.*, 1995, **117**, 4742.
125   H.P. Wessel and G. Englert, *J. Carbohydr. Chem.*, 1995, **14**, 179.
126   J. Alzeer and A. Vasella, *Helv. Chim. Acta*, 1995, **78**, 1219.
127   H.P. Wessel, C.M. Mitchell, C.M. Lobato and G. Schmid, *Angew. Chem. Int. Ed. Engl.*, 1995, **34**, 2712.
128   V. Pozsgay and J.B. Robbins, *Carbohydr. Res.*, 1995, **277**, 51.
129   K. Katsuraya, T. Shoji, K. Inazawa, H. Nakashima, N. Yamamoto and T. Uryu, *Macromolecules*, 1994, **27**, 6695 (*Chem. Abstr.*, 1995, **122**, 265 867).
130   C. Apparu, H. Driguez, G. Williamson and B. Svensson, *Carbohydr. Res.*, 1995, **277**, 313.
131   C. Malet, J.L. Viladot, A. Ochoa, B. Gallégo, C. Brosa and A. Planas, *Carbohydr. Res.*, 1995, **274**, 285.
132   H.P. Wessel and J. Niggemann, *J. Carbohydr. Chem.*, 1995, **14**, 1089.
133   K. Maruta, T. Nakada, M. Kubota, H. Chaen, T. Sugimoto, M. Kurimoto and Y. Tsujisaka, *Biosci. Biotech. Biochem*, 1995, **59**, 1829.
134   U. Nilsson and G. Magnusson, *Carbohydr. Res.*, 1995, **272**, 9.
135   T.K. Park, I.J. Kim and S.J. Danishefsky, *Tetrahedron Lett.*, 1995, **36**, 9089.
136   T. Terada, H. Ishida, M. Kiso and A. Hasegawa, *J. Carbohydr. Chem.*, 1995, **14**, 751.
137   R.K. Jain, X.-G. Liu, S.R. Oruganti, E.V. Chandrasekaran and K.L. Matta, *Carbohydr. Res.*, 1995, **271**, 185.
138   M.S. Motawia, C.E. Olsen, K. Enevoldsen, J. Marcussen and B.L. Møller, *Carbohydr. Res.*, 1995, **277**, 109.
139   T. Stauch, U. Greilich and R.R. Schmidt, *Liebigs Ann. Chem.*, 1995, 2101.
140   A. Endo, M. Iida, S. Fujita, M. Numata, M. Sugimoto and S. Nunomura, *Carbohydr. Res.*, 1995, **270**, C9.
141   F. Reck, E. Meinjohanns, J. Tan, A.A. Grey, H. Paulsen and H. Schachter, *Carbohydr. Res.*, 1995, **275**, 221.
142   P.R. Scudder, K. Shailubhai, K.L. Duffin, P.R. Streeter and G.S. Jacob, *Glycobiology*, 1994, **4**, 929 (*Chem. Abstr.*, 1995, **122**, 240 285).
143   A. Hasegawa, A. Uchimura, H. Ishida and M. Kiso, *Biosci. Biotech. Biochem*, 1995, **59**, 1091.
144   S.A. DeFrees, W. Kosch, W. Way, J.C. Paulson, S. Sabesan, R.L. Halcomb, D.-H. Huang, Y. Ichikawa and C.-H. Wong, *J. Am. Chem. Soc.*, 1995, **117**, 66.
145   T.L. Lowary, E. Eichler and D.R. Bundle, *J. Org. Chem.*, 1995, **60**, 7316.
146   S. Ikeshita, Y. Nakahara and T. Ogawa, *Carbohydr. Res.*, 1995, **266**, C1.
147   R. Niemelä, J. Räbinä, A. Leppänen, H. Maaheimo, C.E. Costello and O. Renkonen, *Carbohydr. Res.*, 1995, **279**, 331.
148   A. Dan, Y. Ito and T. Ogawa, *Tetrahedron Lett.*, 1995, **36**, 7487.
149   A. Dan, Y. Ito and T. Ogawa, *J. Org. Chem.*, 1995, **60**, 4680.
150   M. Yoshikawa, T. Murakami, T. Ueno, M. Kadoya, H. Matsuda, J. Yamohara and N. Murakami, *Chem. Pharm. Bull.*, 1995, **43**, 2115.
151   K. Ekelöf and S. Oscarson, *Carbohydr. Res.*, 1995, **278**, 289.
152   H.B. Mereyala and B.R. Gaddam, *Proc. - Indian Acad. Sci., Chem. Sci.*, 1994, **106**, 1225 (*Chem. Abstr.*, 1995, **123**, 83 876).
153   R. Madsen, U.E. Udodong, C. Roberts, D.R. Mootoo, P. Konradsson and B. Fraser-Reid, *J. Am. Chem. Soc.*, 1995, **117**, 1554.
154   H.P. Wessel, R. Minder and G. Englert, *J. Carbohydr. Chem.*, 1995, **14**, 1101.

155 M.T. Bilodeau, T.K. Park, S. Hu, J.T. Randolph, S.J. Danishefsky, P.O. Livingston and S. Zhang, *J. Am. Chem. Soc.*, 1995, **117**, 7840.

156 I.J. Kim, T.K. Park, S. Hu, K. Abrampah, S. Zhang, P.O. Livingston and S.J. Danishefsky, *J. Org. Chem.*, 1995, **60**, 7716.

157 J.M. Lassaletta and R.R. Schmidt, *Tetrahedron Lett.*, 1995, **36**, 4209.

158 E. Katzenellenbogen, E. Romanowska, N.A. Kocharova, A.S. Shashkov, Y.A. Knirel and N.K. Kochetkov, *Carbohydr. Res.*, 1995, **273**, 187.

159 Y. Ogawa, P.-s. Lei and P. Kováč, *Bioorg. Med. Chem. Lett.*, 1995, **5**, 2283.

160 H. Yamada, T. Harada, H. Miyazaki, H. Tsukamoto, T. Kato and T. Takahashi, *Tennen Yuki Kagobutsu Toronkai Koen Yoshishu*, 1994, **36**, 368 (*Chem. Abstr.*, 1995, **123**, 144 415).

161 T. Terada, T. Toyoda, H. Ishida, M. Kiso and A. Hasegawa, *J. Carbohydr. Chem.*, 1995, **14**, 769.

162 K. Hotta, K. Itoh, A. Kameyama, H. Ishida, M. Kiso and A. Hasegawa, *J. Carbohydr. Chem.*, 1995, **14**, 115.

163 S.J. Danishefsky, V. Behar, J.T. Randolph and K.O. Lloyd, *J. Am. Chem. Soc.*, 1995, **117**, 5701.

164 S.B. Levery, M.E.K. Salyan, S.J. Steele, R. Kannagi, S. Dasgupta, J.-L. Chien, E.L. Hogan, H. van Halbeek and S.-i. Hakomori, *Arch. Biochem. Biophys.*, 1994, **312**, 125 (*Chem. Abstr.*, 1995, **122**, 10 386).

165 J. Natunen, R. Niemela, L. Penttila, A. Seppo, T. Ruohtula and O. Renkonen, *Glycobiology*, 1994, **4**, 577 (*Chem. Abstr.*, 1995, **122**, 133 607).

166 A.S. Cambell and B. Fraser-Reid, *J. Am. Chem. Soc.*, 1995, **117**, 10387.

167 Y. Hasegawa, T. Kawada, Y. Kikkawa and T. Sakuno, *Mokuzai Gakkaishi*, 1995, **41**, 83 (*Chem. Abstr.*, 1995, **123**, 112 571).

168 K. Hotta, H. Ishida, M. Kiso and A. Hasegawa, *J. Carbohydr. Chem.*, 1995, **14**, 491.

169 K. Hotta, T. Kawase, H. Ishida, M. Kiso and A. Hasegawa, *J. Carbohydr. Chem.*, 1995, **14**, 961.

170 W.M. Macindoe, Y. Nakahara and T. Ogawa, *Carbohydr. Res.*, 1995, **271**, 207.

171 M. Iida, A. Endo, S. Fujita, M. Numata, Y. Matsuzaki, M. Sugimoto, S. Nunomura and T. Ogawa, *Carbohydr. Res.*, 1995, **270**, C15.

172 A. Kameyama, T. Ehara, Y. Yamada, H. Ishida, M. Kiso and A. Hasegawa, *J. Carbohydr. Chem.*, 1995, **14**, 507.

173 Y.-M. Zhang, A. Brodzky, P. Sinäy, G. Saint-Marcoux and B. Perly, *Tetrahedron: Asymm.*, 1995, **6**, 1195.

174 S. Koto, H. Haigoh, S. Shichi, M. Hirooka, T. Nakamura, C. Maru, M. Fujita, A. Goto, T. Sato, M. Okada, S. Zen, K. Yago and F. Tomonaga, *Bull. Chem. Soc. Jpn.*, 1995, **68**, 2331.

175 V. Pozsgay, *J. Am. Chem. Soc.*, 1995, **117**, 6673.

176 A. Seppo, L. Penttila, R. Niemela, H. Maaheimo, O. Renkonen and A. Keane, *Biochemistry*, 1995, **34**, 4655 (*Chem. Abstr.*, 1995, **122**, 291 382).

177 D.R.J. Palmer, E. Buncel and G.R.J. Thatcher, *Minutes Int. Symp. Cyclodextrins, 6th*, 1992, 86 (*Chem. Abstr.*, 1995, **122**, 133 571).

178 N. Nakamura, *Methods Carbohydr. Chem.*, 1994, **10**, 269 (*Chem. Abstr.*, 1995, **122**, 240 207).

179 T. Endo, H. Ueda, S. Kobayashi and T. Nagai, *Carbohydr. Res.*, 1995, **269**, 369.

180 K. Fujita, H. Shimada, K. Ohta, Y. Nogami, K. Nasu and T. Koga, *Angew. Chem. Int. Ed. Engl.*, 1995, **34**, 1621.

181 H. Driguez and J.-P. Utille, *Carbohydr. Lett.*, 1995, **1**, 125.

182   N. Sakairi, K. Matsui and H. Kuzuhara, *Carbohydr. Res.*, 1995, **266**, 263.
183   N. Sakairi, L.-X. Wang and H. Kuzuhara, *J. Chem. Soc., Perkin Trans. 1*, 1995, 437.
184   P. Bako, L. Fenichel, L. Toke, L. Szente and J. Szejtli, *J. Inclusion Phenom. Mol. Recognit. Chem.*, 1994, **18**, 307 (*Chem. Abstr.*, 1995, **122**, 187 936).
185   G. Uccello-Barretta, F. Balzano, A.M. Caporusso, A. Iodice and P. Salvadori, *J. Org. Chem.*, 1995, **60**, 2227.
186   P.R. Ashton, S.E. Boyd, G. Gattuso, E.Y. Hartwell, R. Königer, N. Spencer and J.F. Stoddart, *J. Org. Chem.*, 1995, **60**, 3898.
187   H. Yamamura, H. Nagaoka, M. Kawai, Y. Butsugan and K. Fujita, *Tetrahedron Lett.*, 1995, **36**, 1093.
188   C. Bicchi, A. D'Amato, V. Manzin, A. Galli and M. Galli, *J. High Resolut. Chromatogr.*, 1995, **18**, 295 (*Chem. Abstr.*, 1995, **123**, 340 596).
189   S. Hanessian, A. Benalil and C. Laferrière, *J. Org. Chem.*, 1995, **60**, 4786.
190   E. van Dienst, B.H.M. Snellink, I. von Piekartz, J.F.J. Engbersen and D.N. Reinhoudt, *J. Chem. Soc., Chem. Commun.*, 1995, 1151.
191   R. Deschenaux, M.M. Harding, T. Ruch and R. Ziessel, *Minutes Int. Symp.Cyclodextrins, 6th*, 1992, 101 (*Chem. Abstr.*, 1995, **122**, 161 093).
192   S. Hanessian, A. Benalil and M.T.P. Viet, *Tetrahedron*, 1995, **51**, 10131.
193   F. Hirayama, M. Yamanaka, T. Horikawa and K. Uekama, *Chem. Pharm. Bull.*, 1995, **43**, 130.
194   R. Breslow, S. Halfon and B. Zhang, *Tetrahedron*, 1995, **51**, 377.
195   Y. Kuroda, Y. Sukuki, J. He, T. Kawabata, A. Shibukawa, H. Wada, H. Fujima, Y. Go-oh, E. Imai and T. Nakagawa, *J. Chem. Soc., Perkin Trans. 2*, 1995, 1749.
196   J. Yoon, S. Hong, K.A. Martin and A.W. Czarnik, *J. Org. Chem.*, 1995, **60**, 2792.
197   K. Fujita, T. Tahara, K. Ohta, Y. Nogami, T. Koga and M. Yamaguchi, *J. Org. Chem.*, 1995, **60**, 3643.
198   A.Y. Hao, L.H. Tong, F.S. Zhang and X.M. Gao, *Carbohydr. Res.*, 1995, **277**, 333.
199   R. Deschenaux, T. Ruch, P.-F. Deschenaux, A. Juris and R. Zeissel, *Helv. Chim. Acta*, 1995, **78**, 619.
200   P.J. Simms, A.T. Hotchkiss, P.L. Irwin and K.B. Hicks, *Carbohydr. Res.*, 1995, **278**, 1.
201   X.-M. Gao, L.-H. Tong, Y. Inoue and A. Tai, *Synth. Commun.*, 1995, **25**, 703.
202   F. Djedaini-Pilard, N. Azaroual-Bellanger, M. Gosnat, D. Vernet and B. Perly, *J. Chem. Soc., Perkin Trans. 2*, 1995, 723.
203   C.J. Easton, S.F. Lincoln and D.M. Schliebs, *J. Chem. Soc., Chem. Commun.*, 1995, 1167.
204   E. Leray, H. Parrot-Lopez, C. Augé, A.W. Coleman, C. Finance and R. Bonaly, *J. Chem. Soc., Chem. Commun.*, 1995, 1019.
205   T. Ikeda, K. Yoshida and H.-J. Schneider, *J. Am. Chem. Soc.*, 1995, **117**, 1453.
206   K.A. Martin, M.A. Mortellard, R.W. Sweger, L.E. Fikes, D.T. Winn, S. Clary, M.P. Johnson and A.W. Czarnik, *J. Am. Chem. Soc.*, 1995, **117**, 10443.
207   K. Hamasaki and A. Veno, *Chem. Lett.*, 1995, 859.
208   P.R. Ashton, E.Y. Hartwell, D. Philp, N. Spencer and J.F. Stoddart, *J. Chem. Soc., Perkin Trans. 2*, 1995, 1263.
209   T. Jiang, M. Li and D.S. Lawrence, *J. Org. Chem.*, 1995, **60**, 7293.
210   Y. Okabe, H. Yamamura, K.-i. Obe, K. Ohta, M. Kawai and K. Fujita, *J. Chem. Soc., Chem. Commun.*, 1995, 581.
211   V. Lainé, A. Coste-Sarguet, A. Gadelle, J. Defaye, B. Perly and F. Djedaini-Pilard, *J. Chem. Soc., Perkin Trans. 2*, 1995, 1479.
212   S. Kitahata, H. Hashimoto and K. Koizumi, *Oyo Toshitsu Kagaku*, 1994, **41**, 449 (*Chem. Abstr.*, 1995, **122**, 291 354).

213   K. Koizumi, T. Tanimoto, Y. Okada, K. Hara, K. Fujita, H. Hashimoto and S. Kitahata, *Carbohydr. Res.*, 1995, **278**, 129.

214   T. Tanimoto, T. Sakaki and K. Kaizumi, *Carbohydr. Res.*, 1995, **267**, 27.

215   H. Parrot-Lopez, E. Leray and A.W. Coleman, *Supramol. Chem.*, 1993, **3**, 37 (*Chem. Abstr.*, 1995, **122**, 240 200).

216   D.W. Jennings and R.W. Rousseau, *Carbohydr. Res.*, 1995, **273**, 243.

5
Ethers and Anhydro-sugars

1 Ethers

1.1 Methyl Ethers. – Several polyhydroxy carbohydrate derivatives have been efficiently O-methylated using the system 50% aq Na OH, DMSO, MeI.[1] A number of mono-O-methyl ethers of the trimannose oligosaccharide α-D-Man$p$-(1→3)-[α-D-Man p-(1→6)]-β-D-Man$p$-O-C$_8$H$_{17}$ have been prepared.[2] Allyl 2-acetamido-2-deoxy-3-O-β-D-galactopyranosyl-6-O-methyl-α-D-galactopyranoside has been synthesized as a potential acceptor for a sulfotransferase,[3] while mono-2-O-methyl-cyclomaltoheptaose was formed by methylation of cyclo-maltoheptaose in dilute aqueous alkali.[4]

1.2 Other Alkyl and Aryl Ethers. – Diisopropylidene derivatives of D-glucose and D-fructose have been O-alkylated, hydrolysed and reduced affording 3-O-alkyl-D-glucitol and -D-mannitol derivatives.[5] Methyl 3′,4′-O-isopropylidene-β-D-lactoside has been converted into methyl 3′,4′-di-O-n-hexyl- and n-octyl-β-D-lactosides and these ethers were tested for their lyotropic phase behaviour,[6] while 6′-O-alkyl ethers of 6-O-α-D-galactopyranosyl-D-glucopyranose have been prepared and their liquid-crystal properties determined.[7]

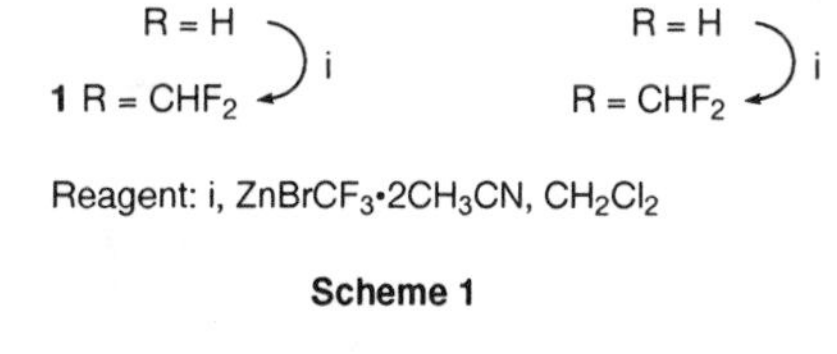

Reagent: i, ZnBrCF$_3$·2CH$_3$CN, CH$_2$Cl$_2$

Scheme 1

Some difluoromethyl ethers have been prepared (Scheme 1) and the difluoro-methyl ether moiety was found to be stable to aqueous trifluoroacetic acid under the conditions required for the deprotection of **1**.[8] A 6′-O-(2-naphthylmethyl) derivative of lactose has been synthesized as a fluorescence labelled substrate for ceramide glycanase.[9] A new route to methyl 4-*O*-benzyl-2-*O*-methyl-α-L-fucopyr-anoside has employed DIBAL reductive opening of a 3,4-*O*-benzylidene deriva-tive.[10] New 'bolaamphiphile' carbohydrate derivatives **2** have been prepared (Scheme 2), and they were deprotected by acid hydrolysis.[11] Electrochemically induced formation of *O*-ethers of some glycals has been studied. The intermediate sugar anions generated afford equilibrium mixtures of regioisomers.[12] Some mono-2-*O*-alkyl-cyclomaltoheptaose derivatives have been prepared by alkyla-tion of β-cyclodextrin in aqueous alkali.[4] The preparation and utilization as nucleoside *O*-protecting groups of *m*- or *p*- *N*-hydroxysuccinimide substituted phenyl-bis(4-methoxyphenyl)methyl ethers and the 1,1-dianisyl-2,2,2-trichlor-oethyl ether groups are covered in Chapter 20.

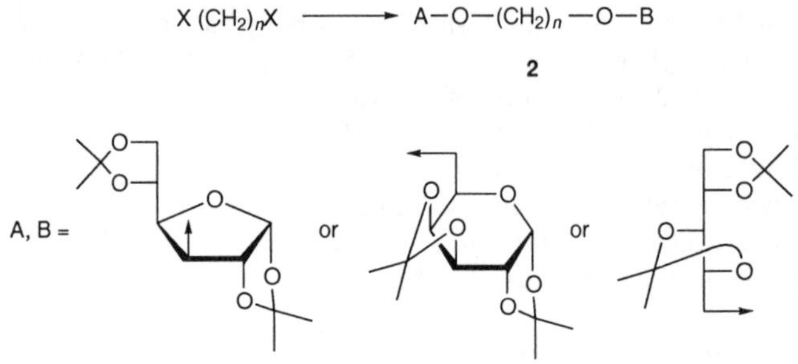

$$X\,(CH_2)_nX \longrightarrow A{-}O{-}(CH_2)_n{-}O{-}B$$

2

A, B =

Scheme 2

Benzylation of the 5-amino-5-deoxy-pentoside **3** *via* its 2,3-*O*-dibutylstannylene derivative afforded a 1:1 mixture of the 2-*O*- and 3-*O*-benzyl ethers, whereas tritylation and silylation under the same conditions generated predominantly the 2-*O*-protected derivative.[13] Similarly, benzylation of the D-erythronolactone **4** *via* its *O*-dibutylstannylene derivative gave predominantly the 2-*O*-benzyl ether **5** while reductive opening (TiCl$_4$, HSiEt$_3$) of the corresponding 2,3-*O*-benzylidene compound afforded the 3-*O*-benzyl ether **6**.[14] Partial benzylation (0.9 eq NaH, DMF, BnBr) of sucrose has afforded 42% of 2-*O*-benzyl-sucrose, isolated as its 3,4,6,1',3',4',6'-heptaacetate. The molecular electrostatic potential profile of sucrose apparently predicts that the 2-OH group is the most electropositive of the eight hydroxy-groups.[15]

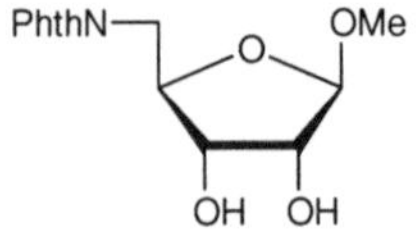

3

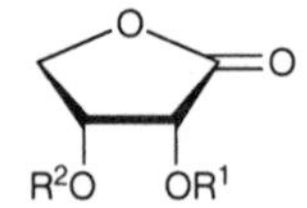

4 R^1 = R^2 = H
5 R^1 = Bn, R^2 = H
6 R^1 = H, R^2 = Bn

Benzyl ethers have been removed in the presence of 4-bromobenzoates without debromination using anhydrous ferric chloride in dichloromethane[16] and TFA in dichloromethane has been used to effect removal of O-4-methoxybenzyl protecting groups.[17]

2-O-(2-Iodoethyl)-D-glucose has been synthesized as a stable iodinated analogue of 2-deoxy-2-fluoro-D-glucose.[18] Epoxidation of methyl 2-O-allyl-4,6-O-benzylidene-α-D-glucopyranoside followed by base treatment gave the cyclic ethers **7** and **8**, which could be deprotected to the parent 4,6-diols.[19] O-Cyanoethylated sugars have been prepared by addition of acrylonitrile to alkaline solutions of semi-protected or unprotected non-reducing sugars such as sucrose, methyl α-D-glucopyranoside, methyl β-D-fructopyranoside and methyl β-D-fructofuranoside. The O-cyanoethyl products were treated with methanolic HCl to give 2-(methoxycarbonyl)ethyl derivatives which were saponified (KOH) to the potassium salts of the carboxylic acids.[20,21] The O-(diethyl phosphonoylmethyl) derivative **9** has been prepared,[22] and the (E) and (Z) but-2-enoic acid-2-yl UDP-GlcNAc derivatives **10** and **11** have been synthesized to study one of the first steps in bacterial cell wall biosynthesis.[23]

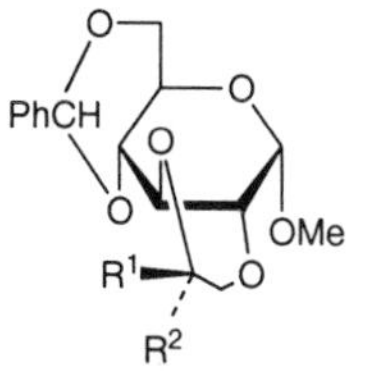

7 R¹ = CH₂OH, R² = H
8 R¹ = H, R² = CH₂OH

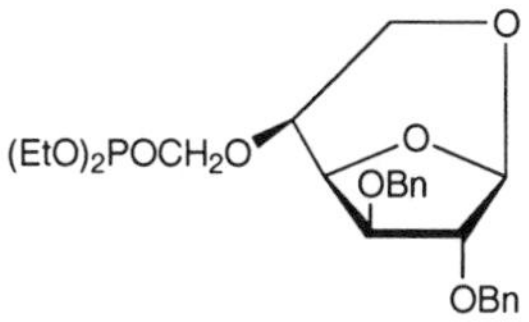

9

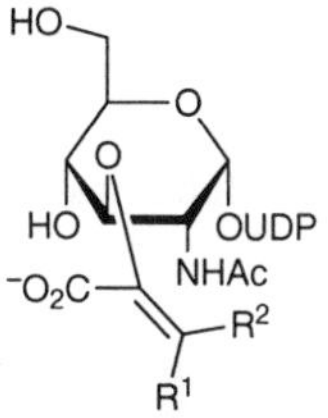

10 R¹ = Me, R² = H
11 R¹ = H, R² = Me

The association constants between the synthetic receptor **12** and a series of *p*-nitrophenyl α- and β-glycosides of D-glucose, D-mannose, D-galactose, D-xylose, L-fucose and L-arabinose have been examined in order to probe carbohydrate -

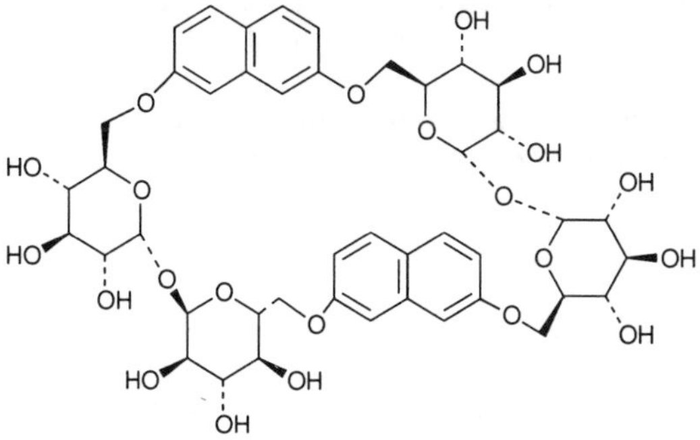

12

carbohydrate interactions in aqueous solution.[24] Chiral crown ethers **13** and **14** have been prepared from D-glucose.[25]

Pyridinium salts of some sugar sulfates can be desulfated and silylated under certain silylating conditions.[26]

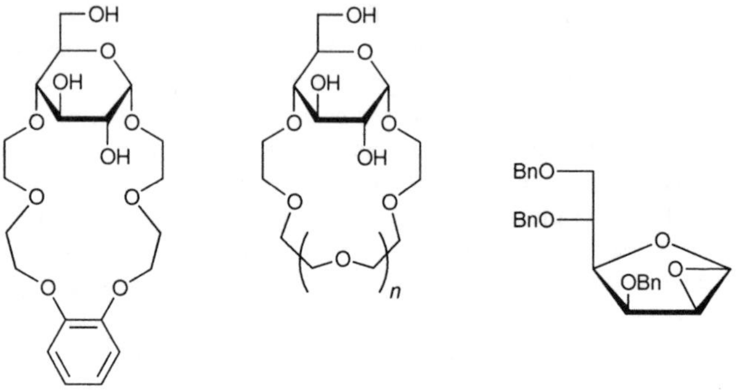

13   14 *n* = 1, 2   15

2   Intramolecular Ethers (Anhydro-sugars)

2.1   Oxirans. – The preparation, and applications in synthesis, of a series of 1,6:2,3- and 1,6:3,4-dianhydro-β-D-hexopyranoses has been reviewed.[27] Some isopropylidene derivatives of D-glucose and D-mannose have been converted into 5,6-anhydro-derivatives by treatment with TsCl in the presence of KOH,[28] and other 5,6-anhydro compounds have been prepared *via* the corresponding 5,6-thionocarbonates, which were opened with methyl iodide to give the 6-deoxy-6-iodo-5-*O*-methylthiocarbonates followed by treatment with base.[29]

Some 1,2-anhydro-aldofuranose derivatives (*e.g.* **15**) have been prepared by base treatment of 2-*O*-tosylates.[30] D-Mannose has been converted into the glycosyl chloride **16** from which the 1,2-anhydro derivative **17** was obtained.[31] Similarly, the D-fucosyl chloride **18** was converted to the 1,2-anhydride **19** *via* the corresponding β-fluoride.[32] Tri-*O*-acetyl-D-glucal and -galactal have been epoxidized (Scheme 3) to give 1,2-anhydro-compounds **20** and **21** respectively as the major products,[33] and use of a 1,2-anhydro-hexopyranose in the synthesis of an antifungal furanone glycoside is covered in Chapter 3.

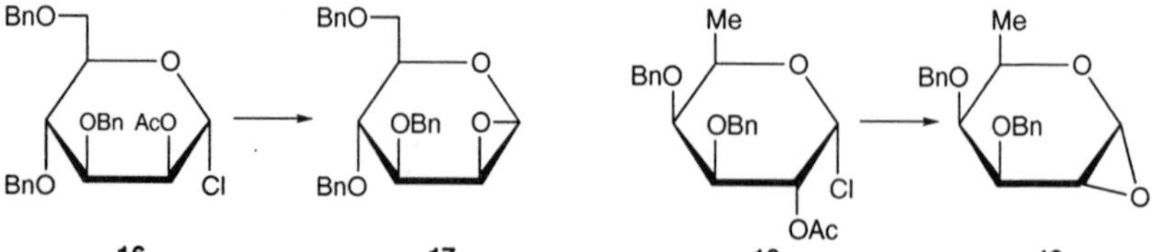

16   17   18   19

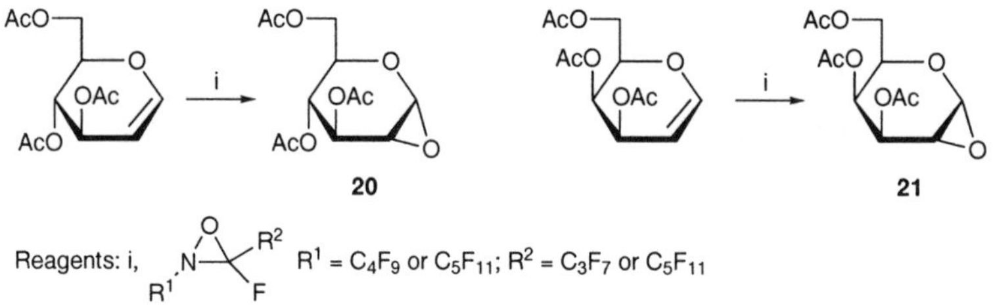

Reagents: i, [structure] $R^1 = C_4F_9$ or C_5F_{11}; $R^2 = C_3F_7$ or C_5F_{11}

Scheme 3

The two bis-epoxides 1,2:5,6-dianhydro-3,4-di-*O*-benzyl-L-iditol and -D-mannitol, on treatment with Na_2S, have afforded mixtures of thiepans and tetrahydrothiopyrans (Scheme 4),[34] and the same compounds when treated with sodium azide (silica gel, CH_3CN, reflux) gave 2,5-anhydro-1-azido-1-deoxy, *i.e.* the furanoid, derivatives.[35]

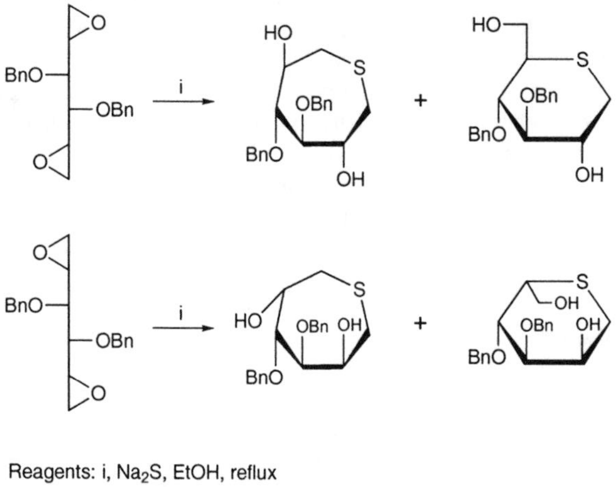

Reagents: i, Na_2S, EtOH, reflux

Scheme 4

When the peroxy compound **22** was allowed to react with nucleophiles (X^-) in the presence of added base the *C*-4 substituted epoxides **23** were formed.[36] The calculated potential surface of methyl 2,3-anhydro-4-deoxy-α-D,L-*ribo*-hexopyranoside has shown the possible existence of seven low energy conformers of the pyranose ring.[37]

The influence of Li^+ or Mg^{2+} chelation on the regioselective azide ion-induced epoxide opening of 2,3-anhydro-tetrose derivatives is covered in Chapter 10, and azide ion opening of a 1,6:3,4-dianhydrohexopyranose is detailed in Chapter 9.

AcO⌐

Bu^tOO

OMe

NO$_2$

22

AcO⌐

X

OMe

NO$_2$

23 X = H, D, OMe, STol, CH(Ac)$_2$

Me   H   OH

OH

H   Me

24

2.2 Other Anhydrides. – Stereoselective chemical and chemoenzymatic approaches to the preparation of 1,6-anhydrohexopyranoses have been reviewed,[38] and base treatment of pentabromophenyl glycosides has been investigated as an improved procedure for the synthesis of 1,6-anhydro sugars.[39] Treatment of *O*-benzylated-3-*O*-acetyl-glycosyl chlorides with base (KO^{t}Bu, THF) has allowed syntheses of 1,3-anhydro-2,4,6-tri-*O*-benzyl-β-D-talopyranose,[40] 1,3-anhydro-2,4-di-*O*-benzyl-α-L-arabinopyranose[41] and 1,3-anhydro-2,4-di-*O*-benzyl-6-deoxy-β-L-talopyranose.[42] The theoretical solution conformation of methyl 3,6-anhydro-α-D-galactopyranoside has been studied using *ab initio* calculations. The computed conformation was similar to that adopted in the crystal (X-ray analysis).[43]

In acid solution, 1-deoxy-D-*threo*-pentulose forms the dimeric **24**,[44] and the stereoselective synthesis of di-β-D-fructopyranose 1,2′:2,1′-dianhydride has been achieved by way of a fructopyranosyl fluoride.[45] The Michael addition of thiols, alcohols and *C*-nucleophiles to levoglucosenone has been used in the synthesis of analogues of the herbicide 1,6-anhydro-4-*O*-benzyl-3-deoxy-2-*O*-methyl-β-D-*ribo*-hexopyranose.[46]

References

1   H. Wang, L. Sun, S. Glazebnik and K. Zhao, *Tetrahedron Lett.*, 1995, **36**, 2953.
2   H. Paulsen, M. Springer, F. Reck, E. Meinjohanns, I. Brockhausen and H. Schachter, *Liebigs Ann. Chem.*, 1995, 53.
3   R.K. Jain, C.F. Piskorz and K.L. Matta, *Carbohydr. Res.*, 1995, **275**, 231.
4   J. Jindrich, J. Pitha, B. Lindberg, P. Seffers and K. Harata, *Carbohydr. Res.*, 1995, **266**, 75.
5   H.W.C. Raaijmakers, E.G. Arnouts, B. Zwanenburg, G.J.F. Chittenden and H.A. van Doren, *Recl. Trav. Chim. Pays-Bas*, 1995, **114**, 301.
6   V. Langlois and J.M. Williams, *J. Chem. Soc., Perkins Trans 1*, 1995, 1611.
7   L.M. Wingert, G.A. Jeffrey, D. Cabaret and M. Wakselman, *Carbohydr. Res.*, 1995, **275**, 25.
8   R. Miethchen, M. Hein, D M Naumann and W. Tyrra, *Liebigs Ann. Chem.*, 1995, 1717.
9   K. Matsuoka, S.-I. Nishimura and Y.C. Lee, *Carbohydr. Res.*, 1995, **276** 31.
10   K. Fujiwara, S. Amano and A. Murai, *Chem. Lett.*, 1995, 191.
11   P. Goueth, A. Ramiz, G. Ronco, G. MacKenzie and P. Villa, *Carbohydr. Res.*, 1995, **266**, 171.

12   S. Fischer and C.H. Harmann, *J. Carbohydr. Chem.*, 1995, **14**, 327.

13   K.S. Kim, J.W. Lim, Y.H. Joo, K.T. Kim, I.H. Cho and Y.H. Ahn, *J. Carbohydr. Chem.*, 1995, **14**, 439.

14   M. Flasche and H.-D. Scherf, *Tetrahedron: Asymm.*, 1995, **6**, 1543.

15   F.W. Lichtenthaler, S. Immel and P. Pokinski, *Liebigs Ann. Chem.*, 1995, 1939.

16   J.I. Padron and J.T. Vazquez, *Tetrahedron:Asymm.*, 1995, **6**, 857.

17   L. Yan and D. Kahne, *Synlett*, 1995, 523.

18   G. Bignan, C. Morin and M. Vidal, *Carbohydr. Res.*, 1995, **271**, 125.

19   L. Holmberg, B. Lindberg and B. Lindqvist, *Carbohydr. Res.*, 1995, **268**, 47.

20   H. Bazin, A. Bouchu and G. Descotes, *J. Carbohydr. Chem.*, 1995, **14**, 1187.

21   H. Bazin, A. Bouchu, G. Descotes and M. Petit-Ramel, *Can. J. Chem.*, 1995, **73**, 1338.

22   A.R.P.M. Valentijn, H.G.H. Broxterman, G.A. van der Marel, L.H. Cohen and J.H. van Boom, *J. Carbohydr. Chem.*, 1995, **14**, 737.

23   W.J. Lees and C.T. Walsh, *J. Am. Chem. Soc.*, 1995, **117**, 7329.

24   J. Jiménez-Barbero, E. Junquera, M. Martin-Pastor, S. Sharma, C. Vicent and S. Penadéz, *J. Am. Chem. Soc.*, 1995, **117**, 11198.

25   P.P. Kanakamma, N.S. Mani, U. Maitra and V. Nair, *J. Chem. Soc., Perkins Trans, 1*, 1995, 2339.

26   R. Takano, T. Kanda, K. Hayashi, K. Yoshida and S. Hara, *J. Carbohydr. Chem.*, 1995, **14**, 885.

27   M. Cerny, *Front. Biomed. Biotechnol.*, 1994, **2**, 121 (*Chem. Abstr.*, 1995, **122**, 240 206).

28   P.Y. Goueth, M.A. Fauvin, M. Mashoudi, A. Ramiz, G.L. Ronco and P.J. Villa, *J. Nat.*, 1994, **6**, 3 (*Chem. Abstr.*, 1995, **123**, 314 270).

29   M. Adiyaman, S.P. Khanapure, S.W. Hwang and J. Rokach, *Tetrahedron Lett.*, 1995, **36**, 7367.

30   Y. Du and F. Kong, *Tetrahedron Lett.*, 1995, **36**, 427.

31   Y. Du and F. Kong, *J. Carbohydr. Chem.*, 1995, **14**, 341.

32   Y. Du and F. Kong, *Carbohydr. Res.*, 1995, **275**, 413.

33   M. Cavicchioli, A. Mele, V. Montanari and G. Resnati, *J. Chem. Soc., Chem. Commun.*, 1995, 901.

34   M. Fuzier, Y. Le Merrer and J.-C. Depezay, *Tetrahedron Lett.*, 1995, **36**, 6443.

35   L. Poitant, Y. LeMerrer and J.-C. Depezay, *Tetrahedron Lett.*, 1995, **36**, 6887.

36   A. Seta, K. Tokuda and T. Sakakibara, *Carbohydr. Res.*, 1995, **268**, 107.

37   V.M. Andrianov, S.G. Kirilova and R.G. Zhbankov, *Zh. Prikl. Spektrosk.*, 1995, **62**, 62 (*Chem. Abstr.*, 1995, **123**, 340 570).

38   Z.J. Witczak, *Front. Biomed. Biotechnol.*, 1994, **2**, 165 (*Chem. Abstr.*, 1995, **122**, 214 333).

39   G.-J. Boons, S. Isles and P. Setala, *Synlett.*, 1995, 755.

40   Z. Gan and F. Kong, *Carbohydr. Lett.*, 1995, **1**, 27.

41   Y. Du and F. Kong, *Carbohydr. Res.*, 1995, **275**, 259.

42   Z. Gan and F. Kong, *Carbohydr. Res.*, 1995, **270**, 211.

43   S.E. Schafer, E.S. Stevens and M.K. Dowd, *Carbohydr. Res.*, 1995, **270**, 217.

44   I.A. Kennedy, T. Hemscheidt, J.F. Britten and I.D. Spenser, *Can. J. Chem.*, 1995, **73**, 1329.

45   J.M. García Fernandez, D.-R. Schnelle and J. Defaye, *Tetrahedron:Asymm.*, 1995, **6**, 307.

46   R. Blattner, R.H. Furneaux, J.M. Mason and P.C. Tyler, *Front. Biomed. Biotechnol.*, 1994, **2**, 43 (*Chem. Abstr.*, 1995, **122**, 240 189).

6
Acetals

1 Methylene and Isopropylidene Acetals

Improved conditions for the synthesis of methylene acetals of D-glucose, D-galactose, D-mannose and D-fructose by transacetalation from dimethoxy-methane have been described.[1] The new 3,4-O-isopropylidene derivatives **1** and **2** of methyl α-D-glucopyranoside, which involve diequatorial diols, have been prepared in excellent yields by conducting the reactions with 2-methoxypropene and catalytic p-TsOH in THF rather than dichloromethane.[2] Use of Zeolite HY as mild acid catalyst in the acetonation of unprotected sugars (D-glucose, D-galactose, L-arabinose, L-sorbose, D-glucofuranurono-6,3-lactone, *etc.*) with acetone gave furanose derivatives, *e.g.* 1,2:5,6-di-O-isopropylidene-α-D-galacto-furanose, as the major products.[3] Exposure of D-ribose-, D-allose- and D-altrose-propane 1,3-diyl-dithioacetals to a variety of isopropylidenation reagents furnished complex mixtures of mono- and di-acetonides; conditions for the selective formation of some mono-acetonides were devised.[4] These studies have been extended to the trimethylenedithioacetals of D-arabinose, D-glucose and D-mannose.[5] Treatment of aldosulose bis(phenylhydrazones) with acetone and catalytic p-TsOH gave mono- and di-O-isopropylidene derivatives, mostly new compounds, such as **3** and **4**.[6] The preparation of acetonides of acid-sensitive substrates is referred to below (Ref. 10).

2 Other Acetals

The isobutylidenation of acyclic polyols with methyl ethyl ketone, either in DMF with catalytic H_2SO_4 or without cosolvent and with p-TsOH as catalyst, has been examined.[7] Highly crystalline 1,2:3,5-di-O-cyclohexylidene-α-D-xylofuranose has been isolated in high yield on treatment of the crude xylose syrup obtained from corncobs with cyclohexanone and sulfuric acid in diethyl ether.[8] Hydroxyethyli-dene acetals of D-threitol, used as building blocks in the synthesis of supramole-cular host systems, are referred to in Chapter 24.

A protocol for the reliable and reproducible preparation of 4,6-O-benzylidene-D-glucopyranose in 72% yield, employing benzaldehyde dimethylacetal and TsOH in DMF, and new, corrected physical data for this compound and its crystalline 1-O-sodio derivative (thought to be the β-anomer) have been re-ported.[9] Rapid and efficient benzylidenation of acid sensitive substrates has been

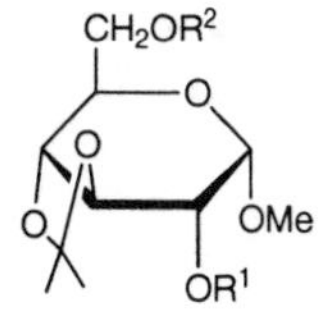

1 R^1 = R^2 = Bz
2 R^1 = H, R^2 = Tbdms

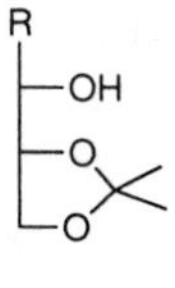

3

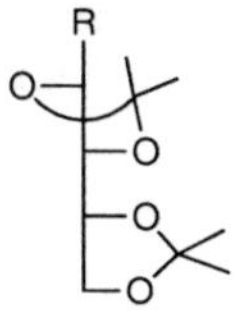

4

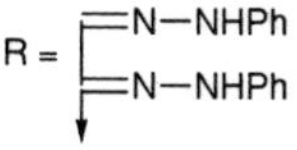

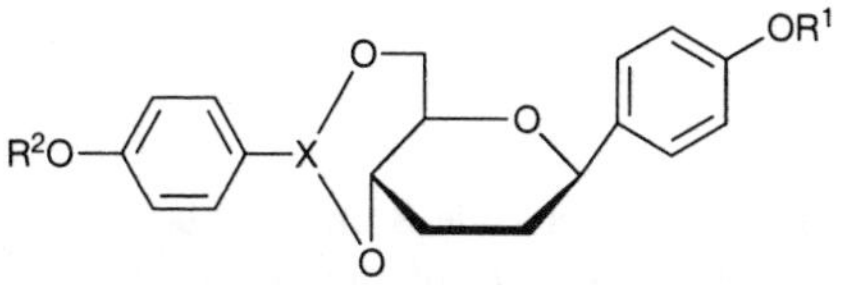

R = $=$N—NHPh
$=$N—NHPh

R^1 = (CH$_2$)$_n$ Me; *n* = 5,7,9,11
R^2 = (CH$_2$)$_m$ Me; *m* = 4, 6, 8

5 X = CH
6 X = B

achieved by halonium ion-induced acetal transfer from benzaldehyde di(pent-4-enyl) acetal; an example is shown in Scheme 1. The method is adaptable to the preparation of other acetals, *e.g.* acetonides.[10] An improved procedure for the synthesis of 1,3:4,6-di-*O*-(*p*-methoxybenzylidene)-D-mannitol, employing *p*-methoxybenzaldehyde, catalytic sulfuric acid and trimethylorthoformate has been published.[11] (*p*-Substituted benzylidene) acetals **5** and their boron analogues **6** have been examined for liquid crystal properties.[12]

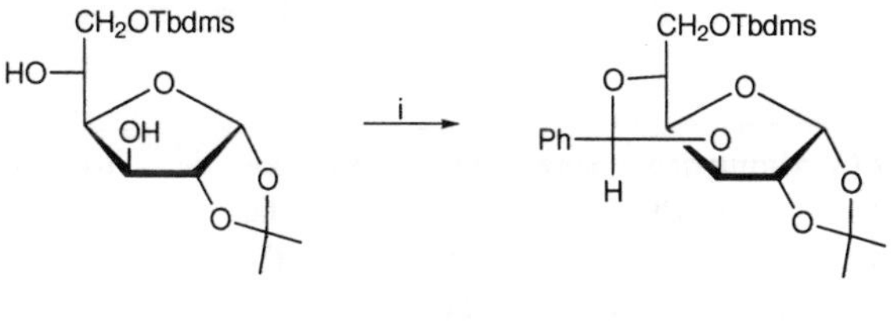

Reagents: i, PhCH[O(CH$_2$)$_3$ CH=CH$_2$]$_2$, NIS, CSA or BF$_3$ ·OEt$_2$

Scheme 1

Cyclitols and symmetrical acyclic polyols have been desymmetrized by formation of dispoke adducts such as *myo*-inositol derivative **7**.[13]

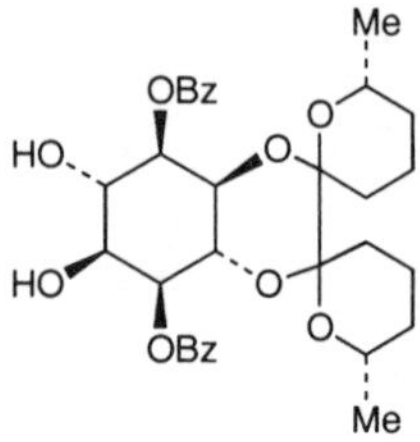

7

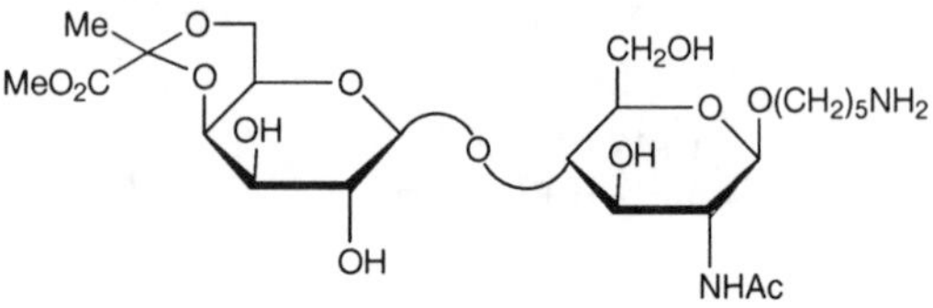

8

The multi-step syntheses of disaccharide **8** and similar pyruvated saccharide fragments related to the aggregation factor of a marine sponge involved introduction of the pyruvate acetal groups without glycosidic bond cleavage by use of methyl pyruvate in the presence of BF_3 etherate without added solvent.[14]

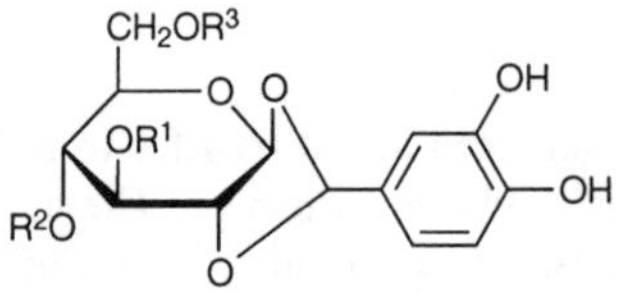

9 $R^1 = R^2 = H$, $R^3 =$

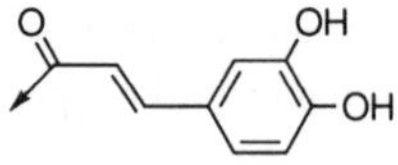

10 $R^1 = \alpha$-L-Rha$p$, $R^2 =$ 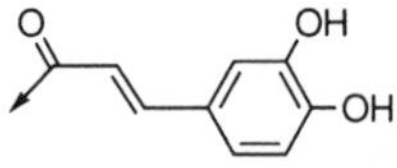   , $R^3 = H$

Two 1,2-*O*-(3′,4′-dihydroxybenzylidene)-β-D-glucose derivatives, plantanoside (**9**) and orobanchoside (**10**), were isolated from the medicinal plant *Plantago asiatica*.[15] Caeruleoside A and a related compound, found in the leaves of the Japanese plant *Lonicera caerulea*, comprise two iridoid moieties joined through an acetal linkage.[16] The first examples of natural products containing L-arabinose acetal moieties, anemoclemosides A (**11**) and B (**12**), were extracted from the roots of *Anemoclema glaucifolium*.[17]

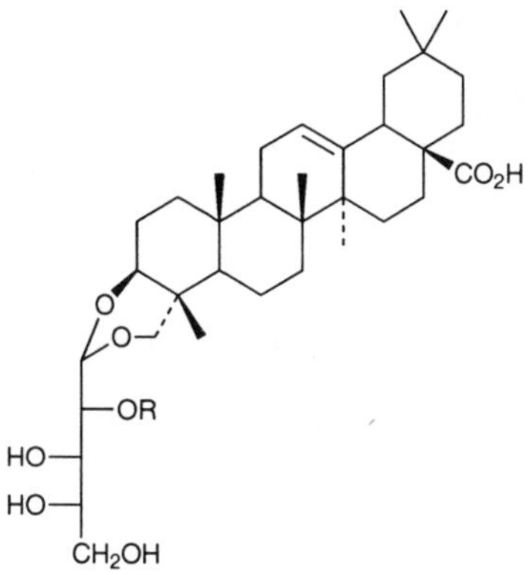

11 R = H
12 R = α-L-Rha*p*

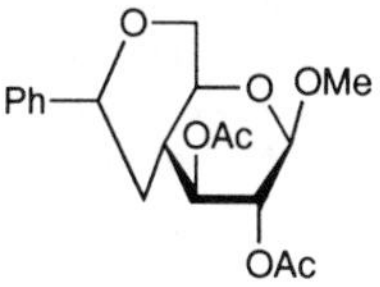

3   Reactions of Acetals

Carbohydrate isopropylidene acetals have been cleaved efficiently with catalytic quantities of DDQ in aqueous acetonitrile.[18] Benzylidene acetal **13** was reduced to the 4-hydroxy-6-*O*-benzyl ether with high efficiency by treatment with triethylsilane and TFA. The corresponding D-galactose acetal was stable under these conditions.[19]

References

1   R. Nouguier, V. Mignon and J.-L. Gras, *Carbohydr. Res.*, 1995, **277**, 339.
2   J. Cai, B.E. Davison, C.R. Ganellin and S. Thaisrivongs, *Tetrahedron Lett.*, 1995, **36**, 6535.
3   A.P. Rauter, F. Ramoa-Ribeiro, A.C. Fernandes and J.A. Figueiredo, *Tetrahedron*, 1995, **51**, 6529.
4   O. Kölln and H. Redlich, *Synthesis*, 1995, 1376.
5   O. Kölln, H. Redlich and H. Frank, *Synthesis*, 1995, 1383.
6   E.S.H. El Ashry, N. Rashed, A. Monsaad and M. El Habrouk, *Carbohydr. Res.*, 1995, **269**, 349.
7   E.S.H. El Ashry, H.A. Hamid and M. El Habrouk, *Carbohydr. Res.*, 1995, **267**, 177.
8   M. Popsavin, V. Popsavin, N. Yukojevic and D. Miljkovic, *Collect. Czech. Cem. Commun.*, 1995, **59**, 1884 (*Chem. Abstr.*, 1995, **122**, 10 392).
9   P.L. Barili, G. Berti, G. Catelani, C. Cini, F. D'Andrea and E. Mastrorilli, *Carbohydr. Res.*, 1995, **278**, 43.
10   R. Madsen and B. Fraser-Reid, *J. Org. Chem.*, 1995, **60**, 772.
11   M.A. Rampy, A.N. Pinehuk, J.P. Weichert, R.W.S. Skinner, S.-J. Fischer, R.L. Wahl, M.D. Gross and R.E. Counsell, *J. Med. Chem.*, 1995, **38**, 3156.
12   V. Vill and H.-W. Tunger, *Liebigs Ann. Chem.*, 1995, 1055.

13   R. Downham, P.J. Edwards, D.A. Entwistle, A.B. Hughes, K.S. Kim and S.V. Ley, *Tetrahedron: Asymm.*, 1995, **6**, 2403.

14   T. Ziegler, *Liebigs Ann. Chem.*, 1995, 949.

15   S. Nishibe, Y. Tamayama, M. Sasahara and C. Andary, *Phytochemistry*, 1995, **38**, 741.

16   K. Machida, J. Asano and M. Kikuchi, *Phytochemistry*, 1995, **39**, 111.

17   X.-C. Li, C.-R. Yang, Y.-Q. Liu, R. Kasai, K. Ohtani, K. Yamasaki, K. Miyahara and K. Shingu, *Phytochemistry*, 1995, **39**, 1175.

18   J.M. García Fernández, C. Ortiz Mellet, A. Moreno Marín and J. Fuentes, *Carbohydr. Res.*, 1995, **274**, 263.

19   M.P. De Ninno, J.B. Etienne and K.C. Duplantier, *Tetrahedron Lett.*, 1995, **36**, 669.

7
Esters

1 Carboxylic Esters

1.1 Synthesis. – Following the recent introduction of the 2-(chloroacetoxymethyl)benzoyl (CAMB) group (Vol. 28, Chapter 7, Ref. 3), the advantages of the 2-(2-chloroacetoxyethyl)benzoyl (CAEB) group for 1,2-*trans*-directing temporary protection of position 2 in glycosyl donors have now been described.[1] Continuing their efforts to provide standards for use in the determination of the primary structures of polysaccharides by the reductive cleavage method, G.R. Gray and co-workers have prepared all positional isomers of partially methylated/acetylated or benzoylated 1,5-anhydro-D-glucitol,[2] 1,5-anhydro-D-galactitol,[2a] 1,5-anhydro-D-mannitol,[3] 1,4-anhydro-L-fucitol,[4] 1,4-anhydro-D-ribitol,[5] and 1,4-anhydro-D-xylitol[6] (see Vol 27, Chapter 7, Ref. 6). A new strategy for the preparation of per-*O*-acetylated-6-amino-6-deoxyhexopyranosides and 1-aminoalditols without acetyl migration from O to N is referred to in Chapter 9.

A detailed study on the conventional benzoylation and pivaloylation of D-fructose has been undertaken; following consideration of factors such as equilibration rates and the distribution of tautomeric forms at given temperatures, practical procedures for the selective preparation of either pyranose, furanose, or acyclic peresters have been worked out.[7] The selectivities in electrochemically induced esterifications of D-glycals have been investigated and compared with selectivities in chemically induced ester formations.[8]

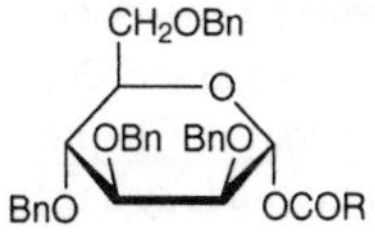

CH_2OBn

OBn BnO

BnO   OCOR

R = Me, Bn, Ph,
α-furyl,
α-naphthyl, *etc.*

1

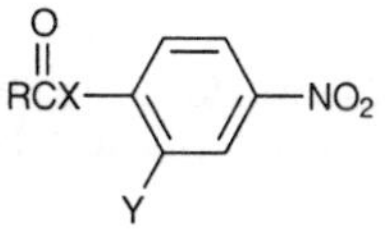

RCX—$\overset{O}{\overset{\|}{}}$—$NO_2$

Y

R = Me(CH$_2$)$_n$;
n = 10,12,14,16
2 X = S, Y = H
3 X = O, Y = NO$_2$

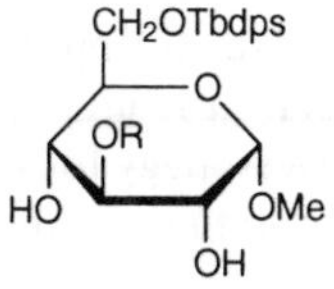

$CH_2OTbdps$

OR

HO   OMe

OH

4 R = H
5 R = Bz

2,3,4,6-Tetra-*O*-benzyl-1-*O*-trimethylsilyl-α-D-mannopyranose reacted with a series of carboxylic acids in the presence of BF$_3$.OEt$_2$ to give α-esters **1** in good yields.[9] Good primary selectivity, with isolated yields of 6-esters of 50-70%, has been achieved in the reactions of methyl α-D-gluco-, α-D-galacto-, and α-D-manno-pyranoside with acylating agents **2** and **3**.[10] The effect of microwave irradiation on the rate and selectivity of the tin-mediated benzoylation of triol **4** has been studied; at relatively low power, with 1 molar equivalent of benzoyl chloride in either toluene or acetonitrile and with a reaction time of 9 min, for example, the mono-ester **5** was preferentially formed in 35-57% yield.[11] Esterification of benzyl α-L-rhamnopyranoside by use of the dibutyltin oxide method gave the 3-esters **6** predominantly.[12] Fluoroacetyl imidazolide (**7**) has been reported to protect the benzyl α-glycoside of *N*-acetylneuraminic acid at O-4 and O-9 selectively.[13]

R = Me(CH₂)ₙ;
 n = 10, 16 *etc.*

6

7

	R¹	R²	R³	R⁴
8	Ac	Ac	Ac	Cl
9	Ac	Ac	H	SEt
10	Ac	H	Ac	SEt
11	H	Ac	Ac	SEt

The synthesis of a glycosyl donor by treatment of tri-*O*-acetyl-α-L-rhamnosyl chloride (**8**) with EtSNa in ethanethiol was accompanied by deacetylation, furnishing a 1:1:1 mixture of the diacetates **9-11**.[14] Radical reduction of bromide **12** with the slow hydrogen donor tris(trimethylsilyl)silane was accompanied by 1→2 migration of the acyloxy group resulting in the formation of the 2'-*O*-acylated α-ribonucleoside derivative **13** as the main product.[15] Sodium methoxide in methanol at −86 °C removed the anomeric acetyl groups from per-*O*-acetylated-2-acetamido-2-deoxy-D-hexopyranoses selectively in yields of >85%,[16] and sodium methoxide in methanol/toluene at ambient temperature removed the primary ester groups from 2-deoxy-3,6-diesters **14** preferentially.[17] The hydrolysis of orthoacetates to give *O*-acetates is referred to in Part 5 below (Refs. 94, 95). The formation and use of halogenated ketene acetals derived from 3-*O*-acyl-1,2:5,6-di-*O*-isopropylidene-α-D-glucofuranose is covered in Chapter 24, the crystal structures of some methyl 6-*O*-acyl-α-D-galactopyranosides in Chapter 22, and a strategy for the preparation of per-*O*-acetylated-6-amino-6-deoxy sugars is referred to in Chapter 9.

12   **13**   **14** R = Ac or Bz

A study of the molecular electrostatic potential profile of sucrose (see Chapter 2, Ref. 3a) in polar, aprotic solvents provided the basis for a convenient procedure for the preparation of sucrose 3,4,6,1',3',4',6'-heptaacetate in 42% yield *via* the 2-*O*-benzyl ether.[18] Ionic complexes of sucrose with various metal ions (*e.g.* Co^{2+} or Mn^{2+}) in DMF reacted with benzoic acid anhydride at low temperatures to form mainly the 3'-esters; the unchelated sucrose anion reacted predominantly at O-2 with acyl-migration to O-6 at moderate temperatures and transannular migration to the fructose ring at elevated temperatures.[19] Acylation of sucrose with 3-acyl-thiazolidine-2-thiones **15** or 3-acyl-5-methyl-1,3,4-thiadiazolo-2(3*H*)-thiones **16** in the presence of triethylamine gave 2-esters in good yields (see Vol. 27, Chapter 7, Ref. 23); with DBU or DBN as base, however, efficient acyl migration took place to furnish 6-esters as the only products.[20] Partially acylated methyl β-lactoside derivatives have been prepared by standard acetylation (2.1-3.0 molar equivalent of Ac_2O or AcCl in pyridine) of the 3',4'-di-*O*-hexyl- or di-*O*-octyl-ethers (see Chapter 5).[21] Malonyl-, succinyl-, and phthaloyl-tethered disaccharides, *e.g.*, compound **17**, have been prepared in connection with a novel, intramolecular glycosylation strategy (see Chapter 3).[22,23]

R = Me(CH₂)ₙ ; *n* = 6,10,12,16

15   **16**   **17**

Out of 16 commercial lipases and proteases tested, *Pseudomonas cepacia* lipase and a *Bacillus* sp. protease in pyridine had the highest activity when applied to the 6-acylation of D-glucose with various straight-chain and branched-chain acyl donors.[24] Immobilized lipases of the *Candida antarctica* type gave the best results in the lipase-catalysed primary acylation of alkyl D-glucosides, D-galactosides, and D-fructosides.[25] Regioselective esterification of 3-*O*-methyl D-glucose and methyl α-D-glucopyranoside with methacryloylaminoundecanoic acid under *Candida antarctica*-catalysis gave the polymerizable 6-esters **18** and **19**, respectively.[26] 6-*O*-Decanoyl- and -dodecanoyl-D-glucono-1,5-lactone, obtained by use of porcine pancreas lipase in pyridine, were the first members of a new class of sugar ester surfactants.[27,28]

Alkyl α- and β-hexopyranosides with protected primary hydroxyl groups were, as a rule, selectively acetylated at O-2 and O-3, respectively, by vinyl acetate and *Pseudomonas cepacia* lipase; the effects of variations in the size and hydrophobicity of the anomeric and primary substituents on the regioselectivity of these reactions have been discussed in terms of enzyme-substrate binding.[29] Methyl β-D-ribopyranoside, methyl β-D-arabinopyranoside, methyl β-D-xylopyranoside, and methyl-α-L-rhamnopyranoside were all acetylated at O-4 by lipases from three different sources, whereas methyl α-D-arabinopyranoside reacted predominantly at the 2-position, and methyl β-L-rhamnopyranoside gave the 2,4-diester under identical conditions.[30] The regioselectivities of the lipase-mediated acylations of 1,6-anhydro-β-D-hexopyranoses with ethyl butanoate[31] or vinyl acetate[32] as acyl donors have been investigated; the glucose and mannose derivatives reacted preferentially at O-4 and the galactose derivative at O-2.

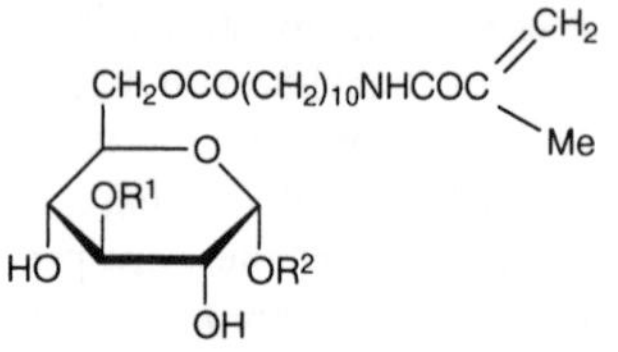

18 R^1 = Me, R^2 = H
19 R^1 = H, R^2 = Me

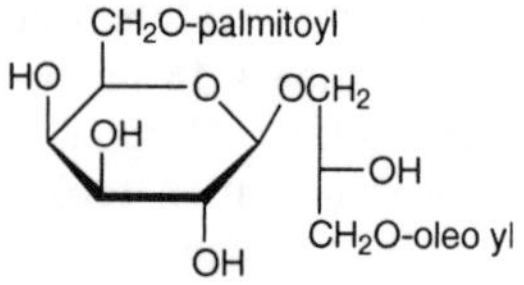

20

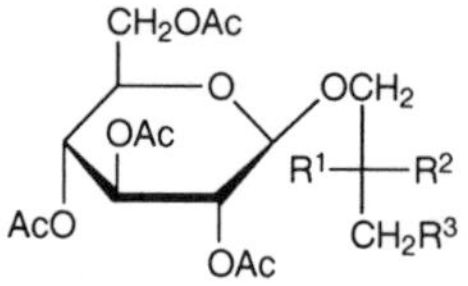

	R^1	R^2	R^3
21	H	OAc	OAc
22	OAc	H	OH
23	OH	H	OAc

A two-step procedure employing two different lipases, the first one in pyridine and the second one in dichloromethane, allowed the acylation of 3-*O*-β-D-galactopyranosyl-*sn*-glycerol consecutively at O-1 and O-6' to give galactolipids, such as diester **20**.[33] *Pseudomonas fluorescens* lipase preferentially hydrolysed the (*R*)-isomer of (2'*RS*)-glycerol β-D-glucoside peracetate to give a mixture of unreacted (2*S*)-isomer **21** and monohydroxy compounds **22** and **23**. Peracetylated DL-erythritol glucoside behaved similarly.[34] 1-*O*-Decanoyl-2,3,4,6-tetra-*O*-acetyl-β-D-glucopyranose (**24**) has been prepared efficiently by a chemo-enzymic procedure.[35]

Sucrose 1'-methacrylate, a highly desirable starting material for making new polymers, has been obtained by treatment of sucrose with vinyl methacrylate in the presence of subtilisin.[36] A theoretical study of the subtilisin-promoted esterification of sucrose with vinyl esters in organic solvents showed that with increasing hydrophobicity of the solvent and chain-length of the acyl group the preference for reaction at the 1'-position decreased in favour of reaction at O-6.[37]

A single acyl group has been introduced into β-cyclodextrin, either at a 2- or a 3-position, by exposure to benzoyl- or α- or β-naphthoyl chloride in alkaline aqueous acetonitrile.[38] A Lipid A disaccharide esterified with (*S*)-3-hydroxytetradecanoic acid and a Lipid A monosaccharide carrying a fluorinated *N*-acyl group are referred to in Chapters 3 and 9, respectively.

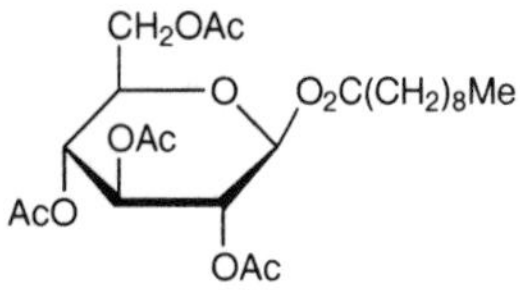

24

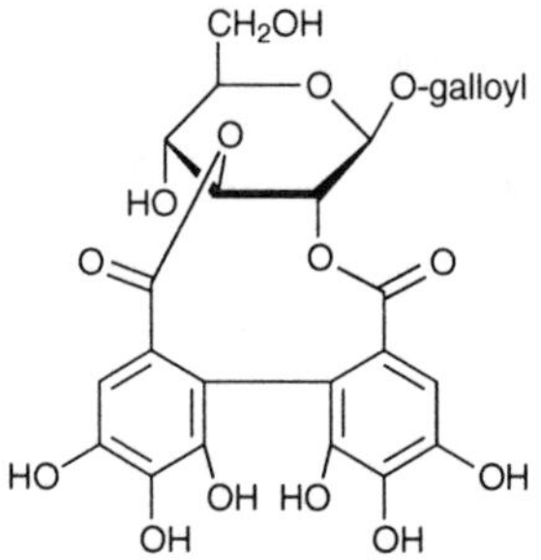

25

1.2 Natural Products. – Two novel tannins, camelliatannins C and E, which contain an elagitannin as well as a flavan-3-ol component, have been isolated from the leaves of *Camellia japonica*; in contrast to known tannins of similar composition they lack a C-C bond between C-1 of the glucose moiety and the aroyl group attached to C-2.[39] Plausible structures have been proposed for vescalagin and castalagin, the C-glucosidic elagitannins of oak, based on ^{1}H-NMR spectroscopy, mass spectroscopy and molecular mechanics calculations.[40] The total synthesis of the elagitannin sanguiin H-5 (**25**) followed a biomimetic route and involved, as the crucial step, the Pb(OAc)$_4$-mediated diastereoselective formation of a biphenyl C-C bond between the galloyl moieties at O-2 and O-3 of the glucose core.[41] Axially chiral hexamethoxydiphenic acid has been resolved *via* its diester with methyl 4,6-*O*-benzylidene α-D-glucopyranoside.[42] The tendencies of penta-*O*-galloyl-β-D-glucopyranose and poly-*O*-galloylated methyl α-D-glucosides to precipitate with protein (*e.g.* bovine serum albumin) have been investigated.[43] Hydrolysable tannins with protein-kinase inhibitory activity are referred to in Chapter 19.

A novel resin glycoside, isolated from *Merremia hungaiensis* roots and named merremin, contains two tetrasaccharide moieties linked by an ester bridge.[44]

2 Phosphates and Related Esters

Considerable attention is given to phosphate ester and related chemistry in Chapter 20. The aldolase-catalysed synthesis of monosaccharides by way of phosphate intermediates is covered in Chapter 2.

A procedure for the preparation of α-D-glucose 1-phosphate by α-glucanphosphorylase-catalysed degradation of starch has been developed.[45] Tri-*O*-acetyl-α-L-rhamnopyranosyl dialkylphosphates **26** and similar, *O*-benzyl protected deri-

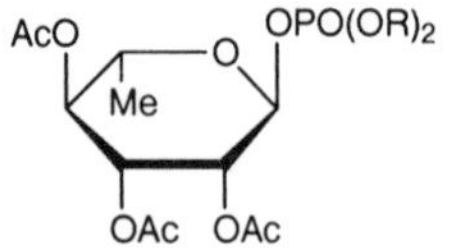

26 R = Et, Pr, Pr^i

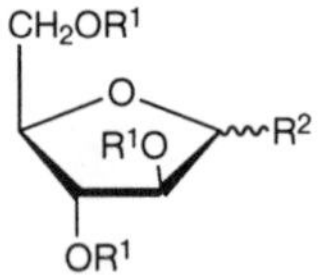

27 R^1 = Tbdms, R^2 = α-OTbdms
28 R^1 = Tbdms, R^2 = α,β-OH
29 R^1 = H, R^2 = β-O-decaprenyl

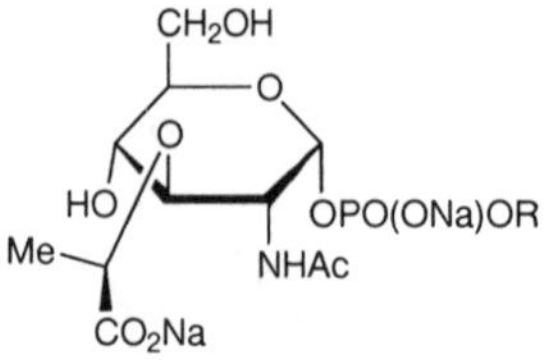

30 R = C$_{16}$H$_{33}$ or

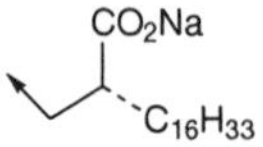

vatives have been obtained by displacement of an anomeric trifluoroacetyl group with the appropriate dialkyl phosphates.[46] The synthesis of the [1-^{14}C]-labelled mycobacterial arabinose donor **29** employed a new protection strategy involving the tetra-silylated derivative **27** which was selectively deprotected at C-1 (TFA in CH$_2$Cl$_2$) to give tri-*O*-silyl ether **28** in 77% yield; the decaprenylphosphate group was then introduced following standard procedures.[47] A series of potential transglycosylase inhibitors, compounds **30-32**, have been synthesized; the 1-alkylphosphates **30** of muramic acid were obtained by use of the trichloroacetimidate method, whereas the (α-D-glucopyranosyl)-methanephosphonates **31** were prepared from propenyl tri-*O*-acetyl-*C*-glucopyranoside, and a similar route involving the corresponing allyl *C*-glucoside led to the ethanephosphonate analogues **32**, as shown in Scheme 1.[48] The introduction of anomeric phosphono-methoxy groups by a Ferrier reaction has been explored using di-O-acetyl-D-arabinal and -xylal as substrates; an example is shown in Scheme 2.[49]

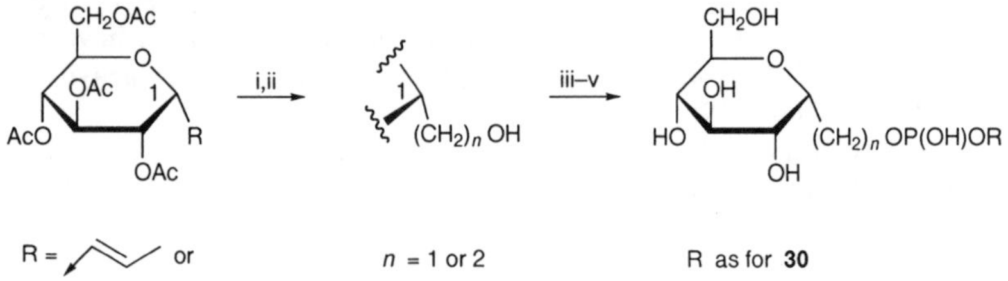

Reagents: i, O$_3$; ii, NaAcOBH$_3$; iii, MsCl, Py; iv, PO(OEt)$_3$; v, TmsBr, CH$_2$Cl$_2$,H$_2$O

Scheme 1

Phosphoramidon (**33**), an inhibitor of the endothelin-converting enzyme,[50] and the glycosylated phosphatidylcholines **34**, which are closely related to a newly dicovered, natural fungicide,[51] have been synthesized by use of the phosphorodi-chloridate method.

Reagents: i, (Pr^{i}O)$_2$ POCH$_2$OH, TmsOTf ; ii, NH$_3$, MeOH

Scheme 2

A convenient synthesis of glucose 1,6-bisphosphates with^{3}H-, ^{14}C-, ^{13}C-, ^{32}P- or ^{33}P-labels is based on the hexokinase-catalysed formation of glucose 1-phosphate from appropriately labelled D-glucose and ATP, followed by phospho-glucomutase-mediated equilibration to the bisphosphate and D-glucose.[52]

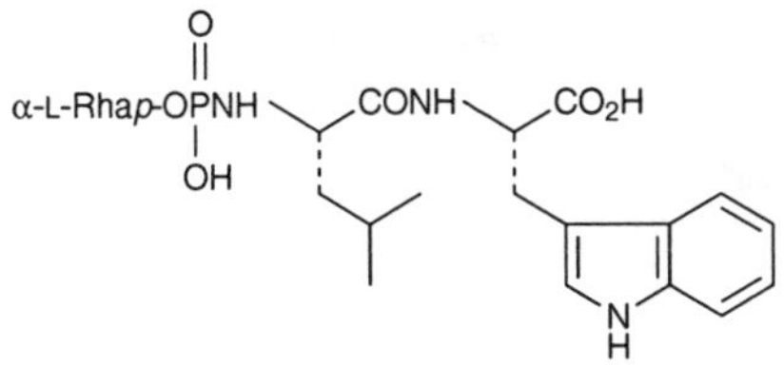

33 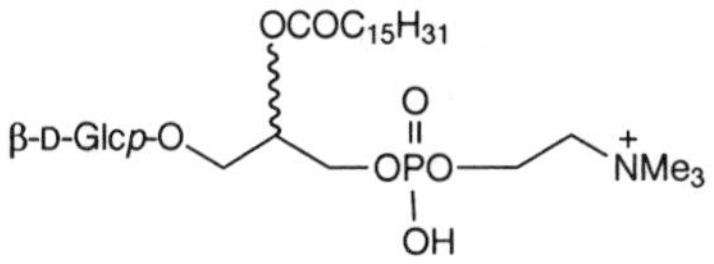   **34**

2-*O*-Phosphorylated furanose derivatives, such as compounds **36**, were obtained in high yields by exposure of alcohol **35** to dialkyl phosphorochloridates in the presence of 1-methylimidazole.[53] Triphosgene-mediated coupling of phosphonic acid with suitably protected nucleosides produced nucleoside-3′-H-phosphonate monoesters, *e.g.* compound **37**, in good to excellent yields.[54] 6-Chloro-6-deoxy-1,2-*O*-isopropylidene-α-D-glucofuranose 3,5-cyclophosphorochloridate (**38**) has been converted to the novel cyclophosphates **39**.[55]

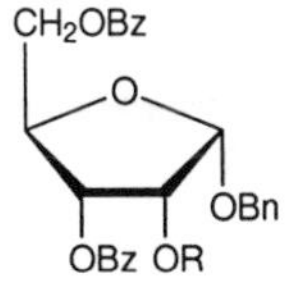

35 R = H
36 R = POX$_2$;
    X = OPh, OPr^i , Ph,
    NMe$_2$ *etc.*

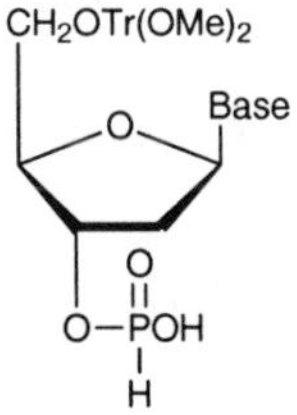

37

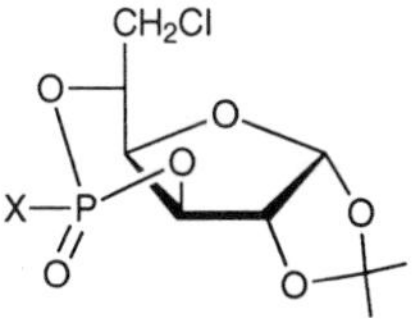

38 X = Cl
39 X = ONa, OEt, OBu,
    O(CH$_2$)$_2$CHMe$_2$

Primary (difluoromethylene)phosphonates **40** have been prepared as phosphate mimics from the corresponding 6-triflates.[56] The bis(galactopyranose 6-*O*-thiophosphoryl)disulfide **41**, made by reaction of diacetone-D-galactose with PS_5 in the presence of triethylamine, formed solvation and inclusion complexes with benzene, hexane and chloroform, which were subjected to detailed NMR spectroscopic studies.[57] An improved synthesis of L-ascorbate 2-polyphosphate involved phosphorylation of L-ascorbate with sodium trimetaphosphate in the presence of bivalent metal ions, especially Ca^{2+}, at pH 10.5.[58] The stereoselective phosphorylation of diacetonides of D-glucose, D-galactose and D-mannose, with free OH groups at C-3, C-6 and C-1, respectively, with the diastereomeric phosphonites **42** has been investigated.[59]

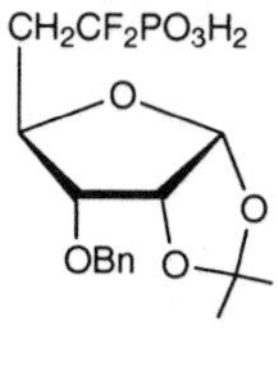

40 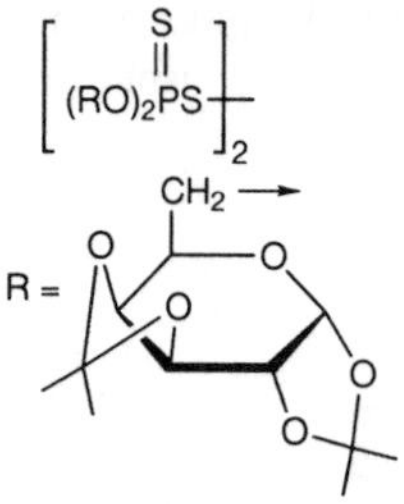   41 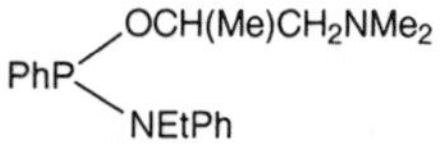   42

The carbohydrate mimics **43** and **44** of inositol 1,4,5-trisphosphate have been synthesized in eight steps from allyl α- or β-D-xylopyranoside[60] and allyl α-D-glucopyranoside,[61] respectively, and methyl 2,3,4-trisphospho-α-D-mannopyranoside has been prepared in three steps from methyl 6-*O*-trityl-α-D-mannopyranoside as a mimic of 1D-*myo*-inositol-trisphosphate.[62] The synthesis of inositol phosphates is covered in Chapter 18.

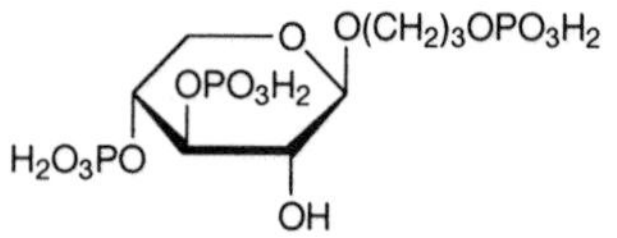

43 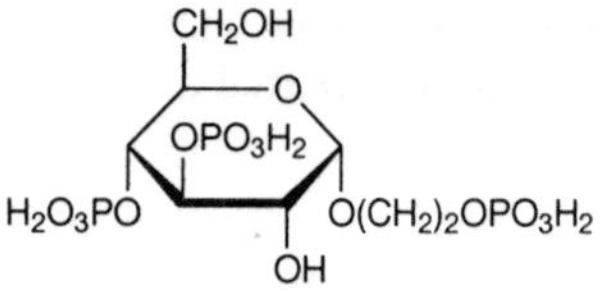   44

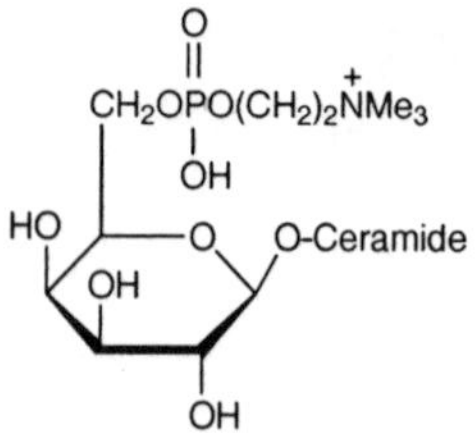

45

Six amphoteric galactocerebrosides **45**, varying in the ceramide moiety, have been isolated from the leech *Hirudo nipponica*.[63]

When moderate heat was applied to solutions of sucrose in sodium orthophosphate buffer during the final stages of freeze drying, mixtures of sucrose monophosphates were formed in appreciable quantities.[64] Treatment of 2,1′:4,6-di-*O*-isopropylidenesucrose with POCl$_3$ and pyridine in aqueous acetontrile, followed by acid hydrolysis, furnished sucrose 6′-phosphate in 15% yield.[65] For the preparation of the 4-phosphate, the 4′-phosphate and the 4,4′-bisphosphate of methyl 3-*O*-(L-*glycero*-α-D-*manno*-heptopyranosyl)-L-*glycero*-α-D-*manno*-heptopyranoside, suitably protected disaccharide precursors were exposed sequentially to phosphorus triimidazolate, benzyl alcohol and MCPBA.[66]

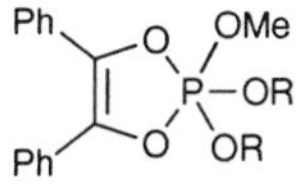

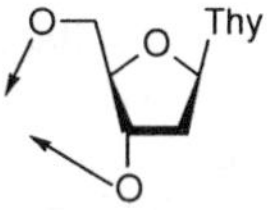

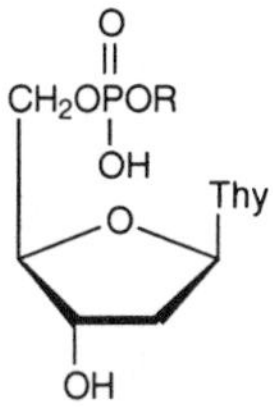

Ester exchange of oxyphosphorane **46** with thymidine gave the bicyclic phosphorane **47**, which on hydrolysis furnished products **48-50**.[67] Synthesis of adenophostin A (**51**), a potent inositol trisphosphate agonist, involved selective phosphorylation of the basic 3′-*O*-(α-D-glucopyranosyl)adenosine skeleton.[68] Phosphodi- and -tri-esters, *e.g.*, compound **52**, incorporating sugars and complexing ligands were synthesized and evaluated as antivirals.[69] The aminated GDP-fucose analogue **53**, suitable for further derivatization, has been synthesized by reaction of 6-*O*-allyl-β-L-galactopyranosyl phosphate with GMP morpholidate, followed by irradiation in the presence of 2-aminoethanethiol.[70] The syntheses of the trisubstrate analogue inhibitor **54** of α-(1→3)-fucosyl transferase[71] and of a tetrasaccharide containing (1→6)-phosphonomethyl-linked β-D-glucopyranosyl residues[72] have been described. The enzymatic synthesis of 3′-phosphoadenosine-5′-phosphosulfate (PAPS) is referred to in Part 3 below, phosphono-sialyl-Lewis X analogues are referred to in Chapter 4, and the preparation of thymidine diphospho-6-deoxy-α-D-*ribo*-3-hexulose is covered in Chapter 20.

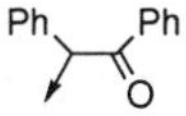

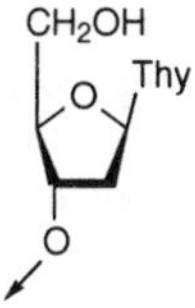

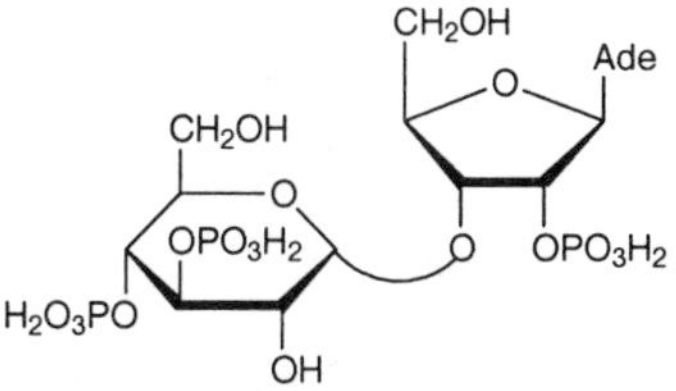

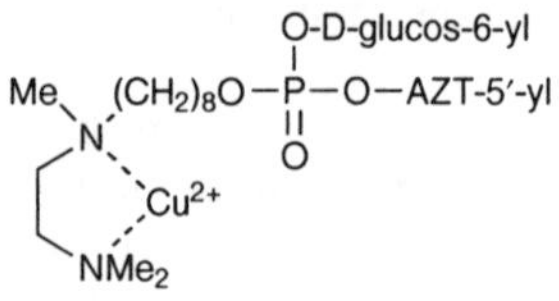

52 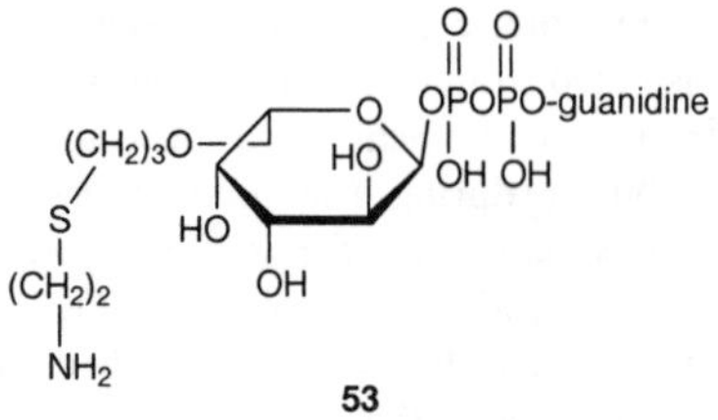   **53**

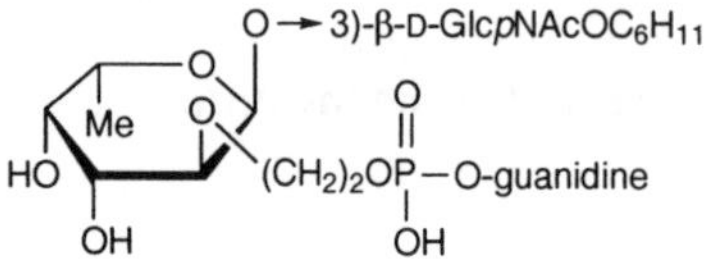

54 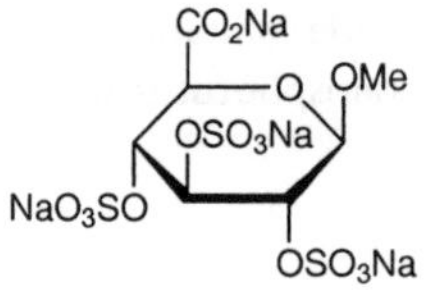   **55**

3   Sulfates

The trisulfated glucuronic acid derivative **55**,[73] all six mono-*O*-sulfates of 8-methoxycarbonyloct-1-yl β-lactosamine,[74] the 3'-, 4'-, and 6'-sulfates of di-saccharide **56**,[75] and several polysulfates of disaccharide **57**[76] have been prepared by standard procedures either for conformational or enzyme inhibition studies. Neuraminic acid has been converted to *N*-glycolyl-8-*O*-sulfoneuraminic acid by use of conventional protecting group methodology and sulfation with $Me_3N.SO_3$ in DMF.[77] The same sulfating agent was employed in the persulfa-tion of β-D-glucosides, β-D-galactosides and lactosides to obtain ceramide mimics.[78] 3'-Phosphoadenosine 5'-phosphosulfate (PAPS) has been prepared from ATP and inorganic sulfate in two enzyme-catalysed reaction steps; it was then used as sulfate donor in the sulfotransferase-catalysed 6-sulfation of *N,N'*-diacetylchitobiose.[79] The synthesis of the 5-amino-5-deoxypentose 2-sulfate component of novel lipid-containing nucleoside antibiotics is covered in Chapters 10 and 20, sulfated sialyl-Lewis X epitope analogues in Chapter 4, and the desulfation of pyridinium salts of sugar sulfates by silylating agents is referred to in Chapter 5.

β-D-GlcpNAc-(1→3)-β-D-Galp-OMe   β-D-Galp-(1→3)-α-D-GalpNAc-OAll
56   **57**

Spirostanol glycosides, isolated from the herb *Peliosanthes sinica*, contain 4-*O*-sulfo-α-L-arabinopyranosyl- and 4-*O*-sulfo-β-D-fucopyranosyl-moieties.[80]

4   Sulfonates

Treatment of diethyl D-galactarate with 2 molar equivalents of TsCl in pyridine gave the tetratosylate as the main product in 37% yield; use of 4.4 molar equivalent of TsCl furnished the ditosylate diene **58** rather than the mono-alkene **59**, as had been reported earlier (R.S. Tipson and M.A. Clapp, *J. Org. Chem.*, 1953, **18**, 952).[81] Mesylation (2.2 molar equivalents MsCl, Py) of benzyl β-D-glucoside was non-selective, furnishing a mixture of dimesylates as well as the 3,4,6-trimesylate.[82] 6-*O*-Benzoyl-1,2-*O*-cyclohexylidene-3-*O*-tosyl- and -triflyl-α-D-xylofuranose have been synthesized as precursors of 3-azido-3-deoxy-D-ribonucleosides.[83] Sucrose analogues modified at the 4-position were accessible by double inversion *via* 4-*O*-triflyl-1,2,3,3′,4′,6,6′-heptapivaloates.[84] On irradiation in methanol in the presence of triethylamine, the pentaflate **60** rearranged to the sulfite **61**.[85]

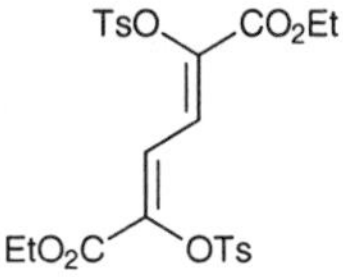

58    **59** 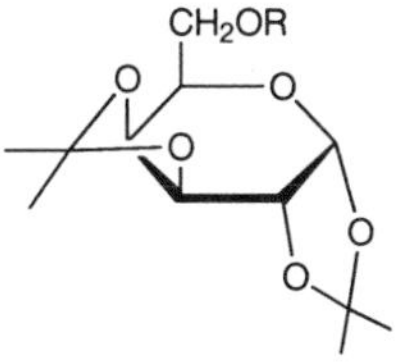   **60** R = $\overset{\overset{\displaystyle O}{\|}}{\underset{\underset{\displaystyle O}{\|}}{S}}$—$C_6F_5$

61 R = $\overset{\overset{\displaystyle O}{\|}}{\underset{\underset{\displaystyle O}{\|}}{SOC_6F_5}}$

5   Other Esters

Cyclic carbonates used as diol protecting groups in cyclitol chemistry are covered in Chapter 18. Allyloxycarbonyl protection of sugar hydroxyl groups has been effected with allyl chloroformate or allyl 1-benzotriazolyl carbonate; use of the reagent system Pd(0)/PBu$_3$/HCO$_2$H/Et$_3$N allowed selective removal of this group.[86] Glucuronyl carbamates **63**, to be used in the preparation of β-glucuronidase-sensitive pro-drugs, have been obtained by treatment of the 2,3,4,6-protected free sugar **62** with the appropriate spacer isocyanates.[87] The previously unreported 7-*O*-carbamoyl-L-*glycero*-D-*manno*-heptose has been isolated from the lipopolysaccharide fraction of several *Pseudomonas* strains.[88]

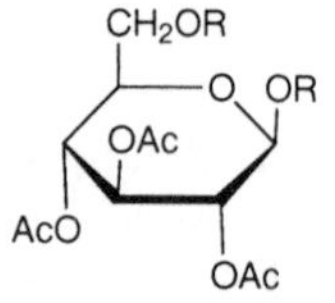
CH2OR
O OR
OAc
AcO
OAc

62 R = H
63 R = CONH-spacer-CO_2All

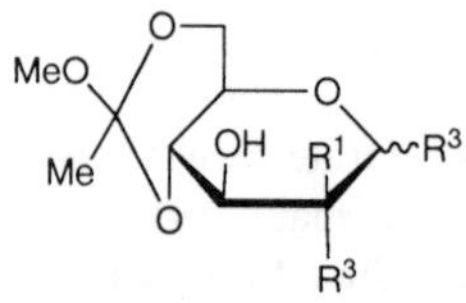
MeO
Me
O
OH
R1
R3

	R^1	R^2	R^3
64	H	OH	α, β-OH
65	OH	H	α, β-OH
66	H	OH	α-OMe

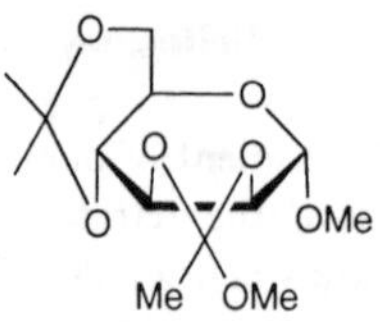
O
OMe
Me OMe

67

The complexation of aldoses with cholesterol-derived boronic acids has been reviewed,[89] and the changes in UV absorption and fluorescence intensity of a stilbene-type boronic acid on complexation with aldoses, especially D-fructose, have been studied with a view to their use in sugar-detection.[90] Similar work has been carried out with diboronic acids derived from biphenyl[91] and a *N*-functionalized diaza-18-C-6 crown ether.[92] A carbohydrate boronic acid derivative with liquid crystal properties referred to in Chapter 6, and further reports on boronate esters are noted in Chapter 17.

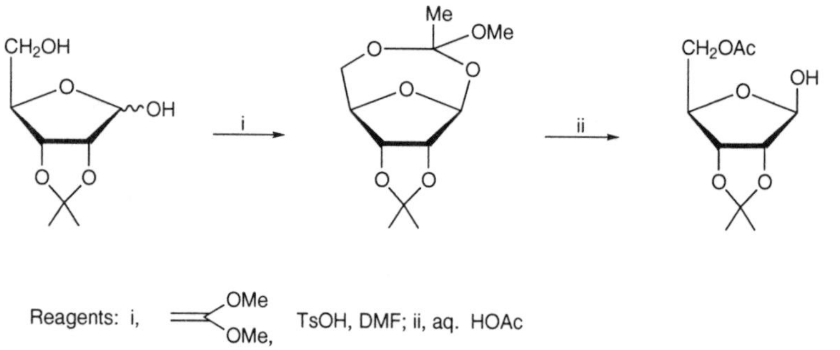
CH2OH
OH
i
Me
OMe
ii
CH2OAc
OH
Reagents: i, OMe OMe, TsOH, DMF; ii, aq. HOAc

Scheme 3

Exposure to mercury(II) bromide and 2,4-collidine converted acetylated glycosyl halides to 1,2-orthoesters.[93] Formation of orthoesters under kinetic control (ketene dimethylacetal, *p*TsOH, DMF) gave 4,6-derivatives **64-66** from D-glucose, D-mannose and methyl α-D-glucopyranoside, respectively, and product **67** from methyl 4,6-*O*-isopropylidene-α-D-mannopyranoside. Mild acid hydrolysis of **67** gave specifically the 2-*O*-acetate.[94] This method has been applied to the selective acetylation of monoisopropylidenated furanoses; an example is given in Scheme 3.[95]

References

1   T. Ziegler and G. Pantkowski, *Tetrahedron Lett.*, 1995, **36**, 5727.
2   L.E. Elvebak II and G.R. Gray, *Carbohydr. Res.*, 1995, **274**, 85.
2a   L.E. Elvebak II, C. Abbott, S. Wall and G.R. Gray, *Carbohydr. Res.*, 1995, **269**, 1.
3   L.E. Elvebak II, H.J. Cha, P. McNally and G.R. Gray, *Carbohydr. Res.*, 1995, **274**, 71.
4   N. Wang, L.E. Elvebak II and G.R. Gray, *Carbohydr. Res.*, 1995, **274**, 59.
5   C.R. Rozanas, N. Wang, K. Vidlock and G.R. Gray, *Carbohydr. Res.*, 1995, **274**, 99.
6   N. Wang and G.R. Gray, *Carbohydr. Res.*, 1995, **274**, 45.
7   F.W. Lichtenthaler, J. Klotz and F.-J. Flath, *Liebigs Ann. Chem.*, 1995, 2069.
8   S. Fischer and C.H. Harmann, *J. Carbohydr. Chem.*, 1995, **14**, 327.
9   H.Q. Huang, Z.J. Li and M.S. Cai, *Chin. Chem. Lett.*, 1995, **6**, 275 (*Chem. Abstr.*, 1995, **123**, 144 390).
10   J. Xia and X. Hui, *Synth. Commun.*, 1995, **25**, 2235.
11   B. Herradón, A. Morcuende and S. Valverde, *Synlett.*, 1995, 455.
12   A.K.M.S. Kabir and M.M. Matin, *J. Bangladesh. Chem. Soc.*, 1994, **7**, 73 (*Chem. Abstr.*, 1995, **122**, 240 179).
13   S. Sepulveda-Boza and U. Stather, *Bol. Soc. Chil. Quim.*, 1994, **39**, 299 (*Chem. Abstr.*, 1995, **122**, 161 131).
14   G. Dekany, P. Ward and I. Toth, *J. Carbohydr. Chem.*, 1995, **14**, 227.
15   T. Gimisis, G. Ialongo, M. Zamboni and C. Chatgilialoglu, *Tetrahedron Lett.*, 1995, **36**, 6781.
16   S.S. Pertel and V.Ya. Chirva, *Khim. Prir. Soedin,* 1994, 177 (*Chem. Abstr.*, 1995, **122**, 314 948).
17   E. Petrakova and C.P.J. Glaudemans, *Carbohydr. Res.*, 1995, **268**, 135.
18   T. Ziegler, G. Lemanski and A. Rakoczy, *Tetrahedron Lett.*, 1995, **36**, 8973.
19   R. Lau, G. Schüle, U. Schwaneberg and T. Ziegler, *Liebigs Ann. Chem.*, 1995, 1745.
20   F.W. Lichtenthaler, S. Immel and P. Pokinskyi, *Liebigs Ann. Chem.*, 1995, 1939.
21   J.L. Navia, R.A. Roberts and R.E. Wingard, Jr., *J. Carbohydr. Chem.*, 1995, **14**, 465.
22   K. Baczko, C. Nugier-Chauvin, J. Banoub, P. Thibault and D. Plusquellec, *Carbohydr. Res.*, 1995, **269**, 79.
23   V. Langlois and J.M. Williams, *J. Chem. Soc., Perkin Trans. 1,* 1995, 1611.
24   T. Watanabe, R. Matsue, Y. Hona and M. Kuwahara, *Carbohydr. Res.*, 1995, **275**, 215.
25   A.T.J.W. de Goede, M. van Oosterom, M.P.J. van Deurzen, R.A. Sheldon. H.van Bekkum and F.van Rantwijk, *Biocatalysis,* 1994, **9**, 145 (*Chem. Abstr.*, 1995, **123**, 228 667).
26   U. Geyer, D. Klemm, K. Pavel and H. Ritter, *Macromol. Rapid Commun.*, 1995, **16**, 337 (*Chem. Abstr.*, 1995, **123**, 112 557).
27   D. Kwoh, D.J. Pocalyko, A.J. Carchi, B. Harichian, L.O. Hargiss and T.C. Wong, *Carbohydr. Res.*, 1995, **274**, 111.
28   D.J. Pocalyko, A.J. Carchi and B. Harichian, *J. Carbohydr. Chem.*, 1995, **14**, 265.
29   D.A MacManus and E.N. Vulfson, *Carbohydr. Res.*, 1995, **279**, 281.
30   N.B. Bashir, S.J. Phythian, A.J. Reason and S.M. Roberts, *J. Chem. Soc., Perkin Trans. 1,* 1995, 2203.
31   M. Woudenberg-van Oosterom, C. Vitry, J.M.A. Baas, F. van Rantwijk and R.A. Sheldon, *J. Carbohydr. Chem.*, 1995, **14**, 237.
32   N. Junot, J.C. Meslin and C. Rabiller, *Tetrahedron: Asymm.*, 1995, **6**, 1387.
33   T. Morimoto, A. Nagatsu, N. Murakami and J. Sakakibara, *Tetrahedron*, 1995, **51**, 6443.

34   A. Soriente, M. de Rosa, A. Trincone and G. Sodano, *Bioorg. Med. Chem. Lett.*, 1995, **5**, 2321.

35   O. Kirk, M.W. Christensen and T. Damhus, *Biocatal. Biotransform.*, 1995, **12**, 91 (*Chem. Abstr.*, 1995, **123**, 314 271).

36   A. W.-Y. Chan and B. Ganem, *Biocatalysis*, 1993, **8**, 163 (*Chem. Abstr.*, 1995, **123**, 56 432).

37   J. O. Rich, B.A. Bedell and J.S. Dordick, *Biotechnol. Bioeng.*, 1995, **45**, 426 (*Chem. Abstr.*, 1995, **122**, 240 214).

38   A.Y. Hao, L.H. Tong, F.S. Zhang and X.M. Gao, *Carbohydr. Res.*, 1995, **277**, 333.

39   T. Hatano, L. Han, S. Taniguchi, T. Shingu, T. Okuda and T. Yoshida, *Chem. Pharm. Bull.*, 1995 **43**, 1629.

40   N. Vivas, M. Laguerre, Y. Glories, G. Bourgeois and C. Vitry, *Phytochemistry*, 1995, **39**, 1193.

41   K.S. Feldman and A. Sambandam, *J. Org. Chem.*, 1995, **60**, 8171.

42   T. Itoh and J.-i. Chika, *J. Org. Chem.*, 1995, **60**, 4968.

43   H. Kawamoto, F. Nakatsubo and K. Murakami, *Phytochemistry*, 1995, **40**, 1503.

44   N. Noda, K. Tsuji, T. Kawasaki, K. Miyahara, H. Hanazono and C.R. Yang, *Chem. Pharm. Bull.*, 1995, **43**, 1061.

45   B. Nidetzki, A. Weinhäuser, R. Griessler and K.D. Kulbe, *J. Carbohydr. Chem.*, 1995, **14**, 1017.

46   Z.-J. Li, Z.-Z. Li and W.-K. Li, *Synth. Commun.*, 1995, **25**, 2545.

47   R.E. Lee, K. Mikusova, P.J. Brennan and G.S. Besra, *J. Am. Chem. Soc.*, 1995, **117**, 11829.

48   G. Brooks, P.D. Edwards, J.D.I. Hatto, T.C. Smale and R. Southgate, *Tetrahedron*, 1995, **51**, 7999.

49   M.-J. Perez-Perez, B. Doboszewski, J. Rozenski and P. Herdewijn, *Tetrahedron: Asymm.*, 1995, **6**, 973.

50   G. De Nanteuil, A. Benoist, G. Rémond, J.-J. Descombes, V. Barou and T.J. Verbeuren, *Tetrahedron Lett.*, 1995, **36**, 1435.

51   D.E. Bierer, L.B. Dubenko, J. Litvak, R.E. Gerber, J. Chu, D.LK. Thai, M.S. Tempesta and T.Y. Truong, *J. Org. Chem.*, 1995, **60**, 7646.

52   T.H. Nielsen, B. Wischmann and B.L. Moller, *J. Labelled Copd. Radiopharm.*, 1995, **36**, 679, (*Chem. Abstr.*, 1995, **123**, 340 565).

53   C. Lamberth, *Org. Prep. Proced. Int.*, 1994, **26**, 595 (*Chem. Abstr.*, 1995, **122**, 65 318).

54   N.N. Bhongle and J.Y. Tan, *Tetrahedron Lett.*, 1995, **36**, 6803.

55   S.B. Khrebtova, M.P. Koroteev and E.E. Nifant'ev, *Zh. Obshch. Khim.*, 1994, **64**, 1846 (*Chem. Abstr.*, 1995, **123**, 144 389).

56   D.B. Berkowitz and D.G. Sloss, *J. Org. Chem.*, 1995, **60**, 7047.

57   M.J. Potrzebowski, J. Blaszczyk and M.W. Wieczorek, *J. Org. Chem.*, 1995, **60**, 2549.

58   X. Wang, W.-W. Qian and P.A. Seib, *J. Carbohydr. Chem.*, 1995, **14**, 53.

59   E.E. Nifantyev and M.A. Gratchev, *Tetrahedron Lett.*, 1995, **36**, 1727.

60   N. Moitessier, F. Chrétien, Y. Chapleur and C. Humeau, *Tetrahedron Lett.*, 1995, **36**, 8023.

61   D.J. Jenkins and B.V.L. Potter, *J. Chem. Soc., Chem. Commun.*, 1995, 1169.

62   M. Malmberg and N. Rehnberg, *Tetrahedron Lett.*, 1995, **36**, 8879.

63   N. Noda, R. Tanaka, K. Tsujino, M. Miura, K. Miyahara and J. Hayakawa, *Chem. Pharm. Bull.*, 1995, **43**, 567.

64   E. Tarelli and S.F. Wheeler, *Carbohydr. Res.*, 1995, **269**, 359.

65   K.B. Kim and E.J. Behrman, *Carbohydr. Res.*, 1995, **270**, 71.
66   K. Ekelöf and S. Oscarson, *J. Carbohydr. Chem.*, 1995, **14**, 299.
67   X. Chen and Y.F. Zhao, *Synth. Commun.*, 1995, **25**, 3691.
68   H. Hotoda, M. Takahashi, K. Tanzawa, S. Takahashi and M. Kaneko, *Tetrahedron Lett.*, 1995, **36**, 5037.
69   C. Desseaux, C. Gouyette, Y. Henin and T. Huynh-Dinh, *Tetrahedron*, 1995, **51**, 6739.
70   C. Hällgren and O. Hindsgaul, *J. Carbohydr. Chem.*, 1995, **14**, 453.
71   B.M. Heskamp, G.A. van der Marel and J.H. van Boom, *J. Carbohydr. Chem.*, 1995, **14**, 1265.
72   K.C. Nicolaou, H. Florke, M.G. Egan, T. Barth and V.A. Estevez, *Tetrahedron Lett.*, 1995, **36**, 1775.
73   H.P. Wessel and S. Bartsch, *Carbohydr. Res.*, 1995, **274**, 1.
74   R.A. Field, A. Otter. W. Fu and O. Hindsgaul, *Carbohydr. Res.*, 1995, **276**, 347.
75   R.K. Jain, X.G. Liu and K.L. Matta, *Carbohydr. Res.*, 1995, **268**, 279.
76   R.K. Jain, C.F. Piskorz and K.L. Matta, *Carbohydr. Res.*, 1995, **275**, 231.
77   M. Tanaka, T. Kai, X.-L. Sun, H. Takayanagi and K. Furuhata, *Chem. Pharm. Bull.*, 1995, **43**, 2095.
78   H. Yoshida, K. Ikeda, K. Achiwa and H. Hoshino, *Chem. Pharm. Bull.*, 1995, **43**, 594.
79   C.-H. Lin, G.-G. Shen, E. García-Junceda and C.-H. Wong, *J. Am. Chem. Soc.*, 1995, **117**, 8031.
80   X.-C. Li, C.-R. Yang, T. Nohara, R. Kasai and K. Yamasaki, *Chem. Pharm. Bull.*, 1995, **43**, 631.
81   T.J. van Saarlos, H. Regeling, B. Zwanenburg and G.J.F. Chittenden, *J. Carbohydr. Chem.*, 1995, **14**, 1007.
82   C.F. Carvalho, C.H.L. Kennard, D.E. Lynch, G. Smith and A. Wong, *Aust. J. Chem.*, 1995, **48**, 1767.
83   M. Popsavin, S. Lajsic, G. Cetkovic, V. Popsavin and D. Miljkovic, *J. Serb. Chem. Soc.*, 1993, **58**, 1011 (*Chem. Abstr.*, 1995, **122**, 10 414).
84   C. Simiand and H. Driguez, *J. Carbohydr. Chem.*, 1995, **14**, 977.
85   S. Duan, E.R. Binkley and R.W. Binkley, *J. Carbohydr. Chem.*, 1995, **14**, 1029.
86   T. Harada, H. Yamada, H. Tsukamoto and T. Takahashi, *J. Carbohydr. Chem.*, 1995, **14**, 165.
87   R.G.G. Leenders, K.A.A. Gerrits, R. Ruijtenbeek and H.W. Scheeren, *Tetrahedron Lett.*, 1995, **36**, 1701.
88   F. Beckmann, H. Moll, K.-E. Jager and U. Zahringer, *Carbohydr. Res.*, 1995, **267**, C3.
89   T.D. James, H. Kawabata, R. Ludwig, K. Marata and S. Shinkai, *Tetrahedron*, 1995, **51**, 555.
90   H. Shinmori, M. Takeuchi and S. Shinkai, *Tetrahedron*, 1995, **51**, 1893.
91   M. Mikami and S. Shinkai, *Chem. Lett.*, 1995, 603.
92   K. Nakashima and S. Shikai, *Chem. Lett.*, 1995, 443.
93   K. Matsuoka, S.-I. Nishimura and Y.C. Lee, *Bull. Chem. Soc. Jpn.*, 1995, **68**, 1715.
94   M. Bouchra, P. Calinaud and J. Gelas, *Carbohydr. Res.*, 1995, **267**, 227.
95   M. Bouchra, P. Calinaud and J. Gelas, *Synthesis*, 1995, 561.

8
Halogeno-sugars

1 Fluoro-sugars

Anomerically unprotected aldopyranoses have been converted into 1-*O*-tetrazole derivatives which undergo displacement by fluoride ions (HF.Py) to afford glycosyl fluorides.[1] Epoxidation of *O*-acetylated glycals has afforded 1,2-anhydro compounds which, on treatment with tetrabutylammonium fluoride, give glycosyl fluorides,[2] and 2,3:4,5-di-*O*-isopropylidene-β-D-fructopyranose in the presence of hydrogen fluoride has given 1,3:4,5-di-*O*-isopropylidene-β-D-fructopyranosyl fluoride in a reaction that involved acetal migration.[3] 2-*O*-(2-Acetoxyethyl)-3,4,6-tri-*O*-methyL-α-D-glucopyranosyl fluoride was obtained by cleavage (HF, CH$_3$NO$_2$, Ac$_2$O) of the corresponding 1,2-*O*-ethanediyl-β-D-glucopyranose derivative. The same medium converted 1,2-*O*-isopropylidene-3,5,6-tri-*O*-methyl-α-D-glucofuranose into the 2-*O*-acetyL-glucofuranosyl fluoride. With 2-*O*-benzyl protected sugars the benzyl ether participated in an intramolecular reaction to form cyclic *C*-glycosides.[4]

UDP-6-Deoxy-6-fluoro-D-galactose has been synthesized and then used as a donor substrate with a β-(1→4)-galactosyltransferase and *N*-acetylglucosamine to give 6'-deoxy-6'-fluoro-*N*-acetyllactosamine.[5] 5-Azido-3,5,6-trideoxy-3,6-difluoro-D-glucose has been prepared and was converted by hydrogenolysis into 1,3,6-trideoxy-3,6-difluoronojirimycin.[6] Sucrose has been converted *via* the 1',2,3,3',4',6,6'-heptapivalate into 4-deoxy-4-fluoro-sucrose[7] and D-ribono-1,4-lactone has been transformed into 1-*O*-acetyl-5-*O*-*tert*-butyldiphenylsilyl-2,3-dideoxy-2-fluoro-D-ribose and hence into nucleoside analogues.[8] Some other deoxyfluoro nucleoside analogues are covered in Chapter 20, and the synthesis of a number of deoxyfluoro-inositols is mentioned in Chapter 18.

A number of 2,6-dideoxy-6,6,6-trifluorohexoses were prepared from a six-carbon acetylenic precursor *via* selective hydroxylation of the derived alkene diastereomers,[9,10] and differences in stereoselectivity of reduction of a trifluoromethyl ketone compared with the corresponding methyl ketone are outlined in Chapter 18. Oxidation of methyl 5-*O*-benzyl-3(2)-deoxy-3(2)-fluoro-α-D-pentofuranosides (DMSO/TFAA) was accompanied by epimerization at the fluorinated carbon atom α- to the ketone resulting in formation of the corresponding 2-(or 3-) keto derivatives as mixtures of two epimers. Reduction then afforded various mixtures of 2- and 3- fluoro compounds.[11]

The effect of K$_2$CO$_3$, KHCO$_3$ and KF on the base-mediated decomposition of 1,3,4,6-tetra-*O*-acetyl-2-*O*-triflyl-β-D-mannopyranose during the synthesis of 2-

deoxy-2-[¹⁸F]-fluoro-D-glucose has been investigated using ¹⁹F NMR spectroscopy. It was shown that impurities result from the elimination of triflic acid and that the substitution of triflate by ¹⁸F is 90% complete after 1 minute when a dry source of fluoride is used.[12] Fluorination of a glucal derivative has been utilized in another synthesis of 2-deoxy-2-[¹⁸F]-fluoro-D-glucose.[13]

2   Chloro-, Bromo-, and Iodo-sugars

Bromination (Ph₃P, CBr₄, Py) of α-D-glucopyranosyl fluoride has afforded the corresponding 6-bromo-6-deoxy-derivative which was treated with sodium azide to give 6-azido-6-deoxy-α-D-glucopyranosyl fluoride.[14] Normal conditions used for the chlorination of chitin have been adapted for bromination (Ph₃P, NBS, LiBr in dimethylacetamide) affording a product with 94% of the 6-OH groups replaced by Br.[15] The bromination (Ph₃P, tribromoimidazole) of methyl β-D-glucopyranoside has been studied in order to establish the likely by-products when cellulose is treated under the same conditions. Minor amounts of methyl 3,6-dibromo-3,6-dideoxy-β-D-gluco- and -allopyranosides were obtained in addition to methyl 6-bromo-6-deoxy-β-D-glucopyranoside.[16]

Acetylated 6^{III}-deoxy-6^{III}-iodo-maltotriose and 6^{IV}-deoxy-6^{IV}-iodo-maltotetraose have been prepared and the iodide has been displaced with the thio group of another 1-thio-sugar.[17] 2,5-Anhydro-D-mannitol has been converted into 3,4-di-O-acetyl-2,5-anhydro-1,6-dideoxy-1,6-diiodo-D-mannitol by standard methods,[18] and 4-deoxy-4-iodo-daunosamine has been prepared as an analogue of daunosamine.[19] 3,6-Dideoxy-6-iodo-1,2-O-isopropylidene-α-D-glucofuranose has been prepared in high yield by treatment of the corresponding 3-deoxy-5,6-thionocarbonate with methyl iodide.[20]

The 2-deoxy-2-fluoro-2-iodo-hexopyranoses **1** and **2** have been prepared from 2-fluoro-D-glucal (Scheme 1).[21] Haloetherification of glycal **3** (NBS, ROH) gave the β-glycosides **4**, which underwent radical cyclisation on abstraction of the bromine atom (see Chapter 14).[22] Free radical bromination (NBS, CCl₄, hv) of 2,3,4,6-tetra-O-acetyl-β-D-glucopyranosylbenzene afforded little discrete material, but when the reaction was conducted in moist CCl₄ a product of replacement of the anomeric proton by OH was obtained. A benzylic C-1 bromo compound was a presumed intermediate.[23]

Bromination of unsaturated lactone **5** (Br₂,CH₂Cl₂) gave only the dibromo compound **6**, whereas unsaturated lactone **7** afforded **8** under the same conditions.[24] 2-Acetamido-4-bromo-2,4-dideoxy-talopyranose derivatives have been prepared *via* 2-acetamido-1,6:3,4-dianhydro-2-deoxy-β-D-talopyranose.[25] The synthesis of deoxyiminoalditols from brominated aldonolactones is covered in Chapter 18.

In a study of the sequential use of 2-deoxyribose 5-phosphate aldolase and fructose 1,6-diphosphate aldolase for the synthesis of hept-2-ulose derivatives some brominated and chlorinated substrates were successfully employed.[26,27] A review on enzymatic oxidoreductions in organic synthesis features a number of halogenated carbohydrate examples.[28]

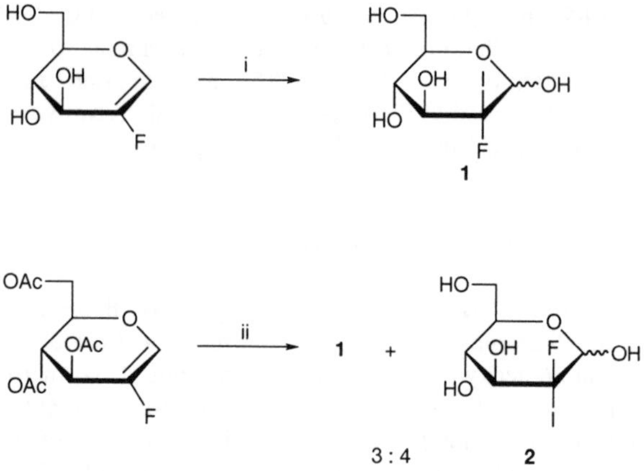

Reagents: i, KI, aq H_2O_2; ii, AgNO$_3$, I$_2$, aq acetone, then NaOMe, MeOH

Scheme 1

The radical reduction (Bu$_3$SnH, AIBN) of tetra-*O*-acetyl-β-D-glucopyranosyl bromide is known to afford predominantly the product of acetyl migration i.e. 1,3,4,6-tetra-*O*-acetyl-2-deoxy-α-D-glucopyranose. In the presence of diphenyl diselenide (or PhSeH) this migration is inhibited and the direct reduction product is present in >95%.[29] Treatment of dibromoalkene 9 with butyl lithium generated the corresponding alkynyl lithium which was successfully added to a cyclohexanone derivative.[30]

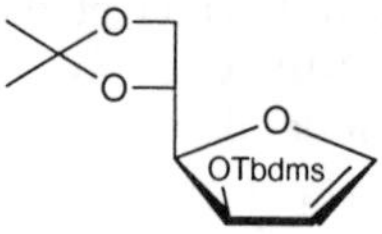

3 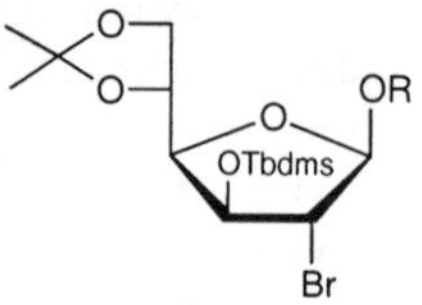   4
R = CH$_2$C≡CH,
CMe$_2$C≡CH,
CH$_2$CH=CH$_2$,
CH$_2$CMe=CH$_2$

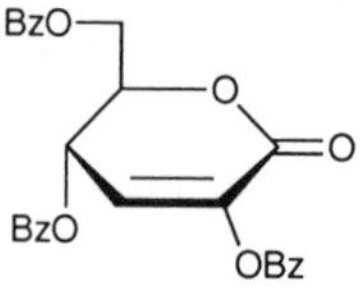

5 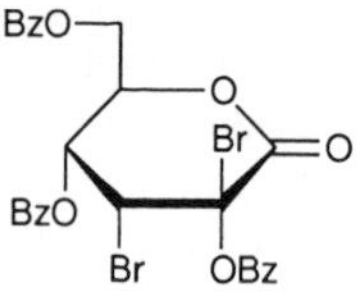   6

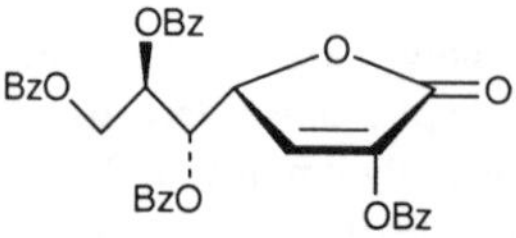

7 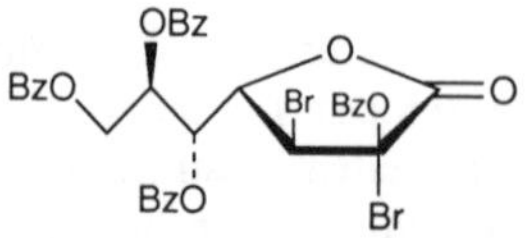   8 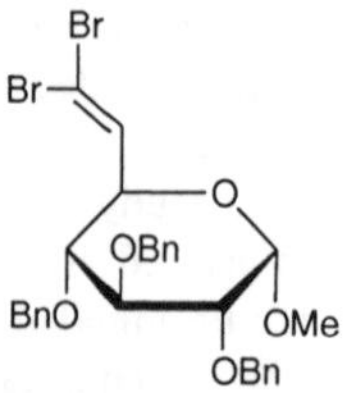   9

References

1   M. Palme and A. Vasella, *Helv. Chim. Acta.*, 1995, **28**, 959.
2   M. Cavicchioli, A. Mele, V. Montanari and G. Resnati, *J. Chem. Soc., Chem. Commun.*, 1995, 901.
3   J.M. García Fernandez, R.-R. Schnelle and J. Defaye, *Tetrahedron: Asymm.*, 1995, **6**, 307.
4   R. Miethchen and G. Torsten, *J. Fluorine Chem.*, 1994, **67**, 11 (*Chem. Abstr.*, 1995, **122**, 10 376).
5   Y. Kajihara, T. Endo, H. Ogasawara, H. Kodama and H. Hashimoto, *Carbohydr. Res.*, 1995, **269**, 273.
6   C.-K. Lee and H. Jiang, *J. Carbohydr. Chem.*, 1995, **14**, 407.
7   C. Simiand and H. Driguez, *J. Carbohydr. Chem.*, 1995, **14**, 977.
8   S. Niihata, T. Ebata, H. Kawakami and H. Matsushita, *Bull. Chem. Soc. Jpn.*, 1995, **68**, 1509.
9   T. Yamazaki, K. Mizutani and T. Kitazume, *J. Org. Chem.*, 1995, **60**, 6046.
10 K. Mizutani, T. Yamazaki and T. Kitazume, *J. Chem. Soc., Chem. Commun.*, 1995, 51.
11 I.A. Mikhailopulo, G.G. Sivets, N.E. Poopeiko and N.B. Khripach. *Carbohydr. Res.*, 1995, **278**, 71.
12 R. Chirakal, B. McCarry, M. Lonergan, G. Firnau and S. Garnett, *Appl. Radiat. Isot.*, 1995, **46**, 149 (*Chem. Abstr.*, 1995, **123**, 56 399).
13 F. Fuechtner, J. Steinbach, R. Luecke, R. Scholz and K. Neubert, *Forschungszent. Rossendorf (Ber.)*, 1995, FZR-73 (*Chem. Abstr.*, 1995, **123**, 112 528).
14 A.M. Horneman and I. Lundt, *J. Carbohydr. Chem.*, 1995, **14**, 1.
15 H. Tseng, K. Furuhata and M. Sakamoto, *Carbohydr. Res.*, 1995, **270**, 149.
16 K. Furuhata, N. Aoki, S. Suzuki, N. Arai, H. Ishida, Y. Saegusa, S. Nakamura and M. Sokamoto, *Carbohyd. Res.*, 1995, **275**, 17.
17 C. Apparu, H. Driguez, G. Williamson and B. Svensson, *Carbohyd. Res.*, 1995, **277**, 313.
18 M.A. Shalaby, F.R. Fronczek, Y. Lee and E.S. Younathan, *Carbohydr. Res.*, 1995, **269**, 191.
19 Y. St-Denis, J.F. Lavallée, D. Nguyen and G. Attardo, *Synlett*, 1995, 272.
20 M. Adiyaman, S.P. Khanapure, S.W. Hwang and J. Rokach, *Tetrahedron Lett.*, 1995, **36**, 7367.
21 J.D. McCarter, M.J. Adam and S.G. Withers, *Carbohydr. Res.*, 1995, **266**, 273.
22 G.V.M. Sharma and K. Krishnudu, *Carbohydr. Res.*, 1995, **268**, 287.
23 P. Cettour, G. Descotes and J.-P. Praly, *J. Carbohydr. Chem.*, 1995, **14**, 451.
24 C. Di Nardo, O. Varela, R.M. deLederkremer, R.F. Baggio, D.R. Vega and M.T. Garland, *Carbohydr. Res.*, 1995, **269**, 99.
25 R. Thomson and M. von Itzstein, *Carbohydr. Res.*, 1995, **274**, 29.
26 M.J.M. Gijsen and C.-H. Wong, *J. Am. Chem. Soc.*, 1995, **117**, 2947.
27 C.-H. Wong, E. García-Junceda, L. Chen, O. Blanco, H.J.M. Gijsen and D.H. Steensma, *J. Am. Chem. Soc.*, 1995, **117**, 3333.
28 J.-M. Fang, C.-H. Lin, C.W. Bradshaw and C.-H. Wong, *J. Chem. Soc., Perkin Trans. 1*, 1995, 967.
29 D. Crich and Q. Yao, *J. Org. Chem.*, 1995, **60**, 84.
30 C. Barband, M. Bols, I. Lundt and M.R. Sierks, *Tetrahedron*, 1995, **51**, 9063.

9
Amino-sugars

1   Natural Products

Chrysopine 1, the spiro-lactone of N^α-(1-deoxy-D-fructos-1-yl)-L-glutamine, and thus a product of the Amadori reaction of glucose with glutamine, has been isolated from crown gall tumours induced on tobacco plants by *Agrobacterium tumefaciens*. Its anomeric configuration was not determined.[1]

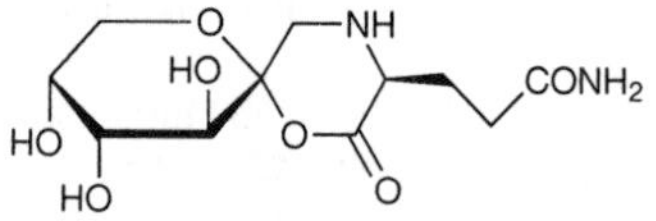

1

2   Syntheses

Syntheses covered in this section are grouped according to the method used for introducing the amino-functionality.

2.1   By Chain Extension. – The popular route to 2-amino-sugars, aza-sugars and higher amino-sugars that involves the reaction of a C-2 metalated thiazole with the nitrone derivative of a sugar aldehyde, has been reviewed (79 refs).[2] Reviews covering the biological recognition and enzyme synthesis of monosaccharides,[3] and the synthetic applications of furan-based materials,[4] have included amino-sugar examples. The α-amino-acid **3**, a constituent of the antibiotic miharamycin A, was obtained by chain extension from C-6 of dialdose derivative **2** (Scheme 1).[5] Chain extension reactions applied to chiral non-carbohydrate compounds are covered in section 2.7.

2.2   By Epoxide Ring Opening. – Over 20 N-alkyl- and N,N-dialkyl-2-amino-2-deoxy-D-altropyranosides were obtained with high selectivity by lithium perchlorate-catalysed *trans*-diaxial ring opening of methyl 2,3-anhydro-4,6-O-benzylidene-α-D-allopyranoside with primary and secondary amines, respectively.[6] 2-Acetamido-2-deoxy-5-thio-D-altropyranose **4** was similarly synthesized by treating the corresponding methyl 2,3-anhydro-4,6-O-

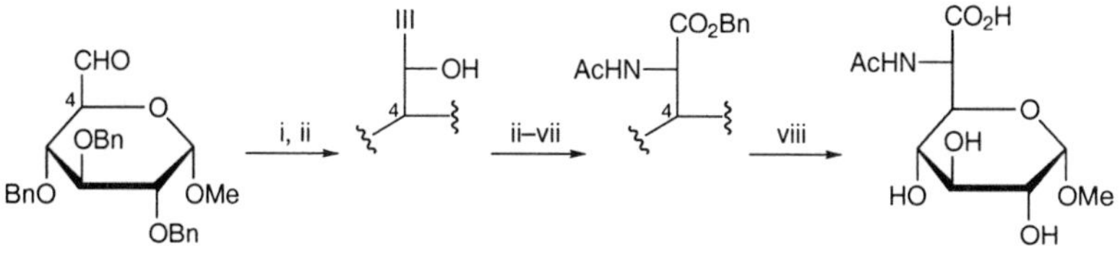

Reagents: i, TmsC≡CMgBr; ii, Bu$_4$NF; iii, Zn(N$_3$)$_2$, Ph$_3$P, Pr^{i}O$_2$CN═NCO$_2$Pr^i;
iv, HS⌒⌒SH, Et$_3$N, MeOH; v, Ac$_2$O–Py; vi, OsO$_4$, NaIO$_4$, THF, H$_2$O;
vii, PhCHN$_2$; viii, H$_2$, Pd/C

Scheme 1

isopropylidene-5-thio-α-D-altropyranoside with ammonia in methanol, followed by *N*-acetylation and acid hydrolysis.[7] Intramolecular reactions of *N*-benzoylcarbamate derivatives of *trans*-epoxyalcohols has featured in various 3-amino-3-deoxy-pentoside syntheses, as exemplified for the 3-amino-3-deoxy-α-D-arabinoside **5** in Scheme 2, where concomitant N→O migration of the benzoyl group occurs.[8,9]

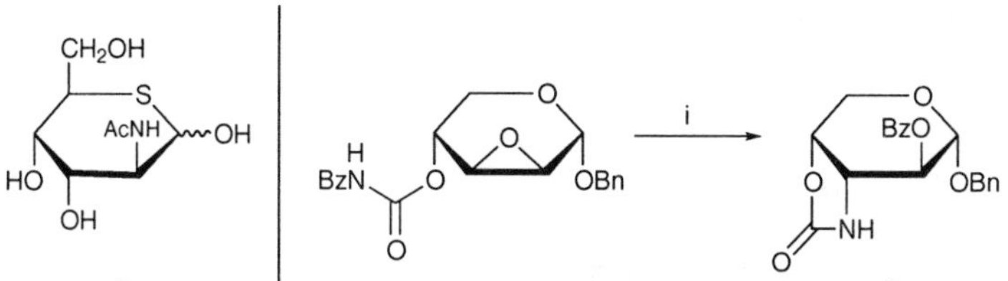

Reagent: i, NaH

Scheme 2

Various *N*-substituted 6-amino-6-deoxy-D-glucose derivatives, e.g. **6-8**, were synthesized by reaction of the corresponding 5,6-anhydro-D-glucose derivative with secondary amines, and shown to be useful as non-ionic surfactants capable of forming reverse micelles for solubilization of amino acids in hexane.[10] Reaction of such tertiary amines with methyl iodide provided quaternary ammonium-sugar derivatives.[11] The 6-amino-2,5-anhydro-6-deoxy-D-gluconate derivative **10**, a potential dipeptide isostere, was obtained from the C$_2$-symmetric, D-mannitol derived bis-epoxide **9** following silica-assisted azidolysis (Scheme 3). Its enantiomer was obtained similarly from an L-iditol bis-epoxide.[12]

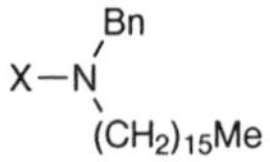

6 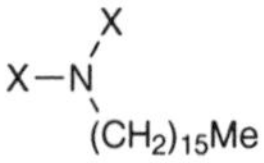   **7** 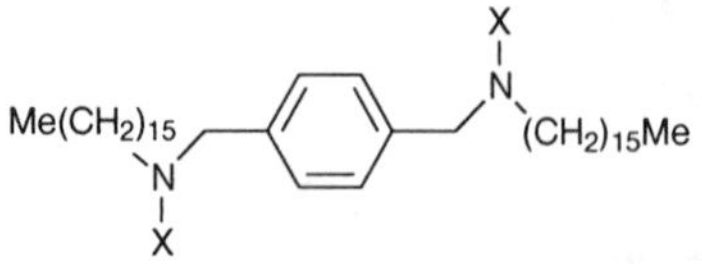   **8**

$$X =$$

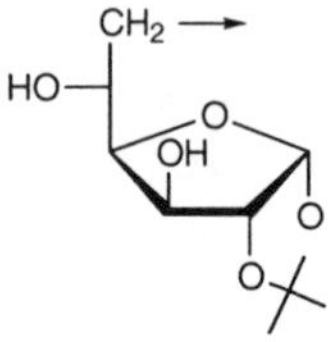

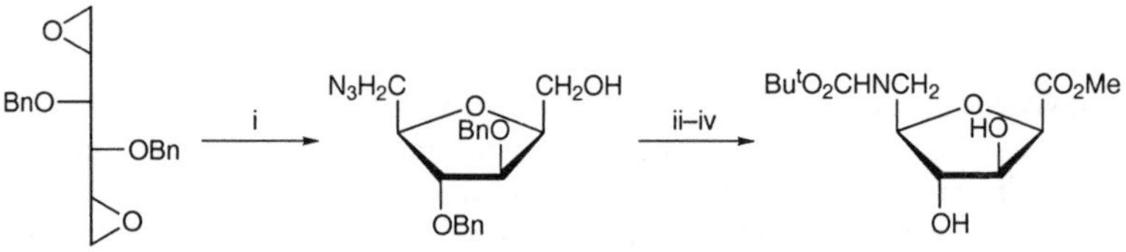

9   **10**

Reagents: i, NaN$_3$, SiO$_2$, MeCN; ii, Na$_2$Cr$_2$O$_7$, EtOH, H$_2$O then
CH$_2$N$_2$; iii, H$_2$, Pd/C, (Bu^{t}O$_2$C)$_2$O; iv, H$_2$, Pd/C, AcOH

Scheme 3

2.3   By Nucleophilic Displacement. – Benzyl 2-acetamido-4-O-acetyl-2-deoxy-3,6-di-O-pivaloyl-β-D-mannopyranoside was obtained from benzyl β-D-galactopyranoside by sequential introduction then displacement of triflate groups with inversion from C-4 (by acetate ion) then C-2 (by azide ion, using Bu$_4$N.N$_3$ in benzene with ultrasound).[13] Six isosteric NADH mimics bearing N-(3-deoxy-D-glucofuranos-3-yl) moieties, e.g. **11** and **12**, were synthesized by displacement of the triflate group from C-3 of the 1,2:5,6-di-O-isopropylidene-D-allofuranose ester. Compound **12** showed stereoselectivity (80% e.e.) in the reduction of methyl benzoylformate to methyl mandelate.[14]

3-Amino-3-deoxy-sucrose has been synthesized from sucrose by sequential oxidation at C-3 by *Agrobacterium tumefaciens* (40% yield), peracetylation, reduction of the C-3 ketone to an *allo*-configured product (H$_2$, Pt), trifluoromethanesulfonylation, displacement by azide with inversion, deacetylation and hydrogenation.[15] 4-Amino-4-deoxy-sucrose was obtained from a sucrose heptapivaloate by double inversion at C-4 by sequential triflate displacement reactions, first with nitrite then azide ion.[16] Various 4-alkylamino-2,4-dideoxy-L-*threo*-pentopyranosides **15**, components of the calicheamicins, were synthesized from

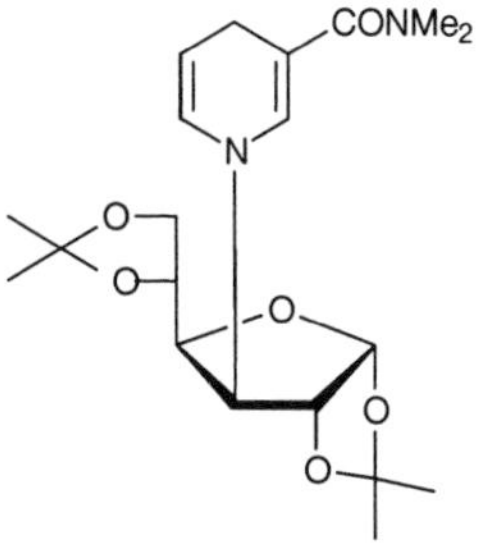

11 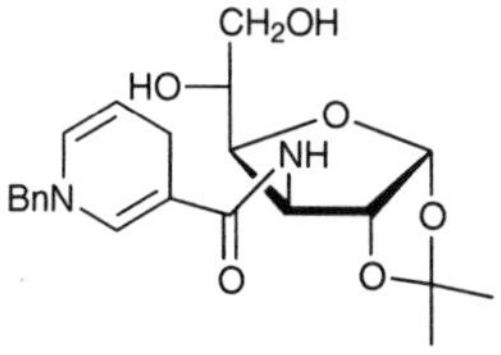   **12**

methyl 2-deoxy-β-D-*erythro*-pentopyranoside by reaction of its tosylate derivative **13** with azide ion to give **14** (Scheme 4).[17] The 4-amino-4-deoxy-D-galactose

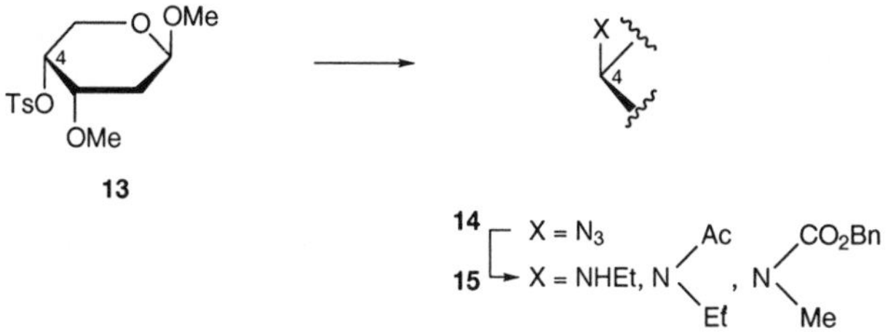

13

14 ⎡ X = N₃
15 ⎣ X = NHEt, N...

Scheme 4

derivative **18** was obtained by hydrolysis of the bicyclic derivative **17**, formed by intramolecular sulfonate displacement reaction of the (4-O-mesylglucosyl)ena-mine **16** (Scheme 5). 4-Amino-4-deoxy-L-arabino- and L-lyxo-pyranose deriva-tives were obtained similarly. The 1,6-imine **19** was obtained from the corresponding D-glucosylenamine 2,3,4,6-tetramesylate, rather than a 1,4-imine.[18] The synthesis of 4-acylamino-2,6-anhydro-2,3,4-trideoxy-non-2-enoic acids is covered in Chapter 16.

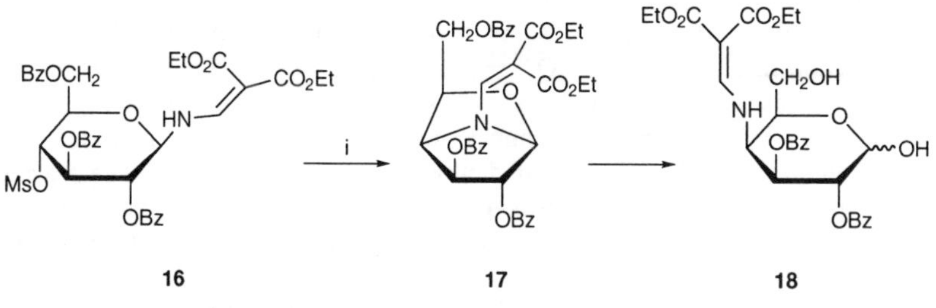

16   **17**   **18**

Reagents: i, NaOMe; ii, H₃O⁺

Scheme 5

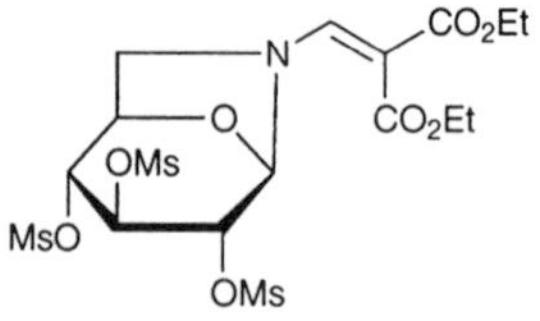

19

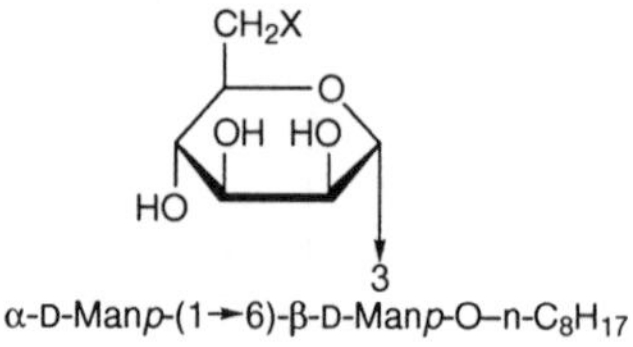

20 X = NHCOCH$_2$I or NH$_2$·HCl

α-D-Man*p*-(1→6)-β-D-Man*p*-O-n-C$_8$H$_{17}$

The branched mannose trisaccharides **20**, bearing an amino-group at C-6 of the (1→3)-linked residue, were synthesized as potential inhibitors of *N*-acetylglucosaminyltransferase I, the nitrogen-functionality being introduced by displacement of a 6-tosylate with azide ion.[19] α-D-Glucopyranosyl fluoride could be converted into the corresponding 6-acetamido-6-deoxy analogue without *O*-protection (i, Ph$_3$P, CBr$_4$; ii, NaN$_3$, DMF; iii, H$_2$, Pd/C; iv, Ac$_2$O, MeOH).[20] Formation of the 6-*N*-alkyl-6 trifluoroacetamido-derivative **21** on reaction of the corresponding 6-triflate with 5-trifluoroacetamidopentanol and sodium hydride in THF has been shown to be the result of selective mono-deprotonation of the trifluoroacetamido-group with this reagent. Use of the corresponding 5-acetamidopentanol, or the combination potassium hydride and 18-crown-6 (to effect double deprotonation), led to *O*-linked ethers.[21] The Amadori rearrangement products **22** were synthesized from D-fructose by

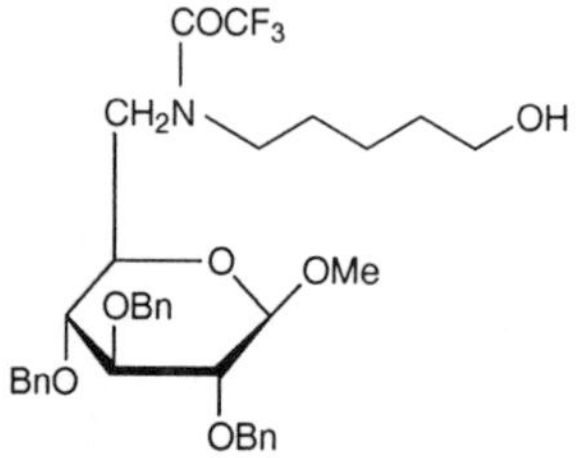

21

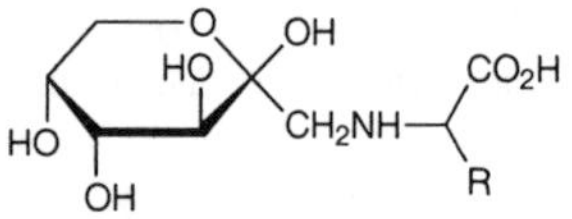

22 R = H, Me *etc.*

reaction of its 2,3:4,5-di-*O*-isopropylidene-α-pyranose 1-triflate with the benzyl ester derivatives of glycine, alanine, valine, leucine or lysine followed by deprotection (i, H$_2$, Pd/C; ii, H$_3$O$^+$).[22] A variety of 5-deoxy-5-dialkylamino-1,2-*O*-isopropylidene-α-D-xylopyranoses were synthesized by reaction of the corresponding 5-tosylate with dialkylamines, and used to induce enantioselectivity (up to 90% e.e.) in the addition of diethylzinc to benzaldehyde.[23] The 5-azide **23**, prepared from D-glucose, was converted to the 1,5-imino-sugar sulfite adduct **24** and then to the 1-cyano-1-deoxy-derivative **25** (Scheme 6) or to the corresponding 1-deoxy-derivative.[24]

The 3′-amino-3′-deoxy-derivative **26** of *N*-acetyl-lactosamine was synthesized using a known 3′-azido-3′-deoxy-galactosyl donor for construction of the disaccharide. It was an inhibitor (K$_i$=104 μM) of the glycosyltransferase that

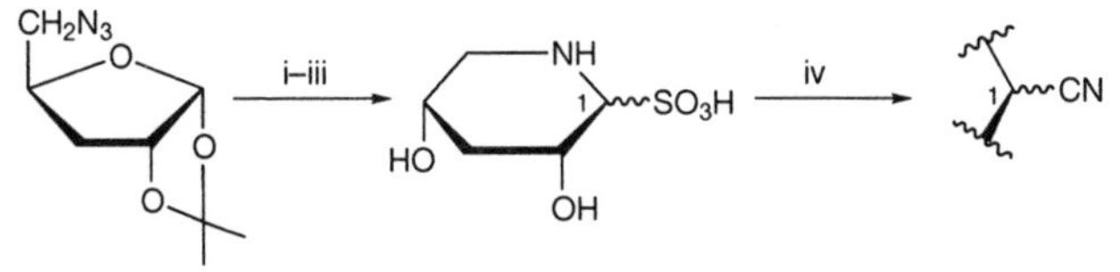

Reagents: i, H_2, Pd/C; ii, Me_3SiCl, DMF; iii, SO_2, H_2O; iv, $Ba(OH)_2$, NaCN

Scheme 6

transfers a β-galactosyl residue to the 3-hydroxy-group of *N*-acetyllactosamine.[25] GDP-3-acetamido- and 3-azido-3-deoxy-α-D-mannopyranose were prepared enzymically from 3-azido-3-deoxy-D-mannose,[26] while related thymidine diphospho-derivatives of 3-acetamido-3-deoxy- and 4-amino-4-deoxy-α-D-glucopyranose were synthesized chemically from the corresponding peracylated azidodeoxyglucoses.[27] The 4-amino-2,4-dideoxy-analogue was similarly prepared following a Barton deoxygenation step.[28] Sugar urea derivatives such as the *N*-(galactos-6-yl)-urea **27** were prepared by reaction of sugar azides with a combination of dialkylamine, carbon dioxide and triphenylphosphine, phosphinimines being formed as intermediates.[29] A five-step synthesis of 1,5-dideoxy-1,5-imino-D-xylono-1,5-lactam from 5-azido-5-deoxy-1,2-*O*-isopropylidene-α-D-xylofuranose involved Lewis acid-catalysed intramolecular cyclization of an acyclic 5-azido-*aldehydo*-sugar derivative.[30] 6-Acetamido-3,4-di-*O*-acetyl-6-deoxy-D-glucal was synthesized from the corresponding 6-azide and its reactions under a variety of conditions were studied (see also Chapters 10 and 13).[31]

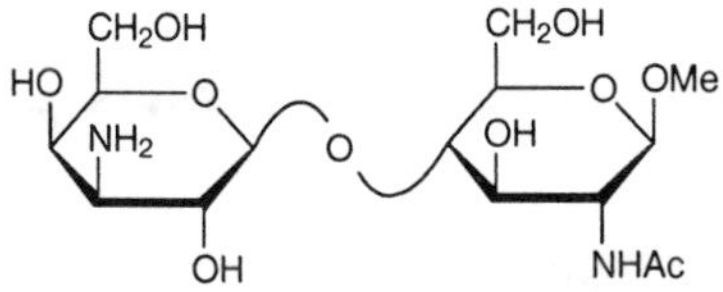

2.4 From Unsaturated Sugars. – A 'trisubstrate analogue' **30** has been synthesized (Scheme 7) as a potential inhibitor of the α-(1→3)-fucosyltransferase that transfers fucose from GDP-fucose onto O-3 of a 2-acetamido-2-deoxy-β-D-glucose residue. The glycosyl donor **29** was synthesized from the L-fuconolactone derivative **28** by methylenation and anti-Markovnikov azido-phenylselenation,

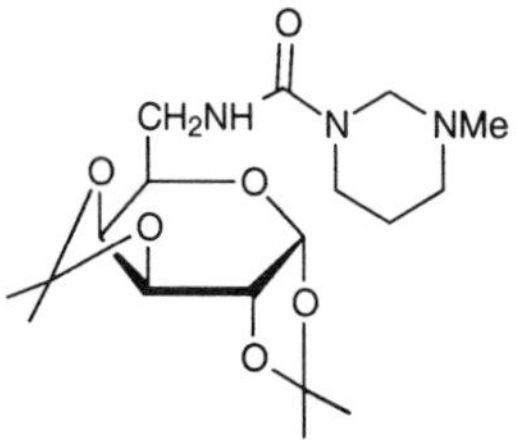

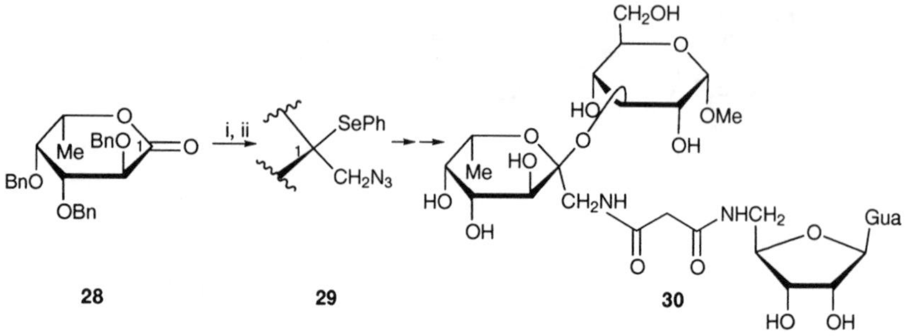

Reagents: i, Cp₂TiMe₂; ii, NaN₃, (PhSe)₂, PhI(OAc)₂

Scheme 7

then used to glycosylate O-3 of a glucoside acceptor. The azido-group was converted to an amine and coupled *via* a malonic acid spacer to 5'-amino-5'-deoxyguanosine. The malondiamide group is a mimic of the pyrophosphate group. Attempts to prepare the analogue of **30** incorporating a 3-O-substituted 2-acetamido-2-deoxy-D-glucoside moiety were frustrated by cleavage of the sensitive α-L-fucoheptulopyranosyl bond during removal of a 4,6-O-benzylidene protecting group on hydrogenolysis in acetic acid.[32]

Analogues of daunosamine have been prepared from L-rhamnal diacetate **31** *via* the known 3-azido-2,3-dideoxy-sugar **32** (Scheme 8). Inversion of configura-

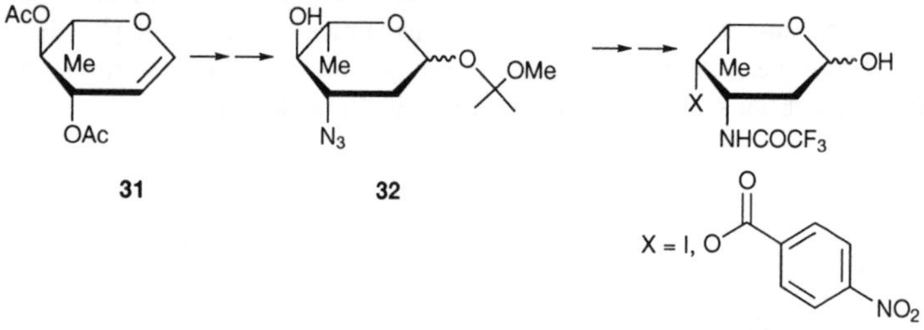

Scheme 8

tion at C-4 to provide analogues **33** was achieved by way of displacements of triflate with iodide or acetate ion.[33] Conjugate addition of benzylamine to 5-hydroxymethyl-2-(5H)-furanone **34**, available from D-mannitol (cf. Vol.21, p.155), led to the aminolactone **35**, which could be elaborated into branched-chain derivatives **36** (Scheme 9), β-amino-esters or β-lactams.[34] Oxyamination reactions of the silyl ether derivative **37** (R=Tbdps) derived from levogluco-senone, gave 4-amino-4-deoxy-D-altrose derivatives **38** or **39**, whereas the corresponding pivalate ester **37** (R=COCMe₃) gave mainly the 3-amino-3-deoxy-

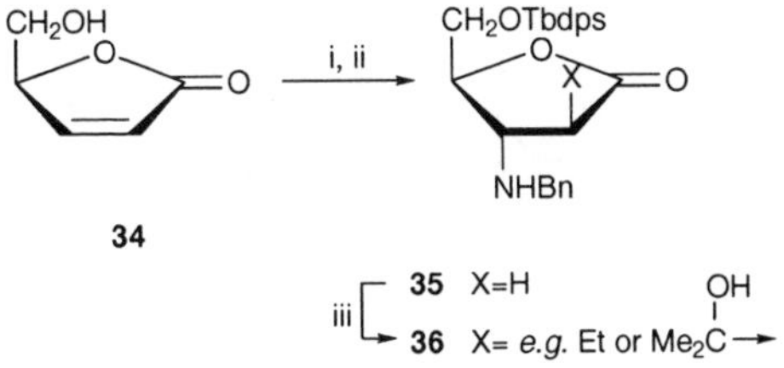

Reagents: i, BnNH$_2$, MeOH; ii, TbdpsCl, imidazole, DMF; iii, LiN(SiMe$_3$)$_2$ then EtI or Me$_2$CO

Scheme 9

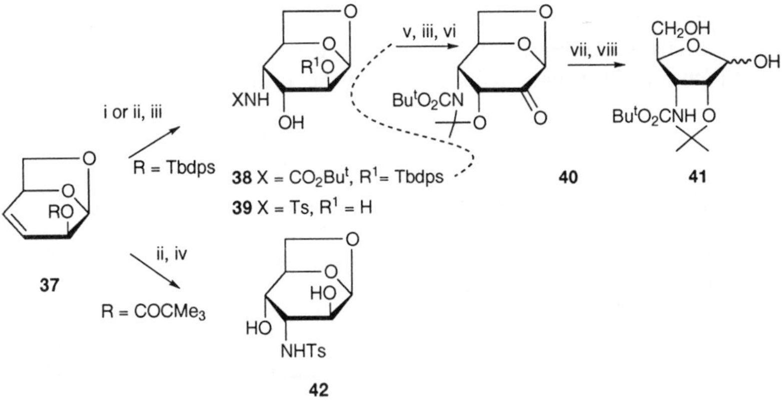

Reagents: i, Bu^{t}O$_2$CClNa, AgNO$_3$, OsO$_4$, MeCN; ii, OsO$_4$, chloramine T, Bu^{t}OH, H$_2$O; iii, Bu$_4$NF;
iv, NaOH; v, Me$_2$C(OMe)$_2$, HOTs; vi, (COCl)$_2$, DMSO then Et$_3$N; vii, Mg monoperoxy-
phthalate, MeOH; viii, Bu$^i_2$AlH

Scheme 10

derivative **42** (Scheme 10). The *N*-tosyl groups of **39** and **42** could be removed by
photolysis in the presence of sodium borohydride and 1,5-dimethoxynaphthalene.
Compound **38** was converted to the 3-amino-3-deoxy-D-ribose derivative **41**, and
thence into 3′-amino-3′-deoxy-adenosine. Chain cleavage was achieved by
Baeyer-Villiger oxidation of the 2-ulose **40**.[35,36]

2.5   From Alduloses. – The phosphonomethyl analogue **45** of 2-acetamido-2-
deoxy-α-D-mannopyranosyl phosphate was obtained from the D-arabinose
derivative **43** (Scheme 11). Reduction of the oxime **44** gave the axial amine,

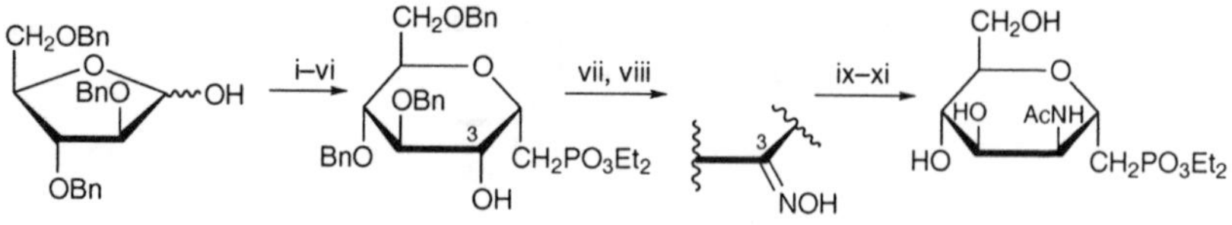

43   **44**   **45**

Reagents: i, Zn(⌒)₂; ii, Hg(OAc)₂, THF then KCl; iii, I₂; iv, TbdmsCl; v, P(OEt)₃; vi, H₃O⁺;
vii, Ac₂O, DMSO; viii, NH₂OH; ix, H₂, Raney Ni; x, Ac₂O, Py; xi, H₂, Pd/C

Scheme 11

whereas the equatorial product is usually obtained in the case of α-*O*-glycosides.[37] Conditions for forming α- or β-glycosides, **47** or **49** respectively, from the lactose-derived 2-benzoyloxyimino-glycosyl bromide **46** have been found (Scheme 12). Borane reduction led to the α-*gluco*- (**48**) or β-*manno*- (**50**)

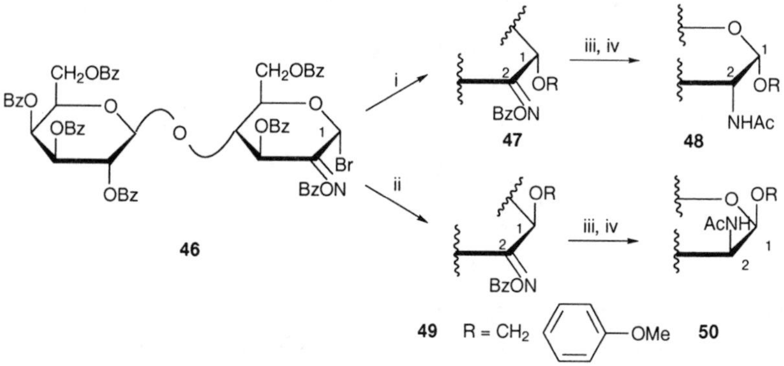

46

47   **48**

49 R = CH₂ ⟨phenyl⟩—OMe **50**

Reagents: i, ROH, collidine, I₂, dioxane; ii, ROH, Ag₂CO₃, I₂, CH₂Cl₂; iii, BH₃·THF; iv, Ac₂O

Scheme 12

products, respectively. These reactions can be performed without isolation of intermediates, so that *N*-acetyl-lactosamine could be obtained in 35% overall yield from lactose without chromatography.[38] This approach was used for the preparation of the good lactosamine donor, *N*-trichloroethoxycarbonyl-β-D-lactosaminyl fluoride.[39] Mixtures of 3-acetamido-3,6-dideoxy-α-L-gulo- and α-L-galacto-pyranosides (epimeric at C-3) resulted from reductions (LiAlH₄) of either 2,4-diester or 2,4-diether derivatives of methyl 3-(*O*-methyloximino)-α-L-*xylo*-pyranos-3-uloside. Similar results were obtained with a β-anomer.[40] In contrast, the 3-amino-sugar derivatives **51** and **52** were the only isomers formed on reduction (NaBH₄) of *N*-(diphenylphosphinyl)-oxime derivatives of the corresponding 3-keto-sugars.[41] The Amadori compound *N*-(1-deoxy-2,3:4,5-di-

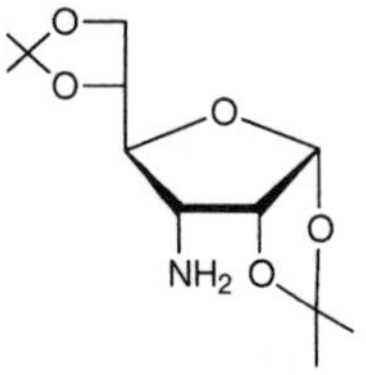

51 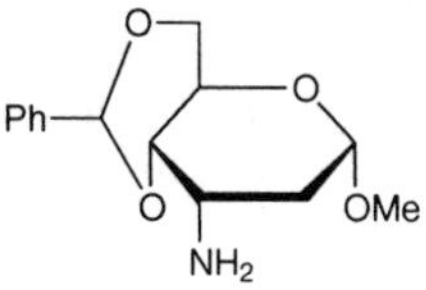 **52**

O-isopropylidene-β-D-fructopyranos-1-yl)-L-tyrosine benzyl ester, was prepared by reductive amination of a 1-*aldehydo*-D-fructose derivative.[41a]

2.6 From Aminoacids. – (+)-Galactostatin **54** has been synthesized by one- then two-carbon chain extensions of the D-serine derivative **53**, using different thiazole reagents (Scheme 13);[42] see Scheme 16 for an alternative synthesis of galactostatin. The calicheamycin constituent sugar **56** was obtained by asymmetric allylboration of the L-serinal derivative **55** (Scheme 14), as part of a comprehensive study of similar reactions.[43]

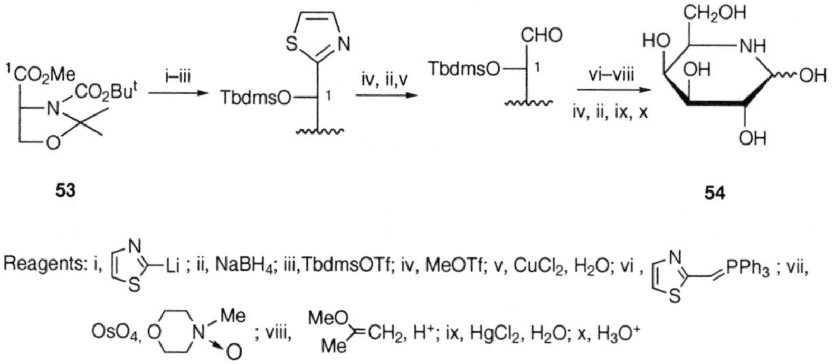

53 **54**

Reagents: i, [thiazole]-Li ; ii, NaBH₄; iii, TbdmsOTf; iv, MeOTf; v, CuCl₂, H₂O; vi, [thiazole]-PPh₃ ; vii,

OsO₄, [morpholine-Me-O] ; viii, $\frac{MeO}{Me}{>}=CH_2$, H⁺; ix, HgCl₂, H₂O; x, H₃O⁺

Scheme 13

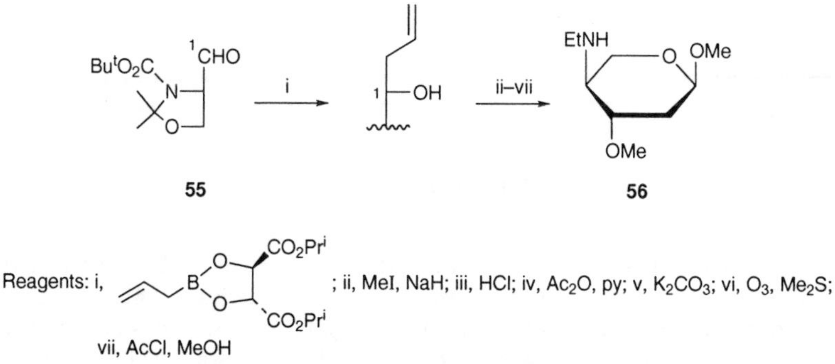

55 **56**

Reagents: i, [allyl boronate CO₂Prⁱ] ; ii, MeI, NaH; iii, HCl; iv, Ac₂O, py; v, K₂CO₃; vi, O₃, Me₂S;

vii, AcCl, MeOH

Scheme 14

2.7   From Chiral Non-carbohydrates. – 2-Amino-2-deoxy-D-threono- and D-erythrono-1,4-lactones were obtained as the minor and major components, respectively, of a separable mixture from 2,3-O-isopropylidene-D-glyceraldehyde in a reinvestigation of the Strecker synthesis (i, KCN, BnNH$_2$; ii, H$_3$O$^+$; iii, H$_2$, Pd(OH)$_2$) first reported by Kuhn and Fisher (*Liebigs Ann. Chem.*, 1961, 152). These were separately reduced to give 2-amino-2-deoxy-D-threitol and -D-erythritol tetracetate.[44] Epoxy-amide **57**, prepared by way of a reaction of 2,3-O-isopropylidene-D-glyceraldehyde with a stabilized ylid (Me$_2$S=CHCONMe$_2$), (cf. Vol.27, p.186), gave the 2-amino- or 2-azido-2-deoxy-D-ribonamides **58** on reaction with excess ammonia, amine or azide ion, but afforded the dimer **59** on reaction with just two equivalents of ammonia (Scheme 15).[45]

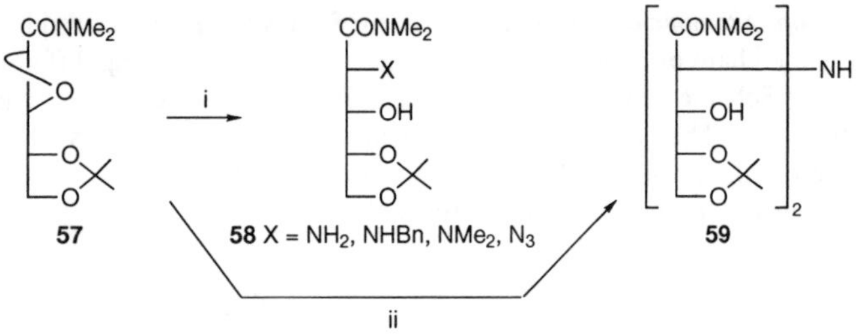

Reagents: i, XH or Mg(N$_3$)$_2$; ii, NH$_4$OH (2 mol. equiv.)

Scheme 15

In a formal total synthesis of (+)-galactostatin **54** (5-amino-5-deoxy-D-galactose; Scheme 16), the L-threose derivative **60**, prepared by a known method

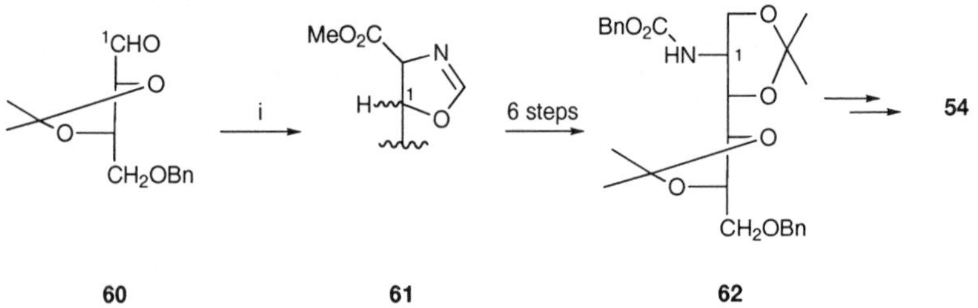

Reagents: i, MeO$_2$CCH$_2$NC, NaCN

Scheme 16

from diethyl (+)-tartrate, was converted to a separable diastereoisomeric mixture containing the 2-amino-2-deoxy-L-galactitol derivative **62**, previously used in the synthesis of **54** (Vol.25, p.206, ref.39). In the first step, stereoselective aldol-like condensation between **60** and methyl isocyanoacetate gave the unstable diastereomeric mixture **61**.[46] For another synthesis of galactostatin see

Scheme 13. Chain extension of a similar tartaric acid-derived L-threose compound **63**, led to the β-hydroxy-α-amino acid, polyoxamic acid **64** (Scheme 17), and the same strategy led to the 5-amino-5-deoxy-L-taluronoside **65** from

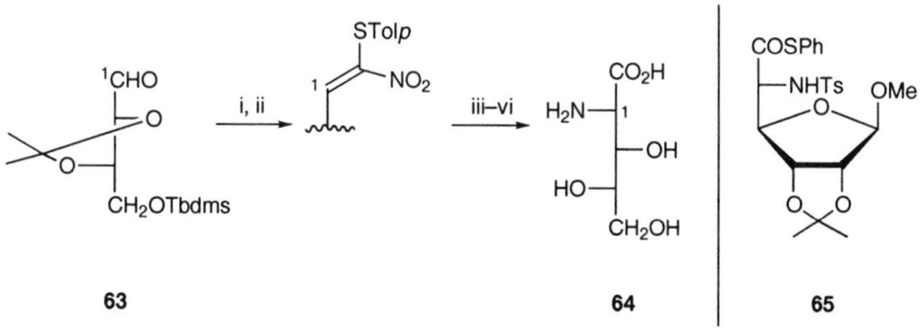

63   **64**   **65**

Reagents: i, *p*-TolSCH$_2$NO$_2$, Bu^{t}OK; ii, MsCl, Pr$^i_2$NEt; iii, Bu^{t}OOK; iv, NH$_3$; v, (Bu^{t}O$_2$C)$_2$O; vi, H$_3$O$^+$

Scheme 17

the corresponding 5-aldehyde.[47] The chlorobenzene microbial oxidation product **66** was converted in a multi-step procedure to the 4-acetamido-4-deoxy-D-mannose derivative **67** (Scheme 18). In a related fashion, the bromobenzene-

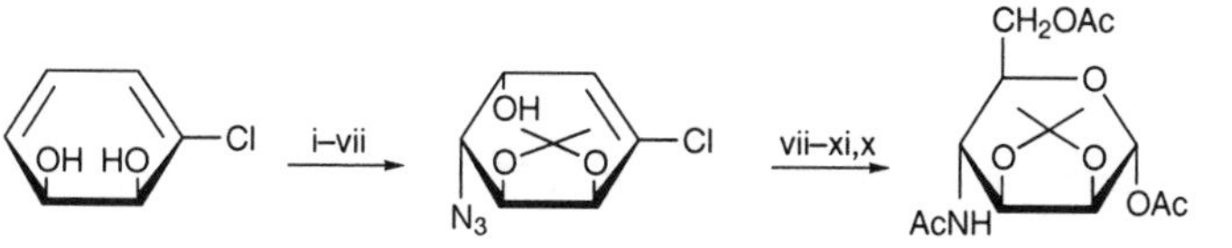

66   **67**

Reagents: i, Me$_2$C(OMe)$_2$, H$^+$; ii, NBS, H$_2$O; iii, NaOH, Bu$_4$NHSO$_4$, CH$_2$Cl$_2$, H$_2$O; iv, TmsCl; v, Tf$_2$O, Py; vi, NaN$_3$; vii, Bu$_4$NF; viii, O$_3$, MeOH, NaHCO$_3$; ix, NaBH$_4$; x, Ac$_2$O, Py, DMAP; xi, H$_2$, Pd/C

Scheme 18

derived analogue was converted to a 2-acetamido-2-deoxy-α,β-D-glucopyranose tetraacetate.[48]

2.8 From Achiral Non-carbohydrates. – Racemic methyl 2-acetamido-2-deoxy-threofuranoside **69** was the major epimer formed as indicated in Scheme 19 from the dihydroisoxazole **68**, prepared by condensation of nitromethane and chloroacetaldehyde.[49] 2-Amino-2-deoxy-L-erythrono-1,4-lactone **73** was synthesized by enzymatic aldol condensation of **70** and **71** to give a 92:8 mixture of *erythro*- and *threo*-adducts, from which **72** was obtained by crystallization (Scheme 20).[50] The lactone **74**, an intermediate in previous syntheses of N-trifluoroacetyl-L-acosamine and -L-daunosamine (Vol.14, p.72, ref.14), was prepared from methyl sorbate as before, but by a rather inefficient route.[51]

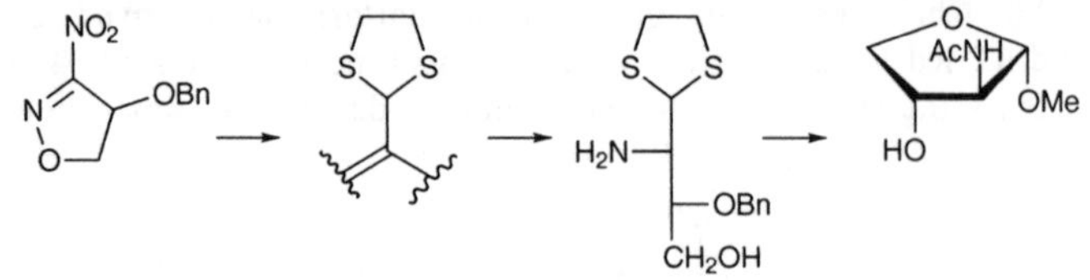

(±)-**68**   (±)-**69**

Scheme 19

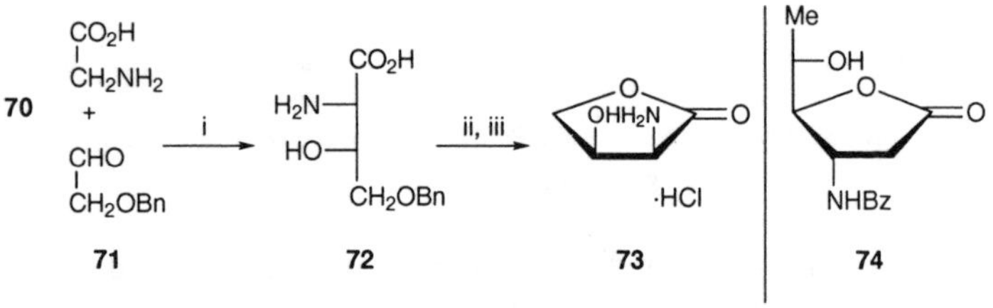

Reagents: i, L-Threonine aldolase; ii, H_2, Pd/C; iii, EtOH, HCl

Scheme 20

The racemic mannonojirimycin derivative **77** was synthesized from furylglycine *via* the furylamide derivative **75**, which formed enone **76** on oxidation (Scheme 21); the *cis*-2,3-diol was introduced by ruthenium-catalysed *cis*-hydroxylation of

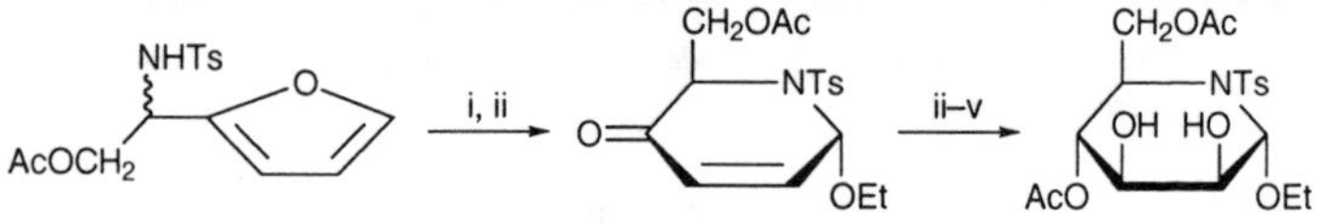

(±)-**75**   (±)-**76**   (±)-**77**

Reagents: i, MCPBA; ii, EtOH, Ce$(NH_4)_2(NO_3)_6$; iii, $NaBH_4$, CeCl$_3$; iv, Ac$_2$O, Et$_3$N, DMAP;
v, RuCl$_3$, NaIO$_4$, MeCN, H_2O, EtOAc

Scheme 21

the 2,3-double bond.[52] The cycloadduct **78**, produced by a highly enantioselective hetero-Diels-Alder condensation between sorbaldehyde dimethyl acetal and 2,3:5,6-di-*O*-isopropylidene-1-nitroso-D-mannofuranosyl chloride, followed by *N*-protection, was converted into 6-deoxy-D-*allo*-nojirimycin or D-fuconojiri-mycin **80** and thence by reduction into their 1-deoxy-analogues. Formation of **80** involved displacement with inversion at both hydroxylated carbon atoms of intermediate **79** (Scheme 22).[53] Further details on the application of this approach

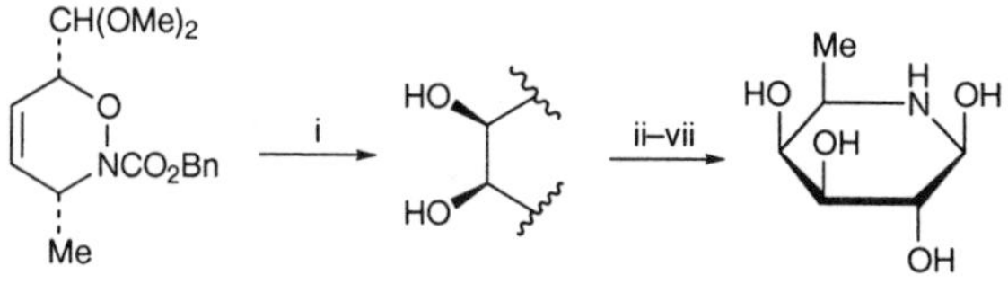

78   **79**   **80**

Reagents: i, OsO$_4$, NMNO; ii, Tf$_2$O, Py; iii, NaOBz; iv, Na$_2$CO$_3$, MeOH; v, H$_2$, Pd/C; vi, SO$_2$, H$_2$O; vii, Ba(OH)$_2$

Scheme 22

to the synthesis of 4-amino-4-deoxy-D-L-erythrose (cf. Vol.28, p.128, ref.40), 4-amino-4,5-dideoxy-D-L-ribose **81** and its C-1 carboxylate derivative **82**, have been published.[54]

3   Properties and Reactions

3.1   Conformational Studies. – Conformation analysis on various 2,3,6-trideoxy-3-amino-α-L-hexopyranoses and their *N*-alkylated derivatives revealed that the $^1C_4$-conformation was favoured.[55]

3.2   *N*-Acyl Derivatives. – Oxazolines have been prepared in near quantitative yield by treatment of peracetylated 2-amino-2-deoxy-glycosyl halides with mercuric bromide and 2,4-collidine.[56] *C*-Silyated byproducts **83** were formed in

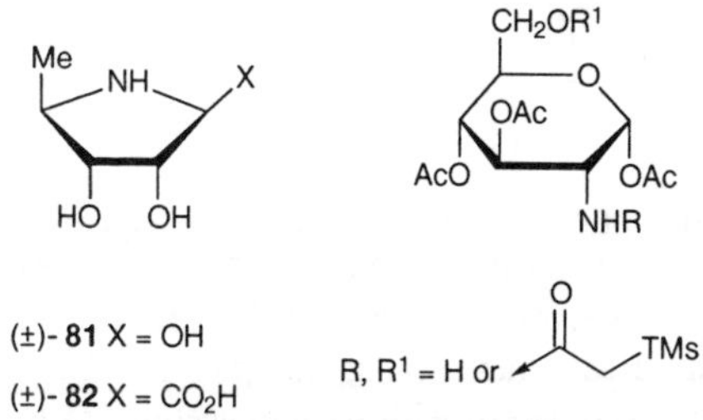

(±)- **81** X = OH

(±)- **82** X = CO$_2$H

R, R^1 = H or

83

the synthesis of the D-*gluco*-oxazoline **84** from peracetylated 2-amino-2-deoxy-α-D-glucopyranosyl acetate using trimethylsilyl triflate and triethylamine.[57] Hydrolysis of **84** proceeds to give the amine salt **85** as the kinetic product (which precipitates from acetonitrile), and this then slowly rearranges to the *N*-acetate **86** by O→N acetyl migration and anomerization (Scheme 23). This is a convenient method for effecting either 2-*N*- or 1-*O*-deacetylation of peracetylated 2-amino-2-deoxy-D-glucose *via* the oxazoline derivative. The free amino-group in **85** can be acylated to provide novel glycoconjugates such as the aspartic acid derivative

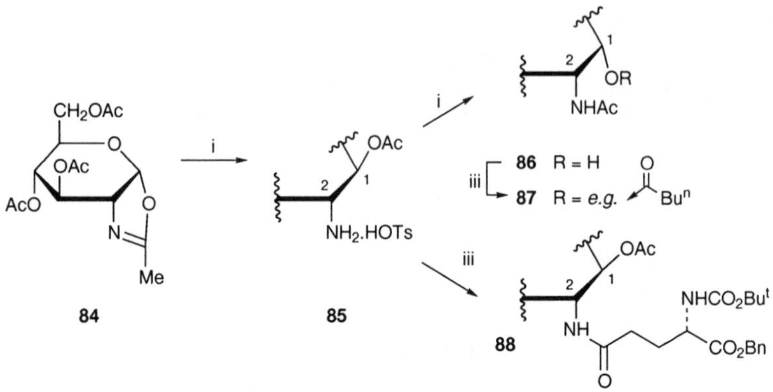

Reagents: i, HOTs, H₂O, MeCN; ii, acylation; iii, $HO_2C\underset{CO_2Bn}{\overset{NHCO_2Bu^t}{\wedge}}$, EEDQ, Et₃N

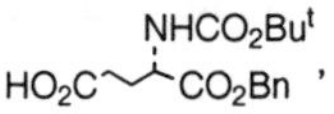

Scheme 23

88.[58] A similar approach has led to the synthesis of various α-glycosyl ester derivatives, e.g. **87** and an intermediate **89** for new lipidic disaccharide analogues

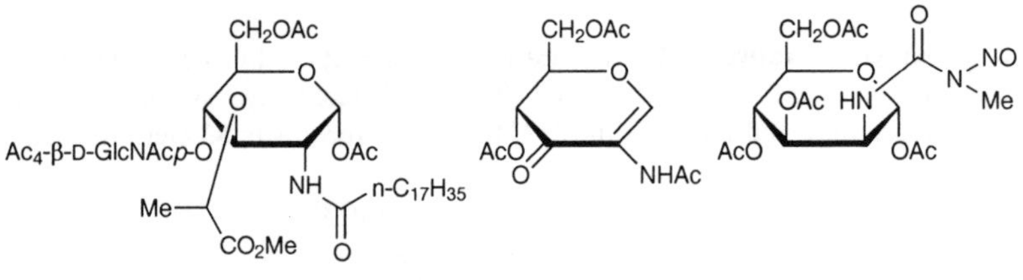

of muramoyl dipeptide.[59] 1,3,4,6-Tetra-*O*-acetyl-2-amino-2-deoxy-α,β-D-mannopyranose was synthesized from 2-amino-2-deoxy-D-mannose *via* either the *N*-(benzyloxycarbonyl)-derivative or the Schiff base formed with 2-hydroxy-1-naphthaldehyde, the free amine being trapped as the stable oxalate salt. Neutralization of this oxalate salt with sodium acetate led to 2-acetamido-1,4,6-tri-*O*-acetyl-2-deoxy-α-D-mannopyranose by O→N acetyl migration. Oxidation of this product at O-3 with concomitant elimination of acetic acid led to enone **90**. A variety of *N*-acyl-derivatives of 2-amino-2-deoxy-D-mannose were synthesized either by acylation of this sugar directly (NHBz, NHCO₂Me, *N*-dansyl) or its tetra-*O*-acetate (NHCONMe, NHSO₂Me, NHCOCF₃, NHSO₂CF₃). The antileukemic activity of these compounds was evaluated, with the acetylated analogue **91** of 2-*epi*-streptozotocin being highly protective in mice.[60]

The utility of several new *N*-protecting groups has been described. The ready cleavage of the *N*-pent-4-enoyl group with iodine in aqueous THF, and the utility

92   93   94

of the 2-deoxy-2-pentenamido-α-D-glucosyl chloride **92** in glycoside synthesis has been described.[61,62] Syntheses of various *N*-tetrachlorophthaloylated 2-amino-2-deoxy-D-glucose derivatives, the use of these derivatives in glycosidation reactions e.g. in combination with the trichloroacetimidate method, and *N*-deprotection using either ethylenediamine or by reduction (NaBH$_4$, HOAc, pH 5) of one of the carbonyl groups and intramolecular phthalide formation have been detailed by two groups.[62,63] A new strategy for *O*-acetyl protection of methyl 6-amino-6-deoxy-hexopyranosides and 1-amino-1-deoxy-alditols which does not cause the commonly observed O→N acetyl migration has been developed. It involved temporary *N*-protection as the *N*-2′,2′-di(ethoxycarbonyl)vinyl group, formed by reaction of the free amine with diethyl (ethoxymethylene)malonate, per-*O*-acetylation, and cleavage by treatment with chlorine in moist chloroform to yield the amine as its hydrochloride salt.[64] A new base-labile 2-(methanesulfonyloxy)-ethoxycarbonyl group has been used for 3′-*N*-protection during the synthesis of 4′-*O*-phosphate and 4′-*O*-sulfate derivatives of daunomycin; 3′-*N*-(β-D-glucopyranosyl- and -glucopyranuronosyl-oxycarbonyl)-derivatives of daunomycin were reported.[65]

Methyl 1-thio-β-glycosides of *N*,*N*-diacetyl-2-amino-2-deoxy-D-glucose and -galactose, readily synthesized by *N*-acetylation of the mono-*N*-acetates (with AcCl, EtNPr$^r_2$), were effective glycosyl donors (with DMTST activation) and the products could be reconverted to mono-*N*-acetates by saponification (NaOMe, MeOH).[66]

Polymerization of *N*-methacroyl derivatives of 2-amino-2-deoxy-D-glucose and -galactose and 5-amino-5-deoxy-D-ribose yielded polymers with MW up to 10^6 daltons. These polymers were covalently linked to various proteases by reductive amination (NaBH$_3$CN). The number of aldehydo-groups in these polymers available for coupling with the lysine amino-groups in the proteases could be enhanced by prior periodate oxidation. Bonding of such a polymer to α-chymotrypsin yielded a significantly stabilized protease, capable of operating in acetonitrile solution.[67]

The novel oligosaccharide mimic **93** and an analogous tetramer, which have peptide-like inter-residue linkages, were formed by amide coupling between a 2-amino-sugar and the acidic group of a 3-*O*-(carboxymethyl)-sugar derivative.[68] Solid-phase techniques were used in the stepwise construction of the alternatively linked oligosaccharide mimic **94**, using *N*-(fluorenylmethoxycarbonyl)-protected benzyl 2-amino-2-deoxy-α-D-glucopyranosiduronic acid as the building block.[69]

Methyl α-glycosides **95-97**, required for the construction of mimics of the O-specific polysaccharide of *Vibrio cholera* O:1, and a 3-deoxy-analogue, were

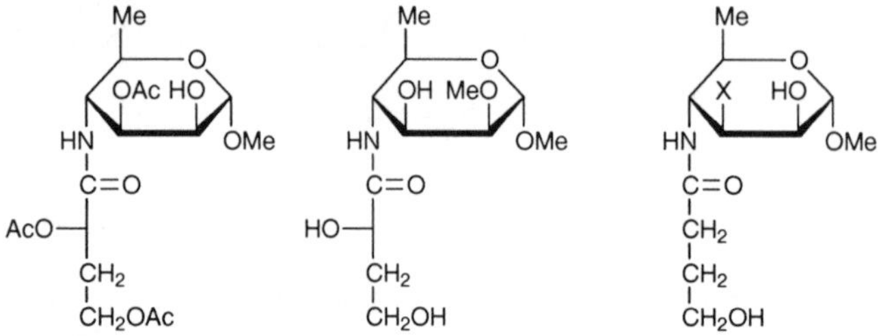

95   **96**   **97** X = H, OH

synthesized by reaction of the corresponding free amino-sugars (from azido-sugar precursors) with 3-deoxy-L-*glycero*-tetrono-1,4-lactone or γ-butyrolactone.[70–72] Reaction of the benzyl α-glycoside of *N*-acetylneuraminic acid with fluoroacetyl imidazolide provided a mixture of 4-*O*-, 5-*N*- and 4,9-di-*O*-fluoroacetate derivatives, while the use of trifluoroacetyl imidazolide allowed an efficient conversion of the 5-*N*-acetyl group in the starting material to a 5-*N*-trifluoro-acetyl group.[73]

3.3   Isothiocyanates and Related Compounds. – A variety of di- and oligo-saccharides containing one isothiocyanato-group per sugar residue have been synthesized from the corresponding amines by reaction with thiophosgene.[74,75] All were stable in the absence of base, but 6-deoxy-6-isothiocyanato-glucose residues unsubstituted at 4-OH underwent cyclization in the presence of triethylamine to give cyclic thiocarbamates.[75] The reaction of 4- and 6-(deoxyisothiocyanato)-sugar derivatives with an amino-cyclitol features in the synthesis of trehalozin analogues (see Chapter 19).[76]

3.4   N-Alkyl Derivatives. – Isomeric mixtures of *N*-(hexosid-2-yl)-pyrrolidine derivatives, e.g. **98**, were formed along with some *N*-demethylated amine **99** on treatment of the *N*-oxide **100** with lithium diisopropylamide in the presence of various dienes, e.g. 2-methyl-1,4-butadiene.[77] The synthesis of carbocyclic analogues of a mannose trisaccharide, linked *via* nitrogen atoms, is covered in Chapter 18.

3.5   Lipid A Analogues. – The analogues **101** of Lipid A, incorporating only 3-hydroxytetradecanoic acid and its homologues as lipid moieties, were synthesized from 2-amino-2-deoxy-D-glucose and shown to have significant mitogenic activity.[78] The analogue **102**, which incorporates a *N*-(2,2-difluorotetradecanoyl) group and a carboxylalkanoyl aglycone, showed LPS-agonist activity towards macrophages, whereas analogues with fewer methylene groups in the aglycon did not.[79]

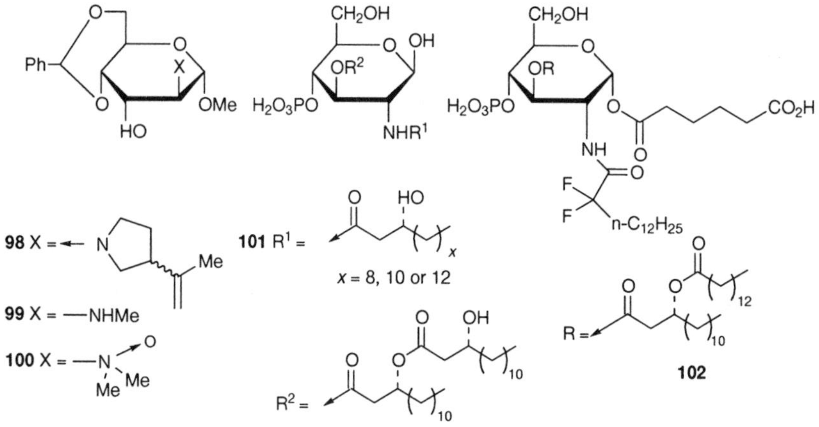

3.6 Amidine and Guanidine Derivatives. – The amidine analogue **103** of methyl 6-*O*-mannosyl-α-D-mannoside has been synthesized and is an effective inhibitor of α-mannosidase (K_i=2.6 μM).[80] Four (deoxy)monosaccharide *N*-benzylamidine analogues were synthesized and their glycosidase inhibitory properties compared to those of **103**, to establish the relative roles of the conformation of the amidine moiety and presence of an aglycone residue in determining potency and selectivity.[80a] Mono- and di-saccharide analogues containing a cyclic guanidinium ion, e.g. **104**, were synthesized by coupling isothiocyanates with 1,2-diamino-1,2-dideoxy-tetritol derivatives, yellow lead oxide being used to effect cyclization of the thiourea adducts formed initially.[81]

3.7 Assorted Derivatives. – Selective 1-*O*-deacetylation of peracetylated 2-amino-2-deoxy-D-hexosides was accomplished with NaOMe in methanol at –96 °C.[82] *N*-Allyloxycarbonyl-protected 2-amino-2-deoxy-D-glucose derivatives served as glycosyl donors in syntheses of 2-acetamido-2-deoxy-β-D-glucosides of hydrophobic alcohols[83] and α,ω-diols,[84] of interest as surfactants. The mixture **105** of methyl 4-amino-4-deoxy-5'-thio-α-maltoside and -α-cellobioside was obtained by condensation of 5-thio-D-glucose with methyl 4-amino-4-deoxy-α-D-glucopyranoside in methanol containing acetic acid, and shown to be a competitive inhibitor of glucoamylase G2. The equivalent reaction involving methyl 2-amino-2-deoxy-β-D-glucopyranoside as the amino-sugar component yielded a mixture of methyl 2-amino-2-deoxy-5'-thio-β-kojibioside and -β-sophoroside.[85]

The reactions of the aziridines **106** with various nucleophiles have been studied. Soft nucleophiles (i.e. RSH, AcOH or Br⁻) and a Lewis acid effected ring opening at C-2. Alcohols and a Lewis acid, on the other hand, initially cleaved the lactone ring to give esters. For aziridine **106** (R=CO₂Bn), the intermediate esters readily converted into lactones **107** in which the aziridine ring had been opened at C-3. For aziridine **106** (R=Ac), reaction with alcohols or benzylamine led to *N*-deacetylation as well as lactone ring cleavage, e.g. to give amide **108**.[86]

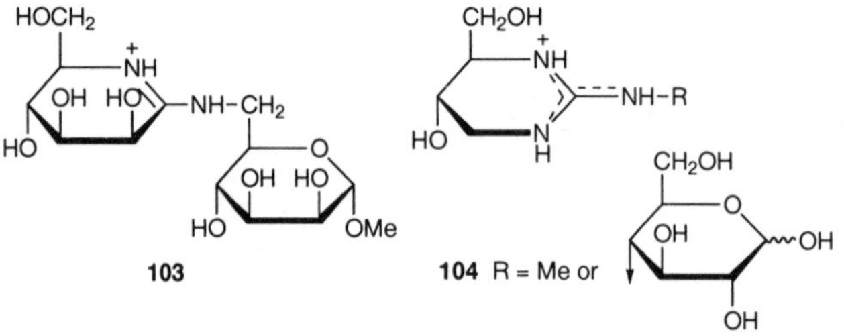

103   **104** R = Me or 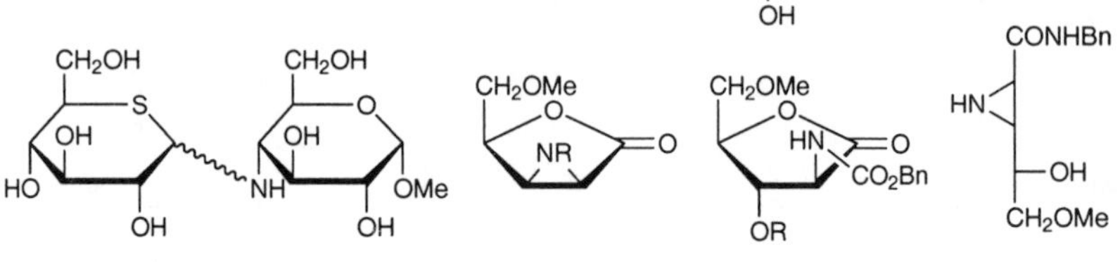   **105**   **106** R = Ac or CO₂Bn   **107** R = Me or Bn   **108**

Synthesis of amino-sugar glycosides is covered in Chapter 3, the formation of 2-acetamido-2-deoxy-D-glucosides from diazirines via glycosyl carbenes in Chapter 10, and the conversion of a deoxynojirimycin derivative into 2,6-dideoxy-2,6-imino-L-gulonic acid in Chapter 18.

4 Diamino-sugars

4.1 Synthesis by Introduction of Two Amino-groups. – The 2,4-diamino-2,4,6-trideoxy-L-altroside **111** was synthesized in 10 steps from L-rhamnal diacetate,

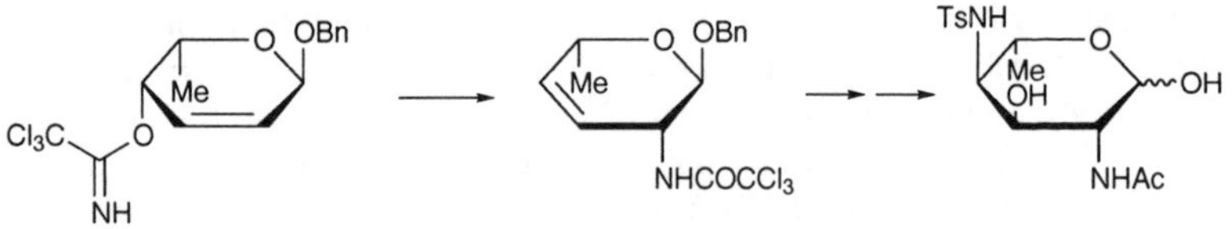

109   **110**   **111**

Scheme 24

key steps being the Overman rearrangement of the allylic trichloroacetimidate **109**, and oxyamination of **110** (Scheme 24).[87] Negamycin lactone **112** was synthesized in 11 steps from 6-*O*-acetyl-2,3,4-trideoxy-D-*glycero*-hex-2-enono-1,5-lactone, with the amino-groups being introduced at C-3 by Michael addition of *N*-benzylhydroxylamine to the unsaturated lactone, and at C-6 by sulfonate displacement, and epimerization at C-5 being effected by intramolecular sulfonate displacement from C-5 by the oxygen atom of a C-2-N(Bn)OH group.[88]

4.2 Synthesis from Amino-sugars. – The 2,4-diamino-2,4,6-trideoxy-D-guloside

115 was synthesized by stereoselective *trans*-diaxial ring opening with azide of epoxide **114**, obtained by several routes, including that from the 2-amino-2-deoxy-D-glucose derivative **113** (R=Bn) *via* formation of the 6-bromide, 6-deoxygenation and 3-*O*-mesylation (Scheme 25).[87] The D-purpurosamine-type glycosyl donors **117** (X=OAc or N₃) were synthesized from the methyl glycoside **113** (R=Me) through lipase-catalysed 6-*O*-acetylation, 3,4-di-*O*-mesylation and elimination (NaI) reactions to give alkene **116**. After catalytic hydrogenation, the 6-azido-group could be introduced by displacement of a 6-mesylate (Scheme 25).

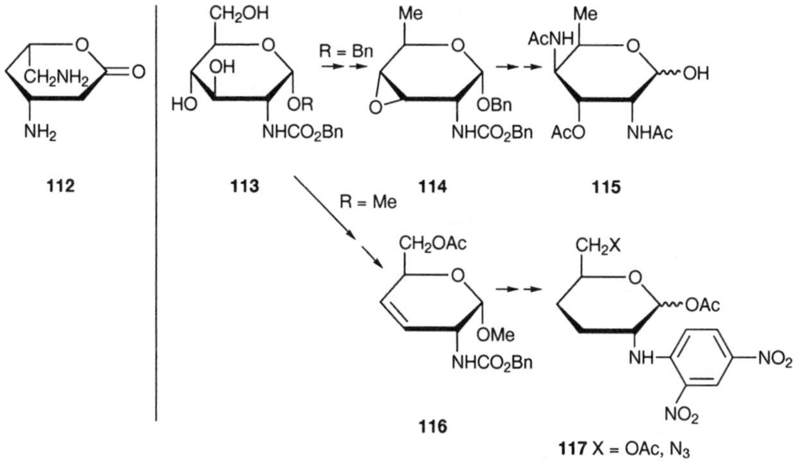

Scheme 25

The L-sugar analogue **118** was prepared by C-5 inversion (sulfonate displacement) in an open chain intermediate en route to **117** (X=N₃).[89] A precursor for purpurosamine synthesis, 6-acetamido-3,4,6-trideoxy-D-glucal, was obtained in 63% e.e. by lipase-catalysed enantioselective *N*-acetylation of the corresponding racemic acrolein-derived amine.[90] Syntheses of various 4-deoxy-4-guanadino-analogues of *N*-acetylneuraminic acid and its 2-deoxy-2,3-unsaturated derivative, are covered in Chapter 16.

4.3 Reactions. – 4-Deoxy-4-guanidino-*N*-acetylneuraminic acid methyl glycoside **120**, a potent inhibitor of influenza virus haemagglutinin, was synthesized from the known azide **119** (Scheme 26).[91] Use of the C-4 epimer of azide **119** in syntheses of 4-acetamido-4-deoxy-4-*epi*-neuraminic acids is covered in Chapter 16. Analogues of the disaccharide **121** modified at C-1 and C-6 of the galactose residue, and at C-4 of the neuraminic acid residue were prepared by standard methods from azide **119** in order to develop a structure-activity relationship for the influenza virus neuraminidase. While analogues bearing azido- or acetamido-groups at C-4 of the NeuAc residue were no longer substrates for the enzyme, nor were they inhibitors. The 4-amino-analogue on the other hand was a potent inhibitor.[92]

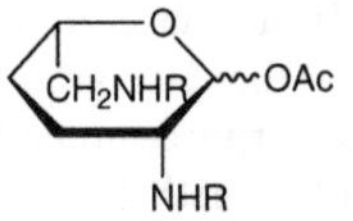

118 R = 2,4-dinitrophenyl 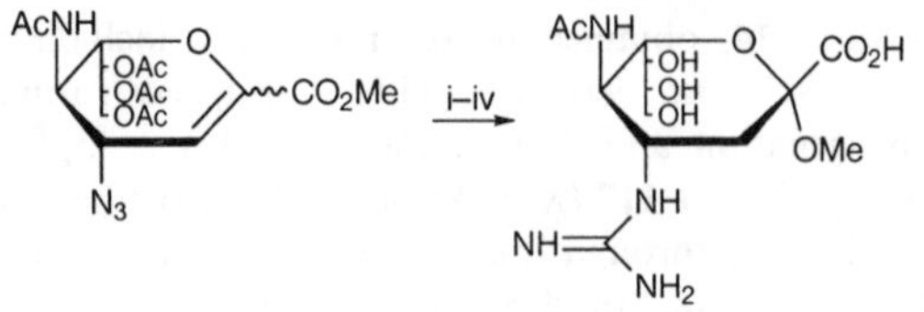   **119**   **120**

Reagents: i, NBS, MeOH; ii, Bu$_3$SnH; iii, MeONa, MeOH;

iv,   NH$_2$·HCl, imidazole, H$_2$O

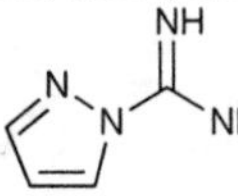

Scheme 26

α-NeuAc-(2→6)-β-D-Gal-OR

121

References

1   W.S. Chilton, A.M. Stomp, V. Beringue, H. Bouzar, V. Vaudequin-Dransart, A. Pettit and Y. Dessaux, *Phytochemistry*, 1995, **40**, 619.

2   A. Dondoni, S. Franco. F. Junquera, F.L. Merchán, P. Merino, T. Tejero and V. Bertolasi, *Chem. Eur. J.*, 1995, **1**, 505.

3   C.-H. Wong, R.L. Halcomb, Y. Ichikawa and T. Kajimoto, *Angew. Chem. Int. Ed. Engl.*, 1995, **34**, 412.

4   G. Casiraghi and G. Rassu, *Synthesis*, 1995, 607.

5   S. Czernecki, S. Horns and J.-M. Valery, *J. Org. Chem.*, 1995, **60**, 650.

6   J.M. Vega-Perez, J.I. Candela, M. Vega and F. Iglesias-Guerra, *Carbohydr. Res.*, 1995, **279**, C5.

7   N.A.L. Al-Masoudi, N.A. Hughes and N.J. Tooma, *Carbohydr. Res.*, 1995, **272**, 111.

8   T.H. Al-Tel, R.A. Al-Qawasmeh, C. Schröder and W. Voelter, *Tetrahedron*, 1995, **51**, 3141.

9   M.H.A. Zarga, T.H. Al-Tel and W. Voelter, *Naturforsch., B: Chem. Sci.*, 1995, **50**, 697 (*Chem. Abstr.*, 1995, **123**, 170 025).

10   L. Sharma and S. Singh, *Carbohydr. Res.*, 1995, **270**, 43.

11   L. Sharma and S. Singh, *Indian J. Chem., Sect. B: Org. Chem. Incl. Med. Chem.*, 1994, **33B**, 851 (*Chem. Abstr.*, 1995, **122**, 106 310)

12   L. Poitout, Y. Le Merrer and J.-C. Depezay, *Tetrahedron Lett.*, 1995, **36**, 6887.

13   K. Sato and A. Yoshitomo, *Chem. Lett.*, 1995, 39.

14   Y. Toyooka, T. Matsuzawa, T. Eguchi and K. Kakinuma, *Tetrahedron*, 1995, **51**, 6459.

15   C. Simiand, E. Samain, O.R. Martin and H. Driguez, *Carbohydr. Res.*, 1995, **267**, 1.

16   C. Simiand and H. Driguez, *J. Carbohydr. Chem.*, 1995, **14**, 977.

17   E.A. Mash, S.K. Nimkar and S.M. De Moss, *J. Carbohydr. Chem.*, 1995, **14**, 1369.

18   M.A. Pradera, D. Olano and J. Fuentes, *Tetrahedron Lett.*, 1995, **36**, 8653.
19   H. Paulsen, M. Springer, F. Reck, I. Brockhausen and H. Schachter, *Liebigs Ann. Chem.*, 1995, 67.
20   A.M. Horneman and I. Lundt, *J. Carbohydr. Chem.*, 1995, **14**, 1.
21   R. Hirschrann, J. Hynes, P.G. Spoors and A.B. Smith, *Tetrahedron Lett.*, 1995, **36**, 2373.
22   S.N. Noomen, G.J. Breel and C. Winkel, *Recl. Trav. Chim. Pays-Bas.*, 1995, **114**, 321.
23   B.T. Cho and N. Kim, *Synth. Commun.*, 1995, **25**, 167.
24   N. Bentley, C.S. Dowdeswell and G. Singh, *Heterocycles*, 1995, **41**, 2499.
25   A.-C. Helland, O. Hindsgaul, M.M. Palcic, L.M. Stults and B.A. Macher, *Carbohydr. Res.*, 1995 **276**, 91-98.
26   W. Klaffke, *Carbohydr. Res.*, 1995, **266**, 285.
27   S. Liemann and W. Klaffke, *Liebigs Ann. Chem.*, 1995, 1779.
28   Y. Ogawa, P. Lei and P. Kovac, *Carbohydr. Res.*, 1995, **277**, 327.
29   I. Pinter, J. Kovács and G. Toth, *Carbohydr. Res.*, 1995, **273**, 99.
30   P. Norris, D. Horton and B.R. Levine, *Tetrahedron Lett.*, 1995, **36**, 7811.
31   W.B. Mathews and W.W. Zajac, Jr., *J. Carbohydr. Chem.*, 1995, **14**, 287.
32   B.M. Heskamp, G.H. Veeneman, G.A. van der Marel, C.A.A. van Boeckel and J.H. van Boom, *Tetrahedron*, 1995, **51**, 8397.
33   Y. St. Denis, J.-F. Lavallée, D. Nguyen and G. Attardo, *Synlett.*, 1995, 272.
34   M.P. Collis, D.C.R. Hockless and P. Perlmutter, *Tetrahedron Lett.*, 1995, **36**, 7133.
35   H. Kuno, S. Niihata, T. Ebata and H. Matsushita, *Heterocycles*, 1995, **41**, 523.
36   K. Matsumoto, T. Ebata and H. Matsushita, *Carbohydr. Res.*, 1995, **267**, 187.
37   L. Cipolla, L. Lay, F. Nicotra, L. Panza and G. Russo, *J. Chem. Soc., Chem. Commun.*, 1995, 1993.
38   E. Kaji and F.W. Lichtenthaler, *J. Carbohydr. Chem.*, 1995, **14**, 791.
39   E. Kaji, F.W. Lichtenthaler, Y. Osa and S. Zen, *Bull. Chem. Soc. Jpn.*, 1995, **68**, 1172.
40   M. Adinolfi, G. Barone, M.M. Corsaro, R. Lanzetta, L. Mangoni and P. Monaco, *J. Carbohydr. Chem.*, 1995, **14**, 913.
41   R.M. Giuliano, V.E. Manetta and G.R. Smith, *Carbohydr. Res.*, 1995, **278**, 345.
41a   B. Kojic-Prodic, V. Milinkovic, J. Kidric, P. Pristovsek, S. Horvat and A. Jakas, *Carbohydr. Res.*, 1995, **279**, 21.
42   A. Dondoni and D. Perrone, *J. Org. Chem.*, 1995, **60**, 4749.
43   W.R. Roush and J.A. Hunt, *J. Org. Chem.*, 1995, **60**, 798.
44   P.A. Wade and S.G. D'Ambrosio, *J. Carbohydr. Chem.*, 1995, **14**, 1329.
45   M. Valpuesta, P. Durante and F.J. López-Herrera, *Tetrahedron Lett.*, 1995, **36**, 4681.
46   M. Kirihata, Y. Nakao, M. Mori and I. Ichimoto, *Heterocycles*, 2996, **41**, 2271.
47   R.F.W. Jackson, N.J. Palmer, M.J. Wythes, W. Clegg and M.R.J. Elsegood, *J. Org. Chem.*, 1995, **60**, 6431.
48   K. Pitzer and T. Hudlicky, *Synlett.*, 1995, 803.
49   P.A. Wade, S.G. D'Ambrosio and D.T. Price, *J. Org. Chem.*, 1995, **60**, 6302.
50   V.P. Vassilev, T. Uchiyama, T. Kajimoto and C.-H. Wong, *Tetrahedron Lett.*, 1995, **36**, 5063.
51   S. Nagumo, I. Umezawa, J. Akiyama and H. Akita, *Chem. Pharm. Bull.*, 1995, **43**, 171.
52   H.-J. Altenbach and R. Wischnat, *Tetrahedron Lett.*, 1995, **36**, 4983.
53   A. Defoin, H. Sarazin and J. Streith, *Synlett.*, 1995, 1187.

54   J.-B. Behr, A. Defoin, N. Mahmood and J. Streith, *Helv. Chim. Acta.*, 1995, **78**, 1161.
55   R. El Bergmi and J.M. Molina, *J. Chem. Res.*, 1995, (S) 484; (M) 2947.
56   K. Matsuoka, S.-I. Nishimura and Y.C. Lee, *Bull. Chem. Soc. Jpn.*, 1995, **68**, 1715.
57   R.M. Werner, M. Barwick and J.T. Davis, *Tetrahedron Lett.*, 1995, **36**, 7395.
58   R. Jha and J.T. Davis, *Carbohydr. Res.*, 1995, **277**, 125.
59   S.S. Pertel, A.L. Kadun and V.Ya. Chirva, *Bioorg. Khim.*, 1995, **212**, 226 (*Chem. Abstr.*, 1995, **123**, 340 564).
60   N.J. Angelino, R.J. Bernacki, M. Sharma, O. Dodson-Simmons and W. Korytnyk, *Carbohydr. Res.*, 1995, **276**, 99.
61   R. Madsen, C. Roberts and B. Fraser-Reid, *J. Org. Chem.*, 1995, **60**, 7920.
62   J.S. Debenham, R. Madsen, C. Roberts and B. Fraser-Reid, *J. Am. Chem. Soc.*, 1995, **117**, 3302.
63   J.C. Castro-Palomino and R.R. Schmidt, *Tetrahedron Lett.*, 1995, **36**, 5343.
64   C. Ortiz Mellet, J.L. Jiménez Blanco, J.M. García Fernández and J. Fuentes, *J. Carbohydr. Chem.*, 1995, **14**, 1133.
65   R.G.G. Leenders, H.W. Scheeren, P.H.J. Houba, E. Boven and H.J. Haisma, *Bioorg. Med. Chem. Lett.*, 1995, **5**, 2975.
66   J.C. Castro-Palomino and R.R. Schmidt, *Tetrahedron Lett.*, 1995, **36**, 6871.
67   C.A. Wartchow, P. Wang, M.D. Bednarski and M.R. Callstrom, *J. Org. Chem.*, 1995, **60**, 2216.
68   H.P. Wessel, C.M. Mitchell, C.M. Lobato and G. Schmid, *Angew. Chem., Int. Ed. Engl.*, 1995, **34**, 2712.
69   C. Müller, E. Kitas and H.P. Wessel, *J. Chem. Soc., Chem. Commun.*, 1995, 2425.
70   P. Lei, Y. Ogawa and P. Kovac, *Carbohydr. Res.*, 1995, **279**, 117.
71   P. Lei, Y. Ogawa, J.L. Flippen-Anderson and P. Kovác, *Carbohydr. Res.*, 1995, **275**, 117.
72   M. Gotoh and P. Kovác, *Carbohydr. Res.*, 1995, **268**, 73.
73   S. Sepulveda-Boza and U. Stather, *Bol. Soc. Chil. Quim.*, 1994, **39**, 299 (*Chem. Abstr.*, 1995, **122** 161 131).
74   M.A. Pradera, J.L. Molin and J. Fuentes, *Tetrahedron*, 1995, **51**, 923.
75   J.M.G. Fernandez, C.O. Mellet, J.L.J. Blanco, J.F. Mota, A. Gadelle, A. Coste-Sarguet and J. Defaye, *Carbohydr. Res.*, 1995, **268**, 57.
76   Y. Kobayasih, M. Shiozaki and O. Ando, *J. Org. Chem.*, 1995, **60**, 2570.
77   J. Chastanet, G. Roussi and G. Negron, *Carbohydr. Res.*, 1995, **268**, 301.
78   H. Ishida, Y. Fujishima, Y. Ogawa, Y. Kumazawa, M. Kizo and A. Hasegawa, *Biosci. Biotech. Biochem.*, 1995, **59**, 1790.
79   M. Shiozaki, H. Miyazaki, M. Arai, T. Hiraoka, S. Kurakata, T. Tatsuta, J. Ogawa, M. Nishijima and Y. Akamatsu, *Biosci. Biotech. Biochem.*, 1995, **59**, 501.
80   Y. Bleriot, T. Dintinger, N. Guillo and C. Tellier, *Tetrahedron Lett.*, 1995, **36**, 5175.
80a   Y. Bleriot, T. Dintinger, A. Genre-Grandpierre, M. Padrines and C. Tellier, *Bioorg. Med. Chem. Lett.*, 1995, **5**, 2655.
81   J. Lehmann, B. Rob, H.-A. Wagenknecht, *Carbohydr. Res.*, 1995, **278**, 167.
82   S.S. Pertel and V. Ya. Chirva, *Khim. Prir. Soedin.*, 1994, 177 (*Chem. Abstr.*, 1995, **122**, 314 948).
83   P. Boullanger, Y. Chevalier, M.-C. Croizier, D. Lafont and M.-R. Sancho, *Carbohydr. Res.*, 1995, **278**, 91.
84   D. Lafont, P. Boullanger and Y. Chevalier, *J. Carbohydr. Chem.*, 1995, **14**, 533.
85   J.S. Andrews, T. Weimar, T.P. Frandsen, B. Svensson and B.M. Pinto, *J. Am. Chem. Soc.*, 1995, **117**, 10799.

86   P, Dauban, L. Dubois, M.E.T.H. Dau and R.H. Dodd, *J. Org. Chem.*, 1995, **60**, 2035.
87   A. Banaszek, Z. Pakulski and A. Zamojski, *Carbohydr. Res.*, 1995, **279**, 173.
88   D. Socha, M. Jurczak and M. Chmielewski, *Tetrahedron Lett.*, 1995, **36**, 135.
89   X. Liang, R. Krieger and H. Prinzbach, *Tetrahedron Lett.*, 1995, **36**, 6433.
90   F. Yang, C. Hoenke and H. Prinzbach, *Tetrahedron Lett.*, 1995, **36**, 5151.
91   S. Ciccotosto and M. von Itzstein, *Tetrahedron Lett.*, 1995, **36**, 5405.
92   S. Sabesan, S. Neira and Z. Wasserman, *Carbohydr. Res.*, 1995, **267**, 239.

10
Miscellaneous Nitrogen Derivatives

1 Glycosylamines and Related Glycosyl-*N*-bonded Compounds

1.1 Glycosylamines. – *N*-(β-D-Glucopyranosyl)nicotinic acid was identified as a major metabolite from niacin in cultured tobacco cells.[1]

β-Glycopyranosylamines were obtained in quantitative yields by treating aqueous solutions of free sugars (D-Glc, D-Gal, lactose, cellobiose, maltose) with concentrated aqueous ammonia and one equivalent of NH_4HCO_3 at 42 °C for 36 hours followed by lyophilization. These were converted into various *N*-acylglycosylamines (*N*-octanoyl, *N*-decanoyl, *N*-myristoyl or *N*-lauroyl) by reaction with acyl chlorides in aqueous ethanol.[2] In a study on the inhibition of the Maillard reaction by aminoguanidine, the product **1** was isolated from reaction of D-glucose with aminoguanidine at pH 7, and converted to the crystalline heptaacetate **2** by peracetylation.[3] The inhibition of β-glucosidases by di-*N*-(β-D-glucopyranosyl)amine and *N*-(α-D-glucopyranosyl)-*N*-(β-D-glucopyranosyl)amine has been investigated.[4]

Formation of separable, diastereoisomeric *N*-glucosyl derivatives (e.g. **3**) of racemic 2,2′-indolylindolines, by direct condensation of the bases with free sugar

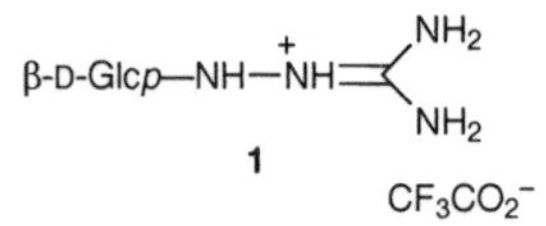

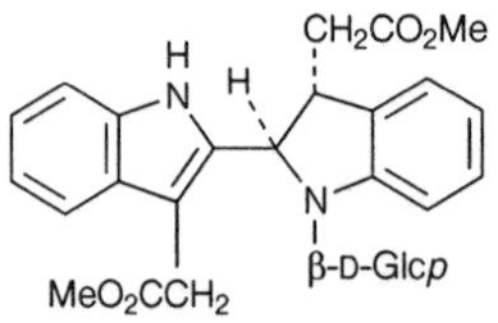

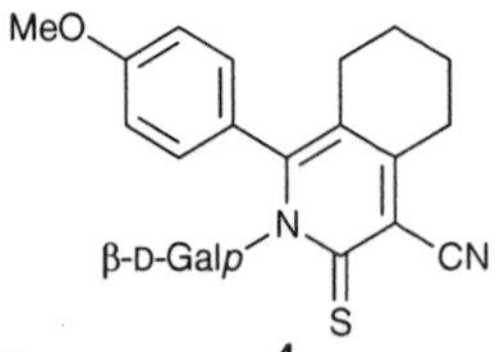

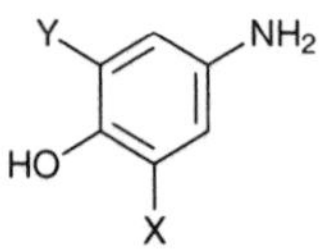

146

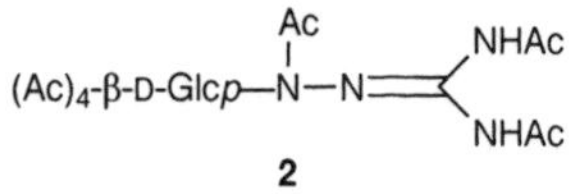

in ethanol under reflux, provided a dual solution to the problems of glycosidation and desymmetrization in the synthesis of indolo[2,3-*a*]carbazole glycosides.[5] Various *N*-glycosyl-pyridinethione derivatives, e.g. **4**, were produced by reaction of the corresponding heterocycles with *O*-acetylated α-halogenosugars.[6] Reaction of 2-acetamido-3,4,6-tri-*O*-acetyl-α-D-glucopyranosyl chloride with the dihalogenated aminophenols **5** in acetonitrile containing triethylamine gave only the *O*-glycosides, whereas reaction with the aminophenols **6** and **7** gave *N*-glycosides as the major products.[7] Regioselective *O*-glycosylations of unprotected *N*-[2,2'-di(ethoxycarbonyl)vinyl]-β-L-fuco- and -rhamno-pyranosylamines are reported in Chapter 3, while intramolecular reactions of *N*-[2,2'-di(ethoxycarbonyl)vinyl]-glycosylamine *O*-mesylates to produce amino-sugar derivatives are covered in Chapter 9.

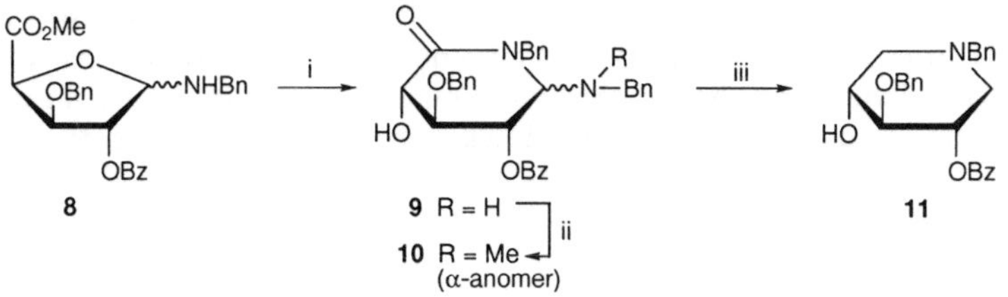

Reagents: i, BnNH$_2$, Me$_3$Al, ii, NaBH$_3$CN, HCHO, AcOH, iii, NaBH$_4$, CF$_3$SO$_2$H, (CH$_2$O)$_n$

Scheme 1

The xyluronosylamine **8** gave the ring expanded benzylamine adduct **9**, that could be *N*-methylated to **10** or reduced to the 1,5-imino-alditol **11** as indicated in Scheme 1.[8] The amidine based pseudo-disaccharide **12** has been synthesized and shown to be a competitive inhibitor of α-mannosidase (K$_i$ = 2.6μM).[9] Glucosylamidines such as **13** were obtained from the corresponding glucosylthioamides such as **14** on reaction with amines and mercury(II) oxide.[10]

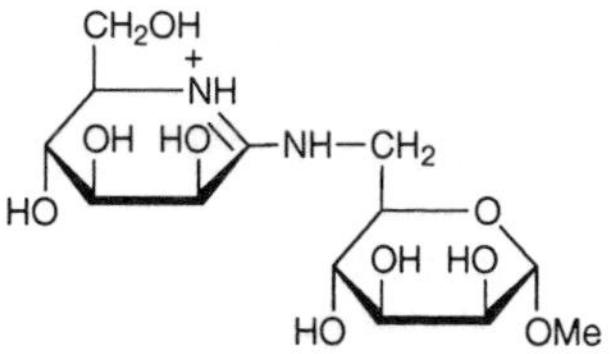

Ac$_4$-β-D-Glc*p*-X

13 X =

or

14 X =

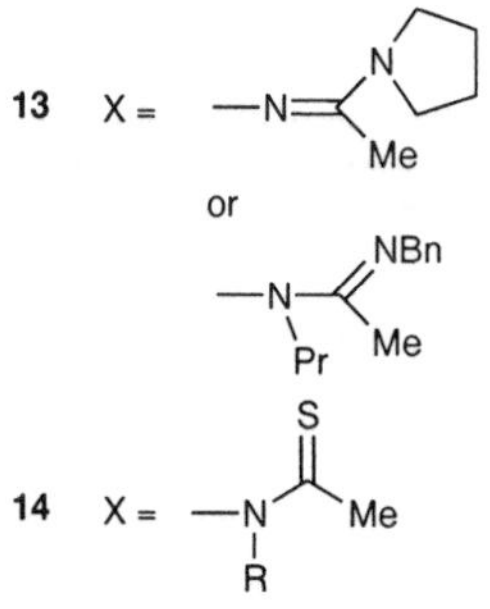

R = H or Pr

An important review entitled 'Reverse Anomeric Effect: Fact or Fiction?' has discounted this as a dominant electronic effect controlling the conformational behaviour of glycosyl species bearing a cationic aglycon. The early evidence was largely from studies of glycosyl-pyridinium and -imidazolium ions, in which there was a preference for the cationic substituent to be axial. There were uncertainties about steric effects in these original examples, however, and reinvestigations reveal no such reverse anomeric effect for glycosylammonium ions, nor does theory support it.[11] The kinetics of reactions of *N*-(2-deoxy-β-D-*arabino*-hexopyranosyl)-3-carboxyamidopyridinium, -isoquinolinium and -4-bromoisoquinolinium bromides with anionic nucleophiles (AcO⁻, Cl⁻, Br⁻, N₃⁻) were studied to probe the influence of the C-2 hydroxy group on the reactivity of glycosyl pyridinium salts.[12]

The fluorescent cyclic derivative **15** was isolated in 4.5% yield on reaction of D-ribose with the *N*^{α}-(benzyloxycarbonyl)-derivative of an arginine-lysine dipeptide. From this it was suggested that intra-molecular reactions may compete with inter-molecular cross-linking reactions during Maillard reactions of sugars with proteins, accounting for the observed lack of protein oligomerisation.[13]

15   16

1.2   Glycosylamides Including *N*-Glycopeptides. – The glucosylamide **16** was the major product (30% yield) obtained on quenching with water the reaction of penta-*O*-benzoyl-α-D-glucopyranose with 2-acetamidoacetonitrile and tin(IV) chloride, along with three minor products.[14] *N*-(2-Deoxy-5-*O*-methoxytrityl-β-D-ribofuranosyl)-formamide and the corresponding α,β-mixture of 2-deoxy-ribofuranosyl-urea derivatives were synthesized by oxidative degradations of thymidine, and the former was incorporated into oligonucleotides.[15]

Photolysis of tri-*O*-methyl or -Tbdms ether derivatives of *N*-(2-deoxy-α-D-*arabino*-hexopyranosyl)succinimide gave bicyclic derivatives **17** following intra-molecular abstraction of H-2, or tricyclic derivatives (see Vol.28, p.142, Scheme 7) following initial intramolecular abstraction of H-5. On further photolysis of bicyclic **17** intramolecular abstraction of H-5 led to the tricyclic derivatives **18** (Scheme 2).[16]

A sialyl-Lewis^x conjugate with the tripeptide Arg-Gly-Asp has been prepared by way of a tetrasaccharide glycosylazide using conventional glycosidation reactions, reduction to the glycosamine and coupling with a C-terminal free peptide. In a cell adhesion test, this conjugate was the best competitive inhibitor

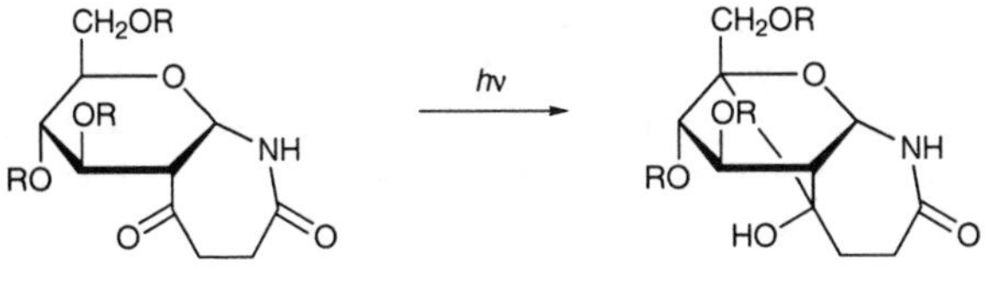

17   18

R = Me or Tbdms

Scheme 2

of P-selectin binding found so far (IC_{50} = 26 μM), suggesting that the binding of glycoproteins to P-selectin involves recognition of both the sugar and its peptide linking region.[17] A versatile solid-phase synthesis of *N*-linked glycopeptides was demonstrated with the coupling of 18 different glycosylamines with various pentafluorophenyl ester-activated peptides on a super-acid sensitive polystyrene resin.[18] *N*-Linked glycopeptides have been constructed in which the sugar moiety is present as part of an *N*-linked substituent on a poly-glycine chain which has no chiral centres, as shown in fragment **19**.[19] The trivalent sialyl-Lewis^x derivative **20** (where SLe^x is α-NeuAc(2→3)-β-Gal(1→4)[α-Fuc(1→3)]-β-GlcNAc), synthesized conventionally from a known tetrasaccharide derivative, had improved *in vitro* binding to E- and P-selectins compared with monovalent derivatives, but this was not the case *in vivo*.[20] Novel *N*-glycosylated dendrimers bearing 2, 4 or 8 β-lactosyl or β-*N*-acetyl-lactosaminyl residues were synthesized by treating *N*-(3-mercaptopropanoyl)glycosylamines with polyl-(*N*-chloroacetylated diglycine)-substituted dendritic polylysines.[21]

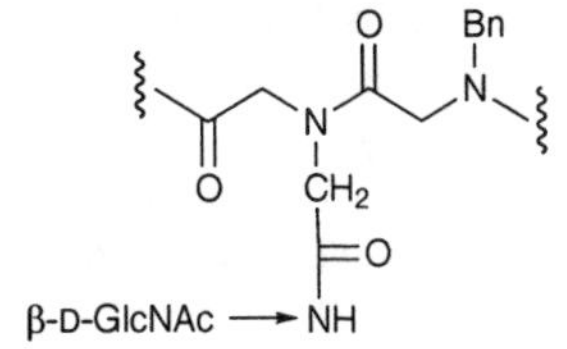

β-D-GlcNAc → NH

19

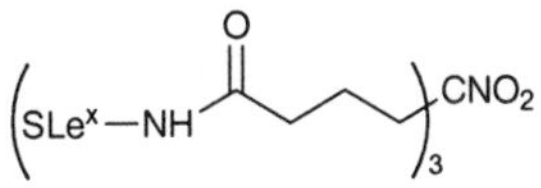

20

β-D-Glc*p*-NH—NPr$^i_2$

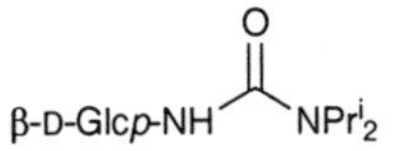

21

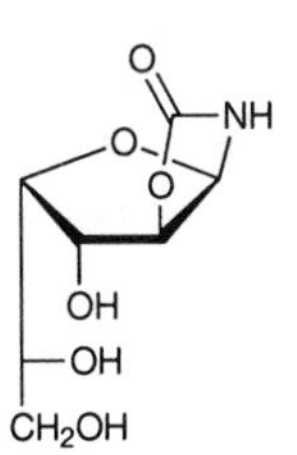

22

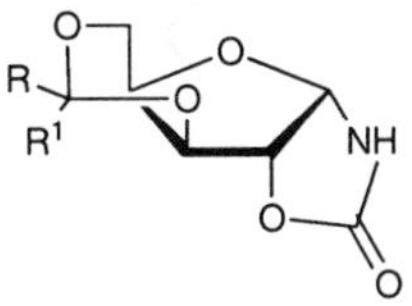

23 R = R^1 = Me or
R = H, R^1 = Ph

1.3 *N*-Glycosyl-carbamates, -isothiocyanates, -thioureas and Related Compounds. – *N*-Glucosyl-ureas such as **21** were synthesized from tetra-*O*-acetyl-β-D-glucopyranosyl azide by reaction with triphenylphosphine and dialkylamines in tetrahydrofuran saturated with carbon dioxide, then *O*-deacetylation. The reaction proceeds by way a phosphinimine intermediate.[22] In the completion of a study on the reaction of hexoses with potassium cyanate in a weakly acidic buffer, it was shown that a single cyclic carbamate, e.g. the D-idofuranosylamine derivative **22**, resulted when the configuration at C-2 and C-4 was the same. When the configuration was different, however, as with D-altrose, mixtures of pyranosyl and furanosyl cyclic carbamates were formed.[23] The chiral oxazolidin-2-ones **23**, produced from D-xylose by reaction with KOCN then acetalation, were evaluated as derivatization reagents for the resolution of racemic carboxylic and sulfonic acids.[24]

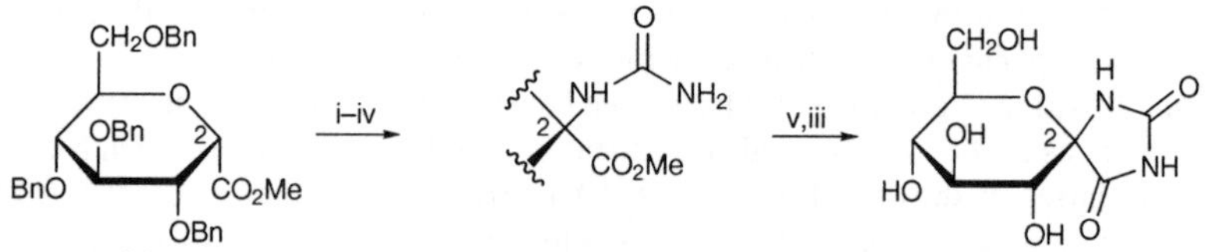

Reagents: i, (Tms)$_2$NLi then CBr$_4$; ii, NaN$_3$; iii, H$_2$, Pd/C; iv, KOCN, AcOH; v, KOBu^t

Scheme 3

Analogues of hydantocidin have been popular targets. The glucopyranose analogue **26** and its C-2 epimer were synthesized from the 2,6-anhydroheptonic acid derivative **24**, *via* a separable mixture of **25** and its C-2 epimer (Scheme 3). Compound **26** was a potent inhibitor of glycogen phosphorylase.[25] The corresponding glucofuranose analogues were made similarly from a 2,5-anhydroheptonic acid derivative, or alternatively from a heptono-1,6-lactone derivative.[26] Full details (cf. Vol.27, p.237) on the synthesis of 5-*epi*-hydantocidin **29** (pre-

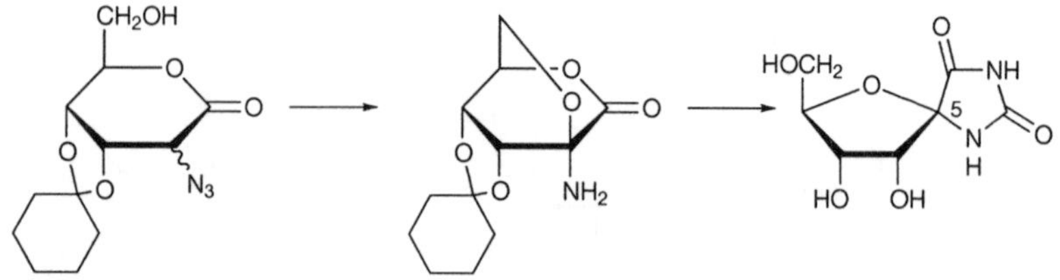

Reagents: i, Pr$^n_4$N$^+$ perruthenate, NMNO

Scheme 4

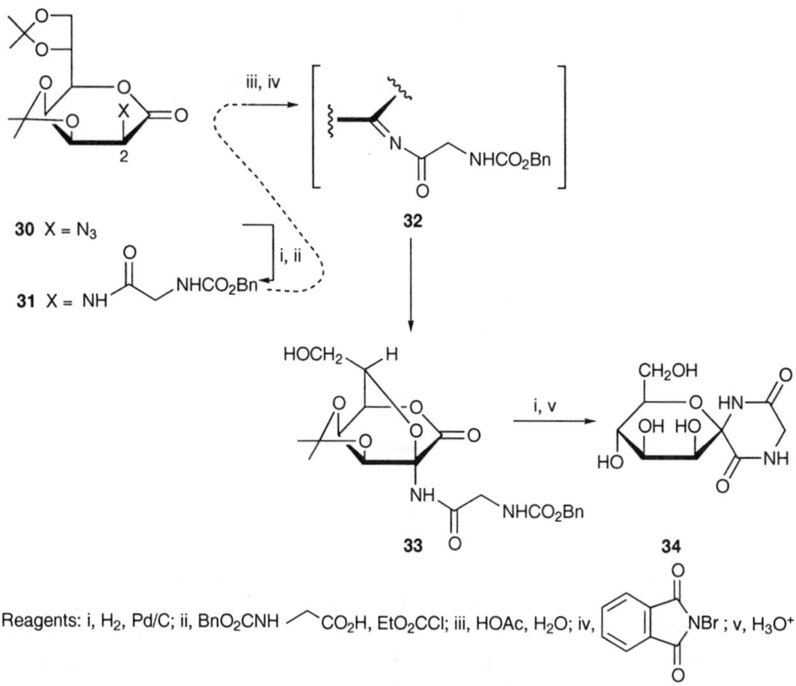
iii, iv
NHCO$_2$Bn
32
30 X = N$_3$
31 X = NH
NHCO$_2$Bn
i, ii
HOCH$_2$
i, v
CH$_2$OH
NHCO$_2$Bn
33
34
Reagents: i, H$_2$, Pd/C; ii, BnO$_2$CNH CO$_2$H, EtO$_2$CCl; iii, HOAc, H$_2$O; iv, NBr ; v, H$_3$O$^+$

Scheme 5

viously incorrectly named as 1-*epi*-hydantocidin) from D-ribose *via* an unanticipated oxidative transformation of the α-azido-lactone **27** to the bicyclic amine **28** (Scheme 4) have been published.[27] Syntheses of spirodiketopiperazines such as the spiro-β-D-mannopyranosyl derivative **34** (from **30**, Scheme 5) and its spiro-β-D-glucopyranosyl analogue made deliberate use of similar oxidative transformations (e.g. **31**→**32**→**33**).[28,29] On treatment with base, compound **34** underwent ring contraction to the more stable spiro-β-D-mannofuranosyl analogue.[28] Hydantocidin analogues **35** and **36** in which one of the carbonyl groups is replaced by a methylene group or a thiocarbonyl group, and **37** which has a spiroimadazolinone ring, were synthesized. While **35** is no longer herbicidal, **36** and its C-2 epimer (sugar numbering) were potent herbicides, and **37** was selectively herbicidal towards dicotyledenous weeds.[30,31] A carba-hydantocidin analogue is covered in Chapter 19.

Solvent-free syntheses of various peracetylated mono- and di-saccharide glycosyl isothiocyanates from the corresponding glycosyl bromides in 41-74% yields were achieved using potassium thiocyanate melts at 190 °C.[32] The α-linked disaccharide **38** and its β-linked isomer, which contain isothiocyanate substituents at both glycosyl and ring positions, were constructed by synthesis of an *N*-protected amino-sugar disaccharide, followed by *N*-deprotection and reaction with thiophosgene.[33] *N*-[5-(*N*,*N*-Dimethylaminopropylcarbamoyl)-1,3,4-oxa-

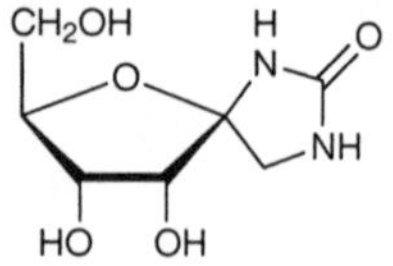

35

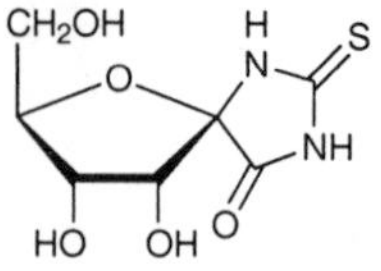

36

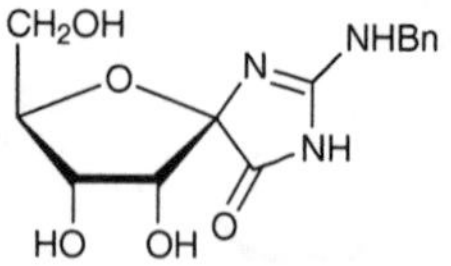

37

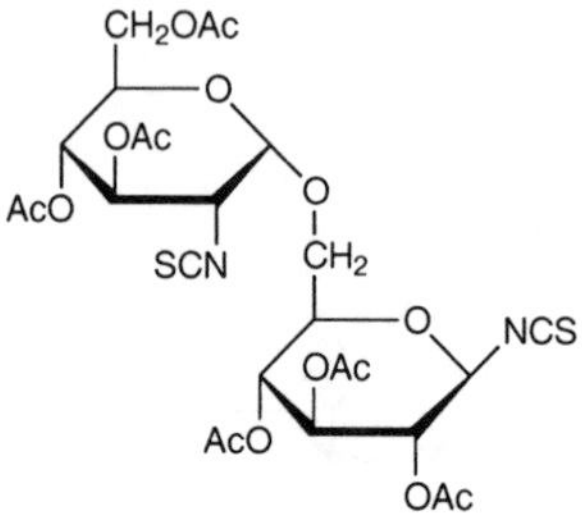

38

diazol-2-yl]-β-D- and L-xylopyranosylamines and -β-D-galactopyranosylamine were synthesized from the corresponding peracetylated glycosyl isothiocyanates by condensation with the requisite oxalyl hydrazide derivative, cyclization (HgO) and deprotection (NH₃, MeOH).[34,35] 5-Amino-6-aryl-3-(N-glycopyranosyl)-tetra-hydro-2-thioxo-4H-1,3-thiazin-4-ones have also been synthesized from glycosyl isothiocyanates (see Chapter 20).[36]

Uchida and co-workers have further exemplified their synthesis (cf. Vol.28, p.233) of cyclic isoureas, e.g. 5'a-carbatrehazolin **39**, by condensing glycosyl isothiocyanates with the cyclitolamine trehazoline, and cyclizing the resulting thiourea adducts (with HgO); for further details see Chapter 19.[37] *cis*-Fused 1,2-oxazolidine-2-thiones such as **40** were the exclusive products from reactions of 1,2-O-sulfinyl-α-D-gluco-furanose or -pyranose derivatives (i.e. 1,2-cyclic sulfites) with sodium thiocyanate.[38]

2   Azido-sugars

Tetra-O-benzyl-α-D-glucopyranosyl azide was obtained by nucleophilic displace-ment reaction of the corresponding α-glucosyl phosphate with azide ion under phase-transfer catalysis conditions.[39] Oxidation of unprotected D-glycopyranosyl azides (with NaOCl, NaHCO₃, catalytic TEMPO, H₂O) afforded the corre-sponding D-glycopyranuronosyl azides, which were isolated as the O-acetylated methyl ester derivatives in good yields.[40]

An approach to the production of small combinatorial libraries has been exemplified by the reaction of tri-O-methyl-D-glucal (with Me₃SiN₃, BnSSBn, SbCl₅ and alcohols) to give mixtures containing structures **41** and **42**.[41] Azido-phenylselenation of tri-O-acetyl- or tri-O-benzyl-D-glucal [with PhI(OAc)₂,

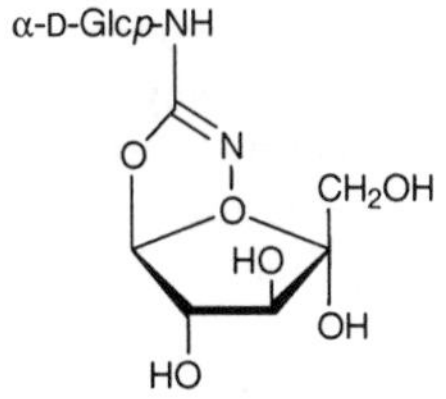

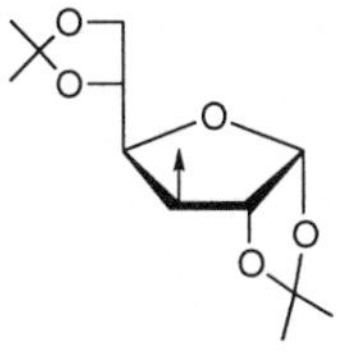

(PhSe)$_2$, NaN$_3$ or Me$_3$SiN$_3$, Bu$_4$NF, *N*-phenylselenophthalimide, respectively]
gave mixtures of the phenyl 2-azido-2-deoxy-1-seleno-D-gluco- and D-manno-
pyranosides, whereas the corresponding D-galactal derivatives gave only the α-D-
galacto-products, e.g. **43**. The seleno-glycosides were readily hydrolysed to the
free sugars [with Hg(OCOCF$_3$)$_2$ or NIS in wet THF].[42] Anti-Markovnikov azido-
phenylselenation of a 1-*exo*-methylene-sugar derivative, to yield a phenyl 1-azido-
1-deoxy-2-seleno-heptulosyl derivative, featured in a synthesis of a potential
glycosyltransferase inhibitor (see also Chapter 9, ref. 32).[43] Nucleophilic ring
opening of the epoxide **45**, obtained from the corresponding 1-*exo*-methylene
sugar **44**, on the other hand led to heptulosyl azide, cyanide or thiophenyl
derivatives **46** (Scheme 6).[44] Allylic azidation of 3-deoxyglycals is covered in
Chapter 13.

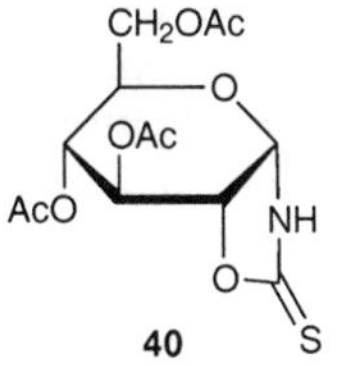

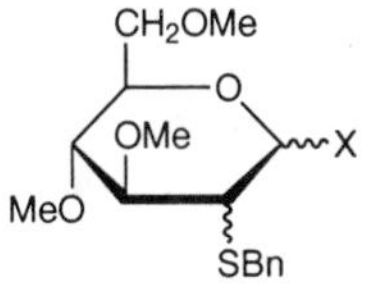

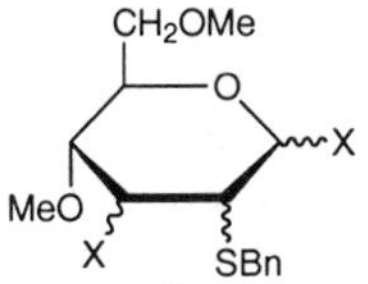

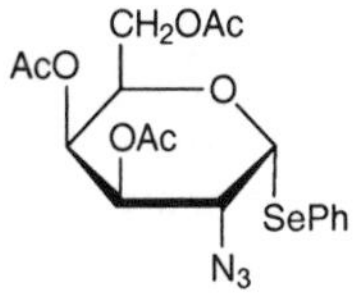

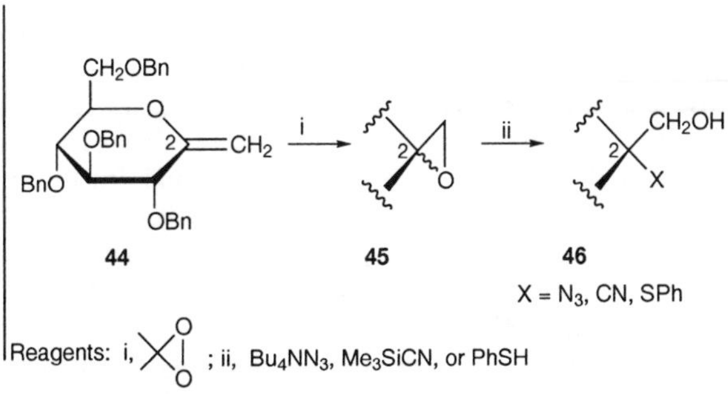

Reagents: i, ; ii, Bu$_4$NN$_3$, Me$_3$SiCN, or PhSH

Scheme 6

Metal ions (Li$^+$ or Mg^{2+}) have a dramatic effect on C2 *vs* C3 regioselectivity in the epoxide ring opening by azide ion of racemic methyl 2,3-anhydro-4-*O*-benzyl-erythronate, with the 2-azido-product being formed almost exclusively in their presence.[45] Only a weak effect is detected for these ions in the analogous reactions involving 2,3-anhydrotetritol derivatives.[46] 2-Acetamido-1,3,6-tri-*O*-acetyl-4-azido-2,4-dideoxy-D-mannopyranose was synthesized from 2-acetamido-2-deoxy-D-mannose by formation of a 1,6:3,4-dianhydrosugar, *trans*-diaxial ring opening of the epoxide with azide ion, and of the 1,6-anhydride ring by acetolysis.[47]

4,4'-Diazido-4,4'-dideoxy-*galacto*-trehalose was synthesized from a known partially benzoylated trehalose derivative by formation and azide ion displacement of a 4,4'-ditriflate, and was claimed to have antimicrobial properties.[48]

2-Azido-3,4,6-tri-*O*-benzyl-D-galactose **49** was synthesized from tri-*O*-benzyl-D-galactal **47** by the short, efficient route shown in Scheme 7, the key step being the reaction of the anion formed from the 2-deoxy-lactone **48** with triisopropyl-phenylsulfonyl azide. 2-Azido-3,4,6-tri-*O*-benzyl-D-mannose was similarly obtained as the sole product from tri-*O*-benzyl-D-glucal. In both cases the azide is introduced *trans* to the substituent at C-4.[49]

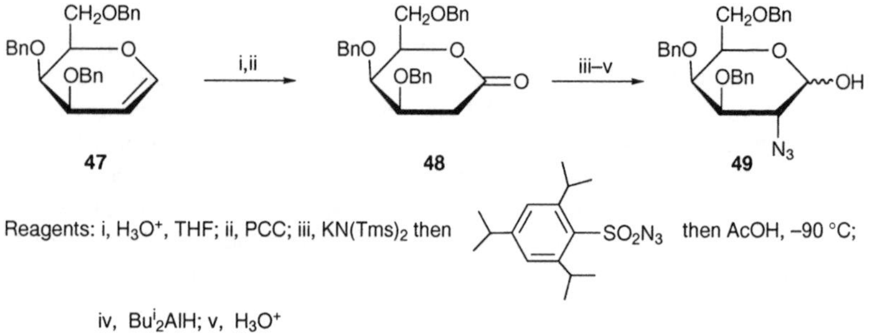

Reagents: i, H$_3$O$^+$, THF; ii, PCC; iii, KN(Tms)$_2$ then [structure] SO$_2$N$_3$ then AcOH, −90 °C;

iv, Bu$^i_2$AlH; v, H$_3$O$^+$

Scheme 7

6-Azido-2,4,6-trideoxy-D-*erythro*-hexose was obtained by 2-deoxyribose-5-phosphate aldolase-catalysed condensation of 2-azidoacetaldehyde with acetaldehyde,[50] while a mixture containing mainly 7-azido-5,7-dideoxy-D-*arabino*-heptulose **50** along with lesser amounts of its C-4 and C-6 epimers, was obtained by sequential one-pot condensation of these aldehydes with dihydroxyacetone monophosphate catalysed by this enzyme and fructose-1,6-diphosphate aldolase.[51] The enzymic oxidation of a D-glucopyranosyl nucleotide, a key step in the biosynthesis of bacterial deoxyoligosaccharides, has been studied; the 3-azido-3-deoxy- and 3-deoxy-analogues were poorer substrates, but still gave the corresponding 4-uloses.[52]

A new route to 1,5-dideoxy-1,5-imino-D-xylonolactam involved Lewis acid catalysed intramolecular reaction of the unstable 5-azido-*aldehydo*-sugar derivative **51** to give **52** (Scheme 8).[53]

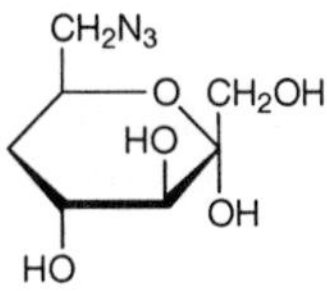

Reagent: i, TiCl$_4$

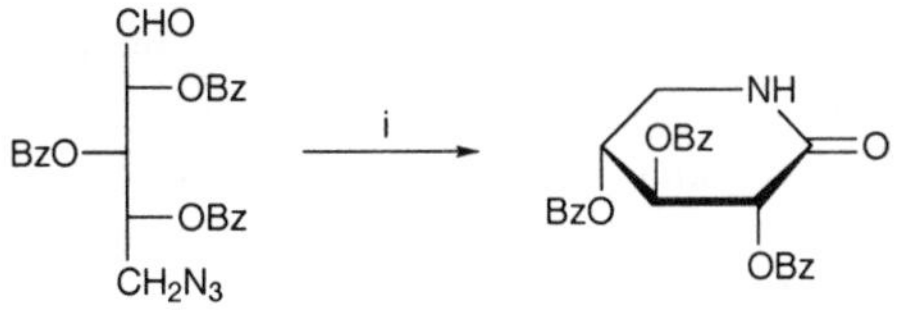

Scheme 8

3   Nitro-sugars

The presence of the ring oxygen atom facilitates Michael addition of nucleophiles (TolS$^-$, MeO$^-$, N$_3$$^-$) to the unsaturated nitro-sugar **53** relative to the unsaturated nitro-cyclitol **54**, but does not affect the stereoselectivity of the reaction; the nature of the solvent does affect the stereoselectivity.[54] A full account (cf. Vol.26, p.129, ref.47) of the Diels-Alder cycloaddition of 1-(trimethylsilyloxy)- or 1-acetoxy-1,3-butadiene to 1,2-dideoxy-1-nitro-D-*galacto*- and D-*manno*-hept 1-enitol tetraacetates has been published.[55] Cycloadditions of substituted cyclopentadienone to the D-*gluco*- and D-*manno*-configured nitrohep-tenitol peracetates, and the conversion of one of the resulting pair of norbornenones into 1-C-arylpentitol derivatives, is covered in Chapter 18. The synthesis of nitro-analogues of the aminoglycosides neamine and kanamycin is covered in Chapter 19. The antibiotic everninomycin contains the nitro-sugar evernitrose **55**, but commercial fermentation produces it in a mixture with analogues having either a hydroxylamino- (**56**) or nitroso-sugar unit, which can be oxidized to the desired nitro-sugar by use of H$_2$O$_2$ and a peroxidase enzyme.[56]

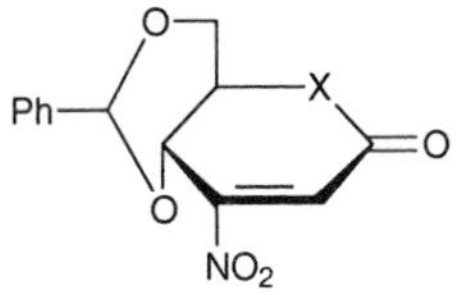

53 X = O
54 X = CH$_2$

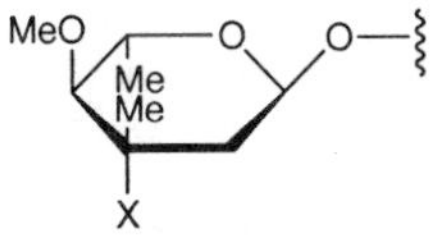

55 X = NO$_2$
56 X = NHOH

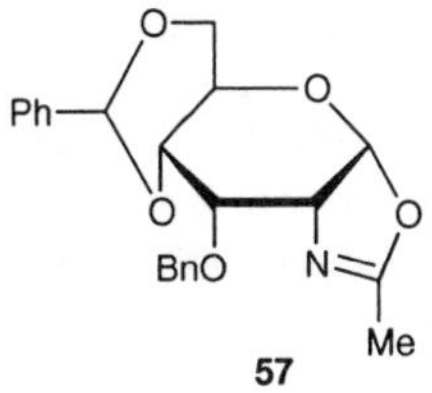

57

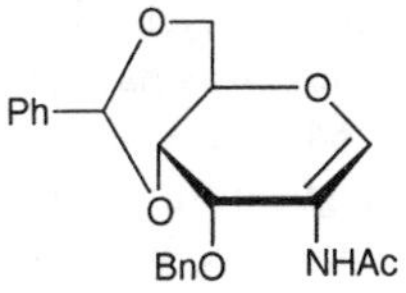

58

4   Diazirino-derivatives

A review on chemical reagents in photoaffinity labelling has reviewed the role of diazirino- ('azi-') sugar derivatives.[57] Disodium 3'-azibutyl α-D-mannopyranoside 6-phosphate and tetrasodium 2-azi-1,10-bis-(α-D-mannopyranosyloxy 6-phosphate)-decane have been synthesized as photoaffinity probes for binding to mannose-6-phosphate receptors.[58]

2-Acetamido-2-deoxy-α-D-glucopyranosides were the predominant products, accompanied in some cases by the β-anomers and a 1,2-fused oxazoline, on reaction of the corresponding 2-acetamido-1,2-dideoxy-1,1-diazirino-sugar with an alcohol or phenol, the reaction proceeding by way of a glycosyl carbene. A somewhat lower α-selectivity was observed for the corresponding 1,1-diazirino-derivative of 2-acetamido-3-*O*-benzyl-4,6-*O*-benzylidene-2-deoxy-D-allose. The oxazoline byproduct in this case, **57**, was show not to be an intermediate, since with a limited quantity of alcohol it gave the alkene **58**, or with an excess of alcohol, the β-glycoside.[59]

5   Oximes, Hydroxylamines, Nitriles and Imines

The *O*-dinitrophenyloximinolactone derivative **59**, a good or modest inhibitor of α- or β-glucosidase, respectively, was synthesized from the corresponding per-*O*-acetylated oxime derivative of D-glucono-1,5-lactone by selective deacetylation of the oxime acetate (with NH_2NH_2), reaction with 1-fluoro-2,4-dinitrobenzene, and deacetylation (with Et_3N, MeOH, H_2O).[60] The Lemieux-type adduct **60** (which exists as a dimer), derived from glucuronic acid, gave α-glycosides **61** or the anomeric glycosylamines **62** as shown in Scheme 9.[61] A synthesis of the *O*-methyloxime derivative of the 4-keto-derivative of *N*-acetylneuraminic acid has been reported.[62]

$$CH_2OH$$

59

In the reaction of chiral aldoximes with divinyl sulfone, which involves tandem nitrone generation-cycloaddition, the best diastereoselectivity was attained when the α-substituent was part of a ring system as it is in the L-threose derivative **63** which gave **64** as a single product (Scheme 10).[63] The 3'-deoxy-3'-(methoxyamino)-analogue of thymidine and its 3'-epimer were synthesized from a 3'-keto-thymidine derivative by *O*-methyloxime formation and reduction ($NaBH_3CN$). 3'-Deoxy-3'-(*N*-benzylhydroxyamino)-thymidine analogues were similarly produced.[64] Hydroxyamino-sugars, such as 3-deoxy-3-hydroxyamino-1,2:5,6-di-*O*-

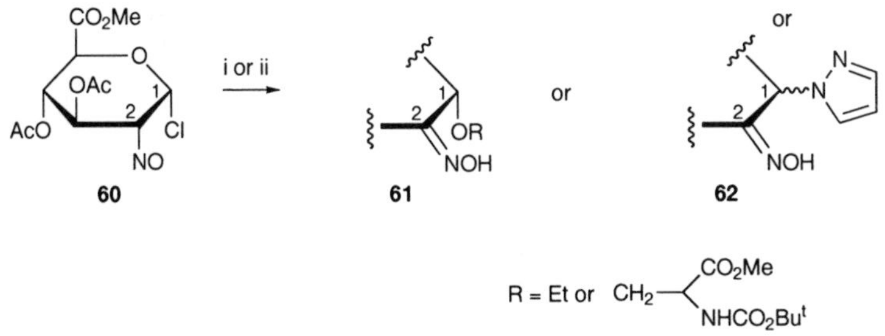

R = Et or $CH_2 - \begin{array}{c} CO_2Me \\ NHCO_2Bu^t \end{array}$

Reagents: i, ROH, DMF; ii, pyrazole, DMF

Scheme 9

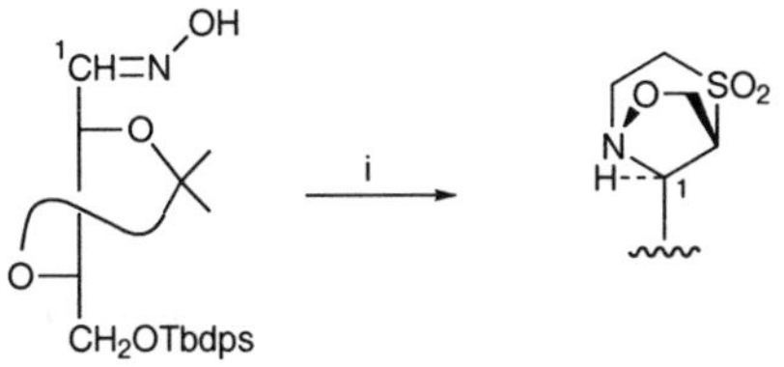

Reagent: i, $\left(\underset{2}{\diagup\!\!\!\diagup} \right) SO_2$, PhMe, Δ

Scheme 10

isopropylidene-α-D-allofuranose, have been converted into derivatives in which the hydroxyamino-group is *O*-acylated or *O,N*-diacylated.[65]

Peracetylated glycopyranosyloxysuccinimides of β-D-Glc, β-D-Gal, β-D-GlcNAc (i.e. **65**), β-lactose, β-maltose and α-sialic acid, were synthesized by condensation of glycosyl halides with *N*-hydroxysuccinimide under phase-transfer catalysis conditions. *O*-Deacetylation with methoxide unavoidably opened the succinimide ring, but this observation permitted the synthesis of acid **66** and the incorporation of such residues into a divalent glycoprobe **67** with a fluorescein tag (Scheme 11). β-D-Galactopyranosyloxyamine was synthesized from the D-galactose analogue of **65** by reaction with hydrazine in ethanol.[66] The *N*-hydroxy-analogues **68** and **69** of hydantocidin were synthesized from 2,3-*O*-isopropylidene-D-erythronolactone, but neither were herbicidal.[67]

In a reinvestigation of the synthesis of aldononitrile peracetates from sugars by oximation and acetylation, it was shown that: (a) peracetylated acyclic oximes are present in the product mixtures, but are converted to the nitriles by elimination of acetic acid in the heated injection port during GLC analysis; and (b) less of the peracetylated *N*-hydroxyglycosylamine byproduct is formed using 1-methylimidazole as a catalyst.[68] The fate of 2,3:5,6-di-*O*-isopropylidene-D-mannose oxime, which can exist as its cyclic *N*-hydroxy-α-D-mannofuranosylamine tautomer, on

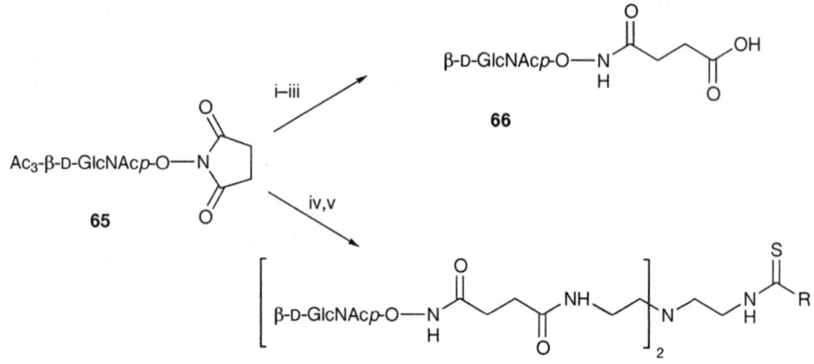

65 **66** **67** R = fluorescamine

Reagents: i, MeONa, MeOH; ii, NaOH, H_2O; iii, H_3O^+; iv, $(H_2NCH_2CH_2)_3N$; v, fluorescein isothiocyanate

Scheme 11

reaction with various dehydrating reagents (Ac_2O, PhNCO, or P_2O_5) has been investigated.[69]

Reaction of trimethylsilyl cyanide with the D-glyceraldehyde-derived nitrone **70** gave the D-*threo*-adduct **71** (Scheme 12), whereas diethylaluminium cyanide gave a mixture of D-*erythro*- and D-*threo*-adducts in a 3:7 ratio.[70]

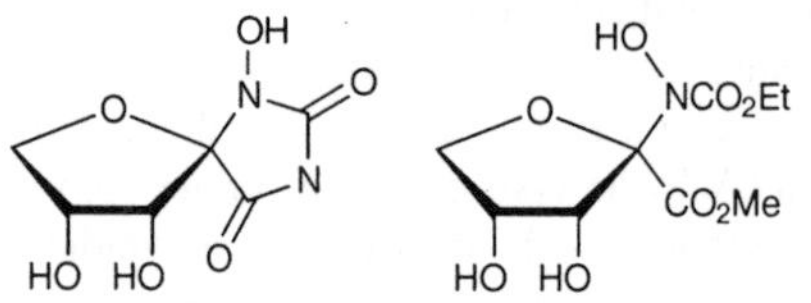

68 **69**

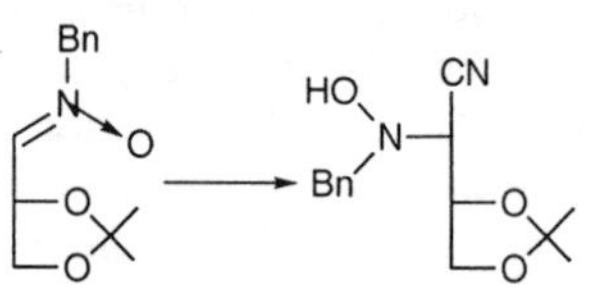

70 **71**

Scheme 12

Further examples (cf. Vol.26, p.130, refs. 6 and 54) have been reported of the conversion of glycosyl azides to aldonic acid *N*-bromoiminolactones on photobromination.[71] Reaction of 2′-deoxyguanosine with glucose and propylamine in phosphate buffer at pH 7 at 40 °C for several days produces the adduct **72** which is considered to arise by condensation of the amino-group of guanosine with the 2-keto-group of 3-deoxy-D-*erythro*-hexos-2-ulose (a glucose degradation product), and molecular rearrangement of the resulting imine.[72]

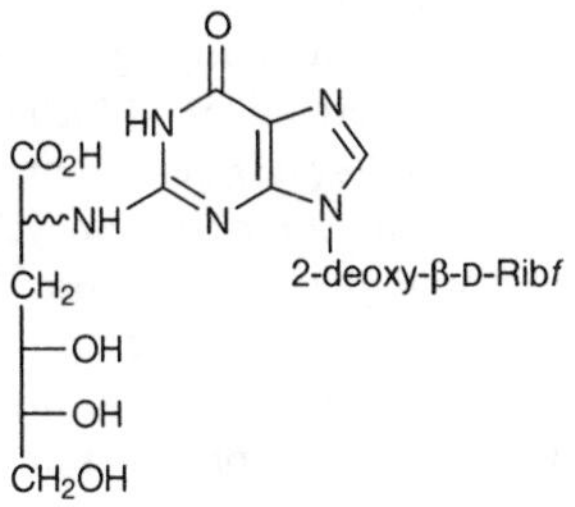

72

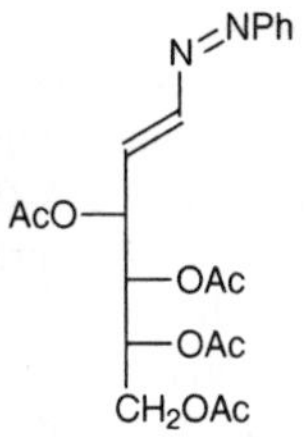

73

6   Hydrazones, Hydrazines and Related Heterocycles

Further details have been reported (cf. Vol.25, p.130) on the formation of acyclic diazabutadienes such as **73** from aldoses (D-Man, D-Gal, D-Rib) on sequential reaction with an arylhydrazine then acetylation (with Ac$_2$O, Py at 60-80 °C).[73] Full details (cf. Vol.28, p.44, ref.229) have been reported on the development of a method for sequential removal of sugar units from the reducing end of an oligosaccharide. Anhydrous hydrazine specifically cleaves glycosidic bonds adjacent to aldehydic or ketonic groups. This is the situation for 1,2-linked disaccharides and 1,3-linked hexosyl-ketoses, and with them hydrazine treatment cleaved the reducing sugars and converted them to various degradation products, e.g. the hydrazones of 2,5-anhydroaldoses. The non-reducing sugars were released intact. This method can be applied to oligosaccharides with other linkages, if they are first reduced to oligosaccharide-alditols and oxidatively cleaved specifically at the alditol moiety [Pb(OAc)$_4$ at low temperature], prior to hydrazinolysis.[74] The method is thus akin to the Barry degradation procedure which uses phenylhydrazine.

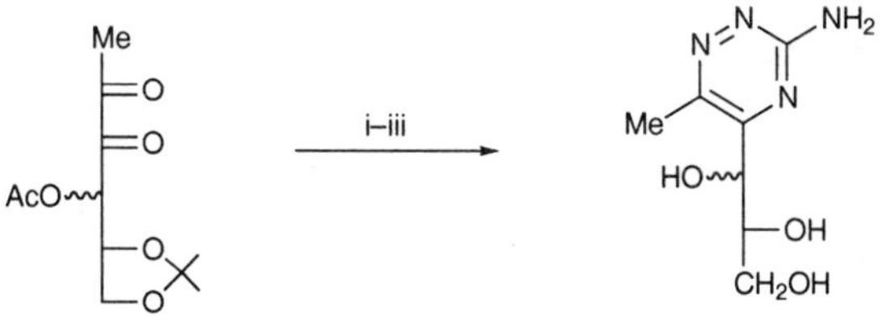

74

75

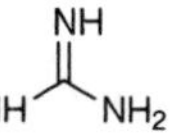

Reagents: i, H$_2$NNH(C=NH)NH$_2$; ii, NaOMe, MeOH; iii, HCl, MeOH

Scheme 13

Aminoguanidine is a known inhibitor of the Maillard reaction. The triazine epimers **75** were synthesized as standards for use in studying this reaction by condensing aminoguanidine with the dicarbonyl derivatives **74** (Scheme 13).[75] 6-Substituted 2-aryl-3-*N*-(D-glycopyranosylamino)-4(3H)-quinazolinones were synthesized either by direct condensation of the amino-heterocycles with D-glucose or D-galactose,[76] or by use of acetobromo-glucose or -xylose.[77]

Aldono-1,4-lactone *N*-tosyl- or *N*-naphthalene-2-sulfonylhydrazone derivatives such as **77** were prepared from the corresponding free sugars (e.g. **76**), and shown to act as glycofuranosylidene carbene precursors on deprotonation and photolysis (Scheme 14). The carbenes could be trapped with alkenes to give adducts such as **78**, or phenols to give glycosides e.g. **79**.[78]

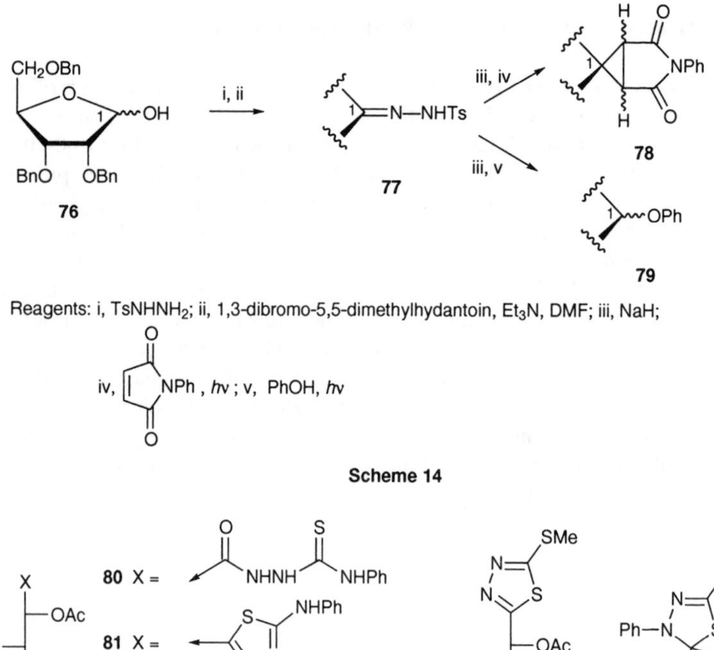

Reagents: i, TsNHNH₂; ii, 1,3-dibromo-5,5-dimethylhydantoin, Et₃N, DMF; iii, NaH; iv, [structure], hν; v, PhOH, hν

Scheme 14

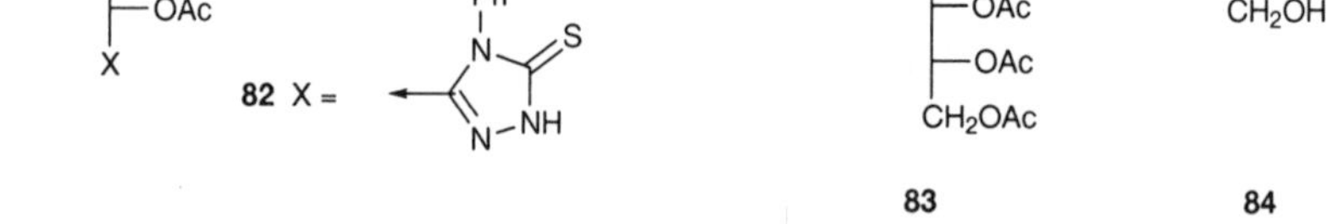

Dehydrative cyclization of the bis-arylthiosemicarbazide derivatives of galactaric acid, e.g. **80**, gave different heterocyclic derivatives depending upon the reagent used, e.g. **81** with POCl₃ or **82** with NaOEt then acetylation.[79] The 1,3,4-

thiadiazole derivative **83** was obtained by reaction of D-glucono-1,5-lactone with *S*-methyl hydrazinecarbodithioate and dehydrative cyclization (with Ac$_2$O).[80] 1,3,4-Thiadiazolidines derivatives **84** were identified with the aid of an X-ray crystal structure as the products from the reactions of aldoses with 2-phenylthio-benzhydrazide, confirming structures proposed more than 65 years ago.[81]

7 Other Heterocycles

Intramolecular cyclization of the iminyl radical generated from the 2,3-unsaturated α-*C*-glycoside **85** gave the bicycle **86**, that readily aromatized to the pyrrole **87** (Scheme 15).[82]

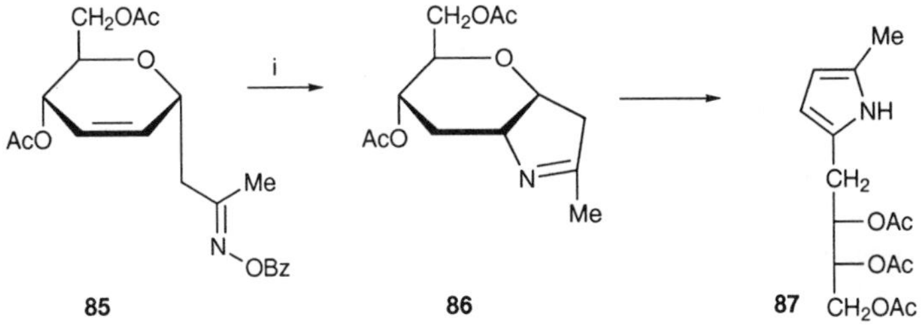

Reagents: i, Bu$_3$SnH, AIBN

Scheme 15

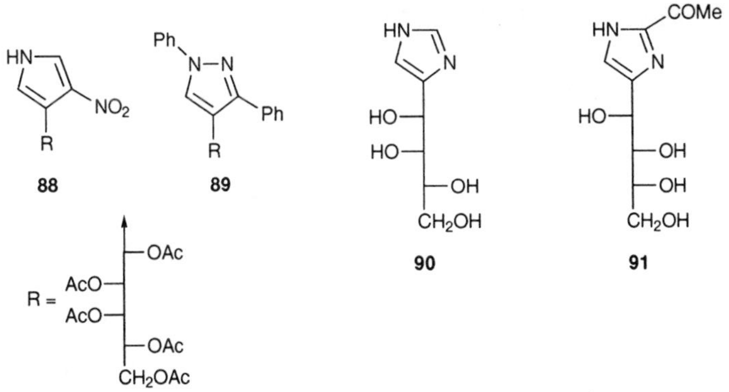

Alditol-1-yl substituted pyrroles or pyrazoles (e.g. **88** and **89**) were obtained by addition to per-*O*-acetylated 1,2-dideoxy-1-nitro-D-*galacto*- and D-*manno*-1-heptenitols of the sodium salt of tosylmethyl isocyanide or the zwitterionic diaryl nitrile imines (e.g. PhC≡N$^+$-N$^-$Ph), respectively.[83] A general one-pot synthesis of alditol-1-yl substituted imidazoles (e.g. **90**) involved reaction of aldoses with formamidine acetate and ammonia (75 °C, 40 atm., 18 h).[84] 2-Acetyl-4-(D-*arabino*-tetritol-1-yl)imidazole **91**, a component of Caramel Colour III that has amongst other biological properties the ability to depress rodent blood lymphocyte counts, was synthesized in five steps from a vinyl stannane derived from 2,3-*O*-isopropylidene-D-glyceraldehyde, the key steps being Stille coupling with an imidazole derivative and asymmetric dihydroxylation of an alkene intermediate.[85]

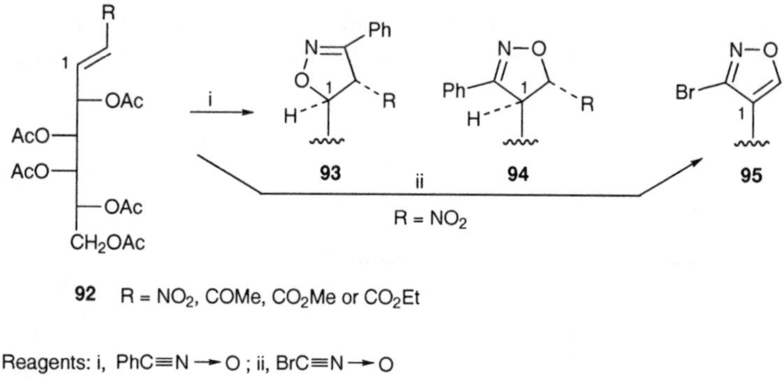

92   R = NO₂, COMe, CO₂Me or CO₂Et

Reagents: i, PhC≡N → O ; ii, BrC≡N → O

Scheme 16

Dipolar cycloadditions of nitrile oxides to the various unsaturated alditol derivatives **92** led to pairs of isoxazolidine regio-isomers, e.g. **93** and **94**, each as predominantly one stereoisomer (Scheme 16). The isoxazole **95** was the major product from addition of bromoformonitrile oxide to the nitro-alkene.[86] The nitrile moiety of D-mannononitrile pentabenzoate was transformed into either a 3-substituted 5-methyl-1,2,4-oxadiazole moiety (with NH_2OH then Ac_2O) or a 5-substituted 2-methyl-1,3,4-oxadiazole moiety (with NH_4N_3 then Ac_2O, Py); cf. Vol.24, p.135.[87] The synthesis of 2-(alditol-1-yl) substituted 2-aza-3,7-dioxabicyclo[3.3.0]octanes by intramolecular dipolar cyclization of 2-*O*-allyl-*aldehydo*-sugar nitrones is covered in Chapter 24.

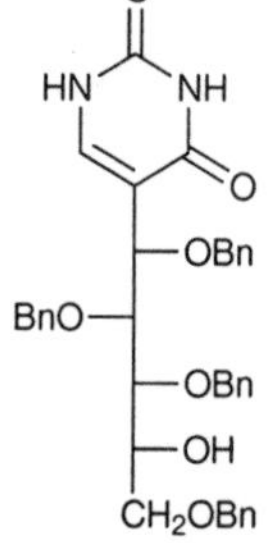

96 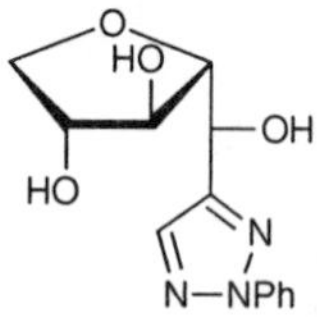   **97**

The D-*gluco*-pentitol-1-yl substituted uracil **96** and its D-*galacto*-isomer were synthesized from tri-*O*-benzyl-D-glucal and -D-galactal, respectively, by [2+2]cycloaddition of trichloroacetyl or chlorosulfonyl isocyanate, cleavage of the β-lactam rings in the products with methanol after *N*-carbamoylation, and cyclization of resulting glycosylureas.[88] Tosylation of 4-(D-*galacto*-pentitol-1-yl)-2-phenyl-2*H*-1,2,3-triazole led to the 3,6-anhydride **97** and two partially tosylated derivatives.[89]

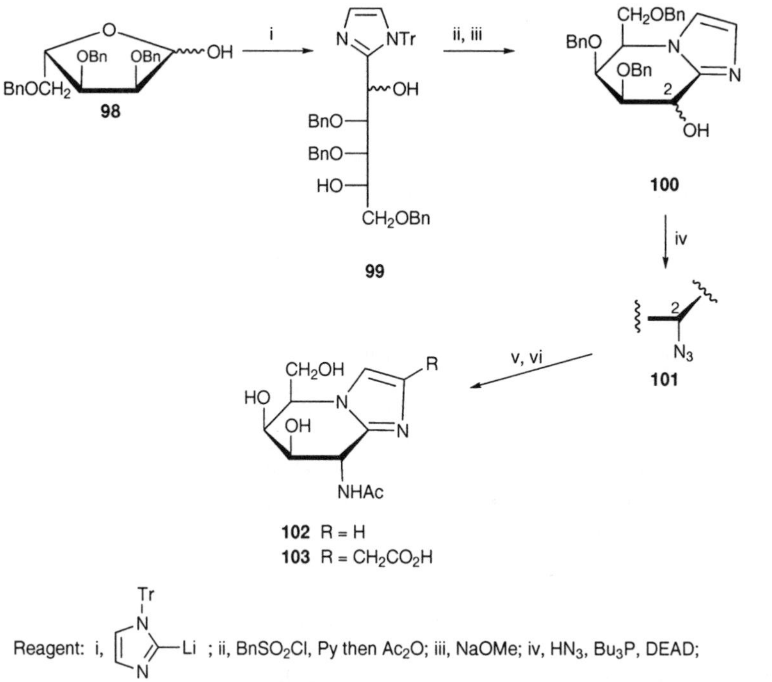

Reagent: i, [imidazole-Tr]—Li ; ii, BnSO₂Cl, Py then Ac₂O; iii, NaOMe; iv, HN₃, Bu₃P, DEAD;

v, H₂, Pd/C, AcOH; vi, Ac₂O, MeOH

Scheme 17

Imidazole- and tetrazole-fused deoxynojirimycin analogues continue to be popular targets for synthesis. The analogue **102** of nagstatin **103** was synthesized from the L-ribose derivative **98** (Scheme 17). The epimers **99** were separately converted to **102**, but both epimers of **100** gave the same azide **101**. The D-*gluco*-, D-*manno*- and 2-acetamido-2-deoxy-D-*gluco*-analogues of **102** were similarly synthesized from an L-xylofuranose derivative. Strong β-glycosidase inhibitory activity was seen with isosteric inhibitors, e.g. the 2-acetamido-2-deoxy-D-*gluco*-isomer had a nanomolar IC_{50} value for β-N-acetylglucosaminidase.[90] Nagstatin

103 itself was synthesized in 12 steps from the epimers **100**.[91] The alternatively fused imidazole derivative **105** was synthesized by intramolecular cyclization of the partially protected acyclic imidazole derivative **104** (Scheme 18).[92]

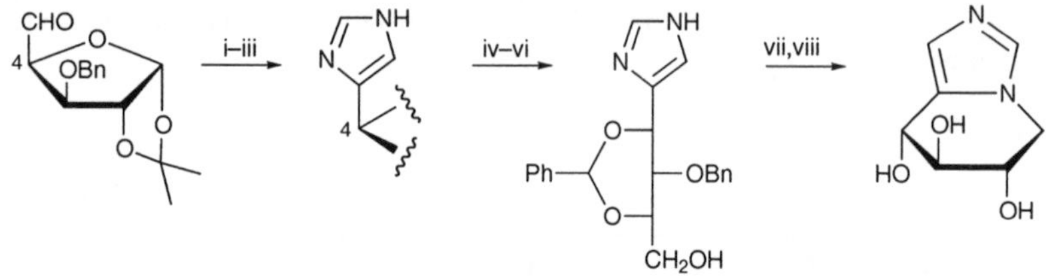

104   **105**

Reagents: i, Ts⌒NC,Bu^{t}OK; ii, POCl$_3$, Et$_3$N; iii, NH$_3$, MeOH; iv, H$_3$O$^+$; v, NaBH$_4$; vi, PhCHO, ZnCl$_2$; vii, Tf$_2$O, Py; viii, H$_2$, Pd(OH)$_2$/C, AcOH

Scheme 18

The fused-tetrazole **106**, with the D-*galacto*-configuration, and its 2-acetamido-2-deoxy-D-*gluco*-analogue **107**, were synthesized from D-galactose and 2-acetamido-2-deoxy-D-glucose, respectively. The method, applied before to the synthesis of other isomers (cf. Vol.25, p.131 and Vol.27, p.137), involved a double inversion at C-5 using sequential oxidation, reduction, and sulfonate displacement with azide ion in an otherwise protected aldononitrile derivative. Compound **106** was a competitive inhibitor of two β-galactosidases (K$_i$ ~ 1 µM), while **107** inhibited bovine kidney N-acetyl-β-D-glucosaminidase (K$_i$ = 0.2 µM). An O-protected derivative of **106** was converted to 1-deoxygalactonojirimycin, the tetrazole ring being reductively cleaved (with LiAlH$_4$).[93] In a variation of this approach, 4-azido-4-deoxy-hexono-1,5-lactone and 5-azido-5-deoxy-hexono-1,4-lactone derivatives were synthesized, converted into the corresponding azido-aldononitriles, and thermally cyclized to yield the tetrazole-fused D-mannofuranose **108**, D-mannopyranose **109** and their D-rhamnose analogues. Inhibitory effects on α- and β-mannosidases and -fucosidases were reported.[94,95]

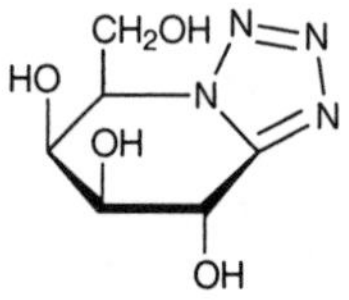

106 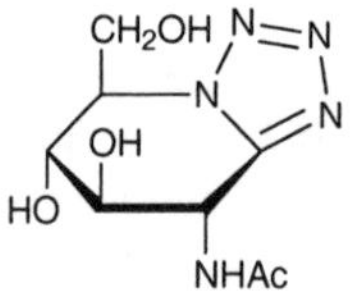   **107** 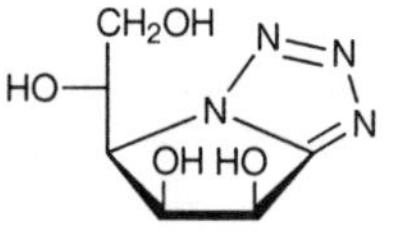   **108** 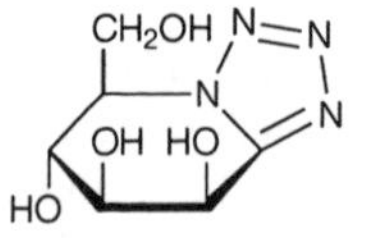   **109**

Photolysis of 2,3,4,6-tetra-*O*-acetyl-1-*C*-cyano-β-D-galactopyranosyl azide gave the oxazepine derivative **110** in 50% yield; a similar product was obtained from the D-arabinopyranosyl analogue.[96] Full details have been reported on the formation of various tetrazole derivatives by photolysis and thermolysis of 1,1-diazido-sugar derivatives (cf. Vol.27, p.138; Vol.28, p.154).[97] Compound **111** was the unexpected product from reaction of 2,3,4,6-tetra-*O*-acetyl-D-glucopyranosyl-idene 1,1-diazide with triphenylphosphine.[98]

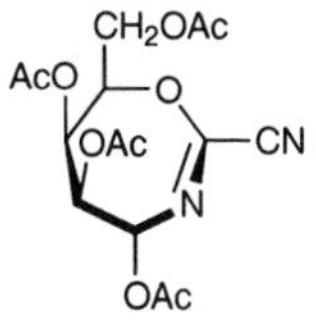

110

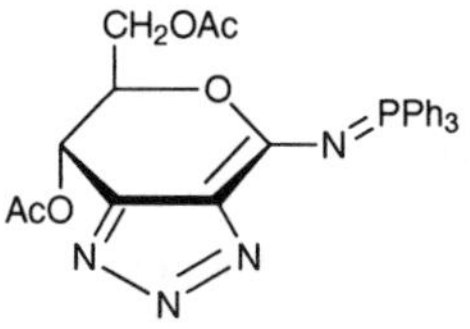

111

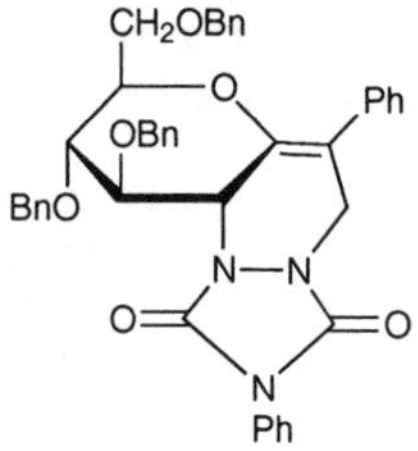

112

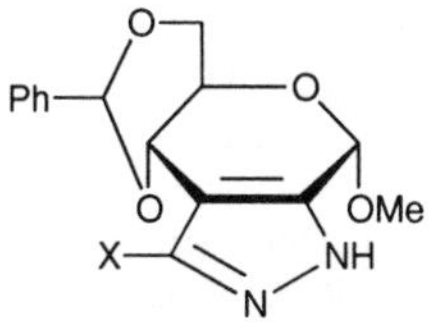

113 X = H or SMe

Tricyclic systems such as **112** have been obtained by trapping 1-*C*-(substituted vinyl)-D-glucal derivatives.[99] The pyrazole-fused pyranosides **113** were formed by reaction of a 3-[methylthio- or bis(methylthio)-methylene]-2-uloside derivative with hydrazine.[100]

5-Substituted 1-(β-D-apiofuranosyl)-1,2,3-triazoles were prepared by reaction of an *O*-protected β-D-apiofuranosyl azide with various Wittig reagents [e.g.

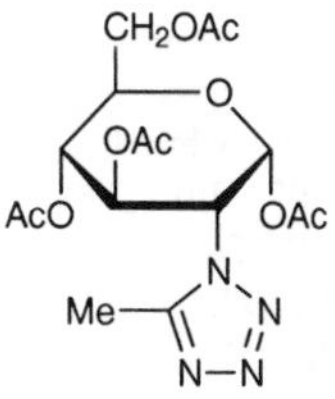

114

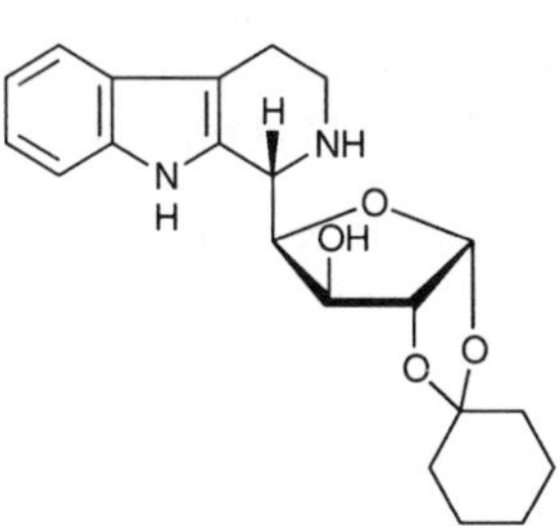

115

Ph$_3$P=CHC(O)Me].[101] 2-Acetamido-1,3,4,6-tetra-*O*-acetyl-2-deoxy-α-D-glucopyr-
anose was converted into the tetrazole derivative **114** by reaction with Lawesson's
reagent then trimethylsilyl azide and tin(IV) tetrachloride.[102]

Condensation of tryptamine with 1,2-*O*-cyclohexylidene-α-D-*xylo*-pentodi-
aldose gave adduct **115**, which on hydrolysis provided tetracyclic aromatic
compounds.[103]

Syntheses of various hydantocidin analogues are covered in Section 1.3 of this
Chapter and in Chapter 19, and cyclic guanadinium ion sugar mimics are covered
in Chapter 9.

References

1   H. Nishitani, Y. Yamada, N. Ohshima, K. Okumura and H. Taguchi, *Biosci.*
     Biotech. Biochem., 1995, **59**, 1336.

2   A. Lubineau, J. Augé and B. Drouillat, *Carbohydr. Res.*, 1995, **266**, 211.

3   J. Hirsch, E. Petrakova, M.S. Feather and C.L. Barnes, *Carbohydr. Res.*, 1995, **267**,
     17.

4   N. Kolarova, R. Trgina, K. Linek and V. Farkas, *Carbohydr. Res.*, 1995, **273**, 109.

5   J.D. Chisholm and D.L. Van Vranken, *J. Org. Chem.*, 1995, **60**, 6672.

6   G.E.H. Elgemeie and A.M.E. Attia, *Sulfur Silicon Relat. Elem.*, 1994, **92**, 95 (*Chem.*
     Abstr., 1995, **122**, 214 431).

7   K. Kasai, K. Okada and N. Yamaji, *Chem. Pharm. Bull.*, 1995, **43**, 266.

8   F.-D. Boyer, A. Pancrazi and Y.-L. Lallemand, *Synth. Commun.*, 1995, **25**, 1099.

9   Y. Bleriot, T. Dintinger, N. Guillo and C. Tellier, *Tetrahedron Lett.*, 1995, **36**, 5175.

10   M. Avalos, R. Babiano, P. Cintas, C.J. Duran, J.L. Jimenez and J.C. Palacios,
     Tetrahedron, 1995, **51**, 8043.

11   C.L. Perrin, *Tetrahedron*, 1995, **51**, 11901.

12   X. Huang, C. Surry, T. Hiebert and A.J. Bennet, *J. Am. Chem. Soc.*, 1995, **117**,
     10614.

13   Y. Al-Abed, P. Ulrich, A. Kapurniotu, E. Lolis and R. Bucala, *Bioorg. Med. Chem.*
     Lett., 1995, **5**, 2929.

14   C. Elias, M.E. Gelpi and R.A. Cadenas, *J. Carbohydr. Chem.*, 1995, **14**, 1209.

15   S. Baillet and J.-P. Behr, *Tetrahedron Lett.*, 1995, **36**, 8981.

16   C.E. Sowa, J. Kopf and J. Thiem, *J. Chem. Soc., Chem. Commun.*, 1995, 211.

17   U. Sprengard, G. Kretzschmar, E. Bartnik, C. Hüls and H. Kunz, *Angew. Chem.,*
     Int. Ed. Engl., 1995, **34**, 990.

18   D. Vetter, D. Tumelty, S.K. Singh and M.A. Gallop, *Angew. Chem., Int. Ed. Engl.*,
     1995, **34**, 60.

19   U.K. Saha and R. Roy, *Tetrahedron Lett.*, 1995, **36**, 3635.

20   G. Kretzschmar, U. Sprengard, H. Kunz, E. Bartnik, W. Schmidt, A. Toepfer,
     B. Horsch, M. Krausse and D. Seiffge, *Tetrahedron*, 1995, **51**, 13015.

21   D. Zanini, W.K.C. Park and R. Roy, *Tetrahedron Lett.*, 1995, **36**, 7383.

22   I. Pinter, J. Kovacs and G. Toth, *Carbohydr. Res.*, 1995, **273**, 99.

23   J. Kovacs, I. Pinter and P. Köll, *Carbohydr. Res.*, 1995, **272**, 255.

24   P. Köll and A. Lutzen, *Tetrahedron Asymm.*, 1995, **6**, 43.

25   C.J.F. Bichard, E.P. Mitchell, M.R. Wormald, K.A. Watson, L.N. Johnson, S.E.
     Zographos, D.D.Koutra, N.G. Oikonomakos and G.W.J. Fleet, *Tetrahedron Lett.*,
     1995, **36**, 2145.

26   T.W. Brandstetter, Y.-ha. Kim, J.C. Son, H.M. Taylor, P.M. de Q. Lilley, D.J. Watkin, L.N. Johnson, N.G. Oikonomakos and G.W.J. Fleet, *Tetrahedron Lett.,* 1995, **36**, 2149.

27   A.J. Fairbanks and G.W.J. Fleet, *Tetrahedron,* 1995, **51**, 3881.

28   J.C. Estevez, D.D. Long, M.R. Wormald, R.A. Dwek and G.W.J. Fleet, *Tetrahedron Lett.,* 1995, **36**, 8287.

29   T.M. Krülle, K.A. Watson, M. Gregoriou, L.N. Johnson, S. Crook, D.J. Watkin, R.C. Griffiths, R.J. Nash, K.A. Tsitsanou, S.E. Zographos, N.G. Oikonomakos and G.W.J. Fleet, *Tetrahedron Lett.,* 1995, **36**, 8291.

30   H. Sano, S. Mio, M. Hamura, J. Kitagawa, M. Shindou, T. Honma and S. Sugai, *Biosci. Biotech. Biochem.,* 1995, **59**, 2247.

31   H. Sano, S. Mio, J. Kitagawa, M. Shindou, T. Honma and S. Sugai, *Tetrahedron,* 1995, **51**, 12563.

32   T.K. Lindhorst and C. Kieburg, *Synthesis,* 1995, 1228.

33   M.A. Pradera, J.L. Molina and J. Fuentes, *Tetrahedron,* 1995, **51**, 923.

34   M. Wojtowicz, *Acta Pol. Pharm.,* 1994, **51** 505 (*Chem. Abstr.,* 1995, **123**, 170 031).

35   M. Wojtowicz, *Acta Pol. Pharm.,* 1994, **51**, 161 (*Chem. Abstr.,* 1995, **122**, 214 389).

36   L.D.S. Yadav and D.S. Yadav, *Liebigs Ann.,* 1995, 2231.

37   C. Uchida and S. Ogawa, *Carbohydr. Lett.,* 1995, **10**, 77.

38   D. Beaupere, A.El. Mesloutti, Ph. Lelievre and R. Uzan, *Tetrahedron Lett.,* 1995, **36**, 5347.

39   J. Bogusiak and W. Szeja, *Pol. J. Chem.,* 1994, **68**, 2309 (*Chem. Abstr.,* 1995, **122**, 187 903).

40   Z. Gyorgydeak and J. Thiem, *Carbohydr. Res.,* 1995, **268**, 85.

41   M. Goebel and I. Ugi, *Tetrahedron Lett.,* 1994, **36**, 6043.

42   S. Czernecki and E. Ayadi, *Can. J. Chem.,* 1995, **73**, 343.

43   B.M. Heskamp, G.H. Veeneman, G.A. van der Marel, C.A.A. van Boeckel and J.H. van Boom, *Tetrahedron,* 1995, **51**, 8397.

44   L. Lay, F. Nicotra, L. Panza and G. Russo, *Synlett,* 1995, 167.

45   F. Azzena, P. Crotti, L. Favero and M. Pineschi, *Tetrahedron,* 1995, **51**, 13409.

46   F. Azzena, F. Calvani, P. Crotti, C. Gardelli, G. Macchia and M. Pineschi, *Tetrahedron,* 1995, **51**, 10601.

47   R. Thomson and M. von Itzstein, *Carbohydr. Res.,* 1995, **274**, 29.

48   R.H. Youssef, R.W. Bassily, A.N. Asaad, R.I. El-Sokkary and M.A. Nashed, *Carbohydr. Res.,* 1995, **277**, 347.

49   F.-Y. Dupradeau, S. Hakomori and T. Toyokuni, *J. Chem. Soc., Chem. Commun.,* 1995, 221.

50   C.-H. Wong, E. Garcia-Junceda, L. Chen, O. Blanco, H.J.M. Gijsen and D.H. Steensma, *J. Am. Chem. Soc.,* 1995, **117**, 3333.

51   H.J.M. Gijsen and C.-H. Wong, *J. Am. Chem. Soc.,* 1995, **117**, 2947.

52   A. Stein, M.-R. Kula, L. Elling, S. Verseck and W. Klaffke, *Angew Chem., Int. Ed. Engl.,* 1995, **34**, 1748.

53   P. Norris, D. Horton and B.R. Levine, *Tetrahedron Lett.,* 1995, **36**, 7811.

54   A. Seta, S. Ito, K. Tokuda, T. Tamura, Y. Kondo and T. Sakakibara, *Carbohydr. Res.,* 1995, **267**, 217.

55   J.A. Serrano, L.E. Cáceres and E. Román, *J. Chem. Soc., Perkin Trans. 1,* 1995, 1863.

56   S. Kalliney and A. Zaks, *Tetrahedron Lett.,* 1995, **36**, 4163.

57   S.A. Fleming, *Tetrahedron,* 1995, **51**, 12479.

58   J. Lehmann, F. Schweizer and U.P. Weitzel, *Carbohydr. Res.,* 1995, **270**, 181.

59   A. Vasella and C. Witzig, *Helv. Chim. Acta.,* 1995, **78**, 1971.

60   M. Therisod, H. Therisod and A. Lubineau, *Bioorg. Med. Chem. Lett.*, 1995, **5**, 2055.
61   Z. Smiatacz, I. Chrzczanowicz, H. Myszka and P. Dokurno, *J. Carbohydr. Chem.*, 1995, **14**, 723.
62   D.R. Groves, J.C. Wilson and M. von Itzstein, *Aust. J. Chem.*, 1995, **48**, 1217.
63   M. Frederickson, R. Grigg, Z. Rankovic, M. Thornton-Pett, J. Redpath and R. Crossley, *Tetrahedron*, 1995, **51**, 6835.
64   J.M.J. Tronchet, M. Zsély, O. Lassout, F. Barbalat-Rey, I. Komaromi and M. Geoffroy, *J. Carbohydr. Chem.*, 1995, **14**, 575.
65   G. Zosimo-Landolfo, J.M.J. Tronchet and F. Habashi, *J. Prakt. Chem./Chem.-Ztg.*, 1994, **336**, 273 (*Chem. Abstr.*, 1995, **122**, 240 245).
66   S. Cao, F.D. Tropper and R. Roy, *Tetrahedron*, 1995, **51**, 6679.
67   S. Hanessian, J.-Y. Sanceau and P. Chemla, *Tetrahedron*, 1995, **51**, 6669.
68   I.B. Niederer, G.G.G. Manzardo and R. Amado, *Carbohydr. Res.*, 1995, **278**, 181.
69   M. Koos and J. Alfoldi, *Chem. Pap.*, 1995, **49**, 32 (*Chem. Abstr.*, 1995, **123**, 286 409).
70   F.L. Merchan, P. Merino and T. Tejero, *Tetrahedron Lett.*, 1995, **36**, 6949.
71   J.-P. Praly, D. Senni, R. Faure and G. Descotes, *Tetrahedron*, 1995, **51**, 1697.
72   S. Ochs and T. Severin, *Carbohydr. Res.*, 1995, **266**, 87.
73   M. Avalos, R. Babiano, P. Cintas and J.L. Jiménez, *Tetrahedron: Asymm.*, 1995, **6**, 945.
74   B. Bendiak, M.E. Salyan and M. Pantoja, *J. Org. Chem.*, 1995, **60**, 8245.
75   J. Hirsch, E. Petrakova and M.S. Feather, *J. Carbohydr. Chem.*, 1995, **14**, 1179.
76   M. Anwar Abdo, M. Farghaly Abdel-Megeed, M. Atia Saleh and G. Abdel-Rahman El-Hiti, *Pol. J. Chem.*, 1995, **69**, 583 (*Chem. Abstr.*, 1995, **123**, 314 312).
77   M.A. Saleh, *Rev. Roum. Chem.*, 1994, **39**, 659 (*Chem. Abstr.*, 1995, **122**, 56 329).
78   S.E. Mangholz and A. Vasella, *Helv. Chim. Acta.*, 1995, **78**, 1020.
79   M.A.E. Shaban, A.Z. Nasr and M.A.M. Taha, *J. Carbohydr. Chem.*, 1995, **14**, 985.
80   M.A.M. Taha, *Alexandria J. Pharm. Sci.*, 1993, **7**, 79 (*Chem. Abstr.*, 1995, **122**, 56 387).
81   G. Argay, R. Csuk, Z. Gyorgydeak, A. Kalman and G. Snatzke, *Tetrahedron*, 1995, **51**, 12911.
82   J. Boivin, A.-C. Callier-Dublanchet, B. Quiclet-Sire, A.-M. Schiano and S.Z. Zard, *Tetrahedron*, 1995, **51**, 6517.
83   J.L. Del Valle, C. Polo, T. Torroba and S. Marcaccini, *J. Heterocycl. Chem.*, 1995, **32**, 899.
84   J. Streith, A. Boiron, A. Frankowski, D. Le Nouen, H. Rudyk and T. Tschamber, *Synthesis*, 1995, 944.
85   M.D. Cliff and S.G. Pyne, *Tetrahedron Lett.*, 1995, **36**, 5969.
86   M. Mancera, I. Roffe and J.A. Galbis, *Tetrahedron*, 1995, **51**, 6349.
87   M.L. Fascio and N.B. D'Accorso, *J. Heterocycl. Chem.*, 1995, **32**, 815.
88   M. Mostowicz and M. Chmielewski, *Carbohydr. Lett.*, 1995, **1**, 95.
89   M.A.E. Sallam, L.B. Townsend and W. Butler, *J. Chem. Res.*, 1995, (S)54; μM) 0467.
90   K. Tatsuta, S. Miura, S. Ohta and H. Gunji, *J. Antibiot.*, 1995, **48**, 286.
91   K. Tatsuta and S. Miura, *Tetrahedron Lett.*, 1995, **36**, 6721.
92   A. Frankowski, D. Deredas, D. Le Nouen, T. Tschamber and J. Streith, *Helv. Chim. Acta*, 1995, **78**, 1837.
93   T.D. Heightman, P. Ermert, D. Klein and A. Vasella, *Helv. Chim. Acta.*, 1995, **78**, 515.
94   B. Davis, T.W. Brandstetter, C. Smith, L. Hackett, B.G. Winchester and G.W.J. Fleet, *Tetrahedron Lett.*, 1995, **36**, 7507.

95 T.W. Brandstetter, B. Davis, D. Hyett, C. Smith, L. Hackett, B.W. Winchester and G.W.J. Fleet, *Tetrahedron Lett.*, 1995, **36**, 7511.

96 J.-P. Praly, C. Di Stéfano, G. Descotes, R. Faure, L. Somsák and I. Eperjesi, *Tetrahedron Lett.*, 1995, **36**, 3329.

97 M. Yokoyama, S. Hirano, M. Matsushita, T. Hachiya, N. Kobayashi, M. Kubo, H. Togo and H. Seki, *J. Chem. Soc., Perkin Trans. 1*, 1995, 1747.

98 J. Kovacs, I. Pinter, M. Kajtar-Peredy, J.-P. Praly and G. Descotes, *Carbohydr. Res.*, 1995, **279**, C1.

99 A. Abas, R.L. Beddoes, J.C. Conway, P. Quayle and C.J. Urch, *Synlett*, 1995, 1264.

100 K. Peseke, G. Thiele and M. Michalik, *Liebigs Ann. Chem.*, 1995, 1633.

101 F. Hammerschmidt, J.-P. Polsterer and E. Zbiral, *Synthesis*, 1995, 415.

102 S. Lehnhoff and I. Ugi, *Heterocycles*, 1995, **40**, 801.

103 R.K. Manna, P. Jaisankar and V.S. Giri, *Synth. Commun.*, 1995, **25**, 3027.

11
Thio- and Seleno-sugars

(1,1'-Binaphthalene)-2,2'-dithiol has been resolved *via* the diastereomeric sugar dithioacetals **1**. [Only the (*R*)-isomer is shown].[1] D-Glucose diethyldithioacetal was cleaved to the free sugar by heating with DDQ (2 molar equivalents) in aqueous acetonitrile; cyclic dithioacetals were stable under these reaction conditions.[2]

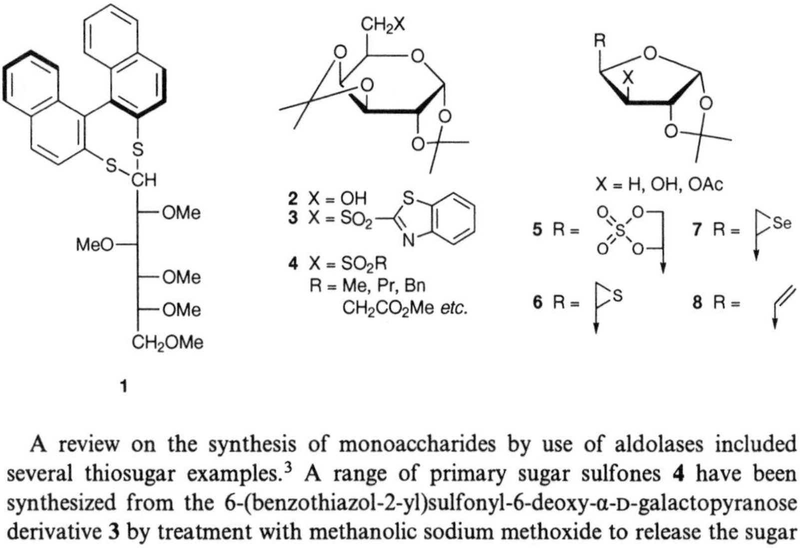

A review on the synthesis of monoaccharides by use of aldolases included several thiosugar examples.[3] A range of primary sugar sulfones **4** have been synthesized from the 6-(benzothiazol-2-yl)sulfonyl-6-deoxy-α-D-galactopyranose derivative **3** by treatment with methanolic sodium methoxide to release the sugar sulfinate and quenching with alkylating agents. Compound **3** was obtained from alcohol **2** by Mitsunobu introduction of the 2-thiobenzothiazole group and subsequent MCPBA oxidation.[4] The cyclic sulfates **5** were transformed to 5,6-episulfides **6** in 'one pot' by nucleophilic opening with potassium thioacetate or thiourea, followed by exposure to sodium methoxide. Reaction of **5** with selenocyanate followed by sodium borohydride gave alkenes **8** by way of the unstable selenirans **7**.[5]

Standard methods have been employed to prepare *N*-acetyl-9-thio- and *N*-acetyl-9-*S*-acetyl-9-thio-neuraminic acid and their methyl α-glycosides.[6] Sulfoquinovosyl diacylglycerols **9**, isolated from the edible brown alga *Hizikia fusiforme*, are inhibitors of yeast α-glucosidase.[7]

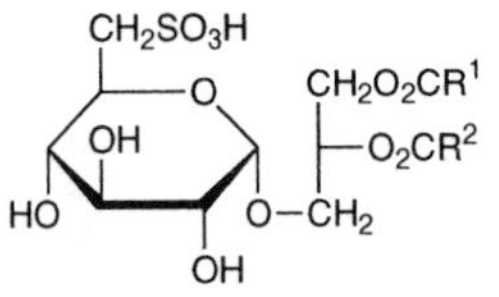

9 R¹, R² = alkyl 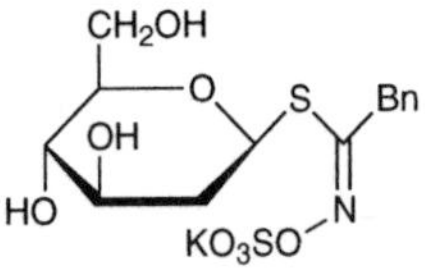   **10**

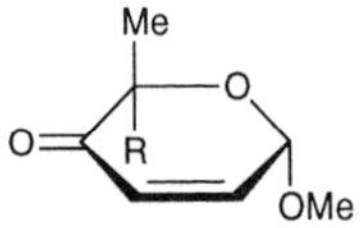

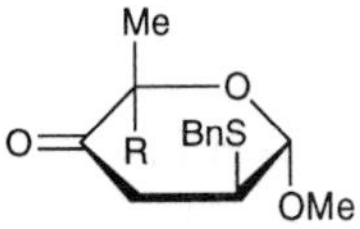

R = C₆H₄SO₂Ph

11   **12**

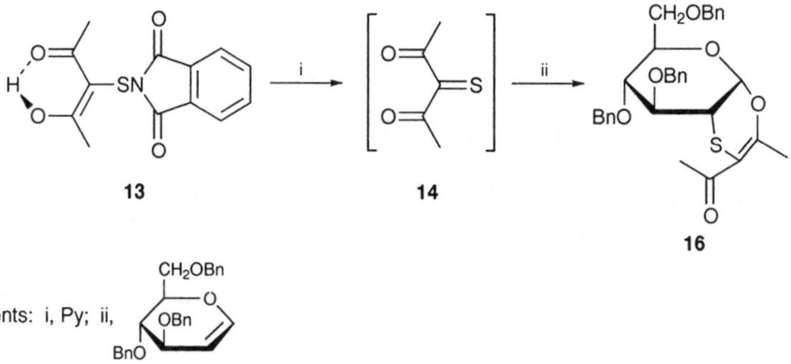

13   **14**   **16**

Reagents: i, Py; ii,

Scheme 1

The four monodeoxy analogues of 1-thio-β-D-glucopyranoses have been synthesized, as their tri-*O*-acetates or -benzoates, by conventional methods. They were converted to the corresponding deoxy analogues of glucosinolates, such as compound **10**.[8] A detailed account on the use of 2,6-anhydro-2-thiosugars for the stereocontrolled synthesis of 2,6-dideoxy-α- and -β-glycosides has been published (see Vol. 28, Chapter 11, Ref. 26).[9] Michael addition of thiols or inorganic sulfur anions to 5-*C*-substituted methyl 2,3,6-trideoxy-hex-2-enopyranos-4-ulosides gave 2-thiosugars stereoselectively (*e.g.*, **11**→**12**).[10] α,α′-Dioxothiones **14**, generated by treatment of α,α′-dioxothiophthalimides **13** with pyridine, reacted slowly, *in situ*, with glycal **15** to furnish cycloadduct **16**, as shown in Scheme 1.[11]

3-*S*-Alkyl-3-thio derivatives of D-glucose **18** and **19**, obtained from diacetone D-glucose *via* the *allo*-configured 3-iodides **17**, have been evaluated as surfactants.[12] A full paper on synthetic routes to methyl 2,6-dideoxy-4-*S*-methyl-4-thio-α-D-*ribo*-hexopyranoside from propargylic alcohol dimer and D-fucal has been presented (see Vol. 27, Chapter 11, Ref. 11 for a preliminary account).[13] Precursors **20** of 4-thio-analogues of AZT were prepared from D-xylose as outlined in Scheme 2, using standard procedures.[14]

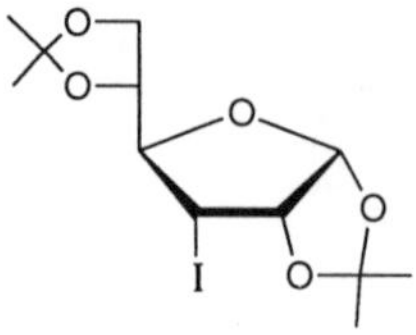

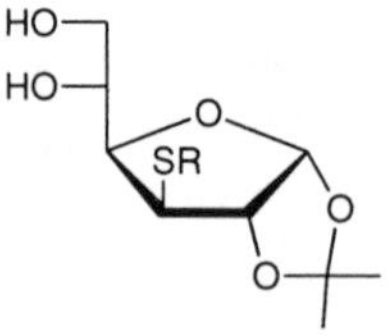

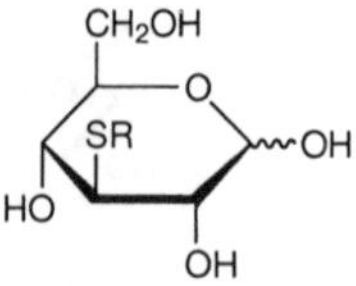

R = C_nH_{2n+1}; $n = 1–10$ or
C_nH_{2n}Ph; $n = 1–3$, *etc.*

17   **18**   **19**

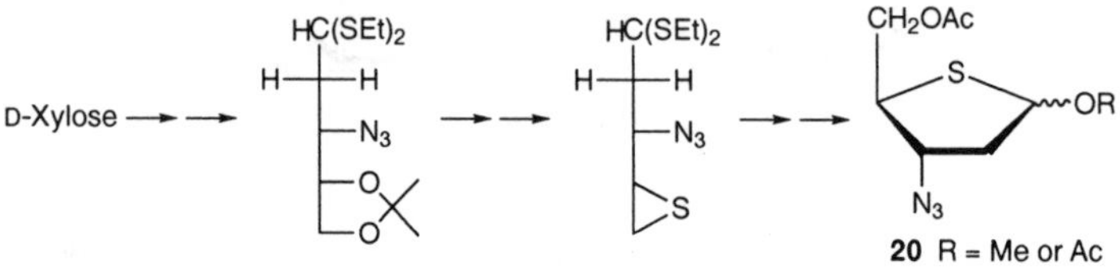

D-Xylose →→

20 R = Me or Ac

Scheme 2

A six-step synthesis of 2-acetamido-1,3,6-tri-*O*-acetyl-4-*S*-acetyl-2-deoxy-4-thio-D-mannopyranose (**22**) from 2-acetamido-2-deoxy-D-mannopyranose used the 1,6:3,4-dianhydrosugar **21** as key-intermediate.[15] The 4-*O*-triflyl-D-galactopyranose derivative **23** was the starting compound for the preparation of methyl 4-*S*-(3-hydroxy-2-pyridyl)-4-thio-β-D-glucopyranosides **24** as potential β-glucosidase inhibitors.[16] A conformational study on L-thiohexofuranoses is referred to in Chapter 21.

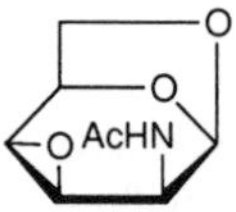

21 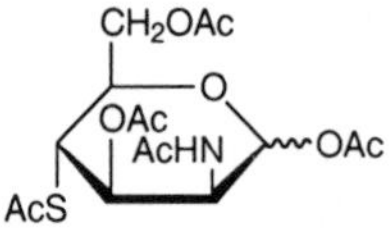   **22** 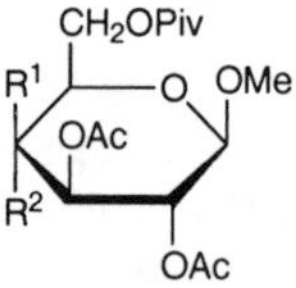   **23** R¹ = OTf, R² = H   **25**

24 R¹ = H, R² =

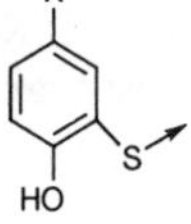

X = H, Me or CH₂OH

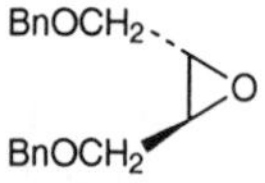

Two new routes to the sugar moiety of 2′,3′-dideoxy-3′-*C*-(hydroxymethyl)-4′-thionucleosides **27**, starting from the (*S,S*)-tartaric acid-derived epoxide **25** or from the unsaturated lactone **26** have been developed; the latter route is outlined in Scheme 3.[17] Compound **28** and its enantiomer, obtained in six steps from (*S*)- and (*R*)-glycidol, respectively, were the key-intermediates for the synthesis of various 5-membered 2′,3′-dideoxy- and 2′,3′-didehydro-2′,3′-dideoxy-4′-thionucleosides.[18]

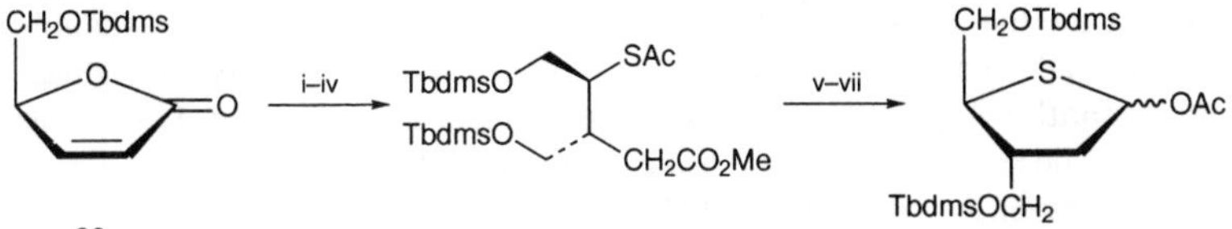

26 → i–iv → → v–vii → **27**

Reagents: i, MeOH, *hv*; ii, TbdmsCl, Im; iii, Me$_2$SO$_4$, ⁻NaOH; vi, I$_2$, PPh$_3$; v, AcSH, Bu$_4$N⁺OH⁻;

vi, DIBAL; vii, Ac$_2$O

Scheme 3

Opening of the known epoxide **29** with methanolic ammonia followed by *N*-acetylation and de-*O*-acetalation gave methyl 2-acetamido-2-deoxy-5-thio-α-D-altropyranoside **30**.[19]

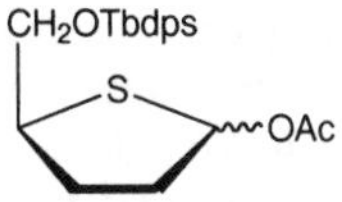

28 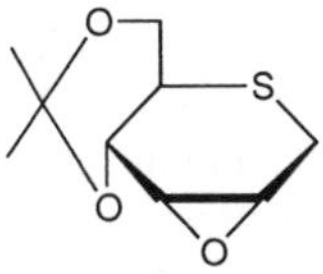   **29** 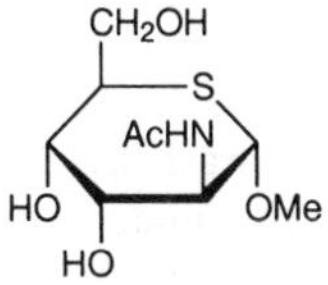   **30** 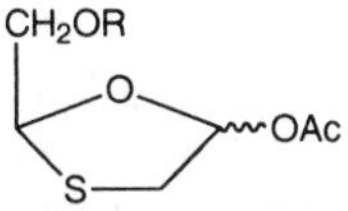   **31** R = Bz or Tbdps

D-1,3-Oxathiolanyl acetates **31** have been prepared from D-glucose and D-mannose.[20] On treatment with sulfide ions, bisepoxide **32** underwent thioheterocyclization to furnish thiepan **33**. Ring-contraction under the conditions indicated in Scheme 4 led to tetrahydrothiopyrans **34** and tetrahydrothiophenes **35**.[21]

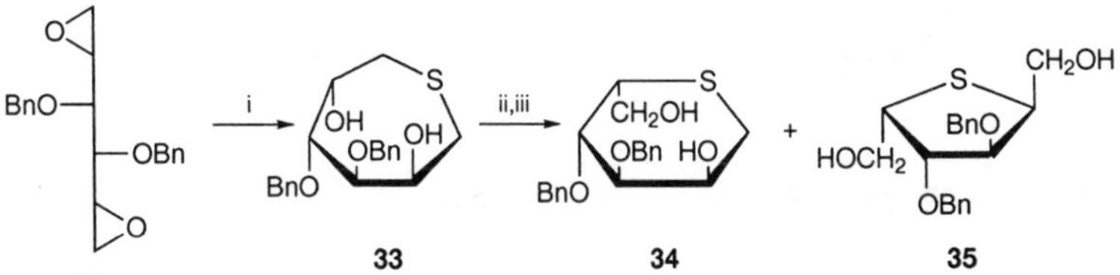

32 → i → **33** → ii,iii → **34** + **35**

Reagents: i, Na$_2$S, EtOH; ii, PPh$_3$ DEAD, BzOH, THF; iii, K$_2$CO$_3$, MeOH

Scheme 4

The 3-thio analogue of sucrose has been obtained from sucrose by way of 3-ketosucrose (see Chapter 15) which gave mainly the *allo*-product on hydrogenation.[22] The preparation of 4-thiosucrose is referred to in Chapter 7. The imidate method was used for the synthesis of sulfur-linked propyl β-kojibioside[23] and of methyl α-maltoside analogues **36** containing sulfur in the non-reducing ring and either oxygen, sulfur, or selenium in the interglycosidic linkage.[24] A mixture **37** of

N-linked methyl 5′-thio-α-maltoside and -cellobioside was readily available by coupling of 5-thio-D-glucopyranose with methyl 4-amino-4-deoxy-α-D-glucopyranoside (both compounds unprotected) in methanolic acetic acid. Use of methyl 2-amino-2-deoxy-α-D-glucopyranoside as acceptor furnished similarly a mixture of the corresponding dihetero methyl kojibioside and sophoroside.[25]

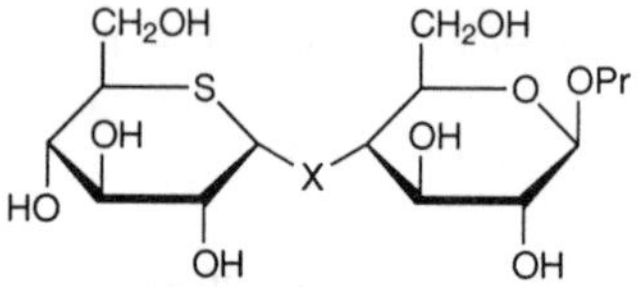

36 X = O, S or Se 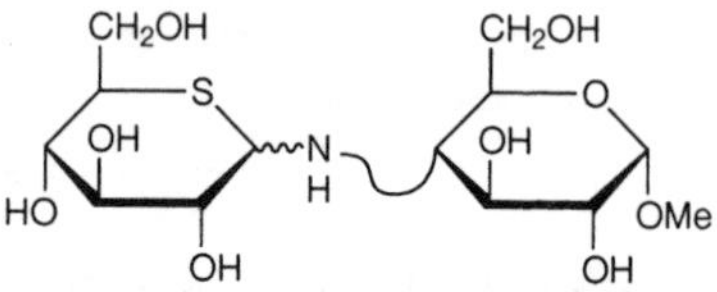   **37**

Reaction of the per-*O*-acetylated methyl ester of 2-thio-*N*-acetylneuraminic acid with bromide **38** in the presence of diethylamine, followed by deprotection, furnished the *S*-sialylnucleoside **39** in 58 % overall yield.[26] Under catalysis by cyclodextrin glucosyltransferase [EC 3.2.1.19] and glucoamylase, 6-deoxy-6-iodo-β-cyclodextrin and D-glucose underwent consecutive coupling and degradation to give mainly 6^{III}-deoxy-6^{III}-iodomaltotriose and 6^{IV}-deoxy-6^{IV}-iodomaltotetraose. On displacement of the iodine atoms by thiosugar derivative **40**, thiooligosaccharides **41** were formed in high yields.[27]

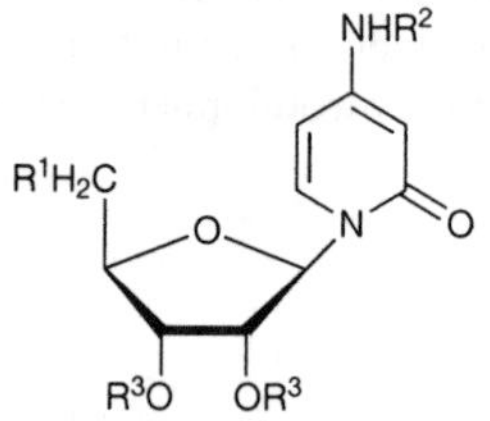

38 R^1 = Br, R^2 = Bz, R^3R^3 =

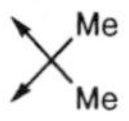

39 R^1 = 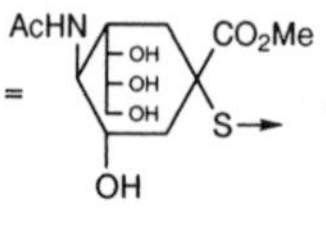 , R^2 = R^3 = H

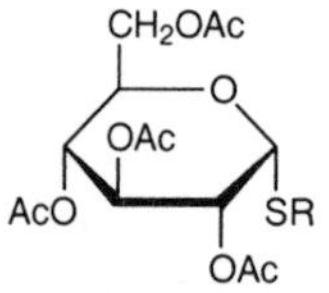

41 R = H

42 R = —→ 6-α-D-Glc*p*-[(1 —→4)-α-D-Glc*p*]$_n$- (1 —→4)-α-D-Glc*p*
n = 1 or 2

References

1   U.K. Bandaragh, G.F. Painter and R.A.J. Smith, *Tetrahedron: Asymm.*, 1995, **6**, 295.

2   J.M. García Fernández, C. Ortiz Mellet, A. Moreno Marín and J. Fuentes, *Carbohydr. Res.*, 1995, **274**, 263.

3   C.-H. Wong, R.L. Halcomb, Y. Ichikawa and T. Kajimoto, *Angew. Chem., Int. Ed. Engl.*, 1995, **34**, 412.

4   C. Lorin, C. Marot, V. Gardon and P. Rollin, *Tetrahedron Lett.*, 1995, **36**, 4437.

5   F. Santoyo-Gonzáles, F. García-Calvo-Flores, P. García-Mendoza, F. Hernandez-Mateo, J. Isac-García and M.D. Pérez-Alvarez, *J. Chem. Soc., Chem. Commun.*, 1995, 461.

6   R. Isecke and R. Bossmer, *Carbohydr. Res.*, 1995, **274**, 303.

7   P. Kurihara, J. Ando, M. Hatano and J. Kawabata, *Bioorg. Med. Chem. Lett.*, 1995, **5**, 1241.

8   H. Streicher, L. Latxague, T. Wiemann, P. Rollin and J. Thiem, *Carbohydr. Res.*, 1995, **278**, 257.

9   K. Toshima, Y. Nozaki, S. Mukayama, T. Tama, M. Nakata, K. Tatsuta and M. Kinoshita, *J. Am. Chem. Soc.*, 1995, **117**, 3717.

10   C.D. Apostopoulos, E.A. Couladouras and M.P. Georgiadis, *Polish J. Chem.*, 1994, **68**, 1725 (*Chem. Abstr.*, 1995, **122**, 133 563).

11   G. Capozzi, R.W. Franck, M. Mattioli, S. Menighetti, C. Nativi and G. Valle, *J. Org. Chem.*, 1995, **60**, 6416.

12   D. Postel, P. Vanlemmens, P. Godé, G. Ronco and P. Villa, *Carbohydr. Res.*, 1995, **271**, 227.

13   F.-Y. Dupradeau, J. Prandi and J.-M. Beau, *Tetrahedron*, 1995, **51**, 3205.

14   B. Tber, N.-E. Fami, G. Ronco, P. Villa, D.F. Ewing and G. Mackenzie, *Carbohydr. Res.*, 1995, **267**, 203.

15   R. Thomson and M. von Itzstein, *Carbohydr. Res.*, 1995, **274**, 29.

16   S. Knapp, Y. Dong, K. Rupitz and S.G. Withers, *Bioorg. Med. Chem. Lett.*, 1995, **5**, 763.

17   J. Mann, A.J. Tench, A.C. Weymouth-Wilson, S. Shaw-Ponter and R.J. Young, *J. Chem., Perkin Trans. 1*, 1995, 677.

18   R.J. Young, S. Shaw-Ponter, J.B. Thomson, J.A. Miller, J.G. Cumming, A.W. Pugh and P. Rider, *Bioorg. Med. Chem. Lett.*, 1995, **5**, 2599.

19   N.A.L. Al-Masoudi, N.A. Hughes and N.J. Tooma, *Carbohydr. Res.*, 1995, **272**, 111.

20   J.H. Hong, Y.C. Xu, M. Cha, E.S. Shin, J.H. Kim, W.-K. Chung and M.W. Chun, *Arch. Pharmacol. Res.*, 1994, **17**, 383 (*Chem. Abstr.*, 1995, **122**, 133 540).

21   M. Fuzier, Y. Le Merrer and J.-C. Depezay, *Tetrahedron Lett.*, 1995, **36**, 6443.

22   C. Simiand, E. Samain, O.R. Martin and H. Driguez, *Carbohydr. Res.*, 1995, **267**, 1.

23   J.S. Andrews and B.M. Pinto, *Carbohydr. Res.*, 1995, **270**, 51.

24   S. Metha, J.S. Andrews, B.D. Johnston, B. Svensson and B.M. Pinto, *J. Am. Chem. Soc.*, 1995, **117**, 9783.

25   J.S. Andrews and B.M. Pinto, *J. Am. Chem. Soc.*, 1995, **117**, 10799.

26   B. Smalec and M. van Itzstein, *Carbohydr. Res.*, 1995, **266**, 269.

27   C. Apparu, H. Driguez, G. Williamson and B. Svensson, *Carbohydr. Res.*, 1995, **277**, 313.

12
Deoxy-sugars

The formation of deoxysugars from small achiral molecules by aldolase-catalysed condensations is covered in Chapter 2 (Refs. 7-9), and the conversion of D-galactose to L-fucose is referred to in Chapter 18.

Two efficient new syntheses of 1-deoxy-D-*threo*-pentulose (**2**), both starting from known D-threitol derivatives and employing a Grignard addition to 4-*O*-benzyl-2,3-*O*-isopropylidene-D-threose (**1**), have been published.[1,2] Compound **2** exists mostly as the acyclic ketone in neutral aqueous solution, but forms a dimeric anhydride in acidic media (see Chapter 22 for crystal structure).[2] [2,3-¹³C₂]-Labelled **2**, which was similarly obtained from [1,2-¹³C₂]-labelled **1**, has been used for investigating the biosynthesis of vitamins B₁ and B₆.[2,3]

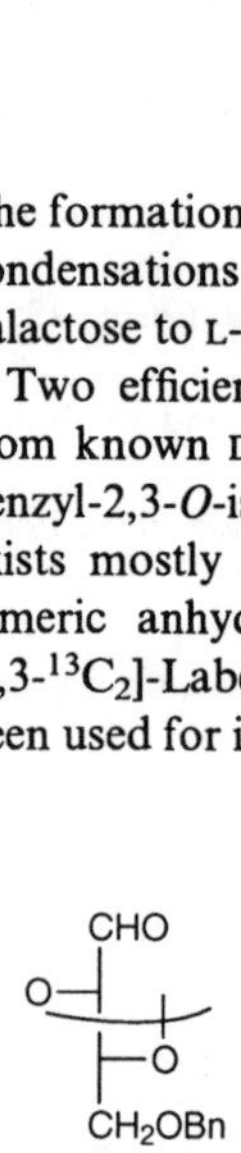

1 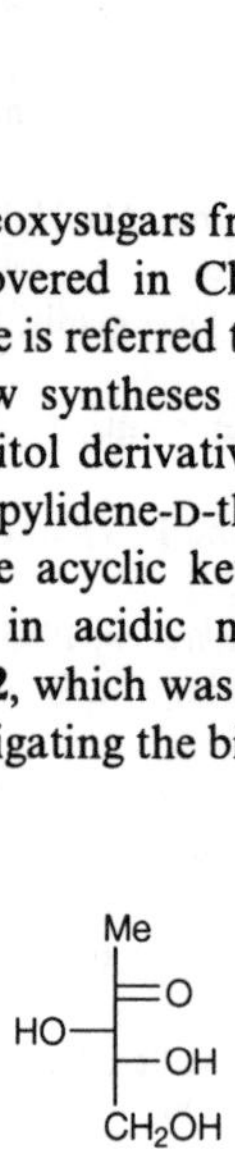 **2** 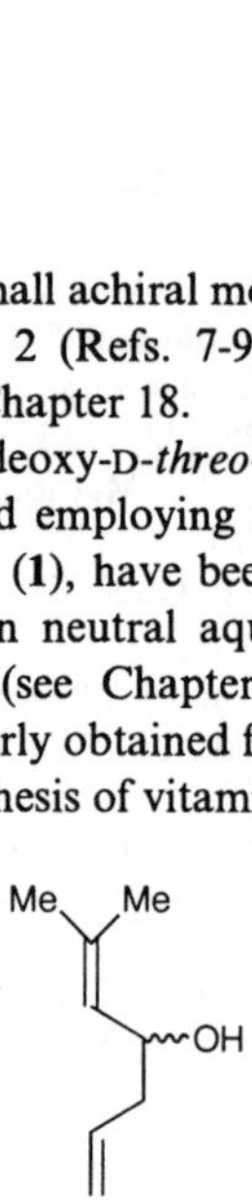 **3** 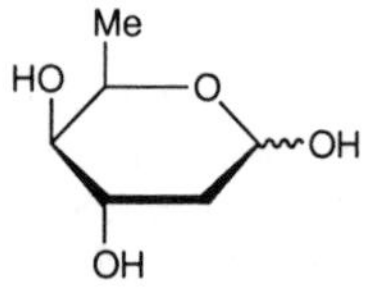 **4** only one enantiomer shown

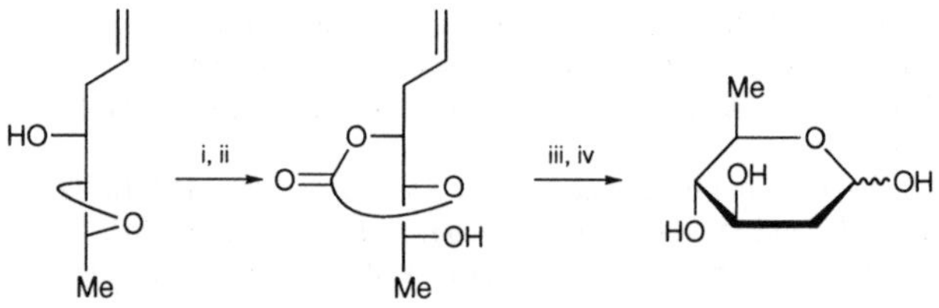

5 **6**

Reagents: i, PhNCO; ii, Et₂AlCl; iii, MeO⁻, MeOH; iv, O₃

Scheme 1

A short synthesis involving a chemo-, regio- and diastereo-selective ene reaction of the trisubstituted double bond of hydroxydiene **3** with singlet oxygen, followed by ozonolysis and hydride reduction, gave 2,6-dideoxy-DL-*xylo*-hexose (boivinose **4**) in 30% overall yield.[4] Its D-*arabino* isomer, D-olivose (**6**), was obtained from epoxyalcohol **5** by application of known methodology (see W.R. Roush *et al., J. Org. Chem.*, 1983, **48**, 5093), as shown in Scheme 1.[5] 2,6-Dideoxy-6,6,6-trifluoro-2,3-*O*-isopropylidene-L-*lyxo*-hexose (**10**) was formed in a four-step reaction sequence from the simple pentynyl alcohol **7** *via* (*E*)-alkene **8** and triol **9**, as detailed in Scheme 2. By use of somewhat different reaction conditions, the D-*xylo*-, D-*ribo*- and L-*arabino*-isomers of **10** were produced from the same starting compound **7**.[6] Diol **11**, available by Sharpless asymmetric dihydroxylation of the corresponding alkene, was resolved by lipase-mediated monoesterification to furnish, after deprotection of the aldehyde function, the monoacetates **12** and **13** of 2,3,6-trideoxy-L-*erythro*-hexose (L-rhodinose) as a separable 1:1 mixture.[7]

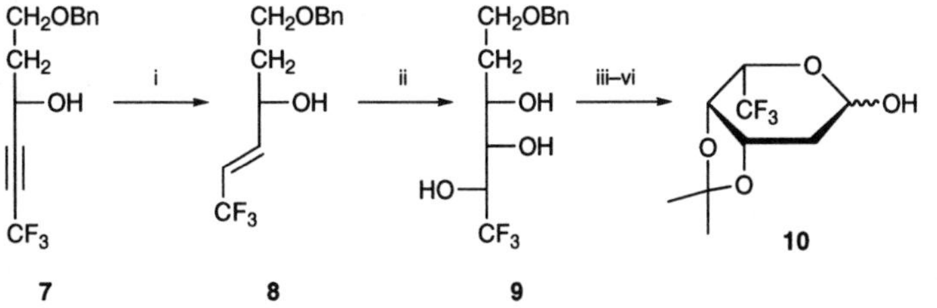

Reagents: i, RedAl; ii, OsO$_4$, NMO; iii, Me$_2$C(OMe)$_2$, H$^+$; iv, Raney Ni, H$_2$; v, PDC; vi, DIBAL

Scheme 2

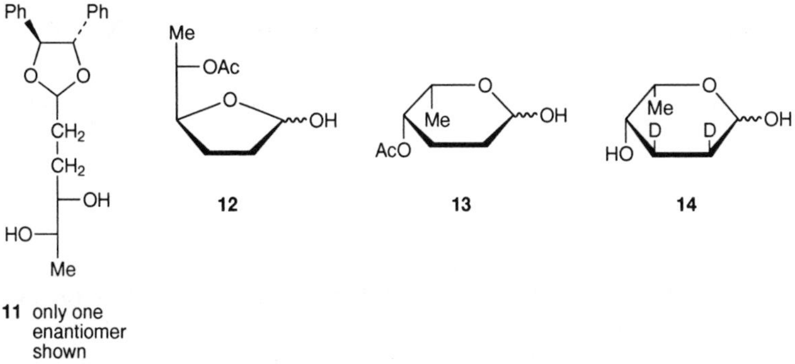

11 only one enantiomer shown

The bis-*C*-deuterated L-rhodinose derivative **14** was obtained by catalytic deuteration of the corresponding 2-enose, which was prepared from L-rhamnal diacetate in 3 steps. [2,3-^{2}H$_4$]D-Rhodinose (**15**) and its C-5 epimer, [2,3-^{2}H$_4$]L-amicetose (**16**), were synthesized from L-threonine, as outlined in Scheme 3.[8] The dideoxygenation of 2-amino-2-deoxy-D-glucose to give the D-purpurosamine analogue **18** and related compounds has been achieved by exposure of dimesylate **17** to sodium iodide in DMF, followed by hydrogenation of the resulting double bond and removal of the *N*-protecting group.[9] The addition of organometallic reagents (MeMgCl, MeLi, MeCeCl$_2$) to methyl β-pentodialdoside **19** gave preferentially the product of non-chelation control **20**, whereas the α-anomer **21** furnished mainly **22** under similar reaction conditions.[10]

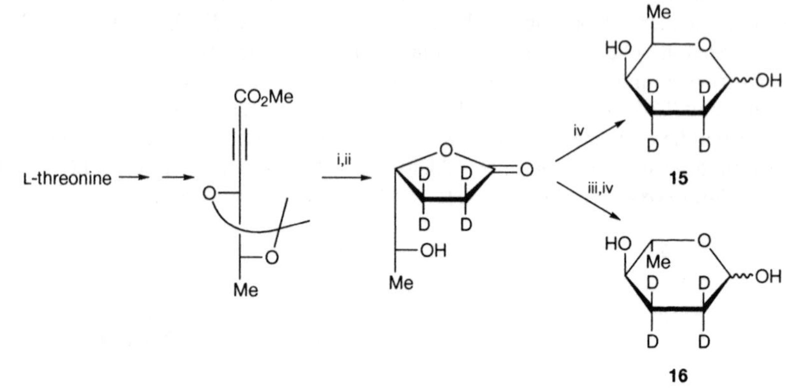

Reagents: i, D$_2$, Pd/C; ii, H$_3$O$^+$; iii, TPP, DEAD, BzOH; iv, DiBAL

Scheme 3

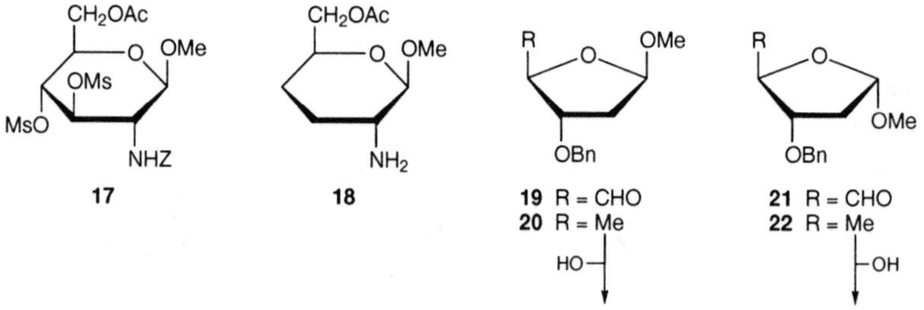

Montmorillonite K-10, an inexpensive, reusable and environmentally acceptable catalyst, proved very efficient in the addition of alcohols to glycals, furnishing, for example, alkyl 2,6-dideoxy-L-hexopyranosides **23** from di-*O*-acetyl-L-rhamnal in short reaction times and high yields, with moderate to good α-selectivity.[11] The 2,3-dideoxy-*C*-glycosides **24**, precursors of carbohydrate-derived liquid crystals (see also Chapters 3 and 6), were prepared from tri-*O*-acetyl-D-glucal by tin tetrachloride-catalysed allylic rearrangement/*C*-glycosylation and subsequent hydrogenation of the double bond.[12] The known addition product of tri-*O*-acetyl-D-glucal to di-*O*-acetyl-2,3-dideoxy-D-*erythro*-hex-2-eno-pyranose under BF$_3$.OEt$_2$-catalysis (see Vol. 18, Chapter 3, Ref.51) has been hydrogenated to give the *C*-linked dideoxy-disaccharide **25**, which was used in the synthesis of a carba-tetrasaccharide (see Chapters 4 and 21).[13] The totally regio- and stereo-selective, inverse electron-demand [4+2] cycloaddition of *ortho*-thioquinones, generated *in situ* from *o*-hydroxythiophthalimides, to substituted

D-glucals represents an effective new route to aryl 2-deoxy-α-D-glucopyranosides. An example is given in Scheme 4.[14]

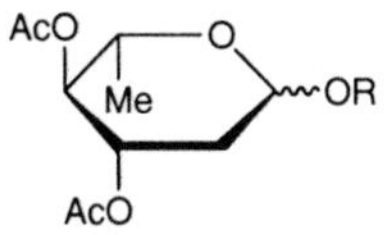

R = (CH₂)₇Me, CHMe₂, cyclohexyl, *etc.*

23

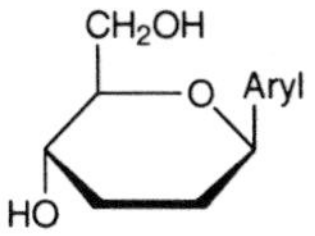

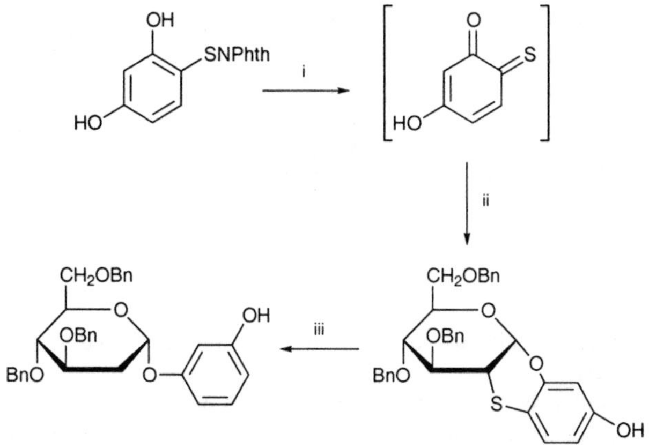

Reagents: i, Py, 60 °C; ii, tri-*O*-benzyl-D-glucal; iii, Raney Ni

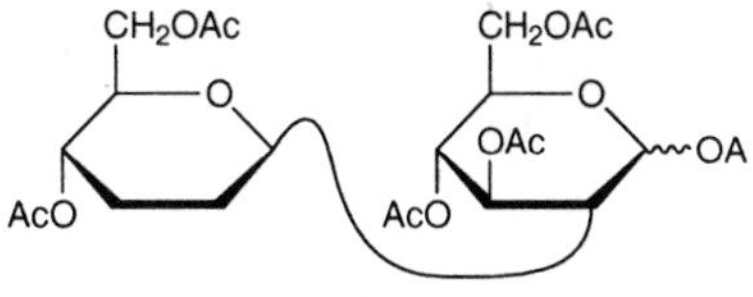

Scheme 4

5,6-Dihydrosilanthrene (**26**) in the presence of AIBN is a mild and efficient deoxygenator of thiocarbamates (*e.g.*, **27→28**, 91%).[15] Deoxygenation *via* thiocarbamate **29** was the key-step in the preparation of the 4,6-dideoxy-α-L-*lyxo*-hexopyranosyl donor **30** from methyl α-L-rhamnopyranoside.[16] The preparation of the 2-deoxy analogue **32** of methyl *N*-acyl-α-D-perosaminide **31**, a constituent of the O-polysaccharide of *Vibrio cholerae*, involved deoxygenation *via* thionocarbonate **33**.[17] Deoxygenation at C-4 of methyl α-D-glucopyranoside derivative **34** by LiEt₃H-reduction of mesylate **35**, *i.e.*, by a nucleophilic process, gave the 4-deoxy sugar derivative **36** in 62% overall yield. This compared well with the correponding radical deoxygenation *via* thionocarbonate **37** (64% overall yield).[18] The triflate group in compound **38** was efficiently removed by ultrasonication in benzene in the presence of tetrabutylammonium hydride.[19] A detailed investigation into the migration of β-(phosphatoxy)alkyl groups during radical debromination (see Vol. 27, Chapter 12, Ref. 3) has been published.[20]

The syntheses of thymidine diphosphates of 2- and 4-deoxy-D-glucose and of 2-

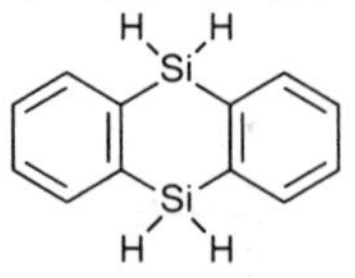

26

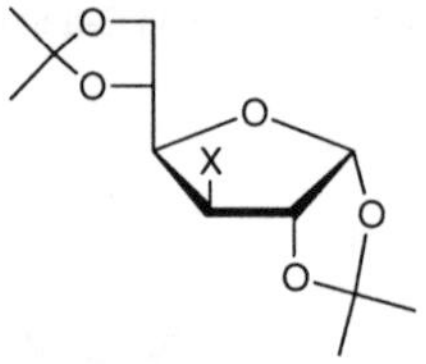

27 X = OCOPh
       ‖
       S
28 X = H

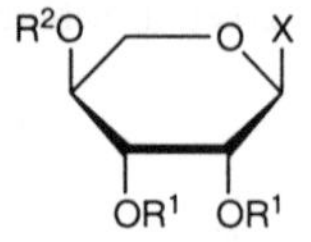

29 X = OMe, R¹ = ⟨Me/Me⟩, R² = OCIm
       ‖
       S
30 X = Cl, R¹ = Bz, R² = H

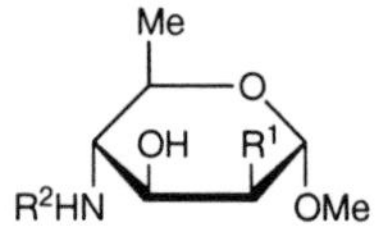

31 R¹ = OH ⎱ R² = ⎡ CO
32 R¹ = H ⎰   ⎢—OH
                      ⎢ CH₂
                      ⎣ CH₂OH

33 R¹ = OCOPh, R² = Boc
       ‖
       S

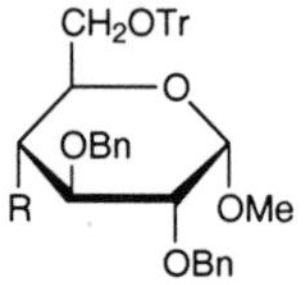

34 R = OH
35 R = OMs
36 R = H
37 R = OCOPh
       ‖
       S

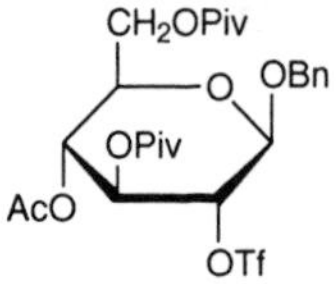

38

deoxy-4-*epi*-neuraminic acid are referred to in Chapters 7 and 9, and in Chapter 16, respectively.

The silyl ether **40**, available by conventional procedures from the known 2-deoxy precursor **39**, was used as glycosyl donor in the preparation of methyl α-isomaltoside and methyl α-isomaltotrioside analogues deoxygenated at C-2 in one or two of the glucose moieties.[21] 4'-Deoxy-α-maltosyl- and 4''-deoxy-α-maltotriosyl fluoride have been synthesized by standard methodology as mechanisitic probes for α-glucosyltransferases.[22] A convenient synthesis of 6'-deoxy-*N*-acetyllactosamine was based on β-(1→4)-galactosyltransferase-mediated coupling of UDP-6-deoxy-D-galactose with *N*-acetylglucosamine, although the transfer rates were very low.[23]

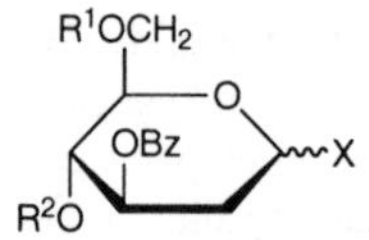

39 X = α-OMe, R¹, R² = ⟩—Ph
40 X = β-OTbdms, R¹ = Ac, R² = Bn

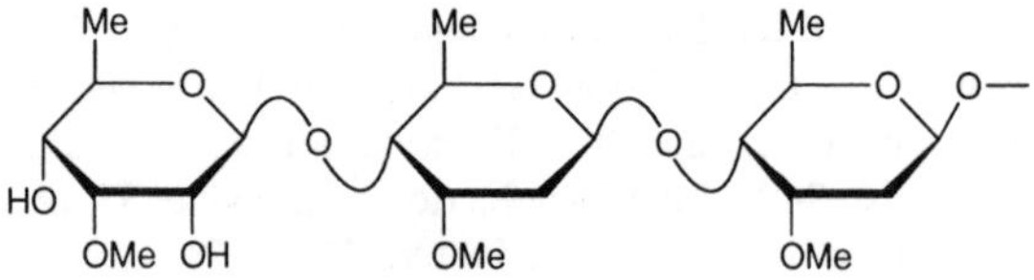

41

As part of a study on the structural features responsible for the intense sweetness of D-fructose, 3-deoxy-D-fructose, prepared by a literature procedure (see Vol. 15, Chapter 11, Ref. 7), has been evaluated; in aqueous solutions at ambient temperature the proportion of open-chain form present is 7.5%, increasing to 47% at 97 °C.[24] 6-Deoxy-β-D-allopyranoside residues have been found in a cardenolide glycoside from the plant *Gomphocarpus sinaicus*.[25] Trisaccharide moieties, such as **41**, containing 6-deoxy-3-O-methyl-β-D-allopyranosyl- and 2,6-dideoxy-3-O-methyl β-D-*ribo*-hexopyranosyl- (cymaropyranosyl-) residues were isolated from aerial parts of the succulent *Stapelia variegata*.[26] Caryophyllose, a new, natural, branched trideoxy sugar is referred to in Chapters 2, 14 and 21.

References

1   A.D. Backstrom, R.A.S. McMordie and T.P. Begley, *J. Carbohydr. Chem.*, 1995, **14**, 177.

2   I.A. Kennedy, T. Hemscheidt, J.F. Britten and I.D. Spenser, *Can. J. Chem.*, 1995, **73**, 1329.

3   I.A. Kennedy, R.E. Hill, R.M. Pauloski, B.G. Sayer and I.D. Spenser, *J. Am. Chem. Soc.*, 1995, **117**, 1661.

4   W.-F. Lu and B.-J. Uang, *J. Chin. Chem. Soc. (Taipei)*, 1994, **41**, 829 (*Chem. Abstr.*, 1995, **122**, 187 883).

5   I. Paterson and M.D. McLeod, *Tetrahedron Lett.*, 1995, **36**, 9065.

6   K. Mizutani, T. Yamazaki and T. Kitazume, *J. Chem. Soc., Chem. Commun.*, 1995, 51.

7   A. Sobti and G.A. Sulikowski, *Tetrahedron Lett.*, 1995, **36**, 4193.

8   A. Kirschning, U. Hary and M. Ries, *Tetrahedron*, 1995, **51**, 2297.

9   X. Liang, R. Krieger and H. Prinzbach, *Tetrahedron Lett.*, 1995, **36**, 6433.

10 R.M. Giuliano and F.J. Villani, *J. Org. Chem.*, 1995, **60**, 202.

11 K. Toshima, I. Ishizuka, G. Matsuo and M. Nakata, *Synlett.*, 1995, 307.

12 V. Vill and H.-W. Tunger, *Liebigs Ann. Chem.*, 1995, 1055.

13 H.P. Wessel and G. Englert, *J. Carbohydr. Chem.*, 1995, **14**, 179.

14 G. Capozzi, C. Falciani, S. Menichetti, C. Nativi and R.W. Franck, *Tetrahedron Lett.*, 1995, **36**, 6755.

15 T. Gimisis, M. Ballestri, C. Ferrari, C. Chatgilialoglu, R. Boukherroub and G. Manuel, *Tetrahedron Lett.*, 1995, **36**, 3897.

16 L.A. Mulard and C.P.J. Glaudemans, *Carbohydr. Res.*, 1995, **274**, 209.

17 Y. Ogawa, P.-s. Lei and P. Kovac, *Carbohydr. Res.*, 1995, **277**, 327.

18 S. Czernecki, S. Horns and J.-M. Valéry, *J. Carbohydr. Chem.*, 1995, **14**, 157.

19 K. Sato and A. Yoshitomo, *Chem. Lett.*, 1995, 39.

20 D. Crich, Q. Yao and G.F. Filzen, *J. Am. Chem. Soc.*, 1995, **117**, 11455.

21 E. Petrakova and C.P.J. Glaudemans, *Carbohydr. Res.*, 1995, **268**, 35.

22 T.K. Lindhorst, C. Brown and S.G. Withers, *Carbohydr. Res.*, 1995, **268**, 93.

23 Y. Kajihara, T. Endo, H. Ogasawa, H. Kodama and H. Hashimoto, *Carbohydr. Res.*, 1995, **269**, 273.

24 W.A. Szarek, R.J. Rafka, T.-F. Yang and D.R. Martin, *Can. J. Chem.*, 1995, **73**, 1639.

25 H. El-Askary, J. Hölzl, S. Hilal and E. El-Kashoury, *Phytochemistry*, 1995, **39**, 943.

26 K.A. El Sayed, A.F. Halim, A.M. Zaghloul, J.D. McChesney, M.P. Stone, M. Voehler and K. Hayashi, *Phytochemistry*, 1995, **39**, 395.

13
Unsaturated Derivatives

1   Glycals

1.1   Syntheses of Glycals. – Further details (see Vol. 27, p. 153, ref. 5 for earlier work) on the unexpected formation of tri-*O*-acetyl-D-glucal from treatment of *S*-phenyl-3,4,6-tri-*O*-acetyl-2-*O*-(diphenoxyphosphoryl)-1-thio-β-D-glucopyrano-side with tributyltin hydride-AIBN have been reported. The reaction was expected to give a glycal 1-phosphonate by way of a β(phosphatoxy)alkyl radical migration.[1] A high yielding synthesis of tri-*O*-acetyl-D-glucal has been achieved by treating tetra-*O*-acetyl-D-glucosyl- or D-mannosyl-bromide with bis(titanocene chloride) (Cp$_2$TiCl)$_2$, the reaction occurring by way of glycosyl radicals.[2]

The D-allal derivative **1** was formed by treating the unsaturated 1-thiophenyl glycoside **2** first with MCPBA then with diethylamine to effect a [2,3] sigmatropic rearrangement of the intermediate sulfoxide. Product **1** was incorporated into the core trisaccharide unit of esparamicin and the aryl tetrasaccharide unit of calicheamicin.[3]

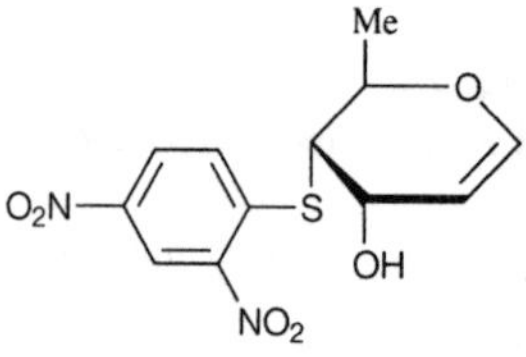

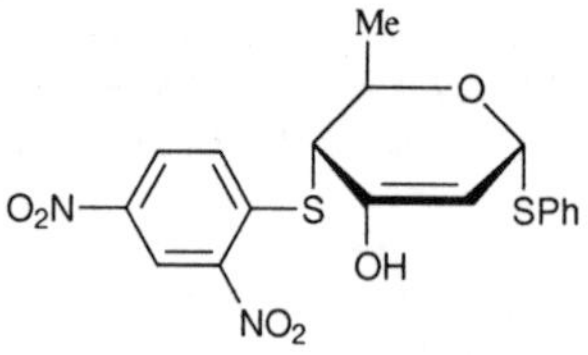

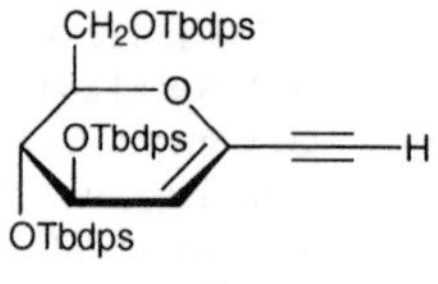

3,4,6-Tri-*tert*-butyldiphenylsilyl-2-deoxy-D-*arabino*-hexonolactone, on sequen-tial reaction with bromomagnesium(trimethylsilylacetylide), phosphorus oxy-chloride-pyridine then sodium hydroxide, gave glycal derivative **3**. The acetylene moiety was further elaborated into an anthracyclinone system representing the core unit found in vineomycinone B.[4]

A convenient synthesis of 'L-2-sorbal' (2,6-anhydro-3-deoxy-L-*threo*-hex-2-enitol) starting from 2,5-anhydro-D-galactitol has been reported. Thus treatment of the latter with 2,2-dimethoxypropane (to produce a mixture of di- and tri-acetals) then potassium *tert*-butoxide in DMSO (to effect elimination of acetone) gave glycal **4**. Its reaction with methanolic acetic acid yielded 'L-2-sorbal'. In addition, reaction of **4** with MCPBA then methanol followed by reaction with acetic acid afforded L-tagatose.[5] The 'D-2-fructal' (2,6-anhydro-3-deoxy-D-*erythro*-hex-2-enitol) derivative **5** is prepared by treating the 2-bromo-3-mesyl compound **6** with a zinc-copper couple. In the absence of a good leaving group in the 3-position a 1,2-elimination takes place instead to produce '*exo* glycals' such as **7**.[6]

4

5

6

7 R^1 = OBz, R^2 = H
R^1 = H, R^2 = OBz

8

9

10

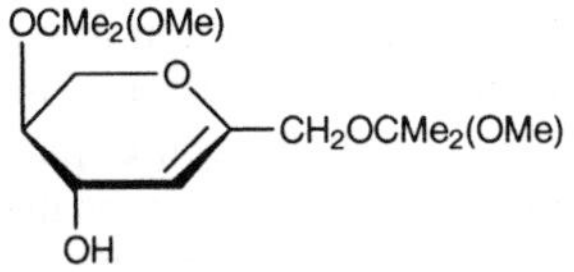

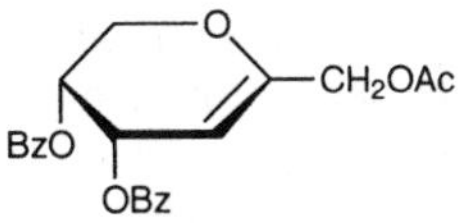

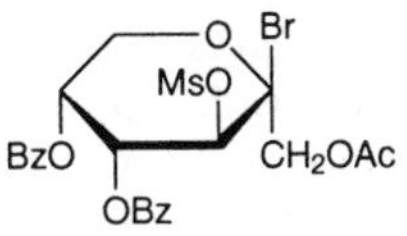

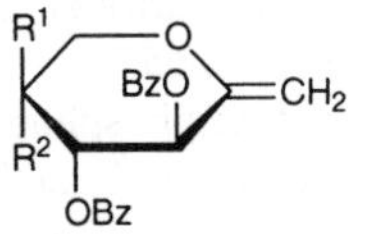

Alkyl 4-*O*-(2-bromoallyl)-2,3-dideoxy-α-D-*erythro*-hex-2-enosyl derivatives **8** (R = OEt, O^{t}Bu, $OC_6H_4{}^tBu$) undergo a palladium-mediated Heck-type reaction to give mainly the bicyclic glycal **9** when the base used is triethylamine. The glycal **10** is formed when **8** (R = H) is treated with the base sodium carbonate.[7]

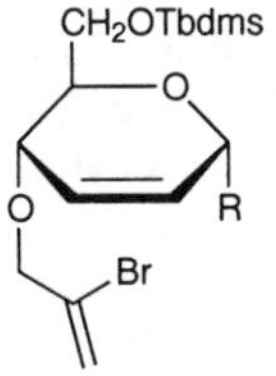

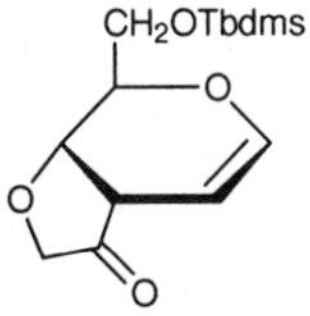

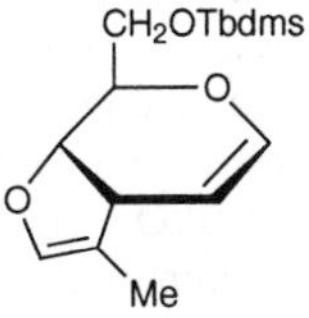

The reaction of 2,3-*O*-isopropylidene-5-*O*-trityl-β-D-ribofuranosyl chloride with the lithioimidazole reagent **11**, then TBAF, affords access to glycal **12**.[8]

Treatment of 2,3,4,6-tetra-*O*-acetyl-D-glucopyranosylidene 1,1-diazide with triphenylphosphine unexpectedly gave the unusual glycal derivative **13**.[9]

11

12

13

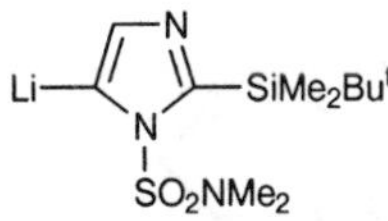

The photoreaction of 2′-deoxy-2′-iododeoxyuridine has been demonstrated to yield nine identifiable products including 1,4-anhydro-2-deoxy-D-*glycero*-pent-1-en-3-ulose; see Chapter 20 for details.

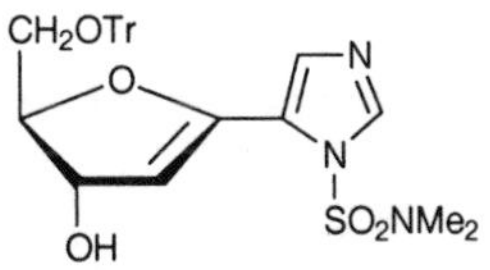

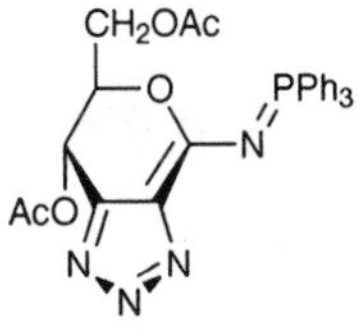

1.2   Allylic Rearrangements of Glycals. – Phenol or 4-methoxyphenol reacts with acetylated glycals in toluene containing a catalytic amount of boron trifluoride etherate at low temperature to yield *O*-2,3-unsaturated glycosides with high α-selectivity (see Vol. 28, p. 174, ref. 17). A change of solvent to dichloromethane gave predominately *C*-2,3-unsaturated glycosides as α/β-mixtures in the case of 4-methoxyphenol. It is further suggested that the *C*-2,3-glycosides originate by rearrangement of the initially formed *O*-2,3-glycosides.[10]

Refluxing acetylated glycals with phenols in chlorobenzene without any acid catalyst gives the usual *O*-linked rearranged compounds as α/β-mixtures. Use of phenols with NO_2 and ^{t}Bu substituents allows for easy crystallization of pure α-anomers and consequently multigram quantities of these products can be made. The products are readily deacetylated and converted to *O*-benzyl ether derivatives.[11]

Treatment of 4,6-*O*-isopropylidene-D-glucal with a variety of phenols, cyclohexane-1,3-dione, 4,6-*O*-isopropylidene-D-glucal and phthalimide under Mitsunobu conditions (triphenylphosphine, diethyl azodicarboxylate) affords 2,3-unsaturated *O*- or *N*-glycosides with α:β ratios varying from 90:10 to 65:35.[12]

Lewis acid-catalysed reaction of diisopropyl (hydroxymethyl)phosphonate [$({}^iPrO)_2P(O)CH_2OH$] with di-*O*-acetyl-D-xylal, followed by reaction with methanolic ammonia, affords **14** as the major isomer; a brief description of the diastereoselectivity of this reaction is also given. Compound **14** on reaction with 2-amino-6-chloropurine under Mitsunobu conditions affords access to dihydropyranyl nucleoside phosphonate analogues, which on reduction of the double bond lead to the corresponding tetrahydropyranyl derivatives. Di-*O*-acetyl-L-arabinal was also used as a starting material.[13,14]

Standard reaction of tri-*O*-acetyl-D-glucal with 4-penten-1-ol and Lewis acid affords the expected glycoside **15**. Treatment of **15** with iodonium dicollidine perchlorate (IDCP) in the presence of primary or secondary alcohols gives the corresponding α-*O*-glycosides as the main products. The use of the pent-4-enoyl glycal **16** as starting material in the presence of alcohol and IDCP affords a 'one-pot' procedure for effecting the glycal ester rearrangement under neutral conditions.[15]

16 R = $\overset{O}{\underset{\|}{C}}CH_2CH_2$CH=CH$_2$

The mechanism of the reaction depicted in Scheme 1 has been studied by chemical means and modelling using AM1 calculations.[16]

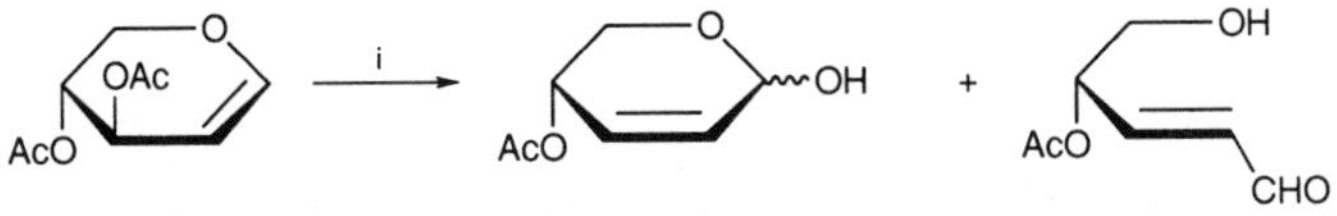

Reagents: i, H$_2$O, Δ

Scheme 1

The reaction of 3,4-di-*O*-acetyl-6-deoxy-L-glucal with titanium tetrachloride-trimethylaluminium followed by reaction with methanolic ammonia affords unsaturated derivative **17**, together with the β-anomer, which was used as an intermediate for the synthesis of (+)-hongconin, thereby establishing the absolute stereochemistry of the natural product as its enantiomer.[17]

For further examples of the Ferrier rearrangement as applied at the start of a synthesis of a thromboxane B$_2$ precursor or of highly oxygenated *cis*-decalinic structures derived from carbohydrates, see Chapter 24. The same reaction applied to the preparation of carbohydrate-based liquid crystals derived from a boronate ester is mentioned in Chapter 17.

1.3 Other Reactions of Glycals. – 3-*O*-Propargylic hexofuranoid glycals undergo a [2,3] Wittig rearrangement on treatment with *n*-butyllithium effecting the preparation of annonaceous acetogenins (Scheme 2). The glycal illustrated is derived from D-mannose and *erythro*-products predominate. The paper also describes the analogous use of a glycal starting material derived from L-gulonic-γ-lactone. However, a pentofuranoid glycal derived from D-ribonic-γ-lactone underwent a [1,2] Wittig rearrangement process instead under similar conditions.[18]

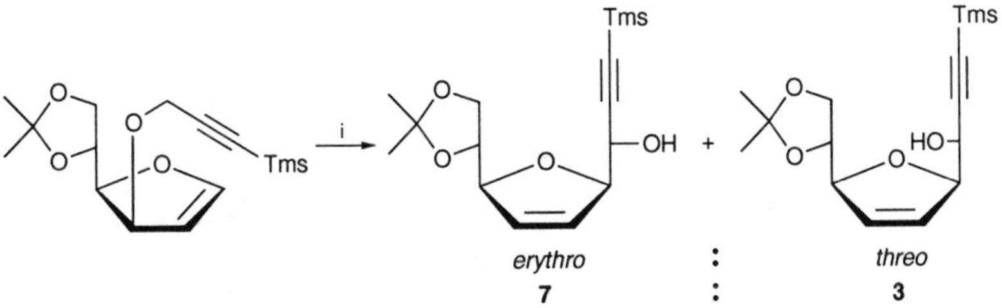

| *erythro* | : | *threo* |
| **7** | : | **3** |

Reagents: i, Bu^{n}Li, –5 °C, THF

Scheme 2

A facile aza-Claisen [3,3] sigmatropic rearrangement takes place on treating furanoid glycal **18** with sodium hydride-trichloroacetonitrile at 0 °C to give **19** as a useful intermediate for preparing dideoxynucleotides.[19]

The silyl enol ether **20** has been reported to undergo a facile rearrangement in the presence of Lewis acids to give levoglucosenone as the major product.[20]

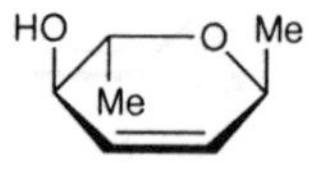

17 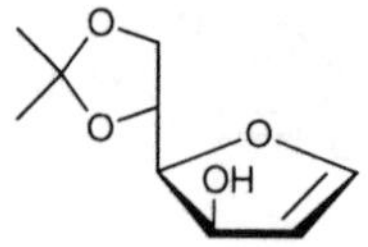   **18** 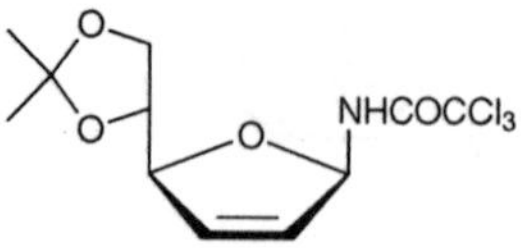   **19**

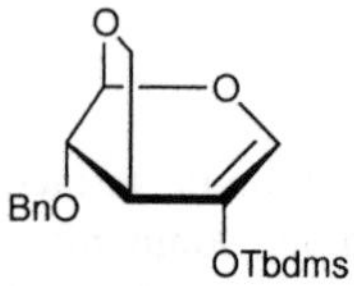

20 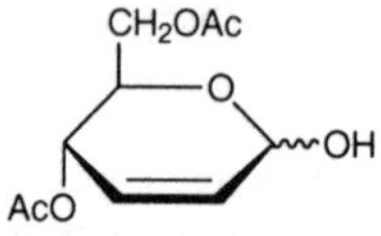   **21** 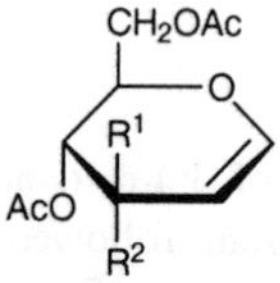   **22** R^1 = N$_3$, R^2 = H
                               23 R^1 = H, R^2 = N$_3$

The hypervalent iodine reagent [hydroxy(tosyloxy)iodo]benzene [PhI(O-H)(OTs)] selectively oxidizes *O*-acyl protected pyranoid glycals in the 3-position affording 4,6-diacyl-1,2-dideoxyhex-1-en-3-uloses as products. The method is applicable to glycals with variable configurations and acyl groups. Glycals without an acyloxy substituent in the 3-position, for example 4,6-di-*O*-acetyl-3-deoxy-D-glucal, produce the Ferrier-type rearrangement product **21** instead.[21] (See also Chapter 5 for a different course for the reaction when *O*-silyl groups are in the 3-position).

3-Deoxyglycals undergo an iodine(III)-promoted azide transfer reaction when treated with the reagent PhI(N$_3$)$_2$. Thus 4,6-di-*O*-acetyl-3-deoxy-D-glucal affords a 1:8 mixture of **22:23**.[22]

Glycals undergo a cycloaddition reaction with α,α'-dioxothiones which are in turn prepared from α,α'-dioxothiophthalimides as illustrated in Scheme 3. The reaction is carried out in 'one-pot' fashion.[23]

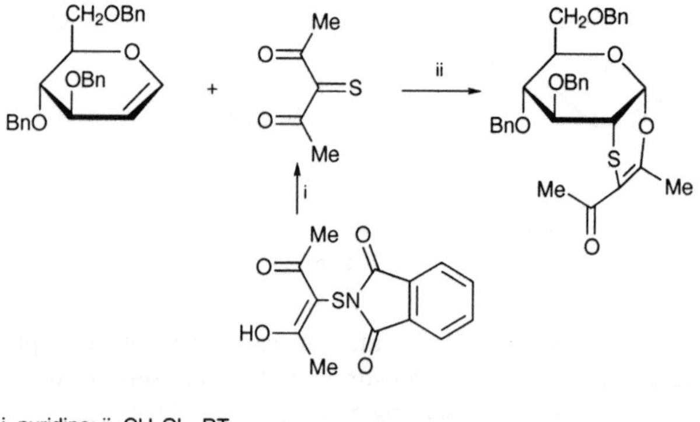

Reagents: i, pyridine; ii, CH$_2$Cl$_2$, RT

Scheme 3

O-Protected and partially O-protected glycals (but not 3,4,6-tri-O-*tert*-butyldi-methylsilyl-D-glucal) can be readily cyclopropanated with diiodomethane-diethyl-zinc to afford intermediate 1,2-C-methylene derivatives. Treatment of these latter compounds with Lewis acids was expected to produce intermediate ring-expanded oxonium ions **24** which should be readily trapped with a nucleophile to give oxepane derivatives. This proved to be the case when cyclopropane derivative **25** was treated with trimethylsilyl triflate in the presence of trimethylsilyl cyanide as internal nucleophile to give oxepane **26**. Cyclopropane derivative **27**, under similar conditions but with allyltrimethylsilane as nucleophile, reacted intramolecularly to oxepane **28**.[24]

The reaction of 1-(tributylstannyl)-3,4,6-tris-O-(triisopropylsilyl)-D-glucal with *n*-butyllithium then triflate **29** afforded the protected undecose backbone **30** found in the herbicidins.[25] The preparation of a carbon-bridged analogue of *N*-acetyllactosamine by reaction of a 1-C-lithiated glycal with a 4-C-formyl protected glucosamine derivative is covered in Chapter 3.

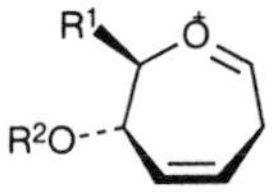

24 R[1] = CH$_2$OAc or Me
    R[2] = Ac

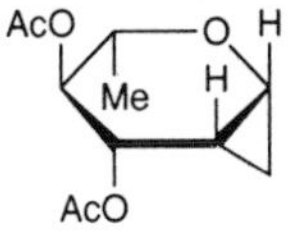

25

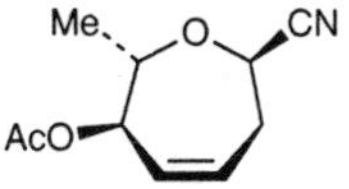

26

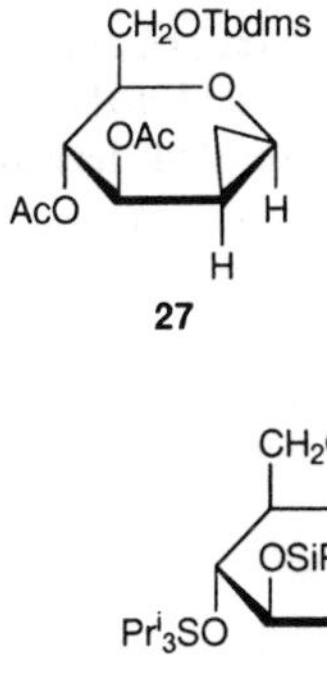

27

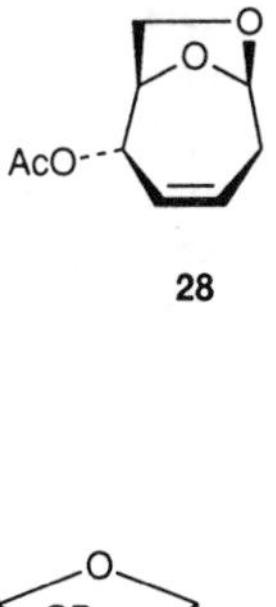

28

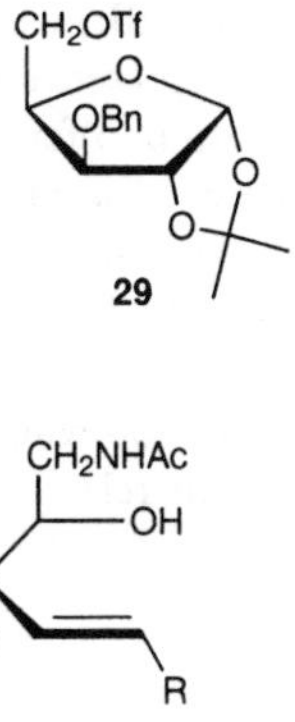

29

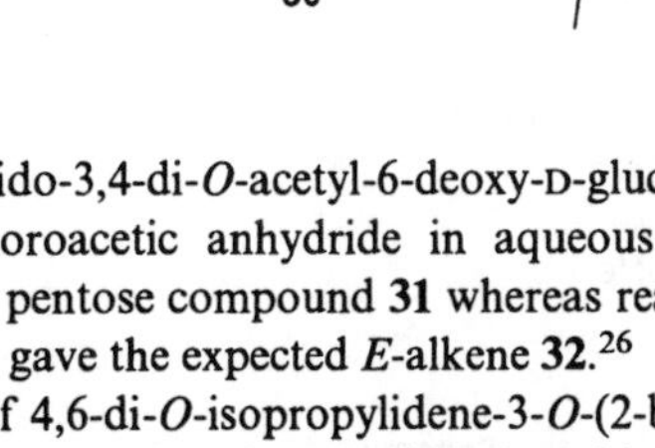

30

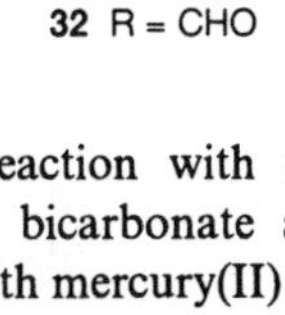

31 R = NO$_2$
32 R = CHO

6-Acetamido-3,4-di-O-acetyl-6-deoxy-D-glucal on reaction with ammonium nitrate-trifluoroacetic anhydride in aqueous sodium bicarbonate affords the unsaturated pentose compound **31** whereas reaction with mercury(II) sulfate and sulfuric acid gave the expected E-alkene **32**.[26]

The use of 4,6-di-O-isopropylidene-3-O-(2-bromo-1-ethoxyethyl)-D-glucal as a precursor to highly functionalized C-glycosides is mentioned in Chapter 3 and the use of polymer supported glycals as precursors to digitalis saponin and neoglyco-conjugates of Le^y and Le^b are mentioned in Chapter 4. Glycal starting materials directed towards the total synthesis of staurosporine and *ent*-staurosporine are covered in Chapter 24.

2 Other Unsaturated Derivatives

1,2-Diol cyclic sulfates on reaction with telluride ion (prepared by reduction of elemental tellurium) afford alkenes as products. Thus D-ribofuranoside derivative **33** gives **34** in 72% yield.[27]

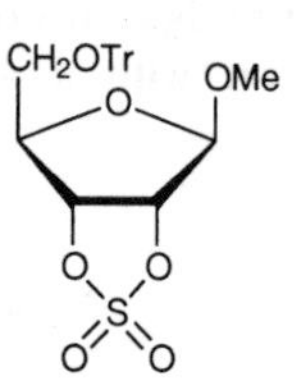

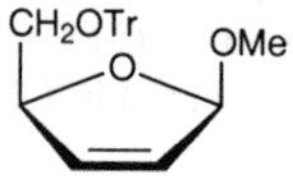

p-tert-Butylphenyl   4,6-di-*O*-benzyl-2,3-dideoxy-α-D-*erythro*-hex-2-enopyrano-side (see ref. 11 above for preparation details) reacts with a variety of Grignard reagents (RMgBr, where R = 4-MeOC$_6$H$_4$, 2-MeOC$_6$H$_4$, 4-MeC$_6$H$_4$, Ph, PHCH$_2$ *etc.*) in the presence of a catalytic amount of [palladium dichloro{1,1'-bis(di-phenylphosphino)ferrocene}] [PdCl$_2$(dppf)] to afford unsaturated derivative **35** with exclusive formation of the α-anomer. In contrast, the same reaction using a catalytic   amount   of   [nickel   dichloro{1,2-bis(diphenylphosphino)ethane}] [NiCl$_2$(dppe)] afforded the β-anomer of **35** exclusively.[28]

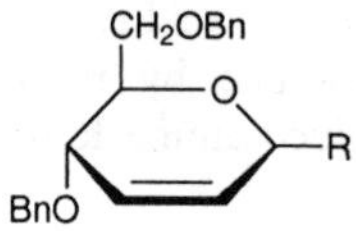

A general synthesis of bicyclic *cis*-fused dihydrofuran derivatives by intramole-cular Mitsunobu reaction is illustrated in Scheme 4. The reaction is also effective with 1,4-*trans*-related starting materials.[29]

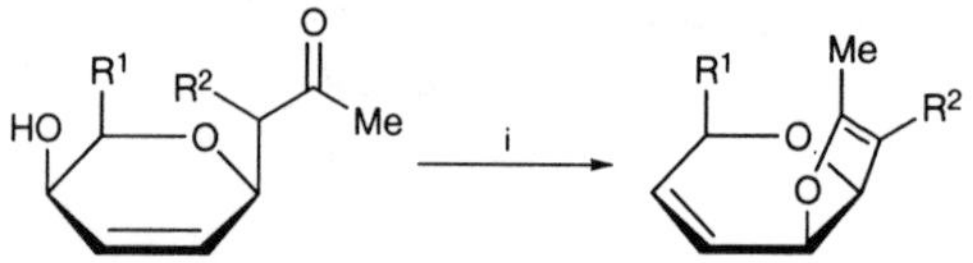

R^1 = CH$_2$OTbdms, R^2 = COMe
R^1 = Me, R^2 = COMe
R^1 = H, R^2 = COMe
R^1 = H, R^2 = CO$_2$Me

Reagents: i, PBu$_3$, DEAD

Scheme 4

Reaction of 1,2:5,6-di-*O*-isopropylidene-3-*O*-triflyl-α-D-glucofuranose with methyllithium gives the corresponding glycal (double bond in the 3,4-position) by an apparent E-2 elimination. However, deuterium labelling experiments suggest the mechanism first involves deprotonation at C-3 followed by a 1,2-hydride shift from C-4→C-3 and hence to product.[30]

A mild conversion of carbonyl derivatives, including lactones, into alkenyl silanes has been reported. For example reaction of 5-*O*-tert-butyldimethylsilyl-2,3-*O*-isopropylidene-D-1,4-ribonolactone with tris(trimethylsilyl)titanacyclobutane gave product **36**.[31]

Heating methyl 2-*O*-benzyl-4,6-*O*-benzylidene-3-*O*-triflyl-β-D-glucopyranoside with water in pyridine and 1,1,1,3,3,3-hexafluoropropan-2-ol affords a low yield of unsaturated compound **37**. The main products from this reaction are ring-contracted bicyclic sugars and are covered in Chapter 14.[32]

Reacting D-mannose with arylhydrazines followed by acetic anhydride in pyridine afforded the diazabutadiene **38**. A similar reaction takes place with D-galactose and D-ribose.[33]

An improved synthesis of 3-cyano-3-deoxy- and 3-deoxy-3-formyl-hex-2-eno-pyranosides has been described. Thus treating methyl 2,3-anhydro-4,6-*O*-benzyl-idene-α-D-galactopyranoside with diethylaluminium cyanide (epoxide ring opening with CN) then elimination of water with diisopropylamine, lithium perchlorate and tosyl chloride afforded **39** which on reduction with DIBAL gave **40**.[34]

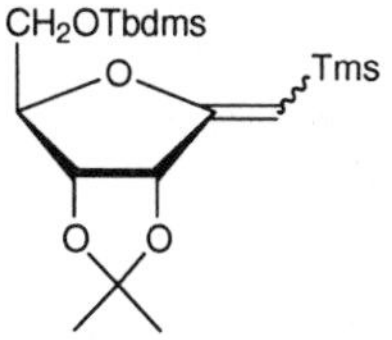

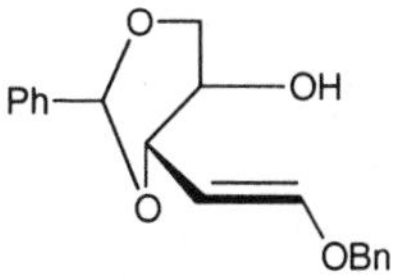

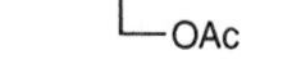

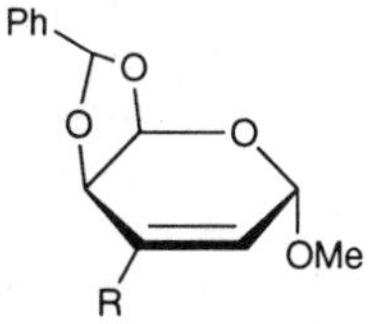

A method for preparing alkenyl triflates from lactones and one for preparing α,β-unsaturated aldonic acids by chain extending D-glyceraldehyde derivatives are covered in Chapter 16, and the preparation of sugar vinyltin derivatives from sugar acetylenes is mentioned in Chapter 17.

References

1   D. Crich, Q. Yao and G.F. Filzen, *J. Am. Chem. Soc.*, 1995, **117**, 11455.
2   C.L. Cavallaro and J. Schwartz, *J. Org. Chem.*, 1995, **60**, 7055.
3   R.L. Halcomb, S.H. Boyer, M.D. Wittman, S.H. Olson, D.J. Denhart, K.K.C. Liu and S.J. Danishefsky, *J. Am. Chem. Soc.*, 1995, **117**, 5720.

4   F.E. McDonald, H.Y.H. Zhu and C.R. Holmquist, *J. Am. Chem. Soc.*, 1995, **117**, 6605.

5   P.L. Barili, G. Berti, G. Catelani, F.D(Andrea and A. Gaudiosi, *Gazzetta Chimica Italiana*, 1995, **124**, 57 (*Chem. Abstr.*, 1995, **122**, 187 941).

6   F.W. Lichtenthaler, S. Hahn and F.-J. Flath, *Liebigs Ann. Chem.*, 1995, 2081.

7   J.-F. Nguefack, V. Bolitt and D. Sinou, *J. Chem. Soc., Chem. Commun.*, 1995, 1893.

8   S. Harusawa, M. Kawabata, Y. Murai, R. Yoneda and T. Kurihara, *Chem. Pharm. Bull.*, 1995, **43**, 152.

9   J. Kovacs, I. Pinter, M. Kajtar-Peredy, J.-P. Praly and G. Descotes, *Carbohydr. Res.*, 1995, **279**, C1.

10 T. Noshita, T. Sugiyama, Y. Kitazumi and T. Oritani, *Biosci. Biotech. Biochem.*, 1995, **59**, 2052.

11 I. Frappa and D. Sinou, *Synth. Commun.*, 1995, **25**, 2941.

12 N.G. Ramesh and K.K. Balasubramanian, *Tetrahedron*, 1995, **51**, 255.

13 M.-J. Pérez-Pérez, B. Doboszewski, J. Rozenski and P. Herdewijn, *Tetrahedron: Asymm.*, 1995, **6**, 973.

14 M.-J. Pérez-Pérez, J. Balzarini, J. Rozenski, E. De Clercq and P. Herdewijn, *Bioorg. Med. Chem. Lett.*, 1995, **5**, 1115.

15 J.C. López, A.M. Gómez, S. Valverde and B. Fraser-Reid, *J. Org. Chem.*, 1995, **60**, 3851.

16 J. Madaj, J. Rak, E. Skorupowa, A. Lopacinska, J. Sokolowski and A. Wisniewski, *J. Chem. Soc., Perkin Trans. 2*, 1995, 569.

17 P.P. Deshpande, K.N. Price and D.C. Baker, *Bioorg. Med. Chem. Lett.*, 1995, **5**, 1059.

18 P. Bertrand, J.-P. Gesson, B. Renoux and I. Tranoy, *Tetrahedron Lett.*, 1995, **36**, 4073.

19 D.L. Armstrong, I.C. Coull, A.T. Hewson and M.J. Slater, *Tetrahedron Lett.*, 1995, **36**, 4311.

20 A. Griffin, N.J. Newcombe and T. Gallagher, *Front. Biomed. Biotechnol.*, 1994, **2** (levoglucosenone and levoglucosan), 23 (*Chem. Abstr.*, 1995, **122**, 265 790).

21 A. Kirschning, *J. Org. Chem.*, 1995, **60**, 1228.

22 A. Kirschning, S. Domann, G. Dräger and L. Rose, *Synlett.*, 1995, 767.

23 G. Capozzi, R.W. Franck, M. Mattioli, S. Menichetti, C. Natiui and G. Valle, *J. Org. Chem.*, 1995, **60**, 6416.

24 J.O. Hoberg and J.J. Bozell, *Tetrahedron Lett.*, 1996, **36**, 6831.

25 J.R. Bearder, M.L. Dewis and D.A. Whiting, *J. Chem. Soc., Perkin Trans. 1*, 1995, 227.

26 W.B. Mathews and W.W. Zajac, Jr., *J. Carbohydr. Chem.*, 1995, **14**, 287.

27 B. Chao, K.C. McNulty and D.C. Dittmer, *Tetrahedron Lett.*, 1995, **36**, 7209.

28 C. Moineau, V. Bolitt and D. Sinou, *J. Chem. Soc., Chem. Commun.*, 1995, 1103.

29 A. Tenaglia, J.-Y. Le Brazidec and F. Souchon, *Tetrahedron Lett.*, 1995, **36**, 4241.

30 A. El Nemr and T. Tsuchiya, *Tetrahedron Lett.*, 1995, **36**, 7665.

31 N.A. Petasis, J.P. Staszewski and D.-K. Fu, *Tetrahedron Lett.*, 1995, **36**, 3619.

32 M. Kassou and S. Castillón, *J. Org. Chem.*, 1995, **60**, 4353.

33 M. Avalos, R. Babiano, P. Cintas and J.L. Jiménez, *Tetrahedron: Asymm.*, 1995, **6**, 945.

34 R.A. Spanevello and S.C. Pellegrinet, *Synth. Commun.*, 1995, **25**, 3663.

14
Branched-chain Sugars

1   Compounds with a $\underset{\overset{|}{O}}{\overset{\overset{R}{|}}{C\text{—}C\text{—}C}}$ Branch-point

1.1   Branch at C-2. – Full details (see Vol. 27, p. 10, ref. 49 and Vol. 26, p. 11, ref. 56) of the preparation of branched C-2-(hydroxymethyl)pentoses from four different ketohexoses (D-psicose, D-fructose, L-sorbose and D-tagatose) by an interesting isomerization reaction catalysed by Ni(II)-N,N'-dialkyl cyclohexanediamine have been reported.[1]

Dess-Martin periodane oxidation of 1,3,5-tri-O-benzoyl-α-D-ribofuranose affords the corresponding 2-keto derivative which reacts with 'MeTiCl₃' (formed *in situ* from titanium tetrachloride and methylmagnesium bromide) to afford the C-2-branched derivative **1** together with the benzoate migration product **2**. Perbenzoylation of a mixture of **1** and **2** produced an intermediate useful for making C-2′-methylribonucleosides by way of standard Vorbrügen coupling procedures with a number of nucleobases.[2]

In studies directed towards the synthesis of pseurotin A, the unsaturated acetal **3** has been used as a common starting material to prepare two different C-2-branched intermediates. In one route, **3** was O-benzylated, the aldehyde unmasked with hydrochloric acid and oxidized to its carboxylic acid derivative. Esterification, dihydroxylation of the alkene then dimethylsulfoxide-thionyl chloride oxidation, addition of ethylmagnesium chloride and removal of the isopropylidene group gave compounds **4**.[3] In an alternative route, **3** was first epoxidized then treated sequentially with hydroxide, benzyl bromide-base and hydrochloric acid to give compounds **5**.[4]

The synthesis of polyhydroxy pyrrolidines containing a hydroxymethyl branching group, starting from 3-deoxy-2-C-hydroxymethyl-D-*erythro*-pentono-1,4-lactone, is mentioned in Chapter 18 and the X-ray crystal structures of the same compound and of 3-deoxy-2-C-hydroxymethyl-D/L-tetrono-1,4-lactone are noted in Chapter 22.

1.2   Branch at C-3. – An interesting entry to the bicyclic moiety found in the nucleoside antibiotics miharamycins has been reported. Methyl-4,6-O-benzylidene-α-D-glucopyranoside can be propargylated in the 2-O-position and the remaining 3-OH oxidized to the ketone. Reaction of this compound with

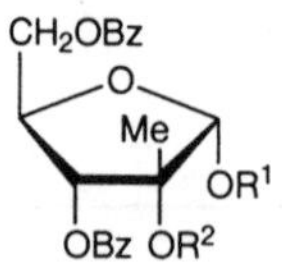

1 R^1 = Bz, R^2= H
2 R^1 = H, R^2= Bz

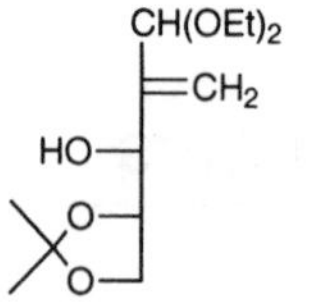

3

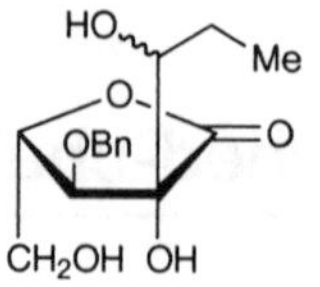

4

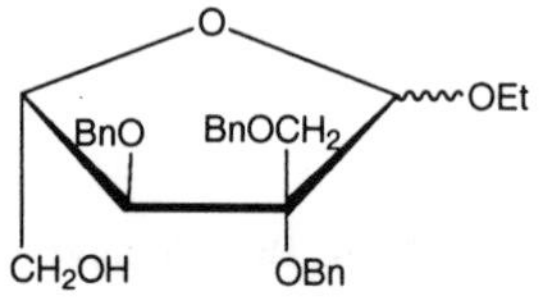

5

samarium diiodide in HMPA-^{t}BuOH-THF afforded a high yield of compound **6** which, on subsequent ozonolysis and sodium borohydride reduction, gave the corresponding key diol intermediate.[5]

Treating 1,2:5,6-di-*O*-isopropylidene-3-*O*-triflyl-α-D-allofuranose with excess butyllithium affords 3-*C*-butyl-1,2:5,6-di-*O*-isopropylidene-α-D-allofuranose in a new type of reaction. Investigation of this reaction through deuterium-labelled derivatives indicates that removal of the C-3 hydrogen (α to the triflate) generates a carbanion which collapses to an intermediate 3-keto-derivative (with expulsion of CF$_3$SO$_2$Li) and then reacts with butyllithium by addition. The corresponding *gluco*-configured triflate undergoes an apparent E-2 elimination to give a 3,4-unsaturated derivative; see Chapter 13.[6]

Lombardo methylenation (zinc-dibromomethane-titanium tetrachloride) of methyl 2-deoxy-5-*O*-(4-phenylbenzoyl)-β-D-*glycero*-pent-3-uloside, followed by dihydroxylation, gave the 3-*C*-hydroxymethyl derivative **7** (R = 4-phenylbenzoyl) exclusively. A range of other 2-deoxypentuloses was also subjected to a similar course of reactions, but generally resulted in the formation of epimers during the dihydroxylation step. The hydroxymethyl group was further transformed first with tosyl chloride to the tosylate which was displaced with adenine or thymine to yield nucleoside analogues.[7]

The addition of dichloromethyllithium to methyl 4,6-*O*-benzylidene-2-*O*-benzoyl- and 2-*O*-methyl-α-D-*ribo*-hexopyranosid-3-ulose affords branched derivatives **8** which, on treatment with caesium acetate-crown ether, affords compounds **9**. The latter derivatives are thought to be formed *via* an intermediate *spiro*-chloroepoxide in a similar way to that described in Vol. 25, p. 162, ref. 9.[8] In a somewhat analogous reaction, *spiro*-tosyl epoxide **10** is formed when the anion derived from chlorotosyl methane (TsCH$_2$Cl) reacts with methyl 4,6-*O*-benzylidene-α-D-*erythro*-hexopyranosid-3-ulose. Compound **10** can be caused to react with a wide range of nucleophiles affording a wide range of branched-derivatives.[9] The synthesis of 1-(β-D-apiofuranosyl)-1,2,3-triazoles is mentioned in Chapters 10 and 20.

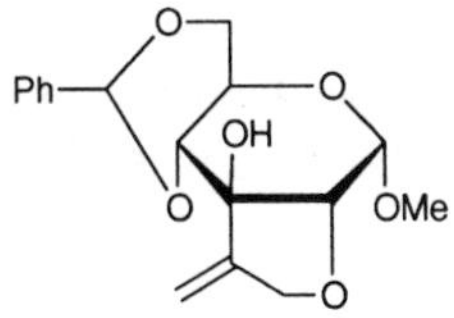

6 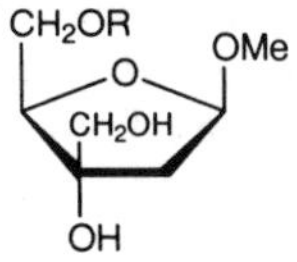 7 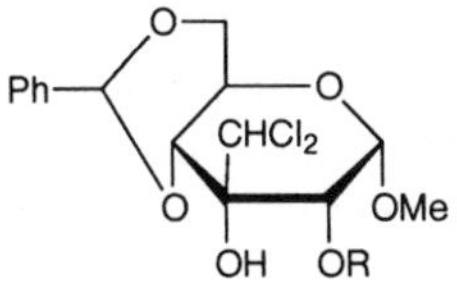 8 R = Me or Bz

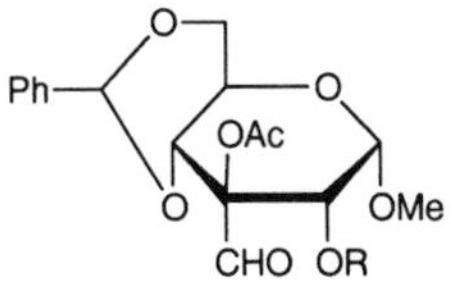

9 R = Me or Bz 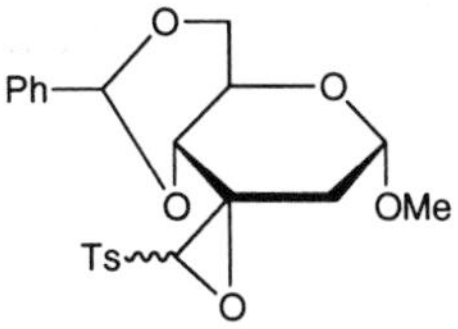 10 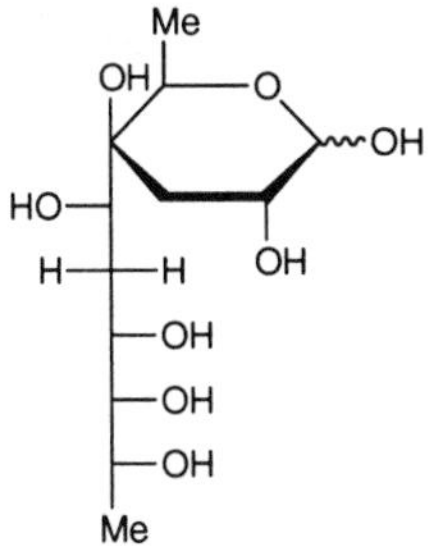 11

1.3 Branch at C-4. – The novel branched-sugar **11**, named caryophyllose has been isolated from the lipopolysaccharide of the bacterium *Pseudomonas caryophylli*[10] and its relative and absolute stereochemistry determined.[11]

The synthesis of the methyl-branched compound **12**, a building block for the synthesis of moenuronamide phosphoglycolipid antibiotics has been described. A stereoselective Grignard reaction with methylmagnesium chloride on a 4-ulose derivative is the key step.[12]

Addition of benzodiynyl derivative **13** (R = thexyldimethylsilyl) to a 4-ulose compound gave **14** (R = thexyldimethylsilyl) an intermediate required to prepare an oxabicyclo[7.3.1] analogue of the aglycone part of esperamicin and calicheamicin; see also Chapter 24.[13]

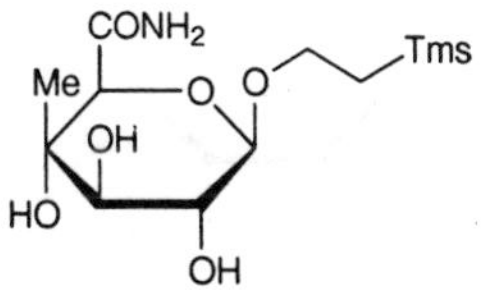

12 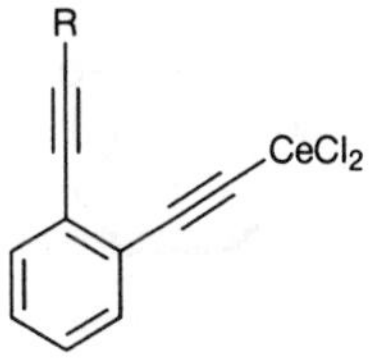 13 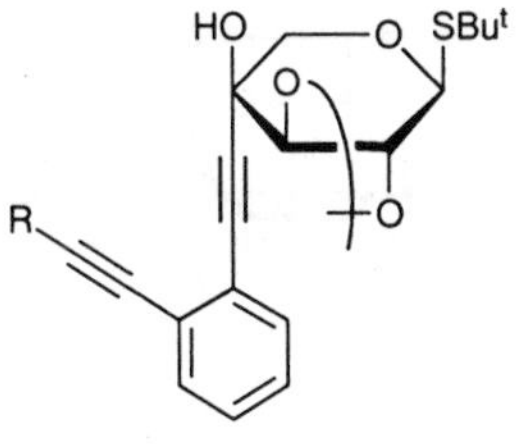 14

The addition of lithium *tert*-butyl acetate to benzyl 2,3-anhydro-α-L-*erythro*-pentopyranosid-4-ulose gave compound **15** which, on treatment with trifluoroacetic acid, afforded bicyclic derivative **16**. The 4-epimer of **15** was also formed but on reaction with acid gave only the product of de-esterification. A similar series of reactions was also performed using lithium *tert*-butyl(methyl) acetate (LiCH(Me)CO$_2$^{t}Bu).[14] The same α-L-*erythro*-pentopyranosid-4-ulose, on treatment with the anion derived from furan followed by reaction with MCPBA, then benzoylation gave the novel oxaspiro compounds 17.[15]

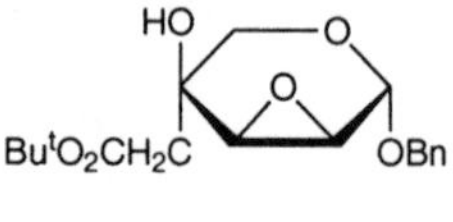

15

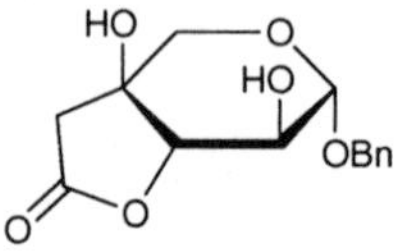

16

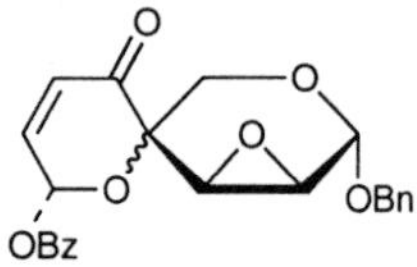

17

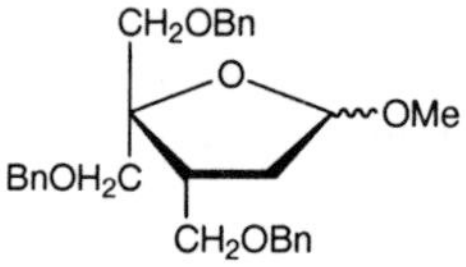

18

3-Deoxy-1,2:5,6-di-*O*-isopropylidene-3-*C*-methylene-α-D-*ribo*-hexofuranose has been converted in several steps to the branched-chain compound 18 as a precursor to 2′,3′-dideoxy-3′,4′-dihydroxymethyl pyrimidine nucleosides. The hydroxymethyl group at C-4 was introduced by way of a crossed aldol condensation of a 5-formyl derivative with formaldehyde followed by a Cannizaro reduction of the intermediate hydroxy aldehyde in a similar manner to that previously described in Vol. 13, p. 126, ref. 16.[16]

It has been observed that acetolysis of the branched 1,6-anhydropyranose sugars shown in Scheme 1, followed by saponification then dehydration affords a mixture of the starting material and a 1,6-anhydrofuranose; the proportion of the latter increases as the electronegativity of the substituent at C-4 increases, as

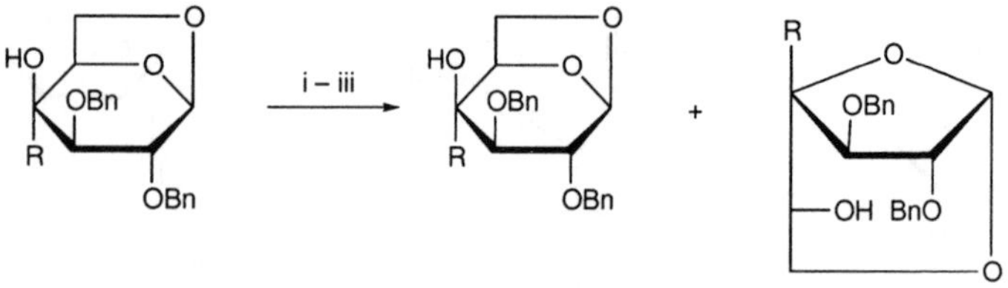

R = Me   95   :   5
R = CO$_2$Et   5   :   95

Reagents: i, TFA, Ac$_2$O; ii, NaOMe, MeOH; iii, TsOH, benzene

Scheme 1

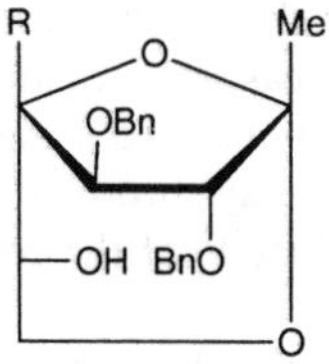

19

shown. This strategy was used to prepare di-*C*-substituted compounds **19** (R = CH$_2$=CHCH$_2$, Ph, CH$_2$OTbdps) for potential use in the synthesis of the zaragozic acids (squalostatins).[17]

2 Compounds with a C—C—C Branch-point
$$\overset{R}{\underset{N}{\mid}}$$

The hydroxylamino-containing evernitrose-derived unit, **20**, of evernitromicin can be oxidized to the nitro-derivative **21** by a peroxidase enzyme.[18]

A synthesis of L-vancosamine derivative **22**, starting from 2,6-dideoxy-β-L-*lyxo*-hexopyranoside, has been reported in which the key step is the stereoselective addition of the *in situ* prepared cerium reagent, CH$_3$CeCl$_2$, to an oxime ether intermediate.[19]

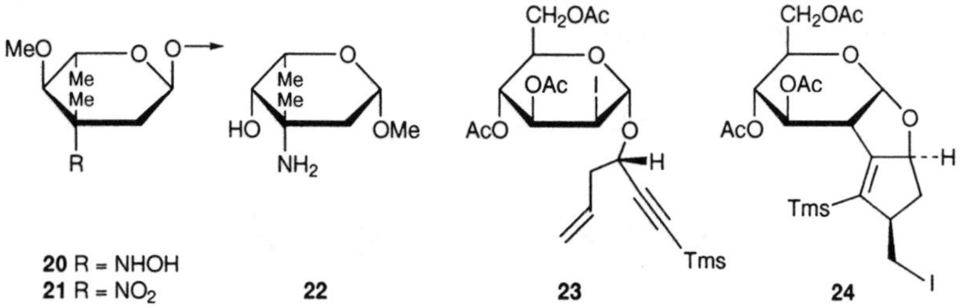

20 R = NHOH
21 R = NO$_2$

22

23

24

3 Compounds with a C—CH—C Branch-point
$$\overset{R}{\mid}$$

A general review has been published on silicon-tethered reactions with parts devoted to the preparation of branched-chain sugars (including nucleosides), through the addition of radicals generated from (bromomethyl)silyl ether tethers to double bonds or by the addition of radicals generated from phenyl selenides onto allylsilyl ether tethers.[20]

3.1   Branch at C-2. – Treating the 1,5-enyne derivative **23** with triethylboron-ethyl iodide induces a radical cascade reaction affording doubly annulated derivative **24**. A similar reaction is observed using the 1,5-diyne analogue of **23**.[21]

The light-induced reaction of tosyl bromide with the diene **25** afforded the bicyclic derivative **26**. A detailed study of the configuration at the newly created chiral centre was reported.[22]

The cyclopentene-annulated sugar derivative **27**, prepared in several steps by ring expansion of cyclobutanone **28** (itself ultimately derived from the addition of dichloroketene to a glycal) has been shown to be a useful precursor directed towards the sythesis of the ABC ring system of forskolin.[23]

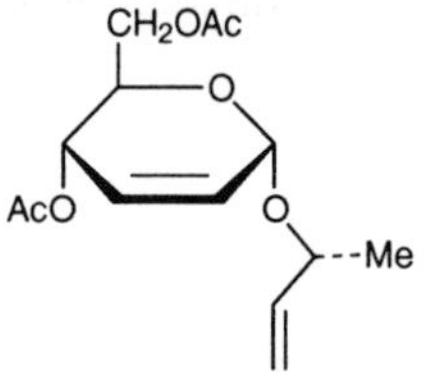

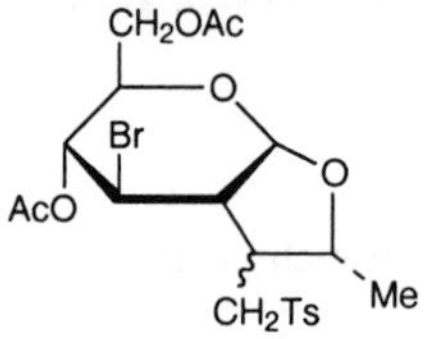

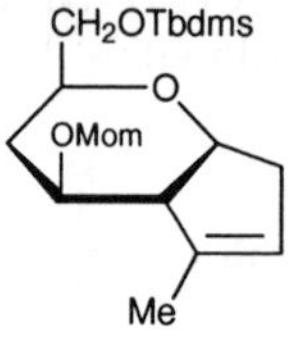

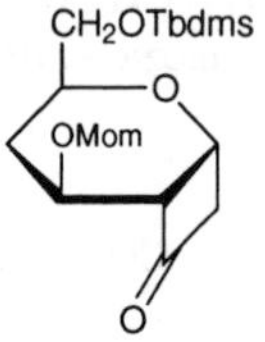

Glycals can be haloetherified (NBS/ROH) to give derivatives such as **29** (R = $CH_2C\equiv CH$, $C(Me_2)C\equiv CH$, $CH_2CH=CH_2$ or $CH_2C(Me)=CH_2$) which, on reaction with tributyltin hydride-AIBN-*tert*-butanol, undergo radical cyclization reactions. For example, **29** (R = $CH_2C\equiv CH$) gave **30**.[24]

The intramolecular [2+3] cycloaddition of the nitrone derivative **31** followed by hydrogenolysis in methanol-hydrochloric acid of the N-O bond in the inter-mediate tricyclic isoxazolidine with concomitant removal of the benzylidene group, furnished compound **32**.[25]

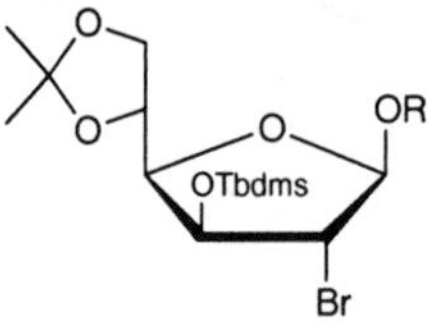

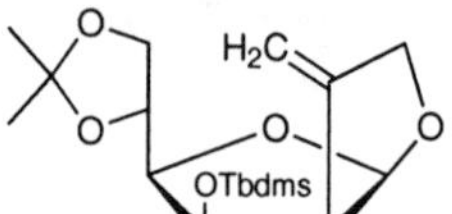

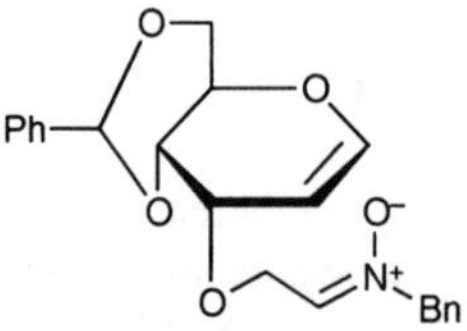

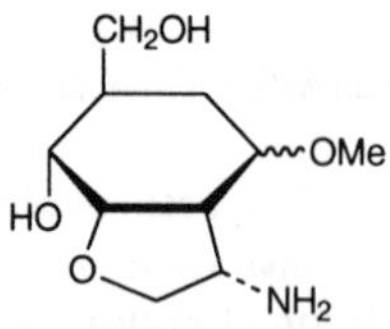

Scheme 2 illustrates a general route to 2-deoxy-2-*C*-vinyl glycosides by way of a cyclopropanation of glycals. Other configurated sugars are described, giving rise to stereochemically different products.[26]

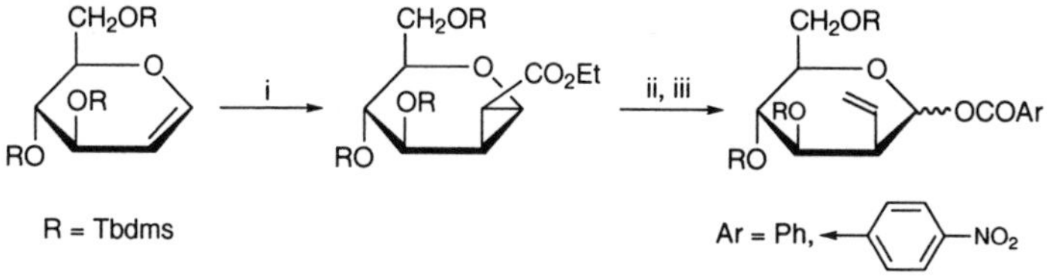

Reagents: i, N_2CHCO_2Et, Cu, MeOBu^t; ii, LAH; iii, Ph$_3$P, DEAD, ArCO$_2$H

Scheme 2

A study on the 1,2-cyclopropanation of pyranoid glycals under two different sets of conditions, zinc-diiodomethane-copper(I) chloride, or chloroform-base then lithium aluminium hydride, has been reported.[27] Similarly, glycals react with diethylzinc-diiodomethane to give cyclopropanation products which can be ring-expanded to oxepane derivatives (see also Chapter 13).[28]

The β-oxy-α-diazoester **33** can be ethylated with triethylboron and then converted through standard chemistry into the lactone **34**.[29]

Thermal cleavage of nitrate ester **35** (made from L-tartaric acid) produces dioxalanyl radicals which undergo stereoselective cyclizations to lactones **36**.[30]

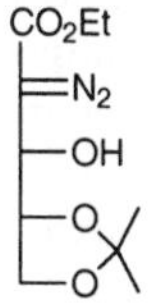

33 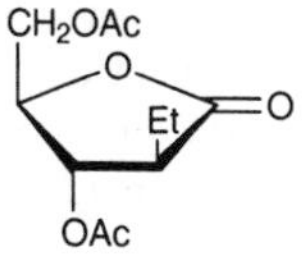   **34** 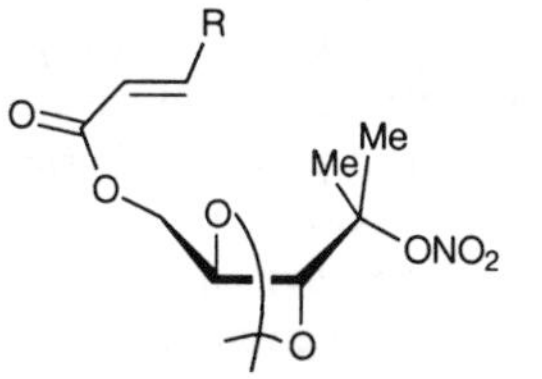   **35** R = Ph or CO$_2$Me   **36** R = Ph or CO$_2$Me

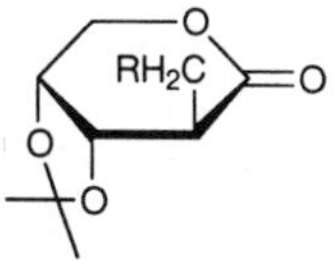

Unsaturated lactones, *e.g.* 2,3-dideoxy-D-*glycero*-1,4-pent-2-enonolactone undergo intermolecular cycloaddition with nitrone **37** to afford products, *e.g.* **38**, by way of *exo*-transition states.[31]

The reaction of benzyl-2-deoxy-2-*C*-formyl-3,4,6-tri-*O*-benzyl-α-D-glucopyranoside (see Vol. 27, p. 170, ref. 27 for its preparation) with a protected glycosyl 2-pyridyl sulfone in the presence of samarium diiodide leads to a high yield of the branched-disaccharide **39** with a 1,2-*trans*-*C*-glycoside linkage.[32] The preparation of *C*-2-branched, highly functionalized *C*-glycosides from 3-*O*-(1-alkoxy-2-bromoethyl)glycals by a radical cascade reaction is covered in Chapter 3.

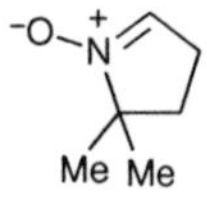

37 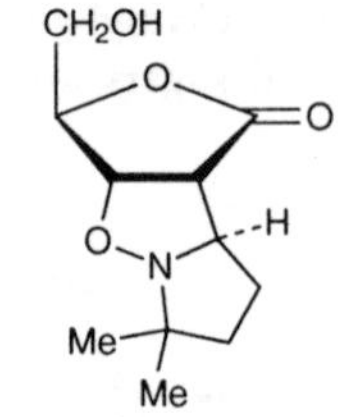   **38** 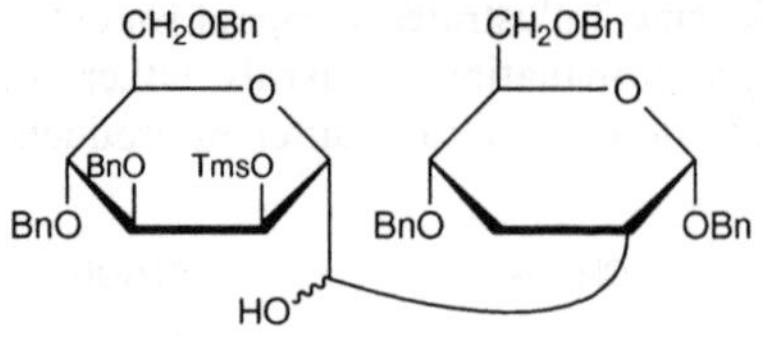   **39**

3.2   Branch at C-3. – A detailed study on the cyclization of tethered radicals **40** (X = SiMe$_2$ or CHOEt) which can react by 6-*exo*-trig or 5-*exo*-trig modes of cyclization to produce C-6- or C-3-branched derivatives, respectively, has been reported. Factors that determine which pathway is followed are discussed.[33]

The regioselective opening of various 2,3-anhydro-pyranoses with trimethyl-silyl cyanide leading to the corresponding 3-cyano-3-deoxy compounds has been described. The cyano group can be further transformed by reduction or controlled hydrolysis.[34] In a similar way reaction of 2,3-anhydropyranose derivatives with the anion derived from *tert*-butyl acetoacetate, followed by treatment with sodium hydride affords bicyclic derivative **41**, for example.[35] (See also ref.44 below). Methyl 2,3-anhydro-5-*O*-benzyl-β-D-ribofuranoside undergoes regioselective epoxide opening in a *trans*-manner with diethylaluminium cyanide to afford methyl-5-*O*-benzyl-3-cyano-3-deoxy-β-D-xylofuranoside which could be epimerized to the D-ribofuranoside with potassium cyanide in pyridine. The same reaction with the α-isomer was not as selective giving 3-cyano-3-deoxy-D-*xylo*- and 2-cyano-2-deoxy-D-*arabino*-products.[36]

Treatment of tri-*O*-acetyl-D-galactal with *p*-substituted phenols in refluxing chlorobenzene containing 5% acetic acid gave aryl 2-deoxy-D-galactopyranosides with no Ferrier rearrangement products. Further treatment of these compounds with three equivalents of boron trifluoride etherate produced the glycosyl benzopyrans **42**. Interestingly, five equivalents of Lewis acid simply caused an O→C rearrangement of the aryl glycosides.[37]

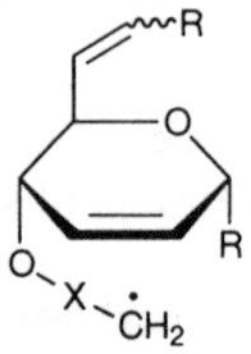

40 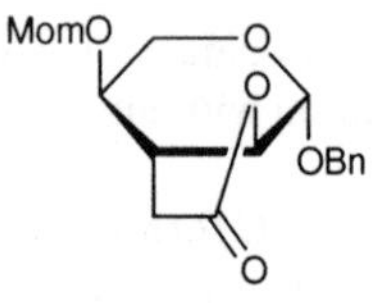   **41** 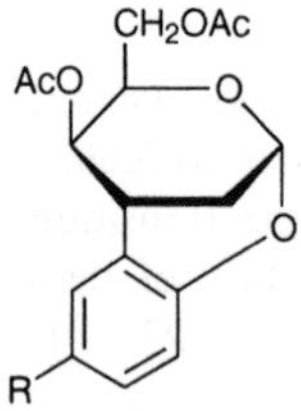   **42**

Analogues of 2-deoxy-D-*erythro*-pentose 3-phosphate have been prepared by way of alkyl radical cyclization onto vinyl phosphonates. Thus compound **43**

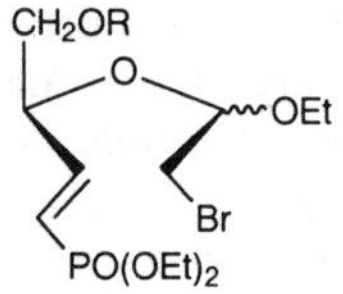

43

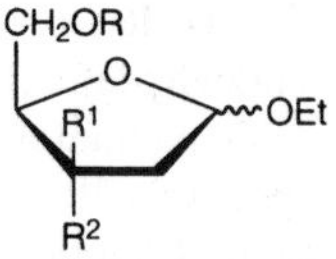

44 R^1 = H, R^2 = $CH_2PO(OEt)_2$
45 R^1 = $CH_2PO(OEt)_2$, R^2 = H

(R = Tbdms or Tbdps, prepared from D-glyceraldehyde), on treatment with tributyltin hydride-AIBN, affords compounds **44** and **45**.[38] An alternative stereospecific synthesis of the phosphonate isostere of 2-deoxy-D-*erythro*-pentose 3-phosphate is illustrated in Scheme 3.[39]

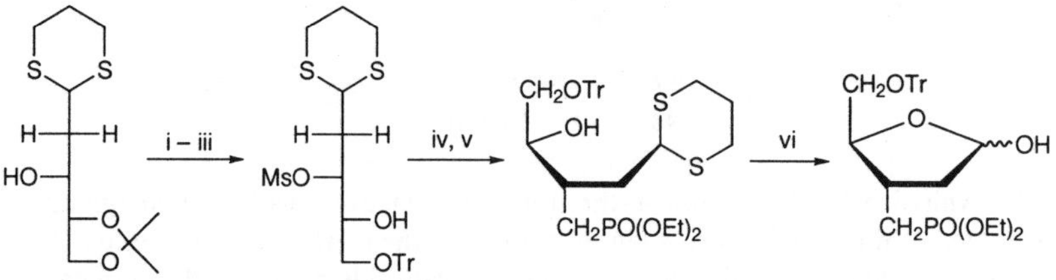

Reagents: i, MsCl, TEA; ii, H$^+$ resin; iii, TrCl, TEA; iv, NaH, THF; v, MePO(OEt)$_2$, BuLi, BF$_3$·OEt$_2$; vi, MeI, CaCO$_3$

Scheme 3

Heating 3-*O*-triflyl-pyranosides in refluxing pyridine-water-toluene gives rise to bicyclic ring-contracted derivatives as shown for methyl 2-*O*-benzyl-4,6-*O*-benzylidene-3-*O*-triflyl-β-D-glucopyranoside in Scheme 4. A higher product yield was obtained when 1,1,1,3,3,3-hexafluoropropan-2-ol was used as solvent. Fragmentation, leading to open-chain unsaturated compounds is a competing side reaction (see Chapter 13).[40]

The stereoselective 1,4-addition of alkyl radicals [$^nC_6H_{13}$, $^cC_6H_{13}$ or Ph(CH$_3$)$_3$]

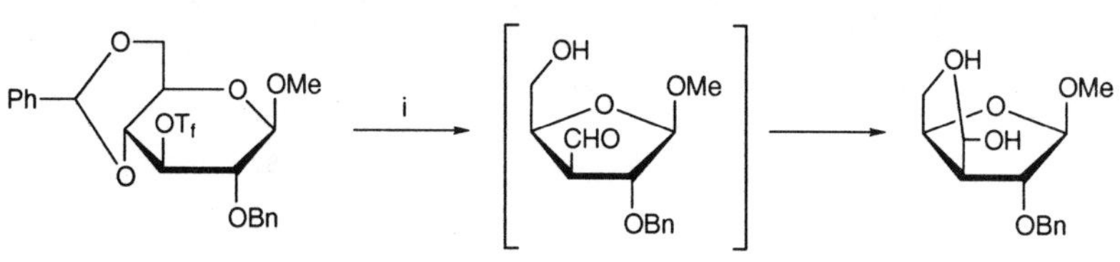

Reagents: i, Py, H$_2$O, toluene, reflux

Scheme 4

to the (Z)-unsaturated ester **46** (R = Me, Et, ^{n}Bu, ^{t}Bu and derived from 1,2-*O*-isopropylidene-D-glyceraldehyde) affords a mixture of *syn*- and *anti*-products which, on treatment with methanolic HCl, affords the D-*threo*- and D-*erythro*-lactones **47** and **48**, respectively.[41]

The synthesis of racemic 1,5-anhydrogalactofuranose **49** (with a substituted *C*-3-(furyl)hydroxymethyl branch) has been achieved starting from 7-oxanorborn-5-en-2-one.[42]

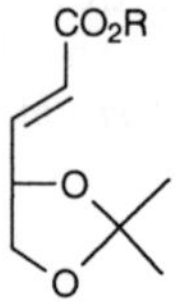

46

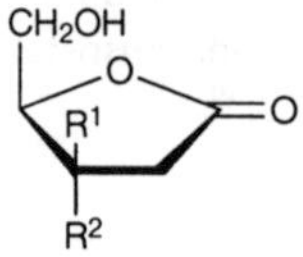

CH₂OH structure

47 R¹ = alkyl, R² = H
48 R¹ = H, R² = alkyl

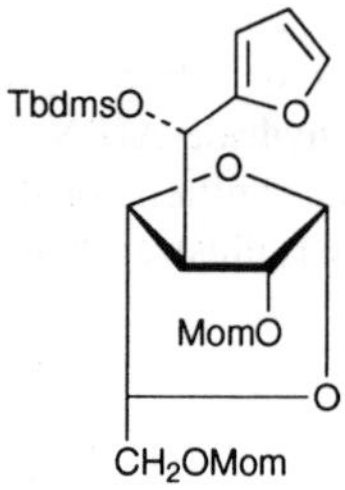

49

2,3-Anhydro-1,4-di-*O*-benzyl-L-threitol on reaction with allylmagnesium bromide, osmium tetroxide, sodium periodate then HCl affords methyl 3-*C*-(benzyloxymethyl)-2,3-dideoxy-L-*threo*-pentofuranoside as a useful intermediate for the synthesis of 2'3'-dideoxy-3'-*C*-(hydroxymethyl)-4'-thionucleosides. In an alternative route to such nucleosides the intermediate branched-sugar **50** was prepared in which the hydroxymethyl group was introduced into the 3-position by a light-induced addition-of-methanol to 6-*O-tert*-butyldimethylsilyl-D-*glycero*-pent-2-eno-1,4-lactone (see Vol. 26, p. 242, ref. 159).[43]

The synthesis of branched-chain glycals by Heck-type reaction of alkyl 4-*O*-(2-bromoalkyl)-2,3-dideoxy-α-D-*erythro*-hex-2-enosyl derivatives is covered in Chapter 13.

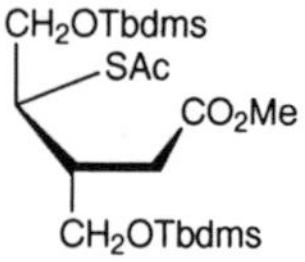

50

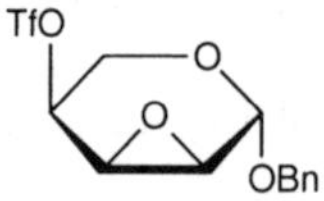

51

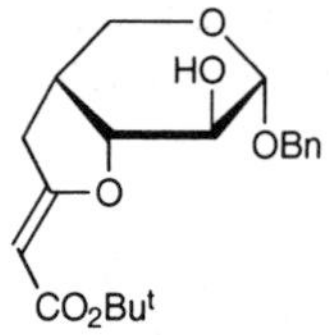

52

3.3 Branch at C-4. – The addition of the anion derived from *tert*-butyl acetoacetate to the epoxy-triflate **51** affords the bicyclic derivative **52** by a process in which the triflate is first displaced with inversion of configuration followed by epoxide ring-opening. With a less reactive leaving group, such as tosylate at position C-4 of **51**, epoxide ring opening at C-3 occurs first, followed by displacement of the tosyloxy group. (See also ref. 35 above).[44]

The unsaturated nitro-peroxy compound **53** reacts with nucleophiles at the 4-position to produce branched-nitro-epoxides **54** [R = D, 4-MeC$_6$H$_4$S, CH(Ac)$_2$].[45] The synthesis of a thromboxane B2 precursor from an intermediate *C*-4-carboxyamidomethyl derivative is covered in Chapter 24 and the preparation of 1→4-carbon linked di- and tri-saccharides is mentioned in Chapter 3.

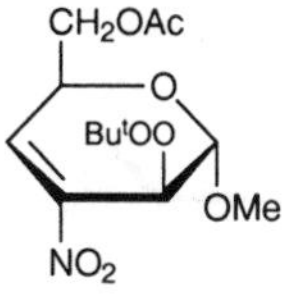

53 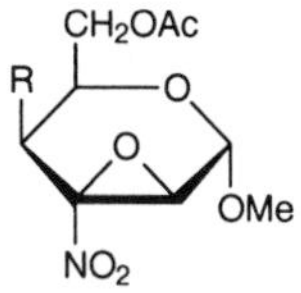   **54** 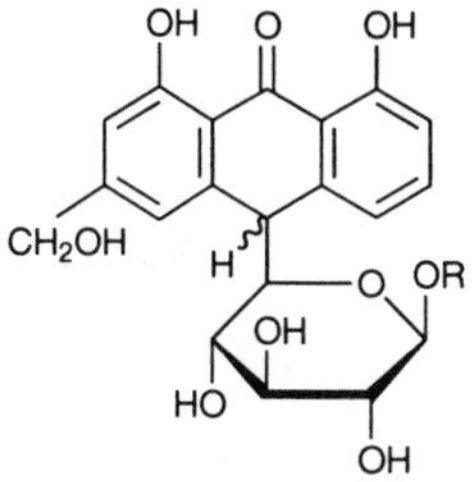   **55**

3.4 Branch at C-6. – The anthroquinone derivatives **55** (R = Ac or Bz) have been isolated from the Central American tree *Picramnia antidesma*. Uncertainties remain as to the absolute stereochemistry of the sugar unit however.[46]

4 Compounds with a $\overset{\text{R}}{\overset{\|}{\text{C—C—C}}}$ **or** $\overset{\text{R}}{\overset{|}{\text{C=C}}}$**—Branch-point**

The 1,4-diketone **56** undergoes an intramolecular aldol cyclopentaannulation reaction on treatment with potassium *tert*-butoxide, to afford **57**.[47]

The reaction of 2-*C*-hydroxymethyl glycals with phenols or phthalimide under Mitsunobu conditions to produce 2-*C*-methylene glycosides has been reported (See Vol. 27, p. 178, ref. 51 for the same reaction under Lewis acid conditions).[48]

The condensation of malononitrile and methyl 4,6-*O*-benzylidene-3-deoxy-α-D-*erythro*-hexopyranoside-2-ulose under basic conditions affords the corresponding methyl 4,6-*O*-benzylidene-2-(dicyanomethylene)-2,3-dideoxy-α-D-*erythro*-hexopyranoside. This latter compound on reaction with 3-benzyl-2-methylthio-2-thiazolinium iodide affords butadiene **58** whilst with 3-benzyl-4,5-dihydro-2-methylthio-1,3-thiazinium iodide, **59** is produced.[49]

Hydrazine adds to methyl 4,6-*O*-benzylidene-3-[bis(methylthio)methylene]-3-deoxy-α-D-*erythro*-hexopyranosid-2-ulose (see Vol. 26, p. 162, ref. 47) and its [mono(methylthio)methylene] analogue to give products **60** and **61**, respectively.[50]

An asymmetric synthesis of ethyl 2,3-dideoxy-4-*C*-methyl-3-*C*-methylene-D-*glycero*-pentofuranosides has been described. Thus, regioselective cyclization of the chiral acetal **62** (derived from 2,3-dimethyl-2-butenal) under acidic conditions,

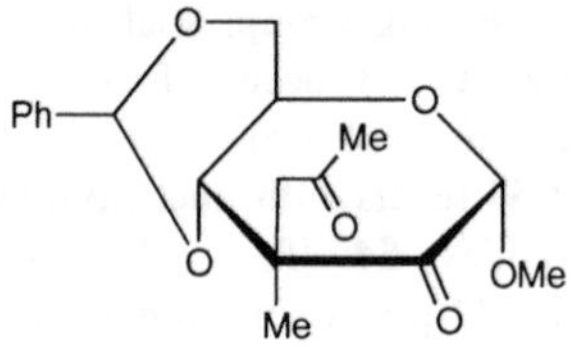

56 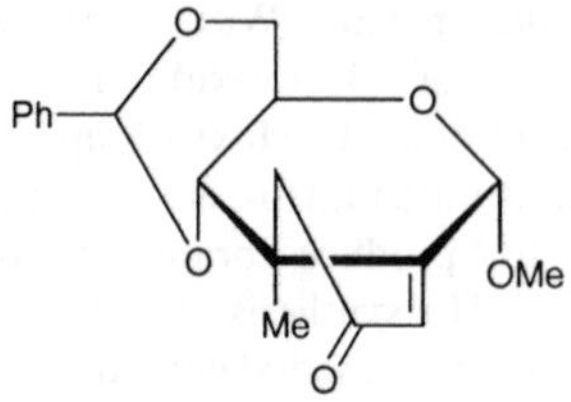   57

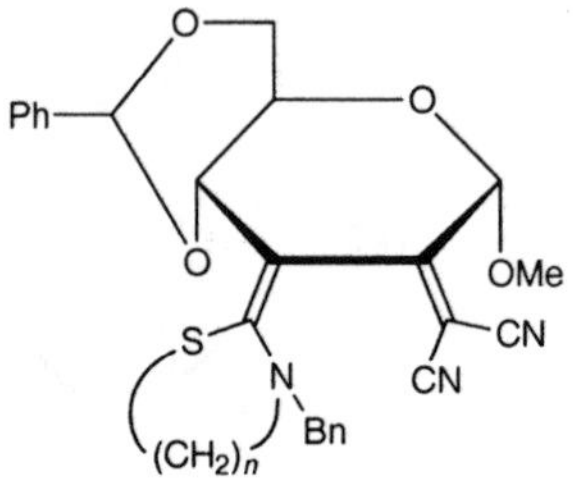

58 *n* = 2   60 R = SMe
59 *n* = 3   61 R = H

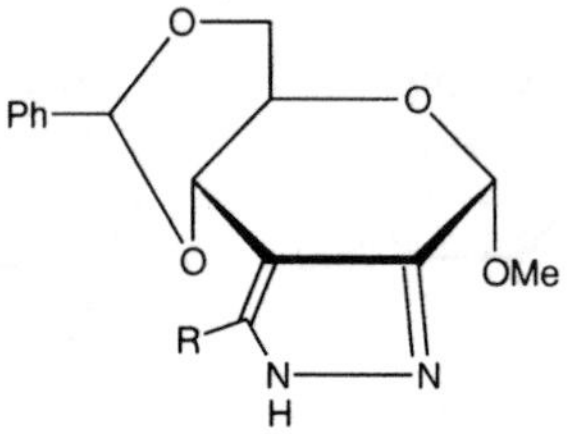

afforded **63** as a 7:3 mixture of α:β anomers.[51] The same group have also prepared 2,3-dideoxy-3-*C*-methylene-D-*glycero*-pentoses from a rearrangement of chiral epoxy-alcohols, available in four steps from methyl derivatives of 2-butenal.[52]

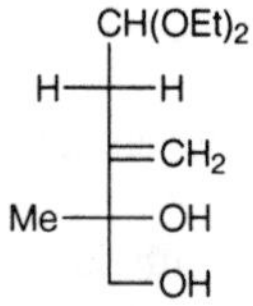

62 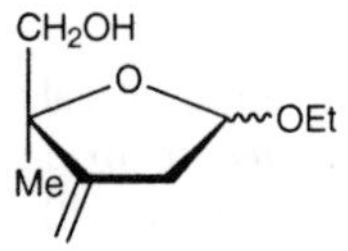   63 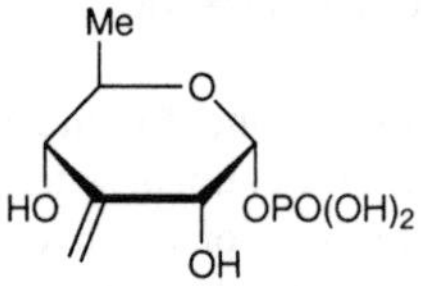   64

The 3-*C*-methylene glycosyl phosphate **64** in which the methylene group is acting as a masked keto group has been prepared in several steps from a 3-*C*-methylene-D-glucose intermediate and used to prepare a nucleoside diphosphate derivative central to the biosynthesis of di- and tri-deoxysugars.[53]

The condensation of 3'-oxo-2',5'-di-*O*-trityluridine with Wittig and related reagents which produces 3-*C*-branched unsaturated nucleosides is mentioned in Chapter 20

An improved synthesis of 3-cyano-3-deoxy- and 3-deoxy-3-formyl-hex-2-eno-pyranosides utilizes the reaction of methyl 2,3-anhydro-4,6-*O*-benzylidene-α-D-galactopyranoside with diethylaluminium cyanide (to effect epoxide ring opening with CN) followed by elimination of water to afford the 3-cyano derivative. The 3-*C*-formyl derivative is then formed from the latter by reduction with DIBAL.[54]

References

1   R. Yanagihara, J. Egashira, S. Yoshikawa and S. Osanai, *Bull. Chem. Soc. Jpn.*, 1995, **68**, 237.
2   M.S. Wolfe and R.E. Harry-O'kuru, *Tetrahedron Lett.*, 1995, **36**, 7611.
3   Z. Su and C. Tamm, *Helv. Chim. Acta*, 1995, **78**, 1278.
4   Z. Su and C. Tamm, *Tetrahedron*, 1995, **51**, 11177.
5   A.J. Fairbanks and P. Sinaÿ, *Synlett.*, 1995, 277.
6   A. El Nemr and T. Tsuchiya, *Tetrahedron Lett.*, 1996, **36**, 7665.
7   C. Scheuer-Larsen, H.M. Pfundheller and J. Wengel, *J. Org. Chem.*, 1995, **60**, 7298.
8   K. Sato, K. Suzuki and Y. Hashimoto, *Chem. Lett.*, 1995, 83.
9   T. Ton-That, *J. Carbohydr. Chem.*, 1995, **14**, 995.
10 M. Adinolfi, M.M. Corsaro, C. De Castro, R. Lanzetta, M. Parrilli, A. Evidente and P. Lavermicocca, *Carbohydr. Res.*, 1995, **267**, 307.
11 M. Adinolfi, M.M. Corsaro, C. De Castro, A. Evidente, R. Lanzetta, L. Mangon and M. Parrilli, *Carbohydr. Res.*, 1995, **274**, 223.
12 T.G. Hansson and N.A. Plobeck, *Tetrahedron*, 1995, **51**, 11319.
13 I. Dancy, T. Skrydstrup, C. Crévisy and J.-M. Beau, *J. Chem. Soc., Chem. Commun.*, 1995, 799.
14 Y. Al-Abed, T.H. Al-Tel and W. Voelter, *Nat. Prod. Lett.*, 1994, **4**, 73 (*Chem. Abstr.*, 1995, **122**, 106 315).
15 T.H. Al-Tel, R.A. Al-Qawasmeh, T. Kaiser and W. Voelter, *Tetrahedron Lett.*, 1995, **36**, 4599.
16 M. Björsne, B. Classon, I. Kers and B. Samuelsson, *Bioorg. Med. Chem. Lett.*, 1995, **5**, 43.
17 S. Caron, A.I. McDonald and C.H. Heathcock, *J. Org. Chem.*, 1995, **60**, 2780.
18 S. Kalliney and A. Zaks, *Tetrahedron Lett.*, 1995, **36**, 4163.
19 R. Greven, P. Jütten and H.-D. Scharf, *Carbohydr. Res.*, 1995, **273**, 83.
20 M. Bols and T. Skrydstrup, *Chem. Rev.*, 1995, **95**, 1253 (*Chem. Abstr.*, 1995, **123**, 83 853).
21 T.J. Woltering and H.M.R. Hoffmann, *Tetrahedron*, 1995, **51**, 7389.
22 M.P. Bertrand, I. De Riggi, C. Lesueur, S. Gastaldi, R. Nouguier, C. Jaime and A. Virgili, *J. Org. Chem.*, 1995, **60**, 6040.
23 J. Pan, I. Hanna and J.-Y. Lallemand, *Bull. Soc. Chim. Fr.*, 1994, **131**, 665 (*Chem Abstr.*, 1995, **122**, 161 066).
24 G.V.M. Sharma and K. Krishnudu, *Carbohydr. Res.*, 1995, **263**, 287.
25 A.T. Hewson, J. Jeffery and N. Szczur, *Tetrahedron Lett.*, 1995, **36**, 7731.
26 K.J. Henry Jr. and B. Fraser-Reid, *Tetrahedron Lett.*, 1995, **36**, 8901.
27 R. Murali, C.V. Ramana and M. Nagarajan, *J. Chem. Soc., Chem. Commun.*, 1995, 217.
28 J.O. Hoberg and J.J. Bozell, *Tetrahedron Lett.*, 1995, **36**, 6831.
29 F.J. López-Herrera and F. Sarabia-García, *Tetrahedron Lett.*, 1995, **36**, 2851.
30 A.S. Batsanov, M.J. Begley, R.J. Fletcher, J.A. Murphy and M.S. Sherburn, *J. Chem. Soc., Perkin Trans. 1*, 1995, 1281.
31 S. Baskaran and G.K.Trivedi, *J. Chem. Res.*, 1995, (S) 308; (M) 1853.
32 D. Mazéas, T. Skrydstrup and J.-M. Beau, *Angew. Chem., Int. Ed. Engl.*, 1995, **34**, 909.
33 J.C. Lopez, A.M. Gomez and B. Fraser-Reid, *Aust. J. Chem.*, 1995, **48**, 333.
34 S.N.-u.-H. Kazmi, Z. Ahmed, A. Malik, N. Afza and W. Voelter, *Z. Naturforsch., B: Chem. Sci.*, 1995, **50**, 294 (*Chem. Abstr.*, 1995, **123**, 9 770).

35   T.H. Al-Tel and W. Voelter, *Tetrahedron Lett.*, 1995, **36**, 523.
36   S. Watanabe, I. Kavai and M. Bobek, *J. Carbohydr. Chem.*, 1995, **14**, 685.
37   C. Booma and K.K. Balasubramanian, *Tetrahedron Lett.*, 1995, **36**, 5807.
38   T. Yokomatsu, T. Shimizu, Y. Ynasa and S. Shibuya, *Synlett.*, 1995 1280.
39   P.V.P. Pragnacharyula and E. Abushanab, *Tetrahedron Lett.*, 1995, **36**, 5507.
40   M. Kassou and S. Castillón, *J. Org. Chem.*, 1995, **60**, 4353.
41   T. Morikawa, Y. Washio, S. Harada, R. Hanai, T. Kayashita, H. Nemoto, M. Shiro
      and T. Taguchi, *J. Chem. Soc., Perkin Trans. 1*, 1995, 271.
42   K. Kraehenbuehl and P. Vogel, *Tetrahedron Lett.*, 1995, **36**, 8595.
43   J. Mann, A.J. Tench, A.C. Weymouth-Wilson, S. Shaw-Ponter and R.J. Young,
      J. Chem. Soc., Perkin Trans. 1, 1995, 677.
44   T.H. Al-Tel, M. Meisenbach and W. Voelter, *Liebigs Ann. Chem.*, 1995, 689.
45   A. Seta, K. Tokuda and T. Sakakibara, *Carbohydr. Res.*, 1995, **268**, 107.
46   P.N. Solis, A.G. Ravelo, A.G. Gonzalez, M.P. Gupta and J.D. Phillipson,
      Phytochemistry, 1995, **38**, 477.
47   A.J. Wood, P.R. Jenkins, J. Fawcett and D.R. Russell, *J. Chem. Soc., Chem.
      Commun.*, 1995, 1567.
48   N.G. Ramesh and K.K. Balasubramanian, *Tetrahedron*, 1995, **51**, 255.
49   K. Peseke, H. Feist, W. Hanefeld, J. Kopf and H. Schulz, *J. Carbohydr. Chem.*,
      1995, **14**, 317.
50   K. Peseke, G. Thiele and M. Michalik, *Liebigs Ann. Chem.*, 1995,1633.
51   Y.E. Raifeld, I.G. Taran, I.E. Mikerin, B.M. Arshava and A.A. Nikitenko, *Zh. Org.
      Khim.*, 1994, **30**, 33 (*Chem. Abstr.*, 1995, **122**, 56 323).
52   I.G. Taran, I.E. Mikerin, B.M. Arshava, G.Y. Vid, A.A. Nikitenko, L.L. Zilberg,
      Y.E. Raifeld and V.I. Shvets, *Bioorg. Khim.*, 1994, **29**, 1013 (*Chem. Abstr.*, 1995, **122**,
      187 888).
53   T. Müller and R.R. Schmidt, *Angew. Chem., Int. Ed. Eng.*, 1995, **34**, 1328.
54   R.A. Spanevello and S.C. Pellegrinet, *Synth. Commun.*, 1995, **25**, 3663.

15
Aldosuloses and Other Dicarbonyl Compounds

1 Aldosuloses

Conditions for the oxidation of sucrose by *Agrobacterium tumefaciens* have been improved and optimized on a molar scale so that 3-ketosucrose is produced in 40% yield.[1] Some 6-deoxy-D-*xylo*-hexos-4-ulose derivatives have been prepared enzymically as part of studies into the biosynthesis of 3,6-dideoxy sugars,[2,3] and the thymidine diphospho-α-D-*ribo*-hex-3-ulose 1 has been synthesized chemically for similar studies.[4] A stereospecific synthesis of methyl 3-*O*-benzoyl-6-*O*-*tert*-butyldiphenylsilyl-2-deoxy-α-D-*erythro*-hexopyranosid-4-ulose, a thromboxane B$_2$ precursor, has been achieved from D-galactose,[5] while a synthesis of D-tagatose from D-galactose proceeds *via* the interesting methyl hexos-2-ulo-2,6-pyranoside intermediate 2.[6] The 6-deoxy-β-D-*arabino*-hexofuranoside 5-ulose 3, a constituent unit of the antibiotic hygromycin, and several related analogues, have been synthesized from 6-deoxy-β-D-glucofuranosides *via* a 2,3-epoxide and selective oxidation at *C*-5.[7]

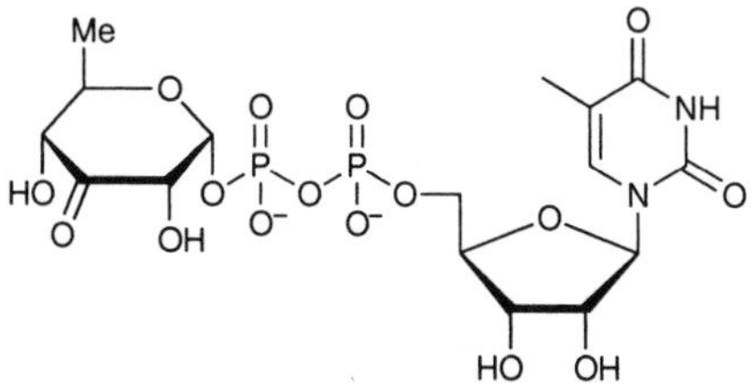

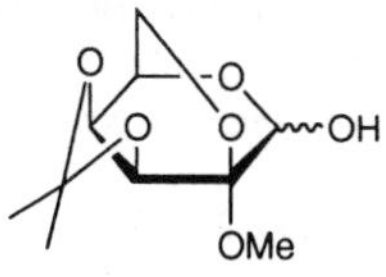

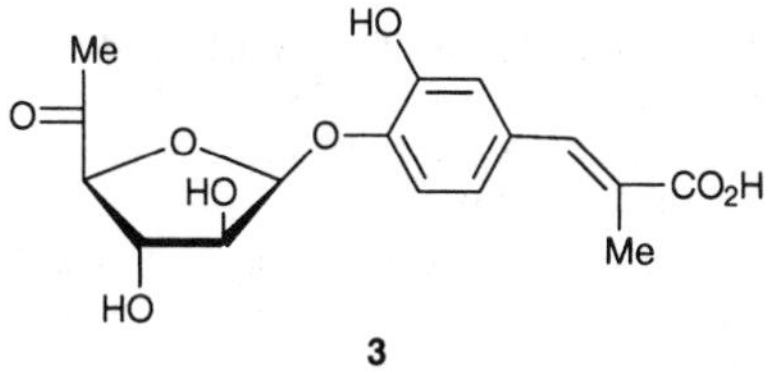

Treatment of the 3-deoxyhexos-2-ulose 4 with trimethylsilyldiazomethane (Scheme 1) has afforded a mixture of epoxides and ring-expanded products.[8] A simple bench-top procedure for the preparation of levoglucosenone has been described.[9]

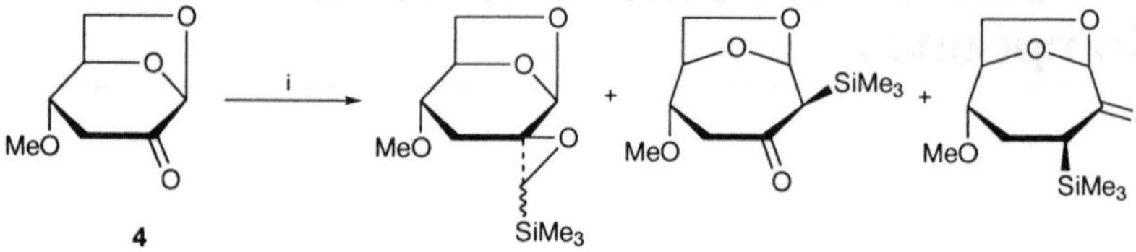

Reagent; i, Me₃SiCHN₂, BF₃·OEt₂

Scheme 1

2   Other Dicarbonyl Compounds

Crude enzyme extracts from the white rot fungus *Oudemansiella mucida* oxidized D-glucose to D-*erythro*-hexos-2,3-diulose,[10] and the diulose phosphonate **5** has been prepared from 2,3:5,6-di-*O*-isopropylidene-D-mannono-1,4-lactone.[11]

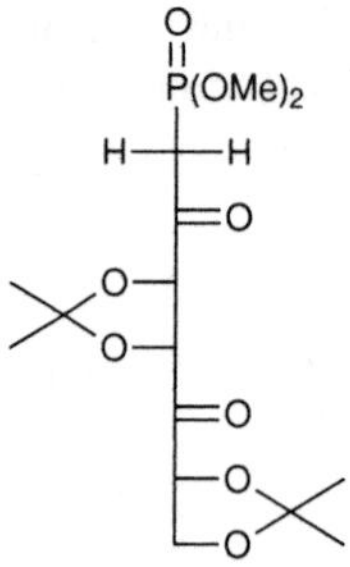

5

References

1   C. Simiand, E. Samain, O.R. Martin and H. Driguez, *Carbohydr. Res.*, 1995, **267**, 1.
2   P.A. Pieper, Z. Guo and H.-W. Liu, *J. Am. Chem. Soc.*, 1995, **117**, 5158.
3   A. Stein, M.-R. Kula, L. Elling, S. Verseck and W. Klaffke, *Angew. Chem., Int. Ed. Engl.*, 1995, **34**, 1748.
4   T. Muller and R.R. Schmidt, *Angew. Chem., Int. Ed. Engl.*, 1995, **34**, 1328.
5   O. Moradei, C. du Mortier, A. Fernández Cirelli and J. Thiem, *J. Carbohydr. Chem.*, 1995, **14**, 525.
6   P.L. Barili, G. Berti, G. Catelani, F.D'Andrea and L. Miarelli, *Carbohydr. Res.*, 1995, **274**, 197.

7   J.G. Buchanan, D.G. Hill, R.H. Wightman, I.K. Boddy and B.D. Hewitt, *Tetrahedron*, 1995, **51**, 6033.

8   T. Kawai, M. Isobe and S.C. Peters, *Aust. J. Chem.*, 1995, **48**, 115.

9   C. Marin, *Front. Biomed. Biotechnol.*, 1994, 17 (*Chem. Abstr.*, 1995, **122**, 240 188).

10 J. Volc, P. Sedmera, V. Havlicek, V. Prikrylova and G. Daniel, *Carbohydr. Res.*, 1995, **278**, 59.

11 R. Csuk and P. Dorr, *J. Carbohydr. Chem.*, 1995, **14**, 35.

16
Sugar Acids and Lactones

1   Aldonic Acids and Lactones

A new convenient and high yielding oxidation of aqueous solutions of aldopentoses to the corresponding aldonolactones utilises Pd/C and oxygen in the presence of one equivalent of magnesium hydroxide.[1] Gluconic acid and glucitol were obtained simultaneously in 90% yields by paired electrolysis of glucose with a lead sheet cathode and a dimensionally stable anode in a press filtration diaphragm.[2] Some phenacyl glycosides (e.g. 1) underwent Norrish type II photochemical reaction to give lactones (e.g. 2) on photolysis with pyrex filtered UV light,[3] while hydrogenation of L-ascorbic acid (Rh/C, H_2) has afforded L-gulono-1,4-lactone.[4]

A rhodium catalyst has been used in the oxidation of free sugars to, initially, the corresponding 1,5-lactones which in DMF as the reaction solvent, isomerize to the 1,4-lactones.[5] The rate of oxidation of 2-deoxy-D-glucose by Cr(VI) reagents in perchloric acid to give 2-deoxy-D-gluconic acid has been studied. Absence of OH at C-2 influences the stability of the chromic ester intermediate and leads to differences in kinetic behaviour between glucose and 2-deoxyglucose.[6] In the hydrothermolysis of hexoses and oligomers thereof (e.g. glucose, cellobiose, cyclodextrin, fructose) small amounts of 3-deoxy-D-hexonic acids 3 are formed.[7] Ozonolysis of the *exo*-glycal 4 derived from L-sorbose has afforded the L-xylonic acid lactone 5.[8]

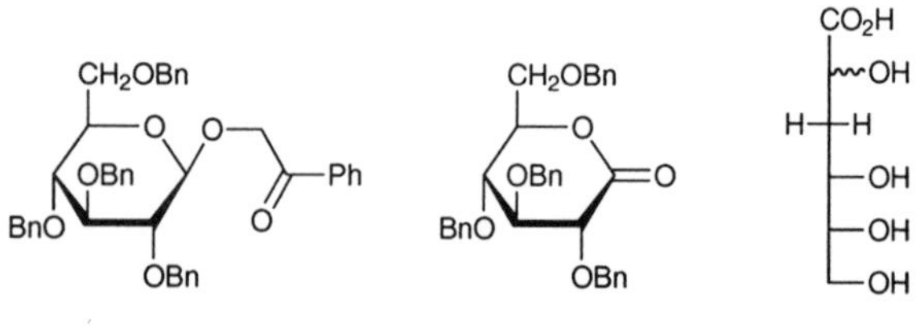

The lactone moiety 6 found in mevinic acids has been synthesized from L-malic acid using a chiral sulfoxide reagent to control the stereochemistry at C-3.[9,10] Synthesis of the negamycin lactone 7 has been reported,[11] while a lactone precursor of L-acosamine and L-daunosamine is discussed in Chapter 9.

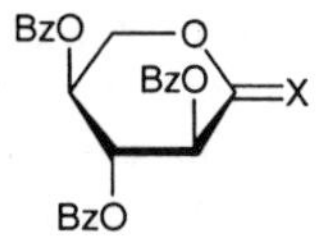

4 X = CH₂
5 X = O

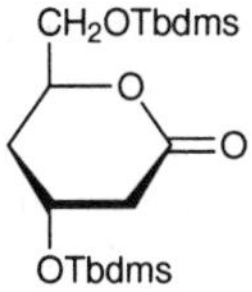

6

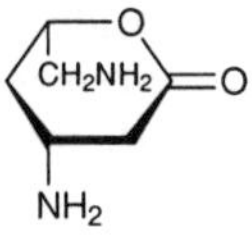

7

The isomeric 2-deoxy-1,4-lactones **8** and **9** have been synthesized in a multi-step procedure from levoglucosenone,[12] while the branched lactones **10** and **11** have been prepared by radical addition [C₆H₁₃I or Ph(CH₂)₃I, Bu₃SnH, AIBN] to an unsaturated aldonic acid ester.[13] The lactone **12**, an intermediate in the synthesis of pseurotin A, has been prepared from D-glyceraldehyde in a multi-step procedure.[14] Thermally-induced cleavage of nitrate ester **13** [derived from

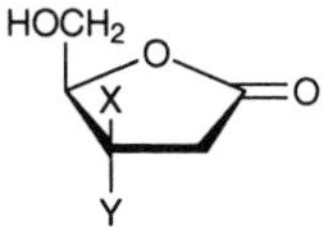

8 X = OH, Y = H
9 X = H, Y = OH
10 X = C₆H₁₃, Ph(CH₂)₃, Y = H
11 X = H, Y = C₆H₁₃, Ph(CH₂)₃

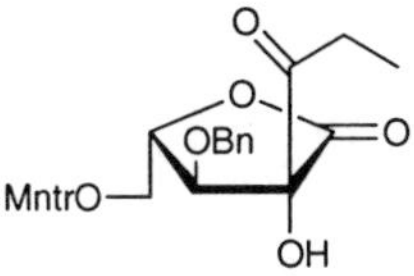

12

(+)-dimethyl tartrate] produced a dioxalanyl radical which underwent stereoselective cyclization to the lactone **14**,[15] while branched lactones **15** have been synthesized by cycloaddition of a chiral 2-alkoxyacrylonitrile to 2,4-dimethylfuran.[16] The synthesis of β-hydroxy-α-aminoacids, such as polyoxamic acid, has been achieved from O-protected acyclic tetroses,[17] and other β-hydroxy-α-amino acid derivatives are covered in Chapters 9 and 10. The ring opening of a 2,3-aziridino-2,3-dideoxy-aldono-1,4-lactone derivative with both hard and soft nucleophiles has been evaluated – with different proportions of regioisomers produced in each case.[18]

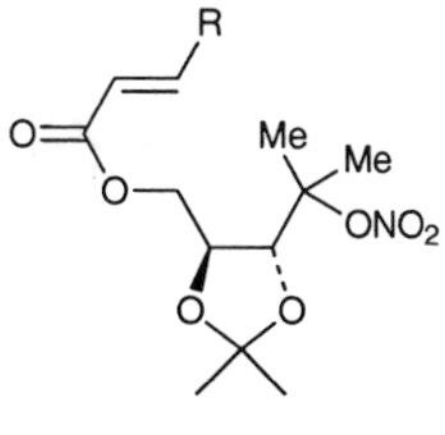

13   R = Ph, CO₂Me   **14** 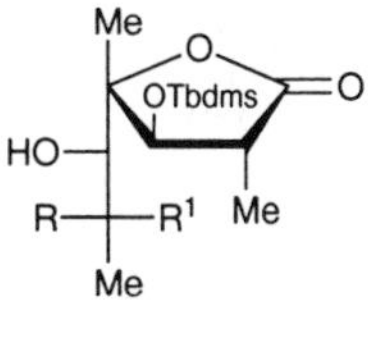   **15** R, R¹ = H, OH

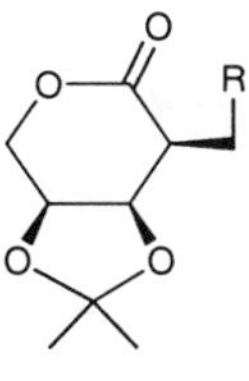

Treatment of 2-deoxy-lactones **16** with base [KN(Tms)$_2$, $-90\,°$C], followed by trimethylsilyl azide generated the 2-azido-2-deoxy-lactones **17**,[19] while 5,6-*O*-isopropylidene-D-mannono-1,4-lactone has afforded, with trifluoromethanesulfonic anhydride, the 2-*O*-monotriflate or, under more forcing conditions, the corresponding 2-*O*-triflyl-3-deoxy-pent-2-enono-1,4-lactone. Both 2-deoxy- and 2,3-dideoxy-derivatives were prepared from these compounds and analogous reactions were performed on D-gulonolactone and L-mannonolactone.[20] Radical oxidative decarboxylation of α-hydroxylactones (e.g. **18**) has been effected by (diacetoxyiodo)benzene to give, in this case, methyl ketone **19**.[21] A new reagent system for the reduction of lactones to lactols (in which the active ingredient is thought to be Cp$_2$TiH) is discussed in Chapter 2.

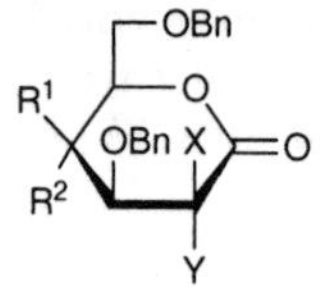

16 R^1 = OBn, R^2 = H } X, Y = H
    R^1 = H, R^2 = OBn }
17 R^1 = OBn, R^2, X = H, Y = N$_3$
    R^2 = OBn, R^1, Y = H, X = N$_3$

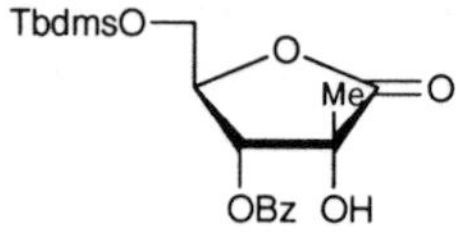

18

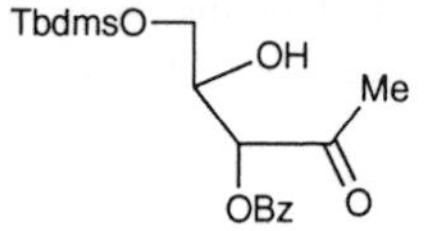

19

A route to 1,5-dideoxy-1,5-imino-D-xylonolactam has utilized a 5-azido-5-deoxy-*aldehydo*-D-xylose derivative,[22] and a route to *C*-glycosides by reaction of aldonolactone derivatives with alkyl lithiums followed by triethylsilane reduction is elaborated in Chapter 3.

2   Anhydroaldonic Acids and Lactones

Some *O*-tosyl-lactones (e.g. **20** and **21**) have been cyclized to tetrahydrofuran derivatives (**22** and **23** respectively) on boiling in dioxane-water[23] and synthesis of 2-deoxy-4-*epi*-neuraminic acid and 2,4-dideoxy-neuraminic acid have been recorded.[24] Liposomes functionalized with neuraminic acid by way of a *C*-

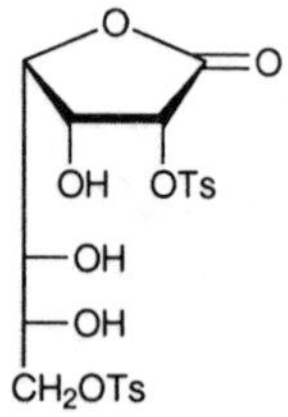

20 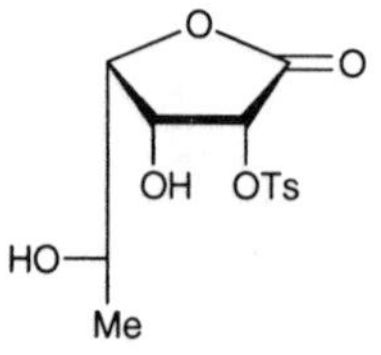   **21** 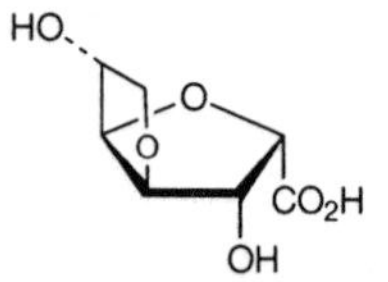   **22** 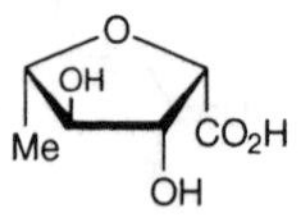   **23**

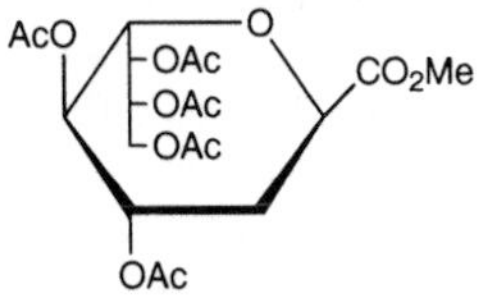

24 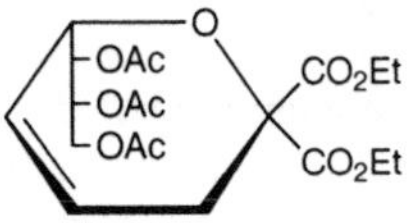 **25** 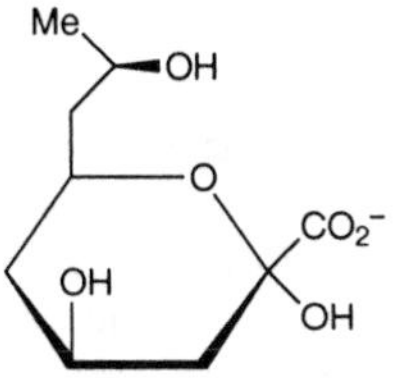 **26**

glycoside linkage have been prepared and tested for viral binding properties.[25] A Diels-Alder reaction has been used as the basis of a multi-step synthesis of Kdn analogue **24** by way of alkene **25**.[26]

3 Ulosonic Acids

A large number of aldoses and derivatives have been tested as substrates for Neu5Ac aldolase which adds pyruvic acid to the aldose. A free hydroxyl group is required at C-3 of the aldose, and aldoses with the *S*-configuration at C-3 lead to products with the *S*-configuration at C-4 in the product.[27] An enzymic method for the conversion of *N*-acetyl-D-mannosamine and D-mannose into neuraminic acid and Kdo respectively has been improved and simplified, and the reactions were carried out on a 10 g scale.[28] *N*-Acetyl-4-azido-4-deoxy-D-mannosamine has been converted with an aldolase into 5-acetamido-7-azido-3,5,7-trideoxy-D-*glycero*-D-*galacto*-2-nonulopyranosonic acid.[29]

Sequential aldol trimerisation of acetaldehyde catalysed by aldolases followed by enzymic addition of pyruvic acid has afforded the tetradeoxy-nonulopyrano-sonic acid **26**,[30] and the preparation of some 3-deoxy-octulosonic acids and 3-deoxy-heptulosonic acids is discussed in Chapter 2.

The base-induced β-elimination of protected aldono-1,4-lactones followed by sequential treatment with acidic methanol and aqueous base has afforded sodium salts of 3-deoxy-ald-2-ulosonic acids (Scheme 1).[31]

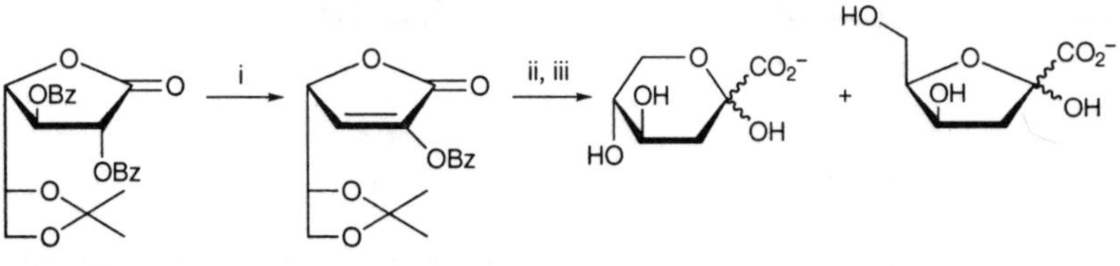

Reagents: i, Et₃N, CH₂Cl₂; ii, HCl, MeOH; iii, aq. NaHCO₃

Scheme 1

A synthesis of 3-deoxy- D-*glycero*-D-*galacto*-2-nonulosonic acid from D-mannose involved use of a keto-thiazole Wittig reagent,[32] while addition of lithiated thiazole to a 2-deoxy-hexono-1,4-lactone featured in a synthesis of 3-deoxy-D-*arabino*-hept-2-ulosonic acid.[33]

In a further example of indium-mediated additions to aldoses, α-(bromomethyl)acrylic acid added to *N*-acetylmannosamine in the presence of indium to give branched derivative **27** and its *C*-4 epimer. Ozonolysis of **27** afforded *N*-acetylneuraminic acid.[34] In a study of the oxidation of L-sorbose (5% Pt/Al$_2$O$_3$, O$_2$) to 2-keto-L-gulonic acid it was found that the reaction rate and selectivity was improved in the presence of certain tertiary amines.[35] The synthesis of a carbocyclic analogue of *N*-acetyl-neuraminic acid is discussed in Chapter 18.

A synthesis of the neuraminidase inhibitor 4-deoxy-4-guanidino-neuraminic acid derivative **28** from neuraminic acid has been effected and a number of *N*-substituted guanidines were also prepared,[36] while further chain-shortened analogues of **28** have been prepared without C-9 then C-8 and then C-7[37]. Other analogues **29** have been prepared from Kdn[38] and the 5-deoxy-derivative **30** has been synthesized from 2-deoxy-D-*arabino*-hexose by way of an aldolase-catalysed reaction with pyruvic acid.[39] Neuraminic acid has also been converted into the 3,4-dideoxy-4-guanidino derivative **31**.[40]

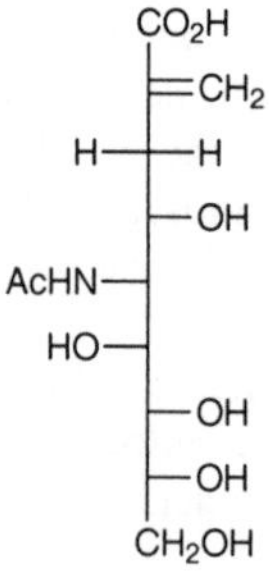

27 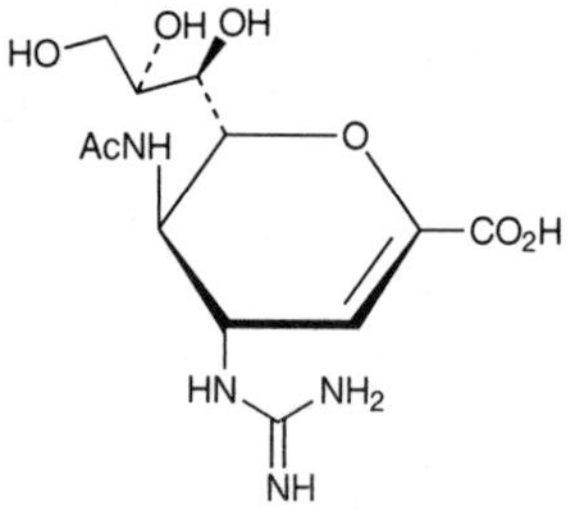   28 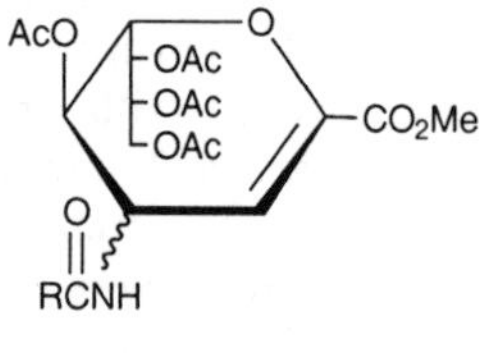   29 R = Me, Ph, Cl$_3$C, Et

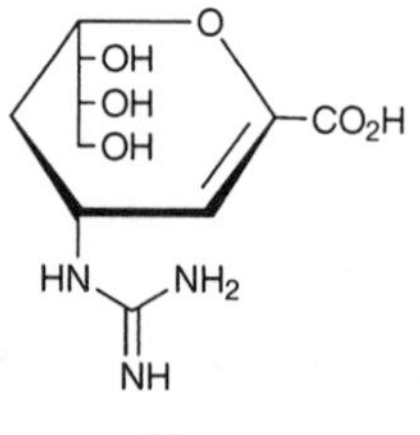

30 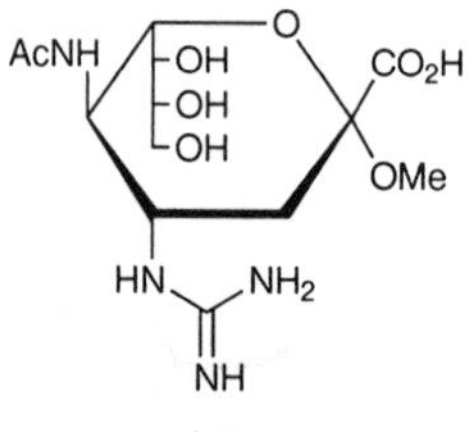   31

Some photoreactive CMP-neuraminic acids **32** have been prepared as substrates for α-(2→6)-sialyltransferases,[41] and synthesis of some thio-umbelliferyl glycosides of ulosonic acids is mentioned in Chapter 3. The ethyl thioglycoside **33** was prepared in 47% overall yield from neuraminic acid and used as a glycosyl donor with good results – selectively forming α-glycosides. A number of sialylated di- and tri-saccharides were synthesized.[42] The 3-deoxy-D-*lyxo*-2-heptulosaric acid derivative **34** has been synthesized from 2,3,4,6-tetra-*O*-acetyl-β-D-galacto-pyranosyl cyanide *via* 1-cyano-D-galactal.[43]

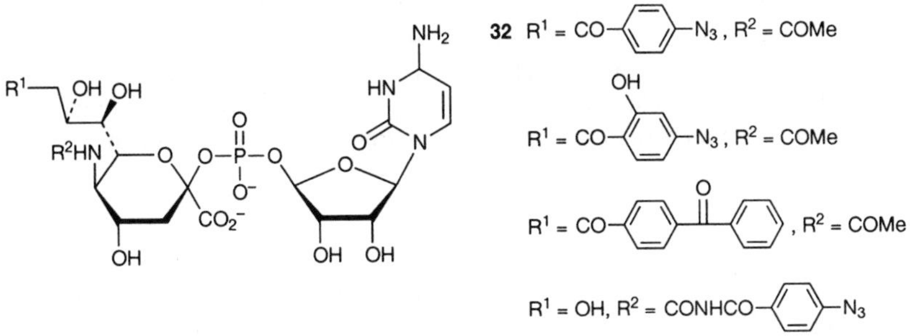

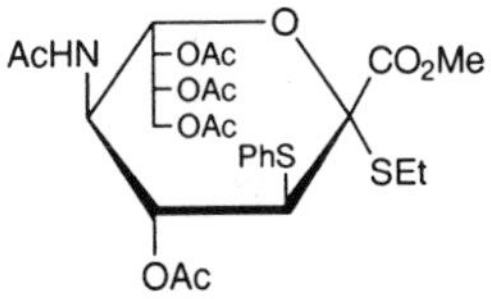

N-Glycolyl-8-*O*-sulfoneuraminic acid, a component of a sea urchin sialosphin-golipid, has been synthesized from neuraminic acid *via* its *O*-benzyl glycoside,[44] and some neuraminic acid lipid conjugates have been prepared as neuritogenic agents by sialylation of lipid alcohols.[45]

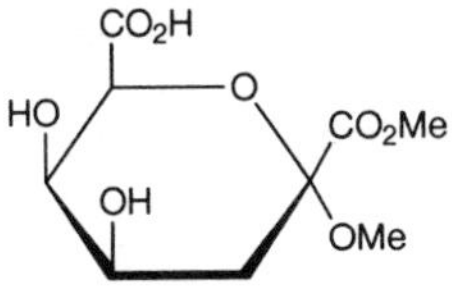

The α- and β-glycopyranosyl phosphates of 3-deoxy-D-*manno*-2-octulosonic acid have been prepared from Kdo as inhibitors of 3-deoxy-D-*manno*-2-octuloso-nate-8-phosphate synthase, which occurs in Gram-negative bacteria.[46] *N*-Acetyl-neuraminic acid has been *O*-alkylated at C-4 by selective alkylation of an 8,9-*O*-isopropylidene derivative,[47] and treatment of the α-benzyl glycoside of *N*-acetylneuraminic acid with fluoroacetyl imidazole has afforded a mixture of 4,5- and 4,9-di-*O*-(fluoroacetyl) derivatives.[48] A 4-acetamido-4-deoxy-4-*epi*-derivative of *N*-acetylneuraminic acid has been prepared using standard techniques.[49]

4   Uronic Acids

Ethyl 3-*O*-benzyl-4,6-*O*-benzylidene-1-thio-β-D-glucopyranoside was a key starting material in the synthesis of methyl (ethyl 2-*O*-acyl-3,4-di-*O*-benzyl-1-thio-β-D-glucopyranosid)uronate, with acyl groups being Ac, Bz, pivaloyl and anisoyl. These compounds were used as glycosyl donors in the synthesis of β-D-glucuronide disaccharides. The 2-*O*-benzoate afforded highest yields.[50] 1,2-*O*-

Isopropylidene-D-glucurono-6,3-lactone has been converted, *via* epimerization at C-5, into the specifically substituted-L-iduronic acid derivative **35**. This compound has potential as a synthon for making glycosaminoglycan fragments[51]

Glycosides of unprotected D-uronic acids have been formed directly in the appropriate alcohol with boron trifluoride etherate as catalyst, affording D-glycosiduronates.[52] A morphine-6-glucuronide analogue has been prepared for use as a hapten in radioimmunoassay, using a glucuronide trichloroacetimidate as glycosyl donor.[53] Synthesis of glucuronides of metronidazole and its hydroxy metabolites has been achieved using the UDP-glucuronosyl transferase activity of rat liver microsomes with disodium UDP-glucuronic acid as glycosyl donor.[54]

35   **36**   **37**

Aldol condensation of an 'acetyl iron' anion with aldehyde-sugar derivatives (e.g. **36**) and decomplexation of the iron led to chain-extended deoxy-uronic acids (e.g. **37** and its *C*-5 epimer),[55] while addition of 2-trimethylsilylethynylmagnesium bromide to the 6-aldehyde derived from methyl 2,3,4-tri-*O*-benzyl-β-D-glucopyranoside followed by a multi-step procedure has afforded the amino-uronic acid **38**, the α-amino-acid on which miharamycin A is based.[56]

A kinetic study of the oxidation of glycosides to glycuronoside sodium salts with the oxidant hypochlorite, bromide and catalytic TEMPO (2,2,6,6-tetramethyl-1-piperidinyloxy radical) at alkaline pH, revealed the *C*-6 aldehyde hydrate as an intermediate which is oxidized rapidly to the carboxylate, and provided a mechanistic interpretation of the selectivity for primary over secondary alcohol groups in the oxidation.[57] Other authors have used the same conditions for a practical synthesis of methyl 4-*O*-methyl-α-D-glucopyranosiduronic acid from methyl 4-*O*-methyl-α-D-glucopyranoside,[58] and yet another group have used these conditions to oxidize unprotected glycopyranosyl azides to the corresponding D-glycopyranuronosyl azide.[59] Highly stable ruthenium-2-(phenyl)azopyridine catalysts effect oxidation of octyl α-D-glucopyranoside in aqueous conditions, with sodium hypobromite as a cooxidant, to octyl α-D-glucopyranosiduronic acid.[60]

Four chiral liquid crystalline compounds, **39** and **40**, were synthesized from D-glucose by standard means. It was determined that **39** (X = OC_7H_{15}) has the properties essential for ferroelectric liquid crystals.[61]

Baeyer-Villiger oxidation of the Ferrier carbocyclization products **41** and **42** derived from glucose, gave isomeric 5-deoxy-hexofuranosiduronic acids **43** and

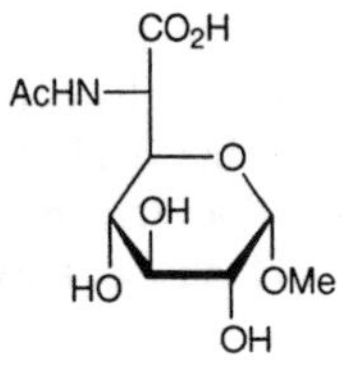

38

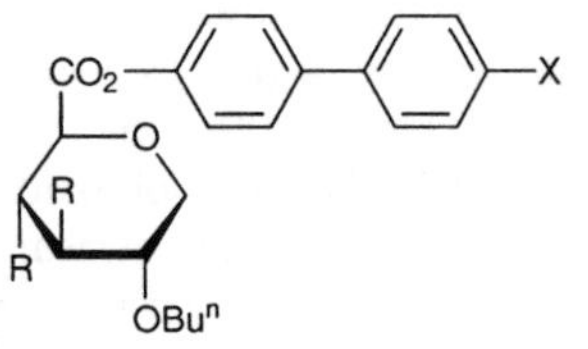

39 R = H, X = OC_7H_{15} or CN
40 R = OMe, X = $OC_{17}H_{15}$ or CN

44 respectively after acid catalysed rearrangement of the ring expanded initial products (Scheme 2).[62] 3-Deoxy-1,2-*O*-isopropylidene-α-D-*erythro*-pentofuran-*N*-methyl-uronamide has been synthesized and transformed into 3'-deoxy-nucleoside derivatives as adenosine receptor antagonists,[63] and synthesis of a 4-*C*-methyl branched glucuronamide is discussed in Chapter 14.

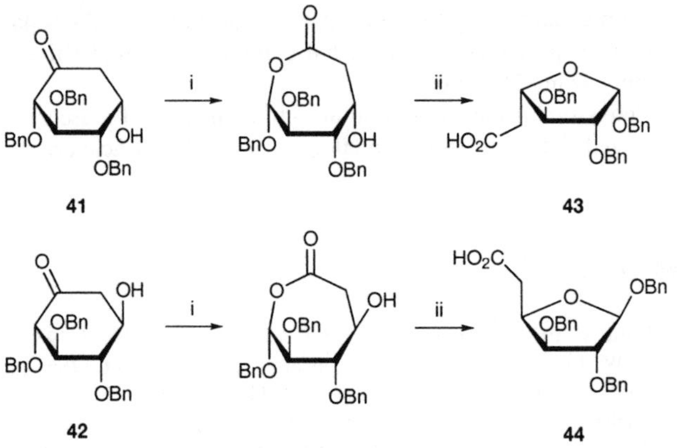

Reagents: i, MCPBA; ii, TsOH, CH_2Cl_2

Scheme 2

5   Ascorbic Acids

5,6-*O*-Cyclic acetals of L-ascorbic acid have been prepared in good yield by the transacetalation of L-ascorbic acid with various ketone dimethylacetals,[64] and a furanyl analogue of dehydro-L-ascorbic acid **45** has been synthesized by condensation of ethyl acetoacetate with D-glucose followed by periodate oxidation.[65]

The kinetics of oxidation of L-ascorbic acid by the mononuclear complex $Co(NH_3)_5H_2O(ClO_3)_3$ have been studied spectrophotometrically. The mechanism involves single electron transfers involving the pentamine hydroxo complex and the ascorbate anions, with subsequent formation of ascorbate radicals and Co(II).[66] Others have studied the kinetics of L-ascorbic acid oxidation by peroxynitrite,[67] and the rates of oxidation of L-ascorbic acid by copper(II)-polyamine complexes have been measured,[68] while the kinetics of L-ascorbic acid oxidation by nitrous acid have been assessed.[69]

45   **46**   **47**

The reaction of 5,6-O-isopropylidene-L-ascorbic acid with superoxide followed by methylation of carboxyl groups has afforded oxidation products **46** and **47**. In contrast, 5,6-O-isopropylidene-3-O-methyl-L-ascorbic acid under the same conditions gave only **47** in 82% yield.[70]

Some 6-deoxy-6-N-trimethylammonium salt analogues of L-ascorbic acid have been synthesized,[71] and carbocyclic analogues of L-ascorbic acid are mentioned in Chapter 18.

References

1   J.F. Witte, R. Frith and R.W. McClard, *Carbohydr. Lett.*, 1994, **1**, 123.
2   H. Li, W. Li, Z. Cuo, D. Gui, S. Cai and A. Fujishima, *Collect. Czech. Chem. Commun.*, 1995, **60**, 928 (*Chem. Abstr.*, 1995, **123**, 286 439).
3   J. Brunckova and D. Crich, *Tetrahedron*, 1995, **51**, 11945.
4   D.S. Soriano, C.A. Meserole and F.M. Mulcahy, *Synth. Commun.*, 1995, **25**, 3263.
5   I. Isaac, I. Stasik, D. Beaupère and R. Uzan, *Tetrahedron Lett.*, 1995, **36**, 383.
6   M. Rizzotto, S. Signorella, M. I. Frascaroli, V. Daier and L.F. Sala, *J. Carbohydr. Chem.*, 1995, **14**, 45.
7   G.C.A. Luijkx, F. van Ratwijk, H. van Bekkum and M.J. Antal Jr., *Carbohydr. Res.*, 1995, **272**, 191.
8   F.W. Lichtenthaler, S. Hahn and E.-J. Flath, *Liebigs Ann. Chem.*, 1995, 2081.
9   J. Tang, I. Brackenridge, S.M. Roberts, J. Beecher and A.J. Willetts, *Tetrahedron*, 1995, **51**, 13217.
10   J. Beecher, I. Brackenridge, S.M. Roberts, J. Tang and A.J. Willetts, *J. Chem. Soc., Perkin Trans. 1*, 1995, 1641.
11   D. Socha, M. Jurczak and M. Chmielewski, *Tetrahedron Lett.*, 1995, **36**, 135.
12   K. Matsumoto, T. Ebata, K. Koseki, K. Orano, H. Kawakami and H. Matsushita, *Bull. Chem. Soc. Jpn.*, 1995, **68**, 670.
13   T. Morikawa, Y. Washio, S. Harada, R. Hanai, T. Kayashita, H. Nemoto, M. Shiro and T.Taguchi, *J. Chem. Soc., Perkin Trans. 1*, 1995, 271.
14   Z. Su and C. Tamm, *Helv. Chim. Acta*, 1995, **78**, 1278.

15 A.S. Batsanov, M.J. Begley, R.J. Fletcher, J.A. Murphy and M.S. Sherburn, *J. Chem. Soc., Perkin Trans. 1*, 1995, 1281.

16 P. Kernen and P. Vogel, *Helv. Chim. Acta*, 1995, **78**, 301.

17 R.F.W. Jackson, N.J. Palmer, M.J. Wythes, W. Clegg and M.R.J. Elsegood, *J. Org. Chem.*, 1995, **60**, 6431.

18 P. Dauban, L. Dubois, M.E.T.H. Dau and R.H. Dodd, *J. Org. Chem.*, 1995, **60**, 2035.

19 F.-Y. Dupradeau, S.-I. Hakomori and T. Toyokuni, *J. Chem. Soc., Chem. Commun.*, 1995, 221.

20 I. Kalwinsh, K.-H. Metten and R. Brückner, *Heterocycles*, 1995, **40**, 939.

21 C.G. Francisco, R. Freire, M.S. Rodriguez and E. Suarez, *Tetrahedron Lett.*, 1995, **36**, 2141.

22 P. Norris, D. Horton and B.R. Levine, *Tetrahedron Lett.*, 1995, **36**, 7811.

23 H. Frank and I. Lundt, *Tetrahedron*, 1995, **51**, 5397.

24 B.P. Bandgar and E. Zbiral, *Carbohydr. Res.*, 1995, **270**, 201.

25 A. Reichert, J.O. Nagy, W. Spevak and D. Charych, *J. Am. Chem. Soc.*, 1995, **117**, 829.

26 A. Lubineau, H. Arcostanzo and Y. Queneau, *J. Carbohydr. Chem.*, 1995, **14**, 1307.

27 W. Fitz, J.-R. Schwark and C.-H. Wong, *J. Org. Chem.*, 1995, **60**, 3663.

28 T. Sugai, A. Kuboki, S. Hiramatsu, H. Okazaki and H. Ohta, *Bull. Chem. Soc. Jpn.*, 1995, **68**, 3581.

29 D.C.M. Kong and M. von Itzstein, *Tetrahedron Lett.*, 1995, **36**, 957.

30 H.J.M. Gijsen and C.-H. Wong, *J. Am. Chem. Soc.*, 1995, **117**, 7585.

31 G. Limberg and J. Thiem, *Carbohydr. Res.*, 1995, **275**, 107.

32 A. Dondoni and A. Marra, *Carbohydr. Lett.*, 1994, **1**, 43.

33 G. Devianne, J.-M. Escudier, M. Baltas and L. Gorrichon, *J. Org. Chem.*, 1995, **60**, 7343.

34 T. H. Chan and M.-C. Lee, *J. Org. Chem.*, 1995, **60**, 4228.

35 C. Brönnimann, T. Mallat and A. Baiker, *J. Chem. Soc., Chem. Commun.*, 1995, 1377.

36 M. Chandler, M.J. Bamford, R. Conroy, B. Lamont, B. Patel, V.K. Patel, I.P. Steeples, R. Storer, N.G. Weir, M. Wright and C. Williamson, *J. Chem. Soc., Perkin Trans. 1*, 1995, 1173.

37 M.J. Bamford, J.C. Pichel, W. Husman, B. Patel, R. Storer and N.G. Weir, *J. Chem. Soc., Perkin Trans. 1*, 1995, 1181.

38 X.-L. Sun, T. Kai, M. Tanaka, H. Takayanagi and K. Furuhata, *Chem. Pharm. Bull.*, 1995, **43**, 1654.

39 I.D. Starkey, M. Mahmoudian, D. Noble, P.W. Smith, P.C. Cherry, P.D. Howes and S.L. Sollis, *Tetrahedron Lett.*, 1995, **36**, 299.

40 S. Ciccotosto and M. von Itzstein, *Tetrahedron Lett.*, 1995, **36**, 5405.

41 P. Mirelis and R. Brossmer, *Bioorg. Med. Chem. Lett.*, 1995, **5**, 2809.

42 T. Ercégovic and G. Magnusson, *J. Org. Chem.*, 1995, **60**, 3378.

43 A. Banaszek, *Tetrahedron*, 1995, **51**, 4231.

44 M. Tanaka, T. Kai, X.-L. Sun, H. Takayanagi and K. Furuhata, *Chem. Pharm. Bull.*, 1995, **43**, 2095.

45 G.H. Veeneman, R.G.A. Van De Hulst, C.A.A. Van Boeckel, R.L.A. Philipsen, G.S.F. Ruigt, J.A.D.M. Tonnaer, T.M.L. Van Delft and P.N.M. Konings, *Bioorg. Med. Chem. Lett.*, 1995, **5**, 9.

46 T. Baasov and A. Kohen, *J. Am. Chem. Soc.*, 1995, **117**, 6165.

47 A. Liav, *Carbohydr. Res.*, 1995, **271**, 241.

48   S. Sepulveda-Boza and U. Stather, *Bol. Soc. Chil. Quim.*, 1994, **39**, 299 (*Chem. Abstr.*, 1995, **122**, 161 131).

49   B.P. Bandgar, S.V. Paul and E. Zbiral, *Carbohydr. Res.*, 1995, **276**, 337.

50   P.J. Garegg, L. Olsson and S. Oscarson, *J. Org. Chem.*, 1995, **60**, 2200.

51   I.R. Vlahov and R.J. Linhardt, *Tetrahedron Lett.*, 1995, **36**, 8379.

52   J.-N. Bertho, V. Ferriéres and D. Plusquellec, *J. Chem. Soc., Chem. Commun.*, 1995, 1391.

53   R.T. Brown, N.E. Carter, K.W. Lumbard and F. Scheinmann, *Tetrahedron Lett.*, 1995, **36**, 8661.

54   U.G. Thomsen, C. Cornett, J. Tjornelund and S.H. Hansen, *J. Chromatogr. A.*, 1995, **697**, 175.

55   Z. Pakulski and A. Zamojski, *Tetrahedron*, 1995, **51**, 871.

56   S. Czernecki, S. Horns and J.-M. Valery, *J. Org. Chem.*, 1995, **60**, 650.

57   A.E.J. de Nooy, A.C. Besemer and H. van Bekkum, *Tetrahedron*, 1995, **51**, 8023.

58   K. Li and R.F. Helm, *Carbohydr. Res.*, 1995, **273**, 249.

59   Z. Gyorgydeak and J. Thiem, *Carbohydr. Res.*, 1995, **268**, 85.

60   A.E.M. Boelrijk and J. Reedijk, *Recl. Trav. Chim. Pays-Bas*, 1994, **113**, 411 (*Chem. Abstr.*, 1995, **122**, 81 781).

61   W.M. Ho, H.N.C. Wong, L. Navailles, C. Destrade, H.T. Nguyen and N. Isaert, *Tetrahedron*, 1995, **51**, 7373.

62   H.B. Mereyala and S. Guntha, *Tetrahedron*, 1995, **51**, 1741.

63   K.A. Jacobson, S.M. Siddiqi, M.E. Olah, X. Ji, N. Melman, K. Bellamkonda, Y. Meshulam, G.L. Stiles and H.O. Kim, *J. Med. Chem.*, 1995, **38**, 1720.

64   T. Kida, A. Masuyama and Y. Nakatsuji, *Yukagaku*, 1994, **43**, 1086 (*Chem. Abstr.*, 1995, **122**, 291 391)

65   H.M. Makhtar M.M. El Sadek and N.B. Zagzoug, *Pak. J. Sci. Ind. Res.*, 1994, **37**, 221 (*Chem. Abstr.*, 1995, **123**, 340 662).

66   D.A. Dixon, N.P. Sadler and T.P. Dasgupta, *Transition Met. Chem. (London)*, 1995, **20**, 295 (*Chem. Abstr.*, 1995, **123**, 340 673).

67   D. Bartlett, D.F. Church, P.L. Bounds and W.H. Keppenol, *Free Radical Biol, Med.*, 1995, **18**, 85 (*Chem. Abstr.*, 1995, **122**, 133 618).

68   S.-D. Kim, J.-E. Park, K.-H. Jang, H.-C. Shin and C.-S. Kim, *J. Korean Chem. Soc.*, 1995, **39**, 29 (*Chem. Abstr.*, 1995, **122**, 187 984).

69   B.D. Beake, R.B. Moodie and D. Smith, *J. Chem. Soc., Perkin Trans. 2*, 1995, 1251.

70   A.A. Frimer and P. Gilinsky-Sharon, *J. Org. Chem.*, 1995, **60**, 2796.

71   J.M. Grisar, G. Marciniàk, F.M. Bolkenius, J. Verne-Mismer and E.R. Wagner, *J. Med. Chem.*, 1995, **38**, 2880.

17
Inorganic Derivatives

1 Carbon-bonded Phosphorus Derivatives

Reaction of methyl 2,3-anhydro-4,6-O-benzylidene -α-D-allo- and manno-pyranosides with LiPPh$_2$ afforded methyl 4,6-O-benzylidene-2-deoxy-2-C-diphenylphosphinyl-α-D-altropyranoside and methyl 4,6-O-benzylidene 3-deoxy-3-C-diphenylphosphinyl-α-D-altropyranoside respectively.[1,2] Disaccharide phosphines, 6,6′-dideoxy-6,6′-bis(diphenylphosphinothioyl)-α,α-trehalose per-O-methyl and -benzyl ethers, have been prepared as chiral ligands for use with Rh(I) hydrogenation catalysts.[3]

The dibromotetritol ether **1** has been converted into the epimeric mixture of phosphinates **2** and **3** (Scheme 1),[4] while phosphinate **4**, on base treatment (NH$_4$OH), released the hypophosphorous acid **5**.[5] Phosphonate **6** has been prepared as a potential inhibitor of fructose-6-phosphate 1-phosphotransferase.[6] Addition of lithiated methyl dimethylphosphonate to 2,3:5,6-di-O-isopropylidene-L-gulonolactone and then oxidation of the product **7** with the Dess-Martin reagent has afforded the diketophosphonate **8** (Scheme 2).[7]

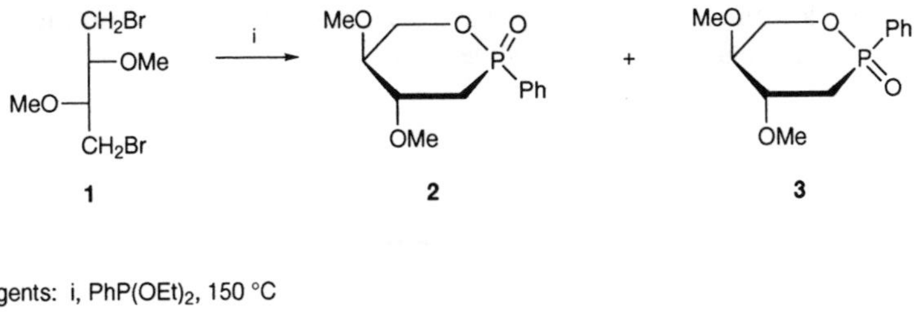

Reagents: i, PhP(OEt)$_2$, 150 °C

Scheme 1

2 Other Carbon-bonded Derivatives

The reaction of methyl 2,3-anhydro-4,6-O-benzylidene-α-D-mannopyranoside with Ph$_2$AsLi gave rise to the corresponding 3-deoxy-3-C-diphenylarsinoaltropyranoside,[8] whereas methyl 2,3-anhydro-4,6-O-benzylidene-α-D-allopyranoside with Ph$_2$AsLi or Ph$_3$SnLi gave the corresponding 2-deoxy-2-C-diphenyl-

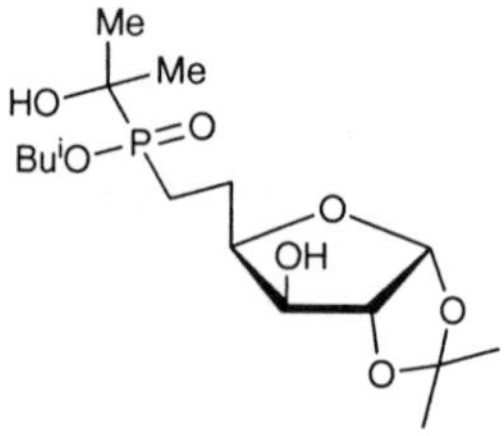

4 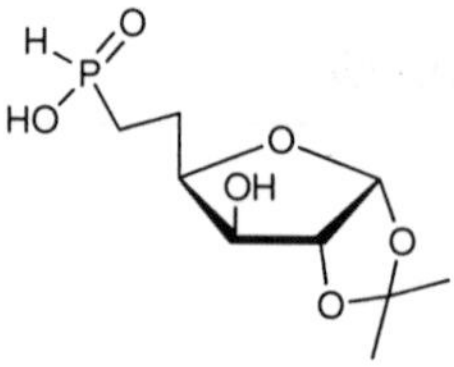   **5** 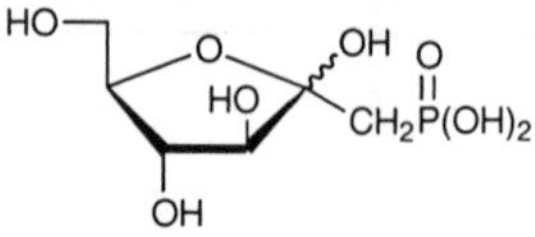   **6**

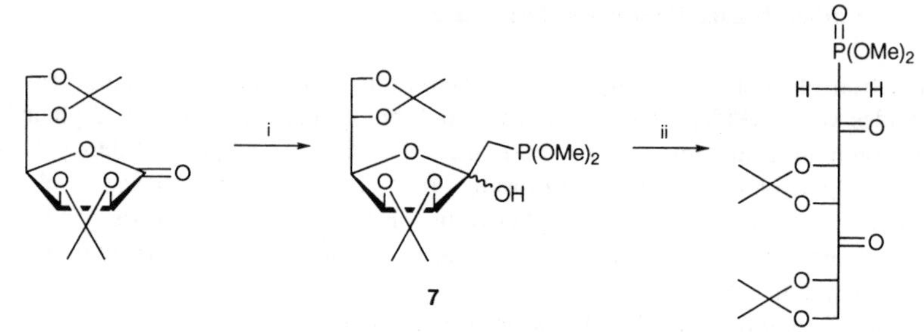

7   **8**

Reagents; i, LiCH$_2$P(O)(OMe)$_2$; ii, Dess Martin reagent

Scheme 2

arsino or -triphenylstannyl-altropyranoside.[2] A number of 5-trimethylarsonio-ribosides **9** with varying R groups have been prepared by quaternization of the appropriate arsines with methyl iodide.[9]

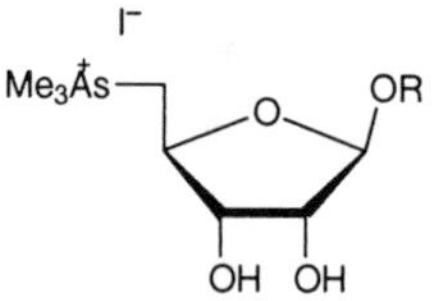

9

Epoxide ring opening of 1,2-anhydro-3,4,6-tri-*O*-benzyl-α-D-glucopyranose with Bu$_3$SnMgMe has afforded the corresponding β-1-*C*-tributylstannyl compound,[10] while lithiation (BuLi, −78 °C) of a similar β-1-*C*-tributylstannyl compound gave the configurationally stable β-carbanion which was trapped by CO$_2$ to give the β-*C*-carboxylate.[11]

Azidophenylselenylation of glycals has generated 2-azido-1,2-dideoxy-1-*C*-phenylselenyl compounds,[12] and a 4-deoxy-4-*C*-selenyl derivative has been used in the synthesis of disaccharides with Se in the interglycosidic linkage.[13] A synthesis of 5'-deuterated ribonucleosides featured the reduction of 5'-acetoxy-5'-*C*-phenylselenyl derivatives (obtained *via* Pummerer rearrangement of 5-deoxy-5-*C*-phenylselenoxides) with Bu_3SnD.[14]

The carbene-complex functionalised sugars **10** and **11** (Scheme 3) have been prepared.[15]

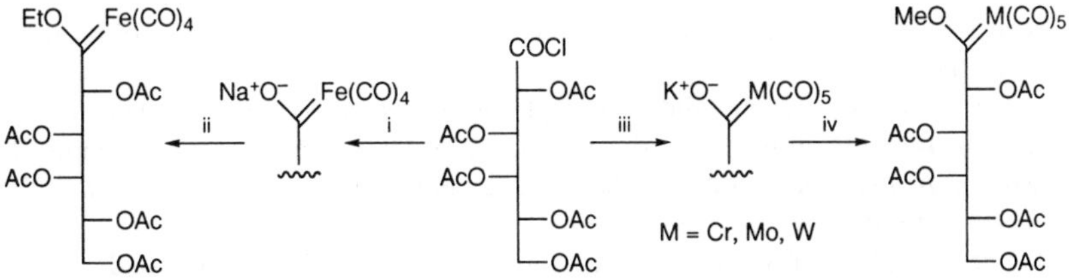

Reagents; i, $Na_2Fe(CO)_4$; ii, EtO_3SCF_3; iii, $K_2M(CO)_5$; iv, Me_3OBF_4

Scheme 3

3   Oxygen-bonded Derivatives

The phosphitylation of D-mannose with phosphorous triazolides has afforded primarily the mannofuranose 2,3,6-*O*-phosphite **12**.[16] The complexes formed separately between D-fructose and L-sorbose with 1,10-phenanthroline and Co(III) ions have been characterised,[17,18] while composites prepared by the reaction of simple sugars with tetraethoxysilane and water have been shown to be capable of resolving racemic cobalt and chromium ion complexes.[19] Sodium alkoxide salts of free sugars have been treated with ferric chloride to give iron-sugar complexes[20] and five-fold deprotonated D-mannose forms dinuclear metalates $[X_2(\beta\text{-}D\text{-}Man\text{-}f)_2]^{4-}$ of trivalent X^{3+} (X = Fe, V, Cr and Al) with *O*-1,2,3,5 and 6 involved in complexation.[21]

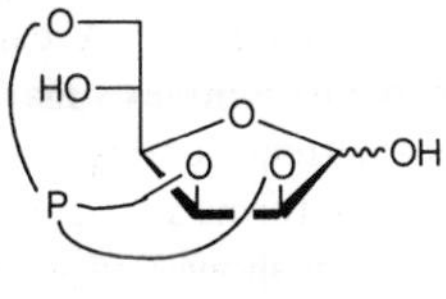

12

A study of the selectivity of metal chelate-directed benzoylation of sucrose dianion, relative to unchelated sucrose anions, was conducted as part of a study on new synthetic approaches to the high potency sweetner sucralose. Ionic complexes of sucrose with various metal ions in DMF reacted at low temperature with benzoic anhydride. Cobalt and manganese salts directed esterification mostly to the 3'-OH of the fructose.[22]

Molecular modelling calculations have been used to study the interactions between D-talopyranose, D-talofuranose and Pb^{2+} and Hg^{2+} ions in the gas phase. In aqueous solutions Pb^{2+} ions form carbohydrate complexes with both forms of the sugar whereas Hg^{2+} ions do not. The calculations implied that the reverse ought to be true.[23]

Some O-carboxymethyl and O-carboxyethyl derivatives of carbohydrates have been prepared and their complexation with Ca^{2+} ions in aqueous solution has been studied.[24]

The tungstate and molybdate complexes of D-*glycero*-D-*manno*-heptitol,[25] D-*glycero*-D-*gulo*-heptonate,[26] and a number of aldoses and ketoses containing the *lyxo* or *manno* configuration[27] have been characterized using ^{13}C- and ^{183}W-NMR methods.

The complexes formed between D-glucose and p-tolylboronic acid in both alkaline and neutral aqueous conditions have been studied using ^{1}H- and ^{13}C-NMR spectroscopy. The sugar always forms a boronate complex in the furanose form, and the complex formed between 2,2'-dimethoxydiphenylmethane-5,5'-diboronic acid and D-glucose has been reassigned.[28] The complex between D-glucose and 3-biphenylboronic acid has been shown to be a stronger inhibitor of α-chymotrypsin than is the specific inhibitor chymostatin.[29] The phenylboronates **13** have been prepared as carbohydrate liquid crystal materials.[30]

13 n = 1, 6, 8, 9, 10, 11, 12, 13

A number of arylboronic acids have been screened in a search for a large fluorescence change on complexing with specific sugars so that they could be used as sugar sensors in analytical systems.[31-33] A calixarene substituted with two arylboronic acids has been shown to complex sugars, the binding being detectable by fluorescence changes.[34] A calixarene bearing two boronic acids on one rim and a crown ether loop on the other rim forms a 1:1 complex with each of D-glucose, D-allose and D-talose. These complexes showed either positive or negative allosterism upon complexation of the crown ether loop with alkali and alkali-earth metal cations.[35] A diaza-18-crown-6 derivative with two arylboronic acid moieties attached has been shown to bind sugars and Ca^{2+} ions,[36] and the complexes formed between D-glucose and another bis-arylboronic acid derivative containing two crown ether moieties has been studied in the presence of various cations. Depending on the cation, some complexes were fluorescent and others not.[37]

L-Fucose forms a polymer with a biarylbisboronic acid derivative in which each fucose forms complexes with two boronic acid groups from different biaryl moieties.[38]

Cholesteryl arylboronic acid derivatives have been used to extract sugars into organic solvents, and the aldose-cholesteryl boronic acid complexation results in a UV-visible absorption shift characteristic of the aldose enantiomer used.[39]

References

1   J.-C. Shi, M.-C. Hong, D.-X.Wu, Q.-T. Liu and B.-S. Kang, *Chem. Lett.*, 1995, 685.
2   M.A. Brown, P.J. Cox, R.A. Howie, O.A. Melvin, O.J. Taylor and J.L. Wardell, *J. Organomet. Chem.*, 1995, **498**, 275 (*Chem. Abstr.*, 1995, **123**, 340 590).
3   S.R. Gilbertson and C.-W.T. Chang, *J. Org. Chem.*, 1995, **60**, 6226.
4   T. Hanaya, A. Akamatsu, S. Kawase and H. Yamamoto, *J. Chem. Res.*, 1995, (S) 194.
5   E.K. Baylis, *Tetrahedron Lett.*, 1995, **36**, 9389.
6   M.S. Chorghade and C.T. Czeke, *Heterocycles*, 1995, **40**, 213.
7   R. Csuk and P. Dorr, *J. Carbohydr. Chem.*, 1995, **14**, 35.
8   M.A. Brown, R.A. Howie, J.L. Wardell, P.J. Cox and O.A. Melvin, *J. Organomet. Chem.*, 1995, **493**, 199 (*Chem. Abstr.*, 1995, **123**, 257 148).
9   K.A. Francesconi, J.S. Edmonds and R.V. Stick, *Appl. Organomet. Chem.*, 1994, **8**, 517 (*Chem. Abstr.*, 1995, **122**, 106 270).
10   Y.Y. Belosludtsev, R.K. Bhatt and J.R. Falck, *Tetrahedron Lett.*, 1995, **36**, 5881.
11   O. Frey, M. Hoffmann and H Kessler, *Angew. Chem., Int. Ed. Engl.*, 1995, **34**, 2026.
12   S. Czernecki and E. Ayadi, *Can. J. Chem.*, 1995, **73**, 343.
13   S. Mehta, J.S. Andrews, B.D. Johnston, B. Svensson and B. M. Pinto, *J. Am. Chem. Soc.*, 1995, **117**, 9783.
14   E. Kawashima, K. Toyama, K. Ohshima, M. Kainosho, Y. Kyogoku and Y. Ishido, *Tetrahedron Lett.*, **1995**, *36*, 6699.
15   K.H. Dotz, W. Straub, R. Ehlenz, K. Peseke and R. Meisel, *Angew. Chem., Int. Ed. Engl.*, 1995, **34**, 1856.
16   S.B. Krebtova, V. Yu. Mishina, M.K. Grachev, A.R. Bekker, M.P. Koroteev and E.E. Nifanteev, *Zh. Obshch. Khim.*, 1994, **64**, 344 (*Chem. Abstr.*, 1995, **122**, 81 761).
17   E. Moraga, S. Bunel, C. Ibarra, A. Blasko and C.A. Bunton, *Carbohydr. Res.*, 1995, **278**, 1.
18   A. Blaskó, C.A. Bunton, E. Moraga, S. Bunel and C. Ibarra, *Carbohydr. Res.*, 1995, **278**, 315.
19   F. Mizukami, H. Izutsu, T. Osaka, Y. Akiyama, N. Uiji, K. Moriya, K. Endo, K. Maeda, Y. Kiyozumi and K. Sakaguchi, *J. Chromatogr. A.*, 1995, **697**, 279.
20   K. Geetha, M.S.S. Raghavan, S.K. Kulshreshtha, R. Sasikala and C.P. Rao, *Carbohydr. Res.*, 1995, **271**, 163.
21   J. Burger, C. Gack and P Klüfers, *Angew. Chem., Int. Ed. Engl.*, 1995, **34**, 2647.
22   J.L. Navia, R.A. Roberts and R.E. Wingard, Jr., *J. Carbohydr. Chem.*, **14**, 465.
23   M. Palma and Y.L. Pascal, *Can. J. Chem.*, 1995, **73**, 22.
24   H. Bazin, A. Bouchu, G. Descotes and M. Petit – Ramel, *Can. J. Chem.*, 1995, **73**, 1338.
25   S. Chapelle and J.-F. Verchére, *Carbohydr. Res.*, 1995, **266**, 161.
26   M. Hlaibi, M. Benaissa, C. Busatto, J.-F. Verchére and S. Chapelle, *Carbohydr. Res.*, 1995, **278**, 227.
27   S. Chapelle and J.-F. Verchére, *Carbohydr. Res.*, 1995, **277**, 39.
28   J.C. Norrild and H. Eggert, *J. Am. Chem. Soc.*, 1995, **117**, 1479.

29   H. Suenaga, M. Mikami, H. Yamamoto, T. Harada and S. Shinkai, *J. Chem. Soc.,*
     Perkin Trans. 1, 1995, 1733.

30   V. Vill and H.-W. Tunger, *J. Chem. Soc., Chem. Commun.*, 1995, 1047.

31   H. Suinaga, M. Mikami, K.R.A.S. Sandanayake and S. Shinkai, *Tetrahedron Lett.*,
     1995, **36**, 4825.

32   T.D. James, K.R.A.S. Sandanayake, R. Iguchi and S. Shinkai, *J. Am. Chem. Soc.*,
     1995, **117**, 8982.

33   H. Shinmori, M. Takeuchi and S. Shinkai, *Tetrahedron*, 1995, **51**, 1893.

34   P. Linnane, T.D. James and S. Shinkai, *J. Chem. Soc., Chem. Commun.*, 1995, 1997.

35   F. Ohseto, H. Yamamoto, H. Matsumoto and S. Shinkai, *Tetrahedron Lett.*, 1995,
     36, 6911.

36   K. Nakashima and S. Shinkai, *Chem. Lett.*, 1995, 443.

37   T.D. James and S. Shinkai, *J. Chem. Soc., Chem. Commun.*, 1995, 1483.

38   M. Mikami and S. Shinkai, *J. Chem. Soc., Chem. Commun.*, 1995, 153.

39   T.D. James, H. Kawabata, R. Ludwig, K. Murata and S. Shinkai, *Tetrahedron*,
     1995, **51**, 555.

18
Alditols and Cyclitols

1 Alditols

1.1 Acyclic Alditols. – A review has appeared on the preparation of long chain-alditols (as well as aldoses and acids) through the use of two-directional synthesis, a process in which both ends of a sugar chain are simultaneously extended.[1]

A mathematical model for determining the number of discrete isomers of deoxyalditols with between 3-10 carbon atom chains has been developed.[2]

The absolute stereochemistry at C-2 of acyclic 1,2,4,-triol systems can be determined by difference CD spectroscopy. The method has been applied to 1,2,4-tri-*O*-benzoyl- and 2,4-di-*O*-benzoyl-1-*O*-pivaloyl-derivatives.[3]

The kinetics and mechanism of the ruthenium(III)-catalysed oxidation of D-glucitol by *N*-bromoacetamide (see also Vol. 26, p. 186, refs. 6 and 7 for related work)[4] and the measurement of thermodynamic parameters concerning the association of divalent and trivalent metal cations with xylitol and glucitol in water[5] have been reported.

1-*O*-(6-*O*-*trans*-Caffeoyl-β-D-glucopyranosides) of erythritol, 3-*C*-methyl-threitol and arabinitol (of undefined absolute stereochemistry) have been isolated from the leaves of *Lonicera gracilipes*.[6]

The binding of tetra-*N*-acetylchitotetraitol [(GLCNAc)$_4$-ol] to hen egg white lysozyme is similar to that of the triose (GLCNAc)$_3$, from which it was concluded that the acyclic moiety does not significantly interact with the enzyme.[7]

The natures of the gels formed from 1,3:2,4-di-*O*-benzylidene-D-glucitol in several organic solvents have been studied by IR, CD and electron microscopy.[8,9]

1-*S*-Alkylthio-D-galactitols in which the alkyl chain is made up of between 6-12 carbon atoms, have been prepared and shown to have liquid crystal properties[10] In addition some 3-*O*-alkyl (*n*-C$_{10}$H$_{21}$ to *n*-C$_{16}$H$_{33}$) derivatives of D-glucitol and of D-mannitol have been prepared by alkylating di-*O*-isopropylidene derivatives of D-glucose and D-mannose, followed by acetal hydrolysis and reduction. The liquid crystalline behaviour of the ethers was studied.[11]

The synthesis of 1,5-dichloro-1,5-dideoxypentitols by reaction of unprotected pentitols with Vilsmeier-Haack iminium salts or Viehe-Janouseck phosgene-based iminium salts, in a similar way to that previously reported (Vol. 26, p. 187, ref. 12 and Vol. 25, p. 202, ref. 20), has been described. A detailed ^{1}H and ^{13}C NMR analysis of the products is also described.[12]

Partial silylation of cyclomaltoheptaose, cellulose and amylose followed by methylation, acid catalysed hydrolysis, reduction and acetylation afforded seven partially methylated glucitol acetates. If reductive cleavage conditions (triethylsilane-trimethylsilyl triflate-boron trifluoride etherate) are used instead of the acid hydrolysis step, 1,5-anhydro-D-glucitol acetates are the products.[13]

A paper has been published concerned with the mechanism of the β-(phosphatoxy)alkyl radical migration on a wide range of compounds (mainly non-carbohydrate) and illustrated with the conversion of *ribo*-diphenylphosphatoxy derivative **1**, into **2**, by treatment with tributyltin hydride-AIBN.[14]

The *cis* dihydroxylation (OsO$_4$-NMO) of 1,5,6-trideoxy-5-(*p*-tosylamido)-*erythro*-hex-3-enitol or its 2-*O*-acetyl derivative proceeds as expected with the osmium complex approaching in a manner *trans* to the O-2 substituent (when drawn in a zig-zag conformation) affording 1,5,6-trideoxy-5-(*p*-tosylamido)-gulitol as the major product. Similarly, 1,5,6-trideoxy-5-(*p*-tosylamido)-*threo*-hex-3-enitol gave the corresponding mannitol derivative as the major diastereomer. The products were further converted to their 3,4-*O*-acetonides and examined by NMR spectroscopy.[15]

The Diels-Alder cycloaddition of highly substituted cyclopentadienones to 1-nitro-ald-1-enitol derivatives give rise to thermally unstable nitro bicyclo[2.2.1]-hept-2-en-7-one cycloadducts which are readily converted to 1-*C*-aryl-1-deoxy-pentitols. Thus **3** can be converted into **4**. A similar reaction is observed for the *gluco*-configured equivalent of **3**.[16]

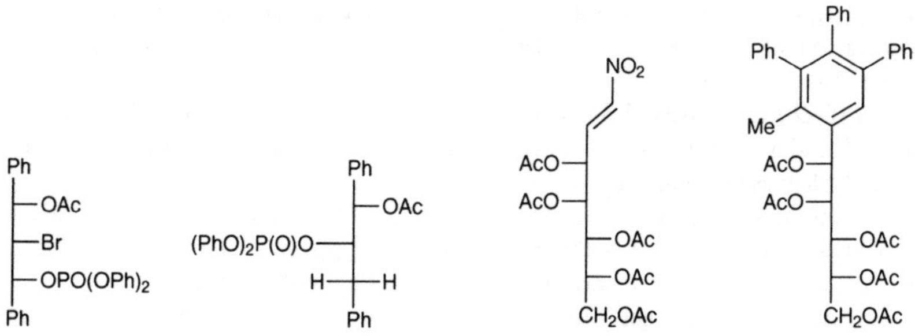

Improved conditions for the reductive desulfurization of the diethyl dithioacetal of D-galactose with Raney nickel-zinc in the presence of sodium hydroxide to afford L-fucitol have been reported. The latter compound was further converted in several easy steps into L-fucose (34% overall from D-galactose).[17]

Calditol, isolated from a bacterial macrocycle, has been presumed to be 4-*C*-(hydroxymethyl)octitol, the acetate of which is shown in Scheme 1. This product has been synthesized and found not to be the natural compound which is now considered to be a cyclopentane derivative.[18]

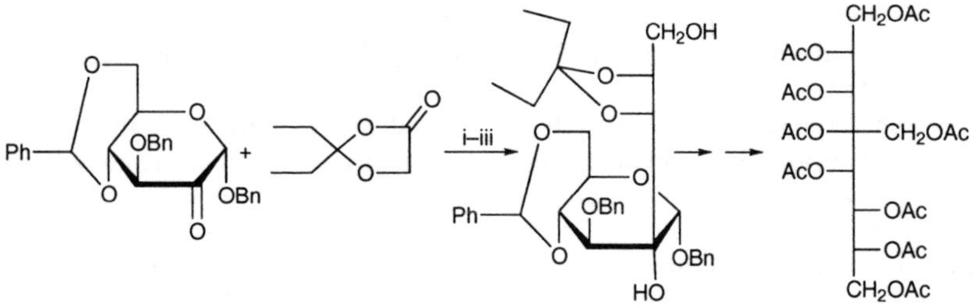

Reagents: i, LDA; ii, Tebbé reagent (Cp$_2$TiCH$_2$AlClMe$_2$); iii, BH$_3$·SMe$_2$

Scheme 1

Reduction of the L-*rhamno*-derived methyl ketone **5** with L-selectride affords the D-*ribo*-pentitol **6**. Under similar conditions, the trifluoromethyl ketone **7** affords the L-*lyxo*-derivative **8**.[19]

The bis 4,4,4-trifluorocrotonyl ester of 1,2:5,6-di-*O*-isopropylidene-D-mannitol shows an interleukin-2 antiproducing effect.[20]

Protection of 1,5-di-*O*-tert-butyldiphenylsilylxylitol with (2*S*,2'*S*)-2,2'-dimethyl -3,3',4,4'-tetrahydro-6,6'-bi-2*H*-pyran affords a 'dispoke' intermediate, which after de-silylation, perbenzylation then acid-catalysed removal of the acetal protecting group gives 1,4,5-tri-*O*-benzyl-D-xylitol in >95% e.e.[21]

Treatment of racemic alcohol **9** with a lipase PS (*Pseudomonas* sp.) and vinyl acetate as acyl donor in isopropyl ether afforded unreacted L-*threo*-compound in >99% e.e. together with the acetate of its enantiomer in >99% e.e. After reaction with an acid resin then Raney nickel, the former afforded 2-deoxy-L-*glycero*-tetritol **10**.[22]

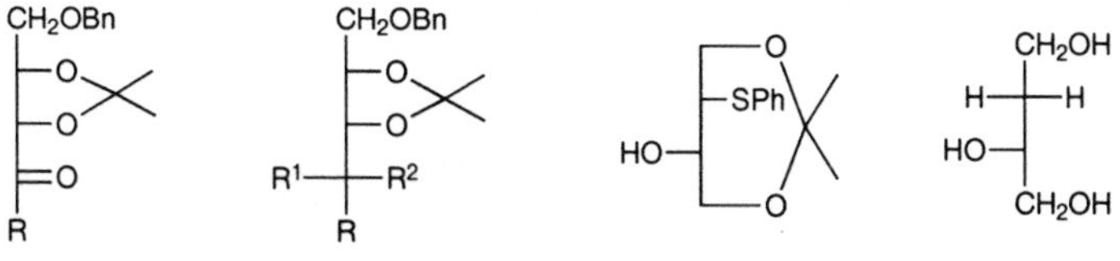

5 R = CH$_3$
7 R = CF$_3$

6 R = CH$_3$, R^1 = H, R^2 = OH
8 R = CF$_3$, R^1 = OH, R^2 = H

9

10

Protection of D-erythrono-1,4-lactone as its 2,3-*O*-benzylidene acetal, followed by regioselective reductive acetal opening (triethylsilane-titanium tetrachloride), then hydride reduction of the lactone function gave 2-*O*-benzyl-L-erythritol. In contrast, tributyltin oxide mediated benzylation of the same 1,4-lactone then hydride reduction of the lactone group gave 2-*O*-benzyl-D-erythritol.[23]

L-Erythrulose has been converted into 2-amino-2-deoxy-L-erythritol in six steps and 35% overall yield, the key step being a stereoselective reduction of the ketoxime orthoformate **11** with K-selectride.[24]

The preparations of 2-acetamido-1,3,4-tri-*O*-acetyl-2-deoxy-D-erythritol and -D-threitol from D-glyceraldehyde by way of Strecker syntheses have been described.[25]

When racemic 2,3-anhydro-erythritol derivative **12** or its *threo* analogue is subjected to azidolysis (sodium azide-ammonium chloride in methanol-water), there is, as expected, little regioselectivity in azide attack at C-2 versus C-3. However, when this reaction is conducted in the presence of metal ions (Li$^+$ or Mg^{2+}) a slight favouring of azide attack at C-2 is seen, reflecting a slight preference for the metal to chelate with the epoxide and the 4-O Bn group. This selectivity is further enhanced by introducing a silyl group at O-1 which discourages metal coordination thereby leading to preparatively useful amounts of 2-azido-2-deoxy-tetritol derivatives. In contrast, the epoxide **13** (R^1 = Bn, triisopropylsilyl, Pmb, R^2 = Bn) shows a slight C-3 preference during azidolysis, in expectation with an electron-withdrawing inductive effect of the 1-O substituent. Again a marked increase in this selectivity is observed in the presence of Li$^+$, leading exclusively in one case (**13**, R^1 = R^2 = Bn) to 3-azido-3,4-dideoxy-*erythro*-pentitols in high yield, indicating selective formation of a 5-membered ring chelate between the epoxide and 1-O Bn group rather than the alternative 6-membered ring chelate formed between the epoxide and 5-O Bn group.[26]

2-Amino-2-deoxy-D-glucitol 6-phosphate, prepared by hydride reduction of D-glucosamine 6-phosphate, is a weak inhibitor of a bacterially derived glucosamine 6-phosphate synthase (Glms), an enzyme that converts glutamine and D-fructose 6-phosphate into D-glucosamine 6-phosphate. Compounds **14** (n = 0,1; m = 1-3) were also prepared as bisubstrate analogues and tested as inhibitors of Glms.[27]

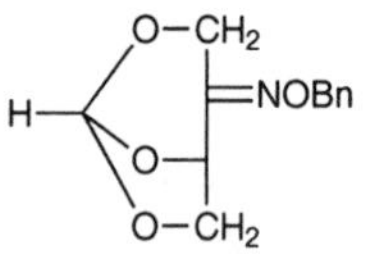

11 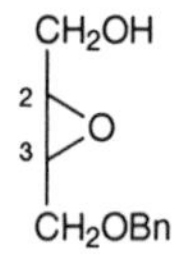   **12** 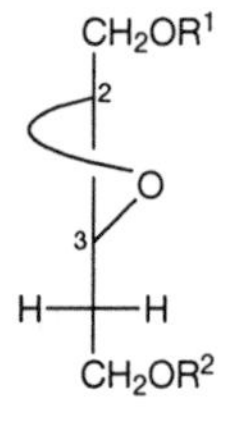   **13**

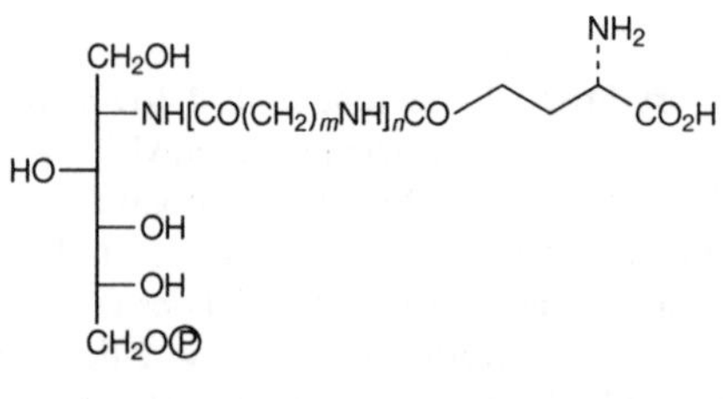

14

The helical polymer **15** has been prepared from the D-mannitol derivative **16** and a diboronic acid. Similar polymers can be obtained using D- and L-threitol derivatives.[28] In a related way condensation of the same diboronic acid with saccharides affords saccharide-containing polymers which display CD spectroscopic properties reflecting the chirality of the saccharides.[29]

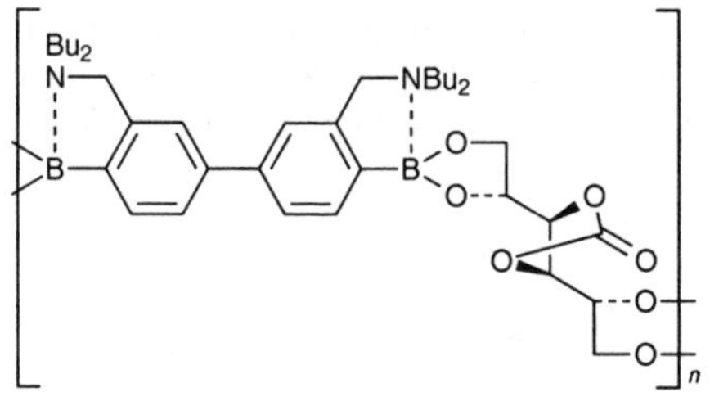

15 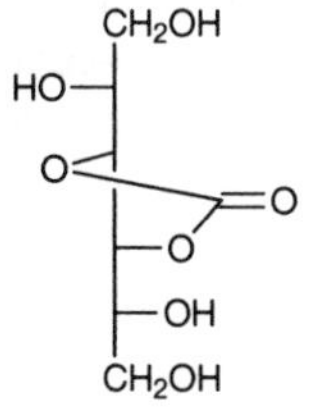   **16**

A hexadecanuclear polyatommetalate of copper(II)-ions and multideprotonated D-glucitol has been prepared and characterized as $Li_8[Cu_{16}(D\text{-glucitol-}H_{-6})_4(D\text{-glucitol-1,2,3,4-}H_{-4})]$ *ca.* $46H_2O$.[30]

1.2   Anhydro-alditols. – A review (in Polish) on the dehydration-cyclization of pentitols and hexitols of plant or mammalian origin in aqueous acid or of *O*-tosyl derivatives under basic conditions has appeared.[31]

The synthesis of all positional isomers of partially methylated-acetylated or methylated-benzoylated derivatives of 1,4-anhydro-D-xylitol,[32] 1,4-anhydro-L-fucitol,[33] 1,4-anhydro-D-ribitol,[34] 1,5-anhydro-D-mannitol,[35] 1,5-anhydro-D-glucitol,[36] and 1,5-anhydro-D-galactitol[37] have been described. The compounds should be useful as standards for determining the primary structure of polysaccharides after reductive cleavage.

1,2:5,6-Di-*O*-isopropylidene-D-mannitol has been converted to the following five anhydro-alditol derivatives using standard chemical transformations: 1,2-anhydro-3-deoxy-5,6-isopropylidene-D-ribitol; 1,2-anhydro-3-deoxy-5,6-isopropylidene-D-arabinitol; 3,4-anhydro-1,2:5,6-di-*O*-isopropylidene-D-mannitol; 3,4-anhydro-1,2:5,6-di-*O*-isopropylidene-D-iditol and 2,3:4,5-dianhydro-1,6-di-*O*-trityl-L-iditol.[38]

The synthesis of 3,4-di-*O*-acetyl-2,5-anhydro-1,6-dideoxy-1,6-diiodo-D-mannitol from 2,5-anhydro-D-mannitol has been described, and differences between the solution and solid state NMR spectra of the former compound have been observed.[39]

A synthesis of (+)-*epiallo*-muscarine from D-glucose has appeared. The key step involves treating 1,2-*O*-isopropylidene-3,5,6-tri-*O*-mesyl-α-D-glucose with ethylene glycol in the presence of tosic acid to give the 2,5-anhydro-L-idose derivative **17**.[40]

Pyranosyl- or furanosyl-1-*O*-acetates are conveniently converted to anhydro-alditol derivatives by reductive cleavage using trimethysilyl triflate-triethylsilane in acetonitrile. Thus 1,2,3,5-tetra-*O*-acetyl-β-D-ribofuranose yields 2,3,5-tri-*O*-acetyl-1,4-anhydro-D-ribitol. Other configured furanoses and pyranoses also undergo this reaction.[41]

During an attempt to α-metalate (^nBuLi-TMEDA) at the carbon bearing the *O*-carbamate group in the 1,4:3,6-dianhydro-D-glucitol derivative **18** in order to introduce bulky substituents with the purpose of obtaining improved chiral auxiliaries, an unexpected reaction took place leading to the formation of the 1,4-anhydro-D-*xylo*-enol carbamate derivative **19**. It is likely that α-metalation occurred but was quickly followed by β-elimination to form a stable alkoxide.[42]

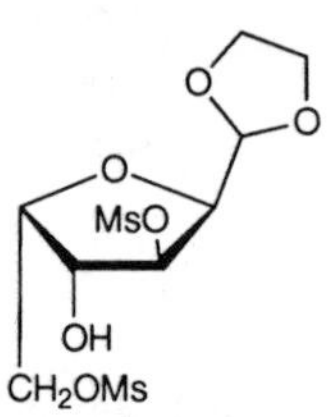

17 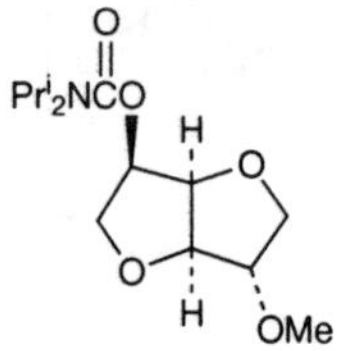   18 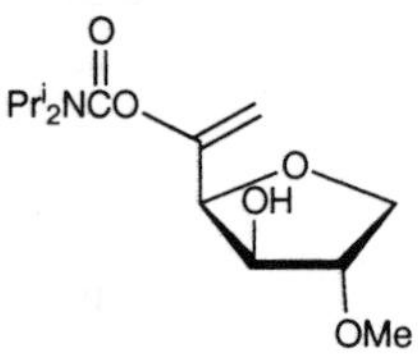   19

The pyranoisoxazolidines **20** and **21** are readily obtained by intramolecular cycloadditions of the nitrones formed by reaction of *N*-benzylhydroxylamine with 3-*O*-allyl-3-*C*-methyl-D-allose and 3-O-allyl-1,2-*O*-isopropylidene-3-*C*-methyl-α-D-*ribo*-pentodialdofuranoside, respectively. The latter two compounds are available from a common precursor, 3-*O*-allyl-1,2:5,6-di-*O*-isopropylidene-3-*C*-methyl-α-D-*allo*-pentofuranose. Compound **20** was further converted to the pyranoid **22**, whilst **21** was converted to the enantiomer of **22** by standard chemistry. Oxepane cycloadducts are also produced when 3-*O*-allyl-3-*C*-methyl-α-D-*xylo*- or 3-*O*-allyl-α-D-*ribo*-pentodialdofuranose is used. (See also Vol. 27, p. 203, ref. 25 for related work).[43]

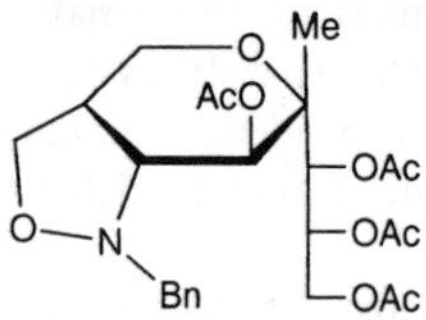

20 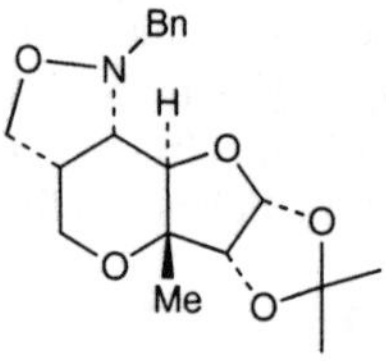   21 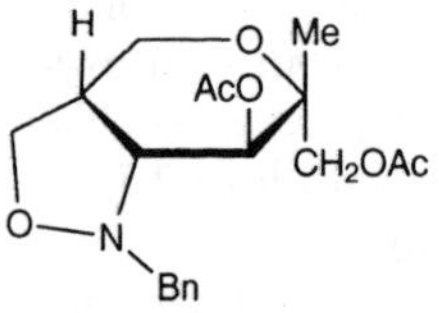   22

The preparation of furanoid *C*-vinyl glycosides from hept-1-enitol derivatives is mentioned in Chapter 3.

1.3 Amino- and Imino-alditols. – A review (253 refs.) on the preparation, antiviral activity and glycoprotein maturation of deoxynojirimycin and its derivatives has appeared.[44]

N-Methyl- and -butyl-derivatives of deoxynojirimycin, 1,4-dideoxy-1,4-imino-
D-arabinitol and 2,5-dideoxy-2,5-imino-D-mannitol have been studied by ^{1}H
NMR spectroscopy to provide data for determining a conformational basis for
the inhibition of glycosidase and HIV-1 replication activity by these types of
compounds.[45]

The 1-amido-1-deoxy-D-glucitol derivative **23** has been prepared and shown to
exhibit liquid crystal properties.[46]

Full details on the preparation of urethane derivatives of 3,4-di-*O*-benzyl-
1,2,5,6-tetradeoxy-1,2:5,6-diimino-L-iditol from D-mannitol as well as the 3,4-
dideoxy-L-*threo* compounds **24** (R = CO$_2$^{t}Bu or CO$_2$Bn), from a hex-3-enitol,
have been reported. (See Vol. 28, p. 230, ref. 66 for an earlier report). In both
cases 1,6-diamino-1,6-dideoxy-2,5-anhydroalditols were minor by-products.[47]
N-Boc-L-*ido*-1,2:5,6-bis-aziridino derivatives have proved to be versatile intermediates for the preparation of imino-alditols. For example, treatment of the 3,4-
di-*O*-benzyl compound with diethylaluminium cyanide affords a 1:2 mixture of
pyrrolidine **25**:piperidine **26**,[48] and with dilithium tetrabromonickelate (Li$_2$NiBr$_4$)
gives bromo-derivative **27** from which displacement of bromide allows further
functionalization.[49]

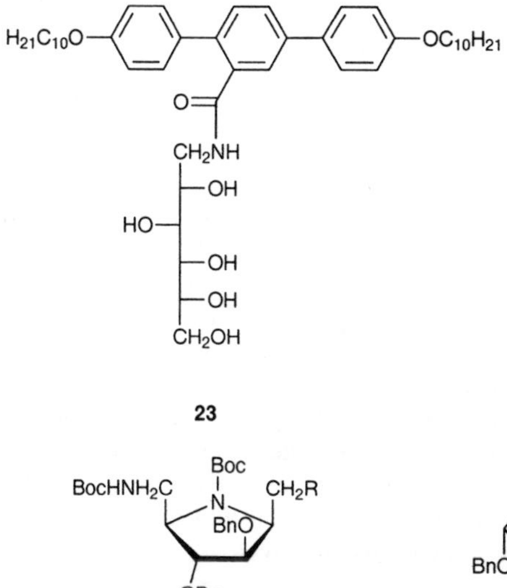

23

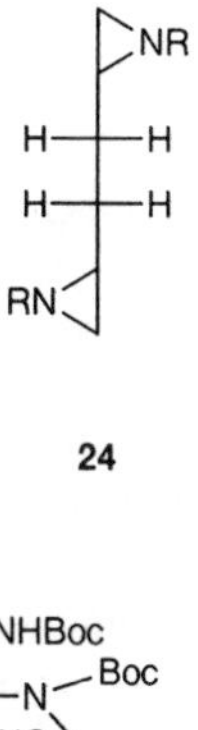

24

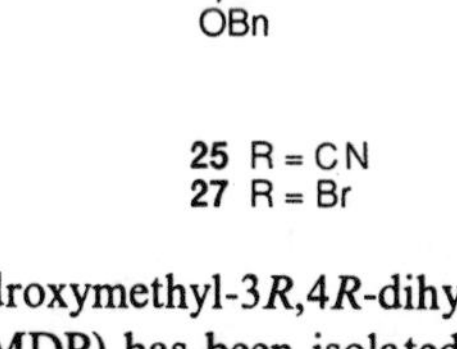

25 R = CN
27 R = Br

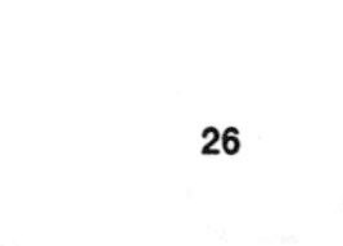

26

2*R*,5*R*-Dihydroxymethyl-3*R*,4*R*-dihydroxypyrrolidine (2,5-dideoxy-2,5-imino-
D-mannitol, DMDP) has been isolated from the fermentation broth of *Streptomyces* sp. KSC-5791 along with minor amounts of deoxynojirimycin and
deoxymannojirimycin. The DMDP was shown to be a potent inhibitor of
trehalases from *Corynebacterium* sp. and the diamondback moth (*Phytella*

xylostella). This is the first time this compound has been isolated from a microorganism.[50] DMDP has also been synthesized by a process covered in Vol. 28, p. 231, ref. 68, designed to act as an inhibitor of pyrophosphate fructose 6-phosphate 1-phosphotransferase and thus function as a herbicide.[51]

A synthesis of 2,5-dideoxy-2,5-imino-D-glucitol (**28**) involves treatment of the D-glucitol derivative **29** first with acid resin in methanol (to remove the silyl and isopropylidene groups) then hydrogenolysis. Cyclization of the so-formed amino-mesylate to **28** surprisingly proceeds with retention of configuration and thus is likely to involve the 5,6-epoxide.[52]

The addition of *n*-butylmagnesium bromide to the lactams **30** (R = ^{n}Bu, Bn or C_9H_{19}), prepared by an oxidation-degradation process in the general way described in Vol. 27, p. 210, ref. 64, then deoxygenation, affords the highly substituted pyrrolidines **31**.[53]

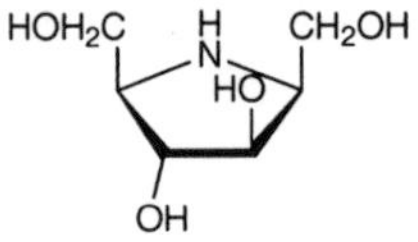

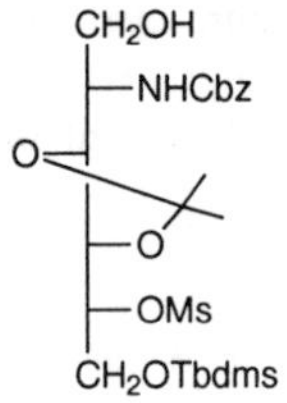

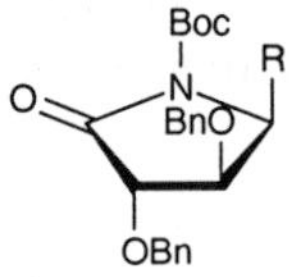

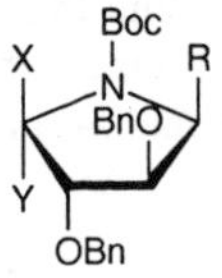

2,3-*O*-Cyclopentylidene-1,4-dideoxy-1,4-*N*-hydroxyimino-D,L-erythritol (a *meso*-pyrrolidine described in Vol. 25, p. 206, ref. 44) can be readily converted into a nitrone with mercury(II) oxide and trapped in a selective manner with organometallic reagents (Grignards, alkyllithiums, *etc.*) to give the racemic *N*-hydroxypyrrolidines **32** (R = Me, Ph, PhC≡C *etc.*). The compounds readily oxidize to aminoxyl free radicals.[54]

The homoazasugars **33** and **34** have been synthesized and shown to be good inhibitors of almond β-glucosidase. The key steps in the syntheses include a two-carbon Wittig chain extension at the anomeric centre of a D-arabinopentose derivative and cyclization by attack of an amino function at an epoxide group (prepared *via* a Sharpless asymmetric epoxidation).[55]

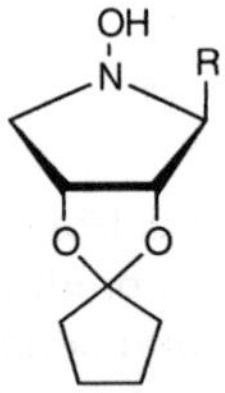

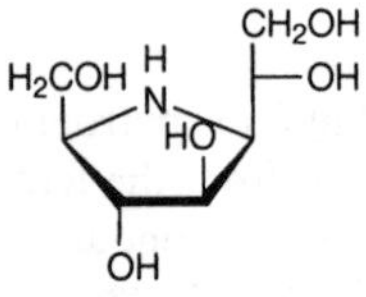

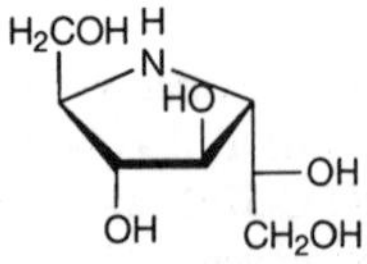

A facile synthesis of both pyrrolidine and piperidine compounds from glycosyl-amines is depicted in Scheme 2.[56]

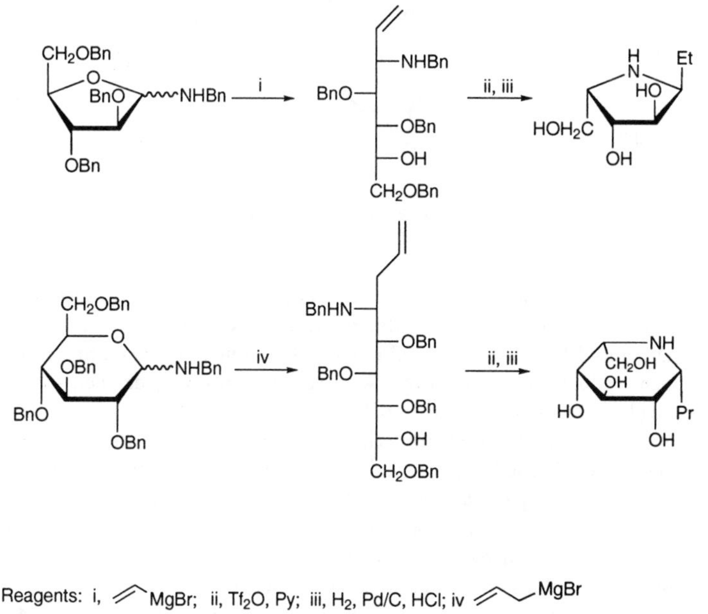

Reagents: i, �•MgBr; ii, Tf₂O, Py; iii, H₂, Pd/C, HCl; iv ⌁MgBr

Scheme 2

A variety of aza-sugars have been prepared and tested as glycosidase inhibitors. These include 1,4,5-trideoxy-1,4-imino-L-lyxitol as an α-fucosidase inhibitor as well as the piperidines **35** and nucleoside analogues **36**.[57]

The branched-chain pyrrolidines **37** (R = Me, Bz) have been reported, the synthesis being accomplished from 3-deoxy-2-C-hydroxymethyl-D-*erythro*-pentono-1,4-lactone (available in two steps from lactose) using standard chemistry.[58]

The β-L-xylosylated pyrrolidines **38** and **39** have been prepared as transition state mimics for glycoside bond cleavage.[59]

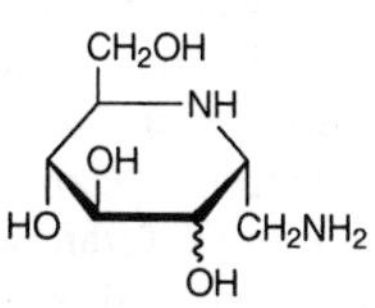

35 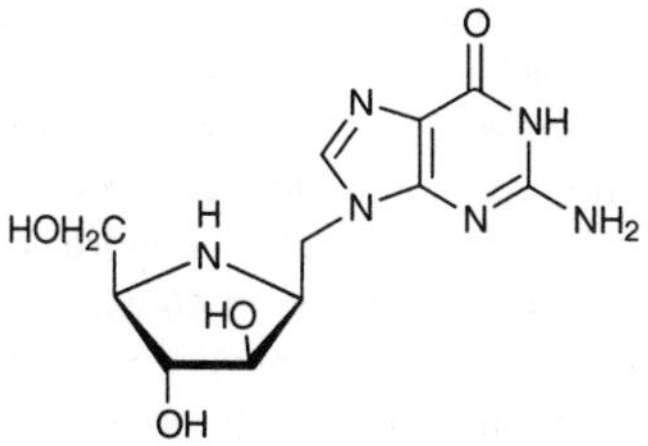   **36**

37

40

38 R = CH₂OH or H, X = OH, Y = H
39 R = H, X = H, Y = OH

Addition of ^{13}C labelled potassium cyanide to 4-azido-2,5-di-*O*-benzyl-4-deoxy-D-arabinose followed by treatment with aqueous sodium bicarbonate gave (1-^{13}C)-5-azido-3,6-di-*O*-benzyl-D-glucono-1,4-lactone plus the mannono lactone. Subsequent reduction of the former with sodium borohydride then hydrogenolysis gave (1-^{13}C)-1-deoxynojirimycin.[60]

The incorporation of oxygen atoms into the alkyl chain of *N*-alkyl-1-deoxynojirimycin derivatives has led to improved α-glucosidase(I)-inhibitory activity relative to toxicity towards Hep 62 cells. An example is illustrated with structure **40**.[61]

The syntheses of 6-amino-2-azido-2,6-dideoxy- and 6-amino-2,6-dideoxy-2-fluoro- analogues of 1-deoxynojirimycin [62] and of 1,3,6-trideoxy-3,6-difluorono-jirimycin [63] by standard chemistry have been described.

Reduction of the cyclopentane-*O*-methylhydroxylamine derivative **41** with lithium aluminium hydride unexpectedly gave tetra-*O*-benzyl-1-deoxynojirimycin through a ring-expansion reaction, in addition to the expected cyclopentylamine.[64]

2,3-Di-*O*-acetyl-4,6-*O*-benzylidene-1,5-*N*-(benzyloxycarbonyl)imino-1,5-dideoxy-D-glucitol (a protected 1-deoxynojirimycin derivative) has been converted into the 1-deoxygalactonojirimycin derivative **42** by a sequence of reactions involving unmasking of the 4-OH group and inversion at C-4 *via* the triflate. Compound **42** as well as deoxynojirimycin derivatives on route to **42** were used to prepare α-sialyl-(2→6) glycosides.[65]

Deoxynojirimycin is readily converted to 2,3-anhydro-4,6-*O*-benzylidene-1,5-*N*-(benzyloxycarbonyl)imino-1,5-dideoxy-D-mannitol in four steps. The reactions of this product with azide and amines were studied.[66]

2,3-Di-*O*-benzyl-1,5-*N*-(benzyloxycarbonyl)imino-6-*O*-*tert*-butyldiphenyl-silyl-1,5-dideoxy-D-glucitol has been converted into the carboxylic acid derivatives **43** and **44** which were tested as glycosidase inhibitors.[67]

A series of imino-sugars with the imino group replacing the anomeric carbon atom have been prepared and tested as glycosidase inhibitors. Compound **45** was made from D-mannose *via* 1,5-anhydro-4-*C*-hydroxymethyl-2,3-*O*-isopropylidene-L-ribofuranose.[68] Compound **46** was also made from D-mannose *via* benzyl 2-*C*-benzyloxymethyl-2,3-isopropylidene-5-triflyl-α,β-D-lyxofuranose by successive treatment with azide, trifluoroacetic acid then hydrogenation-hydro-genolysis.[69]

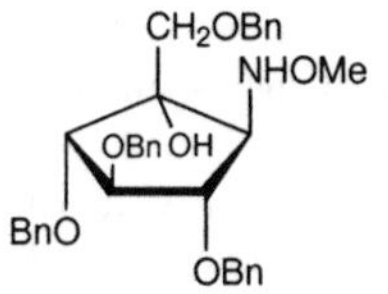

41

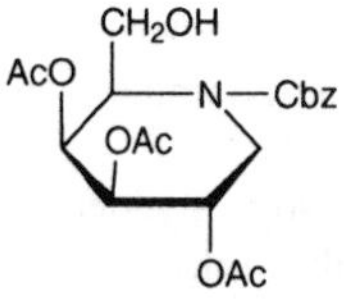

42

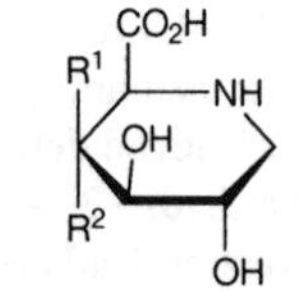

43 $R^1 = H, R^2 = OH$
44 $R^1 = R^2 = H$

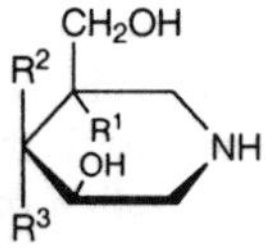

45 $R^1 = OH, R^2 = OH, R^3 = H$
46 $R^1 = OH, R^2 = H, R^3 = OH$
47 $R^1 = H, R^2 = OH, R^3 = H$

Compound **47** was prepared in several steps by standard chemistry from 5-azido-1-*O*-benzoyl-5-deoxy-2,3-isopropylidene-α-D-lyxofuranose.[70]

The guanidine derivative **48** has been prepared by treating 1,5-dideoxy-1,5-imino-xylitol (see Vol. 24, p. 197, ref. 33 for its preparation) with formamidine-sulfonic acid.[71]

The preparations of α-homogalactostatin and of the related substance **49**, in which the imino-cyclitol ring is formed from mercury(II)-assisted ring closure of an amine to a terminal double bond have been described.[72] The synthesis of β-homonojirimycin and compound **50** in which the imino-cyclitol ring was formed by reductive-amination of a 1,5-di-keto derivative has also been described.[73]

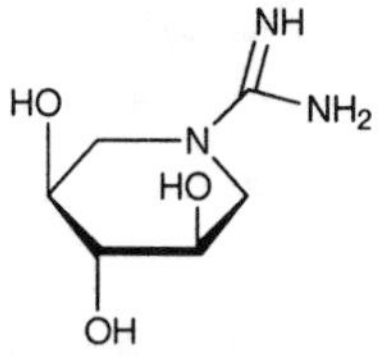

48

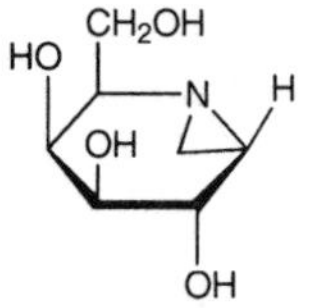

49

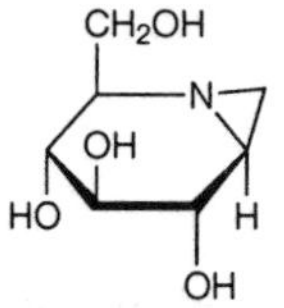

50

The synthesis of 1,5-deoxy-1,5-iminoheptitols from 2-bromo-2-deoxylactones in a related way to that reported in Vol. 27, p. 208, refs. 53 and 54 for the preparation of 1,4-dideoxy-1,4-iminohexitols has been published.[74]

Standard procedures have been used to prepare 1,5-dideoxy-1,5-imino-D-arabinitol which, after linking to aminohexyl agarose, was used to isolate pure α-L-fucosidase from bovine kidney.[75]

A synthesis of a selectively protected 1,5-dideoxy-1,5-imino-D-xylitol has been developed as shown in Scheme 3.[76]

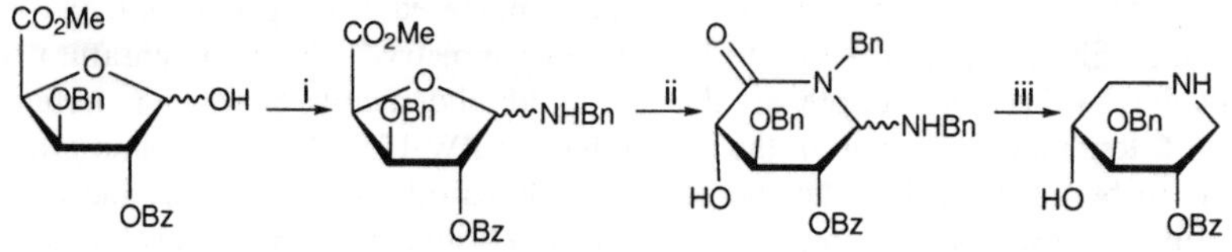

Reagents: i, $BnNH_2$; ii, $BnNH_2$, Me_3Al; iii, TfOH, $(CH_2O)_n$, $NaBH_4$

Scheme 3

Treatment of 2,5-anhydro-3,4-di-O-benzyl-1-deoxy-1-tritylamino-D-glucitol with mesyl chloride, tosic acid-methanol, triethylamine then hydrogen-palladium catalyst affords the 1,6-imino-bridged derivative **51**.[77]

The reaction of 1,2:5,6-dianhydro-3,4-O-isopropylidene-D-mannitol with benzylamine followed by deprotection leads only to the seven-membered azepane **52** in high yield. Other 1,2:5,6-dianhydro-3,4-O-isopropylidene hexitols gave similar results. No six-membered ring products were detected as is the case with 1,2:5,6-dianhydro-3,4-di-O-benzyl-D-mannitol (see Vol. 28, p. 230, ref. 63).[78]

Comound **53** has been prepared as an analogue of ADP ribose and as a potential inhibitor of poly(ADP-ribose)glycohydrolase.[79]

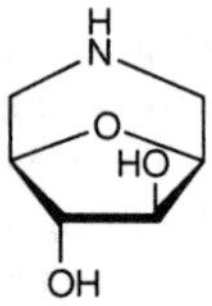

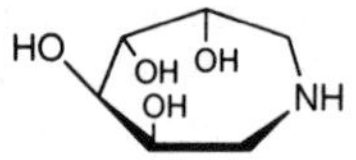

Several imino-alditols have been prepared from non-carbohydrate sources. 2,5-Dideoxy-2,5-imino-D-mannitol has been synthesized from a pyroglutamic acid derivative[80] (see Vol. 27, p. 209, ref. 59 for the use of pyroglutamic acid in the synthesis of other imino-alditols), and 1-deoxy-L-allonojirimycin has been made from a protected L-serine derivative.[81] A protected D-serine aldehyde provided access to a 1,2,4-trideoxy-1,4-imino-D-*erythro*-pentitol derivative for incorporation into DNA,[82] and reduction of the minor natural amino acid, 4-L-hydroxyproline, gave 1,3,4-trideoxy-1,4-imino-D-*erythro*-pentitol which, after N-acylation with a fluorescent probe, was incorporated into oligonucleotides.[83]

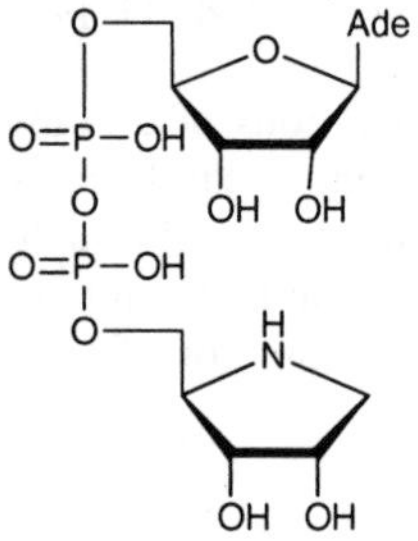

1,4-Dideoxy-1,4-imino-D-lyxitol has been prepared from penta-1,4-dien-3-ol using a Sharpless epoxidation to introduce asymmetry.[84] The same unsaturated alcohol has been used to prepare the amino-alditol derivative **54** which represents the C-terminal component of the renin inhibitor BW-175. The morpholino group has to be introduced at the end of the synthesis otherwise other products can result. For example, treatment of 1-deoxy-2,3-O-isopropylidene-1-morpholino-5-O-tosyl-D-ribitol with base afforded the piperidinium salt **55**.[85]

Asymmetric syntheses of 6-deoxy-D-allonojirimycin and D-fuconojirimycin as well as their 1-deoxy analogues by cylcoaddition of a diene with 1-C-nitroso-D-mannofuranosyl chloride in a similar way to that described in Vol. 27, p. 207, ref. 46 for the preparation of racemic imino-alditols, have been described.[86] (See also

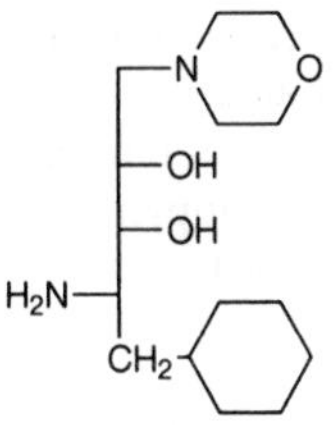

54 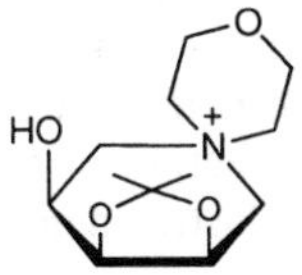   55 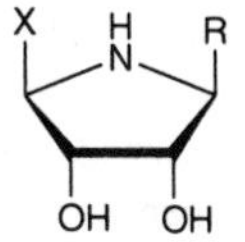   56 X = CO$_2$H   58
             57 X = OH

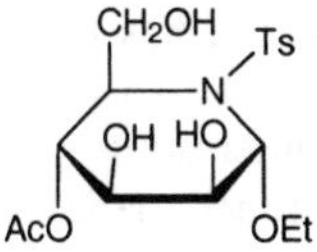

Vol. 28, p. 226, ref. 44 for related work). Similar cycloadditions of dienyl pyrrolidinones [*i.e.* *N*-(penta-1,3-dienyl)- or *N*-(buta-1,3-dienyl)-pyrolidin-2-one] with the *in situ* generated acyl nitroso compound derived from benzyl-*N*-hydroxy-carbamate with periodate have led to the preparations of racemic pyrollidines **56** and **57** (R = H or Me). The latter compound is thought to exist as a dimer.[87] An asymmetric synthesis of 1,5,6-trideoxy-1,5-imino-D-altritol in which the piperidine ring is formed from a pyridinium ring bearing Seebach's oxazolidinone chiral auxiliary has also been described.[88]

1,5,6-Trideoxy-1,5-imino-L-fucitol in protected form has been prepared from a hindered imine derivative of L-alanyl methyl ester,[89] and furylglycine has been utilized as starting material for the synthesis of a 2-hydroxymethyl-dihydropyridine derivative as a building block for preparing aza-sugars including the racemic mannojirimycin derivative **58**.[90]

3,5-Dihydrocyclopentene derivatives (from cyclopentadiene) after kinetic resolution with a lipase are useful precursors for the synthesis of 1-deoxy-D-nojirimycin,[91] and benzene *cis*-diol (from benzene by microbial oxidation) has been used as a precursor to 1-deoxy-D-galactonojirimycin and bicyclic derivative **59**.[92] The preparation of imino-alditol derivatives in which the imino nitrogen atom is part of a fused tetrazole ring is mentioned in Chapter 10 and the synthesis of imino-lactams is covered in Chapters 10 and 16.

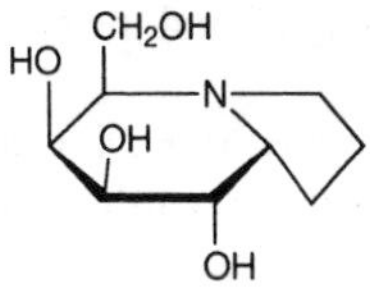

59 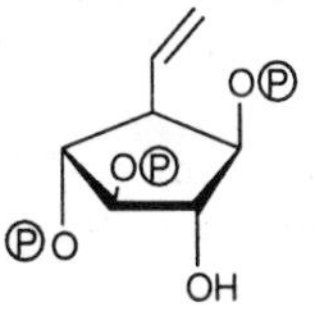   60

2   Cyclitols

2.1   Cyclopentane and Cyclobutane Derivatives. – A review of applications and mechanistic aspects of the 'Cp$_2$Zr' induced direct ring contraction of hept-6-enose and hex-5-enose derivatives to cyclopentane and cyclobutane ring systems with emphasis on the application to oxetanocin synthesis has appeared. (See also Vol.

27, p. 214, ref. 85).[93] A synthesis of the cyclopentane **60** involving such a ring contraction has been reported. The product was designed as a mimic of *myo*-inositol 1,4,5-trisphosphate and found to be an agonist at the natural receptor.[94]

An interesting radical-induced ring contraction of 6-aldehydopyranosides to highly functionalized cyclopentane rings is illustrated in Scheme 4.[95]

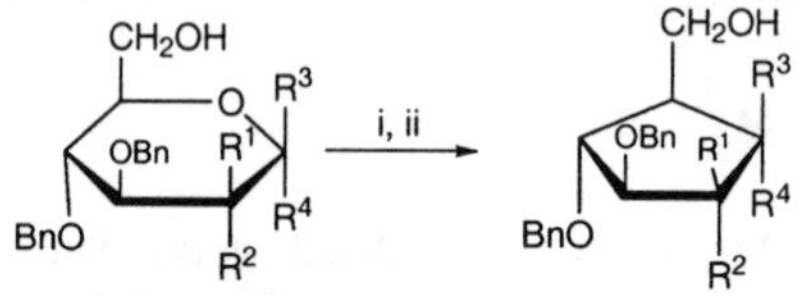

R¹ = R⁴ = H, R² = OBn, R³ = OMe
R¹ = R³ = H, R² = NHCO₂Me, R⁴ = OMe
R¹ = OBn, R² = R³ = H, R⁴ = OMe

Reagents: i, Swern oxidation (modified); ii, SmI₂, Bu^{t}OH, HMPA

Scheme 4

Cyclopentane rings are also formed when the keto-oxime ether **61** (R = Bn) is treated with samarium diiodide, compound **62** (R = Bn) being the only product. Several other examples, usually affording mixtures of isomers are also described.[96] In a similar way treatment of **61** (R = Me) with tributyltin hydride-AIBN gave **62** (R = Me) together with the epimer at the indicated carbon atom. Reductive cleavage of the oxime ether function in **62** (R = Me) with lithium aluminium hydride afforded the expected amine, but surprisingly gave a ring-expanded imino-alditol compound as well.[64]

Cyclopentane rings can also be formed readily by treating 6-deoxy-6-nitro-D-fructose (prepared enzymatically from dihydroacetone phosphate and 3-deoxy-3-nitro-D,L-glyceraldehyde in the presence of an aldolase then phosphatase) with acetic anhydride and boron trifluoride etherate, then subjecting the mixture to chromatography on silica gel thereby affording **63** and the epimer at the indicated carbon atom.[97]

(1*R*,2*S*,4*R*)-4-Amino-2-hydroxy-1-(hydroxymethyl)cyclopentane has been prepared as an intermediate for 2′-deoxynucleoside synthesis from two different

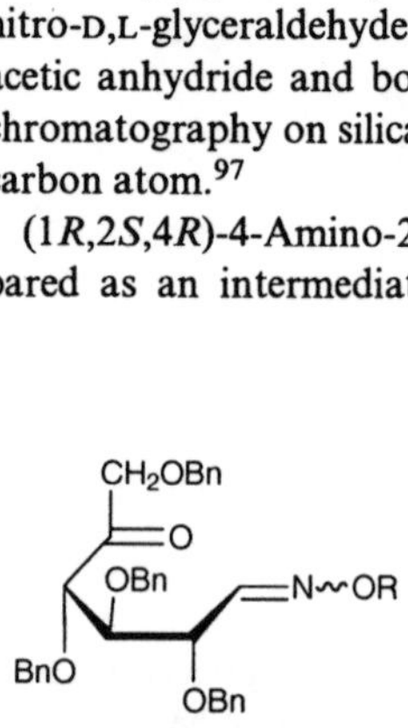

61 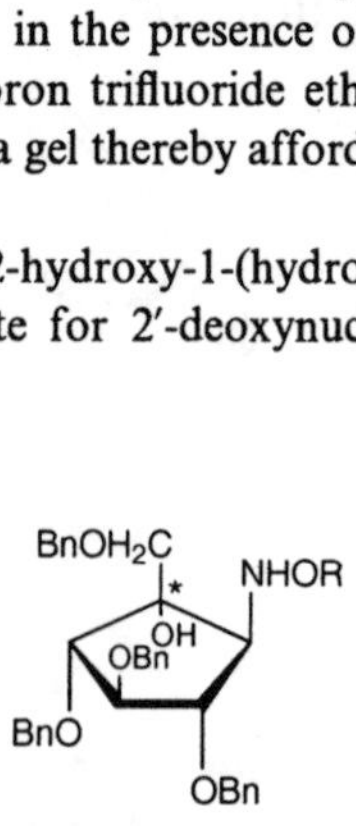   **62** 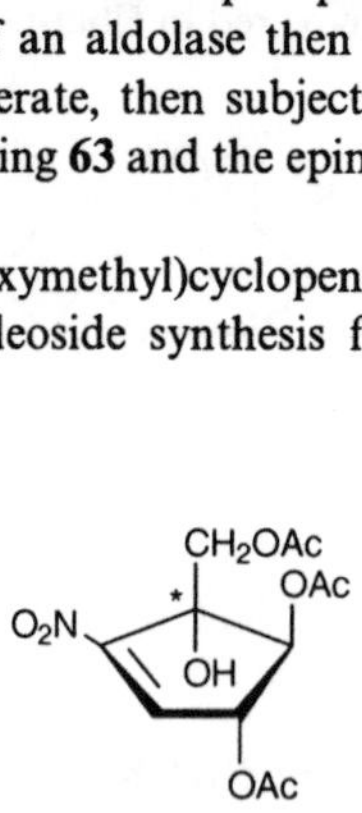   **63** 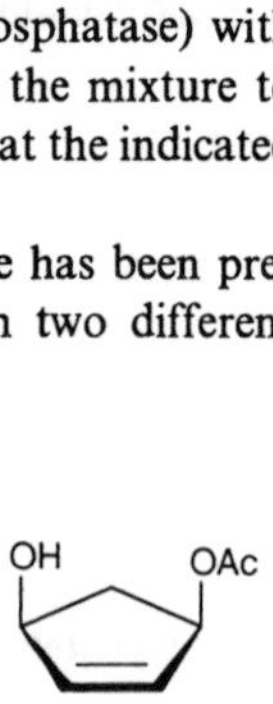   **64**

starting materials; (−)-2-azabicyclo[2.2.1]hept-5-en-3-one; and chiral cyclopentene **64**.[98] (+)-2-Azabicyclo[2.2.1]hept-5-en-3-one, a by-product from a biotransformation process, can be converted into its (−)-enantiomer, in five steps.[99]

A non-carbohydrate route starting from a cyclopentanone has been developed to prepare the carbocyclic analogue of ascorbic and isoascorbic acid,[100] and all possible carbapentofuranoses have been synthesized from norborn-5-en-2-one (optically pure).[101]

Cyclopentane rings as found in mannostatins, trehazoline, aristeromycin, allosamizoline and derivatives are covered in Chapter 19.

2.2 Inositols and Other Cyclohexane Derivatives. – A review, in Japanese, on aspects of the catalytic Ferrier cyclization for producing optically pure cyclohexanones from hex-5-enose derivatives has appeared.[102] The mechanism for carbocyclic ring formation in the synthesis of 2-deoxy-*scyllo*-inosose, involved in the biosynthesis of 2-deoxystreptamine, has been the subject of labelling studies.[103] (See also Vol. 28, p.238, ref. 119 for related work). The syntheses of 6-deoxy-inososes and 6-deoxy-inositols from D-galactose derivatives, by way of the Ferrier cyclization have been reported.[104] See also Chapter 24 for the preparation of a highly functionalized cyclohexane ring as a potential unit for conversion to the C-ring of taxol.

The absolute configuration of (−)- and (+)-1,2:4,5-di-*O*-isopropylidene-*myo*-inositol have been assigned as 1D and 1L, respectively, by transformation of each enantiomer into an established reference compound. These assignments contradict some recent literature reports.[105] (See also Vol. 28, p. 237, ref. 116).

Heating 1,4,5,6-tetra-*O*-benzyl-*myo*-inositol with ethylene carbonate and a trace of sodium bicarbonate affords the 2,3-cyclic carbonate derivative. The reaction fails where there is a *trans*-diol arrangement as with 1,2,3,6-tetra-*O*-benzyl-*myo*-inositol.[106]

The synthesis of (+)-ononitol (**65**) and its (−)-enantiomer from resolved *myo*-inositol has been reported. There appears to be a marked abundance of *O*-methyl inositols present in grains and forage legumes.[107] *Scyllo*-inositol has been per-*O*-methylated under standard conditions and conformationally examined by NMR spectroscopy.[108]

Racemic 4-deoxy-4-fluoro-*myo*-inositol has been made by reaction of **66** with DAST (retention of configuration observed) then hydrogenolysis,[109] and 2-deoxy-2-fluoro-*myo*-inositol, 2-deoxy-2-fluoro-*scyllo*-inositol and 2-deoxy-2,2-difluoro-*myo*-inositol have been synthesized from 3,4,5,6-tetra-*O*-benzyl-*myo*-inositol *via* a selective benzoylation at the equatorial hydroxyl group and subsequent standard chemistry.[110]

Various cyclitols (and acyclic polyols) have been desymmetrized by formation of 'dispoke' intermediates. (See for example, Vol. 28, p. 237, ref. 115).[111] L-*Chiro*-inositol can be converted to the silyl derivative **67** in which the *trans*-diol units are protected on reaction with 1,3-dichloro-1,1,3,3-tetraisopropylidisiloxane (TipsCl). Compound **67** was further converted into conduritol B epoxide and its thioepoxide analogue.[112] The conversion of some tetra-*O*-substituted *myo*-inositols into adipic dialdehyde derivatives is mentioned in Chapter 15.

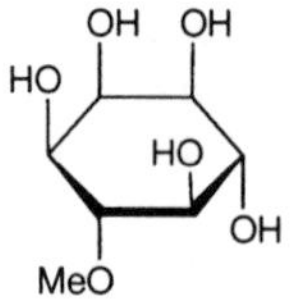

65 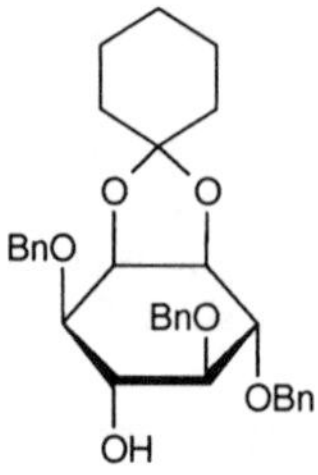   **66** 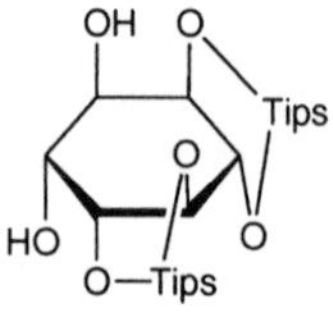   **67** 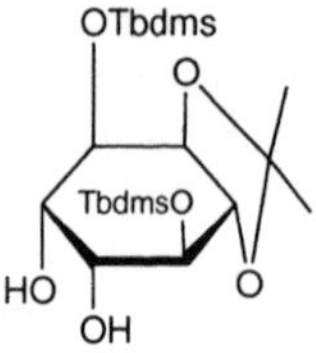   **68**

2,5-Di-*O-tert*-butyldimethylsilyl-3,4-*O*-isopropylidene-D-glucitol on Swern oxidation then treatment with samarium diiodide gives the L-*chiro*-derivative **68** together with a small amount of the *myo*-isomer. Compound **68** was further converted into its hexaacetate for characterization purposes as well as into conduritol F tetraacetate.[113]

The known conduritol **69** (Vol. 25, p. 210, ref. 66) has been used in synthetic studies directed towards the anticancer agent (+)-pancratistatin.[114] See also Chapter 24 for the syntheses of (+)-pancratistatin and (+)-7-deoxypancratistatin involving the use and formation of cyclitols.

The allylic azide **70** has been sythesized in high e.e. by treating the *meso*-carbonate **71** with trimethylsilyl azide in the presence of Pd⁰ and a chiral phosphine ligand as catalyst. Compound **70** was further converted to (+)-conduramine E by use of a reaction involving a [3,3] sigmatropic rearrangement.[115]

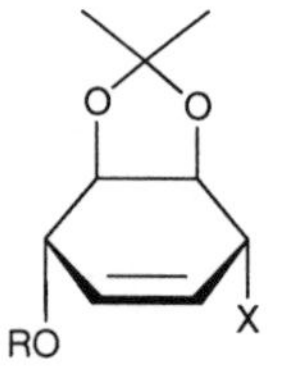

69 R = Tbdms, X = OH
70 R = CO₂Me, X = N₃
71 R = CO₂Me, X = OCO₂Me

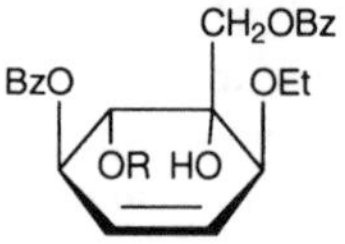

72 R = H
75 R = Ac

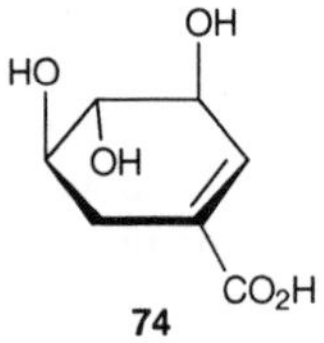

74

Cyclohexene derivatives **72** and **73** containing the unusual ethyl ether group and named uvarigranols C and D, respectively, have been extracted from the plant *Uvaria grandiflora*,[116] and the shikimic acid analogue **74** has been isolated from the plant *Sequoiadendron giganteum*.[117]

A 'one pot' enzymatic synthesis of (6R)- and (6S)-fluoroshikimic acid from erythrose 4-phosphate and both isomers of 3-fluorophosphoenolpyruvate in the presence of 3-deoxy-D-*arabino*-heptulosonic acid 7-phosphate synthase, followed by treatment with 3-dehydroquinate synthase and dehydroquinase simultaneously, then finally shikimate dehydrogenase has been described.[118]

Microbial oxidation of benzene derivatives producing benzene *cis*-diols have been utilized in the preparation of 6-β-hydroxyshikimic acid,[119] polyhydroxylated cyclohexane derivatives related to cyclophellitol[120] and carba-α-L-fucose.[121] It is worth noting here that the *meso*-diol produced by oxidation of benzene itself has been resolved with a lipase giving (1*R*, 2*S*)-1-acetoxy-2-hydroxycyclohexane-3,5-diene in >95% e.e.[122]

Benzene and toluene have also been transformed into cyclohexitols by way of a catalytic photoinduced charge-transfer osmylation in the presence of 0.22 M barium(II) chlorate. Higher concentrations of the chlorate cause a chlorine atom to be incorporated.[123] Hydrogenation of tetrahydroxyquinone in the presence of a large amount of palladium on charcoal produces *cis*-inositol in modest yield.[124]

1,3,5-Trideoxy-1,3,5-tris(dimethylamino)-*cis*-inositol has been shown to be a powerful ligand for hard, highly charged metal ions.[125]

(−)-Quinic acid has been used as starting material to make validamine, 2-*epi*-validamine,[126] valiolamine and several diasteromers of the latter.[127]

The carbocyclic analogues of *N*-acetylneuraminic acid and 3-deoxy-D-*manno*-2-octulosonic acid have been prepared from the Diels-Alder *endo*-adduct formed between furan and acrylic acid,[128] and the guanidino neuraminic acid analogue **75** has been prepared in which the cyclitiol ring was also formed by use of the Diels-Alder reaction.[129]

In an approach to the synthesis of natural, chiral 4-*epi*-sannamine, the *O*-acetyl derivative **76** has been prepared by kinetic resolution of the corresponding racemic alcohol using a lipase and vinyl acetate as acyl donor.[130]

75   76

The syntheses of carba-β-D-glycosylceramides with imino-[131,132,133] ether-[131] or sulfide-[131] linkages have been reported. Scheme 5 illustrates a route to a carba-glucosyl analogue starting from a known carba-sugar derivative (Vol. 26, p. 204, ref. 133).[131]

A synthesis of the carbocyclic analogue of uridine 5'-α-D-galactopyranosyl diphosphate (**77**) as an inhibitor of β-(1→4)-galactosyltransferase has been achieved. The cyclitol ring was formed *via* a Ferrier carbocyclic reaction.[134]

By use of standard chemistry, 1D-3-*O*-benzyl-1,2:4,5-di-*O*-cyclohexylidene-*myo*-inositol has been converted into the carba-disaccharide **78** as a potential intermediate for preparing glycosylphosphatidyl inositols.[135]

The carba-mannosyl trisaccharide **79** has been built up from the addition of

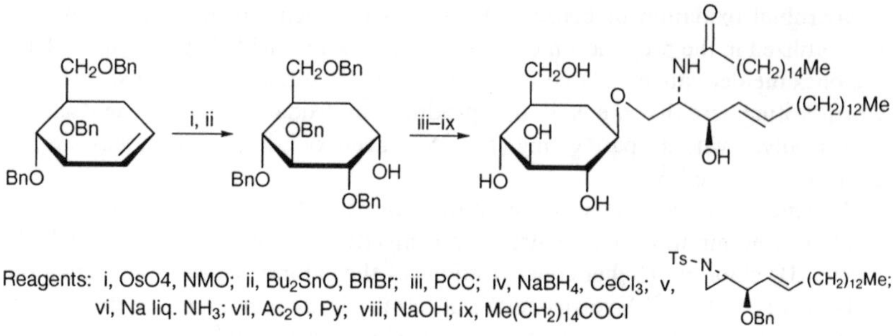

Reagents: i, OsO₄, NMO; ii, Bu₂SnO, BnBr; iii, PCC; iv, NaBH₄, CeCl₃; v, [structure]; vi, Na liq. NH₃; vii, Ac₂O, Py; viii, NaOH; ix, Me(CH₂)₁₄COCl

Scheme 5

methyl 2-*O*-benzyl-4,6-*O*-benzylidene-α-D-mannopyranose to epoxide (see Vol. 27, p. 222, ref. 137) **80** then reaction of another equivalent of **80** with the primary hydroxyl group of the D-mannopyranose sugar after protecting group manipulations.[136] Compound **80** has also been used to prepare β-D-GLC*p*NAc-(1→2)-carba-α-D-Man*p*-(1→6)-β-D-GLC*p*-O(CH₂)₇Me as an acceptor analogue of *N*-acetylglucosaminyl transferase-(V).[137]

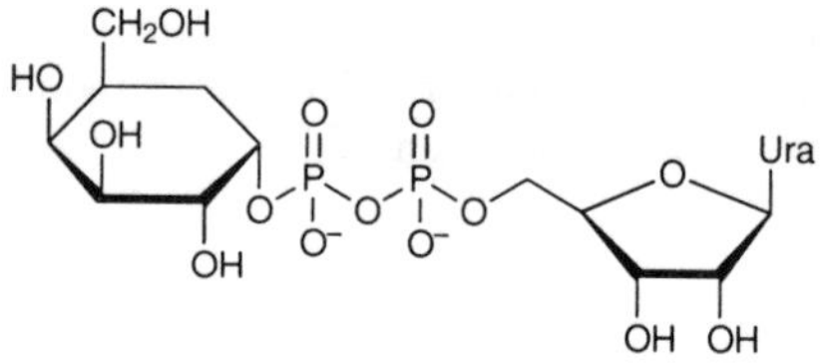

77

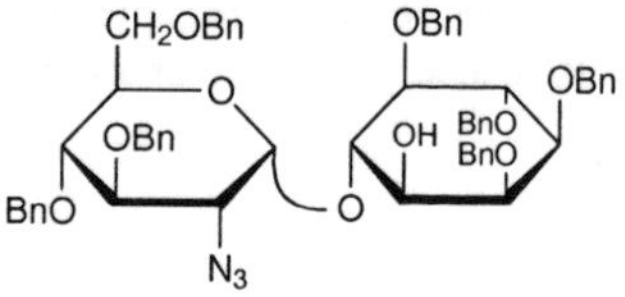

78

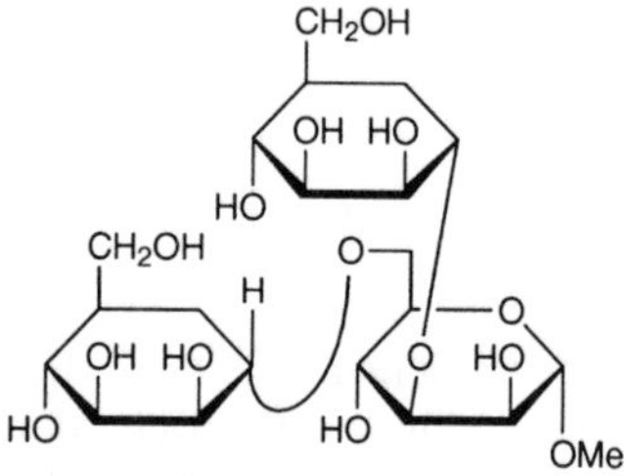

79

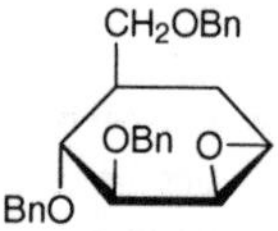

80

The carba-disaccharide **81** has been designed as an antigen for preparing glycoside-bond forming catalytic antibodies with catalytic groups. In its synthesis the addition of an acetylide anion to a carbocyclic ketone, followed by an OH→N₃ replacement, formed the key steps.[138]

The *E. coli* catalysed β-D-galactosidase transglycosylation of 2-nitrophenyl D-galactopyranoside with racemic (3,5/4,6)-3,6-diazido-4,5-dihydroxycyclohexene afforded glycoside **82** which was hydrogenated to diamine **83** or heated in methanol to effect a sigmatropic rearrangement which after hydrogenation gave **84** together with the isomer derived from rearrangement in the alternative sense. Both **83** and **84** were competitive inhibitors of β-D-galactosidase.[139]

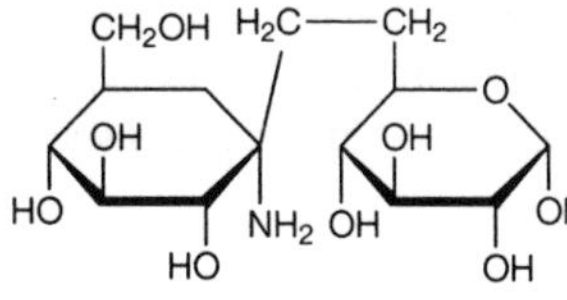

81 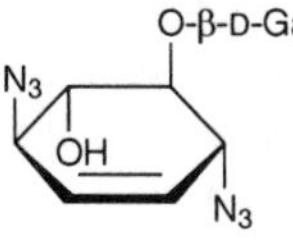　　**82** 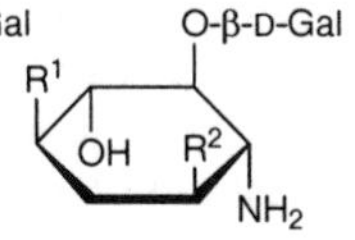　　**83** $R^1 = NH_2$, $R^2 = H$
　　　　　　　　　　84 $R^1 = H$, $R^2 = NH_2$

The α-D-mannosyl-D-*myo*-inositol compound **85** has been reported in which the cyclitol ring was derived from an intermediate α-D-mannosyl oxanorbornane glycoside.[140]

The Ferrier reaction has been utilized in the construction of various amino cyclitols. Thus, starting from *N*-acetylglucosamine, 1D-(1,3,5/2,4)-4-acetamido-5-amino-1,2,3-cyclohexanetriol has been prepared and glycosylated with a D-glucuronic acid derivative to produce an intermediate which, after deprotection, afforded the carba-disaccharide **86**.[141] The cyclitols **87** (R = α-D-arabinofuranosyl or 3-amino-2,3,6-trideoxy-α-L-arabinohexopyranosyl), have been prepared from

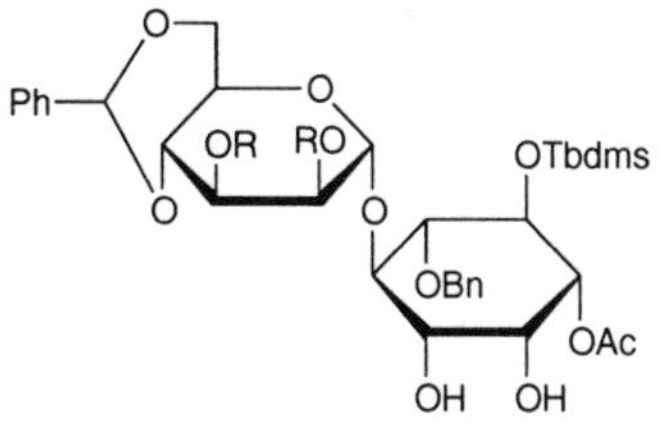

85 R = Et₃Si　　　　　　　　**86**

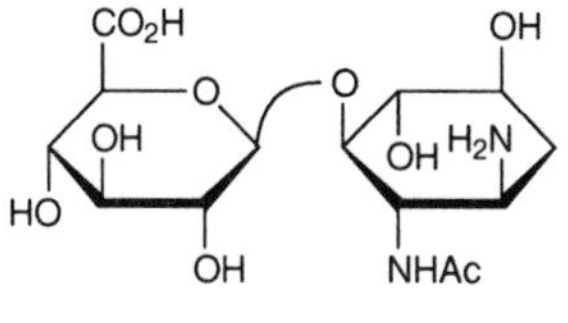

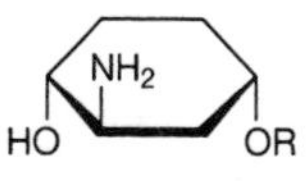

87 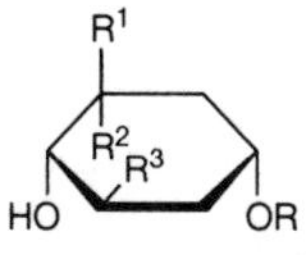　　　**88** $R^1 = R^2 = H$, $R^3 = NH_2$
　　　　　89 $R^1 = R^2 = H$, $R^3 = OH$
　　　　　90 $R^1 = R^2 = NHCOCF_3$, $R^2 = H$
　　　　　91 $R^2 = R^3 = NHCOCF_3$, $R^1 = H$

methyl 3-azido-4-benzoyl-2,3,6-trideoxy-α-D-*erythro*-hex-5-enopyranose by the aforementioned cyclization process.[142]

Carba-disaccharides **88-91** (R = α-D-arabinofuranosyl) have also been synthesized as potential antibiotics.[143]

Several examples of other glycosylated inositols are noted in Chapters 3 and 4. The syntheses of carba-disaccharides related to salbostatin and fortimycins are metioned in Chapter 19, and the syntheses of 6-membered carbocyclic nucleosides and the use of cyclitols as chiral auxilliaries in Diels-Alder reactions are covered in Chapters 20 and 24, respectively.

2.3 Inositol Phosphates and Derivatives. – A review on the recent developments in the synthesis and application of phosphatidylinositols,[144] and one on the biochemistry and chemical synthesis of *myo*-inositol 1,4,5-trisphosphate[145] have appeared.

The oxazoline method and two modifications of the *H*-phosphonate method have been used to obtain glycosyl phosphatidylinositol derivatives.[146]

The tritiated inositol derivatives **92** have been prepared as substrates for *T. brucei* α-D-GLCpNAc-PI de-*N*-acetylase, and derivatives **93** as inhibitors of the enzyme.[147]

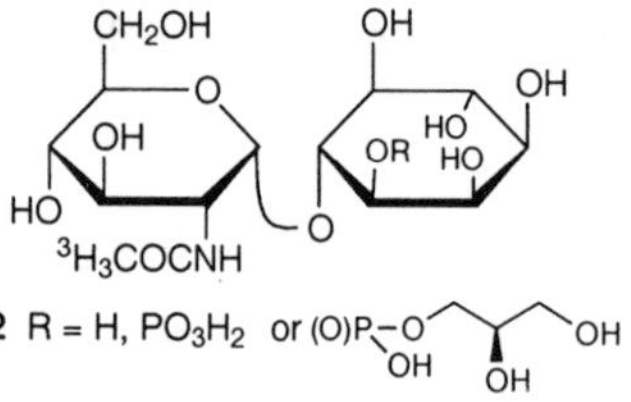

92 R = H, PO₃H₂ or (O)P–O...OH

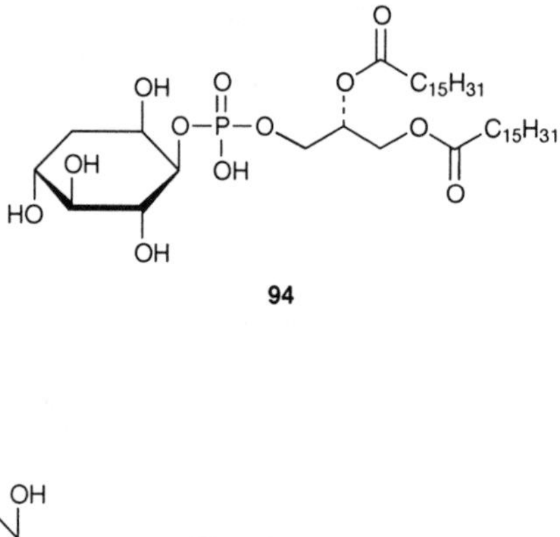

94

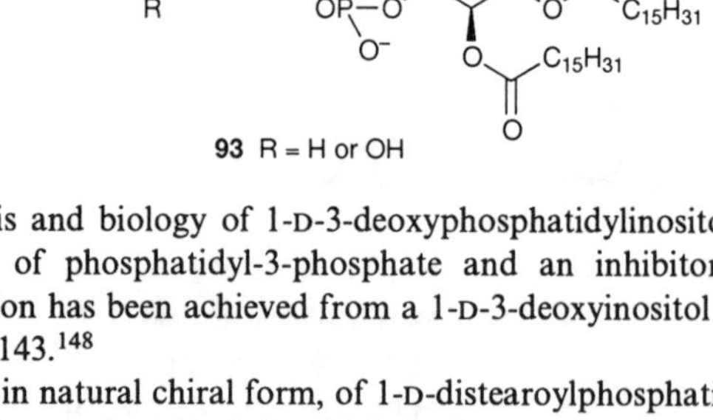

93 R = H or OH

The synthesis and biology of 1-D-3-deoxyphosphatidylinositol **94**, a putative antimetabolite of phosphatidyl-3-phosphate and an inhibitor of cancer cell colony formation has been achieved from a 1-D-3-deoxyinositol reported in Vol. 27, p. 222, ref. 143.[148]

A synthesis, in natural chiral form, of 1-D-distearoylphosphatidyl-*myo*-inositol 3,4,5-tris(dihydrogenphosphate), in which the key step involves a regioselective 1-

O-phosphatidylation of the protected inositol **95** with glycerol derivative **96** by the phoshite-phosphonium salt method has been reported.[149] (See Vol. 28, p. 245, ref. 169 for a racemic synthesis).

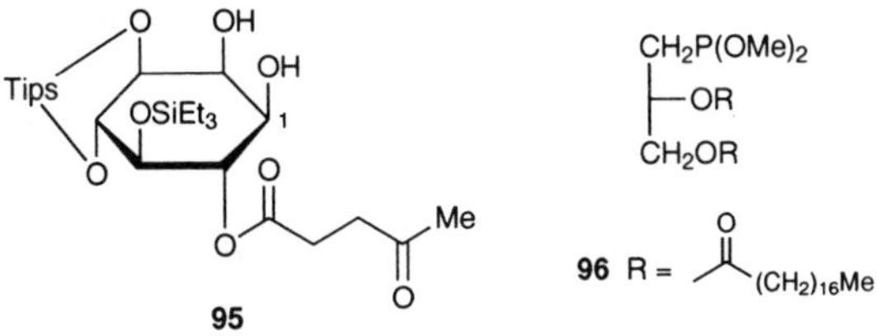

95

96 R =

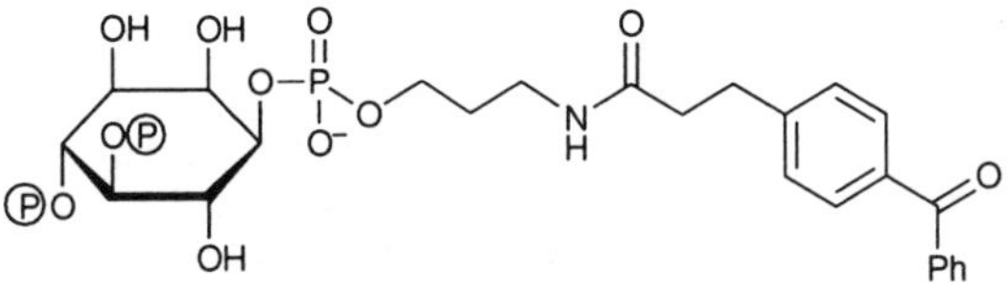

97

An optically pure tethered inositol trisphosphate **97** bearing a selective photo-affinity label for modification of the ligand binding site of inositol trisphosphate receptor proteins has been reported.[150]

A novel phosphorylation method for *myo*-inositol derivatives involving application of the Atherton-Todd reaction under solid-liquid phase transfer conditions has been developed.[151]

The synthesis of 1-*O*-alkyl-(Me, Bn, Oct, dodecyl etc.) and 1-*O*-acyl (acetyl or stearoyl)-*myo*-inositol 3,4,5-trisphosphate as analogues of phosphatidyl *myo*-inositol 3,4,5-trisphosphate[152] and of *myo*-inositol 4,5,6-trisphosphate as an analogue of *myo*-inositol 1,4,5-trisphosphate[153] have been reported.

3-Deoxy-D-*myo*-inositol-1,4,5-trisphosphate and 3-deoxy-D-*muco*-inositol 1,4,5-trisphosphate (the C-4 epimer of D-*myo*-inositol) in which the cyclitol ring is formed by mercury(II)-catalysed reaction of hex-5-enopyranoses bearing allyl protecting groups the latter which did not interfere with the catalysis, have been reported. Biological studies with these compounds indicate that a 4,5-*trans*-relationship of phosphate groups should be preserved for good binding to the receptor of D-*myo*-inositol 1,4,5-trisphosphate.[154]

Standard routes for preparing D-*myo*-inositol 3,4,5,6- and 1,4,5,6-tetrakisphos-phates and their conversion through esterification into membrane permeable derivatives have been described.[155]

Heating readily prepared 1,4-di-*O*-benzoyl-D,L-*myo*-inositol in a pyridine-water mixture effects a migration of the benzoate groups giving rise to all nine possible isomeric dibenzoate derivatives. The mixture could be separated by a combination of fractional crystallization and silica gel chromatography into each pure individual dibenzoate isomer which after subjecting to a phosphorylation-deprotection sequence afforded all nine isomers of *myo*-inositol tetrakisphosphate.[156] A kinetic investigation of the base-catalysed migration reaction of the dibenzoates has also been reported.[157]

L-Quebrachitol has served as a chiral source of 1-D-3-deoxy-3-fluoro-2,4,5-*myo*-inositol trisphosphate, 1-D-3-deoxy-3-fluoro-1,2,4,5-*myo*-inositol tetrakisphosphate and of 1-D-1,2,4,5-*myo*-inositol tetrakisphosphate. The last two compounds are nearly equipotent to 1-D-*myo*-inositol 1,4,5 trisphosphate in both binding and Ca^{2+} release experiments.[158]

The synthesis by established routes of racemic *myo*-inositol 1,4,6-trisphosphorothioate and *myo*-inositol 1,3,4-trisphosphorothioate as low intrinsic partial agonists at the platelet *myo*-inositol 1,4,5-trisphosphate receptor have been reported.[159]

Phytic acid (*myo*-inositol hexaphosphate) can be desymmetrized using phophatases found in Bakers' yeast which allows the isolation of D-*myo*-inositol 1,2,6-trisphosphate. This was further used to prepare 3,4,5-tri-*O*-phenylcarbamoyl D-*myo*-inositol.[160]

myo-2-Inosose 1-phosphate has been synthesized by conventional means as a competitive inhibitor of *myo*-inositol 1-phosphate synthase.[161]

The deoxygenated inositol analogue **98** (R = PO_3^{2-}) is a known potent inhibitor of *myo*-inositol 1-phosphatase (see Vol. 25, p. 218, ref. 131). A set of analogues with less polar groups than phosphate have been prepared: **98** R = $P(O)(O^-)(O^iPr)$, $P(O)(O^-)(Me)$, $P(O)(Me)_2$, SO_3^-, SO_2NH_2 and SO_2NHAc.[162]

The deoxygenated inositol derivatives **99** and **100** have been synthesized in chiral form as mechanism-based inhibitors and probes for *myo*-inositol monophosphatase.[163]

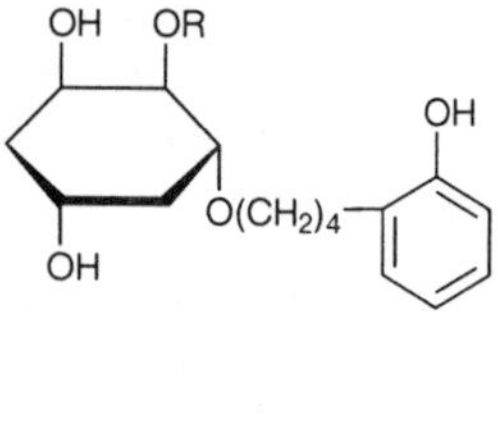

98 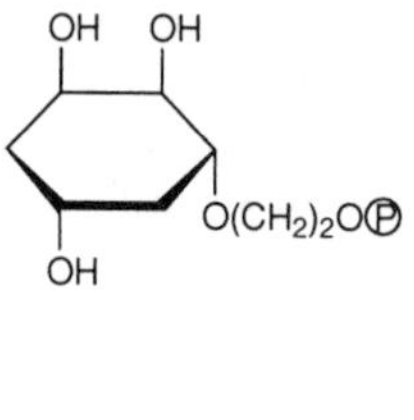   **99** 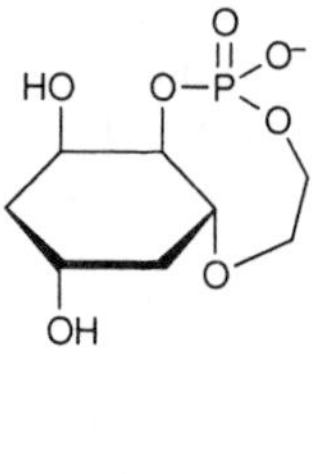   **100**

The synthesis of phosphorylated xylopyranosides and mannopyrannosides as mimics for inositol 1,4,5- and inositol 1,2,6-trisphosphate, respectively are covered in Chapters 3 and 7, respectively.

References

1   S.R. Magnuson, *Tetrahedron*, 1995, **51**, 2167.
2   R.M. Nemba and M. Fah, *Tetrahedron*, 1995, **51**, 3838.

3   Y. Mori and H. Furukawa, *Tetrahedron*, 1995, **51**, 6725.
4   B. Sing, R. Mathur and A. Kumar, *Oxid. Commun.*, 1994, **17**, 24 (*Chem. Abstr.*, 1995, **122**, 10 410).
5   P. Rongere, N. Morel-Desrosiers and J.-P. Morel, *J. Chem. Soc., Faraday Trans.*, 1995, 91 (*Chem. Abstr.*, 1995, **123**, 340 644).
6   N. Matsuda and M. Kikuchi, *Chem. Pharm. Bull.*, 1995, **43**, 1049.
7   T. Fukamizo, T. Ohkawa, Y. Ikeda, T. Torikata and S. Goto, *Carbohydr. Res.*, 1995, **267**, 135.
8   S. Yamasaki and H. Tsutsumi, *Bull. Chem. Soc. Jpn.*, 1995, **68**, 123.
9   S. Yamasaki, Y. Ohashi, H. Tsutsumi and K. Sujii, *Bull. Chem. Soc. Jpn.*, 1995, **68**, 146.
10   W.V. Dahlhoff, K. Radkowski, K. Riehl and P. Zugenmaier, *Z. Naturforsch., B: Chem. Sci.*, 1995, **50**, 1079 (*Chem. Abstr.*, 1995, **123**, 286 428).
11   H.W.C. Raaijmakers, E.G. Arnouts, B. Zwanenburg, G.J.F. Chittenden and H.A. van Doren, *Recl. Trav. Chim. Pays-Bas*, 1995, **114**, 301.
12   M. Benazza, M. Massoui, R. Uzan and G. Demailly, *Carbohydr. Res.*, 1995, **275**, 421.
13   P. Mishnick, M. Lange, M. Gohdes, A. Stein and K. Petzold, *Carbohydr. Res.*, 1995, **277**, 179.
14   D. Crich, Q. Yao and G.F. Filzen, *J. Am. Chem. Soc.*, 1995, **117**, 11455.
15   H. Pettersson-Fasth, A. Gogoll and J.-E. Bäckvall, *J. Org. Chem.*, 1995, **60**, 1848.
16   J.L. Del Valle, T. Torroba, S. Marcaccini, P. Paoli and D.J. Williams, *Tetrahedron*, 1995, **51**, 8259.
17   S. Sarbajna, S.K. Das and N. Roy, *Carbohydr. Res.*, 1995, **270**, 93.
18   A.J. Fairbanks and P. Sinaÿ, *Tetrahedron Lett.*, 1995, **36**, 893.
19   P. Munier, A. Krusinski, D. Picq and D. Anker, *Tetrahedron*, 1995, **51**, 1229.
20   N. Shinohara, T. Yamazaki and T. Kitazume, *Bioorg. Med. Chem. Lett.*, 1995, **5**, 923.
21   P.J. Edwards and S.V. Ley, *Synlett.*, 1995, 898.
22   O. Yamada and K. Ogasawara, *Synthesis*, 1995, 1291.
23   M. Flashe and H.-D. Scharf, *Tetrahedron: Asymm.*, 1995, **6**, 1543.
24   E. Dequeker, F. Compernolle, S. Toppet and G. Hoornaert, *Tetrahedron*, 1995, **51**, 5877.
25   P.A. Wade and S.G. D'Ambrosio, *J. Carbohydr. Chem.*, 1995, **14**, 1329.
26   F. Azzena, F. Calvani, P. Crotti, C. Gardelli, F. Macchia and M. Pineschi, *Tetrahedron*, 1995, **51**, 10601.
27   M.-A. Badet-Denisot, C. Leriche, F. Massière and B. Badet, *Bioorg. Med. Chem. Lett.*, 1995, **5**, 815.
28   M. Mikami and S. Shinkai, *Chem. Lett.*, 1995, 603.
29   M. Mikami and S. Shinkai, *J. Chem. Soc., Chem. Commun.*, 1995, 153.
30   P. Klüfers and J. Schumacher, *Angew. Chem., Int. Ed. Engl.*, 1995, **34**, 2119.
31   A. Wisinewski, *Wiad. Chem.*, 1992, **46**, 847 (*Chem. Abstr.*, 1995, **123**, 169 988).
32   N. Wang and G.R. Gray, *Carbohydr. Res.*, 1995, **274**, 45.
33   N. Wang, L.E. Elvebak (II) and G.R. Gray, *Carbohydr. Res.*, 1995, **274**, 59.
34   C.R. Rozanas, N. Wang, K. Vidlock and G.R. Gray, *Carbohydr. Res.*, 1995, **274**, 99.
35   L.E. Elvebak (II), H.J. Cha, P. McNally and G.R. Gray, *Carbohydr. Res.*, 1995, **274**, 71.
36   L.E. Elvebak (II) and G.R. Gray, *Carbohydr. Res.*, 1995, **274**, 85.
37   L.E. Elvebak (II), C. Abbott, S. Wall and G.R. Gray, *Carbohydr. Res.*, 1995, **269**, 1.
38   J. Mulzer, C. Pietschmann, B. Schöllhorn, J. Buschmann and P. Luger, *Liebigs Ann. Chem.*, 1995, 1433.

39   M.A. Shalaky, F.R. Fronczek, Y. Lee and E.S. Younathan, *Carbohydr. Res.*, 1995, **269**, 191.

40   V. Popsavin, O. Beric, M. Popsavin, J. Csanadi and D. Miljkovic, *Carbohydr. Res.*, 1995, **269**, 343.

41   A. Jeffery and V. Nair, *Tetrahedron Lett.*, 1995, **36**, 3627.

42   R. Tamion, F. Marsals and G. Queguiner, *Tetrahedron Lett.*, 1995, **36**, 2761.

43   A. Bhattacharjee, A. Bhattacharjya and A. Patra, *Tetrahedron Lett.*, 1995, **36**, 4677.

44   A.B. Hughes and A.J. Rudge, *Nat. Prod. Rep.*, 1994, **11**, 135 (*Chem. Abstr.*, 1995, **122**, 81 752).

45   N. Asano, H. Kiso, K. Oseki, E. Tomioka, K. Matsui, M. Okamoto and M. Baba, *J. Med. Chem.*, 1995, **38**, 2349.

46   F. Hildebrandt, J.A. Schröter, C. Tschierske, R. Festag, R. Kleppinger and J.H. Wendorff, *Angew. Chem. Int. Ed. Engl.*, 1995, **34**, 1631.

47   J. Fitremann, A. Duréault and J.-C. Depezay, *Tetrahedron*, 1995, **51**, 9581.

48   J. Fitremann, A. Duréault and J.-C. Depezay, *Synlett.*, 1995, 235.

49   L. Campanini, A. Duréault and J.-C. Depezay, *Tetrahedron Lett.* 1995, **36**, 8015.

50   S. Watanabe, H. Kato, K. Nagayama and H. Abe, *Biosci. Biotech. Biochem.*, 1995, **59**, 936.

51   M.S. Chorghade and C.T. Czeke, *Heterocycles*, 1995, **40**, 213.

52   K.H. Park, *Heterocycles*, 1995, **41**, 1715.

53   H. Yoda, H. Yamazaki, M. Kawauchi and K. Takabe, *Tetrahedron: Asymm.*, 1995, **6**, 2669.

54   J.M.J. Tronchet, B. Balkadjian, G. Zosimo-Landolfo, F. Barbalat-Rey, P. Lichtle, A. Ricca, I. Komaromi, G. Bernardinelli and M. Geoffroy, *J. Carbohydr. Chem.*, 1995, **14**, 17.

55   S. Hiranuma, T. Shimizu, T. Nakata, T. Kajimoto and C.-H. Wong, *Tetrahedron Lett.*, 1995, **36**, 8247.

56   L. Cipola, L. Lay, F. Nicotra, C. Pangrazio and L. Panza, *Tetrahedron* 1995, **51**, 4679.

57   C.-H. Wong, L. Provencher, J.A. Porco, S.-H. Jung, Y.-F. Wang, L. Chen, R. Wang and D.H. Steensma, *J. Org. Chem.*, 1996, **60**, 1492.

58   K. Bennis, J. Gelas and C. Thomassigny, *Carbohydr. Res.*, 1995, **279**, 307.

59   G. Mikkelsen, T.V. Christensen, M. Bols, I. Lundt and M.R. Sierks, *Tetrahedron Lett.*, 1995, **36**, 6541.

60   A. Berger, M. Ebner and A.E. Stütz, *Tetrahedron Lett.*, 1995, **36**, 4989.

61   L.A.G.M. van den Broek, D.J. Vermaas, F.J. van Kemenade, M.C.C.A. Tan, F.T.M. Rotteveel, P. Zandberg, T.D. Butters, F. Miedema, H.L. Ploegh and C.A.A. van Boeckel, *Recl. Trav. Chim. Pays-Bas*, 1994, **113**, 507 (*Chem. Abstr.*, 1995, **122**, 240 250).

62   A. Kilonda, F. Compernolle and G.J. Hoornaert, *J. Org. Chem.*, 1995, **60**, 5820.

63   C.-K. Lee and H. Jiang, *J. Carbohydr. Chem.*, 1995, **14**, 407.

64   T. Kiguchi, K. Tajiri, I. Ninomiya, T. Naito and H. Hiramatsu, *Tetrahedron Lett.*, 1995, **36**, 253.

65   M. Kiso, K. Ando, H. Inagaki, H. Ishida and A. Hasegawa, *Carbohydr. Res.*, 1995, **272**, 159.

66   I.K. Khanna, F.J. Koszyk, M.A. Stealy, R.M. Weier, J. Julien, R.A. Mueller, S.N. Rao, L. Swenton, D.P. Getman, D.A. DeCrescenzo and R.M. Heintz, *J. Carbohydr. Chem.*, 1995, **14**, 843.

67   S. Takahashi and H. Kuzuhara, *Biosci. Biotech. Biochem.*, 1995, **59**, 762.

68   M. Ichikawa and Y. Ichikawa, *Bioorg. Med. Chem.*, 1995, **3**, 161 (*Chem. Abstr.*, 1995, **123**, 56 451).

69   M. Ichikawa, Y. Igarashi and Y. Ichikawa, *Tetrahedron Lett.*, 1995, **36**, 1767.
70   Y. Ishikawa and Y. Igarashi, *Tetrahedron Lett.*, 1995, **36**, 4585.
71   J. Lehmann and B. Rab, *Carbohydr. Res.*, 1995, **272**, C11.
72   O.R. Martin, F. Xie and Li Liu, *Tetrahedron Lett.*, 1995, **36**, 4027.
73   O.R. Martin and O.M. Saavedra, *Tetrahedron Lett.*, 1995, **36**, 799.
74   I. Lundt and R. Madsen, *Synthesis*, 1995, 787.
75   G. Legler, A.E. Stütz and H. Immich, *Carbohydr. Res.*, 1995, **272**, 17.
76   F.-D. Boyer, A. Pancrazi and Y.-L. Lallemand, *Synth. Commun.*, 1995, **25**, 1099.
77   A. Kilonda, E. Dequeker, F. Compernolle, P. Delbeke, S. Toppet, Babady-Bila and
      G.I. Hoornaert, *Tetrahedron*, 1995, **51**, 849.
78   B.B. Lohray, Y. Jayamma and M. Chatterjee, *J. Org. Chem.*, 1995, **60**, 5958.
79   J.T. Slama, N. Aboul-Ela, D.M. Goli, B.V. Cheesman, A.M. Simmons and M.K.
      Jacobson, *J. Med. Chem.*, 1995, **38**, 389.
80   N. Ikota, *Heterocycles*, 1995, **41**, 983.
81   H.-J. Altenbach and K. Himmeldirk, *Tetrahedron: Asymm.*, 1995, **6**, 1077.
82   O.D. Shärer J.-Y. Ortholand, A. Ganesan, K. Ezaz-Nikpay and G.L. Verdine, *J.
      Am. Chem. Soc.*, 1995, **117**, 6623.
83   I.A. Prokhorenko, V.A. Korshun, A.A. Petrov, S.V. Gontarev and Y.A. Berlin,
      Bioorg. Med. Chem. Lett., 1995, **5**, 2081.
84   S. Zhi-cai, Z. Chun-min and L. Guo-qiuang, *Heterocycles*, 1995, **41**, 277.
85   Z. Shi and G. Lin, *Tetrahedron*, 1995, **51**, 2427.
86   A. Defoin, H. Sarazin and J. Streith, *Synlett.*, 1995, 1187.
87   J.-B. Behr, A. Defoin, N. Mahmood and J. Streith, *Helv. Chim. Acta*, 1995, **78**, 1166.
88   J. Streith, A. Boiron, J.-L. Paillaud, E.-M. Rodriguez-Perez, C. Strehler,
      T. Tschamber and M. Zehnder, *Helv. Chim. Acta*, 1995, **78**, 61.
89   D. Sames and R. Polt, *Synlett.*, 1995, 552.
90   H.-J. Altenbach and R. Wischnat, *Tetrahedron Lett.*, 1995, **36**, 4983.
91   C.R. Johnson, B.M. Nerurkar, A. Golebiowski, H. Sundram and J.L. Esker, *J.
      Chem. Soc., Chem. Comm.*, 1995, 1139.
92   C.R. Johnson, A. Golebiowski, H. Sundram, M.W. Miller and R.E. Dwaihy,
      Tetrahedron Lett., 1995, **36**, 653.
93   Y. Hanzawa, H. Ito and T. Taguchi, *Synlett.*, 1995, 299.
94   A.M. Riley, D.J. Jenkins and B.V.L. Potter, *J. Am. Chem. Soc.*, 1995, **117**, 3300.
95   A. Chénedé, P. Pothier, M. Sollogoub, A.J. Fairbanks and P. Sinaÿ, *J. Chem. Soc.,
      Chem. Commun.*, 1995, 1373.
96   J.L. Chiara, J. Marco-Contelles, N. Khiar, P. Gallego, C. Destabel and M. Bernabé,
      J. Org. Chem., 1995, **60**, 6010.
97   W.-C. Chou, C. Fotsch and C.-H. Wong, *J. Org. Chem.*, 1995, **60**, 2916.
98   B.L. Bray, S.C. Dolan, B. Halter, J.W. Lackey, M.B. Schilling and D.J. Tapolczay,
      Tetrahedron Lett., 1995, **36**, 4483.
99   C.F. Palmer and R. McCague, *J. Chem. Soc., Perkin Trans. 1*, 1995, 1201.
100   J. Schachtner, H.-D. Stachel and K. Polborn, *Tetrahedron*, 1995, **51**, 9005.
101   C. Marschner, J. Baumgartner and H. Griengl, *J. Org. Chem.*, 1996, **60**, 5224.
102   C. Noritaka and S. Ogawa, *Yuki Gosei Kagaku Kyokaishi*, 1995, **53**, 858 (*Chem.
      Abstr.*, 1995, **123**, 286 405).
103   N. Yamauchi and K. Kakinuma, *J. Org. Chem.*, 1995, **60**, 5614.
104   J. Cleophax, D. Dubreuil, S.D. Gero, A. Loupy, M. Viera de Almeida, A.D. Da
      Silva, G. Vass, E. Bischoff, E. Perzborn, G. Hecker and O. Lockhoff, *Bioorg. Med.
      Chem. Lett.*, 1995, **5**, 831.
105   R. Aneja, S.G. Aneja and A. Parra, *Tetrahedron: Asymm.*, 1995, **6**, 17.

106   T. Desai, J. Gigg and R. Gigg, *Carbohydr. Res.*, 1995, **277**, C5.

107   K.M. Pietrusiewicz and G.M. Salamonczyk, *Synth. Commun.*, 1995, **25**, 1863.

108   J.E. Anderson, S.J. Angyal and D.C. Craig, *Carbohydr. Res.*, 1995, **272**, 141.

109   S. Ballereau, P. Guédat, B. Spiess, N. Rehnberg and G. Schlewer, *Tetrahedron Lett.*, 1995, **36**, 7449.

110   H.-X. Zhai, P.-S. Lei, J.C. Morris, K. Mensa-Wilmot and T.Y. Shen, *Tetrahedron Lett.*, 1995, **36**, 7403.

111   R. Downham, P.J. Edwards, D.A. Entwistle, A.B. Hughes, K.S. Kim and S.V. Ley, *Tetrahedron: Asymm.*, 1995, **6**, 2403.

112   F. Tagliaferri, S.G. Johnson, T.F. Seiple and D.C. Baker, *Carbohydr. Res.*, 1995, **266**, 301.

113   J.L. Chiara and N. Valle, *Tetrahedron: Asymm.*, 1995, **6**, 1895.

114   T.J. Doyle, D. VanDerveer and J. Haseltine, *Tetrahedron Lett.*, 1995, **36**, 6197.

115   B.M. Trost and S.R. Pulley, *Tetrahedron Lett.*, 1995, **36**, 8737.

116   P. Xi-Ping and Y. De-Quan, *Phytochemistry*, 1995, **40**, 1709.

117   H. Geiger, S. El-Dessouki and T. Seeger, *Phytochemistry*, 1995, **40**, 1705.

118   P.J. Duggan, E. Parker, J. Coggins and C. Abell, *Bioorg. Med. Chem. Lett.*, 1995, **5**, 2347.

119   A.J. Blacker, R.J. Booth, G.M. Davies and J.K. Sutherland, *J. Chem. Soc., Perkin Trans. 1*, 1995, 2861.

120   S.M. Roberts and P.W. Sutton, *J. Chem. Soc., Perkin Trans. 1*, 1995, 1499.

121   H.A.J. Carless and S.S. Malik, *J. Chem. Soc., Chem. Commun.*, 1995, 2447.

122   G. Nicolosi, A. Patti, M. Piattelli and C. Sanfilippo, *Tetrahedron Lett.*, 1995, **36**, 6545.

123   W.B. Motherwell and A.S. Williams, *Angew. Chem., Int. Ed. Engl.*, 1995, **34**, 2031.

124   S.J. Angyal, L. Odier and M.E. Tate, *Carbohydr. Res.*, 1995, **266**, 143.

125   K. Hegetschweiler, T. Kradolfer, V. Gramlich and R.D. Hancock, *Eur. Chem. J.*, 1995, **1**, 74.

126   T.K.M. Shing and V.W.-F. Tai, *J. Org. Chem.*, 1995, **60**, 5332.

127   T.K.M. Shing and L.H. Wan, *Angew. Chem., Int. Ed. Engl.*, 1995, **34**, 1643.

128   S. Ogawa, M. Yoshikawa, T. Taki, S. Yokoi and N. Chida, *Carbohydr. Res.*, 1995, **269**, 53.

129   M. Chandler, R. Conroy, A.W.J. Cooper, R.B. Lamont, J.J. Scicinski, J.E. Smart, R. Storer, N. G. Weir, R.D. Wilson and P.G. Wyatt, *J. Chem. Soc., Perkin Trans.1*, 1995, 1189.

130   F. Yang, C. Hoenke and H. Prinzbach, *Tetrahedron Lett.*, 1995, **36**, 5151.

131   H. Tsunoda and S. Ogawa, *Liebigs Ann. Chem.*, 1995, 267.

132   H. Tsunoda, J. Inokuchi, K. Yamagishi and S. Ogawa, *Liebigs Ann. Chem.*, 1995, 279.

133   S. Ogawa and H. Tsunoda, *Methods Enzymol.*, 1994, **247** (Neoglycoconjugates, Pt. B), 136 (*Chem. Abstr.*, 1995, **122**, 314 994).

134   H. Yuasa, M.M. Palcic and O. Hindsgaul, *Can. J. Chem.*, 1995, **73**, 2190.

135   S. Cottaz, J.S. Brimacombe and M.A.J. Ferguson, *Carbohydr. Res.*, 1995, **270**, 85.

136   S. Ogawa, S. Sasaki and H. Tsunoda, *Carbohyd. Res.*, 1995, **274**, 183.

137   S. Ogawa, T. Furuya, H. Tsunoda, O. Hindsgaul, K. Stangier and M.M. Palcic, *Carbohydr. Res.*, 1995, **271**, 197.

138   C. Barbaud, M. Bols, I. Lundt and M.R. Sierks, *Tetrahedron*, 1995, **51**, 9063.

139   J. Lehmann and B. Rob, *Carbohyd. Res.*, 1995, **276**, 199.

140   O. Arjona, A. de Dios, C. Montero and J. Plumet, *Tetrahedron*, 1995, **51**, 9191.

141   L.-X. Wang, N. Sakairi and H. Kuzuhara, *Carbohydr. Res.*, 1995, **275**, 33.

142   I.F. Pelyvás, M. Mádi-Puskás, Z.G. Tóth, Z. Varga, G. Batta and F. Sztaricskai, *Carbohydr. Res.*, 1995, **272**, C5.

143   I.F. Pelyvás, M. Mádi-Puskás, Z.G. Tóth, Z. Varga, M. Hornyak, G. Batta and F. Sztaricskai, *J. Antibiotics*, 1995, **48**, 683.

144   M.S. Shashidhar, *Proc.-Indian Acad. Sci., Chem. Sci.*, 1994, **106**, 1231 (*Chem. Abstr.*, 1995 **122**, 240 162).

145   B.V.L. Potter and D. Lampe, *Angew. Chem., Int. Ed. Engl.*, 1995, **34**, 1933.

146   N.S. Shastina, L.I. Einisman, A.E. Stepanov, V.I. Shvets, *Bioorg. Khim.*, 1994, **20**, 71 (*Chem. Abstr.* 1995, **122**, 214 359).

147   S. Cottaz, J.S. Brimacombe, M.A.J Ferguson, *J. Chem. Soc., Perkin Trans. 1*, 1995, 1673.

148   A.P. Kozikowski, J.J. Kiddle, T. Frew, M. Berggren and G. Powis, *J. Med. Chem.*, 1995, **38**, 1053.

149   Y. Watanabe, M. Tomioka and S. Ozaki, *Tetrahedron*, 1995, **51**, 8969.

150   G. Dormán, J. Chen and G.D. Prestwich, *Tetrahedron Lett.*, 1995, **36**, 8719.

151   Y.-X. Ding, S.-F. Zhou and J. Shen, *Huaxue Xuebao*, 1995, **53**, 305 (*Chem. Abstr.*, 1995, **123**, 33 539).

152   T. Sawada, R. Shirai, Y. Matsuo, Y. Kabuyama, K. Kimura, Y. Fukui, Y. Hashimoto and S. Iwasaki, *Bioorg. Med. Chem. Lett.*, 1995, **5**, 2263.

153   L. Schmitt, B. Spiess and G. Schlewer, *Bioorg. Med. Chem. Lett.* 1995, **5**, 1225.

154   E. Poirot, H. Bourdon, F. Chrétien, Y. Chapleur, B. Berthon, M Hilly, J.-P. Mauger and G. Guillon, *Bioorg. Med. Chem. Lett.*, 1995, **5**, 569.

155   S. Roemer, M.T. Rudolf, C. Stadler and C. Schultz, *J. Chem. Soc., Chem. Commun.*, 1995, 411.

156   S.-K. Chung and Y.-T. Chang, *J. Chem. Soc., Chem. Commun.*, 1995, 11.

157   S.-K. Chung and Y.-T. Chang, *J. Chem. Soc., Chem. Commun.*, 1995, 13.

158   A.P. Kozikowski, A.H. Fauq, R.A. Wilcox and S.R. Nahorski, *Bioorg. Med. Chem. Lett.*, 1995, **5**, 1295.

159   S.J. Mills, A.M. Reiley, C.T. Murphy, A.J. Bullock, J. Westwick and B.V.L. Potter, *Bioorg. Med. Chem. Lett.*, 1995, **5**, 203.

160   C. Blum, S. Karlsson, G. Schlewer, B. Spiess and N. Rehnberg, *Tetrahedron Lett.*, 1995, **36**, 7239.

161   M.E. Migaud and J.W. Frost, *J. Am. Chem. Soc.*, 1995, **117**, 5154.

162   A.M.P. van Steijn, H.A.M. Willems, Th. De Boer, J.L.T. Guerts and C.A.A. van Boekel, *Bioorg. Med. Chem. Lett.*, 1995, **5**, 469.

163   J. Schulz, J. Wilkie, P. Lightfoot, T. Rutherford and D. Gani, *J. Chem. Soc., Chem. Commun.*, 1995, 2353.

19
Antibiotics

1 Aminoglycosides and Aminocyclitols

An extensive report from Prinzbach's laboratory has described the synthesis of the pseudodisaccharide sannamycin (1), its 2-epimer and various other isomers, including the enantiomers of sannamycin and its 2-epimer. A detailed study was made of the glycosylations necessary to assemble these sytems.[1] Various other novel pseudodisaccharides have also been prepared, including the α-D-arabino-furanosyl systems 2 (R=H, α-NH$_2$, β-NH$_2$), and the pyranose 3;[2,3] these workers also describe a Ferrier carbocyclization at the disaccharide level, using an acid labile 2'-deoxy-disaccharide, to give the aminocyclitol derivative 4.[3]

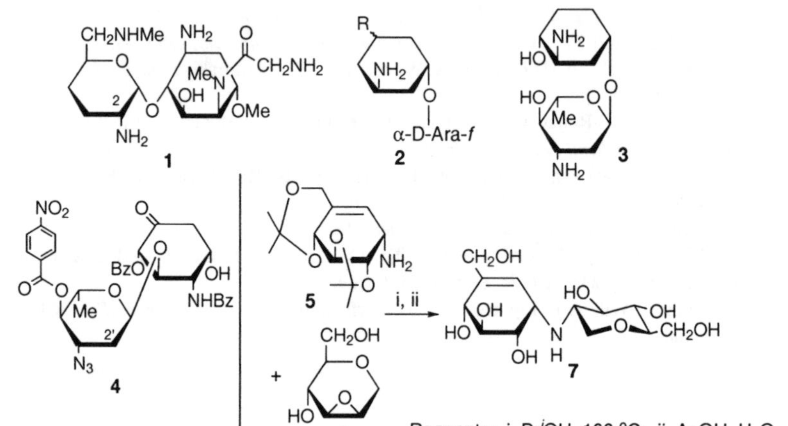

Reagents: i, Pr^{i}OH, 100 °C; ii, AcOH, H$_2$O
Scheme 1

The trehalase inhibitor salbostatin (7) has been synthesized (Scheme 1) *via* interaction of 1,5:2,3-dihydro-D-mannitol (6) with the known α-valienamine derivative 5; some of the product of diaxial opening of the epoxide was also produced, although the desired regioisomer was isolated in 58% yield.[4]

Various aminoacyl and dipeptidyl derivatives of kanamycin A and metilmycin have been prepared by regioselective acylation of Cu(II) complexes.[5] In an approach to overcoming resistance to aminoglycoside antibiotics, thought to be due to enzymes which remove important electrostatic interactions by phosphorylation or acylation, all four analogues of neamine and three of the four possible analogues of kanamycin A which involve deletion of an amino-function have been synthesized.[6] As part of a study on fluorination – toxicity relationships for aminoglycoside antibiotics, 5-deoxy-5-epifluoroarbekacin (**8**), 5-deoxy-5-epifluoroamikacin, and some related fluorinated antibiotics have been prepared. The 5-deoxy-5-epifluoro-species showed acute toxicity values almost identical with the modified antibiotics, in sharp contrast to the toxicities of the 5-deoxy-5-fluoro-compounds, and this was rationalized on the basis of basicity values for the 3-NH_2 groups.[7]

Streptomycin A and neomycin B have found novel use as effective displacer compounds in the ion-exchange displacement chromatographic separation of proteins.[8] A paper on the ^{1}H-nmr assignments of the kanamycins and butirosin A is mentioned in Chapter 21.

Tatsuta and co-workers have reported elegant work on the iminoacid-disaccharide gualamycin (**9**). The structure of gualamycin has been determined by NMR methods together with an X-ray structure of the aglycon methyl ester hydrochloride.[9] The disaccharide unit has been prepared in protected form, as has the aglycon methyl ester hydrochloride. This was made by Wittig chain extension of a known L-hexopyranoside to give **10**, followed by hydroxylation of the alkene, formation of the pyrrolidine by cyclization of an amino group at C-2 onto C-5, and further manipulation.[10] The aglycon could be converted into a bicyclic lactam in which just the required hydroxy group at C-8 was unprotected, and this was linked to a phenylthio glycoside of the disaccharide, benzylated at O-2′, to complete the total synthesis of **9**.[11]

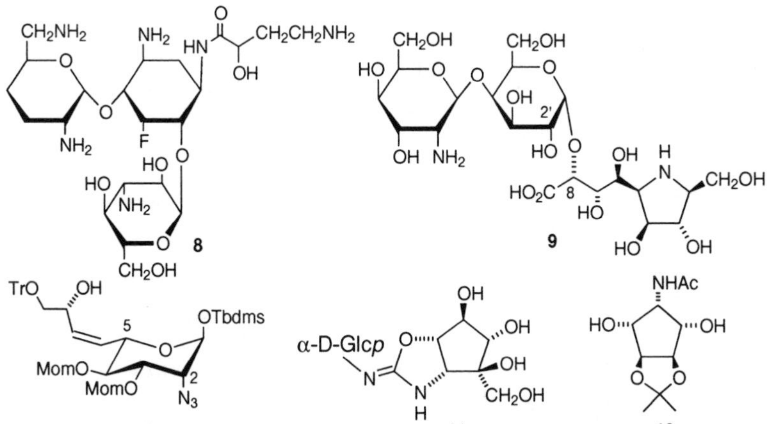

There has been further work on aminocyclopentitol antibiotics. A series of eight analogues of trehazolin (**11**) has been prepared in which the α-D-glucopyranose unit is replaced by other mono- and disaccharides and by carbosugar units,[12] whilst a further synthesis of (+)-trehazolin (11) has been described, in which the aminocyclitol is derived from (R)-epichlorohydrin and cyclopentadiene.[13] A much higher-yielding route has been developed to make the *meso*-compound 12, which has been used previously, with a subsequent resolution, to make (+)-trehazolin.[14] Free radical cyclization of the oxime ether 13 (Scheme 2) gave 14 with high stereocontrol over the orientation of the hydroxylamino substituent, and 14 could be converted to the aminocyclopentitol 15, stereoisomeric with that found in trehazolin.[15] The diastereoisomer 16 was the major product obtained by treatment of the *meso*-diol with (S)-O-acetyl mandelic acid, and 16 could be converted to mannostatin A (17); use of (R)-O-acetyl mandelic acid led to the enantiomer of 17.[16] There has been a further report on radical cyclization of oxime ether dithioacetals as a route to mannostatin analogues (see Vol.28, p. 234),[17] and cyclization of 18 using tributylstannane and AIBN gave 19, which has the stereochemistry found in allosamizoline, as well as other stereoisomers.[18]

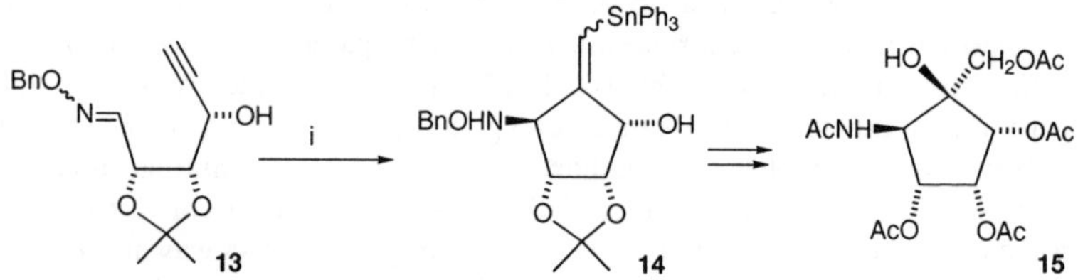

Reagents: i, Ph$_3$SnH, Et$_3$B

Scheme 2

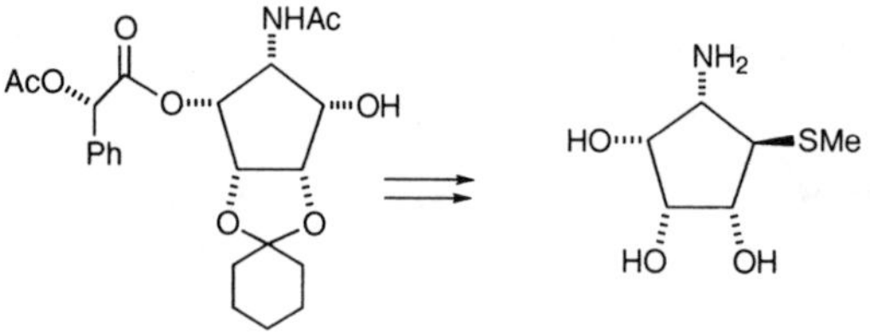

16   17

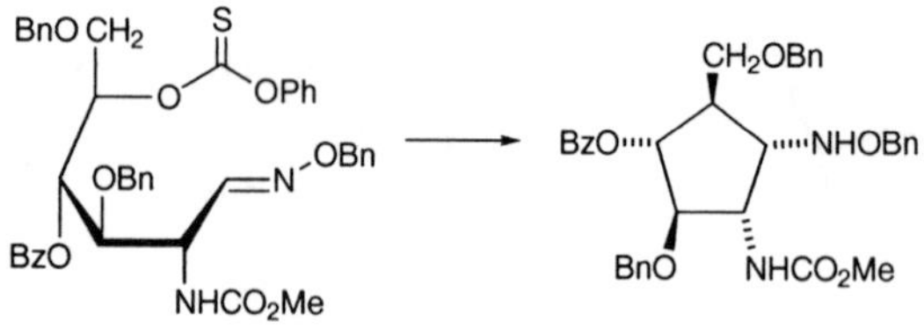

18   19

2   Macrolide Antibiotics

Full ¹H- and ¹³C-NMR assignments have been reported for erythromycin A enol ether, formed by acid treatment of the antibiotic.[19] Erythromycin A has also been degraded to a C(1)-C(9) fragment, still bearing both the sugar units, which is a useful intermediate for analogue synthesis. Two analogues were made by rebuilding the macrolide ring.[20] Detailed conformational analysis has been carried out on erythromycin 9-ketone and the related antibiotics azithromycin and clathromycin, in aqueous solution and when bound to bacterial ribosomes. The cladinose unit at C-3 adopted different conformations in the different antibiotics when bound to ribosomes, unlike other regions of the structure, suggesting that structural variation of the C-3 sugar could be tolerated, whilst desosamine would be required at C-5.[21]

The complete stereostructure of the 30-membered ring macrolide aculeximycin, which contains the trisaccharide aculexitriose (Vol.23, p. 199-200) has now been determined.[22]

3   Anthracyclines and other Glycosylated Polycyclic Antibiotics

Daunomycin has been converted into its 4'-*O*-phosphate and 4'-*O*-sulfate, and into the 3'-*N*-β-D-glucuronylcarbamate, a potential substrate for β-glucuronidase, these compounds being prodrugs for antibody directed enzyme-prodrug therapy.[23] The same team has also made the more complex spacer-linked

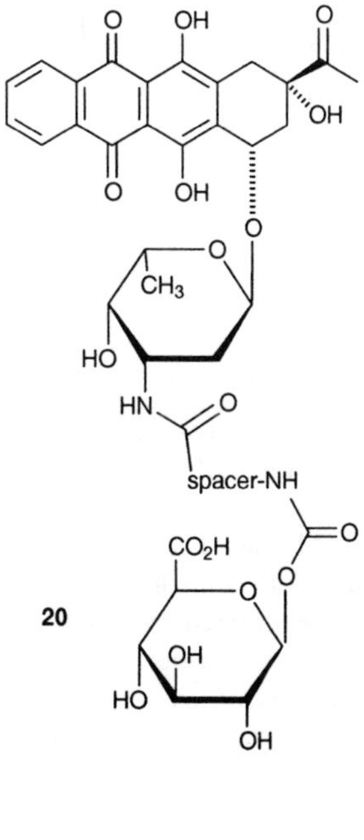

20

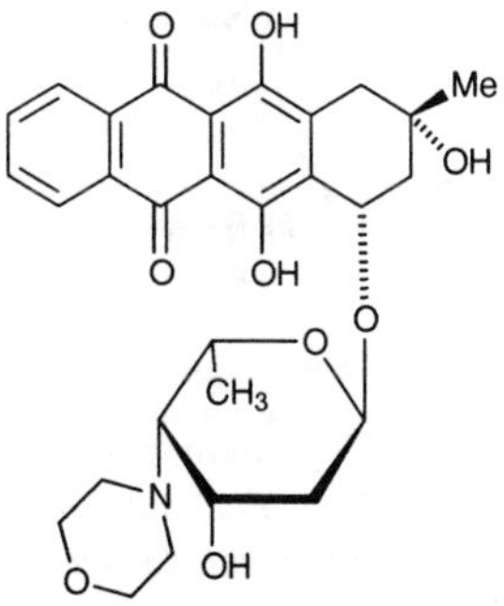

21

glucuronide carbamate **20**, where the spacer is designed such that hydrolysis by β-glucuronidase should be followed by intramolecular cyclization to a lactam, liberating daunomycin.[24] Somewhat similarly, doxorubicin has been linked to carboxymethyl-pullulan via an oligopeptide spacer, in an attempt to improve the therapeutic index of the drug.[25] Various doxorubicin derivatives have been described in which the 3′-amino group is incorporated into enaminomalonyl-β-alanine units, as potential drug-linking domains of an antibody-drug conjugate.[26]

Accounts of presentations at a symposium have described daunorubicin and doxorubicin analogues fluorinated at C-2′,[27] and other anthracyclines with reduced basicity and increased stability of the glycosidic bond, again by introduction of halogens at C-2′.[28] The 4′-morpholino-9-methylanthracycline **21** has been synthesized,[29] and the 3′-amino-function of carminomycin has been incorporated into a morpholine ring which also carries a hypoxanthine unit, by reductive amination with 2′,3′-seco-inosine.[30]

The ¹H- and ¹³C-NMR spectra of aclacinomycin A have been fully assigned, and the results were used in a study of the solution conformation of the trisaccharide **22** of this antibiotic.[31] The related trisaccharide **23** of ditrisarubicin B and betaclacamycin has been synthesized and its DNA-binding was studied. It was found that the DNA binding of the trisaccharide is low, indicating the need for the chromophere to anchor the antibiotic in the DNA.[32]

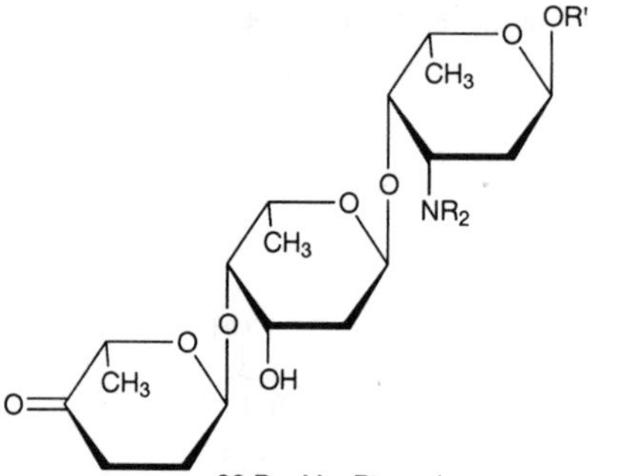

22 R = Me, R′ = aglycon
23 R = H, R′ = Me

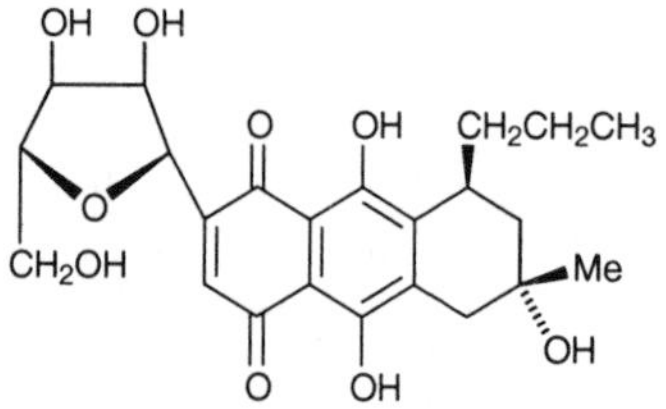

24

The 2,4-di-*O*-*p*-coumaroyl-α-ʟ-rhamnopyranoside of the flavanone kaempferol, present in New Zealand *Epicridaceae* species, has antimicrobial activity against multi-resistant *Staphylococcus aureus*.[33]

The *C*-ribosyl naphthoquinone exfoliamycin (**24**) has been isolated from *Streptomyces exfoliatus*, along with the products of *O*-methylation and of elimination of water in ring C. The structure was determined by X-ray crystallography of a derivative; the absolute configuration was not determined, but is assumed to correspond with ᴅ-ribose.[34] The first synthesis of angucycline antibiotic C104 (**26**) has been accomplished; the key intermediate **25** was made by

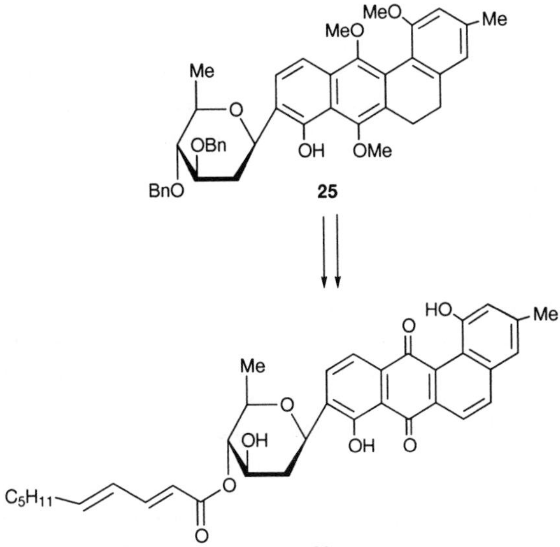

25

26

reaction of the phenol with the glycosyl acetate in the presence of Cp$_2$HfCl$_2$ and AgClO$_4$, a process thought to involve initial *O*-glycosylation followed by O→C rearrangement.[35] The structure of the related amicenomycin A (**27**) has been determined; amicenomycin B was also isolated from the same *Streptomyces* species, and in this the D-ring has undergone oxidative cleavage.[36] A brief review has been given by Suzuki on the formation of aryl *C*-glycoside antibiotics by O→C rearrangement,[37] and a review on the synthesis of gilvocarcin discusses the glycosylation steps.[38]

4'-Deacetyl-griseusin A (**28**) has been isolated from an unidentified actino-mycete, along with 4'-deacetyl griseusin B.[39]

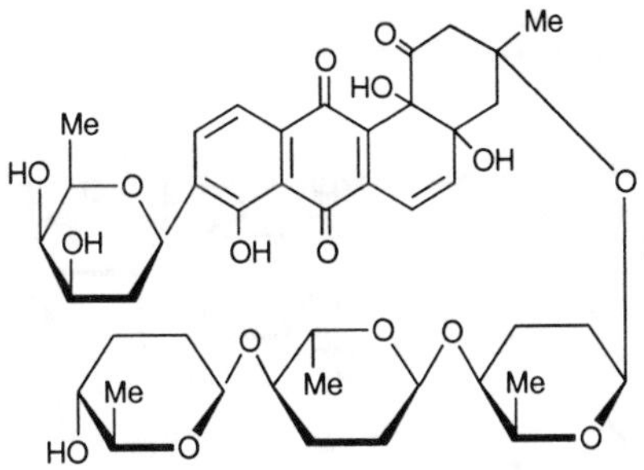

27

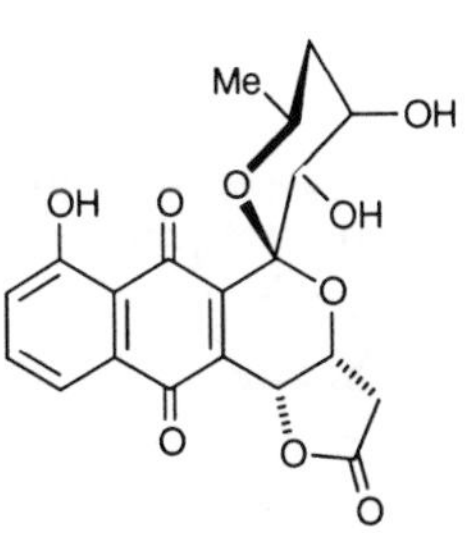

28

4   Nucleoside Antibiotics

The chiral dihydrofuran **29** has been made by a process involving asymmetric epoxidation to introduce chirality, and converted to cordycepin (**30**) by α-selective dihydroxylation followed by Vorbrüggen-type glycosylation.[40] 2'-Deoxy-2'-fluoropuromycin (**31**) has been prepared, the fluorine being introduced into an intermediate of *arabino*-configuration using DAST.[41] 2'-Deoxypuromycin (**32**) has also been made by base-sugar condensation under Vorbrüggen conditions; sodium-salt glycosylation was α-selective, leading to the synthesis of the α-anomer of **32**.[42]

There have been several reports which relate to pyrrolo[2,3-*d*]pyrimidine (7-deazapurine) nucleoside antibiotics. 2'-Deoxy-2'-*ara*-fluorotoyocamycin (**33**, R = CN) has been made by base-sugar condensation in two laboratories,[43-44] and converted to the sangivamycin (**33**, R = CONH$_2$) and thiosangivamycin (**33**, R = CSNH$_2$) analogues.[43, 44] The tubercidin analogue (**33**, R = H) was similarly prepared.[45] The 5'-fluoro- and 5'-amino-5'-deoxy-derivatives of toyocamycin and sangivamycin have also been made by condensation between modified sugar units and a heterocycle.[46] Some 2'-deoxy-pyrrolo[2,3-*d*]pyrimidine nucleosides, including 2'-deoxycadeguomycin (**34**), have been made by coupling with a glycosyl chloride.[47] Tubercidin 5'-*O*-methyl ether has been prepared for evaluation against adenosine receptors.[48]

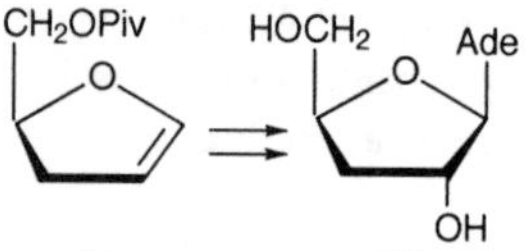

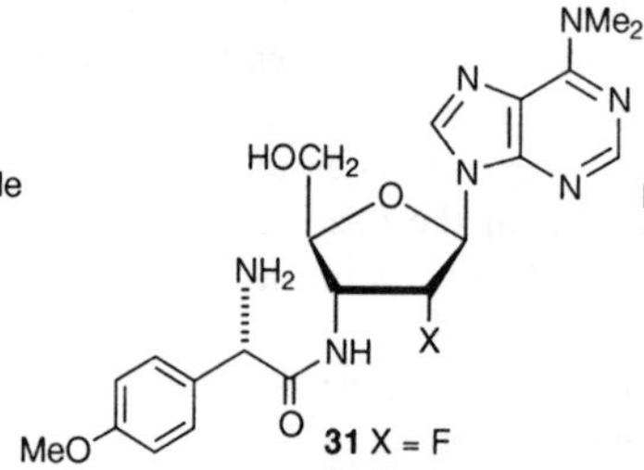

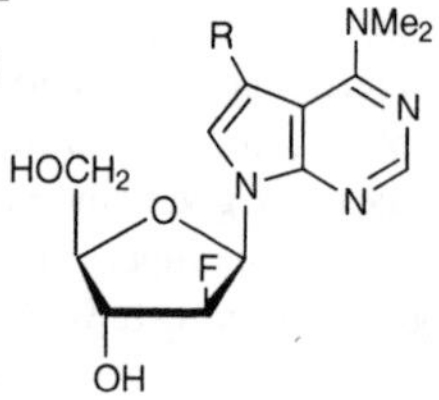

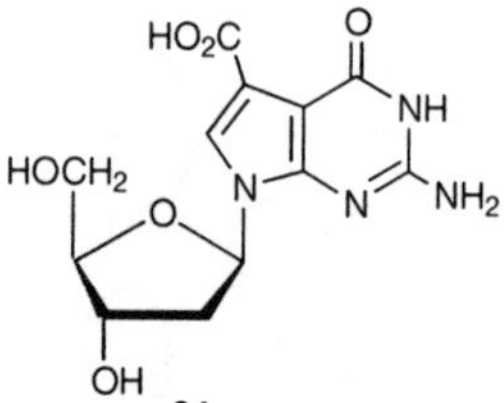

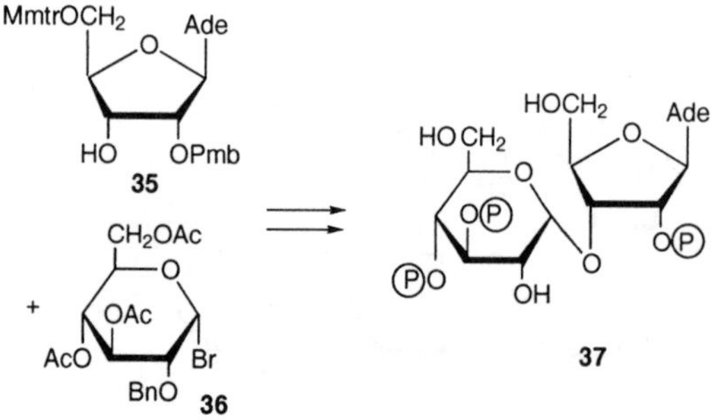

Adenophostin A (**37**) is a naturally-occurring IP$_3$ receptor agonist isolated from strains of *Penicillium brevicompactum*. It has been synthesized in a sequence in which a key step was the α-selective glycosylation of adenosine derivative **35** with the 2-*O*-benzylated glycosyl bromide **36**.[49]

The ribosylated furan **38** can be made by hydroxylation of the previously-reported 2′,3′-ene, and can be converted in two different ways to the *C*-nucleoside antibiotic showdomycin (**39**).[50,51]

In the area of carbocyclic nucleoside antibiotics, hydrolysis of the racemic esters **40** (R= *n*-Bu or *n*-C$_6$H$_{13}$) by the lipase from *Candida rugosa* proceeds with very high enantiomeric selectivity, and from the resolved materials both enantiomers of aristeromycin were made by an established route. The authors report that a previous similar method (Vol.21, p. 182) is not as enantioselective.[52] In a new synthesis of neplanocin A (**43**), the alcohol **41**, derived from D-ribose, was converted to the cyclopentene **42** using an intramolecular insertion reaction of an alkylidene carbene. The new stereocentre in **42** was mostly of the 'wrong' β-configuration, but could be corrected by a process of desilylation, oxidation and borohydride reduction.[53] The biosynthesis of neplanocin A (**43**) and aristeromycin has been reinvestigated, and the cyclopentenone **44** has been proposed as an intermediate, which is converted to aristeromycin *via* neplanocin A without any bifurcation.[54] The 3-deaza-analogue **45** of 5′-*nor*-aristeromycin has been prepared, and the antiviral activity of it and of the 7-deaza-compound (Vol.27, p. 235) are reported.[55]

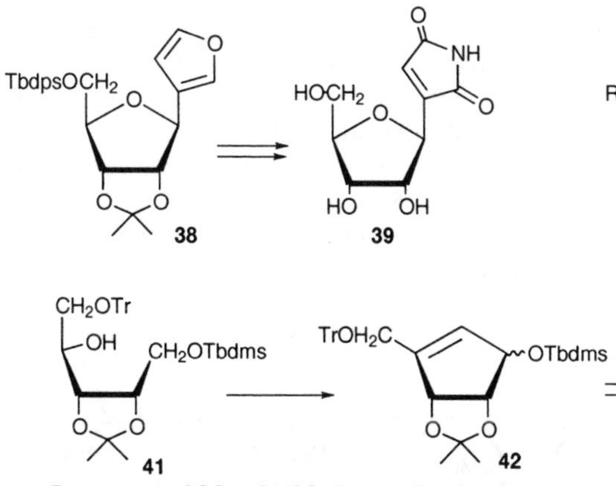

Reagents: i, (COCl)$_2$, DMSO, Et$_3$N; ii, TmsC(N$_2$)Li

Scheme 3

In several nucleoside antibiotics, the nucleoside unit is chain-extended to give an α-amino acid functionality, and there have been further reports of synthetic activity towards such compounds. There has been a study on the synthesis of the 5-aminopentose-2-sulfate unit which is found linked glycosidically in the liposido-mycins; D-ribose has been converted into the compounds **46** (R = Bn, Tr,

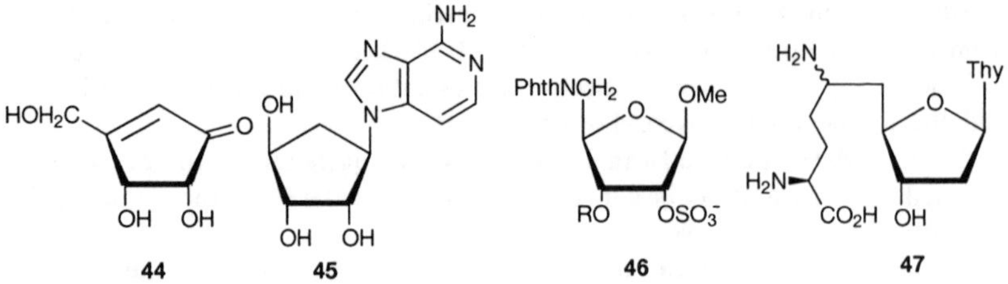

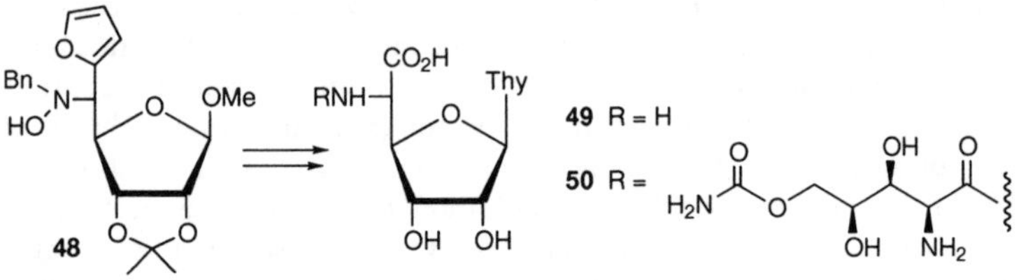

Tbdms), with the regioselectivity of alkylation at O-3 *via* stannylene acetals being investigated.[56] A carbohydrate-based synthesis of the 1,4-dimethyl-1,4-diazepan-2-one unit of the liposidomycins is mentioned in Chapter 24. The thymidine analogue **47** of sinefungin (both epimers) has been made by a method previously used for the antibiotic itself (Vol.25, p. 230).[57] The hydroxylamine **48** can be made diastereoselectively by addition of 2-furyllithium to the appropriate nitrone; it has now been converted into thymine polyoxin C (**49**) by a route which offers a higher overall yield than the one previously reported (Vol.28, p. 254), and thymine polyoxin C has been converted to polyoxin J (**50**) by coupling to the previously-known aminoacid.[58] In a synthesis of the bicyclic moiety of the miharamycins (Scheme 4), a key step was the intramolecular reductive cyclization of the keto-alkyne **51** using SmI_2. Data obtained by these workers are at variance with those reported earlier (Vol.21, p. 158) by others for allegedly the same compounds; the present work is supported by X-ray data and thus probably constitutes the first synthesis of the miharamycin bicycle.[59]

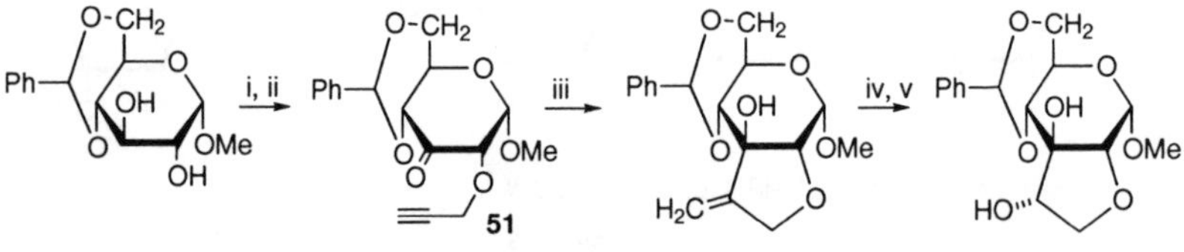

Reagents: i, HC≡CCH$_2$Br, phase transfer; ii, PCC; iii, SmI$_2$, HMPA, 94%; iv, O$_3$, then Me$_2$S; v, NaBH$_4$

Scheme 4

There has been further activity concerning oxetanocin and its analogues. Oxetanocin A (**52**) can be obtained selectively protected at O-4' by either lipase-catalysed hydrolysis of its di-*O*-acetyl derivative or by photolysis of its di-*O*-(2-nitrobenzyl) ether.[60] There has been a further route reported for the synthesis of the carbocyclic oxetanocin analogue 'cyclobut G' (**53**, B = Gua), which has potent antiviral activity. A photochemical [2+2] cycloaddition has been used to make racemic 'cyclobut A' (**53**, B = Ade),[62] whilst a number of similar purines have been prepared via the intermediacy of the amidine **54**.[63] The related structures **55** have also been reported, [64] as have fluorinated species of type **56** (B = Ade, Gua, Hypoxanthyl).[65]

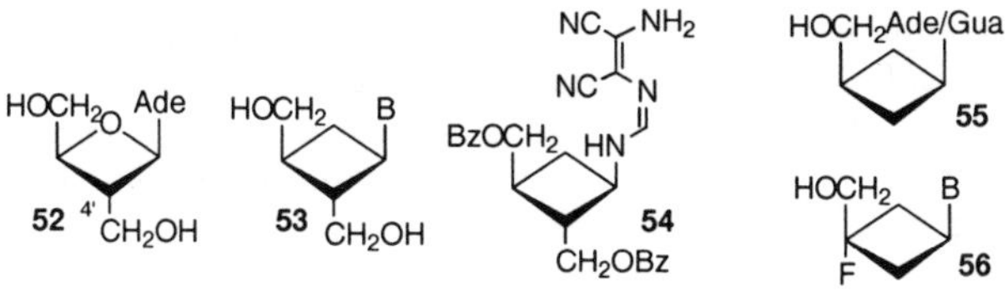

5   Miscellaneous Antibiotics

The chemistry of staurosporine has been reviewed,[66] and 3'-demethoxy-3'-hydroxystaurosporine has been isolated from a blocked mutant.[67]

There have been several reports relating to the spirohydantoin hydantocidin. In a route to 5-*epi*-hydantocidin (**59**) (Scheme 5), the α-azidolactone **57**, made from D-ribose by Kiliani chain extension and displacement of a 2-*O*-sulfonate, underwent an interesting transformation to the bridged bicyclic amine **58** on oxidation with TPAP. In the conversion of **58** to **59**, the pyranose to furanose interconversion produced only the 5-*epi*-isomer.[68] On silylation, followed by treatment with TmsN₃ and TmsOTf, 1,2:3,4-di-*O*-isopropylidene-β-D-psicofuranose gave rise to **60**, convertible to the thionoanalogue **61** of hydantocidin.[69] The carba-analogue **62** was produced in a similar sequence using allyl trimethylsi-

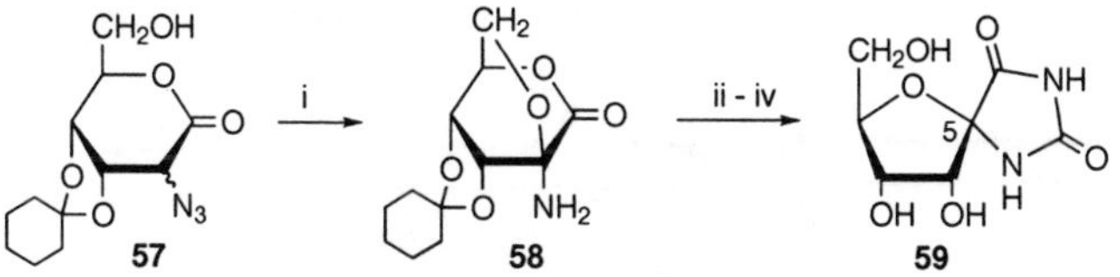

Reagents: i, TPAP, NMO; ii, KCNO, AcOH; iii, KOBu^t, DMF; iv, TFA, H₂O
Scheme 5

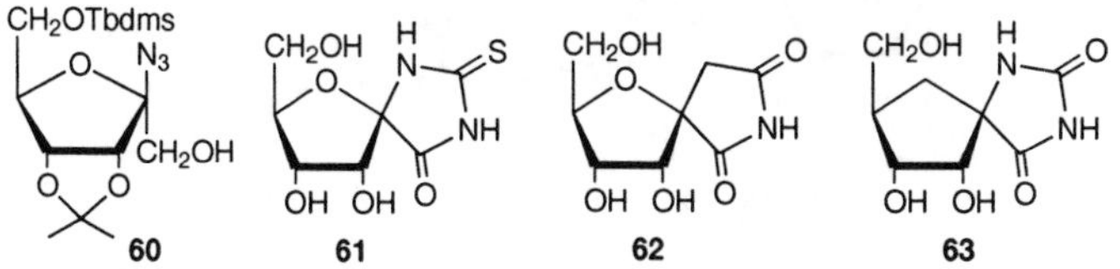

lane.[70] Two hydantocidin analogues lacking the CH_2OH group have been made from 2,3-*O*-isopropylidene-D-erythronolactone,[71] and the carbocyclic analogue **63** of hydantocidin has been made in racemic[72] and optically-active forms.[73]

Structure **64** has been determined for the anti-tumour antibiotic spicamycin, from *Streptomyces elanosinicus;* the *N*-acylglycine unit can be removed by acid hydrolysis, giving 'spicamycin aminonucleoside',[74] which can then be linked to other *N*-acyl aminoacids to give spicamycin analogues.[75] Alternatively, other fatty acids can be introduced, to give in some cases, analogues with higher bioactivity.[76]

The total synthesis of nagstatin (**65**) has been reported, starting from tri-*O*-benzyl-L-ribofuranose (sugar carbons indicated), which was treated with 2-lithio-1-tritylimidazole, followed by cyclization, functionalization of the imidazole ring and final introduction of nitrogen by azide displacement.[77] Nagstatin analogues, lacking the carboxymethyl substituent and with different configurations on the piperidine ring, have been similarly prepared.[78]

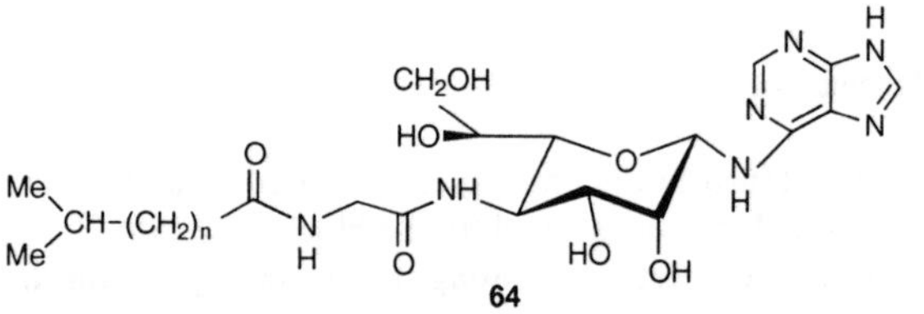

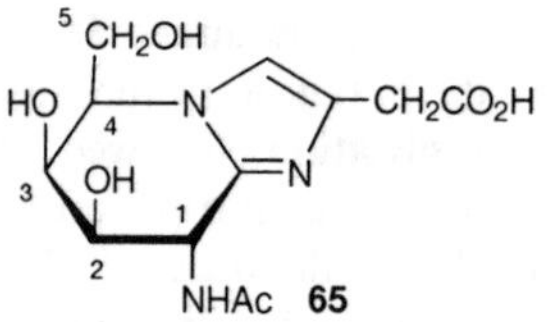

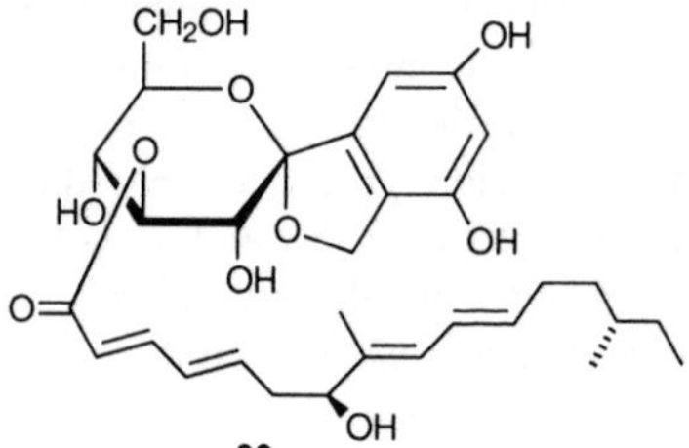

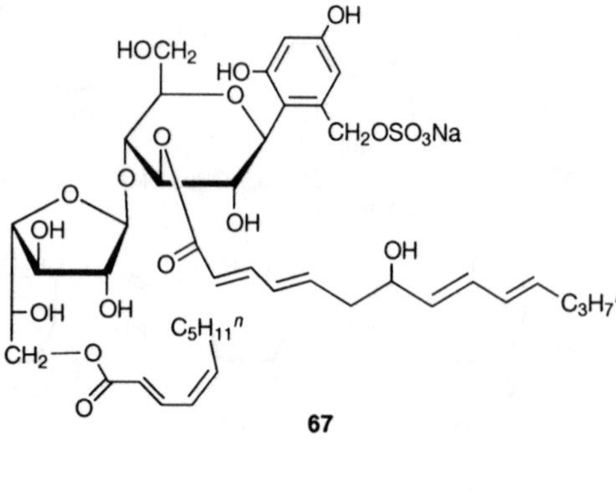

The anti-fungal agent papulacandin D (**66**) has been synthesized; the spiro-cyclic core was made by attack of an aryllithium reagent on tetra-*O*-Tms-D-glucono-1,5-lactone, followed by acid-catalysed spirocyclization.[79] The fungal metabolite lecythophorin has been assigned structure **67**; desulfated lecytho-phorin seems to be the same as chaetiacandin (Vol.19, p. 186), the structure of which would then require revision to a galactofuranose.[80] Fusacandins A and B are related structures of the papulacandin class, containing a β-D-Gal*p*-(1→2)-β-D-Gal*p* unit attached at O-4' of the core unit.[81]

The glycoside **68**, a constituent of the antibiotic hygromycin, has been prepared, the stereochemistry being established by regioselective opening of a 2',3'-D-*manno*-epoxide.[82] Screening has led to the identification of a microorganism, *Streptomyces rimosus*, that can metabolize the naturally-occurring ACE inhibitor A58365 A to the glycoside **69**.[83]

A short review has been given on lipid A analogues as potential candidates for treatment of Gram-negative sepsis.[84] Moenomycin A is a new member of the moenomycin class of phosphoglycolipid antibiotics; it differs from moenomycin A in lacking the C-4 methyl branch on the reducing terminal sugar. Degradation studies revealed that the trisaccharide core was the smallest antibiotically-active unit.[85] A synthesis of the methyl-branched uronamide of moenomycin A is mentioned in Chapter 14.

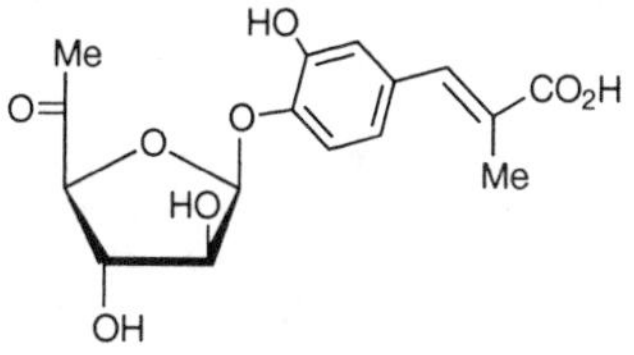

68

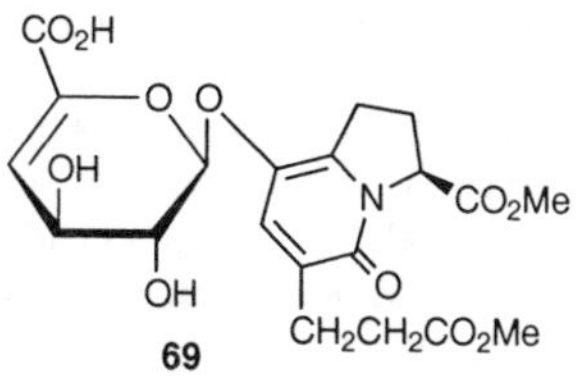

69

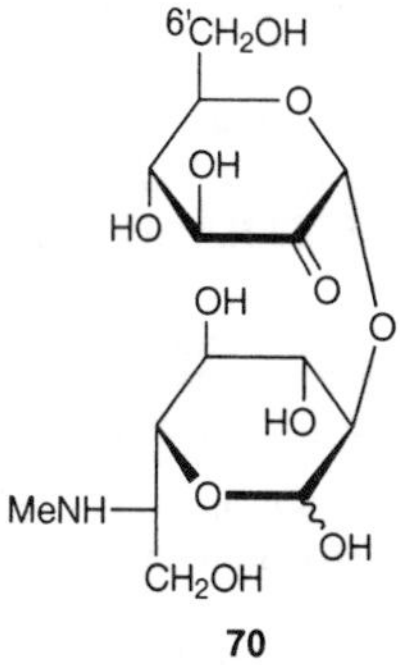

70

By use of ^{1}H-^{13}C correlation techniques, all 66 carbon signals of vancomycin have been assigned.[86]

In the structure of the plant growth regulator calonyctin A, a long-chain hydroxyacid spans the terminal sugar units of a branched tetrasaccharide. The sugar unit has been prepared (Vol.28, p.71) and has now been assembled into the complete structure of calonyctin A.[87]

In the salmycins, from a strain of *Streptomyces violaceus*, the disaccharide unit **70**, or its oxime, is linked via O-6′ to the known trihydroxamate siderophere danoxamin.[88]

The hydroxylamino group of hydroxylaminoeverninomycin can be oxidized to a nitrogroup using horseradish peroxidase.[89]

References

1 C. Ludin, T. Weller, B. Seitz, W. Meier, S. Erbeck, C. Hoenke, R. Krieger, M. Keller, L. Knothe, K. Pelz, A. Wittmer and H. Prinzbach, *Liebigs Ann. Chem.*, 1995, 291.

2 I.F. Pelvyás, M. Mádi-Puskás, Z.G. Tóth, Z. Varga, G. Batta and F. Sztaricskai, *Carbohydr. Res.*, 1995, **272**, C5.

3 I.F. Pelvyás, M. Mádi-Puskás, Z.G. Tóth, Z. Varga, M. Hornyak, G. Batta and F. Sztaricskai, *J. Antibiotics*, 1995, **48**, 683.

4 T. Yamagishi, C. Uchida and S. Ogawa, *Bioorg. Med. Chem. Lett.*, 1995, **5**, 487; *Chem. Eur. J.*, 1995, **1**, 634.

5 S. Kotretsou, M.P. Mingeot-Leclercq, V. Constantinou-Kokotou, R. Brasseur, M.P. Georgiadis and P.M. Tulkens, *J. Med. Chem.*, 1995, **38**, 4710.

6 J. Roestamadj, I. Grapsas and S. Mobashery, *J. Am. Chem. Soc.*, 1995, **117**, 11060.

7 T. Shitara, E. Umemura, T. Tsuchiya and T. Matsuno, *Carbohydr. Res.*, 1995, **276**, 75.

8 A. Kundu, S. Vunnum and S.M. Cramer, *J. Chromatogr. A*, 1995, **707**, 57.

9 K. Tsuchiya, S. Kobayashi, T. Kurokawa, T. Nakagawa, N. Shimada, H. Nakamura, Y. Iitaka, M. Kitagawa and K. Tatsuta, *J. Antibiotics*, 1995, **48**, 630.

10 K. Tatsuta, M. Kitagawa, T. Horiuchi, K. Tsuchiya and N. Shimada, *J. Antibiotics*, 1995, **48**, 741.

11 K. Tatsuta and M. Kitagawa, *Tetrahedron Lett.*, 1995, **36**, 6717.

12 C. Uchida, H. Kitahashi, S. Watanabe and S. Ogawa, *J. Chem. Soc., Perkin Trans. 1*, 1995, 1707.

13 B.E. Ledford and E.M. Carreira, *J. Am. Chem. Soc.*, 1995, **117**, 11811.

14 B.K. Goering, J. Li and B. Ganem, *Tetrahedron Lett.*, 1995, **36**, 8905.

15 J. Marco-Contelles, C. Destabel, J.L. Chiara and M. Bernabe, *Tetrahedron: Asymm.*, 1995, **6**, 1547.

16 S. Ogawa, S. Kimura, C. Uchida and T. Ohashi, *J. Chem. Soc., Perkin Trans. 1*, 1995, 1695.

17 J. Marco-Contelles, C. Destabel and P. Gallego, *J. Carbohydr. Chem.*, 1995, **14**, 1343.

18 S. Takahashi, H. Inoue and H. Kuzuhara, *J. Carbohydr. Chem.*, 1995, **14**, 273.

19 P. Alam, P.C. Buxton, J.A. Parkinson and J. Barber, *J. Chem. Soc., Perkin Trans. 2*, 1995, 1163.

20 A. Nishida, K. Yagi, N. Kawahara, M. Nishida and O. Yonemitsu, *Tetrahedron Lett.*, 1995, **36**, 3215.

21 A. Awan, R.J. Brennan, A.C. Regan and J. Barber, *J. Chem. Soc., Chem. Commun.*, 1995, 1653.

22 H. Murata, I. Ohama, K. Harada, M. Suzuki, T. Ikemoto, T. Shibuya, T. Haneishi, A. Torikata, Y. Itezono and N. Nakayama, *J. Antibiotics*, 1995, **48**, 850.

23   R.G.G. Leenders, H.W. Scheeren, P.H.J. Houba, E. Boven and H.J. Haisma, *Bioorg. Med. Chem. Lett.*, 1995, **5**, 2975.

24   R.G.G. Leenders, K.A.A. Gerrits, R. Ruijtenbeek, H.W. Scheeren, H.J. Haisma and E. Boven, *Tetrahedron Lett.*, 1995, **36**, 1701.

25   H. Nogusa, T. Yano, S. Okuno, H. Hamana and K. Inoue, *Chem. Pharm. Bull.*, 1995, **43**, 1931.

26   D.E. Seitz, J.E. Ezcurra and D.L. Guttman-Carlisle, *Tetrahedron Lett.*, 1995, **36**, 1413.

27   T. Tsuchiya and Y. Takagi, *ACS Symp. Ser.*, 1995, **574** (Anthracycline Antibiotics), 100 (*Chem. Abstr.*, 1995, **122**, 133599).

28   W. Priebe, P. Skibicki, O. Varela, N. Neamati, M. Sznaidman, K. Dziewiszek, G. Grynkiewicz, D. Horton and Y. Zou, *ACS Symp. Ser.*, 1995, **574** (Anthracycline Antibiotics), 14 (*Chem. Abstr.*, 1995, **122**, 133596).

29   J. Nafsiger, G. Averland, E. Bertounesque, G. Gaudel and C. Monneret, *J. Antibiotics*, 1995, **48**, 1185.

30   N.P. Todorova, H.S. Milkov and M.N. Preobrazenskaya, *Dokl. Bulg. Akad. Nauk.*, 1995, **47**, 43 (*Chem Abstr.*, 1995, **123**, 286435).

31   J.A. Parkinson, I.H. Sadler, M.B. Pickup and A.B. Tabor, *Tetrahedron*, 1995, **51**, 7215.

32   C.J. Shelton and M.M. Harding, *J. Chem. Res.*, 1995, (*S*) 158; (*M*) 1201.

33   S.J. Bloor, *Phytochemistry.*, 1995, **38**, 1033.

34   C. Volkmann, A. Zeeck, O. Potterat, H. Zähner, F.-M. Bohnen and R. Herbst-Irmer, *J. Antibiotics*, 1995, **48**, 431.

35   T. Matsumoto, T. Sohma, H. Yamaguchi, S. Kurata and K. Suzuki, *Synlett*, 1995, 263; *Tetrahedron*, 1995, **51**, 7347.

36   N. Kawamura, R. Sawa, Y. Takahashi, T. Sawa, N. Kinoshita, H. Naganawa, M. Hamada and T. Takeuchi, *J. Antibiotics*, 1995, **48**, 1521.

37   K. Suzuki, *Pure Appl. Chem.*, 1994, **66**, 2175 (*Chem. Abstr.*, 1995, **122**, 31760).

38   D.H. Hua and S. Saha, *Recl. Trav. Chim. Pays-Bas*, 1995, **114**, 341.

39   M. Igarashi, W. Chen, T.Tsuchida, M. Umekita, T. Sawa, H. Naganawa, M. Hamada and T. Takeuchi, *J. Antibiotics*, 1995, **48**, 1502.

40   F.E. McDonald and M.M. Gleason, *Angew. Chem., Int. Ed. Engl.*, 1995, **34**, 350.

41   T. Maruyama, K. Utsumi, H. Tomiyoka, M. Kasamoto, Y. Sato, T. Anne and E. De Clercq, *Chem. Pharm. Bull.*, 1995, **43**, 955.

42   M.S. Motawia, M. Meldal, M. Sofan, P. Stein, E.B. Pedersen and C. Nielsen, *Synthesis*, 1995, 265.

43   B.K. Bhattacharya and G.R. Revankar, *J. Chem. Soc., Chem. Commun.*, 1995, 115.

44   B.K. Bhattacharya, T.S. Rao and G.R. Revankar, *J. Chem. Soc., Perkin Trans. 1*, 1995, 1543.

45   S.H. Krawczyk, M.R. Nassiri, L.S. Kucera, E.R. Kern, R.G. Ptak, L.L. Wotring, J.C. Drach and L.B. Townsend, *J. Med. Chem.*, 1995, **38**, 4106.

46   M. Sharma, Y.X. Li, M. Ledvina and M. Bobek, *Nucleosides Nucleotides*, 1995, **14**, 1831.

47   E.D. Edstrom and Y. Wei, *J. Org. Chem.*, 1885, **60**, 5069.

48   C.G. Smith, S.J. Lee and D.L. Marquardt, *J. Med. Chem.*, 1995, **38**, 2259.

49   H. Hotoda, M. Takahashi, K. Tanzawa, S. Takahashi and M. Kaneko, *Tetrahedron Lett.*, 1995, **36**, 5037.

50   S.H. Kong and S.B. Lee, *J. Chem. Soc., Chem. Commun.*, 1995, 1017.

51   S.H. Kong and S.B. Lee, *Tetrahedron Lett.*, 1995, **36**, 4089.

52   R. Czuk and P. Dörr, *Tetrahedron*, 1995, **51**, 5789.

53   S. Ohira, T. Sawamoto and M. Yamato, *Tetrahedron Lett.*, 1995, **36**, 1537.
54   J.M. Hill, G.N. Jenkins, C.P. Rush, N.J. Turner, A.J. Willetts, A.D. Buss, M.J. Dawson and B.A.M. Rudd, *J. Am. Chem. Soc.*, 1995, **117**, 5391.
55   S.M. Siddiqui, X. Chen, J. Rao, S.W. Schneller, S. Ikeda, R. Snoeck, G. Andrei, J. Balzarini and E. De Clercq, *J. Med. Chem.*, 1995, **38**, 1035.
56   K.S. Kim, J.W. Lim, Y.H. Joo, K.T. Kim, I.H. Cho and Y.H. Ahn, *J. Carbohydr. Chem.*, 1995, **14**, 439.
57   D.H.R. Barton, S.D. Gero, G. Negron, B. Quiclet-Sire, M. Samadi and C. Vincent, *Nucleosides Nucleotides*, 1995, **14**, 1619.
58   A. Dondoni, F. Junquera, F.L. Merchan, P. Merino and T. Tejero, *J. Chem. Soc., Chem. Commun.*, 1995, 2127.
59   A.J. Fairbanks and P. Sinaÿ, *Synlett*, 1995, 277.
60   N. Katagiri, M. Makino and C. Kaneko, *Chem. Pharm. Bull.*, 1995, **43**, 884.
61   G.S. Bisacchi, J. Singh, J.D. Godfrey, T.P. Kissick, T. Mitt, M.F. Malley, J.D. Di Marco, J.Z. Gougoutas, R.H. Mueller and R. Zahler, *J. Org. Chem.*, 1995, **60**, 2902.
62   K. Somekawa, R. Hara, K. Kinnami, F. Muraoka, T. Suishu and T. Shimo, *Chem. Lett.*, 1995, 407.
63   B.L. Booth and P.R. Eastwood, *J. Chem. Soc., Perkin Trans. 1*, 1995, 669.
64   V. Kaiwar, C.B. Reese, E.J. Gray and S. Neidle, *J. Chem. Soc., Perkin Trans. 1*, 1995, 2281.
65   T. Gharbaoui, M. Legraverend, O. Ludwig, E. Bisagni, A.-M. Aubertin and L. Chertanova, *Tetrahedron*, 1995, **51**, 1641.
66   S. Omura, Y. Sasaki, Y. Iwai and H. Takeshima, *J. Antibiotics*, 1995, **48**, 535.
67   P. Hoehn, O. Ghisalba, T. Moerker and H.H. Peter, *J. Antibiotics*, 1995, **48**, 300.
68   A.J. Fairbanks and G.W.J. Fleet, *Tetrahedron*, 1995, **51**, 3881.
69   H. Sano, S. Mio, J. Kitagawa, M. Shindai, T. Honma and S. Sugai, *Tetrahedron*, 1995, **51**, 12563.
70   H. Sano, S. Mio, N. Tsukaguchi and S. Sugai, *Tetrahedron*, 1995, **51**, 1387.
71   S. Hanessian, J.-Y. Sanceau and P. Chemla, *Tetrahedron*, 1995, **51**, 6669.
72   H. Sano and S. Sugai, *Tetrahedron*, 1995, **51**, 4635.
73   H. Sano and S. Sugai, *Tetrahedron: Asymm.*, 1995, **6**, 1143.
74   T. Sakai, K. Shindo, A. Odagawa, A. Suzuki, H. Kawai, K. Kobayashi, Y. Hayakawa, H. Seto and N. Otake, *J. Antibiotics*, 1995, **48**, 899.
75   T. Sakai, H. Kawai, M. Kamishohara, A. Odagawa, A. Suzuki, T. Uchida, T. Kawasaki, T. Tsuruo and N. Otake, *J. Antibiotics*, 1995, **48**, 504.
76   T. Sakai, H. Kawai, M. Kamishohara, A. Odagawa, A. Suzuki, T. Uchida, T. Kawasaki, T. Tsuruo and N. Otake, *J. Antibiotics*, 1995, **48**, 1467.
77   K. Tatsuta and S. Miura, *Tetrahedron Lett.*, 1995, **36**, 6721.
78   K. Tatsuta, S. Miura, S. Ohta and H. Gunji, *J. Antibiotics*, 1995, **48**, 286.
79   A.G.M. Barrett, M. Peña and J.A. Willardsen, *J. Chem. Soc., Chem. Commun.*, 1995, 1147.
80   W.A. Ayer and N. Kawahara, *Tetrahedron Lett.*, 1995, **36**, 7953.
81   J.E. Hochlowski, D.N. Whittern, A. Buko, L. Adler and J.B. McAlpine, *J. Antibiotics*, 1995, **48**, 614.
82   J.G. Buchanan, D.G. Hill, R.H. Wightman, I.K. Boddy and B.D. Hewitt, *Tetrahedron*, 1995, **51**, 6033.
83   J.S. Myndersa, D.S. Fukuda and A.H. Hunt, *J. Antibiotics*, 1995, **48**, 425.
84   O. Holst, *Angew. Chem., Int. Ed. Engl.*, 1995, **34**, 2000.
85   A. Donnerstag, S. Marzian, D. Müller, P. Welzel, D. Böttger, A. Stärk, H.-W.

Fehlhaber, A. Markus, Y. van Heijenoort and J. van Heijenoort, *Tetrahedron*, 1995, **51**, 1931, and corrigendum, 5509.

86   C.M. Pearce and D.H. Williams, *J. Chem. Soc., Perkin Trans. 2*, 1995, 153.

87   Z.-H. Jiang, A. Geyer and R.R. Schmidt, *Angew. Chem., Int. Ed. Engl.*, 1995, **34**, 2520.

88   L. Vértezy, W. Aretz, H.-W. Fehlhaber and H. Kogler, *Helv. Chim. Acta*, 1995, **78**, 8286.

89   S. Kalliney and A. Zaks, *Tetrahedron Lett.*, 1995, **36**, 4163.

20
Nucleosides

---

1 General

A concise review on nucleoside synthesis concentrates on convergent approaches and emphasises stereoselectivity, including the synthesis of 2'-deoxy-, 2',3'-dideoxy- and 3'-oxa/thia-nucleosides.[1]

Two groups have independently isolated from sponges a novel, branched and chlorine-containing 5'-deoxy nucleoside 1, named kumusine[2] and trachycladine A;[3] the latter workers also isolated trachycladine B, the corresponding chlorine-free inosine analogue.

1-α-D-Ribofuranosyl-4-pyridone-3-carboxamide (2) has been isolated from the urine of normal and leukaemic patients, and was synthesized in an α-selective way.[4]

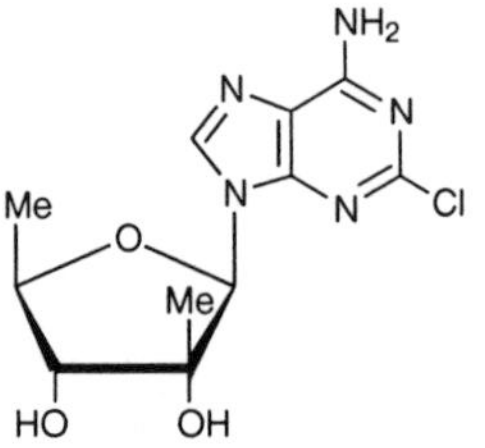

2 Synthesis

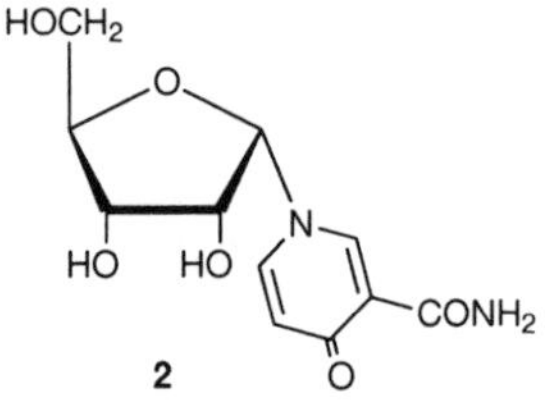

Vorbrüggen and co-workers have developed a new direct method for nucleoside synthesis, illustrated by the case in Scheme 1. The method is β-selective, with the bridged ion shown being postulated as an intermediate, but was not successful for the synthesis of 2'-deoxynucleosides.[5]

Treatment of 3 (see Vol.27, p.248 and also Section 5) with either tris(trimethyl-silyl)silane[6] or tributylstannane – AIBN[7] led to the formation of 4 as major product, via a rearrangement of the initially-formed radical. When similar chemistry was carried out on 5 (the major isomer formed by treatment of the relevant 1',2'-ene with NBS – pivalic acid when Tips, as opposed to Tbdms, protection was used), the β-D-arabino-nucleoside 6 was formed as major product,

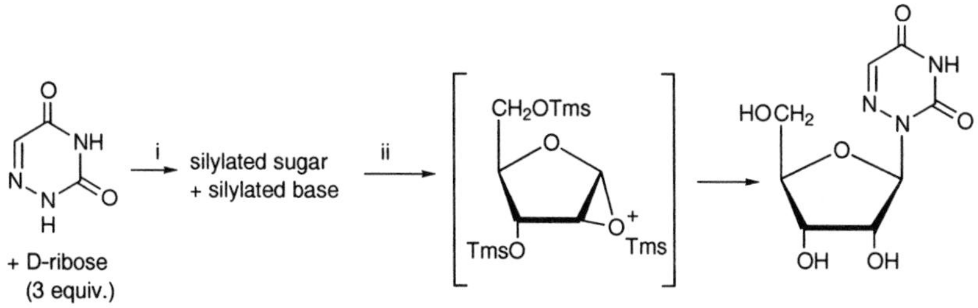

Reagents: i, HMDS, TmsCl; ii, TmsOTf, MeCN

Scheme 1

indicating that the rearranged radical reacts predominantly *trans-* to the pivaloyloxy group at C-2'. Treatment of **5** with allyl tributylstannane – AIBN gave the *C*-allyl analogue **7**.[7]

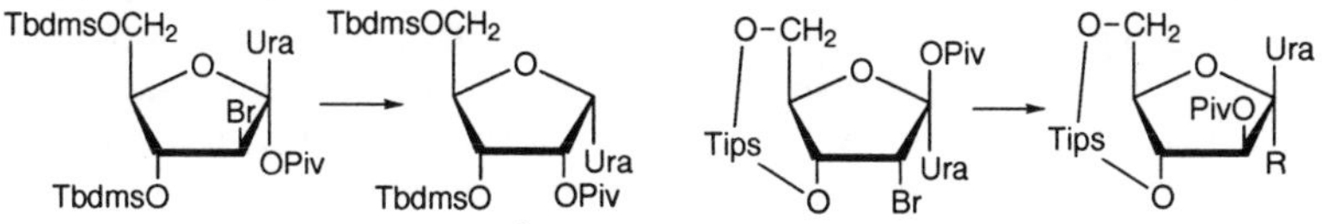

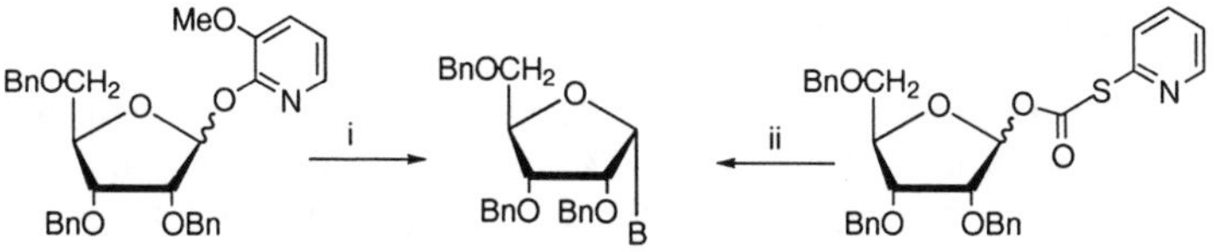

Reagents: i, silylated base, TmsOTf, toluene; ii, silylated base, AgOTf, toluene

Scheme 2

A novel method for the stereocontrolled synthesis of α-ribonucleosides, using either of two pyridine-containing leaving groups, is outlined in Scheme 2. The method was also applied to the synthesis of 1-(β-D-arabinofuranosyl)cytosine (also with 1,2-*cis*-stereochemistry), and the authors discuss the reasons for the selectivity in terms of 'remote activation'.[8]

2-Alkyl-3-hydroxy-4-pyridone ribonucleosides (potential oral iron chelators) have been prepared by conventional base-sugar coupling,[9] and simple routes to spongosine (6-amino-2-methoxyadenosine) and its 8-aza-analogue have been described.[10] 6-Nitro-1-deazapurine undergoes ribosylation with SnCl₄ catalysis to give 60% of the 9-β-product, but some 30% of 3-β-product is also formed, whereas 6-nitro-1,3-dideazapurine (4-nitrobenzimidazole) gives just the 9-β-product (purine numbering),[11,12] which could be used in an improved synthesis of 1,3-dideazaadenosine.[12] Various 2,5,6-trihalogenobenzimidazole ribonucleosides have been described,[13] as has 2-(β-D-ribofuranosyl)indazole, which, as its tri-*O*-

acetyl derivative, rearranges to the 1-substituted isomer.[14] Base-sugar condensation has been used to prepare the thieno[2,3-*d*]pyrimidine **8** and the thieno[3,2-*d*]pyrimidine **9**,[15] the furo[2,3-*d*]pyrimidines **10**[16] and **11** (where the β-D-*arabino*-compound was also prepared via a cyclonucleoside and the β-D-*xylo*-analogue by condensation),[17] and some 1-β-D-ribofuranosyl-5,7-disubstituted pyrido[2,3-*d*]pyrimidines.[18]

There have been reports of various 3-cyano-1-β-D-glucopyranosyl- and -1-β-D-galactopyranosyl-pyridin-2-ones[19-20] and pyridin-2-thiones,[21,22] along with α-L-arabinopyranosyl- and β-D-xylopyranosyl-3-cyano-2-pyridinethiones.[23] 5-Amino-6-aryl-3-(β-D-gluco-or -galacto-pyranosyl)tetrahydro-2-thioxo-4H-1,3-thiazin-4-ones have been made by elaboration of the appropriate tetra-*O*-acetyl-β-D-hexopyranosylisothiocyanates,[24] and 1-(2-acetamido-2-deoxy-β-D-glucopyranosyl)-6-azauracil has been made by base-sugar condensation.[25]

3   Anhydro- and Cyclo-nucleosides

Treatment of **12** (Vol.27, p.228; see also Section 5) with fluoride ion, followed by acetylation, gave the anhydrosystem **13**, and the corresponding di-*O*-benzoyl compound could be converted to the β-D-arabinofuranosyl nucleoside analogue.[26] The dihydrothymidine anhydronucleoside **14** has been prepared by reaction of the D-*arabino*-oxazoline with methyl methacrylate. Dehydrogenation using DDQ gave the anhydrothymidine **15**, whilst interestingly use of MnO$_2$ caused oxidation of only the minor diastereomer of **14**, leaving the major isomer, of 5*S*-chirality, unchanged.[27]

Some β-D-xylofuranosylpyrimidines have been prepared by condensation using 1,2-di-*O*-acetyl-3,5-di-*O*-benzoyl-D-xylofuranose; selective deblocking and activation of O-2' then led to the formation of 2,2'-anhydro-β-D-lyxofuranosylpyrimidines.[28]

Treatment of 2'-deoxyuridine derivatives of type **16** with a polymer-supported fluoride at room temperature gave the 2,3'-anhydrosystems **17** in good yield.[29]

The 8,5'-cyclonucleoside **18** was prepared by intramolecular displacement of an 8-bromosubstituent,[30] and the conformationally fixed uridine cyclic phosphate **19**, and the corresponding 2,5'-compound, have been prepared to study the effect of the conformation on their hydrolysis by ribonuclease.[31]

There have been further studies on the reaction of purine and pyrimidine nucleosides with α-acetoxyisobutanoyl bromide, and the use of the resultant bromoacetates to form 2',3'-anhydroribonucleosides.[32] Branched nucleosides of type **20** (B=Purine) have been prepared by condensation reactions; subsequent selective deacetylation, mesylation and base treatment then gave the 2',3'-anhydro-lyxonucleosides **21**. Similar chemistry was also used to make the D-*ribo*-analogues.[33]

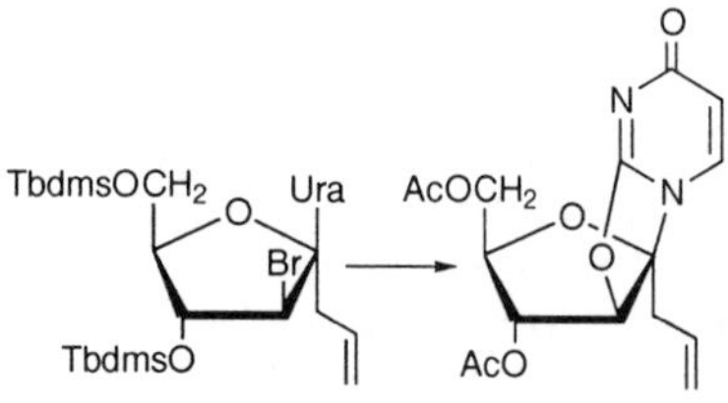

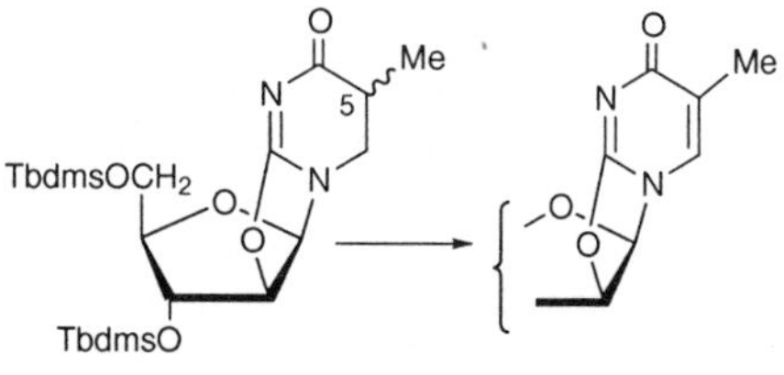

4 Deoxynucleosides

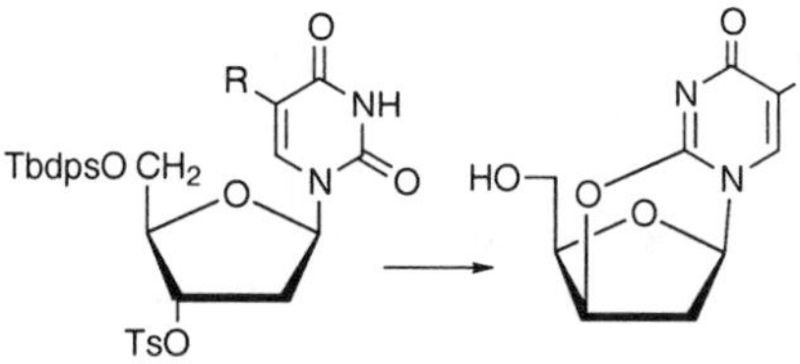

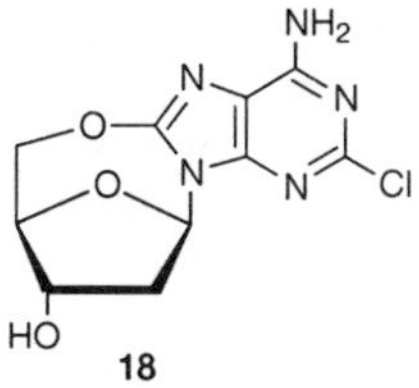

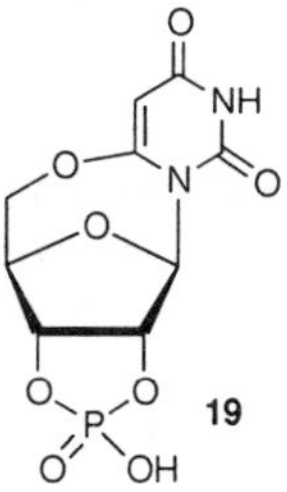

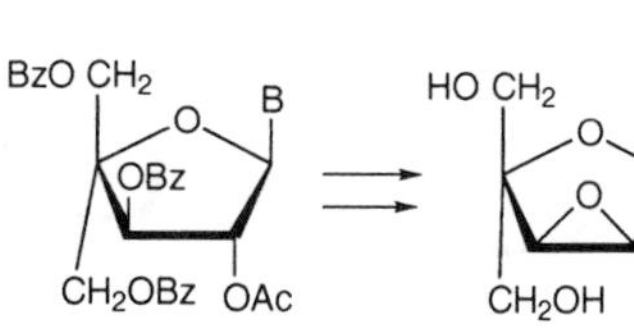

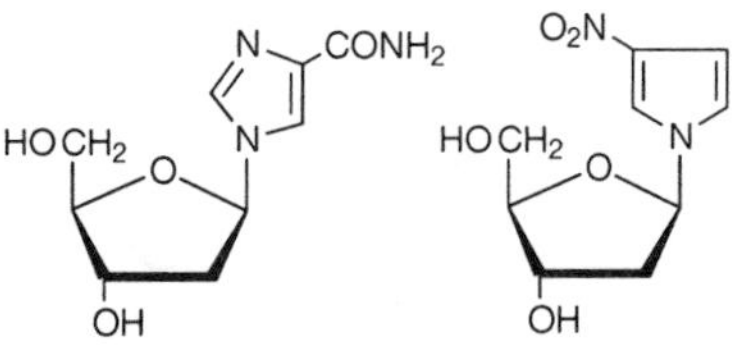

The enzymic propanoylation of 2-deoxy-D-ribose at O-5 is the first step in a new one-pot chemicoenzymatic synthesis of 2'-deoxyribonucleosides.[34] Alginate gel-entrapped cells of an auxotrophic thymine-dependent strain of *E. coli* have been used to catalyse the transfer of the 2-deoxy-D-ribofuranosyl unit from 2'-deoxyuridine to purine and pyrimidine bases, as well as to their aza- and deaza-analogues.[35] Enzymic transglycosylation was also used as a stereoselective alternative to chemical synthesis in the preparation of the imidazole deoxynucleoside **22**.[36]

Base-sugar coupling has been used in the synthesis of the nitropyrrole **23**, designed as a universal replacement for any of the natural deoxynucleosides in DNA sequences,[37] in a practical synthesis of 5-nitro-2'-deoxyuridine,[38] and for the synthesis of 8-substituted-2-chloro-2'-deoxyadenosine derivatives,[39] 2-

anilino-2'-deoxypurine nucleosides,[30] and 1-(2'-deoxy-β-D-ribofuranosyl)-quina-zoline-2,4-dione.[40]

Phase-transfer glycosylation using 2-deoxy-3,5-di-*O*-toluoyl-α-D-*erythro*-pento-furanosyl chloride led to considerable amounts of the N[7]-nucleosides **24** (X = O or S), from which the adenine and hypoxanthine analogues could be made,[41] and other workers have reported the synthesis of the N[7]-regioisomers of 2-chloro-2'-deoxyadenosine[42] and of 2'-deoxyguanosine.[43]

Free-radical deoxygenation procedures have been used to convert 6-chlorogua-nosine into 2-aminopurine-2'-deoxyriboside, Bu_3SnH effecting both deoxygena-tion at C-2' and dechlorination at C-6,[44] and for the synthesis of 2-aza-2'-deoxyinosine (**25**), which was then incorporated into oligonucleotides.[45] Treat-ment of the 3',5'-di-*O*-acetyl analogue of **14** with acetyl chloride in MeCN gave a chlorocompound, reducible with tributylstannane to the dihydrothymidine deri-vative **26**.[27]

A route to 2'-deoxynucleosides stereoselectivity deuteriated at C-2' involves the formation of derivatives **27** (B = Ade, Thy, Ura) by reduction of the corresponding *ribo*-bromides with Bu_3SnH – Et_3B at low temperatures; the stereoselectivity of this method was very good in comparison with related procedures.[46] Similar reduction of α-acetoxyselenides **28**, made by seleno-Pum-merer reactions, gave 5'-deuterio-compounds **29** where the *R/S* ratio was base-dependent, but with the 5'-*S*- isomer always predominant.[47]

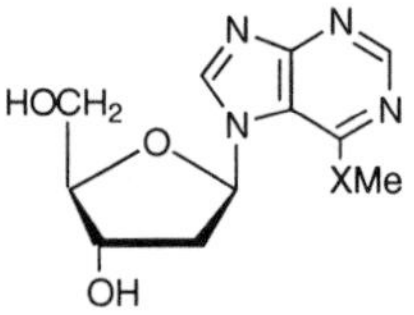

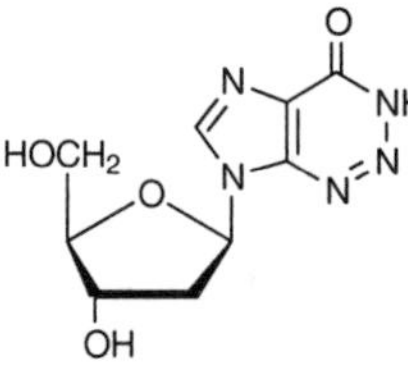

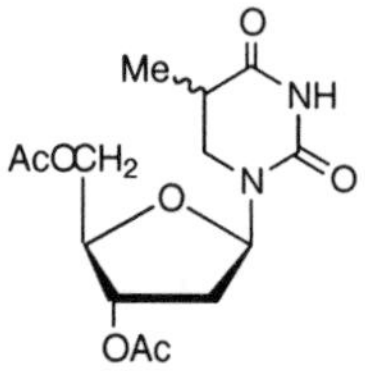

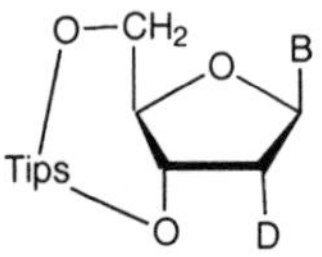

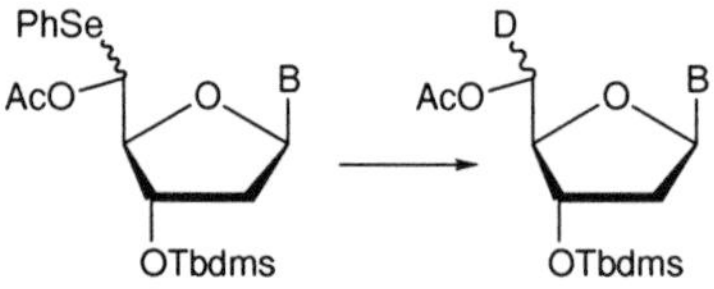

Treatment of 3',5'-di-*O*-toluoylthymidine with acetic anhydride and H_2SO_4 in acetonitrile-dichloromethane causes equilibration at the anomeric centre to give substantial amounts of the α-thymidine derivative.[48]

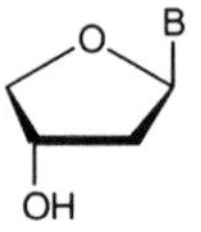

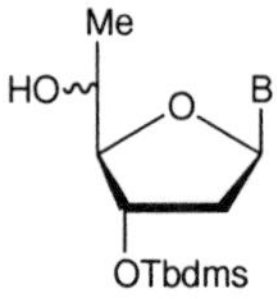

Truncated analogues **30** of 2'-deoxynucleosides have been prepared, along with their α-anomers, by base-sugar coupling, and shown to be inhibitors of uridine phosphorylase.[49] Compounds of type **31** have been synthesized by addition of MeMgBr to deoxynucleoside 5'-aldehydes, and incorporated into oligonucleo-tides using phosphoramidite methodology.[50] The chain-extended nucleoside analogue **32** has been prepared in a sequence in which the 'sugar' was elaborated from non-carbohydrate sources, and related structures epimeric at C-4' were also described.[51]

3'-Deoxycytidine, 3'-deoxyuridine, and their 5-fluoro-derivatives have been prepared using 1-O-acetyl-2,5-di-O-p-chlorobenzoyl-3-deoxy-D-ribofuranose, and normal coupling methods; the sugar unit was in turn derived from the antibiotic cordycepin.[52] 3'-Deoxyguanosine has been synthesized by regioselective opening of a 2',3'-$ribo$-epoxide with LiBHEt$_3$, the epoxide being produced from N^2-dimethylaminomethyleneguanosine by the use of α-acetoxyisobutanoyl bromide and subsequent base treatment (see ref. 32).[53]

In the area of 2',3'-dideoxynucleosides, a report from Reese's laboratory has described a route to pyrimidine 2',3'-dideoxy-2-thionucleosides, involving 2,3'-anhydro-2'-deoxynucleosides as intermediates.[54] 2',3'-Dideoxy-N^6-cycloalkyl-1-deazaadenosines have been prepared by base-sugar coupling, and the paper also reports further examples of related 2'-deoxynucleosides (see Vol.28, p.268).[55] Purine 2',3'-dideoxy-β-L-nucleosides have been prepared by base-'sugar' condensation,[56] as has the 5-azacytidine derivative 33; this latter report also describes the synthesis of 1-(2'-deoxy-β-L-arabinofuranosyl)-5-iodouracil and -5-(2-bromovinyl)uracil from L-arabinose.[57] The chiral building block 34, derived from D-glutamic acid, has been used to prepare dideoxy-L-nucleosides, and gives good β-selectivity (ca. 5:1) with pyrimidine bases.[58]

There is still considerable interest in routes to 2',3'-didehydro-2',3'-dideoxynucleosides (d4 systems). A report from M.J. Robins' laboratory describes, with many examples, the efficient conversion of ribonucleosides to d4 systems by reaction with α-acetoxyisobutanoyl bromide in moist acetonitrile, followed by reductive elimination on the mixed $trans$-bromoacetates.[32] An efficient conversion of thymidine into d4T (36) involved the conversion of thymidine into anhydronucleoside 35 under Mitsunobu conditions followed by manipulation as indicated in Scheme 3.[59] Bromomesylate 37, prepared from 5-methyluridine, has also been employed in a route to d4T,[60] as has the cyclonucleoside 38 (see Vol.28, p.267-8), convertible into d4T (36) in a one-pot procedure (Scheme 4).[61] Treatment of 2'-deoxynucleoside derivatives of type 16 with a polymer-supported fluoride reagent in THF at reflux gives the d4 analogues in moderate to good yield,[29] whilst treatment of the thymidine derivative 39 with potassium t-butoxide followed by an alkyl halide gives 3-alkyl-derivatives of d4T.[62] Thioacetals of type 40 have also been converted into the corresponding d4 systems,[63] and a symposium report has described the use of levoglucosenone as a precursor of d4T.[64] The d4 analogues of thieno[3,2-d]pyrimidine nucleosides have been described,[65] as has the didehydrodideoxy analogue of 9-(β-D-hexofuranosyl)adenine, along with related deoxy- and dideoxy-species.[66]

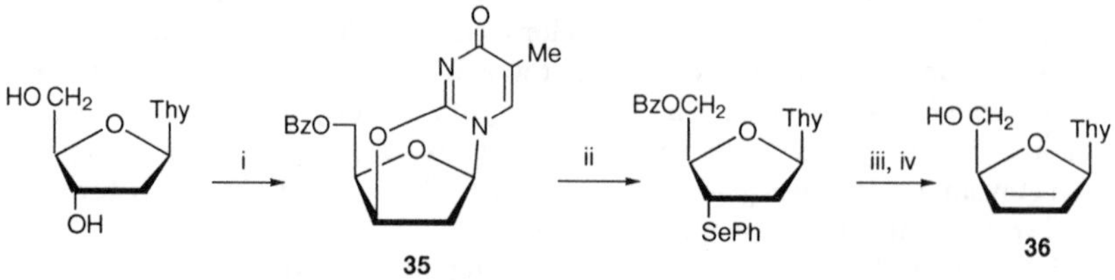

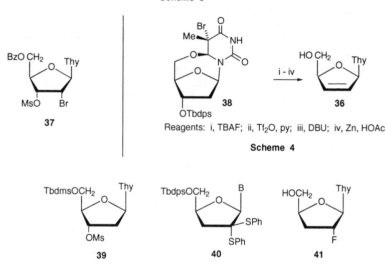

Reagents: i, PhCO₂H, PPh₃, DIAD; ii, PhSeH; iii, H₂O₂, HOAc; iv, NaOMe, MeOH

Scheme 3

Reagents: i, TBAF; ii, Tf₂O, py; iii, DBU; iv, Zn, HOAc

Scheme 4

5   Halogenonucleosides

2-Chloro-2′-deoxy-2′-fluoroadenosine, and its N⁷-regioisomer, have been prepared using base-sugar condensations, in which the regiochemistry of the reaction depended upon the reaction conditions.[67] Vorbrüggen-type coupling was also used to make the 2′-fluoro-compound **41**, and its α-anomer.[68] Sodium-salt glycosylation has been used to prepare 2′-deoxy-2′-fluoro-ara-A (**42**),[69] and its 2-chloroderivative (**43**), separable from its α-anomer in gram quantities by HLPC.[70] 1-(2′-Deoxy-2′-fluoro-β-D-arabinofuranosyl)-5-nitrouracil has also been prepared by condensation,[38] as have pyrrolopyrimidine nucleosides such as **44**, where the use of a standard deoxygenative procedure during the synthesis led to the 3′-deoxycompound **45**.[71] The adenosine analogues **46** (R=Me, Et) have been prepared as lipophilic prodrugs of FddI, to which they are converted by the action of adenosine deaminase.[72]

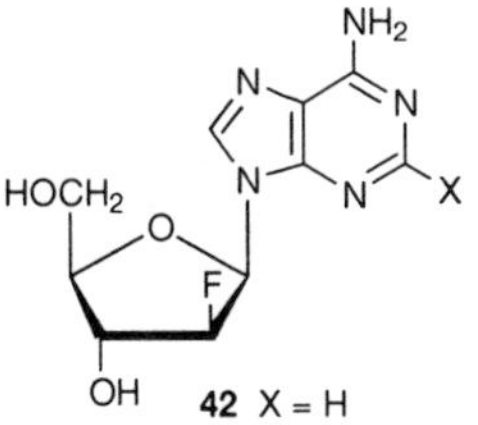

Fluorinated L-nucleosides have also been reported. Compounds **47** (B=Ade, Cyt) have been prepared from L-xylose,[73] and the same team have also made 2',2'-difluorocompounds of type **48**, the sugar being assembled using a Reformatsky reaction between ethyl bromodifluoroacetate and isopropylidene-L-glyceraldehyde. The adenosine analogue showed good anti-HIV activity with no cytotoxicity.[74]

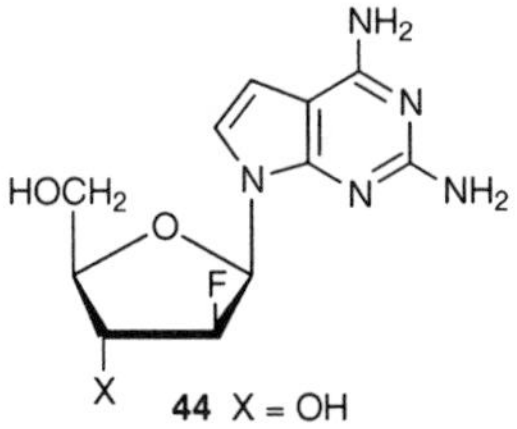

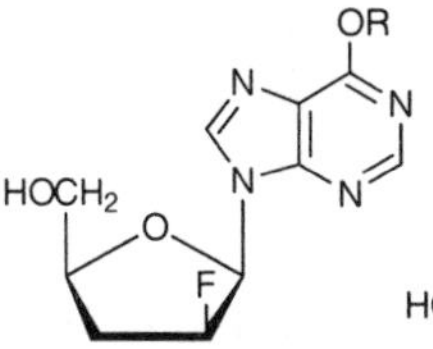

The fluorinated thionucleoside **49** was prepared from the corresponding 5'-O-trityl-2'-alcohol by fluorination by DAST with retention of configuration. Use of the sulfoxide gave fluorination with predominant inversion of stereochemistry.[75]

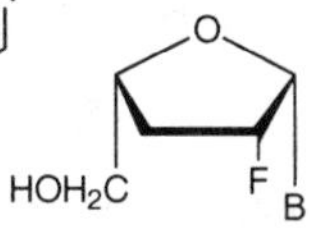

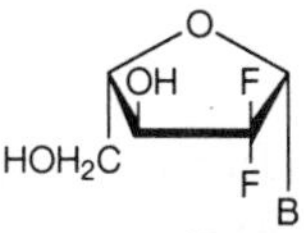

A full account has been given of the synthesis of **3** by addition of NBS and pivalic acid to the 1'-ene, and the subsequent Lewis acid-catalysed reactions with soft nucleophiles to give products **50** (R=allyl, CN, etc.) with a carbon substituent at C-1' (see Vol.27, p.248).[26]

The 4-methylthio analogue[76] and some hydantoin analogues[77] of 3'-fluoro-3'-deoxythymidine (FLT) have been prepared by condensation methods.

The fluorinated xylofuranosyladenine **51** has been prepared by opening of the 2',3'-*ribo*-epoxide using KHF$_2$, and was incorporated into 2-5A analogues.[78] When the alcohol **52** was treated with DAST, the β-D-*lyxo*-difluoride **53** was obtained. This is the first report of this stereochemical pattern for a 2',3'-difluoronucleoside (earlier unsuccessful routes using base-sugar coupling were reported last year), and it was found that the cytidine analogue could be made directly from **53** by treatment with ammonia under mild conditions, the two fluorines making the base much more susceptible to nucleophilic substitution.[79] Treatment of 5'-O-acetyl-2',3'-didehydro-2',3'-dideoxyuridine with fluorine gas (10% in nitrogen) gave the trifluorocompound **54**, together with the product of fluorination in the base only. Treatment of **54** with KOBu^t gave the fluoroalkene **55**, which could be hydrogenated stereoselectively from the α-face.[80]

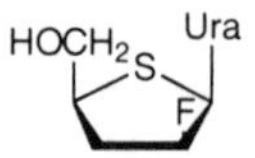

HOCH₂ Ura TbdmsOCH₂ Ura HOCH₂ Ade

49 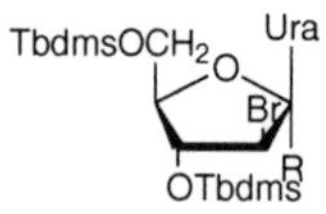   **50** 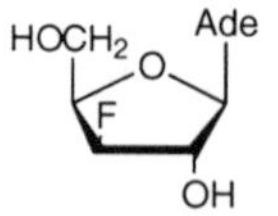   **51**

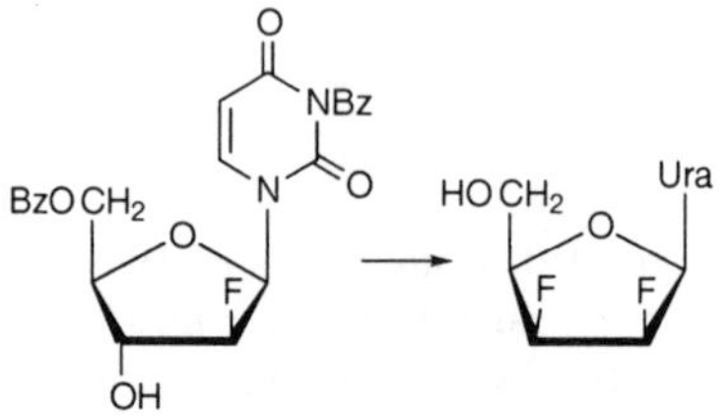

BzOCH₂ ... → HOCH₂ Ura   AcOCH₂ ... → AcOCH₂ ...

52   **53** 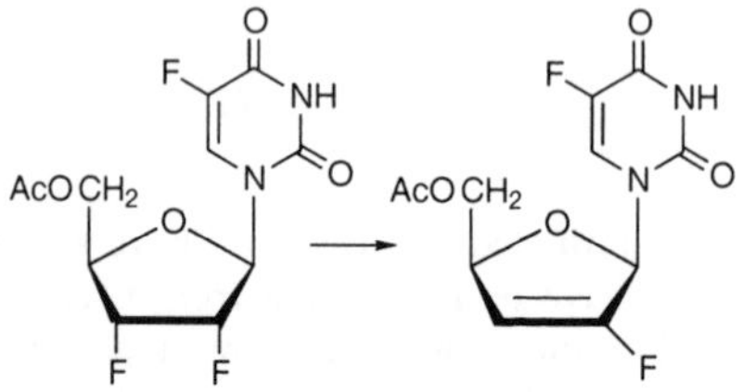   **54**   **55**

PhS—F Ura/Gua

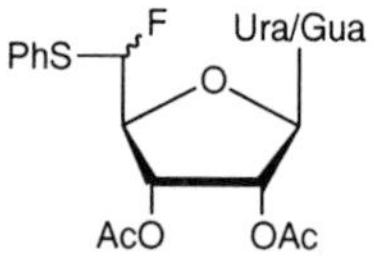

56

4'-Fluoroadenosine has been prepared in unprotected form for the first time, using chemistry similar to that employed by Moffatt and co-workers for the synthesis of nucleocidin (Vol.10, p.160), but with some improvements in protecting groups and reagents. The compound was found to be a time-dependent inhibitor of adenosylhomosyteine hydrolase.[81]

The positive fluorine reagent Selectofluor has been used for the α-fluorination of (arylthio)nucleosides, to give products such as **56**, and analogous products from reactions at C-2' and C-3' were also described.[82]

6   Nucleosides with Nitrogen-substituted Sugars

2'-Amino-2'-deoxyuridine has been coupled at the amino group with various Fmoc-protected aminoacids; the resultant conjugates were incorporated into ribozymes using phosphoramidite methods.[83]

3'-Amino-3'-deoxyadenosine has been prepared by coupling of the base to a sugar synthon derived from levoglucosenone (see Chapter 9).[84]

Reactions of the uridine-derived *lyxo*-epoxide with various amines, including an aza-crown ether, each gave a mixture of the two products from attack of the amine at C-3 and C-2, as exemplified by **57** and **58** in the case of pyrrolidine, with the product of type **57** predominating. This mixture could be converted as indicated in Scheme 5 into the 2,2'-anhydro-system **59**, presumably *via* an aziridinium intermediate. Reaction of **59** with thiophenol gave **60**, whilst base hydrolysis gave the pure *arabino*-compound **57**. When primary amines were used,

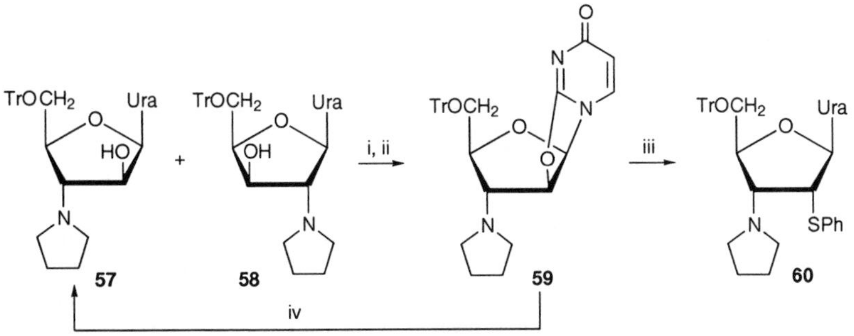

Reagents: i, MsCl, Py; ii, Py, reflux; iii, PhSH, (Me$_2$N)$_2$C=NH; iv, NaOH aq.

Scheme 5

the initial *arabino-/xylo-* mixture could be converted to the 2',3'-aziridine under Mitsunobu conditions.[85]

There has been a further report (see Vol.27, p.250) on the formation of 3'-*N*-alkylamino-3'-deoxythymidines, either by reductive alkylation, or by treatment of AZT with TPP followed by an alkyl halide.[86] 3'-Amino-3'-deoxythymidine has been converted to various *N*-substituted analogues such as **61**, which showed some anti-HIV activity.[87]

A new route to AZT involves bromomesylate **37**; reductive removal of the bromine, followed by treatment with sodium azide and lithium carbonate gave 5'-*O*-benzoyl-AZT, presumably via a 2,3'-anhydronucleoside.[88] 2-Alkylthio-[89] and the 4-methylthio-analogues[76] of AZT have been made by base-sugar condensation, as have 3'-azido-2',3'-dideoxynucleosides of hydantoins.[90] Some computational work on conformations of AZT is mentioned in Chapter 21.

N-Acyl-L-phenylalanyl prodrugs of 3'-amino-2',3'-dideoxycytidine have been prepared from 2'-deoxycytidine, using a double inversion at C-3' to generate an azide intermediate which was reduced and coupled to *N*-acylphenylalanines.[91] A report from Pedersen's laboratory has described the synthesis of thymidine analogues of type **62** (X=CH$_2$, NH, O), either in a non-stereoselective manner from triacetyl-D-glucal using a method based on previous work (Vol.25, p.252) or by ring-opening of a 2,3'-anhydronucleoside to give just the β-anomers. This latter approach with 1,2,4-triazole as nucleophile gave the triazolyl analogue **63**,[92] whilst others have adopted a similar approach to make tetrazoles of type **64**.[93] Reaction of 3'-amino-3'-deoxythymidine, and various similar compounds, with *N*-ethylmaleimide gave Michael adducts involving the 3'-aminogroup.[94]

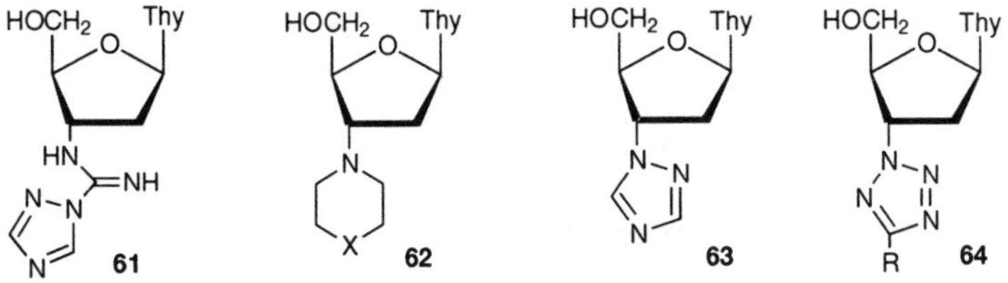

A new synthesis of 3'-deoxy-3'-nitrothymidine involves displacement of an 'up'-iodo-substituent by nitrite ion in DMSO in the presence of phloroglucinol.[95]

The (hydroxylamino)nucleoside **65** has been made by reduction of a C-3' nitrone with $NaBH_3CN$, and related *O*-alkylhydroxylamines were also described, prepared by reduction of *O*-alkyloximes.[96] Similar *O*-alkyloximes at both C-2' and C-3' of uridine have been prepared as potential inhibitors of ribonucleoside diphosphate reductase.[97]

The uncharged analogue **66** of cyclic AMP has been prepared, with 5'-amino-5'-deoxy-2',3'-*O*-isopropylideneadenosine as an intermediate.[98] Tosylates of type **67** (X=Me, I, F, CF$_3$) have been made by Vorbrüggen-type condensations, and used to prepare the 2',3'-*ribo*-epoxides and hence, *via* anhydronucleosides, derivatives of type **68**. The same paper also describes the synthesis of 3',5'-diazido-3',5'-dideoxy-5-methyluridine.[99] 5'-Azido- and 5'-amino-2',5'-dideoxynucleosides of quinazoline-2,4-diones have been made by base-sugar condensation.[100]

Nucleopeptidic conjugates **69** have been described, with the C-N bond being formed by Mitsunobu coupling of the appropriate 2'-deoxynucleoside with an *N*-Boc sulfamide.[101] The *S*-adenosylmethionine analogue **70** has been prepared and its biology studied.[102] The use of 5'-amino-5'-deoxyguanosine in a trisubstrate analogue inhibitor of a fucosyltransferase is mentioned in Chapter 9.

As regards hexopyranosyl systems, 5-substituted-6-azauracil nucleosides of *N*-acetylglucosamine have been prepared,[103] and acetobromoglucose was used as starting material to make the bis(nitroimidazole) **71**, in a search for hypoxia-selective antitumour agents with two bioreductively-activated units.[104]

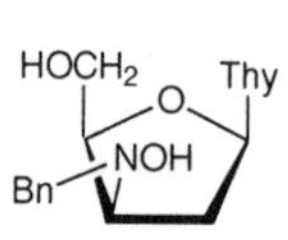

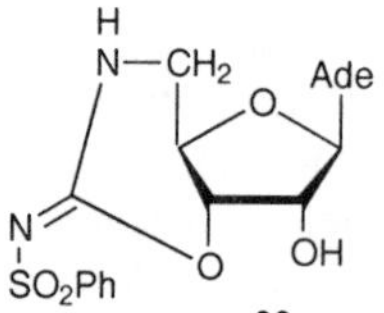

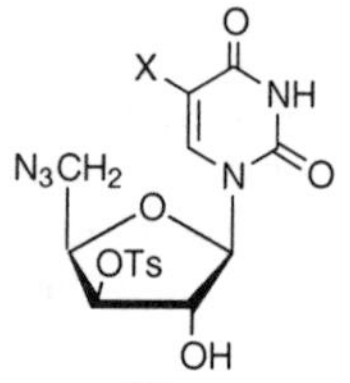

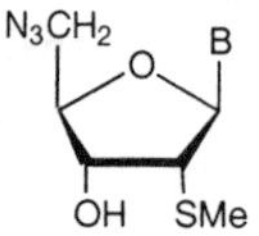

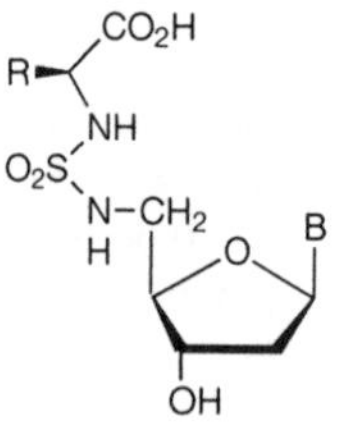

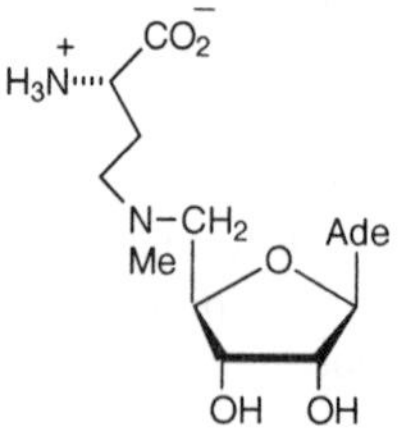

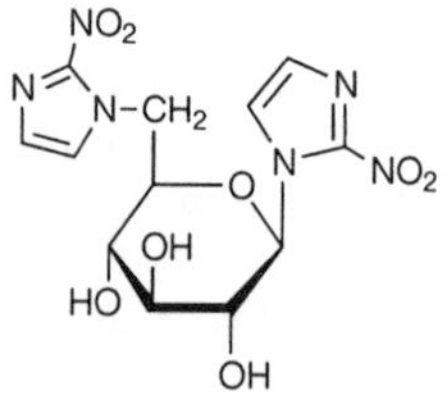

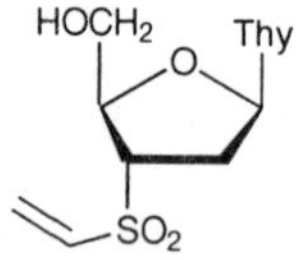

7   Thio- and Seleno-nucleosides

3'-Thiothymidine and 2'-deoxy-3'-thioadenosine have been prepared by condensation methods.[105] The vinyl sulfone **72** has been synthesized as a reactive analogue of AZT, and it was shown to react in a Michael fashion with various C-, N- or S-nucleophiles.[106]

4'-Thionucleosides continue to attract attention. The stereoisomer **73** was the major product of Lewis acid-catalysed reaction between 2-*t*-butyldimethylsilyoxy-thiophene and isopropylidene L-glyceraldehyde, and could be converted to the 4'-thioanalogue **74** of ddC; the enantiomer was also made similarly.[107] In an alternative approach, the thiolactone **75** was prepared from (*S*)-glycidol, and used to make 4'-thio-dideoxynucleosides similar to **74**. The α-phenylselenyl lactone **76** could be made stereoselectively from **75**, and this could be used, after DIBAL reduction and acetylation, in β-selective Vorbrüggen couplings, which led, after selenoxide elimination, to the d4 systems of type **77**.[108] The branched lactone **78**, made by photochemical addition of methanol to the enone, has been used as a precursor for 4'-thionucleosides **79** (R = H or F), in a sequence involving double inversion at C-4'. The same workers also describe an alternative route to the same type of analogue in which the building block **80** was made from (2*S*, 3*S*)-2,3-bis(benzyloxymethyl)oxirane by reaction with allylmagnesium bromide followed by ozonolytic cleavage of the double bond, this approach being used to make uridine analogues.[109]

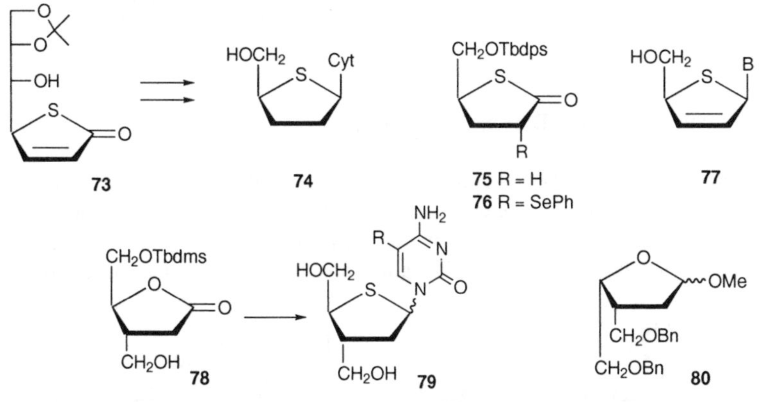

An interesting pyranose-to-thiofuranose rearrangement was used in a route to 2'-deoxy-4'-thionucleosides (Scheme 6), but unfortunately the initial rearrangement of the thionocarbonate **81** was not regioselective.[110]

Purine 2',3'-dideoxy-4'-thio-L-nucleosides have been prepared by base-sugar condensation.[56]

The cyclopentene derivative **82** has been prepared from 2',3'-*O*-isopropylidene-5'-thioadenosine, and was an irreversible inhibitor of *S*-adenosylmethionine decarboxylase.[111]

The sialic acid-cytidine conjugate **83** has been synthesized, the units being linked by displacement of a 5'-bromo-substituent on the nucleoside by a 2-thio-sialyl nucleophile.[112]

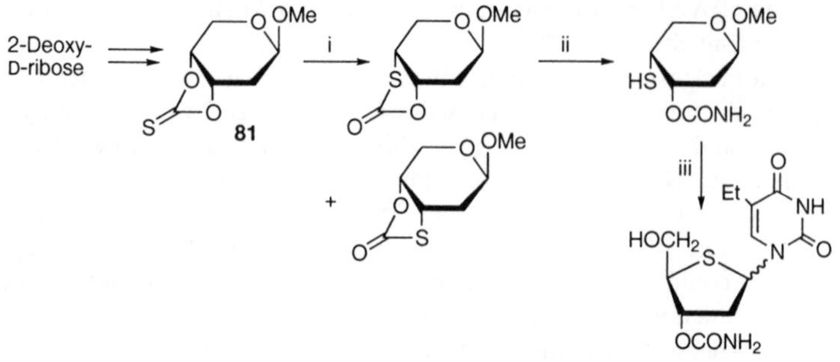

Reagents: i, Bu$_4$NBr, diglyme, 150 °C; ii, NH$_3$; iii, silylated 5-ethyluridine, TmsOTf, HMDS

Scheme 6

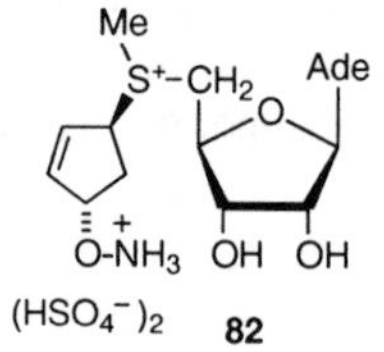

(HSO$_4^-$)$_2$ **82** 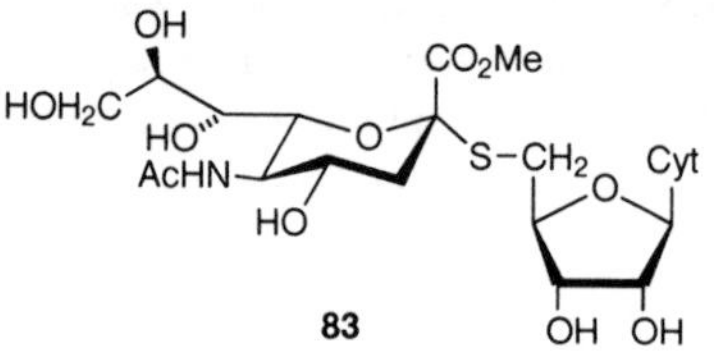   **83**

8   Nucleosides with Branched-chain Sugars

In a novel route to 3'-deoxy-2'-methylenethymidine **85** (Scheme 7), the sulfone **84** was prepared from α-D-isosaccharinolactone; reductive elimination of **84**, in a manner previously described for pyranoid systems (Vol.26, p.148) then gave an intermediate which underwent [3,3]-sigmatropic rearrangement as indicated.[113]

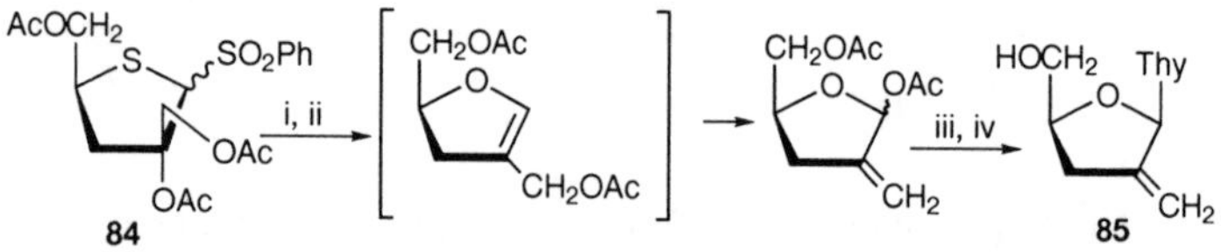

Reagents: i, SmI$_2$, HMPA; ii, silica gel; iii, (Tms)$_2$Thy, Pd$_2$(dba)$_3$, PPh$_3$; iv, NH$_3$, MeOH

Scheme 7

Previously reported intermediates (Vol.27, p.253-4) have been converted into other nucleoside analogues with two-carbon chains at C-2', such as the selectively-protected triol **87**, prepared by reduction of the lactone **86**.[114] 2'-C-Methyl ribonucleosides have been made by Vorbrüggen-type coupling of the appropriate base to a sugar unit (see Chapter 14).[115]

It has been found that the anti-tumour agent CNDAC (**88**) (see Vol.27, p.254 and earlier volumes) undergoes equilibration in base to the D-*ribo*- epimer **89**, degradation to the glycal being a slower process. The *arabino*-isomer has the higher anti-tumour activity and the observed activity of the *ribo*-compound could be due to epimerization.[116]

As regards compounds branched at C-3, 3'-*C*-trifluoromethyl ribonucleosides have been made by base-sugar coupling (see also Chapter 14).[117] Addition of a Ce(III) organometallic to the appropriate 3'-ketonucleoside gave predominantly the *xylo*-products **90**, whilst if O-5' was unprotected, the *ribo*-products were formed with high stereoselectivity, possibly because of coordinative guidance from the hydroxyl function.[118] The 3'-alkyne **91** was prepared by ring-opening of the 2',3'-*lyxo*-epoxide; subsequent deoxygenation at C-2', followed by further chemistry at C-5' using $Ph_3P=CBr_2$ then gave the dialkyne **92** which could be polymerized using Cu(I). Elaboration of the second alkyne was not successful if O-2' was still present.[119]

DmtrOCH₂ Ura DmtrOCH₂ Ura HOCH₂ Cyt HOCH₂ Cyt

OH CH₂CH₂OH OH OH CN
O **86** **87** NC **88** **89**

TbdmsOCH₂ Ade/Ura TrOCH₂ Ura Ura HOCH₂ Thy

OH HO HO CH₂NO₂
OTbdms Tms **90** **91** **92** **93**

HOCH₂ Thy HOCH₂ Thy H₂NOC HOH₂C Thy

C CH₂ CH₂OH CH₂OH CH₂OH OH
94 OH **95** OH **96** CH₂OH **97** HOH₂C **98**

The AZT analogue **93** has been prepared from a 3'-deoxy-3'-iodo compound of the same configuration, by formation of a cobaloxime and photochemical reaction with the nitronate anion.[120] A related iodo-derivative was treated with propargyl triphenylstannane and AIBN to give, after deprotection, the allenyl analogue **94** of AZT.[121] There have been further reports on the synthesis of 3'-(hydroxymethyl)thymidine **95** (see Vol.28, p.278), its incorporation into oligonucleotides via the 3',5'- and 3',3'-CH_2OH positions, and evaluation of their hybridization and enzymatic stability.[122,123] Some β-D-apiofuranosyl triazoles such as **96** have been made by cycloadditions of a glycosyl azide with stabilized phosphoranes.[124] Pyranosyl 3'-hydroxymethyl nucleosides of type **97** have also been prepared, the hydroxymethyl group being delivered onto a 2',3'-ene from O-4' using a silicon tether, followed by inversion of stereochemistry at C-4'. Oligonucleotides were assembled involving units **97** (B=Ade), and these formed duplexes with oligothymidylate as predicted by computer modelling.[125,126] There have been further reports on the synthesis of 4'-(hydroxymethyl)thymidine **98**, which has now been incorporated into oligodeoxynucleotides either in the normal manner, or involving the extra hydroxyl function; branched systems through all three hydroxyls were also prepared.[127,128] The Prague group have also described the synthesis of 2'-deoxy-, 2',3'-dideoxy- and 2',3'-didehydro-dideoxy analogues of 4'-*C*-hydroxymethyladenosine.[129] A paper describing new adenosine analogues as agonists or antagonists at adenosine receptors reports the synthesis of a ribofuranosyl uronamide with a 4'-*C*-methyl branch.[130]

Chattopadhyaya's laboratory has reported further on systems with extensive modification at C-2' and C-3'. The unsaturated bromide **99** was made from a known seconucleoside (Vol.28, p.277) and converted to **100** by treatment with tributylstannane followed by deprotection. In related fashion, **101** was prepared by intramolecular Diels-Alder reaction.[131]

The doubly-branched pyrimidine nucleosides **102** have been prepared by coupling of silylated uracil with a sugar unit (Chapter 14),[132] and branched pyranosyl systems **103** have been reported.[133]

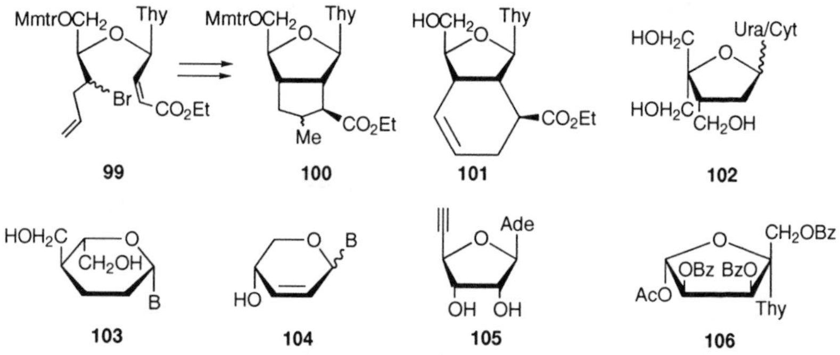

9　Nucleosides of Unsaturated Sugars, Aldosuloses and Uronic Acids

As in previous volumes, 2',3'-didehydro-2',3'-dideoxyfuranosyl derivatives are discussed in Section 4, together with their saturated analogues.

Compounds of type **104** have been prepared by reaction of 3,4-bis-*O*-(*p*-nitrobenzoyl)-D-xylal with nucleobases in refluxing DMF, followed by deacylation. The conformations of these analogues have been studied in some detail.[134,135] An alternative synthesis has been described of the alkyne **105** (see Vol.25, p. 258-9 for the earlier route), involving reaction of a protected 5'-aldehyde with $Ph_3P=CBr_2$.[136]

The thymine nucleoside **106** has been prepared by coupling of silylated thymine with the glycosyl bromide (Vol.18, p.84-85), made by a photobromination procedure.[137]

Some glucuronate nucleosides of barbiturates have been reported,[138] as has the analogue **107** of a thymidine dinucleotide, which was incorporated into oligonucleotides.[139]

10   *C*-Nucleosides

The new tiazofurin analogues 'furanofurin' (**108**, X=O) and 'thiophenfurin' (**108**, X=S) have been prepared by direct reaction between tetra-*O*-acetyl-β-D-ribofuranose and the heterocycles, as their ethyl esters, in the presence of SnCl$_4$. Thiophenfurin had anti-tumour activity and inhibited IMP dehydrogenase, implying the need for the sulfur of tiazofurin, but that the nitrogen was not essential.[140] Pyridazine *C*-nucleosides have been made by cyclizations of tetrazines and ribofuranosyl alkynes, as in the case of **109** and **110**, which gave **111** after deprotection.[141] 1,2,4-Triazines have been similarly prepared by cycloadditions to ribofuranosyl imidates; use of **109** gave **112**.[142] Pyrazolo[1,5-*a*]pyrimidine *C*-nucleosides such as **113** have been prepared by interaction of 3-aminopyrazole with a previously-described intermediate.[143]

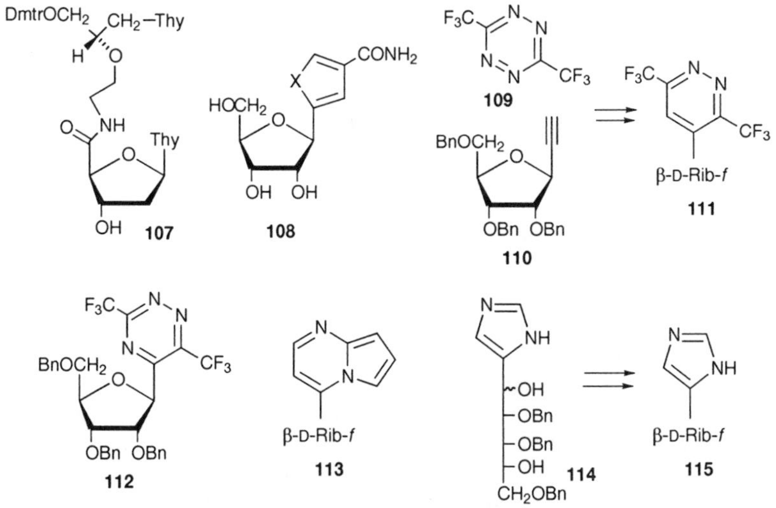

An improved route to 3-(β-D-ribofuranosyl)pyrazole has been described, starting from 2,3-*O*-isopropylidene-D-ribofuranose and involving acetylenic intermediates.[144] A stereocontrolled route to 4(5)-(β-D-ribofuranosyl)imidazole (**115**) has also been developed; the epimeric mixture **114** was obtained *via* addition of a lithiated imidazole to 2,3,5-tri-*O*-benzyl-D-ribofuranose, and cyclization of this under modified Mitsunobu conditions gave selectively the β-product. A mechanistic rationale was presented, and the 2'-deoxycompound was similarly made.[145] Various other 2'-deoxy-*C*-nucleosides of type **116** have been prepared in a non-stereoselective way by reaction of lithiated heterocycles with 3,5-*O*-Tips-2-deoxyribose, followed by acid-catalysed cyclization; compounds made included the 2-furyl, 2-indolyl, 2-pyridyl and benzothiophen-2-yl analogues.[146] A practical synthesis of N^1-methyl-2'-deoxy-ψ-uridine ('ψ-thymidine') from ψ-uridine has been developed, using free radical deoxygenation. The product was incorporated into oligodeoxynucleotides,[147] as were 2'-deoxyformycin,[148] and N[2]-isobutanoyl-2'-deoxy-9-deazaguanosine.[149]

Several papers have reported further uses of palladium-catalysed reactions to make 2'-deoxy-*C*-nucleosides. Illustrative is the synthesis (Scheme 8) of the thymidine analogue **117**,[150] and similar chemistry has been used to prepare 2'-deoxypyrazine[151] and -pyrazolo[1,5-*a*]-1,3,5-triazine *C*-nucleosides.[152]

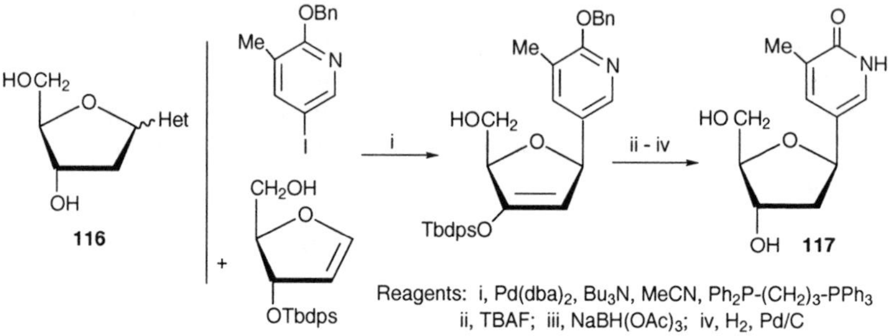

Reagents: i, Pd(dba)$_2$, Bu$_3$N, MeCN, Ph$_2$P-(CH$_2$)$_3$-PPh$_3$
ii, TBAF; iii, NaBH(OAc)$_3$; iv, H$_2$, Pd/C

Scheme 8

11 Carbocyclic Nucleoside Analogues

The carbocyclic analogue **118** of AICAR, an intermediate in purine nucleoside biosynthesis, has been prepared in a sequence which involved elaboration of the imidazole ring from a cyclopentylamine; a similar sequence was used to make the later biosynthetic intermediate SAICAR.[153]

Various 2'-*O*-alkylated derivatives of carbocyclic 5-methyluridine have been prepared, as have 6'-alkoxy and -acyloxy-derivatives of carbocyclic thymidine (**119**), in order to test the effect of the structural alterations on DNA/RNA duplex stability.[154] A short and high-yielding synthesis of carbocyclic bromo-vinyl-deoxyuridine (BVDU) has been described, involving a cyclopentylamine derivative as an intermediate.[155] Carbocyclic BVDU and carbocyclic 2'-deoxy-guanosine have been made by a modification of an earlier route to carbocyclic 2'-deoxynucleosides, involving a microbiological reduction.[156] Racemic difluoro-carbocyclic nucleosides of type **120** have been made via the fluorination of (±)-*N*-Boc-2-azabicyclo[2.2.1]hept-5-en-3-one (see Vol.28, p.284),[157] and the (+)-enantiomer of 2-azabicyclo[2.2.1]hept-5-en-3-one has been converted into the (−)-enomtiomer, which is of considerable use in the synthesis of carbocyclic nucleosides in the D-series.[158]

A new asymmetric synthesis of the 'unnatural' (+)-form of carbovir has been described, involving the cycloaddition of a camphor-derived nitrone with cyclopentadiene,[159] and racemic *trans*-carbovir has been made using a Ramberg-Bäcklund reaction to prepare the cyclopentene unit.[160] (±)-2',3'-Didehydro-2',3'-

dideoxy-carbocyclic nucleosides of 5-substituted uracils and cytosines, where the 5-substituent is a thiophene ring, have been reported,[161] and these have been hydroxylated, to give substantial amounts of the all-*cis* ('β-*lyxo*') products.[162] The carbocyclic 5'-nor-adenosine analogue **121**, its C-4'-epimer, and their enantiomers have been prepared from chiral cyclopentenes, and these compounds were converted to phosphonates such as **122**.[163] Enantiomerically-pure 5'-aza-noraristeromycin analogues have been synthesized by cycloaddition of a chiral acylnitroso dienophile with cyclopentadiene, to give, after further manipulation, the analogue **123** which could be hydroxylated.[164] A similar approach was used to make hydantoin analogues such as **124**.[165] 5'-Nor-aristeromycin analogues, and also a uronamide analogue, have been described as potential agonists/antagonists at adenosine receptors.[130] Branched carbovir analogues of type **125** have been made from a previously-known chiral cyclopentanone derivative.[166]

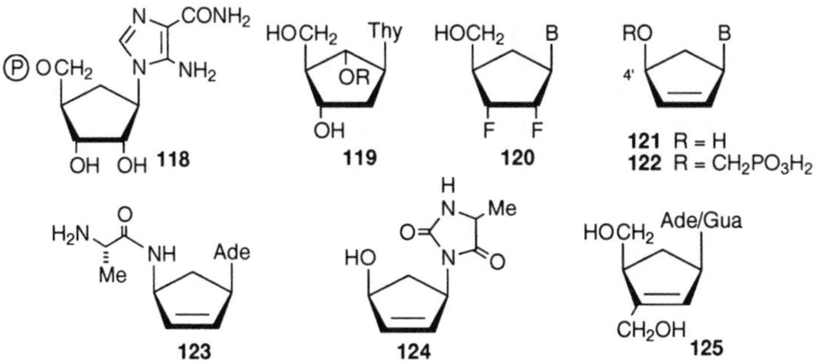

An alternative route has been developed in Marquez' laboratory to prepare the methano-carbocyclic thymidine **127** (see Vol.28, p.286). A key step involved the highly stereoselective formation of **126** by addition of diazomethane to the α,β-unsaturated nitrile, itself accessible from a known intermediate in two steps. Photochemical loss of nitrogen from **126** was followed by a Curtius sequence to establish an amine from which the thymine ring was formed.[167] The aristeromycin analogue **128** has also been prepared, using chemistry reminiscent of earlier work on analogues of this type (Vol.28, p.285-6),[168] and bicyclic compounds of type **129** have been prepared as racemates.[169]

A series of cyclohexenyl nucleoside analogues such as **130**, which can be regarded as a carbovir analogue with an extra methylene group inserted, have been described,[170] as have nucleoside phosphonates of type **131**[171] and the compounds **132** and **133** (B = Gua, Ade, Ura), where the base units were added to the cyclohexyl ring by nucleophilic opening of epoxides.[172]

12 Nucleoside Phosphates and Phosphonates

The Mitsunobu reaction, using dibenzyl phosphate or benzyl methylphosphonate as nucleophile, can be used to convert 2',3'-*O*-isopropylidene purine nucleosides

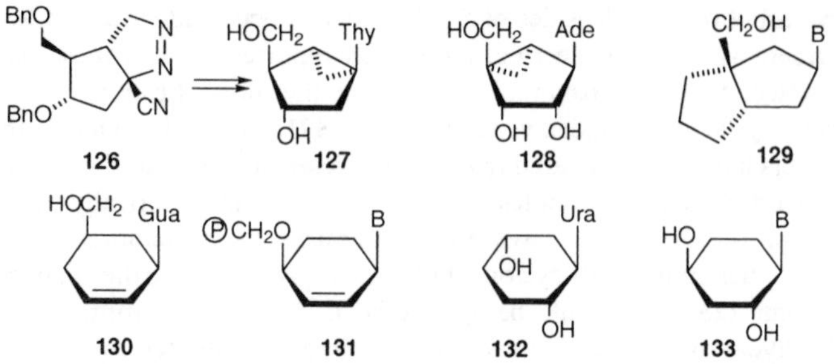

126　　127　　128　　129

130　　131　　132　　133

into their 5'-phosphates or methylphosphonates. The use of pyridine as solvent is important in preventing the formation of 3,5'-cyclonucleosides.[173] Phosphorylation of guanosine with $POCl_3$ is more selective for the 5'-*O*-monophosphate when carried out in triethyl phosphate containing some water.[174] Protected 1-hydroxybenzotriazolyl esters of adenosine 5'-phosphorothioate (134, X = O) and phosphorodithioate (134, X = S) have been prepared; aminolysis of these gives phosphorothioamidates and phosphorodithioamidates, whilst alkaline hydrolysis forms mono- and dithiophosphates accompanied by 3',5'-cyclic species.[175] Various [17]O-labelled thymidine 5'-phosphate triesters , alkylphosphonates, dialkylphosphinates and phosphoramidates have been prepared. The chemical shifts in the ^{17}O-NMR spectra seemed to be useful for structural assignment, but diastereomers at phosphorus did not display discernably different chemical shifts.[176]

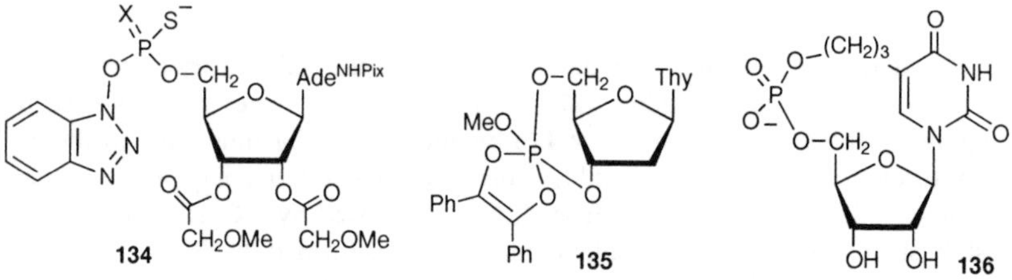

134　　135　　136

The thymidine oxyphosphorane 135 has been prepared by ester exchange, and underwent hydrolysis to give primarily the 5'-dialkyl phosphate.[177] The cyclic analogue 136 of UMP has been synthesized for incorporation into antisense sequences to impart nuclease resistance. Its conformation was found to be similar to that of a nucleotide unit in A-type DNA duplexes.[178]

There continue to be reports of potential prodrugs of biologically-active nucleoside-5'-phosphates. The *S*-acetyl-2-thioethyl ester 137 of d4T has been prepared, and undergoes esterase hydrolysis in the cell.[179] The same team have also made the analogous derivative of AZT.[180] A series of aromatic aminoacid phosphoramidates of AZT have been described[181] as have bis-ketol AZT monophosphates[182] and 4-acyloxybenzyl bis(nucleosid-5'-yl) phosphates of AZT and ddI.[183] Phosphodi- and tri-esters incorporating both AZT and a metal-complexing ligand have been prepared; although they catalysed the cleavage of RNA fragments, they were not effective in antiviral assays.[184] The triester **138** has been described; hydrolysis by esterase followed by elimination of acrolein gives FdUMP,[185] and other workers have developed a similar phosphoramidate system, again as a prodrug of FdUMP.[186]

The 3'-phosphate **139** has been prepared as the first mechanism-based inhibitor of a phosphodiesterase.[187]

New methods have been developed for the synthesis of nucleoside 3'-H-phosphonates, using triphosgene as an activating agent for phosphonic acid,[188] and involving pivaloyl chloride induced coupling of the nucleoside with 9-fluorenemethyl phosphonic acid followed by deprotection using triethylamine in pyridine.[189] Various β-D-arabinofuranosyl nucleoside 3'-H-phosphonates have been prepared.[190] The conversion of nucleoside H-phosphonate diesters into phosphoramidates using the reagent system CCl_4 and n-$BuNH_2$ in acetonitrile has been shown to result in inversion of configuration at phosphorus.[191]

Oxazaphospholidines of type **140** have been advocated as alternatives to the more normal phosphoramidite units used in solid-phase synthesis.[192] The unit **141**, as an isomeric mixture, has been developed as an all-purpose adaptor for synthesis of oligonucleotides on any of the commercial supports. The unit **141** is attached to the support through O-5'. After removal of the Dmtr group, the first nucleoside is attached via a 2'(3')-3' link using a phosphoramidite. The desired oligonucleotide is then conventionally assembled, and disconnected by removal of the *O*-benzoyl group from the adaptor, which is followed by cyclic phosphodiester formation.[193]

Various papers have described novel routes to 3'-phosphorothioates and related species. Reactions of 5'-*O*-pixyl-2'-deoxyribonucleoside 3'-H-phosphonates with *N*-(2-cyanoethylsulfanyl) phthalimide in the presence of TmsCl and *N*-methylmorpholine gives the corresponding 3'-phosphorothioate *S*-(2-cyanoethyl) esters, useful building blocks in oligodeoxyribonucleoside synthesis by the phosphotriester approach.[194] Nucleoside cyanoethyl H-phosphonate diesters have been converted to thiophosphotriesters of type **142** by reaction with TmsCl followed by reagents of the type $ArSO_2SR$,[195] and nucleoside 3'-H-phosphonates can be converted to nucleoside 3'-H-phosphonothioate monoesters by activation with PivCl in pyridine, followed by H_2S in dioxane.[196] The dithio-H-phosphonate **143** could be prepared by use of PCl_3 and a base followed by H_2S, and converted into the dithiophosphate **144** by oxidative coupling with 9-fluorenemethanol, followed by treatment with NH_3; direct oxidation was less successful, and similar chemistry was carried out at O-5'.[197] The use of cyclic phosphitylating agent **145**, followed by sulfur treatment, has been developed for the one-pot synthesis of monothiophosphates of nucleosides, including 2',3'-cyclic monothiophosphates, and for the synthesis of monothio-derivatives of dinucleoside 5'-oligophosphates.[198] Building blocks of type **146** have been developed for use in making oligonucleoside phosphorothioates.[199]

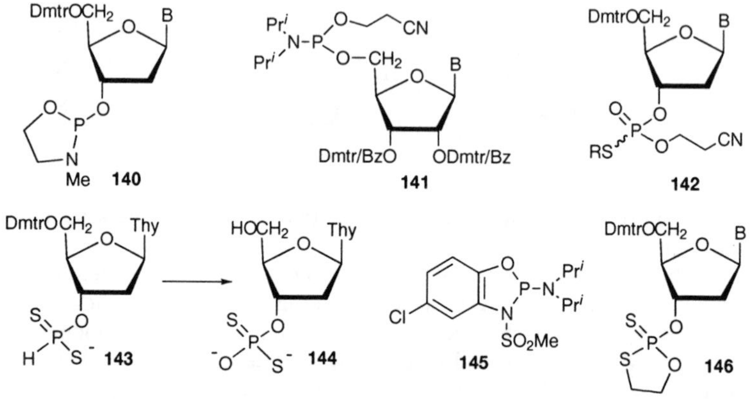

Stable phosphorofluoridites of type **147** have been prepared,[200] and, by reaction with a second nucleoside followed by thionation, gave dinucleosidyl phosphorofluoroidothioates **148**. Isomer separation was possible at the phosphorofluoridite stage.[201] Reactions of **147** with *t*-butanol or β-cyanoethanol gave, after sulfuration, intermediates which could be converted thermally (*t*-butyl esters) or by base treatment (cyanoethyl esters) into nucleosidyl 3'-*O*-phosphorofluoridothioates (**149**, X=S). Oxidation of the P(III) intermediates also gave a convenient route to nucleosidyl phosphorofluoridates (**149**, X=O).[202]

Diphenylmethylsilylethyl esters at the internucleosidic link, as in **150**, can be rapidly deprotected using SiF$_4$ in acetonitrile containing a little water.[203] On treatment with acid, compounds such as **150** undergo migration of the trialkylsilylethyl group from oxygen to sulfur, probably involving an ion pair in which the carbocation is stabilized by the β-silicon effect.[204]

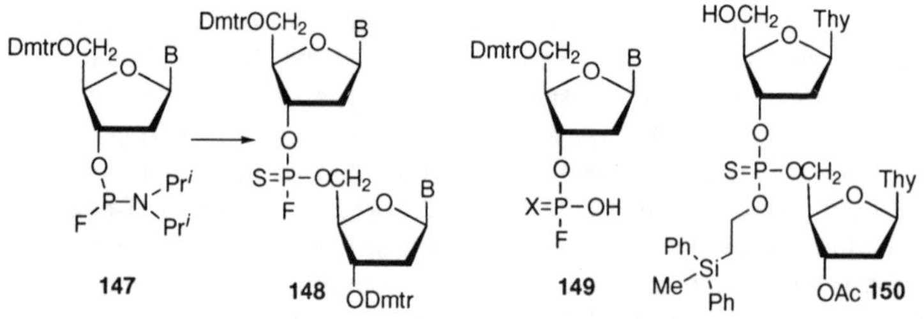

As regards other modifications of the internucleosidic link, a new route to dinucleosidyl phosphorofluoridates is illustrated in Scheme 9. The hydrolytic stability of the product was investigated, and various triesters were prepared from **151**.[205] Triesters have also been prepared by alkylation of dinucleosidyl tributylstannyl phosphates.[206]

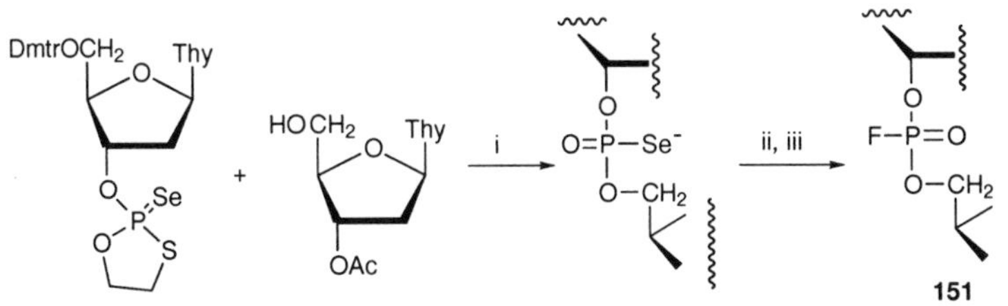

Reagents: i, DBU, MeCN; ii, MeI; iii, AgF or Et₃NH F

Scheme 9

The first dinucleotide phosphoramidimidate **152** derived from an aliphatic amine has been prepared by reaction of the phosphorimidite with the amine in the presence of iodine.[207]

Dideoxynucleoside methylphosphonates have been prepared by reaction of a 5′-*O*-protected deoxynucleoside with MePOF₂ to give a phosphonofluoridate, which was then treated with a 3′-protected nucleoside in the presence of base. The success of the method depends on the second fluoride displacement being a much slower process. Dinucleoside methylphosphonothioates were prepared similarly.[208]

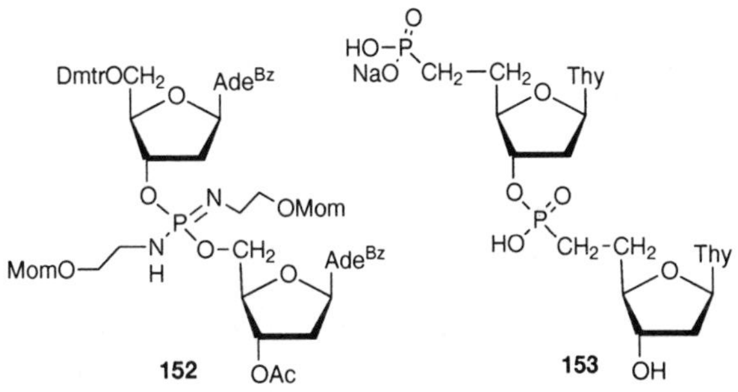

An efficient synthesis of the bis(phosphonate) **153** has been described,[209] and a route to a thymidine dimer containing an internucleoside phosphinite link is outlined in Scheme 10, the dimeric unit being incorporated into oligodeoxynucleotides.[210] The dinucleotide **154** with a novel hydroxymethylphosphonate link was prepared by reaction of a dithymidinyl trimethylsilylphosphite with formaldehyde; the procedure could be carried out on a solid support to give a decathymidylate with all links modified.[211] Also reported is a procedure for the synthesis of dinucleoside trifluoromethylphosphonates (**155**).[212] Various dinucleoside phosphonamidates of antiviral agents, such as **156**, have also been described.[213]

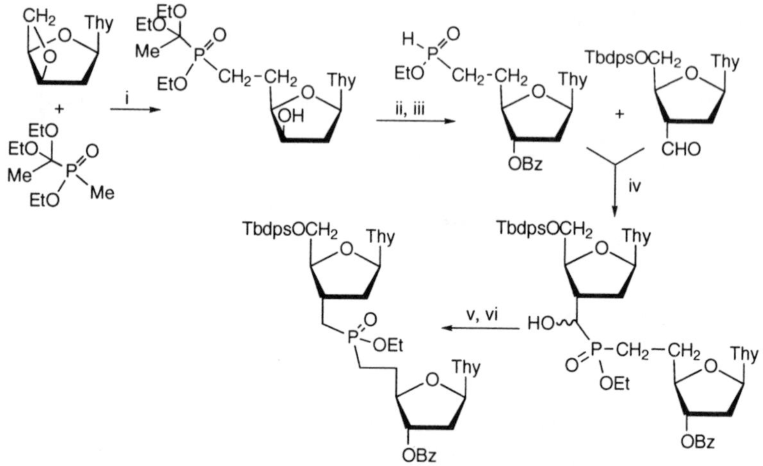

Reagents: i, BuLi, then BF₃.Et₂O; ii, DEAD, Ph₃P, PhCO₂H; iii, TmsCl, CHCl₃, EtOH; iv, DBU; v, *p*-Tol-CSCl, Et₃N; vi, Bu₃SnH, AIBN

Scheme 10

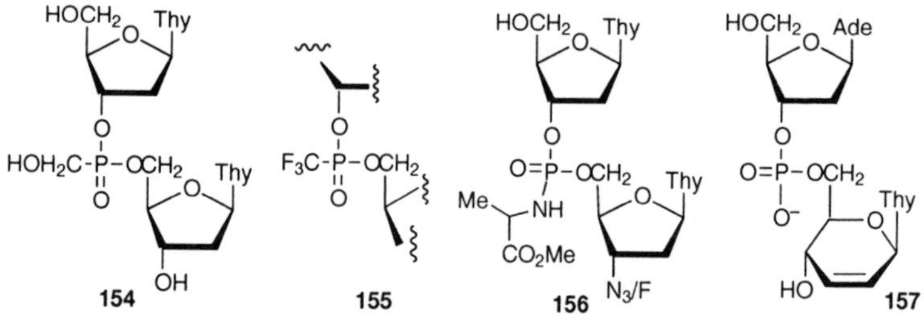

With regard to oligonucleotide analogues with modifications in the sugar unit, there has been a further contribution from Eschenmoser's laboratory concerning 'pyranosyl RNA', where β-D-ribopyranosyl nucleosides are linked by 2′,4′-phosphate units.[214] Other workers have described the dinucleotide analogue **157**, an antisense construct with potentially increased hybridization properties.[215] 3′-Dimethoxytrityl-5′-phosphoramidites of α-deoxynucleosides **158** have been prepared, and these, together with the usual 5′-*O*-Dmtr-derivatives of β-deoxynucleoside 3′-phosphoramidites, have been used to make oligodeoxynucleotides with alternating α,β-linkages, and alternating 3′-3′ and 5′-5′ linkages.[216]

Various new 2′-5′-adenylate trimers have been described, involving a 5′-amino-5′-deoxyribonucleotide unit at the 5′-end and in some cases an acyclonucleoside terminal unit. These had increased resistance to phosphodiesterase.[217]

There have been further papers concerned with nucleoside phosphonates. A full and extended account has been given of the formation of hydroxyphosphonates such as **159** by stereoselective addition of diethyl phosphite anion to the ketonucleoside (see Vol.27, p.263-264) and the deoxygenation of these to give products such as **160**. The deoxygenation gave low stereoselectivity (~2:1) for the β-phosphonate. Similar chemistry can be effected at C-2'.[218] Also reported on more fully are difluoromethylene phosphonates of type **161** (see Vol.28, p.290).[219] 5'-*O*-Phosphonomethyl derivatives of nucleosides have been prepared by alkylation using reagents TsOCH$_2$-P(O)(OH)$_2$ and TsOCH$_2$ -P(O)(OH)(OEt).[220]

In the area of cyclic phosphates, adenosine 3',5'-cyclic phosphorofluoridate has been prepared; it was shown to undergo facile conversion to adenosine 2',3'-cyclic phosphate.[221] A photolabile 3',5'-cyclic phosphotriester of adenosine has been described,[222] as have 3',5'-cyclic phosphotriesters and phosphoramidates of 8-chloroadenosine.[223]

Acidity constants have been measured for AMP and ADP in various mixtures of water and organic solvents using potentiometric titration.[224] The 3'-triphosphate of 2'-deoxyadenosine has been described for the first time, together with 3'-di- and -tri-phosphates of 2',5'-dideoxyadenosine.[225] 5'-Triphosphates of 2'-amino-2'-deoxypyrimidine nucleosides have been reported,[226] as have long-chain acyl nucleoside di- and tri-phosphates derived from AZT and d4T, eg **162**, as potential lipophilic prodrugs.[227] Ara-C has been similarly conjugated to ether lipids *via* a diphosphate link[228] and analogues of thymidine triphosphate and 3'-deoxy-3'-fluorothymidine diphosphate with large hydrophobic substituents at the α-phosphorus (linked as a phosphonate) have been reported.[229] The analogue of AZT triphosphate with two intraphosphorus methylene groups, (**163**, X=PO$_2^-$) has been prepared, along with the trimethylene analogue (**163**, X=CH$_2$).[230] A report from Wong's laboratory describes an enzymatic synthesis of 3'-phospho-adenosine-5'-phosphosulfate (PAPS) on a 100mg-plus scale.[231]

In the area of sugar nucleotides, a rapid synthesis of uridine diphosphosugars has been reported, involving the interaction of a glycosyl bromide with UDP; UDP-Gal, UDP-L-Ara and UDP-L-Fuc were made in this way and shown to act as glycosyl sources with transferases. Although the materials were not anomerically pure, the wrong anomer of the sugar nucleotide did not inhibit the transferases.[232] An efficient route to CMP-NeuNAc and CMP-[α-NeuNAc-(2→8) NeuNAc] has been described, using phosphoramidite methods.[233] Some other CMP-sialic acid derivatives are mentioned in Chapter 16. Syntheses of GDP-3-acetamido-and-azido-3-deoxy-D-mannose have been described, the sugar-nucleotide link being made enzymatically,[234] and TDP-6-deoxy-α-D-*ribo*-3-hex-ulose, a central intermediate in the biosynthesis of di-and tri-deoxy sugars, has also been prepared chemically.[235] The analogue of UDP-Gal with a carbocyclic galactose unit has been described.[236] A paper describing the linkage of mannose and dimannose units to oligodeoxyribonucleotides *via* a spacer area is mentioned in Chapter 3.

Analogues of tiazofurin adenine dinucleotide (TAD) containing 2'-deoxy-2'-fluoroadenosine, 3'-deoxy-3'-fluoroadenosine and 2'-deoxy-2'-fluoro-ara-A, have been synthesized and evaluated as inhibitors of IMP dehydrogenase.[237]

There have been further papers related to the second messenger cyclic adenosine 5'-diphosphate ribose (cADPR). Potter's laboratory has reported that the enzyme which cyclizes NAD to cADPR will also produce analogues of cADPR from NAD analogues with modifications in the adenosine unit (2'- or 3'-deoxycompounds, 8-substituents, and the formycin isostere),[238] whilst Sih and co-workers have shown the enzymic cyclization of NADP to cADPR phosphate; 3'-NADP and 2',3'-cyclic NADP also were converted to the cADPR deriva-tives.[239] The cyclase enzyme has been shown to convert the guanine and hypoxanthine analogues of NAD into products **164** (X=NH$_2$ or H) containing a 7-1″ link, as opposed to the 1-1″ link in cADPR.[240] The same sort of 7-1″ link is formed on enzymic cyclization of 1,N^6-etheno-NAD.[241]

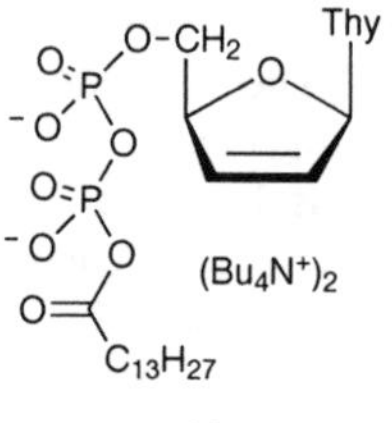

162 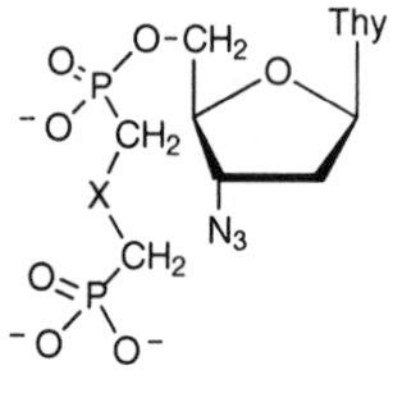   **163** 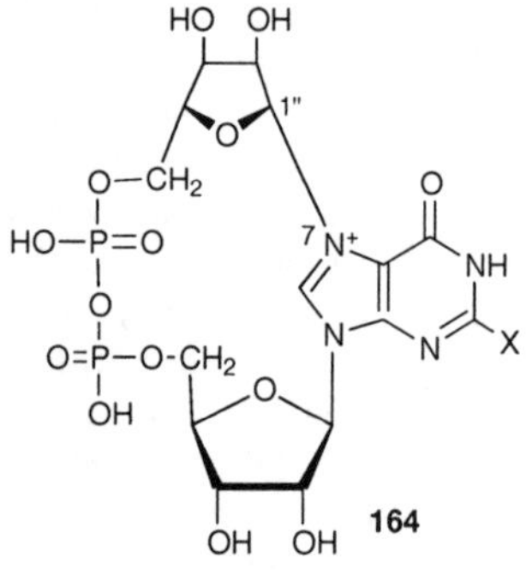   **164**

Some papers on the incorporation of 1,4-iminoalditols into nucleotide analogues are mentioned in Chapter 18, and some protecting-group chemistry of relevance in oligonucleotide synthesis is discussed in Section 14 below.

13 Oligonucleotide Analogues with Phosphorus-free Linkages

A recent Symposium Report entitled 'Carbohydrate Modifications in Antisense Research' gives an excellent overview of the types of analogue discussed in this Section, and also includes contributions concerned with modifications to the sugar unit of oligonucleotides.[242]

5'-Thioformacetal dinucleosides **165** have been prepared by a modified method, which is more suitable for the cases where purine bases are involved,[243] whilst a mild method for making the isomeric 3'-thioformacetals **168** (B=Thy, 5-Me-Cyt, Ade-Bz) involves the formation of phosphinates **166** and their reaction with the 3'-thiothymidine derivative **167** in the presence of DBU.[244] A quantum mechanical study of the substitution of a phosphodiester linkage by the thioform-acetal group has been reported.[245] There have been further examples of dinucleosides linked by the $^{3'}CH_2$-CH_2-S-$CH_2^{5'}$ unit, with additional substituents at C-2' of one of the units.[246]

The all-carbon alkene replacement **169** has been prepared by a Wittig reaction, and could be isomerized to the *trans*-alkene. Both analogues were incorporated into oligodeoxynucleotides and compared with the amide replacements (T_m values). The *trans*-alkene was expected to have the higher T_m, but it was not in fact much higher than for the *cis*-isomer.[247] The 5-atom replacements **170** and **172** have been made and incorporated into decamers for comparison with the previously-reported 4-atom types **171** and **173**; it was found that the 4-atom link gave more stable complexes for the amides, but the reverse was the case for the secondary amines.[248] There has been further work reported on the $^{3'}CH_2NHCH_2CH_2^{5'}$ replacement,[249] and thymidine dimers **174** with a chiral centre in the link have been described (both diastereomers).[250]

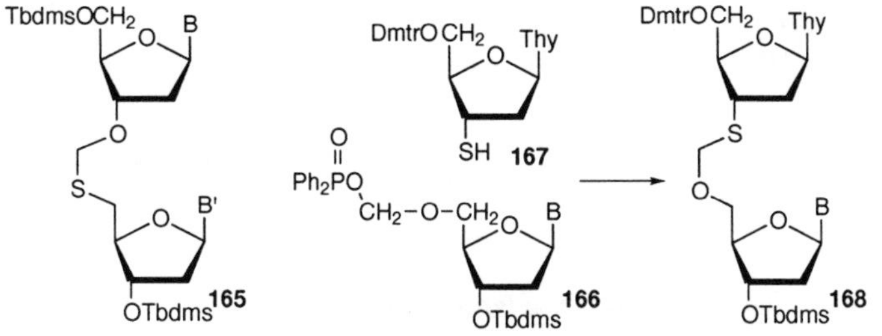

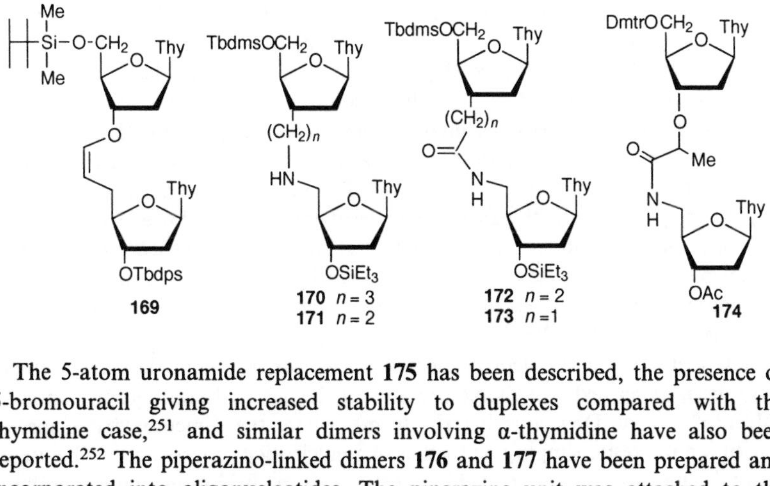

The 5-atom uronamide replacement **175** has been described, the presence of 5-bromouracil giving increased stability to duplexes compared with the thymidine case,[251] and similar dimers involving α-thymidine have also been reported.[252] The piperazino-linked dimers **176** and **177** have been prepared and incorporated into oligonucleotides. The piperazine unit was attached to the 'upper' nucleoside by displacement of a mesylate, and for the the synthesis of **177**, reductive amination was used, with the 5'-aldehyde of the 'bottom' unit.[253] The conformational dynamics of nucleoside dimers linked by *N*-methylhydroxylamine units (Vol.26, p.241-2) have been studied theoretically and by NOESY NMR methods.[254] Reaction of amine **178** with isothiocyanate **179** gave the thiourea-linked dimer **180**; reduction of the azide with H_2S, followed by further reaction with **178** gave an iterative process used for making oligomers with up to four thiourea links. The thiourea units could be converted into internucleosidic guanidine units by treatment with peracetic acid followed by ammonia.[255]

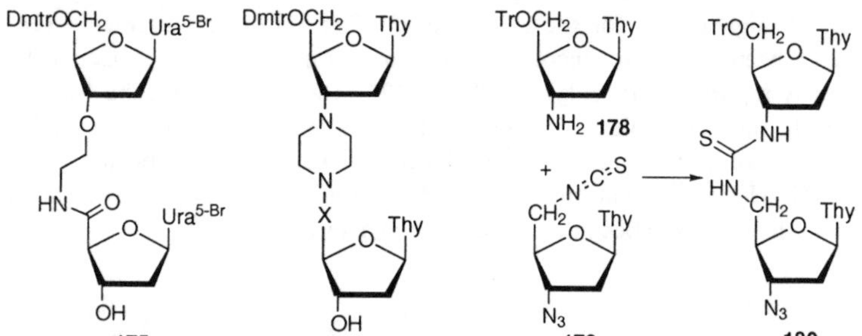

'Dimers' formed by linking a 2'-deoxynucleoside with an acyclonucleoside *via* a carbonate unit have been synthesized.[256]

14   Ethers, Esters and Acetals of Nucleosides

When 5'-*O*-Dmtr-cytidines, protected at N-4, were treated with alkyl iodides in the presence of silver(I) oxide, the 2'-*O*-alkylated compounds were the major product. N^4-*t*-Butylphenoxyacetyl protection gave the best selectivity, and the products were deaminated to 2'-*O*-alkyluridines.[257] An alternative approach to 2'-*O*-alkylated ribonucleosides involves the use of **181**, produced by selective debenzylation of the tris-(2,4-dichlorobenzyl) derivative (see Vol.21, p.49-50), in condensations with silylated bases.[258] 3',5'-Di-*O*-trityluridine has been alkylated at O-2' using 2-chloromethylanthracene; detritylation was followed by incorporation into oligonucleotides which exhibited enhanced fluorescence on binding to complementary RNA segments.[259] The synthesis of 3'-*O*-(2-aminoethyl)-2'-deoxyuridines has been described;[260] these materials were used for the synthesis of 'dimers' of type **175**.[251,252]

New methods for the deprotection of nucleoside 5'-*O*-dimethoxytrityl ethers have been described, involving the use of 1,1,1,3,3,3-hexafluoropropanol as solvent,[261] or of Dowex resin (H^+ form), where the dimethoxytrityl species remain on the resin, simplifying further oligonucleotide processing.[262] The use of triethylsilane, together with dichloroacetic acid in dichloromethane, has been advocated to scavenge dimethoxytrityl cations during the synthesis of deoxyribo-nucleotide phosphorothioates.[263]

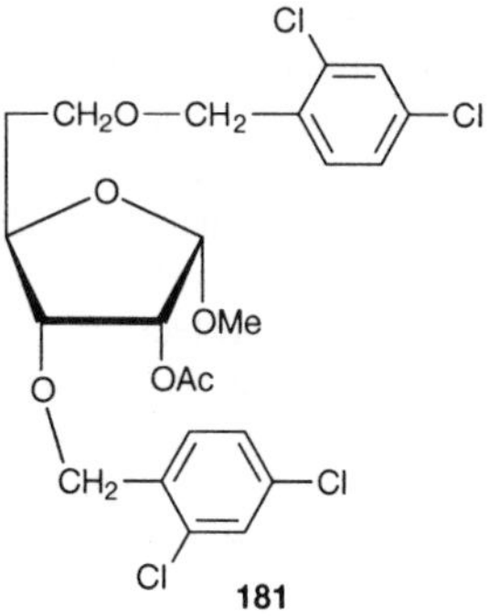

181

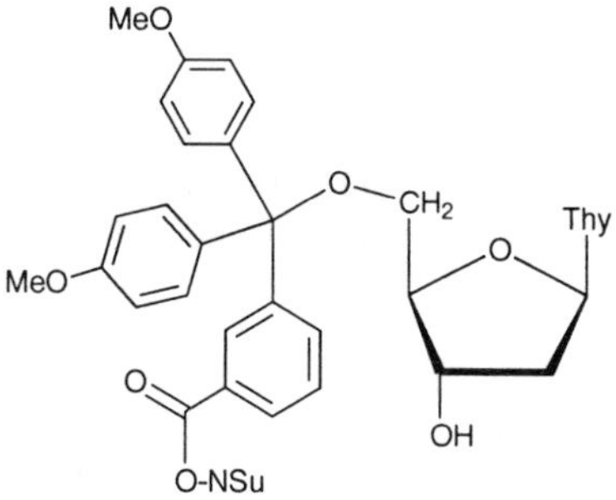

182

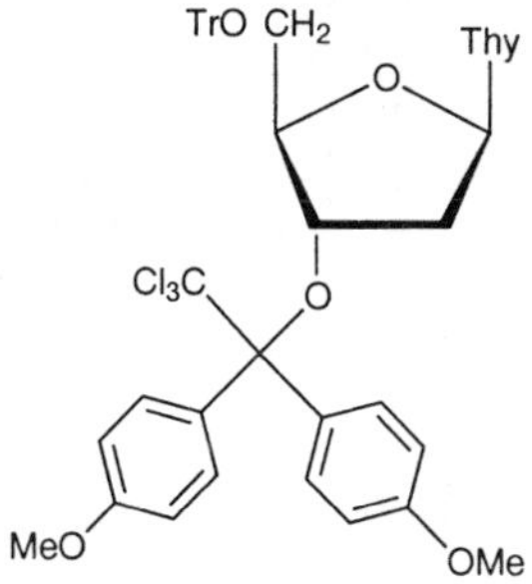

183

Deoxynucleosides have been protected at O-5' by a modified dimethoxytrityl group as in **182**; the *N*-hydroxysuccinimidyl ester permits attachment to resins with pendant amine groups, and the nucleoside can be recovered by mild acid treatment.[264] On the other hand, the 1,1-dianisyl-2,2,2-trichloroethyl ether (DATE) protecting group, as in **183**, is particularly stable towards both acids and bases, but easily removed by reductive fragmentation using the supernucleophilic lithium cobalt(I) phthalocyanine.[265]

The 2'- and 3'-*O*-(6-aminohexyl) ethers of uridine have been prepared by alkylation of the 2',3'-stannylene derivative, and the tether was used to link the uridine unit into various biologically significant conjugates[266] and to make lipophilic derivatives.[267]

When the bis-silyl ether **184** was treated with KOH in ethanol, the 5'-*O*-Tbdms group was hydrolysed; this finding seems to depend on the 'up' 2'-OH group, since 3',5'-di-*O*-Tbdms-thymidine was stable to the conditions.[268]

The ω-aminoacyl derivatives **185** of TMP and dCMP have been prepared and converted to their 5'-triphosphates prior to attachment of a fluorescent probe to the aminohexanoyl group.[269] 5-Fluorouridine has been converted into its 2',3'- and 5'-*O*-(4-carboxybutanoyl) esters by reaction with glutaric anhydride. These were linked to chitosan via an amide link, and the conjugates were slow-release devices for 5-fluorouridine at physiological pH.[270]

2'-Deoxyuridine and 6-azauridine have been converted into esters with nicotinic acid,[271] as have thymidine and trifluorothymidine, where quaternization and reduction gave 1,4-dihydronicotinoyl derivatives.[272]

Full details have been given of the enzyme-catalysed reaction of 3'-and 5'-*O*-vinyloxycarbonyl nucleosides and their reaction with amines to give 3'- and 5'-*O*-carbonates (see Vol.27, p.266),[273] and the same laboratory has described the enzymic alkoxycarbonylation of α-, *xylo*-, anhydro- and *arabino*-nucleosides.[274]

The dimethoxybenzoin carbonate protecting group, as in **186**, has been developed for use in DNA synthesis; the protection is removable photochemically.[275] Intermediates of type **187** have been used in the synthesis of oligoribonucleotides; the 2-(4-nitrophenyl)ethylsulfonyl (Npes) protecting group at O-2' is selectively removable, using DBU, in the presence of various protecting groups on the base.[276]

The sulfamoyl derivative **188** of AZT triphosphate has been synthesized in Vince's laboratory.[277]

Reaction of 3',5'-di-*t*-butylsilanediyl derivatives of nucleosides with 2-(trimethylsilyl) ethoxymethyl chloride (Sem chloride) gives the 2'-*O*-Sem derivatives **189**, from which the *O*-silyl protection can be removed by HF in pyridine.[278]

There have been further reports of the enzymic galactosylation of nucleosides at O-5' using β-galactosidases.[279,280]

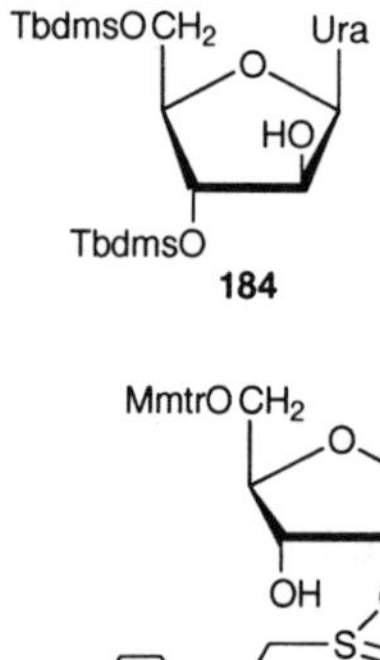

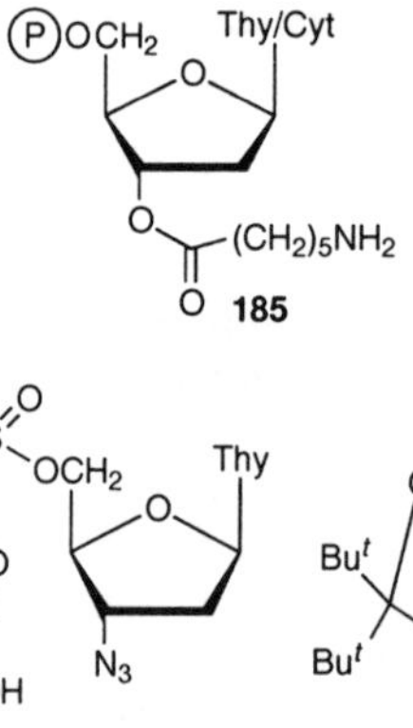

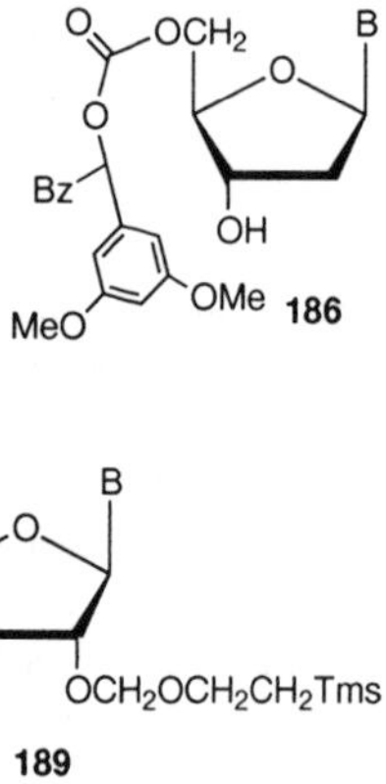

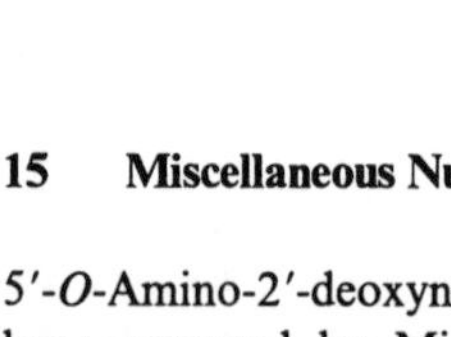

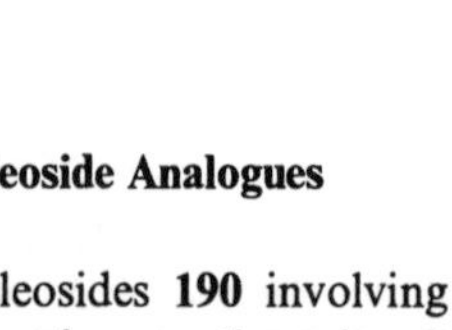

15   Miscellaneous Nucleoside Analogues

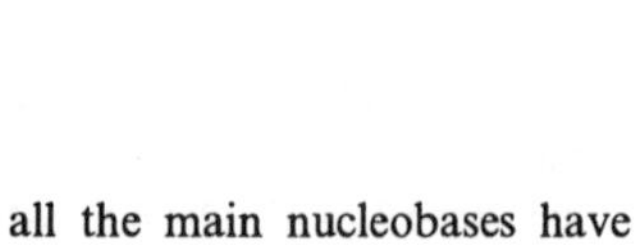

5'-*O*-Amino-2'-deoxynucleosides **190** involving all the main nucleobases have been prepared by Mitsunobu reactions involving *N*-hydroxyphthalimide; the products are of relevance in the synthesis of antisense oligonucleotides (see Vol.26, p.242).[281]

There have been further reports from Leumann's laboratory on the incorporation of bicyclodeoxynucleosides (see Vol.27, p.268-9)[282] and their α-anomers[283] into oligodeoxynucleotides, and the C-5' epimer **191** has been prepared by inversion of stereochemistry; incorporation of this epimer into oligomers led to significant reduction in affinity for complementary sequences.[284]

Further 1,5-anhydrohexitol nucleosides have been described (see Vol.27, p.269 for earlier),[285] and incorporated into oligonucleotides.[286] The phosphonate **192** was prepared from diacetyl-L-arabinal by Ferrier rearrangement followed by deacetylation; subsequent Mitsunobu reaction with 2-amino-6-chloropurine led to the unsaturated nucleotide analogue **193**; similar chemistry on diacetyl-D-xylal gave rise to the enantiomer of **193**.[287,288]

Reports from Nair's laboratory have described the synthesis, from D-glucose, of homologues **194** of 'isodideoxynucleosides',[289] and the incorporation of (S,S)-iso-ddA into a dinucleotide with deoxyadenosine.[290]

There have been more reports on dioxolanyl nucleoside analogues. 5-Substituted analogues of D-(+)-dioxolan C have been made from 1,6-anhydro-D-mannose as for the parent system (Vol.26, p.256),[291] whilst phosphonates of type **195** (B=Thy, Cyt, Gua) have been prepared, along with their *trans*-isomers.[292] Homologated compounds of type **196** have also been described, and the rates of hydrolysis of the dioxolane rings were studied.[293] Chu's group has prepared the dioxolane analogue **197** of tiazofurin,[294] and also analogues such as **198**, with the nucleobase at C-2 of the dioxolan ring; these compounds, were, not surprisingly, of limited stability, but the use of 5-fluorouracil as base helped matters.[295] The *C*-nucleoside system **199** and its enantiomer have also been prepared in the same laboratory.[296]

In the oxathiolane area, a paper from Wellcome Laboratories describes work done there on the preparation of 3-thiacytosine and related compounds,[297] whilst a synthesis of 3TC? involving an enantioselective enzymic hydrolysis has been reported.[298] The difluorophosphonate analogue **200** of FTC has been made as a racemate with other diastereomers,[299] and 1,2,6-thiadiazine dioxide analogues of 3TC? have been reported.[300] A full account has appeared concerning the synthesis of regioisomeric systems such as **201** (see Vol.28, p.297).[301]

The thiazolidinyl analogue **202** of pseudouridine, and its enantiomer, have been made, with chirality originating in D- or L-cysteine,[302] whilst thiazolidinones **203** (B=Cyt, Thy, 5-fluoro-Cyt) have been prepared, using Pummerer-type chemistry.[303] 4'-Azathymidine derivatives such as **204** have been prepared by cycloadditions of nitrones with vinyl acetate, followed by base-'sugar' condensation,[304] and the isoxazolidinyl nucleoside **205**, and its *trans*-isomer, have been made as racemates.[305] The N-in-ring analogue **206** of 3'-deoxythymidine has been synthesized by linkage of the base to 4-hydroxyproline,[306] and the analogues **207** (B = Ade, Gua) of oxetanocin have been made by building up the purine ring from a previously-known chiral *N*-aminoazetidine.[307]

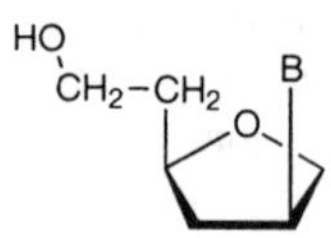

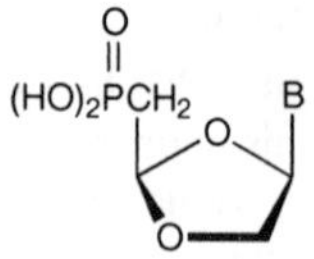

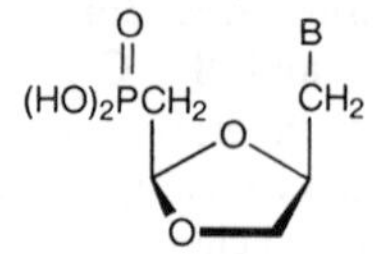

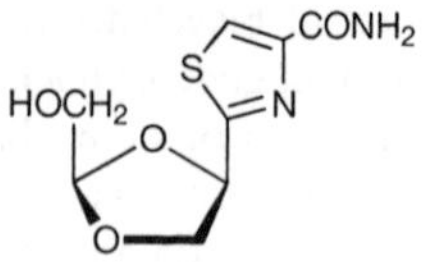

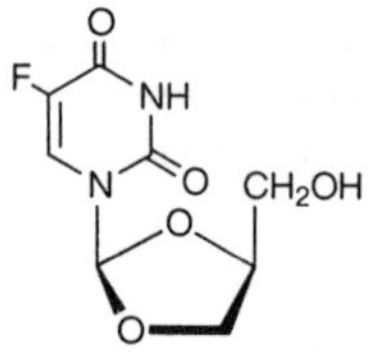

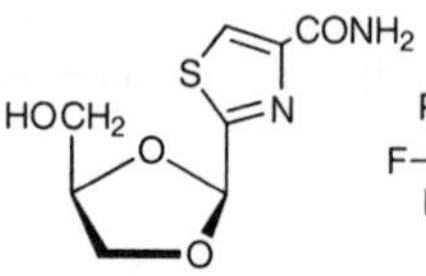

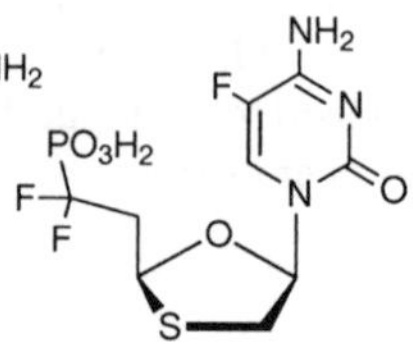

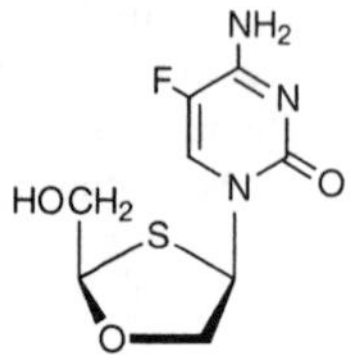

The cyclopropane **208** could be prepared in high diastereomeric excess by treatment of the *cis*-alkene (derived from L-gulonolactone) with Et$_2$Zn/CH$_2$I$_2$. Subsequent manipulation, including a Curtius reaction, then gave the L-nucleoside analogues **209**.[308] There has also been a full account of work first reported last year (Vol.28, p.298) in which similar chemistry is used to prepare structures such as **209**, but in the enantiomeric series.[309] The *meso*-cyclopropane **210** has also been described,[310] as has the racemic spirocompound **211**, which has antiviral and antileukaemic activity.[311]

The Wadsworth-Emmons product **212** has been converted to **213**, from which both 5'- and 3'-phosphoramidites were made, for incorporation into *alt*-DNA sequences.[312]

A Wadsworth-Emmons reaction on an adenosine 5'-aldehyde derivative was a key step in the synthesis of 'adenosylspermidine' (**214**), an inhibitor of spermidine synthase.[313]

2',3'-*O*-Isopropylideneuridine-5'-aldehyde could be used in condensations with pyrrole to make porphyrins containing either two or four nucleoside units attached at the meso-positions.[314]

16   Reactions

A range of purine 2',3'-seconucleosides have been prepared, using periodate cleavage followed by borohydride reduction, both reagents being resin-supported. The products were tested as antimalarials.[315]

Kinetic parameters for the oxidation of various nucleotides and 2'-deoxynucleotides by oxoruthenium (IV) species have been determined. The kinetics and product analyses were consistent with oxidation at C-1', and the increased reactivity of DNA compared with RNA could be explained on the basis of deactivation of the oxidation product by the polar effect of the 2'-hydroxyl group.[316] The Fe(III) complex of 2-aminomethyltetrahydrofuran-*N,N*-diacetic acid, in the presence of H_2O_2 catalyses hydroxylation of deoxyguanosine at the 2'- and 8-positions, thus indicating that *in vivo* conversion of deoxynucleotides to ribonucleotides might be possible.[317] A paper on the oxidative degradation of thymidine is mentioned in Chapter 10.

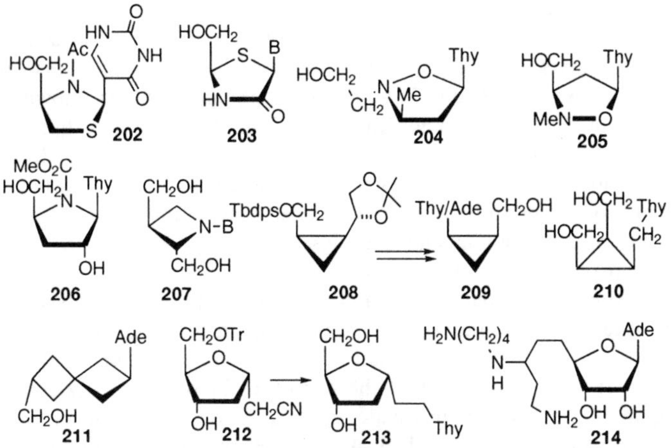

ESR spin trapping experiments were used to study the radicals produced by reaction of $^tBuO^\bullet$ and $PhC(Me)_2O^\bullet$ with nucleic acid components. It was found that H-abstraction from sugars was a minor reaction and may occur by radical transfer from the bases.[318]

The photolysis of 2'-deoxy-2'-iodouridine has been studied, to investigate the fate of C-2' radicals in nucleosides. The results support the idea that initial dissociation into radicals is followed by single electron transfer to give the ion-pair **215** (Scheme 11), which then undergoes 1,2-hydrogen shifts to give the products **216** and **217**, although other products are also formed. Deuterium-labelling studies supported this scheme.[319]

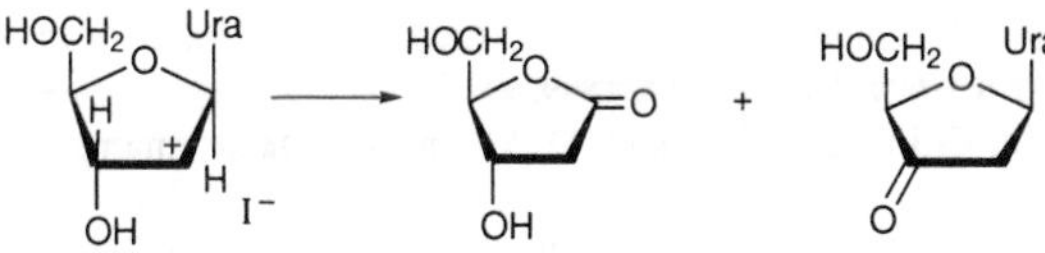

Scheme 11

In the hydrolysis of dinucleotides, bimetallic cooperation of Zn(II) with Sn(IV), In(III), Fe(III) or Al(III) leads to an increased rate of reaction, and a mechanism was proposed for the cooperactivity.[320] The same group have also shown that a dinuclear Zn(II) complex efficiently cleaves ApA at 50 °C and pH 7.[321] The rate of hydrolysis of CpU has been compared, under several conditions, with the hydrolysis of analogues with a methyl substituent at C-5', both diastereomers being investigated.[322] The mechanism of hydrolysis of UpU and related dialkyl phosphates in aqueous morpholine has been shown to differ in detail from the Breslow and Xu mechanism.[323] An investigation has been reported into the kinetics and mechanism of desulfurization, ester hydrolysis and transesterification (to the 2'-5'-esters) of the diastereomeric uridylyl (3'-5') uridine phosphoromonothioates.[324] A detailed kinetic analysis has been carried out on the interconversions and hydrolyses of 5'-*O*-pivaloyluridine 2'- and 3'-dimethylphosphates in the pH range 0-9.[325]

2'-Thiouridine, 2'-thioadenosine and 2'-thiocytidine undergo glycosidic cleavage in aqueous solution at and above pH 6.5, whereas the C-2' epimer of 2'-thioadenosine, and the corresponding disulfides, are stable under mildly basic conditions. It was postulated that the hydrolysis involved the 1,2-episulfide.[326] The degradation of monoamino analogues of 2'- and 3'-deoxyadenosine, and the *ara*-stereoisomers, has been studied. At acid pH, the amino-analogues are more stable than the hydroxy-compounds.[327]

References

1　　L.J. Wilson, M.W. Hager, Y.A. El-Kattan and D.C. Liotta, *Synthesis*, 1995, 1465.

2　　T. Ichiba, Y. Nakao, P.J. Scheuer, N.U. Sata and M. Kelly-Borges, *Tetrahedron Lett.*, 1995, **36**, 3977.

3　　P.A. Searle and T.F. Molinski, *J. Org. Chem.*, 1995, **60**, 4296.

4　　G.B. Chheda, H.B. Patrzyc, H.A. Tworek and S.P. Dutta, *Nucleosides Nucleotides*, 1995, **14**, 1519.

5　　B. Bennua-Skalmowski, K. Krolikiewicz and H. Vorbrüggen, *Tetrahedron Lett.*, 1995, **36**, 7845.

6　　T. Gimisis, G. Ialongo, M. Zamboni and C. Chatgilialoglu, *Tetrahedron Lett.*, 1995, **36**, 6781.

7　　Y Itoh, K. Haraguchi, H. Tanaka, K. Matsumoto, K.T. Nakamura and T. Miyasaka, *Tetrahedron Lett.*, 1995, **36**, 3867.

8　　S. Hanessian, J.L. Condé and B. Lou, *Tetrahedron Lett.*, 1995, **36**, 5865.

9　　G. Liu, F.W. Bruenger, A.M. Barrios and S.C. Miller, *Nucleosides Nucleotides*, 1995, **14**, 1901.

10　　L.M. Ohja, D. Gulati, N. Seth, D.S. Bhakuni, R. Pratap and K.C. Agarwal, *Nucleosides Nucleotides*, 1995, **14**, 1889.

11　　T.A. Devlin, E. Lacrosaz-Rouanet, D. Vo and D.J. Jebaratnam, *Tetrahedron Lett.*, 1995, **36**, 1601.

12　　T.A. Devlin and D.J. Jebaratnam, *Synth. Commun.*, 1995, **25**, 711.

13　　L.B. Townsend, R.V. Devivar, S.R. Turk, M.R. Nassiri and J.C. Drach, *J. Med. Chem.*, 1995, **38**, 4098.

14　　J. Boryski, *Nucleosides Nucleotides*, 1995, **14**, 77.

15　　F. Jourdan, J. Renault, A. Karamat, D. Ladourée and M. Robba, *J. Heterocycl. Chem.*, 1995, **32**, 953.

16　　K. Egar, M. Jalalian and M. Schmidt, *J. Heterocycl. Chem.*, 1995, **32**, 211.

17　　J. Renault, F. Jourdan, D. Ladourée and M. Robba, *Heterocycles*, 1995, **41**, 937.

18   A. Gupta and L. Parkash, *Boll. Chim. Farm.*, 1994, **133**, 163 (*Chem. Abstr.*, 1995, **122**, 56383).

19   A.M.E. Attia and G.E.H. Elgemie, *Nucleosides Nucleotides*, 1995, **14**, 1211.

20   E.S. Ibrahim, G.E.H. Elgemie, M.M. Abbasi, Y.A. Abbas, M.A. Elbadawi and A.M.E. Attia, *Nucleosides Nucleotides*, 1995, **14**, 1415.

21   A.M.E. Attia and G.E.H. Elgemie, *Carbohydr. Res.*, 1995, **268**, 295.

22   A.M.E. Attia, *Sulfur Lett.*, 1994, **17**, 143 (*Chem. Abstr.*, 1995, **122**, 56 379).

23   A.M.E. Attia, E.I. Ibrahim, F.E.A. Hay, M.M.A. Abbasi and H.A.E. Mansour, *Nucleosides Nucleotides*, 1995, **14**, 1581.

24   L.D.S. Yadav and D.S. Yadav, *Liebigs Ann. Chem.*, 1995, 2231.

25   N.A. Al-Masoudi and A.A. Al-Atoom, *Nucleosides Nucleotides*, 1995, **14**, 1341.

26   Y. Itoh, K. Haraguchi, H. Tanaka, E. Gen and T. Miyasaka, *J. Org. Chem.*, 1995, **60**, 656.

27   P.V.P. Pragnacharyulu, C. Vargeese, M. McGregor and E. Abushanab, *J. Org. Chem.*, 1995, **60**, 3096.

28   G.A. Tolstikov, A.G. Mustafin, R.R. Gatullin, L.V. Spirikhin, V.S. Sultanova and I.B. Abdrakhmanov, *Izv. Akad. Nauk., Ser. Khim.*, 1993, 1137 (*Chem. Abstr.*, 1995, **122**, 214423).

29   E. Larsen, T. Kofoed and E.B. Pedersen, *Synthesis*, 1995, 1121.

30   Z. Kazimierczuk, J.A. Vilpo and F. Seela, *Nucleosides Nucleotides*, 1995, **14**, 1403.

31   Y.-Q. Chen and J.-L. Lin, *Chin. Sci Bull.*, 1994, **39**, 730 (*Chem. Abstr.*, 1995, **122**, 31828).

32   M.J. Robins, J.S. Wilson, D. Madej, N.H. Low, F. Hansske and S.F. Wnuk, *J. Org. Chem.*, 1995, **60**, 7902.

33   S.A. Surzhykov, *Bioorg. Khim.*, 1994, **20**, 1114 (*Chem. Abstr.*, 1995, **122**, 188010).

34   A.K. Prasad, M.D. Sørensen, V.S. Parmar and J. Wengel, *Tetrahedron Lett.*, 1995, **36**, 6163.

35   I. Votruba, A Holy, H. Dvorakova, J. Gunter, D. Hockova, H. Hrebabecky, T. Cihlar and M. Masojidkova, *Collect. Czech. Chem. Commun.*, 1994, **59**, 2303 (*Chem. Abstr.*, 1995, **123**, 170044).

36   S. Pochet, L. Dugué, A. Meier and P Marlière, *Bioorg. Med. Chem. Lett.*, 1995, **5**, 1679.

37   D.E. Bergstrom, P. Zhang, P.H. Toma, P.C. Andrews and R. Nichols, *J. Am. Chem. Soc.*, 1995, **117**, 1201.

38   T.S. Rao, R.F. Rando, J.H. Huffmann and G.R. Revankar, *Nucleosides Nucleotides*, 1995, **14**, 1997.

39   M. Medveczky, T.F. Yang, J. Gambino, P. Medveczky and G.E. Wright, *J. Med. Chem.*, 1995, **38**, 1811.

40   B.K. Bhattacharya, M.V. Chari, R.H. Durland and G.R. Revankar, *Nucleosides Nucleotides*, 1995, **14**, 45.

41   F. Seela and H. Winter, *Nucleosides Nucleotides*, 1995, **14**, 129.

42   V.L. Worthington, W. Fraser and C.H. Schwalbe, *Carbohydr. Res.*, 1995, **275**, 275.

43   J. Hunziker, E.S. Priestley, H. Brunar and P.B. Dervan, *J. Am. Chem. Soc.*, 1995, **117**, 2661.

44   S. Schmidt and D. Cech, *Nucleosides Nucleotides*, 1995, **14**, 1445.

45   M. Acedo, E. De Clercq and R. Eritja, *J. Org. Chem.*, 1995, **60**, 6262.

46   E. Kawashima, Y. Aoyama, T. Sekine, M. Miyahara, M.F. Radwan, E. Nakamura, M. Kainosho, Y. Kyogoku, and Y. Ishido, *J. Org. Chem.*, 1995, **60**, 6980.

47   E. Kawashima, K. Toyama, K. Ohshima, M. Kainosho, Y. Kyogoku and Y. Ishido, *Tetrahedron Lett.*, 1995, **36**, 6699.

48   E. Larsen, A.A.H. Abdel Aleem and E.B. Pedersen, *J. Heterocycl. Chem.*, 1995, **32**, 1645.

49   T.-F. Tang, L.P. Kotra, Q. Teng, F.N.M. Naguib, J.-P. Sommadossi, M. el Kouni and C.K. Chu, *Tetrahedron Lett.*, 1995, **36**, 983.

50   A.K. Saha, T.J. Caulfield, C. Hobbs, D.A. Upson, C. Waychunas and A.M. Yawman, *J. Org. Chem.*, 1995, **60**, 788.

51   K. Nacro, J.-M. Escudier, M. Baltas and L. Gorrichon, *Tetrahedron Lett.*, 1995, **36**, 7867.

52   M. Saneyoshi, M. Kohsaka-Ichikawa, A. Yahata, S. Kimura, S. Izuta and T. Yamaguchi, *Chem. Pharm. Bull.*, 1995, **43**, 2005.

53   G.-X. He and N. Bischofsberger, *Tetrahedron Lett.*, 1995, **36**, 6991.

54   B.V. Joshi, C.B. Reese and C.V.N.S. Varaprasad, *Nucleosides Nucleotides*, 1995, **14**, 209.

55   G. Cristalli, S. Vittori, A. Eleuteri, R. Volpini, E. Camaioni, G. Lupidi, N. Mahmood, F. Bevilacqua and G. Palu, *J. Med. Chem.*, 1995, **38**, 4019.

56   T.-S. Lin, M.-Z. Luo, J.-L. Liu, M.-C. Liu, Y.-L. Zhu, G.E. Dutchman and Y.-C. Cheng, *Nucleosides Nucleotides*, 1995, **14**, 1759.

57   T.-S. Lin, M.-Z. Luo and M.-C. Liu, *Tetrahedron*, 1995, **51**, 1055.

58   A. Tse and T.S. Mansour, *Tetrahedron Lett.*, 1995, **36**, 7607.

59   S. Becouarn, S. Czernecki and J.-M. Valéry, *Nucleosides Nucleotides*, 1995, **14**, 1227.

60   B.-C. Chen, S.L. Quinlan, D.R. Stark, J.G. Reid and R.H. Spector, *Tetrahedron Lett.*, 1995, **36**, 7957.

61   B.H. Lipshutz, K.L. Stevens and R.F. Lowe, *Tetrahedron Lett.*, 1995, **36**, 2711.

62   G. Negrón, B. Quiclet-Sire, Y. Diaz, R. Gaviño and R. Cruz, *Nucleosides Nucleotides*, 1995, **14**, 1539.

63   S. Niihata, H. Kuno, T. Ebata and H. Matsushita, *Bull. Chem. Soc. Jpn.*, 1995, **68**, 2327.

64   T. Ebata, H. Kawakami, K. Matsumoto and H. Matsushita, *Front. Biomed. Biotechnol.*, 1994, **2**, 73 (*Chem. Abstr.*, 1995, **122**, 240307).

65   C. Fossey, D. Ladurée and M. Robba, *J. Heterocycl. Chem.*, 1995, **32**, 627.

66   H. Hrebabecky, J. Dockal and A. Holy, *Collect. Czech. Chem. Commun.*, 1994, **59**, 1408 (*Chem. Abstr.*, 1995, **122**, 214400).

67   G.V. Zaitseva, G.G. Sivets, Z. Kazimierczuk, J.A. Vilpo and I.A. Mikhailopulo, *Bioorg. Med. Chem. Lett.*, 1995, **5**, 2999.

68   S. Niihata, T. Ebata, H. Kawakami and H. Matsushita, *Bull. Chem. Soc. Jpn.*, 1995, **68**, 1509.

69   O.D. Schärer and G.L. Verdine, *J. Am. Chem. Soc.*, 1995, **117**, 10781.

70   M.F. Evangelisto, R.E. Adams, W.V. Murray and G.W. Caldwell, *J. Chromatogr. A*, 1995, **695**, 128.

71   B.K. Bhattacharya, J.O. Ojwang, R.F. Rando, J.H. Huffman and G.R. Revankar, *J. Med. Chem.*, 1995, **38**, 3957.

72   H. Ford, Jr., M.A. Siddiqui, J.S. Driscoll, V.E. Marquez, J.A. Kelley, H. Mitsuya and T. Shirasaka, *J. Med. Chem.*, 1995, **38**, 1189.

73   Y. Xiang, S. Cavalcanti, C.K. Chu, R.F. Schinazi, S.B. Pai, Y.-L. Zhu and Y.-C. Cheng, *Bioorg. Med. Chem. Lett.*, 1995, **5**, 877.

74   Y. Xiang, L.P. Kotra, C.K. Chu and R.F. Schinazi, *Bioorg. Med. Chem. Lett.*, 1995, **5**, 743.

75   L.S. Jeong and V.E. Marquez, *Chem. Lett.*, 1995, 301.

76   A.A. H. Abdel-Rahman, H.M. Abdel-Bary, E.B. Pedersen and C Nielsen, *Arch. Pharm. (Weinheim, Ger.)*, 1995, **328**, 67 (*Chem. Abstr.*, 1995, **122**, 240319).

77　H.M. Abdel-Bary, A.A. El-Barbary, A.I. Khodair, A.E.S. Abdel-Majied, E.B. Pedersen and C. Nielsen, *Bull. Soc. Chim. Fr.,* 1995, **132**, 149.

78　E.N. Kalininchenko, T.L. Podkopaeva, N.E. Poopieko, M. Kelve, M. Saarma, I.A. Mikhailopulo, J.E. van den Boogaart and C. Altona, *Recl. Trav. Chim. Pays-Bas,* 1995, **114**, 43.

79　L.S. Jeong and V.E. Marquez, *J. Org. Chem.,* 1995, **60**, 4276.

80　P.L.Coe, R.R. Talekar and R.T. Walker, *J. Fluorine Chem.,* 1994, **69**, 19 (*Chem. Abstr.,* 1995, **122**, 81876).

81　D. Guillerm, M. Muzard, B. Allart and G. Guillerm, *Bioorg. Med. Chem. Lett.,* 1995, **5**, 1455.

82　Q.S. Lal, *Synth. Commun.,* 1995, **25**, 725.

83　J. Matulic-Adamic, L. Beigelman, L.W. Dudycz, C. Gonzalez and N. Usman, *Bioorg. Med. Chem. Lett.,* 1995, **5**, 2721.

84　H. Kuno, S. Niihata, T. Ebata and H. Matsushita, *Heterocycles,* 1995, **41**, 523.

85　S. Bera, T. Pathak and G.J. Langley, *Tetrahedron,* 1995, **51**, 1459.

86　M.V. Jasko, I.I. Fedorov, A.M. Atrazhev, D.Y. Mozzherin, N.A. Novicov, A.V. Bochkarev, G.V. Gurskaya and A.A. Krayevsky, *Nucleosides Nucleotides,* 1995, **14**, 23.

87　C. Pannecouque, R. Busson, J. Balzarini, P. Claes, E. De Clercq and P. Herdewijn, *Tetrahedron,* 1995, **51**, 5369.

88　B.-C. Chen, S.L. Quinlan and J.G. Reid, *Tetrahedron Lett.,* 1995, **36**, 7961.

89　M.A. Zahran, A.E.-S. Abdel-Megied, A.A.-H. Abdel-Rahman, A. El-Emam, E.B. Pedersen and C. Nielsen, *Heterocycles,* 1995, **41**, 2507.

90　A.A. El-Barbary, A.I. Khodair and E.B. Pedersen, *Arch. Pharm. (Weinheim, Ger.),* 1994, **327**, 653 (*Chem. Abstr.,* 1995, **122**, 106349).

91　T. Kawaguchi, H. Sakairi, S. Kimura, T. Yamaguchi and M. Saneyoshi, *Chem. Pharm. Bull.,* 1995, **43**, 501.

92　T. Kofoed, A.A. H. Abdel-Aleem, P.T. Jørgensen, T.R. Pederson and E.B. Pedersen, *Acta Chem. Scand.,* 1995, **49**, 291.

93　V.A. Ostrovskii, E.P. Studentsov, V.S. Poplavskii, N.V. Ivanova, G.V. Gurskaya, V.E. Zavodnik, M.V. Jasko, D.G. Semizarov and A.A. Krayevsky, *Nucleosides Nucleotides,* 1995, **14**, 1289.

94　A.A. Bakhmedova, I.V. Yartseva and S.Y. Mel'nik, *Bioorg Khim.,* 1995, **21**, 45 (*Chem. Abstr.,* 1995, **123**, 340742).

95　J.J. Huang, A. Ragouzeos and J.L. Rideout, *J. Heterocycl. Chem.,* 1995, **32**, 691.

96　J.M.J. Tronchet, M. Zsély, O. Lassout, F. Barbalat-Rey, I. Komaromi and M. Geoffroy, *J. Carbohydr. Chem.,* 1995, **14**, 575.

97　S.P. Auguste and D.W. Young, *J. Chem. Soc., Perkin Trans. 1,* 1995, 395.

98　G. Ceulemans, F. Vandendriessche, J. Rosenski and P. Herdewijn, *Nucleosides Nucleotides,* 1995, **14**, 117.

99　N.A. Al-Massoudi and W. Pfleiderer, *Carbohydr. Res.,* 1995, **275**, 95.

100　A.A. El-Barbary, N.R. El-Brollosy and E.B. Pedersen, *J. Heterocycl. Chem.,* 1995, **32**, 719.

101　M. Criton, G. Dewynter, N. Aouf, J.-L. Montero and J.-L. Imbach, *Nucleosides Nucleotides,* 1995, **14**, 1795.

102　G.M. Blackburn, D.P. Hornby, A, Meekhalfia and P. Shore, *Nucleic Acids Symp. Ser.,* 1994, **31**, 19 (*Chem. Abstr.,* 1995, **122**, 133648).

103　N.A. Al-Massoudi, F.B. Issa, W. Pfleiderer and H.B. Lazrek, *Nucleosides Nucleotides,* 1995, **14**, 1693.

104 M.P. Hay, H.H. Lee, W.R. Wilson, P.B. Roberts and W.A. Denny, *J. Med. Chem.*, 1995, **38**, 1928.

105 A.A. El-Barbary, A.I. Khodair, E.B. Pedersen and C Nielsen, *Monatsh. Chem.*, 1994, **125**, 1017 (*Chem. Abstr.*, 1995, **122**, 31820).

106 S.Bera, K. Sakthivel, T. Pathak and G.J. Langley, *Tetrahedron*, 1995, **51**, 7857.

107 G. Rassu, P. Spanu, L. Pinna, F. Zanardi and G. Casiraghi, *Tetrahedron Lett.*, 1995, **36**, 1941.

108 R.J. Young, S. Shaw-Ponter, J.B. Thomson, J.A. Miller, J.G. Cumming, A.W. Pugh and P. Rider, *Bioorg. Med. Chem. Lett.*, 1995, **5**, 2599.

109 J. Mann, A.J. Tench, A.C. Weymouth-Wilson, S. Shaw-Ponter and R.J. Young, *J. Chem. Soc., Perkin Trans. 1*, 1995, 677.

110 K.S. Jandu and D.L. Selwood, *J. Org. Chem.*, 1995, **60**, 5170.

111 J. Guo, Y.Q. Wu, D. Rattendi, C.J. Bachi and P.M. Woster, *J. Med. Chem.*, 1995, **38**, 1770.

112 B. Smalec and M. von Itzstein, *Carbohydr. Res.*, 1995, **266**, 269.

113 R. Pontikis, J. Wolf, C. Monneret and J.-C. Florent, *Tetrahedron Lett.*, 1995, **36**, 3523.

114 A.J. Lawrence, J.B.J. Pavey, I.A. O'Neil and R. Cosstick, *Tetrahedron Lett.*, 1995, **36**, 6341.

115 M.S. Wolfe and R.E. Harry-O'kuru, *Tetrahedron Lett.*, 1995, **36**, 7611.

116 A. Azuma, K. Hanaoka, A. Kurihara, T. Kobayashi, S. Miyauchi, N. Kamo, M. Tanaka, T. Sasaki and A. Matsuda, *J. Med. Chem.*, 1995, **38**, 3391.

117 C.R. Johnson and D.R. Bhumralkar, *Nucleosides Nucleotides*, 1995, **14**, 185.

118 P.M.J. Jung, A. Burger and J.-F. Biellmann, *Tetrahedron Lett.*, 1995, **36**, 1031.

119 M.A. Amin, H. Stoeckli-Evans and A. Gossauer, *Helv. Chim. Acta*, 1995, **78**, 1879.

120 Y. Takahashi, Y. Honda and T. Tsuchiya, *Carbohydr. Res.*, 1995, **270**, 77.

121 S. Becouarn, S. Czernecki and J.-M. Valéry, *Tetrahedron Lett.*, 1995, **36**, 873.

122 P.N. Jorgensen, M.L. Svendsen, C. Scheuer-Larsen and J. Wengel, *Tetrahedron*, 1995, **51**, 2155.

123 J. Wengel, M.L. Svendsen, P.N. Jorgensen and C. Nielsen, *Nucleosides Nucleotides*, 1995, **14**, 1465.

124 F. Hammerschmidt, J.-P. Polsterer and E. Zbiral, *Synthesis*, 1995, 415.

125 B. Doboszewski, N. Blaton, J. Rozenski, A. De Bruyn and P. Herdewijn, *Tetrahedron*, 1995, **51**, 5381.

126 B. Doboszewski, H. DeWinter, A. Van Aerschot and P. Herdewijn, *Tetrahedron*, 1995, **51**, 12319.

127 J. Fensholdt, H. Thrane and J. Wengel, *Tetrahedron Lett.*, 1995, **36**, 2535.

128 H. Thrane, J. Fensholdt, M. Regner and J. Wengel, *Tetrahedron*, 1995, **51**, 10389.

129 H. Hrebabecky and A. Holy, *Collect. Czech. Chem. Commun.*, 1994, **59**, 1654 (*Chem. Abstr.*, 1995, **122**, 106346).

130 S.M. Siddiqi, K.A. Jacobson, J.L. Esker, M.E.Olah, X. Ji, N. Melman, K.N. Tiwari, J.A. Secrist III, S.W. Schneller, G. Cristalli, G.L. Stiles, C.R. Johnson and A.P. IJzerman, *J. Med. Chem.*, 1995, **38**, 1174.

131 A. Papchikhin, P. Agback, J. Plavec and J. Chattopadhyaya, *Tetrahedron*, 1995, **51**, 329.

132 M. Björsne, B. Classon, I. Kers and B. Samuelsson, *Bioorg. Med. Chem. Lett.*, 1995, **5**, 43.

133 A. Lundquist, I. Kvarnström, S.C.T. Svensson, B. Classon and B. Samuelsson, *Nucleosides Nucleotides*, 1995, **14**, 1493.

134 B. Doboszewski, N. Blaton and P. Herdewijn, *Tetrahedron Lett.*, 1995, **36**, 1321.

135   B. Doboszewski, N. Blaton and P. Herdewijn, *J. Org. Chem.*, 1995, **60**, 7909.
136   A. Matsuda, H. Kosaki, Y. Yoshimura, S. Shuto, N. Ashida, K. Konno and S. Shigeta, *Bioorg. Med. Chem. Lett.*, 1995, **5**, 1685.
137   A. Grouiller, V. Uteza, I. Komaromi and J.M.J. Tronchet, *J. Carbohydr. Chem.*, 1995, **14**, 1387.
138   S.M. Neighbours, W.H. Soine and S.G. Paibir, *Carbohydr. Res.*, 1995, **269**, 259.
139   E. Larsen, K. Danel and E.B. Pedersen, *Nucleosides Nucleotides*, 1995, **14**, 1905.
140   P. Franchetti, L. Cappellacci, M. Grifantini, A. Barzi, G. Nocentini, H. Yang, A. O'Connor, H.N. Jayaram, C. Carrell and B. M. Goldstein, *J. Med. Chem.*, 1995, **38**, 3829.
141   M. Richter and G. Seitz, *Arch. Pharm. (Weinheim, Ger.)*, 1994, **327**, 365 (*Chem. Abstr.*, 1995, **122**, 81854).
142   M. Richter and G. Seitz, *Arch. Pharm. (Weinheim, Ger.)*, 1995, **328**, 175 (*Chem. Abstr.*, 1995, **123**, 228741).
143   I. Maeba, Y. Nishiyama, S. Kanazawa and A. Sato, *Heterocycles*, 1995, **41**, 507.
144   A.D. Rycroft, G. Singh and R.H. Wightman, *J. Chem. Soc., Perkin Trans. 1*, 1995, 2667.
145   S. Harusawa, Y. Murai, H. Moriyama, H. Ohishi, R. Yoneda and T. Kurihara, *Tetrahedron Lett.*, 1995, **36**, 3165.
146   M. Yokoyama, T. Akiba and H. Togo, *Synthesis*, 1995, 638.
147   B.K. Bhattacharya, R.V. Devivar and G.R. Revankar, *Nucleosides Nucleotides*, 1995, **14**, 1269.
148   H. Kuhn, D.P. Smith and S.S. David, *J. Org. Chem.*, 1995, **60**, 7094.
149   T.S. Rao, A.F. Lewis, T.S. Hill and G.R. Revankar, *Nucleosides Nucleotides*, 1995, **14**, 1.
150   H.-P. Hsieh and L.W. McLaughlin, *J. Org. Chem.*, 1995, **60**, 5356.
151   J.J. Chen, J.A. Walker II, W. Liu, D.S. Wise and L.B. Townsend, *Tetrahedron Lett.*, 1995, **36**, 8363.
152   H.-C. Zhang, M. Brakta and G.D. Daves, Jr., *Nucleosides Nucleotides*, 1995, **14**, 105.
153   L. Schmitt and C.A. Caperelli, *Nucleosides Nucleotides*, 1995, **14**, 1929.
154   K.-H. Altmann, M.-O. Bévierre, A. De Mesmaeker and H.E. Moser, *Bioorg. Med. Chem. Lett.*, 1995, **5**, 431.
155   D.G. Wyatt, A.S. Anslow, B.A. Coomber, R.P.C. Cousins, D.N. Evans, V.S. Gilbert, D.C. Humber, I.L. Paternoster, S.L. Sollis, D.J. Tapolczay and G.G. Weingarten, *Nucleosides Nucleotides*, 1995, **14**, 2039.
156   A.D. Borthwick, A.J. Crame, A.M. Exall, G.G. Weingarten and M. Mahmoodian, *Tetrahedron Lett.*, 1995, **36**, 6929.
157   A.Toyota, M. Aizawa, C. Habutami, N. Katagiri and C. Kaneko, *Tetrahedron*, 1995, **51**, 8783.
158   C.F. Palmer and R. McCague, *J. Chem. Soc., Perkin Trans. 1*, 1995, 1201.
159   T. Berranger and Y. Langlois, *Tetrahedron Lett.*, 1995, **36**, 5523.
160   A. Grumann, H. Marley and R.J.K. Taylor, *Tetrahedron Lett.*, 1995, **36**, 7767.
161   A. Popescu, A.-B. Hörnfeldt, S. Gronowitz and N.G. Johansson, *Nucleosides Nucleotides*, 1995, **14**, 1233.
162   A. Popescu, A.-B. Hörnfeldt, S. Gronowitz and N.G. Johansson, *Nucleosides Nucleotides*, 1995, **14**, 1639.
163   N.B. Dyatkina, F. Thiel and M. von Janta-Lipinski, *Tetrahedron*, 1995, **51**, 761.
164   A. Ghosh, A.R. Ritter and M.J. Miller, *J. Org. Chem.*, 1995, **60**, 5808.
165   A. Ghosh and M.J. Miller, *Tetrahedron Lett.*, 1995, **36**, 6399.

166   J. Wachtmeister, B. Classon, B. Samuelsson and I. Kvarnström, *Tetrahedron*, 1995, **51**, 2029.
167   A. Ezzitouni, J.J. Barchi, Jr. and V.E. Marquez, *J. Chem. Soc., Chem. Commun.*, 1995, 1345.
168   L.S. Jeong, V.E. Marquez, C.-S. Yuan and R.T. Borchardt, *Heterocycles*, 1995, **41**, 2651.
169   Q. Chao and V. Nair, *Tetrahedron Lett.*, 1995, **36**, 7375.
170   M.J. Konkel and R. Vince, *Nucleosides Nucleotides*, 1995, **14**, 2061.
171   M.-J. Pérez-Pérez, J. Rozenski, R. Busson and P. Herdewijn, *J. Org. Chem.*, 1995, **60**, 1531.
172   F. Calvani, M. Macchia, A. Rossello, M.R. Gismondo, L. Drago, M.C. Fassina, M. Cisternino and P. Domiano, *Bioorg. Med. Chem. Lett.*, 1995, **5**, 2567.
173   M. Saady, L. Lebeau and C. Mioskowski, *Tetrahedron Lett.*, 1995, **36**, 2239.
174   T. Ikemoto, A. Haze, H. Hatano, Y. Kitamoto, M. Ishida and K. Nara, *Chem. Pharm. Bull.*, 1995, **43**, 210.
175   C.B. Reese, L.H.K. Shek and Z. Zhao, *J. Chem. Soc., Perkin Trans. 1*, 1995, 3077.
176   T.S. Rao, A.E. Sopchik and W.G. Bentrude, *Nucleosides Nucleotides*, 1995, **14**, 1163.
177   X. Chen and Y.-F. Zhao, *Synth. Commun.*, 1995, **25**, 3691.
178   K. Seio, T. Wada, K. Sakamoto, S. Yokoyama and M. Sekine, *Tetrahedron Lett.*, 1995, **36**, 9515.
179   J.-L. Girardet, C. Périgaud, A.-M. Aubertin, G. Gosselin, A. Kirn and J.-L. Imbach, *Bioorg. Med. Chem. Lett.*, 1995, **5**, 2981.
180   I. Lefebvre, C. Périgaud, A. Pompon, A.-M. Aubertin, J.-L. Girardet, A. Kirn, G. Gosselin and J.-L. Imbach, *J. Med. Chem.*, 1995, **38**, 3941.
181   C.R. Wagner, E.J. McIntee, R.F. Schinazi and T.W. Abraham, *Bioorg. Med. Chem.. Lett.*, 1995, **5**, 1819.
182   G.F. Koser, Y. Huang, K. Chen and K.C. Calvo, *J. Chem. Soc., Perkin Trans. 1*, 1995, 299.
183   A. Routledge, I. Walker, S. Freeman, A. Hay and N. Mahmood, *Nucleosides Nucleotides*, 1995, **14**, 1545.
184   C. Desseaux, C. Gouyette, Y. Henin and T. Huynh-Dinh, *Tetrahedron*, 1995, **51**, 6739.
185   D. Farquhar, R. Chen and S. Khan, *J. Med. Chem.*, 1995, **38**, 488.
186   K.M. Fries, C. Joswig and R.F. Borch, *J. Med. Chem.*, 1995, **38**, 2672.
187   J.K. Stowell, T.S. Widlanski, T.G. Kutateladze and R.T. Raines, *J. Org. Chem.*, 1995, **60**, 6930.
188   N.N. Bhongle and J.Y. Tang, *Tetrahedron Lett.*, 1995, **36**, 6803.
189   Z.-W. Yang, Z.-S. Xu, N.-Z. Shen and Z.-Q. Fang, *Nucleosides Nucleotides*, 1995, **14**, 167.
190   E. Rozners and E. Bizdena, *Nucleosides Nucleotides*, 1995, **14**, 2009.
191   I. Tomoskozi, E. Gacs-Baitz and L. Otvos, *Tetrahedron*, 1995, **51**, 6797, and corrigendum, 8407.
192   R.P. Iyer, D. Yu, T. Devlin, N.-H. Ho and S. Agrawal, *J. Org. Chem.*, 1995, **60**, 5388.
193   M.E. Schwartz, R.R. Breaker, G.T. Asteriadis and G.R. Gough, *Tetrahedron Lett.*, 1995, **36**, 27.
194   X. Liu and C.B. Reese, *J. Chem. Soc., Perkin Trans. 1*, 1995, 1685.
195   W.K.-D. Brill, *Tetrahedron Lett.*, 1995, **36**, 703.
196   R. Zain, R. Strömberg and J. Stawinski, *J. Org. Chem.*, 1995, **60**, 8241.
197   P.H. Seeberger, E. Yau and M.H. Caruthers, *J. Am. Chem. Soc.*, 1995, **117**, 1472.

198   N. Puri, S. Hünsch, C. Sund, I. Ugi and J. Chattopadhyaya, *Tetrahedron*, 1995, **51**, 2991.
199   W.J. Stec, A. Grajkowski, A. Kobylanska, B. Karwowski, M. Koziolkiewicz, K. Misiura, A. Okruszek, A. Wilk, P. Guga and M. Boczkowska, *J. Am. Chem. Soc.*, 1995, **117**, 12019.
200   W. Dabkowski and I. Tworowska, *Tetrahedron Lett.*, 1995, **36**, 1095.
201   W. Dabkowski, I. Tworowska, J. Michalski and F. Cramer, *J. Chem. Soc., Chem. Commun.*, 1995, 1435.
202   W. Dabkowski and I. Tworowska, *Chem. Lett.*, 1995, 727.
203   V.T. Ravikumar, *Synth. Commun.*, 1995, **25**, 2161.
204   A.H. Krotz, P. Wheeler and V.T. Ravikumar, *Angew. Chem., Int. Ed. Engl.*, 1995, **34**, 2406.
205   K. Misiura, D. Pietrasiak and W.J. Stec, *J. Chem. Soc., Chem. Commun.*, 1995, 613.
206   H. Ayukawa, S. Ohuchi, M. Ishikawa and T. Hata, *Chem. Lett.*, 1995, 81.
207   R.W. Fisher and M.H. Caruthers, *Tetrahedron Lett.*, 1995, **36**, 6807.
208   W. Dabkowski, I. Tworowska and R. Saiakhov, *Tetrahedron Lett.*, 1995, **36**, 9223.
209   T. Szabo and J. Stawinski, *Tetrahedron*, 1995, **51**, 4145.
210   S.P. Collingwood and A.D. Baxter, *Synlett*, 1995, 703.
211   T. Wada and M. Sekine, *Tetrahedron Lett.*, 1995, **36**, 8845.
212   M. Mayer, I. Uga and W. Richter, *Tetrahedron Lett.*, 1995, **36**, 2047.
213   G.H. Hakimelahi, A.A. Moosavi-Movahedi, M.M. Sadeghi, S.-C. Tsay and J.R. Hwu, *J. Med. Chem.*, 1995, **38**, 4648.
214   S. Pitsch, R. Krishnamurthy, M. Bolli, S. Wendeborn, A. Holzner, M. Minton, C. Lesueur, I Schlönvogt, B. Jaun and A. Eschenmoser, *Helv. Chim. Acta*, 1995, **78**, 1621.
215   L.W. Tari, K.L. Sadana and A.S. Secco, *Nucleosides Nucleotides*, 1995, **14**, 175.
216   M. Koga, A. Wilk, M.F. Moore, C.L. Scremin, L. Zhou and S.L. Beaucage, *J. Org. Chem.*, 1995, **60**, 1520.
217   E.I. Kvasyuk, T.I. Kulak, I.A. Mikhailopulo, R. Charubala and W. Pfleiderer, *Helv. Chim. Acta*, 1995, **78**, 1777.
218   W.L. McEldoon and D.F. Wiemer, *Tetrahedron*, 1995, **51**, 7131.
219   J. Matulic-Adamic, P. Haeberli and N. Usman, *J. Org. Chem.*, 1995, **60**, 2563.
220   M.V. Jasko, N.A. Novikov and N.B. Tarussova, *Bioorg. Khim.*, 1994, **20**, 50 (*Chem. Abstr.*, 1995, **122**, 214409).
221   J. Baraniak, W.J. Stec and G.M. Blackburn, *Tetrahedron Lett.*, 1995, **36**, 8119.
222   T. Furuta, H. Torigai, M. Sugimoto and M. Iwamura, *J. Org. Chem.*, 1995, **60**, 3953.
223   X.-B. Tian and L.-H. Zhang, *Prog. Nat. Sci.*, 1994, **4**, 726 (*Chem. Abstr.*, 1995, **122**, 214414).
224   H.A. Azab, A.M. El-Nady and S.A. El-Shatoury, *Monatsh. Chem.*, 1994, **125**, 1049 (*Chem. Abstr.*, 1995, **122**, 56391).
225   L. Désaubry, I. Shoshani and R.A. Johnson, *Tetrahedron Lett.*, 1995, **36**, 995.
226   D.P.C. McGee, C. Vargeese, Y. Zhai, G.P. Kirschenheuter, A. Settle, C.R. Siedem and W.A. Pieken, *Nucleosides Nucleotides*, 1995, **14**, 1329.
227   D. Bonnaffé, B. Dupraz, J. Ughetto-Monfrin, A. Namane and T.H. Dinh, *Tetrahedron Lett.*, 1995, **36**, 531.
228   C.I. Hong, A. Nechaev, A.J. Kirisits, R. Vig, S.-W. Hui and C.R. West, *J. Med. Chem.*, 1995, **38**, 1629.
229   N. Dyatkina, A. Arzumanov, L. Victorova, M. Kukhanova and A. Krayevsky, *Nucleosides Nucleotides*, 1995, **14**, 91.

230 P. Labataille, H. Pélicano, G. Maury, J.-L. Imbach and G. Gosselin, *Bioorg. Med. Chem. Lett.*, 1995, **5**, 2315.
231 C.-H. Lin, G.-J. Shen, E. Garcia-Junceda and C.-H. Wong, *J. Am. Chem. Soc.*, 1995, **117**, 8031.
232 M. Arlt and O. Hindsgaul, *J. Org. Chem.*, 1995, **60**, 14.
233 Y. Kajihara, T. Ebata, K. Koseki, H. Kodama, H. Matsushita and H. Hashimoto, *J. Org. Chem.*, 1995, **60**, 5732.
234 W. Klaffke, *Carbohydr. Res.*, 1995, **266**, 285.
235 I. Müller and R.R. Schmidt, *Angew. Chem., Int. Ed. Engl.*, 1995, **34**, 1328.
236 H. Yuasa, M.M. Palcic and O. Hindsgaul, *Can. J. Chem.*, 1995, **73**, 2190.
237 A. Zatorski, B.M. Goldstein, T.D. Colby, J.P. Jones and K.W. Pankiewicz, *J. Med. Chem.*, 1995, **38**, 1098.
238 G.A. Ashamu, A. Galione and B.V.L. Potter, *J. Chem. Soc., Chem. Commun.*, 1995, 1359.
239 F.-J. Zhang, Q.-M. Gu, P. Jing and C.J. Sih, *Bioorg. Med. Chem. Lett.*, 1995, **5**, 2267.
240 F.-J. Zhang and C.J. Sih, *Tetrahedron Lett.*, 1995, **36**, 9289.
241 F.-J. Zhang and C.J. Sih, *Bioorg. Med. Chem. Lett.*, 1995, **5**, 1701.
242 Y.S. Sanghvi and P.D. Cook (Eds.), *ACS Symp. Ser.*, 1994, **580** (Carbohydrate Modifications in Antisense Research).
243 Y. Ducharme and K.A. Harrison, *Tetrahedron Lett.*, 1995, **36**, 6643.
244 J. Zhang and M.D. Matteucci, *Tetrahedron Lett.*, 1995, **36**, 8375.
245 J.M. Veal and F.K. Brown, *J. Am. Chem. Soc.*, 1995, **117**, 1873.
246 D. Wang, B. Meng, M.J. Damha and G. Just, *Nucleosides Nucleotides*, 1995, **14**, 1961.
247 S. Wendeborn, R.M. Wolf and A. De Mesmaeker, *Tetrahedron Lett.*, 1995, **36**, 6879.
248 G. Stork, C. Zheng, S. Gryaznov and R. Schultz, *Tetrahedron Lett.*, 1995, **36**, 6387.
249 C.V.C. Prasad, T.J. Caulfield, C.P. Prouty, A.K. Saha, W.C. Schairer, A. Yawman, D. Upson and L.I. Kruse, *Bioorg. Med. Chem. Lett.*, 1995, **5**, 411.
250 L. De Napoli, A. Iadonisi, D. Montesarchio, M. Varra and G. Piccialli, *Bioorg. Med. Chem. Lett.*, 1995, **5**, 1647.
251 A.A.H. Abdel Aleem, E. Larsen and E.B. Pedersen, *Nucleosides Nucleotides*, 1995, **14**, 2027.
252 A.A.H. Abdel Aleem, E. Larsen and E.B. Pedersen, *Tetrahedron*, 1995, **51**, 7867.
253 G.V. Petersen and J. Wengel, *Tetrahedron*, 1995, **51**, 2145.
254 V. Mohan, R.H. Griffey and D.R. Davis, *Tetrahedron*, 1995, **51**, 6855.
255 R.O. Dempcy, K.A. Browne and T.C. Bruice, *J. Am. Chem. Soc.*, 1995, **117**, 6140.
256 S. Obika, Y. Takashima, Y. Matsumoto, K. Kuromaru and T. Imanishi, *Tetrahedron Lett.*, 1995, **36**, 8617.
257 R.P. Hodge and N.D. Sinha, *Tetrahedron Lett.*, 1995, **36**, 2933.
258 P. Martin, *Helv. Chim. Acta*, 1995, **78**, 486.
259 K. Yamana, R. Aota and H. Nagano, *Tetrahedron Lett.*, 1995, **36**, 8427.
260 A.A.H. Abdel-Aleem, E. Larsen, E.B. Pedersen and C. Nielsen, *Acta Chem. Scand.*, 1995, **49**, 609.
261 N.J. Leonard and Neelima, *Tetrahedron Lett.*, 1995, **36**, 7833.
262 R.P. Iyer, Z. Jiang, D. Yu, W. Tan and S. Agrawal, *Synth. Commun.*, 1995, **25**, 3611.
263 V.T. Ravikumar, A.H. Krotz and D.L. Cole, *Tetrahedron Lett.*, 1995, **36**, 6587.
264 E. Leikauf, F. Barnekow and H. Köster, *Tetrahedron*, 1995, **51**, 3793.
265 R.M. Karl, R. Klösel, S. König, S. Lehnhoff and I. Ugi, *Tetrahedron*, 1995, **51**, 3759.

266   M. Manoharan, K.L. Tivel, L.K. Andrade and P.D. Cook, *Tetrahedron Lett.*, 1995, **36**, 3647.

267   M. Manoharan, K.L. Tivel and P.D. Cook, *Tetrahedron Lett.*, 1995, **36**, 3651.

268   L. Le Hir de Fallois, J.-L. Décout and M. Fontecave, *Tetrahedron Lett.*, 1995, **36**, 9479.

269   R.S. Sarfati, T. Berthod, C. Guerreiro and B. Canard, *J. Chem. Soc., Perkin Trans. 1*, 1995, 1163.

270   H. Onishi, Y. Machida and T. Nagai, *Chem. Pharm. Bull.*, 1995, **43**, 340.

271   S.V. Makutnova, I.L. Plikhtyak, I.V. Yartseva, T.P. Ivanova and S.Y. Mel'nik, *Bioorg. Khim.*, 1995, **21**, 289 (*Chem. Abstr.*, 1995, **123**, 340741).

272   A. Miyake and N. Bodor, *Pharm. Sci. Commun.*, 1994, **4**, 231 (*Chem. Abstr.*, 1995, **122**, 214422).

273   L.F. Garcia-Alles and V. Gotor, *Tetrahedron*, 1995, **51**, 307.

274   V. Gotor, F. Moris and L.F. Garcia-Alles, *Biocatalysis*, 1994, **10**, 295 (*Chem. Abstr.*, 1995, **122**, 240306).

275   M.C. Pirrung and J.-C. Bradley, *J. Org. Chem.*, 1995, **60**, 1116.

276   M. Pfister, H. Schirmeister, M. Mohr, S. Farkas, K.-P. Stengele, T. Reiner, M. Dunkel, S. Gokhale, R. Charubala and W. Pfleiderer, *Helv. Chim. Acta*, 1995, **78**, 1705.

277   R. Vince and P.T. Pham, *Nucleosides Nucleotides*, 1995, **14**, 2051.

278   T. Wada, M. Tobe, T. Nagayama, K. Furusawa and M. Sekine, *Tetrahedron Lett.*, 1995, **36**, 1683.

279   W.H. Binder, H. Kählig and W. Schmid, *Tetrahedron: Asymm.*, 1995, **6**, 1703.

280   J.J. Krepinsky, D.M. Whitfield, S.P. Douglas, N. Lupescu, D. Pulleybank and F.L. Moolten, *Methods Enzymol.*, 1994, **247**, 144 (*Chem. Abstr.*, 1995, **122**, 315019).

281   M. Perbost, T. Hoshiko, F. Morvan, E. Swayze, R.H. Griffey and Y.S. Sanghvi, *J. Org. Chem.*, 1995, **60**, 5150.

282   M. Bolli and C. Leumann, *Angew. Chem., Int. Ed. Engl.*, 1995, **34**, 694.

283   M. Bolli, P. Lubini and C. Leumann, *Helv. Chim. Acta*, 1995, **78**, 2077.

284   J.C. Litten, C. Epple and C.J. Leumann, *Bioorg. Med. Chem. Lett.*, 1995, **5**, 1231.

285   I. Verheggen, A. Van Aerschot, L. Van Meervelt, J. Rosenski, L. Wiebe, R. Snoeck, G. Andrei, J. Balzarini, P. Claes, E. De Clercq and P. Herdewijn, *J. Med. Chem.*, 1995, **38**, 826.

286   A. Van Aerschot, I. Verheggen, C. Hendrix and P. Herdewijn, *Angew. Chem., Int Ed. Engl.*, 1995, **34**, 1338.

287   M.-J. Pérez-Pérez, J. Balzarini, J. Rosenski, E. De Clercq and P. Herdewijn, *Bioorg. Med. Chem. Lett.*, 1995, **5**, 1115.

288   M.-J. Pérez-Pérez, B. Doboszewski, J. Rosenski and P. Herdewijn, *Tetrahedron: Asymm.*, 1995, **6**, 973.

289   P. Bolon, T.S. Jahnke and V. Nair, *Tetrahedron*, 1995, **51**, 10443.

290   T.S. Jahnke and V. Nair, *Bioorg. Med. Chem. Lett.*, 1995, **5**, 2235.

291   M. Lee, C.K. Chu, S.B. Pai, Y.-L. Zhu, Y.-C. Cheng, M.W. Chun and W.K. Chung, *Bioorg. Med. Chem. Lett.*, 1995, **5**, 2011.

292   K. Bednarski, D.M. Dixit, T.S. Mansour, S.G. Colman, S.M. Walcott and C. Ashman, *Bioorg. Med. Chem. Lett.*, 1995, **5**, 1741.

293   E.V. Efimtseva, S.N. Mikhailov, S. Meshkov, T. Hankamäki, M. Oivanen and H. Lönnberg, *J. Chem. Soc., Perkin Trans. 1*, 1995, 1409.

294   J. Du, F. Qu, D.-W. Lee, M.G. Newton and C.K. Chu, *Tetrahedron Lett.*, 1995, **36**, 8167.

295   C. Liang, D.-W. Lee, M.G. Newton and C.K. Chu, *J. Org. Chem.*, 1995, **60**, 1546.

296   Y. Xiang, Q. Teng and C.K. Chu, *Tetrahedron Lett.*, 1995, **36**, 3781.
297   J.J. Huang, J.L. Rideout and G.E. Martin, *Nucleosides Nucleotides*, 1995, **14**, 195.
298   J. Milton, S. Brand, M.F. Jones and C.M. Rayner, *Tetrahedron Lett.*, 1995, **36**, 6961.
299   R.E. Austin and D.G. Cleary, *Nucleosides Nucleotides*, 1995, **14**, 1803.
300   T. Breining, A.R. Cimpoia, T.S. Mansour, N. Cammack, P. Hopewell and
       C. Ashman, *Heterocycles*, 1995, **41**, 87.
301   T.S. Mansour, H. Jin, W. Wang, E.U. Hooker, C. Ashman, N. Cammack,
       H. Salomon, A.R. Belmonte and M.A. Wainberg, *J. Med. Chem.*, 1995, **38**, 1.
302   A. Inaba, K. Inami, Y. Kimoto, R. Yanada, Y. Miwa, T. Taga and K. Bessho,
       Chem. Pharm. Bull., 1995, **43**, 1601.
303   J.C. Graciet, P. Faury, M. Camplo, A.S. Charvet, N. Mourier, C. Trabaud,
       V. Niddam, V. Simon and J.L. Kraus, *Nucleosides Nucleotides*, 1995, **14**, 1378.
304   J.M.J. Tronchet, M. Iznaden, F. Barbalat-Rey, I. Komaromi, N. Dolatshahi and
       G. Bernardinelli, *Nucleosides Nucleotides*, 1995, **14**, 1737.
305   Y. Xiang, J. Chen, R.F. Schinazi and K. Zhao, *Tetrahedron Lett.*, 1995, **36**, 7196.
306   L. Pickering, B.S. Malhi, P.L. Coe and R.T. Walker, *Tetrahedron*, 1995, **51**, 2719.
307   S. Nishiyama, Y. Kikuchi, H. Kurata, S. Yamamura, T. Izawa, T. Nagahata,
       R. Ikeda and K. Kato, *Bioorg. Med. Chem. Lett.*, 1995, **5**, 2273.
308   M. Lee, D. Lee, Y. Zhao, M.G. Newton, M.W. Chun and C.K. Chu, *Tetrahedron
       Lett.*, 1995, **36**, 3499.
309   Y. Zhao, T. Yang, M. Lee, D. Lee, M.G. Newton and C.K. Chu, *J. Org. Chem.*,
       1995, **60**, 5236.
310   L. Mévellec and F. Huet, *Tetrahedron Lett.*, 1995, **36**, 7441.
311   B.C.N.M. Jones, J.C. Drach, T.H. Corbett, D. Kessel and J. Zemlicka, *J. Org.
       Chem.*, 1995, **60**, 6277.
312   C.L. Scremin, J.H. Boal, A. Wilk, L.R. Phillips, L. Zhou and S.L. Beaucage,
       Tetrahedron Lett., 1995, **36**, 8953.
313   J.R. Lakanen, A.E. Pegg and J.K. Coward, *J. Med. Chem.*, 1995, **38**, 2714.
314   M. Cornia, S. Binacchi, T. Del Soldato, F. Zanardi and G. Casiraghi, *J. Org. Chem.*,
       1995, **60**, 4964.
315   J.R. Sufrin, A.J. Spiess, C.J. Marasco, Jr., S.L. Croft, D. Snowdon, V. Yardley and
       C.J. Bacchi, *Bioorg. Med. Chem. Lett.*, 1995, **5**, 1961.
316   G.A. Neyhart, C.-C. Cheng and H.H. Thorp, *J. Am. Chem. Soc.*, 1995, **117**, 1463.
317   Y. Nishida and S. Ito, *J. Chem. Soc., Chem. Commun*, 1995, 1211.
318   C. Hazlewood and M.J. Davies, *J. Chem. Soc., Perkin Trans. 2*, 1995, 895.
319   H. Sugiyama, K. Fujimoto and I Saito, *J. Am. Chem. Soc.*, 1995, **117**, 2945.
320   M. Irisawa, N. Takeda and M. Komiyama, *J. Chem. Soc., Chem. Commun*, 1995,
       1221.
321   M. Yashiro, A. Ishikubo and M. Komiyama, *J. Chem. Soc., Chem. Commun*, 1995,
       1793.
322   M. Oivanen, N.Sh. Padyukova, S. Kuusela, S.N. Mikhailov and H. Lönnberg, *Acta
       Chem. Scand.*, 1995, **49**, 307.
323   C.L. Perrin, *J. Org. Chem.*, 1995, **60**, 1239.
324   M. Oivanen, M. Ora, H. Almer, R. Strömberg and H. Lönnberg, *J. Org. Chem.*,
       1995, **60**, 5620.
325   M. Kosonen and H. Lönnberg, *J. Chem. Soc., Perkin Trans. 2*, 1995, 1203.
326   R. Johnson, C.B. Reese and Z. Pei-Zhou, *Tetrahedron*, 1995, **51**, 5093.
327   G. Thoithi, A. Van Schepdael, R. Busson, P. Herdewijn, E. Roets and J. Hoogmar-
       tens, *Nucleosides Nucleotides*, 1995, **14**, 1559.

21
NMR Spectroscopy and Conformational Features

1   General Aspects

Reviews on theoretical and experimental aspects of ^{13}C nuclear magnetic relaxation and motional behaviour of carbohydrate molecules in solution,[1] and on ^{13}C-^1H coupling constants in the conformational analysis of sugar molecules[2] have been published. The spectral quality of the ensembles obtained with the CICADA algorithm, a newly developed conformational search method, and those obtained with MM3 have been compared and tested against experimental NMR and optical rotation data of small carbohydrate molecules in solution.[3] The capacity of three very different molecular mechanics force fields to reproduce a set of experimental spectral data (o.r.d., NOE, NMR coupling constants) has been probed by application to the conformational behaviour of ethyl β-lactoside.[4]

Postacquisition enhancement yields pure-phase spectra with concomitant increase in the signal to noise ratio and improved accuracy in the measurement of long-range heteronuclear coupling constants, as has been shown for $^3J_{C-4,H-6S}$ and $^3J_{C-4,H-6R}$ in the reducing-end glucose residue of isomaltose.[5] A new method for determining long-range ^{13}C-^1H coupling constants accurately, even when they are close to zero, is based on the propagation of magnetization in a multi-stage Hartmann-Hahn experiment involving a chain of coupled protons; the technique has been illustrated with α-D-glucopyranose as the model.[6]

Theoretical conformational analyses of 2-hydroxypiperidine and 2-hydroxy-hexahydropyrimidine indicated that the anomeric effect is due to charge back donation from lone pairs rather than to dipolar repulsion.[7] A molecular orbital study of the reverse anomeric effect in N-pyranosylimidazoles suggested that the effect is not a general phenomenon even for glycosylated quarternary ammonium compounds.[8] The proportions of axial anomers of various glucosylamines and their conjugate acids have been determined by ^1H-NMR spectroscopy. The changes upon N-protonation were small and were accounted for by steric reasons rather than reverse anomeric effect.[9]

Electron-nuclear relaxation studies with 'TEMPOL'-labelled carbohydrate molecules, e.g. β-D-glucopyranose derivative 1, allowed the evaluation of intramolecular interspin distances up to 10-15 Å in length with good accuracy.[10] In a symposium report, hydrogen-bonding in small sugar molecules (threitol, erythritol, β-D-glucopyranose, maltose) as described by the 1992 version of MM3, has been compared with diffraction data obtained by construction and

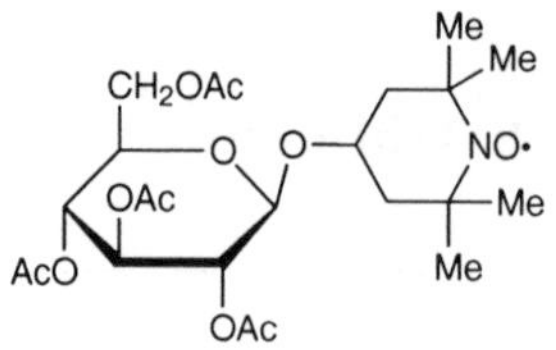

1

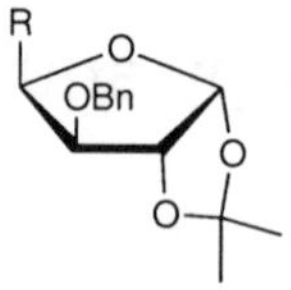

2 R = CHO

3 R = H₂C

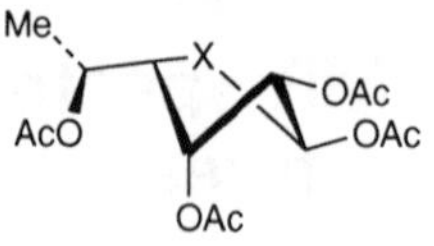

4 X = O
5 X = S

optimization of model miniature crystals (*ca.* 700 atoms).[11] Evidence from molecular modelling and NMR spectroscopic experiments for the entrapment of water in a simple carbohydrate complex has been presented.[12] The electronic properties of glycosylated chromophores (carminic acid, glycosylated oxazines, *etc.*) have been studied by AM1 semiempirical calculations in an effort to understand and predict the static and dynamic characteristics of these compounds, which are used to probe the nucleation and crystallization behaviour of sugar solutions.[13]

2   Furanose Systems

NOE Studies have been used to establish unambiguously the stereochemistry at C-5 of compound **3**, the new chiral centre being formed in the chain-extension of **2** by a Reformatzki reaction.[14] A comparative study of the conformations of furanose and thiofuranose derivatives, in particular compounds **4** and **5**, has been published; the relatively weak anomeric effect in **5** is reflected in the tendency of its anomeric substituent to remain equatorial, as shown, whereas **4** exists as $aT_3/^3T_2$ mixture.[15] An attempt has been made to correlate the biological activity of 1,4-dideoxy-1,4-imino-D–arabinitol derivatives **6** with their conformations which change on *N*-alkylation. The activity also depend on the size of the *N*-alkyl group.[16] According to molecular dynamics calculations the *N*-hydroxypyrrolidine analogues **7** of *C*-glycofuranoside derivatives should prefer *N-exo-C-exo* conformations; ¹H-NMR spectroscopic evidence indicates, however, that in solution the *N-endo* forms with equatorial *N*-hydroxy groups predominate.[17]

An investigation of one-bond ¹³C-¹H spin coupling constants has been performed with the model compound **8** to allow predictions to be made for other furanose rings, especially those of nucleosides.[18] A new program, the 'HETROT' algorithm, has been developed for the conformational analysis of sugar rings in nucleosides and nucleotides; it combines information on ¹H-¹H and ¹³C-¹H coupling constants to calculate all possible values for the five

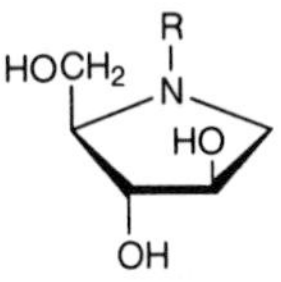

HOCH$_2$ ... R
N
HO
OH

R = H, Me or Bu
6

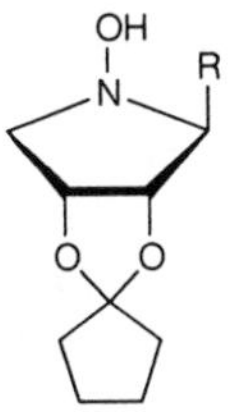

OH
R
N
O   O

R = Me, Ph, C≡CPh,
C$_6$H$_4$OMe *etc.*
7

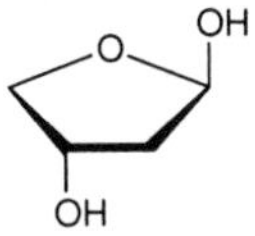

OH
O   OH
OH
8

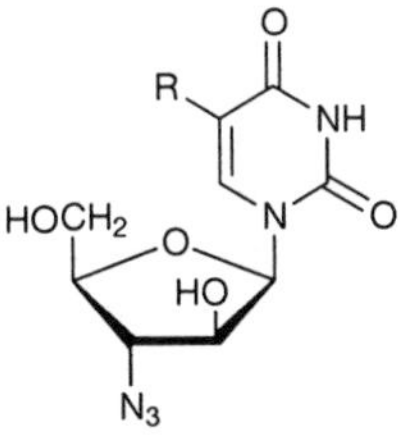

R
NH
N   O
HOCH$_2$ O
HO
N$_3$
9 R = Me
10 R = CH$_2$OH

parameters that describe the conformation of the furanose ring and has been applied to AZT (**9**).[19] Conformational analysis of the 5-hydroxymethyl analogue **10** of AZT by the PROFIT program showed that the presence of the CH$_2$OH group at C-5 changes the conformation of the exocyclic side-chain of the sugar moiety, which might explain why **10** is not phosphorylated by human thymidine kinase.[20] Analysis of ribo- and 2'-deoxyribo-nucleosides **11** and **12** and their 3'-ethylphosphates **13** and **14**, respectively, by ^{1}H-NMR spectroscopy revealed that phosphorylation at O-3' shifts the conformational equilibrium towards the 'South' form; this tendency is enhanced by a specific interaction when the 2'-position is hydroxylated so that **13** adopts a unique (S, ε^-) conformation.[21]

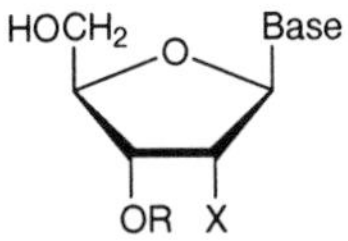

HOCH$_2$ O   Base
OR   X
11 X = OH, R = H
12 X = R = H
13 X = OH, R = PO$_2$H(OEt)
14 X = H, R = PO$_2$H(OEt)

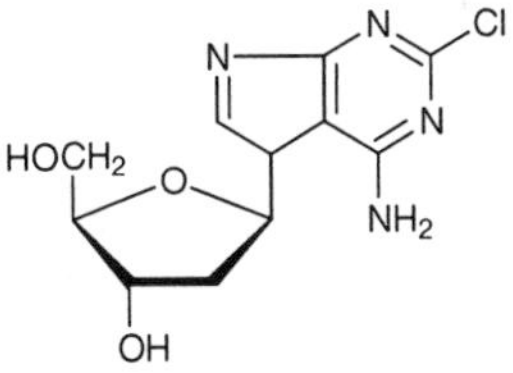

N   Cl
N
HOCH$_2$ O
NH$_2$
OH
15

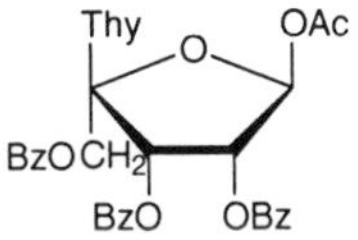

Thy   OAc
O
BzOCH$_2$
BzO   OBz
16

Conformational analyses by NMR spectroscopy and/or theoretical calculations have also been reported for 6-methyl-2'-deoxyuridine,[22] 2',3'-dideoxythymidine analogues with known antiviral activity,[23] 2'-deoxy-2'-fluoroarabinofuranosyluracils,[24] some D-arabino-, D-lyxo- and D-xylo-furanosyl pyridine-C-nucleosides,[25] the N-7 regioisomer **15** of 2-chloro-2'-deoxyadenosine,[26] the 4-thymin-1-yl α-L-lyxofuranose derivative **16**,[27] cyclic ADP ribose,[28] and 1,2-dideoxy-D-ribofuranose, its 3- and 5-phosphates and the trimer **17**.[29]

3   Pyranose and Related Systems

The ^{1}H- and ^{13}C-NMRspectra of a series of partially O-methylated methyl glycopyranosides have been unambiguously assigned by HMBC (heteronuclear multi-bond correlation) experiments, and values for the changes in ^{1}H and ^{13}C chemical shifts at the site of methylation and in the β- and γ-positions have been tabulated.[30] The complete relative stereochemistry of caryophyllose (**18**), a new, branched 12-carbon sugar from the lipopolysaccharide fraction of *Pseudomonas caryophylli*, has been assigned by ^{1}H- NMR spectroscopy; its absolute configuration was independently elucidated by Mosher's empirical method and by Exciton chiral coupling.[31] The configurations at the quarternary carbon atoms of branched glycopyranose derivatives, such as **19**, have been determined.[32]

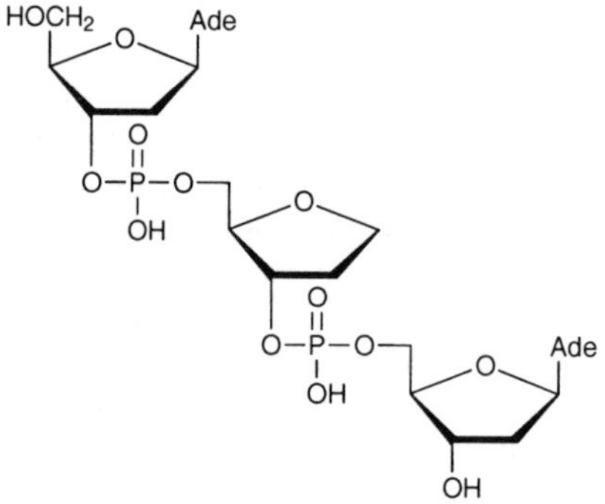

17

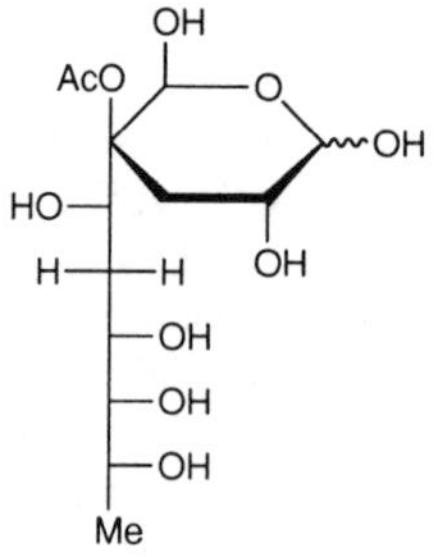

18

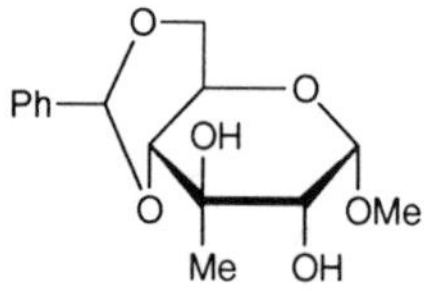

19

The effect of electric fields of various strengths and directions on the conformational behaviour of 2-methoxytetrahydropyran (a simple glycoside model) has been investigated by the *ab initio* molecular orbital method.[33] In another theoretical study (by AM1) the influence of the solvent (H_2O) on the conformational stabilities and the rotational motions of exocyclic groups in the model compounds 2-hydroxy-, 2-hydroxymethyl- and 2,3-dihydroxy-tetrahydro-pyran, as well as in α- and β-D-gluco- and D-galacto-pyranose has been examined.[34] Molecular mechanics and quantum mechanical electronic structure theory have been employed to analyse the factors contributing to the relative energies of two hydroxymethyl conformers for each of the two chair forms ($^4C_1$ and $^1C_4$) of β-D-glucopyranose. Manifestations of the greater steric strain in the $^1C_4$ chair include longer ring bonds, a larger bond angle at the ring oxygen atom and smaller puckering amplitudes.[35] A theoretical study on the dependence of $^3J_{C,H}$ values in 16 hexopyranoses on the conformation of their hydroxymethyl groups resulted in an equation which might be useful as a tool for estimating the conformational properties of CH_2OH groups in other monosaccharides and the corresponding substituted groups in $(1 \rightarrow 6)$-linked disaccharides.[36] A correlation has been established on the basis of CD and NMR data ($^3J_{H-5,H-6R}$ values) between rotamer populations around the C-5−C-6 bonds and electron withdrawing effects of the aglycons of the non-chiral alkyl β-D-glucopyranosides **20**, and between these populations and the absolute configuration of the chiral aglycon carbon atoms in alkyl β-D-glucopyranosides, such as compounds **21**.[37] The rotamers about the C-5−C-6 bond of the C-6-functionalized alkyl α-D-glucopyranoside derivatives **22** have been investigated by ^1H-NMR spectroscopy.[38]

High-resolution solid state ^{13}C-NMR spectroscopy has been applied to examine the crystal structures of 1,2:3,4-di-*O*-isopropylidene-α-D-galactopyranose derivatives.[39] Intramolecular motions in polycrystalline methyl α- and β-D-glucopyranoside and methyl β-D-galactopyranoside have been probed by measuring ^1H spin-lattice relaxation times as well as ^1H second moments over a wide temperature range.[40] A series of D-aldopento- and aldohexo-pyranoses containing ^{13}C enrichment at various single positions have been subjected to extensive NMR spectroscopic analysis in order to refine empirical relationships between ring structure and configuration.[41]

Conformational analysis by MM and MO methods indicated that the $^1C_4$ chair is the most stable form for methyl 3-amino-2,3,6-trideoxy-α-L-*lyxo*-hexopyranoside **23** and its derivatives **24**.[42] ^1H-NMR spectra of 1,6-anhydro-β-D-glucopyranose derivatives and their 3-amino-3-deoxy analogues recorded in DMSO-d_6 and D_2O showed that in solvents more polar than $CDCl_3$ the $B_{O,3}$ forms are populated to an appreciable extent, due to disruption of intramolecular hydrogen-bonding.[43] Because of charge interactions which cause repulsion between the substituents, the polysulfates **25** exist partly in non-chair forms, whereas their α-anomers and the uncharged sulfamoyl analogue **26** are entirely in the $^4C_1$ conformation, as shown by ^{1}H-^{1}H coupling constants.[44] According to MMX (a new version of MM2) calculations, the 1,3-anhydro-α-L-arabinopyr-anose derivative **27**, as well as its D-galactose and D-mannose equivalents, assume near E_2 conformations in solution.[45]

MM2 Calculations performed on a series of stereoisomeric methyl 5,7-*O*-benzylidene-3-deoxy-3-nitro-α-D-heptoseptanosides **28** showed that the lowest energy conformations for these compounds are twist forms derived from $^5C_{1,2}$ and $^0C_{3,4}$ chairs; this was confirmed by ^{1}H-NMR spectroscopic data.[46] A study of the hexamethyl ether of *scyllo*-inositol by ^{1}H-NMR spectroscopy revealed that all six exocyclic C−O bonds are eclipsed to relieve steric strain.[47]

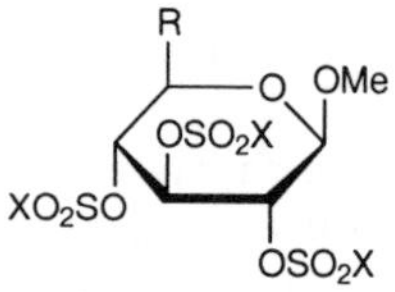

25 X = ONa, R = CO$_2$Na
        or CH$_2$OSO$_3$Na
26 X = NH$_2$, R = CH$_2$OSO$_2$NH$_2$

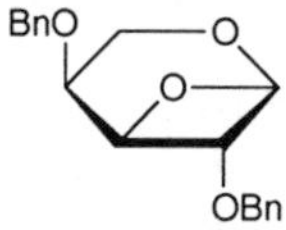

27

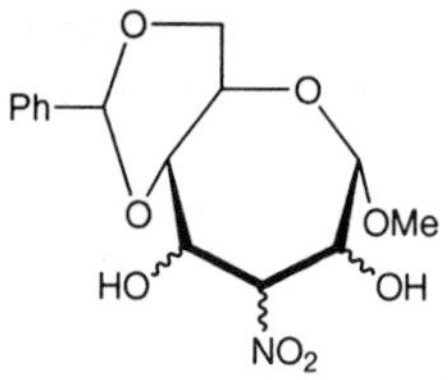

28

4   Disaccharides

A new, quick and convenient method for quantifying α- and β-maltose has been established on the basis of a thorough ^{13}C-NMR investigation considering the influence of factors such as sample concentration, choice of solvent, spin-lattice relaxation times and NOE suppression on the accuracy.[48] The hydration of α-maltose and amylose has been studied by molecular modelling and thermo-dynamic methods.[49] A hydrogen bond between 4-OH and 2′-OH of disaccharide **29**, which manifests itself by the doubling of the ^{1}H-NMR signals for the two hydroxyl hydrogen atoms involved in a partially deuterated sample in DMSO-d_6, has been taken as proof of an 'anti' conformer.[50]

x-D-Gal*p*-(1→3)-x-D-GLC*p*NAc-OMe   α-L-Rha*p*-(1→2)-α-L-Rha*p*-OMe
   29 x = β   **30** x = α   **31**

β-D-Gal*p*-(1→2)-β-D-Xyl*p*-OMe   α-D-Man*p*-(1→n)-α-D-Man*p*-OMe
            32   **33** n = 6   **34** n = 2

Computational and/or NMR spectroscopic methods have been applied to conformational analyses of the following disaccharides: the methyl glycosides **30**-**33**,[51-54] the α-(1→2)-, α-(1→3)- and α-(1→6)-linked mannosyl dimers, as models for the linkages found in asparagine-linked glycoproteins,[55,56] mannobiose and epimelibiose, as model disaccharides of galactomannan,[57] eight derivatives of galabiose with an α-(1→4) diaxial linkage,[58] sucrose,[59] sucrose octasulfate,[60,61] methyl α-thiomaltoside,[62] methyl α-lactoside,[63,64] and ethyl β-lactoside, sucralose, arabinobiose and a galacturonic acid dimer.[65] The optical rotation of disaccharide **34** has been calculated semiempirically as a function of the linkage dihedral angles θ and ψ.[66] Conformational studies on disaccharide fragments of a lectin are mentioned below (Ref. 81).

5 Oligosaccharides

Multiple-field ^{13}C-relaxation data for melezitose have been reported.[67] The ^{13}C-NMR spectra of the central D-GLC*p*-, D-Man*p*-, D-Gal*p*-, L-Fuc*p*- or L-Rha*p*-rings of 46 trisaccharides have been simulated by use of multiple linear regression analysis and neural networks.[68] A technique for assisting the resonance assignment process in ^{1}H-NMR spectra of oligosaccharides containing multiple residues of the same type involving modified HOHAHA experiments with constant-time acquisition t_1 has been described.[69]

Intramolecular dipolar and chemical shift anisotropy interactions in a penta-saccharide fragment of heparin and their effect on ^{13}C-NMR relaxation rates have been discussed.[70] ^{1}H- and ^{13}C-NMR relaxation measurements have been used to study the internal and overall motions of the pentasaccharide corresponding to the antithrombin III binding-site of heparin.[71] A number of tetra- to hexa-oses derived from porcine intestinal heparin[72] and a digalactosylmannopentaose liberated from legume seed galactomannan by β-mannonase[73] have been fully characterized by 2D ^{1}H- and ^{13}C-NMR spectroscopy, and the hexasaccharide fragment **35** of the O-specific polysaccharide of *Hafnia alvei* has been identified by methylation analysis combined with NMR spectroscopy.[74]

α-D-Man*p*-(1→2)-α-D-Man*p*-(1→3)-β-D-GLC*p*NAc-(1→2)-β-D-Qui*p*3NFo-(1→3)-
α-D-GLC*p*NAc-(1→4)-α-D-GLC*p*A

35   (NFo = *N*-formamide)

α-L-Fuc*p*-(1→2)-β-D-Gal*p*-(1→3)-β-D-Gal*p*NAc-(1→3)-α-D-Gal*p*-OPr

36

A review, in Chinese (4 pages, 20 refs.) on experimental and theoretical aspects of oligosaccharide conformations has been published.[75] The conformational space of the terminal tetrasaccharide **36** of tumour-associated antigens has been examined by molecular modelling. A certain rigidity of the first two linkages (Fuc-Gal and Gal-GalNAc) was observed with flexibility in the third one (GalNAc-Gal).[76] An investigation of the pseudo-tetrasaccharide **37** by ^{1}H-NMR spectroscopy including 1D TOCSY and ROESY experiments showed that the two C-linked rings are in the unfavourable[1] C_4C-conformation.[77]

Conformational studies, using experimental and/or theoretical methods, have been undertaken with several high-mannose-type oligosaccharides generated during the biochemical degradation of Man*p*9GLC*p*NAc$_2$ to Man-α, α-trehalos-4-yl-O

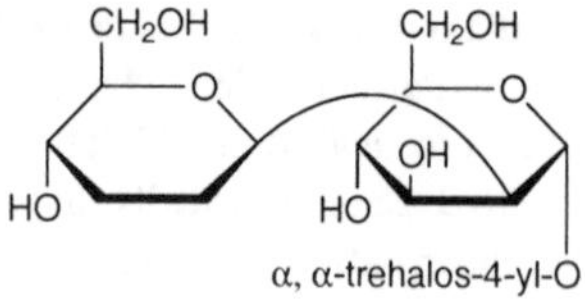

37

p_5GLCpNAc$_2$,[78] solvated Manp$_9$GLCp-NAc$_2$,[79] several arabinoxylan oligomers,[80] the carbohydrate moiety of a lectin, heptasaccharide **38**, a related
hexasaccharide and disaccharide fragments thereof,[81] a series of β-(1→3)-
branched β-(1→6) oligosaccharides involved in immune reactions in plants,[82] the
pentasaccharide of ganglioside GM1,[83,84] the bioactive forms of the carbohydrate
ligands in a sialyl Lewis X-selectin complex,[85] the oligosaccharide moieties of a
steroidal[86] and a triterpenoid[87] saponin, α-, β- and γ-cyclodextrins,[88] 6-deoxy-6-
L-tyrosinylaminocyclomaltoheptaose, a self-complexing β-cyclodextrin derivative,[89] fully benzylated cyclotetraisomaltoside,[90] the cyclic (1→2)-β-glucan
cyclosophoroheptadecaose,[91] and β-'cycloaltrin' (**39**).[92]

β-D-Xyl$p$
1
↓
2
α-L-Man$p$-(1→3) — (β-D-Man$p$-(1→4)-β-D-GlcpNAc- (1→4) — (β-D- GlcpNAc
6   3
↑   ↑
1   1
α-L-Man$p$ α-L-Fuc$p$

38

Gradient-enhanced homonuclear 2D NMR techniques, applied to D-mannose-
containing oligosaccharides, provided improved intensity in correlations between
protons in relatively short experimental times, especially when $^3J_{H,H}$ values were
small.[93]

6   Other Compounds

A Co(III) complex containing L-sorbose and two molecules of 1,10-phenanthro-
line has been investigated by CD, o.r.d. and ^{1}H-NMRspectroscopy, all ^{1}H-NMR
signals have been assigned.[94] The complete ^{1}H-NMR assignments of the
aminoglycoside antibiotics butirosin A and kanamycins A and B, as well as
conformational studies on butirosin A, have been reported.[95] The conformation
of adriamycin has been examined by 500 MHz ^{1}H-NMR spectroscopy.[96] The ^{1}H-
and ^{13}C-NMR spectra of the tumour antibiotic aclacinomycin A have been fully
assigned and its solution conformation analysed.[97] A conformational analysis of

the erythromycin analogues azithromycin and clathromycin in aqueous solution and bound to bacterial ribosomes relied mainly on ROESY and TRNOESY experiments.[98]

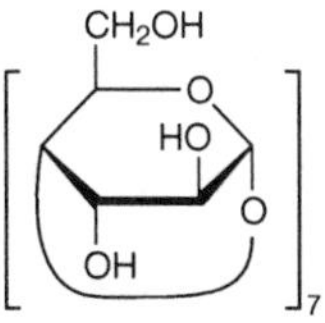

39

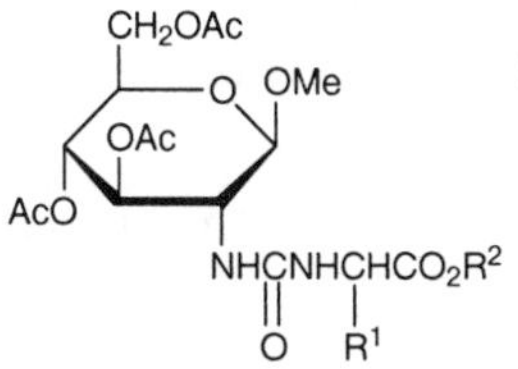

R[1] = H, R[2] = Et(GlyOEt) or
R[1] = Bn, R[2] = Me(L-PheOMe) *etc.*

40

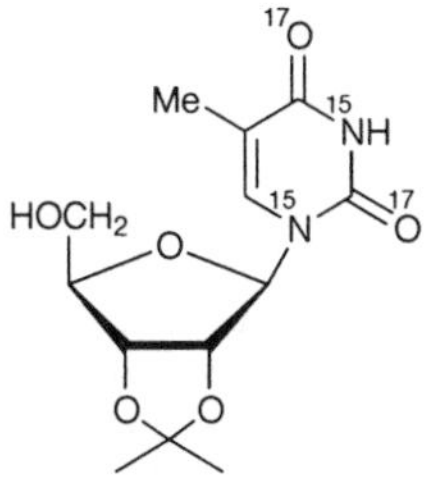

41

7 NMR of Nuclei other than ^{1}H and ^{13}C

A number of ureido sugar derivatives **40** have been examined by ^{1}H-, ^{13}C- and ^{15}N-NMR spectroscopy in solution and by ^{13}CP MAS in the solid state; the NMR data indicated that replacement of one amino acid residue by another has no significant effect on the conformation of the glucopyranose moiety.[99] Thermodynamics calculations of the weakly hydrogen-bonded complexes formed by nucleosides, such as **41**, have been carried out on the basis of information from ^{15}N- and ^{17}O-NMR data.[100] A conformational study of cyclic ADP-ribose made use of ^{1}H-, ^{13}C- and ^{31}P-NMR methods.[28] The spectral spin-diffusion process between the ^{31}P nuclei of dipotassium α-D-glucopyranose 1-phosphate has been studied under magic-angle conditions.[101] ^{13}C- and ^{183}W-NMR experiments have been used in the structural characterization of the tungstate complexes of D-glycero-D-*manno*-heptitol[102] and of a number of other aldoses and ketoses with the *lyxo*-configurations.[103]

References

1   P. Dais, *Adv. Carbohydr. Chem. Biochem.*, 1995, **51**, 63.
2   I. Tvaroska and F. R. Taravel, *Adv. Carbohydr. Chem. Biochem.*, 1995, **51**, 15.
3   S.B. Engelsen, J. Koca, I. Braccini, C. Hervé du Penhoat and S. Pérez, *Carbohydr. Res.*, 1995, **276**, 1.
4   S.B. Engelsen, S. Pérez, I. Braccini and C. Hervé du Penhoat, *J. Comput. Chem.*, 1995, **16**, 1096 (*Chem. Abstr.*, 1995, **123**, 314 283).
5   L. Poppe, S. Sheng and H. van Halbeek, *J. Magn. Reson., Ser. A*, 1994, **111**, 104 (*Chem. Abstr.*, 1995, **122**, 161 095).

6   J.M. Nuzillard and R. Freeman, *J. Magn. Reson., Ser. A*, 1994, **110**, 262 (*Chem. Abstr.*, 1995, **122**, 31 722).

7   U. Salzner, *J. Org. Chem.*, 1995, **60**, 986.

8   S.S.C. Chan, W.A. Szarek and G.R.J. Thatcher, *J. Chem. Soc., Perkin Trans. 2*, 1995, 45.

9   C.L. Perrin, *Pure Appl. Chem.*, 1995, **67**, 719 (*Chem. Abstr.*, 1995, **123**, 286 497).

10   F. Cinget, P.H. Fries, U.Greilich and P.J.A. Vottéro, *Magn. Reson. Chem.*, 1995, **33**, 260 (*Chem. Abstr.*, 1995, **123**, 228 664).

11   A.D. French and D.P. Miller, *A.C.S. Symp. Ser.*, 1994, **569**, 235 (*Chem. Abstr.*, 1995, **122**, 81 767).

12   P. Irwin, G. King, T.F. Kumosinski, P. Pieffer, J. Klein and L. Doner, *A.C.S. Symp. Ser.*, 1994, **576**, 342 (*Chem. Abstr.*, 1995, **122**, 161 143).

13   J.P. Rasimas and G.J. Blanchard, *J. Phys. Chem.*, 1994, **98**, 12949 (*Chem. Abstr.*, 1995, **122**, 10 393).

14   A.P. Rauter, M.J. Ferreira, J. Font, A. Virgili, M. Figueiredo, J.A. Figueiredo, M.I. Ismael and T.L. Canda, *J. Carbohydr. Chem.*, 1995, **14**, 929.

15   P.A. Zunszain and O. Varela, *J. Chem. Res.*, 1995, (S)486; (M)2910.

16   N. Asano, H. Kiso, K. Oseki, E. Tomioka, K. Matsui, M. Okamoto and M. Baba, *J. Med. Chem.*,1995, **38**, 2349.

17   J.M.J. Tronchet, M. Balkadjian, G. Zosimo-Landolfo, F. Barbalat-Rey, P. Lichtle, A. Ricca, I. Komaromi, G. Bernardinelli and M. Geoffroy, *J. Carbohydr. Chem.*, 1995, **14**, 17.

18   A.S. Serianni, J. Wu and I. Carmichael, *J. Am. Chem. Soc.*, 1995, **117**, 8645.

19   D.O. Cicero, G. Barbato and R. Bazzo, *Tetrahedron*, 1995, **51**, 10303.

20   S.V. Gupta, S.V.P. Kumar, A.L. Stuart, R. Shi, K.C. Brown W.M. Zoghaib, J. Li and L.T.J. Delbaere, *Nucleosides Nucleotides*, 1995, **14**, 1675.

21   J. Plavec, C. Thibaudeau, G. Viswanadham, C. Sund, A. Sandström and J. Chatto-padhyaya, *Tetrahedron*, 1995, **51**, 11775.

22   S.N. Rao, *Nucleosides Nucleotides*, 1995, **14**, 1179.

23   D. Galisteo, J.A. Lopez Sastre, H. Martinez García and R. Nunez Miguel, *J. Mol. Struct.*, 1995, **350**,147 (*Chem. Abstr.*, 1995, **123**, 228 754).

24   J. Molina Molina and M. Rodriguez Espinosa, *Struct. Chem.*, 1994, **5**, 155 (*Chem. Abstr.*, 1995, **122**, 10 452).

25   F. Verberckmoes and E.L. Esmans, *Spectrochim. Acta*, 1995, **51A**, 153 (*Chem. Abstr.*, 1995, **122**, 214 421).

26   V.L. Worthington, W. Fraser and C.H. Schwalbe, *Carbohydr. Res.*, 1995, **275**, 275.

27   A. Grouiller, V. Uteza, I. Komaromi and J.M.J. Tronchet, *J. Carbohydr. Chem.*, 1995, **14**, 1387.

28   T. Wada, K. Inageda, K. Aritomo, K. Tokita, H. Nishina, K. Takahashi, T. Katada and M. Sekine, *Nucleosides Nucleotides*, 1995, **14**, 1301.

29   J.E. van den Boogaart, C.J.M. Huige, J.H. Boal, W. Egan and C. Altona, *Recl. Trav. Chim. Pays-Bas*, 1995, **114**, 79.

30   A. De Bruyn, *J. Carbohydr. Chem.*, 1995, **14**, 135.

31   M. Adinolfi, M.M. Corsaro, C. De Castro, A. Evidente, R. Lanzetta, L. Mangoni and M. Parilli, *Carbohydr. Res.*, 1995, **274**, 223.

32   A.U. Rehman and M.Y. Khan, *Sci. Int. (Lahore)*, 1994, **6**, 133 (*Chem. Abstr.*, 1995, **123**, 257 143).

33   I. Tvaroska and J.P. Carver, *J. Phys. Chem.*, 1995, **99**, 6234 (*Chem. Abstr.*, 1995, **122**, 265 806).

34   F. Zuccarello and G. Buemi, *Carbohydr. Res.*, 1995, **273**, 129.

35   S.E. Barrows, F.J. Dulles, C.J. Cramer, A.D. French and D.G. Truhlar, *Carbohydr. Res.*, 1995, **276**, 219.

36   I. Tvaroska and J. Gajdos, *Carbohydr. Res.*, 1995, **271**, 151.

37   E.Q. Morales, J.I. Padrón, M. Trujillo and J.T. Vázquez, *J. Org. Chem.*, 1995, **60**, 2537.

38   D. Mentzafos, M. Polissiou, I. Grapsas, A. Hountas, M. Georgiadis and A. Terzis, *J. Chem. Crystallogr.*, 1995, **25**, 157 (*Chem. Abstr.*, 1995, **123**, 286 411).

39   L.K Prikhodchenko, T.N. Koilosova, R.G. Zhbankov and R.E. Teeyaer, *Zh. Prikl. Spektrosk.*, 1994, **60**, 305 (*Chem. Abstr.*, 1995, **122**, 161 073).

40   L. Lantanowicz, E.C. Reynhardt, R. Utrecht and W. Medycki, *Ber. Bunsen-Ges.*, 1995, **99**, 152 (*Chem. Abstr.*, 1995, **122**, 265 804).

41   C.A. Podlasek, J. Wu, W.A. Stripe, P.B. Bondo and A.S. Serianni, *J. Am. Chem. Soc.*, 1995, **117**, 8635.

42   R. El Bergmi and J.M. Molina, *J. Chem. Res.*, 1995, (S)484; (M)2947.

43   T.B. Grindley, A. Cude, J. Kralovic and R. Thangarasa, *Front. Biomed. Biotechnol.*, 1994, **2**, 147 (*Chem. Abstr.*, 1995, **122**, 265 847).

44   H.P. Wessel and S. Bartsch, *Carbohydr. Res.*, 1995, **274**, 1.

45   Y. Du and F. Kong, *Carbohydr. Res.*, 1995, **275**, 259.

46   J. Molina Molina, D.P. Olea and H.H. Baer, *Carbohydr. Res.*, 1995, **273**, 1.

47   J.E. Andersson, S.J. Angyal and D.C. Craig, *Carbohydr. Res.*, 1995, **272**, 141.

48   J. Tang, R.D. Tan, X.L. Ding, G.J. Tao and S.D. Ding, *Spectrosc. Lett.*, 1995, **28**, 261, (*Chem. Abstr.*, 1995, **122**, 291 359).

49   C. Fringart, I. Tvaroska, K. Mazeau, M. Rinaudo and J. Desbrieres, *Carbohydr. Res.*, 1995, **278**, 27.

50   J. Dabrowski, T. Kozar, H. Grosskurth and N.E. Nifant'ev, *J. Am. Chem. Soc.*, 1995, **117**, 5534.

51   V. Pozsgay and B. Coxon, *Carbohydr. Res.*, 1995, **277**, 171.

52   B.J. Hardy, W. Egan and G. Widmalm, *Int. J. Biol. Macromol.*, 1995, **17**, 149 (*Chem. Abstr.*, 1995, **123**, 340 604).

53   M. Martín-Pastor, J.L. Asensio, R. López and J. Jiménez-Barbero, *J. Chem. Soc., Perkin Trans. 2*, 1995, 713.

54   B.A. Spronk, A. Rivera-Sagredo, J.P. Kamerling and J.F.G. Vliegenthart, *Carbohydr. Res.*, 1995, **273**, 11.

55   M.K. Dowd, A.D. French and P.J. Reilly, *J. Carbohydr. Chem.*, 1995, **14**, 589.

56   R.J. Woods, B. Fraser-Reid, R.A. Dwek and C.J. Edge, *ACS Symposium Ser.*, 1994, **669**, 252 (*Chem. Abstr.*, 1995, **122**, 81 812).

57   J.F. Bergamini, C. Boisset, K. Mazeau, A. Heyraud and F.R. Taravel, *New J. Chem.*, 1995, **19**, 115 (*Chem. Abstr.*, 1995, **122**, 265 820).

58   C. Gouvion, K. Mazeau and I. Tvaroska, *J. Mol. Struct.*, 1995, **344**, 157 (*Chem. Abstr.*, 1995, **122**, 265 873).

59   S. Immel and F.W. Lichtenthaler, *Liebigs Ann. Chem.*, 1995, 1925.

60   J. Shen and L.E. Lerner, *Carbohydr. Res.*, 1995, **273**, 115.

61   U.R. Desai, J.R. Vlahov, A. Pervin and R.J. Linhardt, *Carbohydr. Res.*, 1995, **275**, 391.

62   A. Khatoon, N. Akhtar, I.M. Kidwai and M.A. Haleem, *Pak. J. Sci. Ind. Res.*, 1993, **36**, 517 (*Chem. Abstr.*, 1995, **122**, 187 913).

63   J.L. Asensio and J. Jimenez-Barbero, *Bioploymers*, 1995, **35**, 55 (*Chem. Abstr.*, 1995, **122**, 291 355).

64   J.L. Asensio, M. Martín Pastor and J. Jiménez-Barbero, *Int. J. Biol. Macromol.*, 1995, **17**, 137 (*Chem. Abstr.*, 1995, **123**, 257 157).

65   N. Bouchemal-Chibani, I. Braccini, C. Derouet, C. Hervé du Penhoat and
      V. Michon, *Int. J. Biol. Macromol.*, 1995, **17**, 177 (*Chem. Abstr.*, 1995, **123**, 340 676).

66   E.S. Stevens, *Biopolymers*, 1994, **34**, 1403, (*Chem. Abstr.*, 1995, **122**, 31 789).

67   L. Maeler, J. Lang, G. Widmalm and J. Kowalewski, *Magn. Reson. Chem.*, 1995, **33**,
      541 (*Chem. Abstr.*, 1995, **123**, 257 155).

68   D.L. Clouser and P.C. Jurs, *Carbohydr. Res.*, 1995, **271**, 65.

69   T.J. Rutherford and S.W. Homans, *J. Magn. Reson., Ser. B*, 1995, **106**, 10 (*Chem.
      Abstr.*, 1995, **122**, 291 378).

70   G. Torri, *Chem. Pap.*, 1994, **48**, 211 (*Chem. Abstr.*, 1995, **122**, 240 281).

71   M. Hricovini and G. Torri, *Carbohydr. Res.*, 1995, **268**, 159.

72   W. Chai, E.F. Hounsell, C.J. Bauer and A.M. Lawson, *Carbohydr. Res.*, 1995, **269**,
      139.

73   A.L. Davies, R.A. Hoffmann, A.L. Russell and M. Debet, *Carbohydr. Res.*, 1995,
      271, 43.

74   E. Katzenellenbogen, E. Romanowska, N.A. Kocharova, A.S. Shashkov, Y.A.
      Knirel and N.K. Kochetkov, *Carbohydr. Res.*, 1995, **273**, 187.

75   L. Lai and Y. Yang, *Shengwu Huaxue Yu Shengwu Wuli Jinzhan*, 1995, **22**, 290
      (*Chem. Abstr.*, 1995, **123**, 257 138).

76   L. Toma, D. Colombo, F. Roncchetti, L. Panza and G. Russo, *Helv. Chim. Acta*,
      1995, **78**, 636.

77   H.P. Wessel and G. Englert, *J. Carbohydr. Chem.*, 1995, **14**, 179.

78   P. Balaji, P.K. Qasba, K. Pradman and V.S.R. Rao, *Glycobiology*, 1994, **4**, 497
      (*Chem. Abstr.*, 1995, **122**, 31 800).

79   A.J. Woods, C.J. Edge, M.R. Wormald and R.A. Dwek, *Alfred Benzon Symp.*, 1993
      (Pub. 1994), **36**,15 (*Chem. Abstr.*, 1995, **122**, 161 102).

80   K.S. Novinski and K. Ribka, *Acta Biochim. Pol.*, 1994, **41**, 216 (*Chem. Abstr.*, 1995,
      122, 187 910).

81   P.K. Qasba, P.V. Balaji and V.S. R. Rao, *Glycobiology*, 1994, **4**, 805 (*Chem. Abstr.*,
      1995, **122**, 240 265).

82   M.G. Petukhov, A.K. Mazur and L.A. Elyakova, *Carbohydr. Res.*, 1995, **279**, 41.

83   A. Bernardi and L. Raimondi, *J. Org. Chem.*, 1995, **60**, 3370.

84   J.C. Rodgers and P.S. Portoghese, *Biopolymers*, 1994, **34**, 1311, (*Chem. Abstr.*, 1995,
      122, 31 810).

85   K. Scheffler, B. Ernst, A. Katopodis, J.L. Magnani, W.T. Wang, R. Weisemann and
      T. Peters, *Angew. Chem., Int. Ed. Engl.*, 1995, **34**, 1841.

86   Y. Miyaki, O. Nakamura, Y. Sashida, K. Koike, T. Nikaido, T. Ohmoto,
      A. Nishino, Y. Satoni and H. Nishino, *Chem. Pharm. Bull.*, 1995, **43**, 971.

87   N.P. Sahu, K. Koike, J. Jia and T. Nikaido, *Tetrahedron*, 1995, **51**, 13435.

88   I. Bako and L. Jicsinszki, *J. Inclusion Phenom. Mol. Recognit. Chem.*, 1994, **18**, 275
      (*Chem. Abstr.*, 1995, **122**, 187 934).

89   P. Berthault, D. Duchesne, H. Desvaux and B. Gilquin, *Carbohydr. Res.*, 1995, **276**,
      267.

90   K. Fujita, H. Shimada, K. Ohla, Y. Nogamiu, K. Nasu and T. Koga, *Angew. Chem.,
      Int. Ed. Engl.*, 1995, **34**, 1621.

91   S. Houdier and P.J.A. Vottero, *Carbohydr. Lett.*, 1995, **1**, 13.

92   I. Andre, K. Mazeau, F.R. Taravel and I. Tvaroska, *Int. J. Biol. Macromol.*, 1995,
      17, 1897 (*Chem. Abstr.*, 1995, **123**, 340 621).

93   A. Otter, O. Hindsgaul and D.R. Bundle, *Carbohydr. Res.*, 1995, **275**, 381.

94   A. Blasko, C.A. Bunton, E. Moraga, S. Bunel and C. Ibarra, *Carbohydr. Res.*, 1995,
      278, 315.

95 J.R. Cox and E.H. Serpersu, *Carbohydr. Res.*, 1995, **271**, 55.
96 R. Barthwal, N. Srivastava, U. Sharma and G. Govil, *J. Mol. Struct.*, 1994, **327**, 201 (*Chem. Abstr.*, 1995, **122**, 106 313).
97 J.A. Parkinson, I.H. Sadler, M.B. Pickup and A.B. Tabor, *Tetrahedron*, 1995, **51**, 7215.
98 A. Awan, R.J. Brennan, A.C. Regan and J.Barber, *J. Chem. Soc., Chem. Commun.*, 1995, 1653.
99 I. Waver, B. Piekarska-Bartoszewicz and A. Temeriusz, *Carbohydr. Res.*, 1995, **279**, 83.
100 P. Strazewski, *Helv. Chim. Acta*, 1995, **78**, 1113.
101 R. Challoner, J. Kuemmerle and C.A. McDowell, *Mol. Phys.*, 1994, **83**, 687 (*Chem. Abstr.*, 1995, **122**, 133 546).
102 S. Chapelle and J.-F. Verchère, *Carbohydr. Res.*, 1995, **266**, 161.
103 S. Chapelle and J.-F. Verchère, *Carbohydr. Res.*, 1995, **277**, 39.

22
Other Physical Methods

1   IR Spectroscopy

FTIR reflection spectroscopy has been employed in a structural study of D-glucose, D-fructose and sucrose in aqueous solution. Mono- and disaccharide spectra evidenced inter- and intramolecular interactions, 'dimerization effect' of monosaccharides and conformational equilibria.[1] Molecular mechanics calculations of the vibrational spectra of the conformers of methyl 2,6-di-O-acetyl-3,4-anhydro-α-DL-talopyranoside-6,6-^{2}H$_2$ and its galactopyranoside analogue indicate vibrations in the 400-850 cm^{-1} region which show the largest ring conformation dependent shifts.[2] Experimental IR and Raman spectra agree (to an average error of 4 cm^{-1}) with calculations for N-acetyl-α-D-muramic acid and N-acetyl-β-D-neuraminic acid (in the crystalline state).[3] Molecular mechanics calculations of the IR and Raman spectra of α- and β-D-glucose showed significant differences from experimental spectra.[4]

Several studies of carbonyl group migration through enediol intermediates have been reported. FTIR of reducing sugars revealed two sets of temperature sensitive bands in the carbonyl (1700-1750 cm^{-1}) and alkene regions, the latter assigned to the enediol from acyclic aldehyde and ketone sugars.[5] FTIR of several pentuloses, pentoses and hexuloses showed bands at 1733, 1726 and 1717 cm^{-1}, assigned to 3-keto, 2-keto and aldehyde forms. Tetruloses showed only the bands at 1733 and 1717 cm^{-1}, while the corresponding 2-deoxyaldoses showed only one peak. Sugars exhibiting carbonyl migration showed an alkene band at 1650 cm^{-1} (from enediol intermediate) which, like the carbonyl bands, was also pH and temperature sensitive.[6] Vibrational (Raman) optical activity (ROA) spectra of D-fructose, D-sorbose, D-tagatose and D-psicose in aqueous solution, combined with data from previous studies, provide information about anomeric configuration, exocyclic CH$_2$OH orientation and relative disposition of the ring hydroxyls.[7]

FTIR, along with Mössbauer spectroscopy and thermogravimetry, has been used to indicate that dibenzyltin(IV) dichloride reacts with sugars to give tin(IV) oxide-containing products along with dibenzyltin(IV) sugar complexes.[8]

The use of vibrational spectra in oligosaccharide studies has been furthered this year. For a set of six disaccharides with different types of glycosidic linkages (α,α-trehalose, sophorose, laminaribiose, maltose, cellobiose and gentiobiose) IR and Raman spectra show good agreement with calculated frequencies below 1500 cm^{-1}. No additional terms are needed in the potential energy calculations to express the exo-anomeric effect.[9] Another study of eight oligosaccharides in the crystal state using FTIR in the 1500-100 cm^{-1} range and Raman between 1500-600 cm^{-1}, provided information relating to stereochemistry and the nature of the glycosidic linkages.[10] An analysis of the sulfate region of the IR and Raman spectra of heparin from cattle lung mucus was reported.[11]

IR has been used to characterize the carbohydrate α-amino acids 2-amino-2-deoxy-D-*glycero*-D-*talo*-heptonic acid and 2-amino-2-deoxy-D-*glycero*-L-*gluco*-heptonic acid.[12]

Polarized reflection spectra from two faces of 2'-deoxyadenosine single crystals indicate that the deoxyribose group has little effect on the adenine chromophore.[13] IR and Raman spectroscopy have been used for structural and conformational studies of phosphonylmethyl analogues of diribonucleoside monophophates.[14] The relative humidity-induced structural transitions of guanosine and disodium ATP have been monitored by low frequency Raman spectroscopy. Five well defined states of the former and four of the latter are seen, with stacked bases forming long chains. The transitions are reversible.[15] The use of IR to study polymorphism of erthythromycin has been reported.[16] An IR method for determining the crystal structure of α-D-galactopyranose has been described.[17]

2   Mass Spectrometry

Low resolution EI mass spectral studies of pyrimidine, cytosine and derivatives of cytosine and isocytosine has been reviewed [54 refs].[18]

Fast atom bombardment (FAB) collison-induced dissociation mass spectra (CID-MS) of the [M-H]$^-$ ions of methyl 2-, 3-, 4- and 6-deoxy-D-galactopyranosides and related compounds, indicate that charge localization in fragment ions strongly directs fragmentation reactions.[19] Ion cyclotron resonance (ICR) MS has been used to measure gas phase affinities of Me$_3$Si$^+$ (guest) and a series of permethylated monosaccharides and simple crown ethers (hosts).[20] Gas phase reactions of α- and β-glycosyl fluorides with tetramethylsilane generate [M+SiMe$_3$]$^+$ ions which are more stable than the corresponding [M+H]$^+$ ions.[21] Negative ion FAB-MS allowed the differentiation of three isomeric deoxyfluorinated sugars, the peculiar fragmentations observed also allowing investigation of the role of the F in the decomposition processes.[22] Positive ion FAB-MS can distinguish permethylated aldo- and pseudo-aldobiouronic acids; CID spectra of the [M+H]$^+$ ions and [M+H-MeOH]$^+$ ions allow identification of the type of interglycosidic linkage.[23] FAB-MS of disaccharide analogues of muramyl dipeptide have been reported.[24] Surface-induced dissociation spectra of four isomers of mannosyl-α(1→Y)-mannose (Y=2-4, 6) on a 2-(perfluorooctyl)ethanethiol monolayer gold surface were reported.[25]

This year has seen a number of mass spectrometric applications to oligosaccharide characterizations. Liquid secondary ion mass spectrometry (LSIMS) has been used in characterization of tetra-, hexa- and octasaccharides derived from porcine intestinal heparin.[26] Laser desorption time-of-flight MS of heparin fragments complexed with (Arg-Gly)$_{10}$ and (Arg-Gly)$_{15}$ overcomes the problem of weak signals from the uncomplexed saccharides.[27] For decasaccharides and larger, complexation with the protein angiogenin (MW 14,120) was employed.

Positive ion FAB tandem MS provides a means of differentiating positional isomers of monosulfated disaccharides derived from heparin and heparin sulfate, and was also applied to di- and trisulfated disaccharides.[28] Electrospray MS was used to observed the non-covalent complex between human antithrombin III (which inhibits thrombin and is activated by heparin) and the heparin pentasaccharide recognition determinant.[29] Sequence determinations have been made for trifluoroacetylated plant glycosides.[30] HPLC-MS using positive ion FAB of malto-oligosaccharides (4-10) reductively aminated with 4-hexadecylaniline has been described.[31] Thermospray and continuous FAB HPLC-MS of plant extract glycosides (of 1-8 sugars) provide complimentary information, giving mass, number of sugars, type of aglycone amd sometimes sugar sequence.[32] Electrospray tandem-MS allows structural characterization of underivatized polysialo-gangliosides.[33]

The $1 \rightarrow 2$, $1 \rightarrow 3$ or $1 \rightarrow 4$ linkage type of xylobioses has been distinguished using unimolecular decomposition spectra (MIKE) of the $[M+NH_4]^+$ ions or CID-MS of $[M+MeNH_3]^+$ ions or permethylated derivatives.[34]

CID-MS of permethyl or peracetyl derivatives of the products of three isomeric (linear/branched) pentasaccharides reductively aminated with tri-methyl(*p*-aminophenyl)ammonium chloride (and thus bearing a preformed charge) yields exclusively reducing end containing fragment ions.[35] A combination of mass spectral techniques, positive and negative ion FAB, electrospray and low energy tandem MS have been examined on the underivatized, *N*-acetylated-permethylated and permethylated core oligosaccharide of *Aeromonas hydrophilia* (chemotype III).[36]

A matrix of 2,5-dihydroxybenzoic acid and 1-hydroxyisoquinoline (wt ratio of 3:1) proved most suitable for MALDI-MS of oligosaccharides.[37] A new study of the use of electrospray ionization MS of cyclodextrins and guests shows that some amino acids least likely to form inclusion complexes show the most intense complex ions. This suggests previous data identifying complexes in this way are probably due to electrostatic adducts formed in the electrospray process.[38]

Mass spectrometry has also seen a number of applications to glycoconjugates. Electrospray mass spectra have been used to obtain molecular weight, sequence, linkage and branching data for glycolipids and glycoprotein glycans.[39] Low and high energy CID-LSIMS have been used for structural analysis of mono- and di-sulfated glycosphingolipids. The method can identify the position and number of sulfates in samples of less than 1 nmol.[40] LSIMS and MALDI-TOF have also been used to study glycosphingolipids as native compounds, or permethylated, peracetylated or perbenzoylated derivatives, the latter giving sensitivity advantages of at least two orders of magnitude. Fragmentation of derivatives differed, with permethylated compounds favouring ceramide derived ions, and peracylated compounds enhancing production of carbohydrate derived ions.[41] Electrospray MS of methylated glycosphingolipids provided more structurally diagnostic fragments than high energy tandem MS.[42] A study of a range of matrices for MALDI-TOF analysis of sphingo- and glycosphingolipids was reported.[43]

Mass spectral methods for the structural characetrization of flavonoid glyco-sides have been discussed.[44] Negative ion frit-FAB LC-MS was used to charac-terize the structures of the glycosidic antibiotics of the glykenin family produced by *Basiciomycetes* spp. Molecular weights and the exact acetyl locations in the sugar moieties were provided.[45] CID-MS (He or Argon collision gases) of a number of carbohydrate antibiotics (RMM of 700-1500) indicate that cationized antibiotics give a higher yield of diagnostically useful high mass ions that do protonated antibiotics.[46] The exact positions of acetyl groups of glykenin glycosidic antibiotics (from *Basidiomycetes* spp.) were determined using a MS/MS technique under high energy collision conditions.[47] Fragmentation patterns of *O*-acetylated 4-*O*-methyl- and 6-acetamido- and 3,4-, 2,4- and 2,3-diacet-amido-pyranosides confirm the location of the acetamido groups.[48] Three general fragmentation pathways of 46 new 2-amino-*N*-alkyl-4,6-*O*-benzylidene and *N,N*-dialkyl-2-deoxy-D-hexopyranosides were identified in EI-MS.[49]

Solid SIMS has proved useful in the study of acetylated precursors of a series of glycosylated porphyrins (solid SIMS spectra are not complicated by chemical noise or matrix reactions).[50]

Capillary HPLC-MS employing a new 'in source' thermospray-type interface is suitable for low RMM polar compounds, with adenosine being a demonstration example.[51] A quantitative FAB investigation of four RNA nucleosides indicates H-bonded complexes form primarily at the liquid-matrix interface.[52] Electrospray tandem-MS has proved useful for characterization of AZT and a series of 3'-azido-2',3',4'-trideoxy-5-halogenated-4'-thio-uridine analogues.[53] A study of ion-trap collisional activation of deprotonated deoxymononucleoside monophos-phates and deoxydinucleotide monophosphates shows that loss of base in monophosphates occurs at a lower threshold for 3'- than 5'-phosphates. Loss of charged base from dinucleotides is highly dependent on the identity of both bases.[54] In another study, several mass spectral techniques provide an order of pair stability for H-bonded nucleoside dimers.[55]

There was a report on solid state MS of cyanocobalamine, glucose and raffinose on a disperse oxide surface.[56]

3   X-ray and Neutron Diffraction Crystallography

Specific X-ray crystal structures have been reported as follows (solvent molecules of crystallization are frequently not recorded).

3.1   Free Sugars and Simple Derivatives Thereof. – A comparison of calculated low energy crystal structures with observed structures for α- and β-D-glucose, α- and β-D-galactose, α-D-talose and β-D-allose, provided some examples in which the observed structures were more than 20 kJ mol^{-1} higher in energy than the calculated lowest energy structures.[57] Dehydrated lactose exists in a hygroscopic α_H form whose X-ray structure is the same as that of the monohydrate. A thermally induced transition converts this into a stable α_S form.[58]

3.2   Glycosides, Disaccharides and Derivatives Thereof. – Penta-O-benzoyl-α-D-fructofuranose adopts an E_2 configuration.[59] An X-ray of methyl α-D-arabinofuranoside bound to concanavalin A has been reported.[60]

The O-glycosides: Benzyl 2-acetamido-4-azido-3-O-benzoyl-6-O-(*tert*-butyldiphenylsilyl)-2,4-dideoxy-β-D-glucopyranoside,[61]   2-chlorophenyl   3,4,6-tri-O-benzyl-2-deoxy-2-methylene-β-D-glucopyranoside[62] and benzyl 3,4,6-tri-O-mesyl-β-D-glucopyranoside (see Chap.8 for synthesis).[63] The glycosides bearing long chain C6 O-acyl groups: methyl 6-O-n-octanoyl-β-D-glucopyranoside, methyl 6-O-n-decanoyl-β-D-glucopyranoside[64] and the two α-D-galactopyranoside analogues.[65] X-Ray analysis has contributed to studies of the C5-C6 rotational isomerism of 6-substituted methyl 2,3,4-tri-O-acetyl-α-D-glucopyranosides.[66]

Compounds containing less common sugars: Methyl 2-O-(cyclohexylcarbamoyl)-6-deoxy-3,4-O-hexafluoroisopropylidene-α-L-altropyranoside,   co-crystallized with 1,1,1,7,7,7-hexafluoro-2,6-dihydroxy-2,6-bis(trifluoromethyl)-4-heptanone.[67] X-Ray analysis of methyl 4,6-O-benzylidene-3-O-methyl-α-D-altropyranoside showed $^1C_4$ conformation and a single type of H-bond, OH(2)$\cdots$O(1);[68] methyl   2,2'-anhydro-4,6-O-benzylidene-3-deoxy-3-C-(2-hydroxyethyl)-α-D-allopyranoside;[69] 16-β-acetoxy-3-β-[2,6-dideoxy-3-O-methyl-α-L-*arabino*-hexopyranosyloxy]-14-hydroxycard-20(22)-enolide.[70]

Peracetylated derivatives: the plant hormone-related 2-(indol-3-yl)-ethyl and 2-phenylethyl β-D-xylopyranosides **1** and **2** respectively;[71] 2-(indol-3-yl)ethyl β-D-galactopyranoside tetraacetate, related to the plant hormone auxin. Structural analysis was also assisted by ^{1}H NOE NMR experiments;[72] peracetates of (2S)-butyl 6-O-(β-D-apiofuranosyl)-β-D-glucopyranoside glycosides isolated from cassava (*Manihot esculenta*).[73]

The disaccharides α-D-galactopyranosyl α-D-galactopyranoside,[74] α-D-allopyranosyl   α-D-allopyranoside,[75] 3,3',4,4'-tetra-O-benzoyl-2,2',6,6'-tetradeoxy-α,α-ribotrehalose **3**,[76] 2,3,4,6,1',3',4',6'-octa-O-acetyl-β-sophorose, methyl 2,3,4,6,3',4',6'-hepta-O-acetyl-β-sophoroside, methyl 2,3,4,6,3',4'-hexa-O-acetyl-6'-deoxy-β-sophoroside[77] and methyl 2-O-(2,3,4-tri-O-acetyl-β-D-xylopyranosyl)-3-O-benzyl-4,6-O-benzylidene-α-D-mannopyranoside **4**.[78]

The trehalose derivatives 3-*O*-benzyl-4,6-*O*-benzylidene-2-deoxy-α-D-*ribo*-hexopyranosyl 4',6'-*O*-benzylidene-2'-deoxy-α-D-*ribo*-hexopyranoside[79] and 2,2',3,3'-tetra-*O*-benzoyl-6,6'-dideoxy-4,4'-di-*O*-mesyl-6,6'-dithiocyanato-α,α-trehalose;[80] di-*N*-acetylchitobiose complexed to turkey egg lysozyme.[81]

3.3 Higher Oligosaccharides and *C*-Glycosides.

– Following the previous section, not only di-, but tri- and tetrasaccharide complexes of rainbow trout lysozyme have been characterized by X-ray analysis.[82] The orientation of the (α1→6) mannosyl arm of a heptasaccharide and hexasaccharide (and disaccharide fragments) has been studied by molecular dynamics simulations and compared with protein-carbohydrate complex crystal structures.[83]

The crystal structure of methyl β-D-cellotrioside complexed with water and ethanol shows 4 independent molecules with an extended conformation. All rings are $^4C_1$ and each O6 primary hydroxyl group is in the *gt* conformation. The structure shows many common points with cellulose II and these data can thus be used to provide an improved structure of cellulose II.[84] Crystal structures of 1,6-anhydro-β-maltotriose nonaacetate **5** and 6''-bromo-6''-deoxy-1,6-anhydro-β-maltotriose octaacetate **6** show glycosidic conformations almost the same as other α(1→4) linked oligosaccharides, and weak intermolecular C-H···O H-bonds.[85] The crystal structure of Konjac mannan has been reported.[86] Small angle neutron scattering (along with light scattering) has been used for analysis of succinoglycan in solutions, the unexpected results being explained on the basis of the stereoregularity and semirigid nature of the polysaccharide.[87]

Several cyclodextrins and complexes have been the subject of X-ray studies. The azido and allyl groups of mono-6-azido-6-deoxy-α-CD and mono-2-*O*-allyl-α-CD are included in the cavity of the adjacent unit leading to a helical polymer.[88] The π-system in β-CD-but-2-yne-1,4-diol interacts with sugar ring CHs, while other ring CHs interact with the CH_2OH hydroxyl proton and one CH_2OH methylene proton interacts with an axial oxygen linker between the sugars.[89]

Hexakis-(2,6-di-*O*-methyl)-α-cyclodextrin crystallizes in two different forms when grown from 10% aqueous sodium hydroxide by slow evaporation of water at *ca.* 90 °C.[90] In the one form, the cyclodextrins are arranged in a helically extended polymeric chain formed by inclusion of an O6 methyl into the cavity of the adjacent molecule. The other form shows a cage-like packing, but the macrocycle conformation in both cases is nearly identical. The structure of the inclusion complex of hydroquinone with α-cyclodextrin has been determined.[91] The structure of heptakis-(2,6-di-*O*-methyl)-cyclomaltoheptose (dimethyl β-CD) is stabilized by intramolecular (O3)H···O(2) H-bonds between neighbouring glucose units.[92] The structures have been determined of tetra-*O*-acetylanhydroexfoliamycin **7** from *Streptomyces exfoliatus*;[93] the tetrabenzylated *C*-imidazoglycoside (see Chapter 3) **8** which is a precursor to a β-glycosidase inhibitor;[94] the bisacetylene linked glycoside **9**[95] and the diacetylene containing enediyne analogue **10**;[96] the *C*-linked sugars **11** and **12** derived by cycloaddition to the aldose nitroalkene (see Chapter 18).[97]

7

9

8

10

11

12

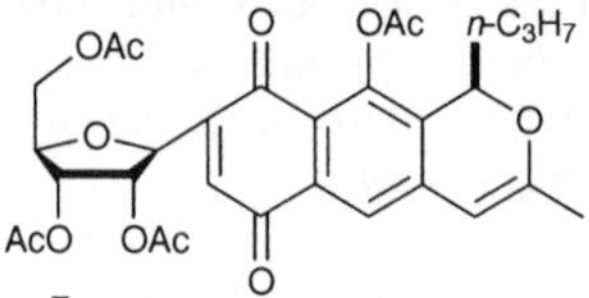

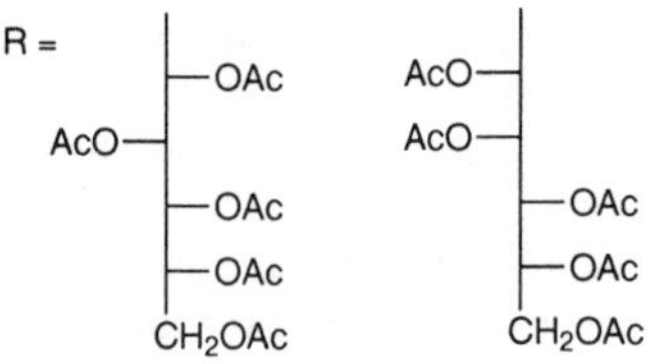

3.4 Anhydro-sugars. – The ^{13}C-labelled dimeric anhydride **13**,[98] 1,6-anhydride **14**,[99] spiroepoxide **15**,[100] 1,6-anhydro-4-*O*-benzyl-3-deoxy-2-*O*-methyl-β-D-*ribo*-hexopyranose and 1,6-anhydro-4-*O*-benzyl-2-*C*-(2-cyanoethyl)-2,3-dideoxy-β-D-*ribo*-hexopyranose.[101]

13 • = ^{13}C

14 R = Tbdps

15

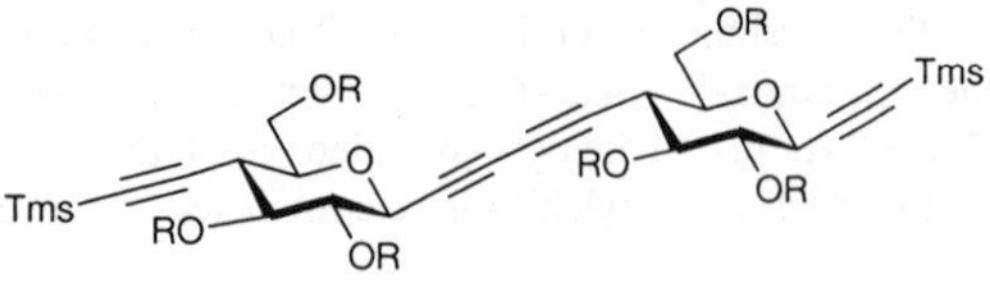

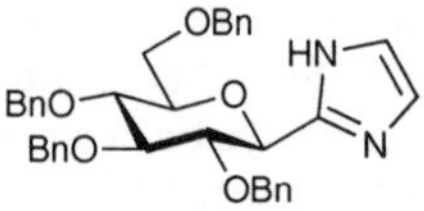

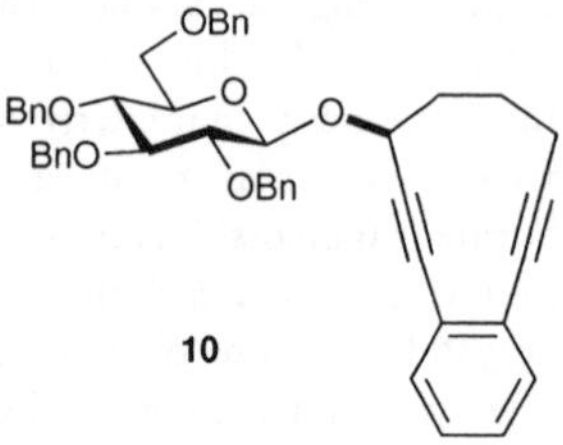

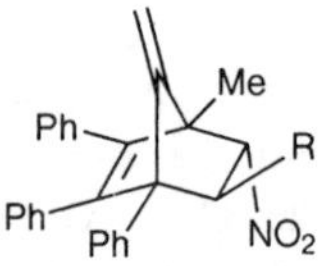

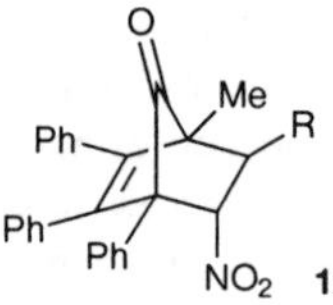

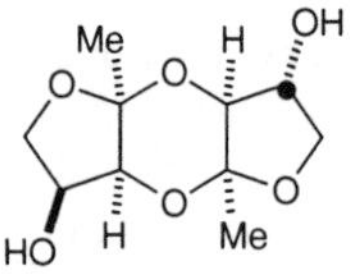

3.5 Nitrogen-, Sulfur- and Selenium-containing Compounds. – The peracetate (**16**) of the product of reaction of D-glucose with aminoguanidine,[102] the phenylsulfenamide **17**,[103] the imidazolidin-2-one **18**,[104] 3-*O*-β-D-glucuronosyl morphine,[105] three *N*-aryl-*N*-pentopyranosylamines *N*-*p*-nitrophenyl-*N*-(2,3,4-tri-*O*-acetyl-β-D-lyxopyranosyl)amine, *N*-acetyl-*N*-*p*-nitrophenyl-*N*-(2,3,4-tri-*O*-acetyl- α-L-arabinopyranosyl)amine and *N*-*p*-nitrophenyl-*N*-(2,3,4-tri-*O*-acetyl-α-L-arabinopyranosyl)amine,[106] the *N*-bromoiminolactone **19**[107] and *Z*-β-D-glucopyranosyloxy-*N*,*N*,*O*-azoxymethane.[108]

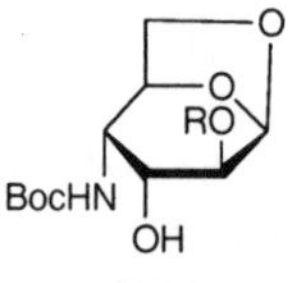

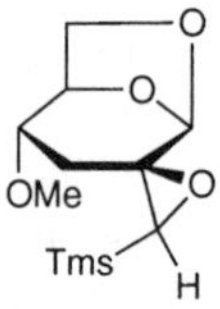

16
17
18
19

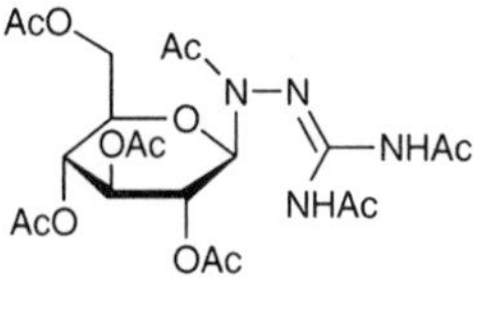

The β-D-fructofuranosylamine-derived oxazolidinone **20** in the ^{5}E configuration,[59] *N*-(*n*-octyl)-6-deoxy-D-gluconamide,[109] *N*-(1-octyl)-D-arabinonamide, *N*-(1-dodecyl)-D-ribonamide,[110] and the amide **21**,[111] the non-2-enoic acid ester **22** which is derived from a KDN derivative (see Chapter 16).[112]

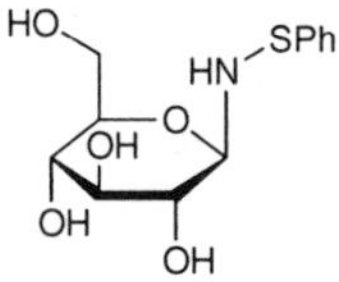

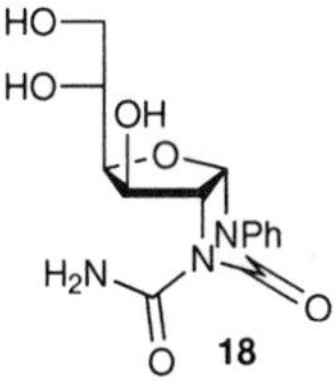

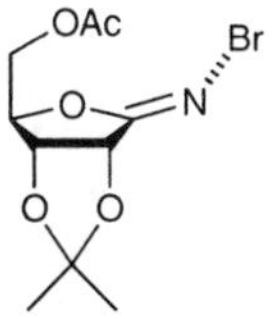

The terminal unit **23** of the *O*-specific polysaccharide of *Vibrio cholorae*, and the corresponding *N*-trifluoroacetate,[113] the *N*-linked L-tyrosine benzyl ester **24**,[114] and *N*-(1-deoxy-β-D-fructopyranos-1-yl)glycine.[115]

20
21
22
23
24
25
26

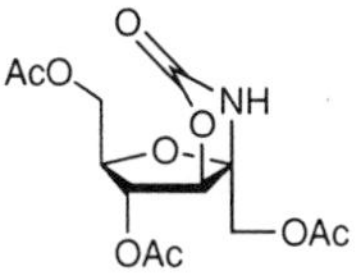

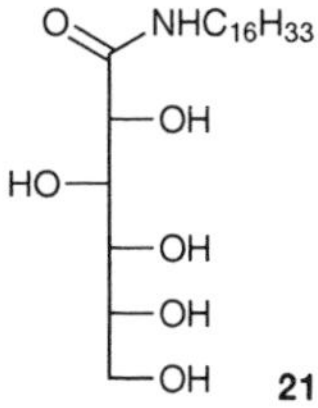

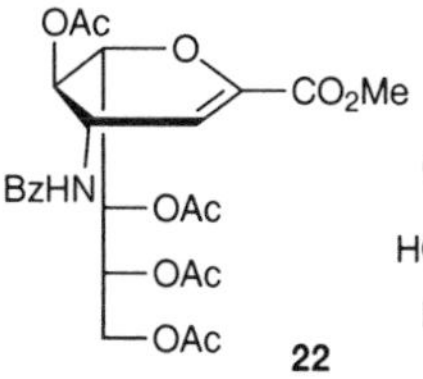

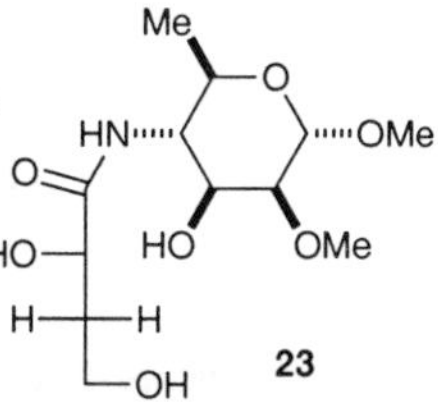

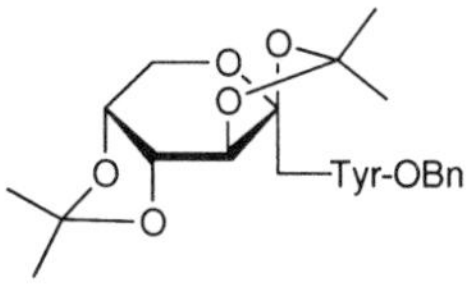

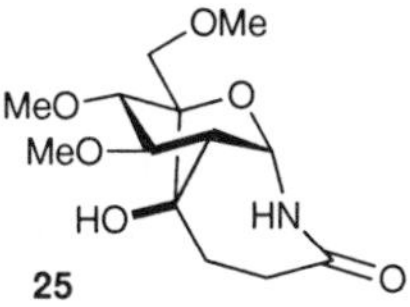

X-ray structures of the unusual tricyclic amido glycoside **25** and azepine **26** were reported.[116] Their synthesis by radical chemistry is described in Chap. 24. The *N*-hydroxypyrrolidine **27**[117] and the imine hetero-Diels-Alder adduct **28** (see Chapter 24).[118]

27
28
29
30

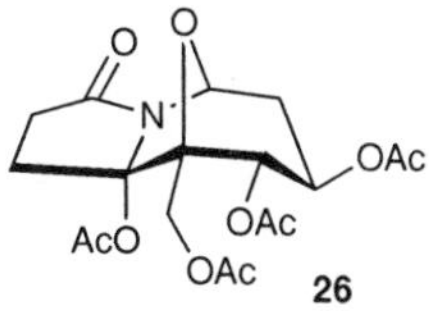

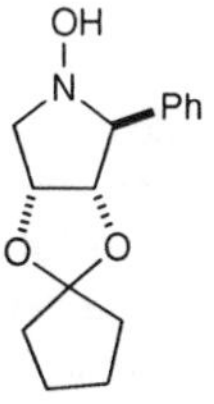

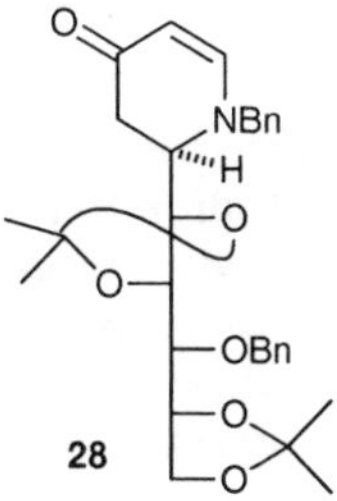

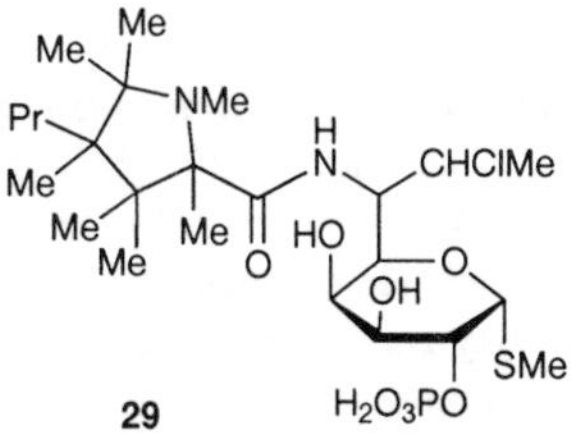

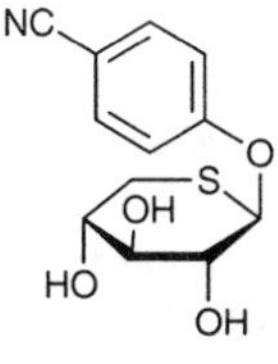

The thioglycoside antibiotic clindamycin-2-phosphate **29** has three separate molecules in the crystal, differing mainly in the orientation of the phosphate group and the pyrrolidinyl moiety.[119] Thiosugar glycoside **30**[120] and bis(2,3,4,6-tetra-*O*-acetyl-β-D-glucopyranosyl)diselenide and disulfide.[121]

The heterocycle-containing sugar derivatives **31**, **32**,[122] *C*-glycosidic thiophene **33**[123] and the thiazolidine pentacetate **34**,[124] this confirming a structure proposed over 65 years ago.

Sugar sulfates: 2-deoxy-2-(sulfamino)-α-D-glucopyranose sodium salt,[125,126] 2-amino-2-deoxy-D-glucopyranose-3-sulfate, 2-amino-2-deoxy-β-D-glucopyranose-6-sulfate[127] and 2-amino-2-deoxy-α,β-D-glucopyranose-3-hydrogensulfate, 2-amino-2-deoxy-α-D-glucopyranose-6-hydrogensulfate.[125]

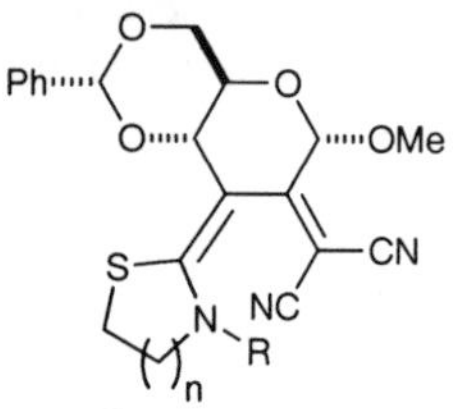

31 n =1 R = Bn
32 n =2 R = Bn

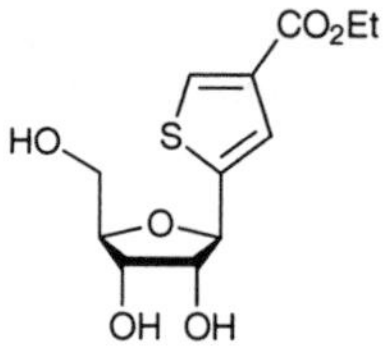

33

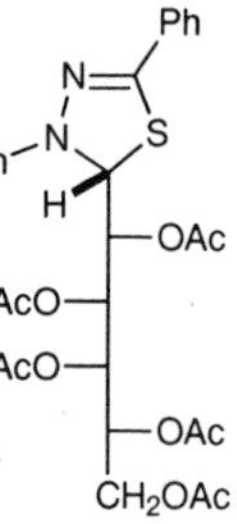

34

3.6   Branched-chain Sugars. – The oxazolidinone **35**,[128] and the novel fused cyclopentenone **36** (see Chapter 24 for synthesis).[129]

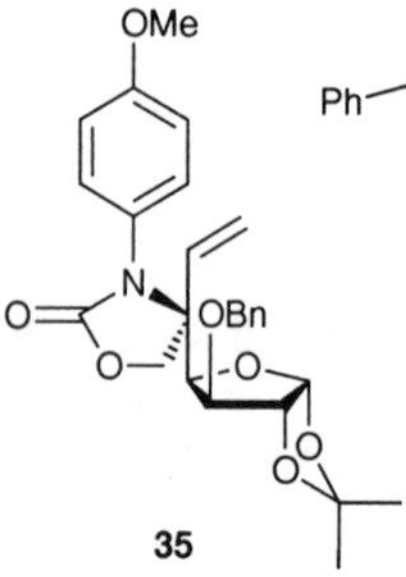

35

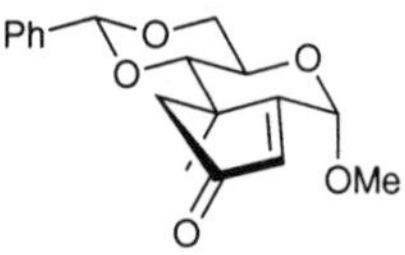

36

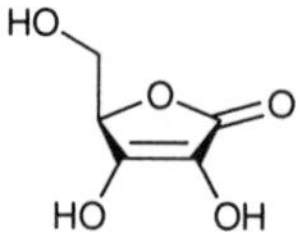

37

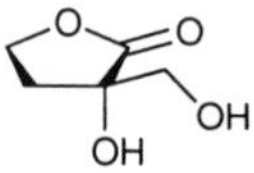

38

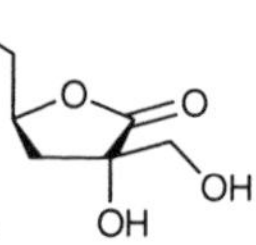

39

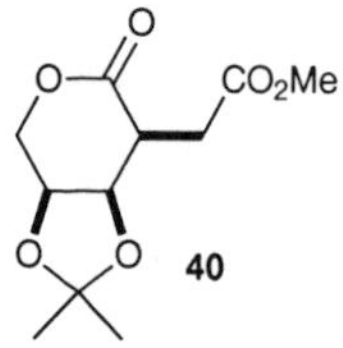

40

3.7   Sugar Acids and their Derivatives. – D-Erythroascorbic acid (2,3-didehydro-D-*glycero*-pentono-1,4-lactone) **37**,[130] 3-deoxy-2-*C*-hydroxymethyl-D/L-*erythro*-tetrono-1,4-lactone **38**, 3-deoxy-2-*C*-hydroxymethyl-D-*erythro*-pentono-1,4-lactone **39**,[131] lactone **40** (see Chapter 16),[132] brominated lactone **41** (see Chapter 8),[133] and 2,3-dideoxy-D-*erythro*-hex-2-enono-1,4-lactone and -1,5-lactone.[134] The neuraminic acid analogue **42**[135] and **43**,[136] an intermediate in a synthesis of a 4-guanidino carba-neuraminic acid analogue.

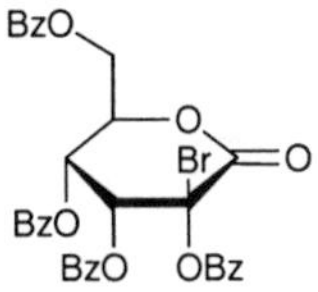

3.8 Inorganic Derivatives. – The X-ray structure of 6-deoxy-1,2-*O*-isopropylidene-6-triphenylstannyl)-α-D-glucofuranose shows tetrahedral tin, ring carbohydrate hydroxyl groups held together by intramolecular H-bonds and an intermolecular contact between one ring OH proton and an isopropylidene ring oxygen atom.[137]

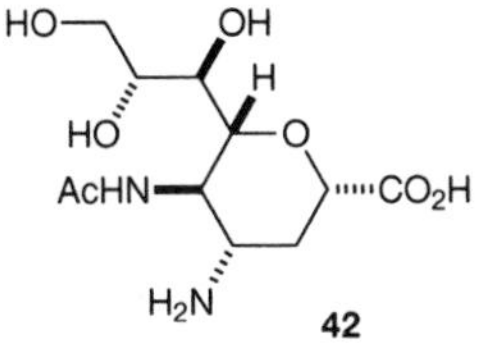

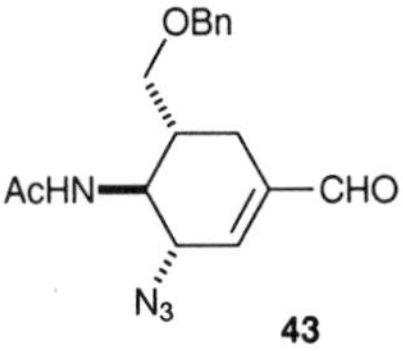

This year has seen continued interest in metal complexes of multiply deprotonated sugars. Five fold deprotonated mannose is a ligand in homoleptic dinuclear metalates of trivalent metals, and the crystal structure of one complex, **44** of formula $[Fe_2(\beta\text{-D-Man}f\text{H}_{-5})_2]^{4-}$ was reported.[138] The structure of the hexadecanuclear polymetalate of Cu(II) and multideprotonated D-sorbitol of formula $Li_8[Cu(D\text{-Sorb}\text{H}_{-6})(D\text{-Sorb1,2,3,4H}_{-4})_4]$ca.46H$_2$O **46** has a toroidal structure.[139] Vertex sharing rings of Cu_3O_3 are a major structural motif. This structure is comparable with the toroidal cyclodextrin lead complexes reported previously, though in that case the toroidal structure was ligand pre-determined. X-Ray structures of a series of $[M(tdci)_2]Cl_n.15H_2O$ (tdci=1,3,5-tri-deoxy-1,3,5-tris(dimethylamino)cis-inositol) complexes (**45**), with M=Al, Fe, Ga and In, show coordination exclusively through trisdeprotonated hydroxyl groups (and not amino groups).[140]

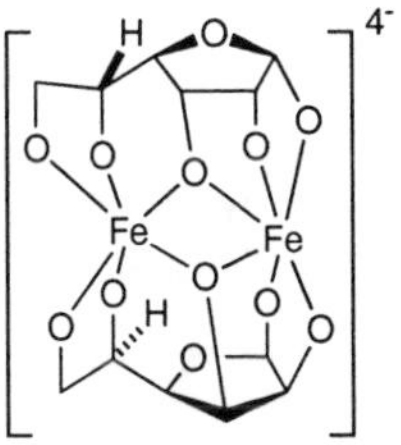

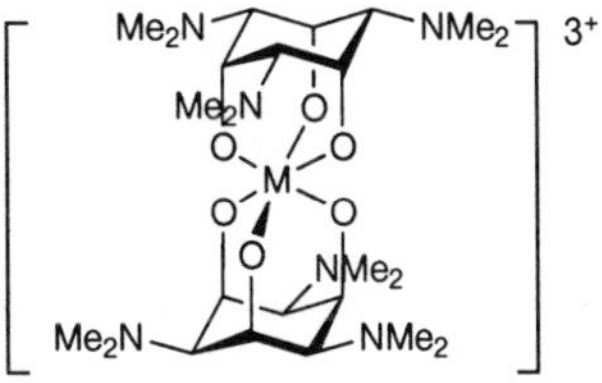

The structure of bis[6-amino-1,3-dimethyl-5-(2′-ethyl)-phenylazoniumuracil]-tris)thiocyanato-*S*)cuprate(I) has been obtained at 193K.[141]

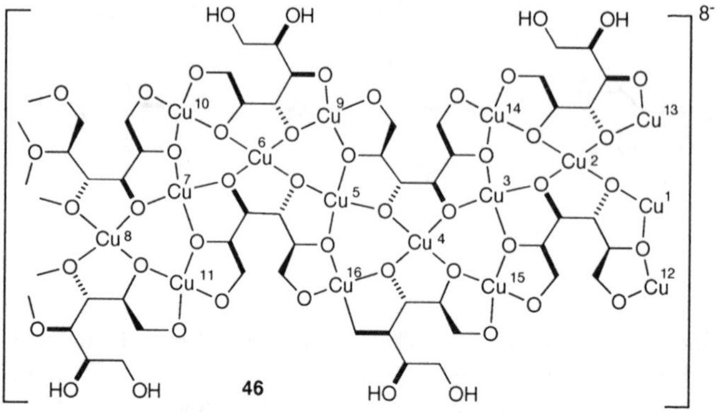

46

3.9 Alditols, Cyclitols and Derivatives Thereof

The structural determination of gualamycin **47** has been assisted by X-ray analysis of the aglycon methyl ester hydrochloride.[142]

3,4-Di-*O*-acetyl-2,5-anhydro-1,6-dideoxy-1,6-diiodo-D-mannitol **48**,[143] the sialyl Lewis X mimetic intermediate **49**,[144] and cyclopentane **50**.[145] The X-ray structure of the hexamethylether of *scyllo*-inositol shows that all six exocyclic C-O bonds are eclipsed with the corresponding ring C-H bonds (the methoxy group is constrained by the two adjacent, equatorial, methoxyl groups).[146] Structures of the glycosidase inhibitors **52** and **52**[147] and of the *N*-Boc-deoxynojirimycin **53** has been determined adopting a $^1C_4$ conformation.[148]

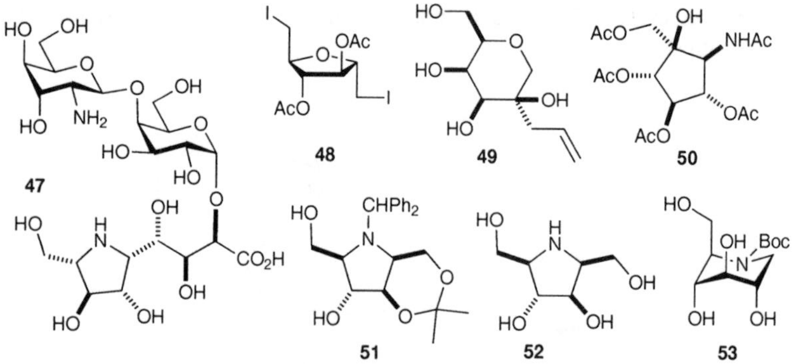

47   **48**   **49**   **50**   **51**   **52**   **53**

3.10 Nucleosides and their Analogues and Derivatives Thereof

5-Acetyl-1-(3,5-*O*-isopropylidene-α-D-xylofuranosyl)uracil monohydrate has the furanose ring in a C2'-endo, C3'-endo conformation and a glycosidic torsion angle of $-31°$.[149] X-ray structures of adenosine, cytidine, uridine and thymidine were interpreted in terms of C-H$\cdots$O interactions, while guanosine lacks these interactions.[150] Another structural analysis of cytidine was reported[151] and studies of crystal engineering of cytidine, 2'-deoxycytidine and their phosphonate salts were undertaken.[152]

Base analogues: The *N*-7 regioisomer of 2-chloro-2'-deoxyadenosine (**54**), *N*-[1-phenyl(2*R*)-prop-2-yl]-2-chloroadenosine,[153] 3-methyluridine, 1-methylinosine,[154] 2'-deoxy-5-ethyluridine,[155] the pentafluorophenyl base analogue **55**,[156] the 8*H*-pyrimido[4,5-*c*][1,2]-oxazin-7(6*H*)-one **56**,[157] spirocyclic bridged system **57**,[158] and 3,3-dimethyl-2-butene cycloadduct **58** (a crystalline mixture of diastereoisomers).[159]

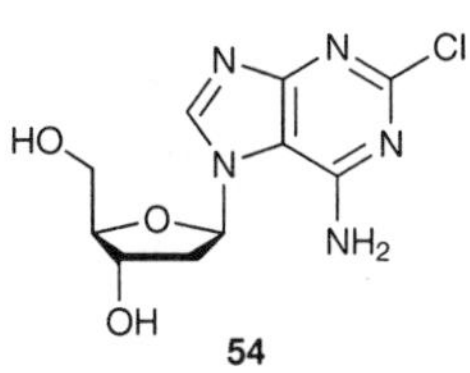

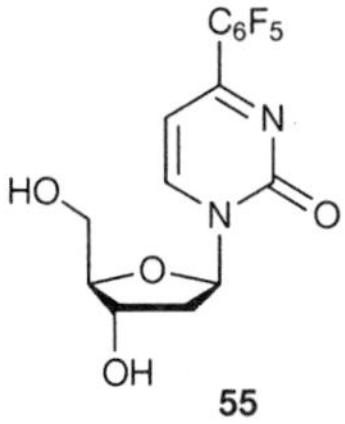

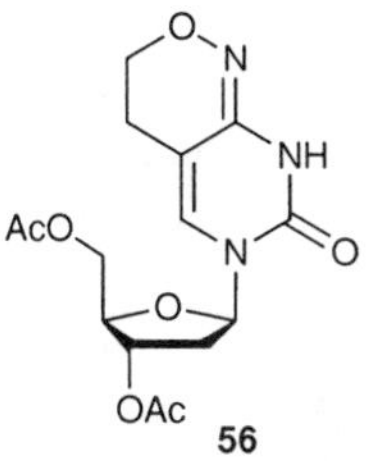

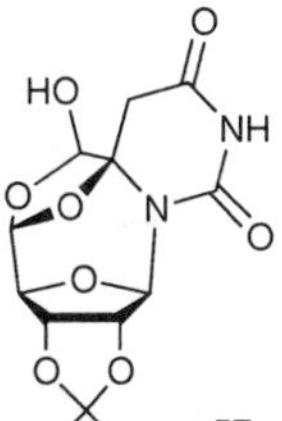

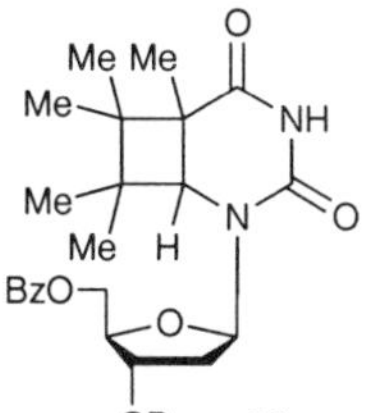

The X-ray of the 4'-azathymidine **59** shows the N-substituent to be α-disposed.[160] The branched nucleoside analogues **60**,[161] the 1,5-anhydrohexitol 5-iodouracil nucleoside **61**,[162] the dioxolane 2',3'-dideoxynucleoside analogues *cis*-4-(hydroxymethyl)-2-(thymin-1-ylmethyl)-1,3-dioxolane, *cis*-2-(hydroxymethyl)-2-(thymin-1-ylmethyl)-1,3-dioxolane and *trans*-4-(hydroxymethyl)-2-(uracil-1-yl-methyl)-1,3-dioxolane,[163] the pyranosyl nucleoside component **62** of antitumour antibiotic spicamycin[164] and the carbocyclic nucleoside analogue precursor **63**.[165]

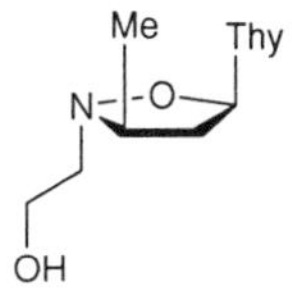

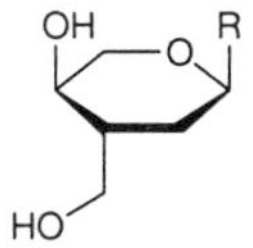

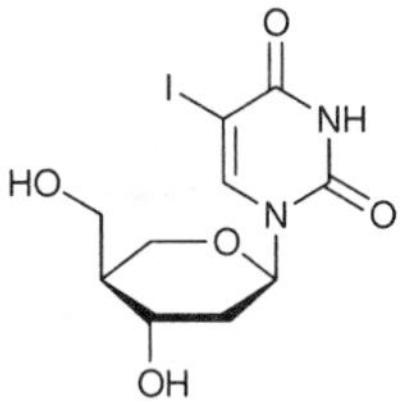

Anhydronucleoside **64**,[166] heterocycle fused nucleoside **65**,[167] 3'-isocyano-2',3'-dideoxyuridine,[168] 5'-azido-2',5'-dideoxythymidine,[169] 2'-amidonucleoside ana-logue **66**,[170] the triazole derivative **67**,[171] the imidazole-4-carboxamide nucleoside **68**,[172] 3'-(tetrazol-2''-yl)-3'-deoxythymidine and its 5''-methyl derivative **69** and **70** respectively,[173] diacetylene nucleoside **71**,[174] base protected α-anomer of the bicyclic cytosine nucleoside analogue **72**.[175]

62

63

64

65

66

67

68

69 R = H
70 R = Me

71

72

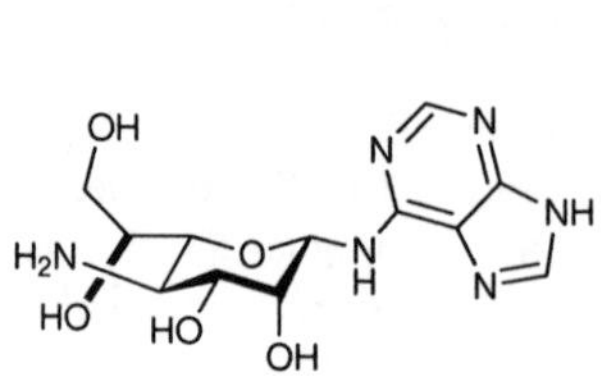

4 ESR Spectroscopy

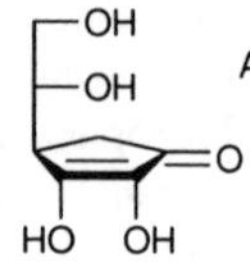

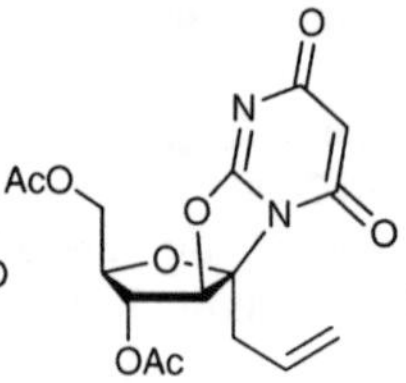

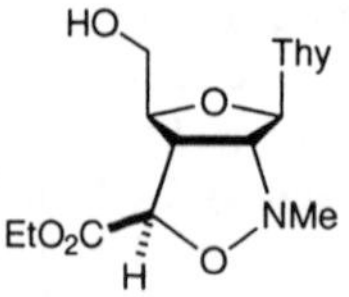

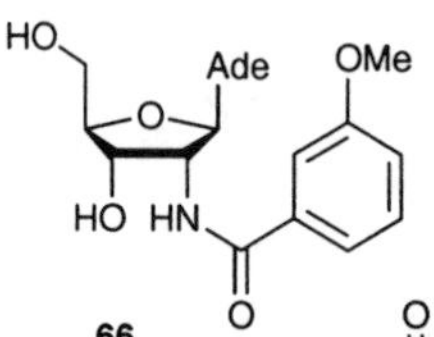

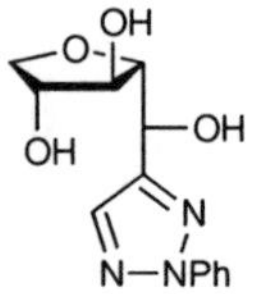

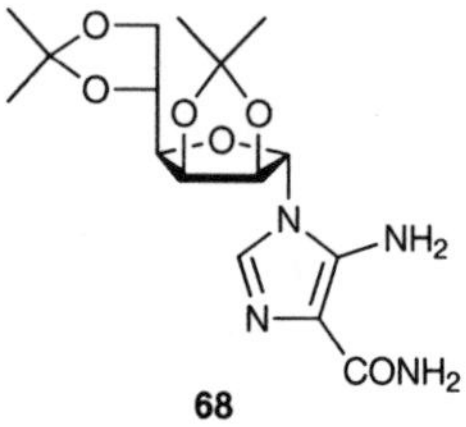

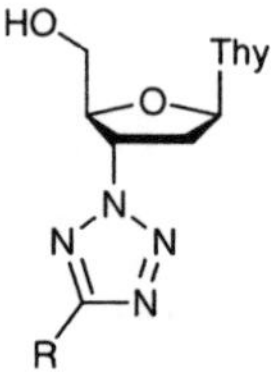

A series of D-glucosyl-based neoglycolipids, both β- and α- anomers of *O*-, *S*- and *C*-glycosides, have been prepared bearing spin labels. The relatively short-lived imino *N*-oxyl group was used for *O*-glycosides, and the *N*-acylamino *N*-oxyl moiety was employed for *S*- and *C*-glycosides. ESR spectra provided conformational information about the lipophilic chain.[176]

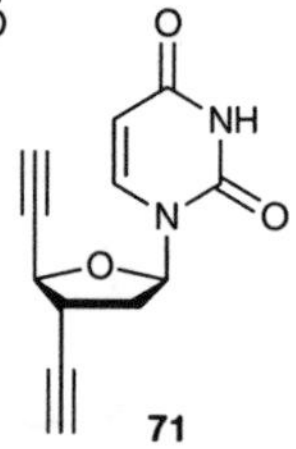

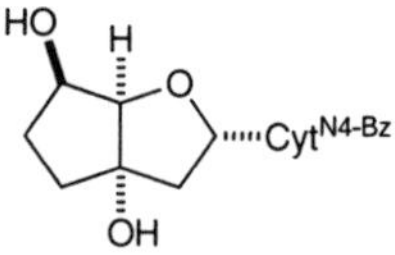

5 Polarimetry, Circular Dichroism, Calorimetry and Related Studies

Theoretical predictions of vibrational CD spectra of D-glyceraldehyde, D-erythrose and D-threose have been made through development of a method to scale calculated force constants to agree with experimental values.[177, 178]

CD spectra of tautomeric and protonated cyclic adenosine derivatives e.g. **73** have been described.[179] A review of the chiroptical properties of cholesterol glycosides and chirality distinction has appeared (e.g. D-glucose and L-glucose lead to red and blue shifts respectively).[180] CD has been used for chirality determination of dideoxy-*glycero*-mannonopyranose-derived **74**.[181] The C2 chirality of acyclic 1,2,4-triols **75** and **76** has been determined from CD difference spectra by the excitol interaction of the terminal 1,2-dibenzoate.[182] Cyclodextrin has been used to determine the effect of 21 solvents on aggregation of amphotericin B. Absence of strong solvent interactions leads to self-association.[183] Cyclodextrin **77** and the porphyrin **78** associate as box and lid. The absorption profile of this complex changes markedly when pentachlorophenol is complexed in the cyclodextrin cavity.[184]

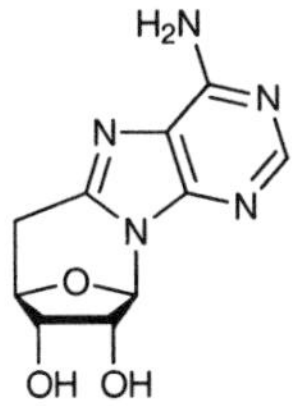

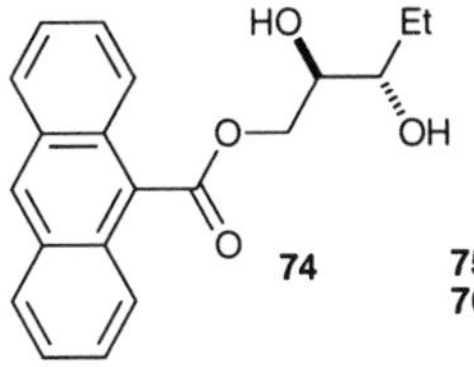

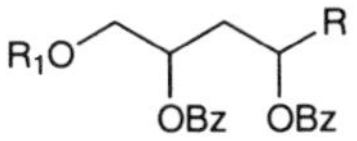

This year has seen several further reports by Shinkai's group on the development of boronic acid based carbohydrate recognition molecules. Boronic acid **79** forms a polymeric complex **80** with L-fucose.[185] The CD spectrum shows one positive maximum and two negative minima. An anthracene-based diboronic acid bearing two 15-crown-5 rings complexes D-glucose to form the CD active fluorescent complex **81**, but metal complexation (Ba^{2+}, Sr^{2+}, K^+, Na^+) of the two 15-crown-5-rings leads to disruption of the complexation to the saccharide leading to the CD inactive non-fluorescent complex **82**.[186] The calixarene scaffold for boronic acid sugar binders **83** forms a 2:1 complex **84** with D-fructose but a 1:1 complex with D-glucose involving O-1, O-2, O-4 and O-6. Binding is detected by fluorescence.[187] CD detection of saccharide recognition with several diboronic acids has also been discussed.[188]

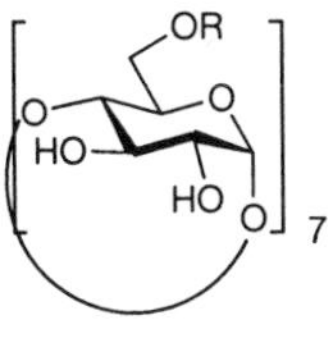

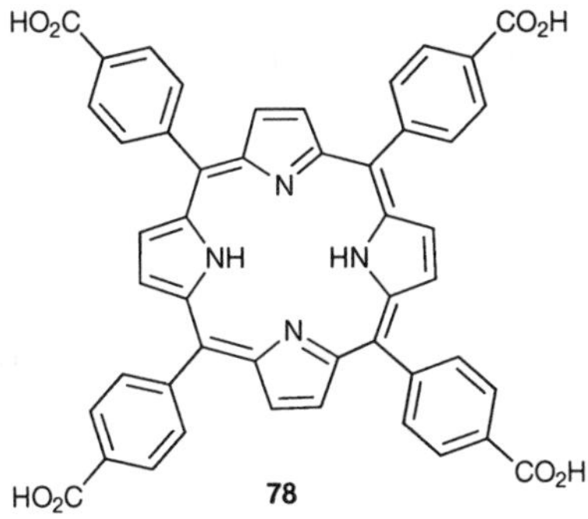

Time domain reflectometry has been used to study the hydration of L-*xylo*- and D-*arabino*-ascorbic acid solutions in water and water ethanol mixtures,[189] differential scanning calorimetry and thermogravimetric analysis have been used to study hydration of α-maltose and amylose,[190] a new physical chemical model describes the thermodynamic properties of binary water-carbohydrate

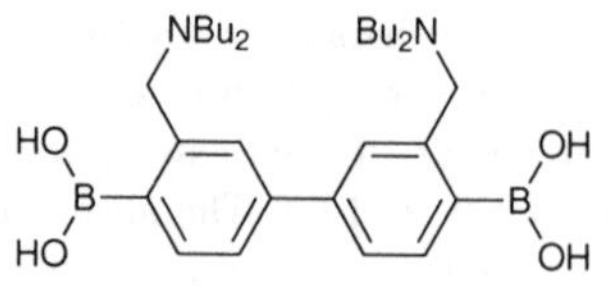

79

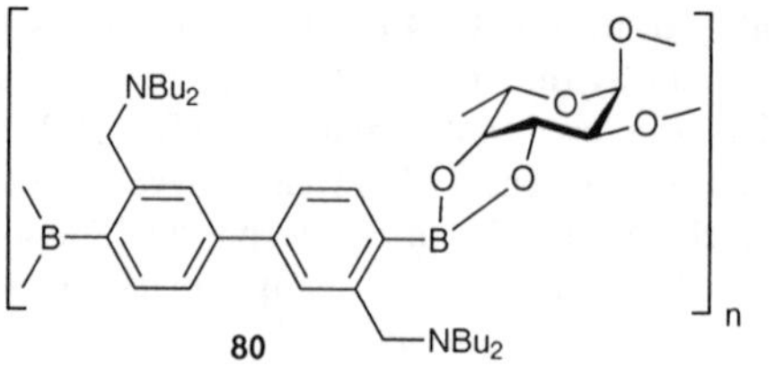

80

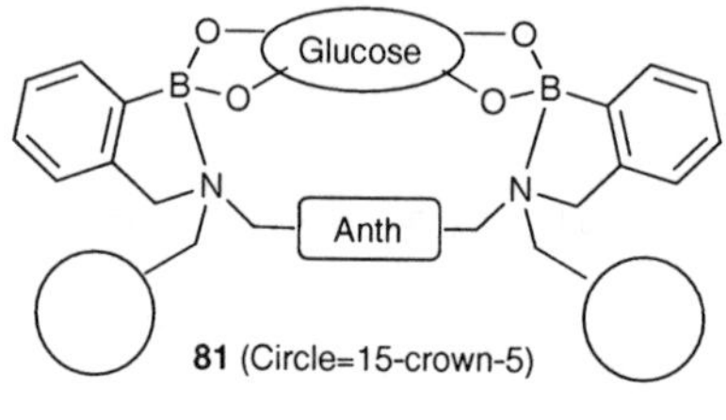

81 (Circle=15-crown-5)

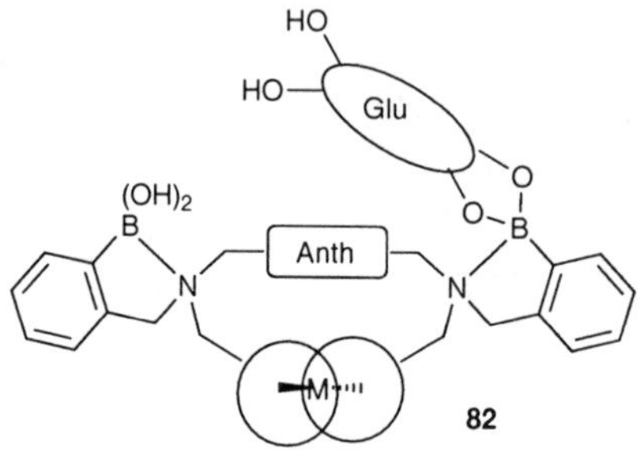

82

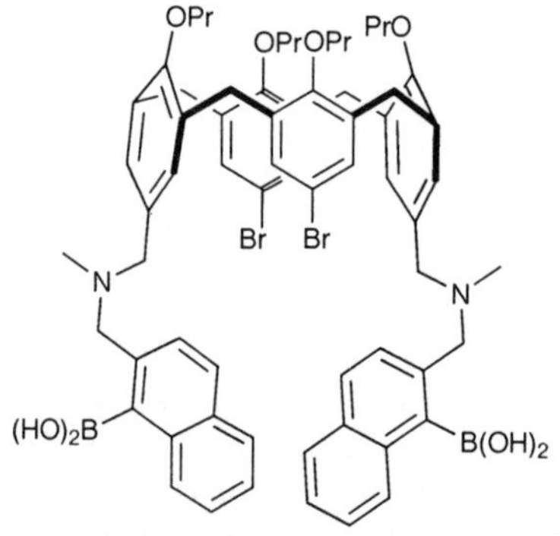

83

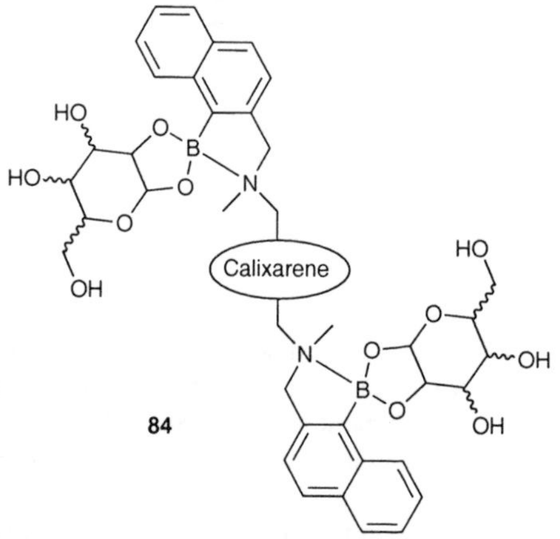

84

mixtures,[191] the surface, interfacial, emulsification, foaming and solubilization properties of rhamnolipids and sophorolipids have been examined.[192]

N-Hydroxysuccinimidyl esters react with glycosylamines to provide glycoconjugates which can be cross-linked to amino-functionalized resins. Lectin immunostaining, or flow cytometry using a fluorescently labelled lectin are used to identify carbohydrate-protein recognition of these sugar-bead conjugates.[193]

References

1   H. Kodad, R. Mokhliddr, E. Davin and G. Mille, *Can. J. Appl. Spectrosc.*, 1994, **39**, 107.
2   V.M. Andrianov and R.G. Zhbankov, *Zh. Struct. Khim.*, 1994, **35**, 35 (*Chem. Abstr.*, 1994, **122**, 265 786).
3   I. Kouach-Alix, J.P. Huvenne, P. Legrand and G. Vergoten, *J. Mol. Struct.*, 1994, **327**, 1.
4   R.G. Zhbankov, V.M. Adrianov, H. Ratjczak, and M. Marchewka, *Zh. Strukt. Khim.*, 1995, **36**, 322 (*Chem. Abstr.*, 1995, **123**, 314 277).
5   V.A. Yaylayan and A.A. Ismail, *Carbohydr. Res.*, 1995, **276**, 253.
6   V.A. Yaylayan and A.A. Ismail, *Spec. Publ. - RSC (Maillard Reactions in Chemistry, Food and Health)*, 1994, **151**, 69 (*Chem. Abstr.*, 1995, **123**, 112 526).
7   A.F. Bell, H. Lutz and L.D. Barron, *Spectrochim. Acta, Part A*, 1995, **51A**, 1367.
8   N. Buzas, M.A. Pujar, L. Nagy, A. Vertes, E. Kuzmann and H. Mehner, *J. Radioanal. Nucl. Chem.*, 1995, **189**, 237 (*Chem. Abstr.*, 1995, **123**, 9803).
9   M. Dauchez, P. Derreumaux, P. Lagant and G. Vergoten, *J. Comput. Chem.*, 1995, **16**, 188 (*Chem. Abstr.*, 1995, **122**, 240 223).
10 M. Sekkal, V. Dincq, P. Legrand and J.P. Huvenne, *J. Mol. Struct.*, 1995, **349**, 349.
11 N.I. Sushko, S.P. Firsov, R.G. Zhbankov and V.M. Tsarenkov, *Zh. Prikl. Spektrosk.*, 1994, **61**, 365 (*Chem. Abstr.*, 1995, **123**, 340 628).
12 M.A. Diaz-Diez, F.J. Garcia-Barros, A. Bernalte-Garcia, and C. Valenzuela-Calahorro, *Thermochim. Acta.*, 1994, **247**, 439 (*Chem. Abstr.*, 1994, **122**, 161 144).
13 L.B. Clark, *J. Phys. Chem.*, 1995, **99**, 4466.
14 P. Mozjes, J. Stepanek, I. Rosenberg, Z. Tocik, M. Burian, M. Pavelcikova, M. Refregiers and H. Videlot, *J. Mol. Struct.*, 1995, **348**, 45.
15 U. Hisako, Y. Sugawara and T. Kasuya, *Phys. Rev. B: Condens. Matter*, 1995, **51**, 5666.
16 L. Lupan, R. Bandula, D. Crisan and V. Popa, *Rev. Roum. Chim.*, 1994, **39**, 1423 (*Chem. Abstr.*, 1995, **123**, 286 436).
17 M.L.C.E. Kouwijzer, B.P. van Eijck, H. Kooijman and J. Kroon, *Acta Crystallogr.*, 1995, **B51**, 209.
18 A.S. Plaziak, *Wiad. Chem.*, 1993, **47**, 51 (*Chem. Abstr.*, 1995, **123**, 169 989).
19 J.W. Dallinga, D.M.F.A. Pachen and J.C.S. Kleinjans, *Biol. Mass. Spectom.*, 1994, **23**, 764 (*Chem. Abstr.*, 1995, **122**, 133 557).
20 M. Sawada, Y. Okumura, Y. Takai, S. Takahashi, M. Mishima and Y. Tsuno, *J. Mass. Spectrom. Soc. Jpn.*, 1994, **42**, 225 (*Chem. Abstr.*, 1995, **123**, 228 656).
21 V.I. Kaderitsev, A.A. Stomakhin and O.S. Chizhov, *Izv. Akad. Nauk, Ser. Khim.*, 1994, **4**, 629 (*Chem. Abstr.*, 1995, **122**, 314 951).
22 M. Coppola, D. Favretto, P. Traldi and G. Resnati, *Org. Mass. Spectrom.*, 1994, **29**, 553.
23 V. Kovacik, J. Hirsch, W. Heerma, C.G. de Koster and J. Haverkamp, *Org. Mass. Spectrom.*, 1994, **29**, 707.
24 T. Vaisar, M. Ledvina and J. Jezek, *J. Mass Spectrom.*, 1995, **30**, 767.
25 A.R. Dongre and V.H. Wysocki, *Org. Mass Spectrom.*, 1994, **29**, 700.
26 W. Chai, E.F. Hounsell and A.M. Lawson, *Carbohydr. Res.*, 1995, **269**, 139.
27 P. Juhasz and K. Biemann, *Carbohydr. Res.*, 1995, **270**, 131.
28 T. Li, M. Kubota, S. Okuda, T. Hirano and M. Ohashi, *Eur. Mass Spectrom.*, 1995, **1**, 11 (*Chem. Abstr.*, 1995, **123**, 144 420).

29   A. Tuong, F. Uzabiaga, M. Petitou, J.C. Lormeau and C. Picard, *Carbohydr. Lett.*, 1995, **1**, 55.

30   D. Chassagne, J. Crouzet, R.L. Baumes, J.-P. Lepourte and C.L. Bayonove, *J. Chromatogr. A.*, 1995, **694**, 441.

31   L. Johansson, H. Karlsson and K.-A. Karlsson, *J. Chromatogr. A.*, 1995, **712**, 149.

32   J.-L. Wolfender, K. Hostettman, F. Abe, T. Nagao, H. Okabe and T. Yamauchi, *J. Chromatogr. A.*, 1995, **712**, 155.

33   T. Ii, Y. Ohashi and Y. Nagai, *Carbohydr. Res.*, 1995, **273**, 27.

34   V. Kovacik, J. Hirsch and V. Patoprsty, *Chem. Pap.*, 1994, **48**, 422 (*Chem. Abstr.*, 1995, **123**, 340 607).

35   B. Domon, D.R. Mueller and W.J. Richter, *Org. Mass Spectrom.*, 1994, **29**, 713.

36   J. Banoub, E. Gentil and D.H. Shaw, *Spectroscopy*, 1994, **12**, 55 (*Chem. Abstr.*, 1995, **123**, 286 440).

37   M.D. Mohr, K.O. Boersen and H.M. Widmer, *Rapid Commun. Mass Spectrom.*, 1995, **9**, 809.

38   J.B. Cunniff and P. Vouros, *J. Am. Mass Spectrom.*, 1995, **6**, 437.

39   V.N. Reinhold, B.B. Reinhold and C.E. Costello, *Anal. Chem.*, 1995, **67**, 1772.

40   K. Tadano-Aritami, H. Kubo, P. Ireland, M. Okudo, T. Kasami, S. Handa and I. Ishizuka, *Carbohydr. Res.*, 1995, **273**, 41.

41   H. Perreault and C.E. Costello, *Org. Mass Spectrom.*, 1994, **29**, 720.

42   B.B. Reinhold, S.-Y. Chan, S. Chan and V.N. Reinhold, *Org. Mass Spectrom.*, 1994, **29**, 736.

43   D.J. Harvey, *J. Mass Spectrom.*, 1995, **30**, 1311.

44   Q. Li, *Diss. Abstr. Int. B.*, 1994, **55**, 1408 (*Chem. Abstr.*, 1995, **122**, 265 791).

45   F. Nishida, K. Masuda, K.-i. Harada, M. Suzuki, V. Meevootisom, T.W. Flegel, Y. Thebtaranonth and S. Intararuangsorn, *J. Mass Spectrom. Soc. Jpn.*, 1995, **43**, 27 (*Chem. Abstr.*, 1995, **123**, 228 674).

46   M.H. Florencio, D. Despeyroux and K.R. Jennings, *Org. Mass Spectrom.*, 1994, **29**, 483.

47   F. Nishida, K.-i. Harada, M. Suzuki, T. Fujita, H. Noki, V. Meevootisom, T. Flegel, Y. Thebtaranonth and S. Intararuangsorn, *J. Mass. Spectrom. Soc. Jpn.*, 1995, **43**, 37 (*Chem. Abstr.*, 1995, **123**, 257 151).

48   Y.N. Elkin, Y.A. Knirel, N.A. Kocharova, E.V. Vinogradov and K. Capek, *Bioorg. Khim.*, 1994, **20**, 790 (*Chem. Abstr.*, 1995, **122**, 187 889).

49   J.M. Vega-Perez, J.L. Espartero, M. Vega, J.I. Candela, F. Iglesias-Guerra and F. Alcudia, *Eur. Mass Spectrom.*, 1995, **1**, 161.

50   M. Spiro, J.C. Blais, G. Bolbach, F. Fournier, J.C. Tabet, K. Driaf, O. Gaud, R. Granet and P. Krausz, *Int.J. Mass Spectrom. Ion Processes*, 1994, **134**, 229 (*Chem. Abstr.*, 1995, **122**, 31 818).

51   A. Carrier, L. Varfalvy and M.J. Bertrand, *J. Chromatogr. A.*, 1995, **705**, 205.

52   S.A. Aksyonov and L.F. Sukhodub, *Rapid Commun. Mass Spectrom.*, 1995, **9**, 775.

53   J. Banoub, E. Gentil, B. Tber, N.-E. Fahmi, G. Ronco, P. Villa and G. Mackenzie, *Spectroscopy*, 1994, **12**, 69.

54   S. Habibi-Goudarzi and S.A. McLuckey, *J. Am. Soc. Mass. Spectrom.*, 1995, **6**, 102.

55   L.F. Sukhodub and S.A. Aksjonov, *Biofizika*, 1995, **40**, 506 (*Chem. Abstr.*, 1995, **123**, 286 488).

56   V.A. Pokrovskiy, N.P. Galagan and L.I. Dembnovetskaya, *Rapid Commun. Mass Spectrom.*, 1995, **9**, 585.

57   B.P. van Eijck, W.T.M. Mooij and J. Kroon, *Acta Crystallogr.*, 1995, **B51**, 99.

58   L.O. Figura and M. Epple, *J. Therm. Anal.*, 1995, **44**, 45 (*Chem Abstr.*, 1995, **122**, 291 353).

59   F.W. Lichtenthaler and F. Klotz, F.-J., *Liebigs Ann. Chem.*, 1995, 2059.

60   A.J. Kalb, F. Frolow, J. Yarin and M. Eisenstein, *Acta Crystallogr.*, 1995, **D51**, 1077.

61   J.C. Barnes and R.A. Field, *Acta Crystallogr.*, 1995, **C51**, 1018.

62   V. Kabaleeswaran, S.S. Rajan, C. Booma and K.K. Balasubramanian, *Acta Crystallogr.*, 1995, **C51**, 758.

63   C.F. Carvalho, C.H.L. Kennard, D.E. Lynch, G. Smith and A. Wong, *Aust.J. Chem.*, 1995, **48**, 1767.

64   Y. Abe, K. Harata, M. Fujiwara and K. Ohbu, *Carbohydr. Res.*, 1995, **269**, 43.

65   Y. Abe, M. Fujiwara, K. Ohbu and K. Harata, *Carbohydr. Res.*, 1995, **275**, 9.

66   D. Mentzafos, M. Polissiou, I. Graspsas A. Hountas, M. Georgiadis, and A. Terzis, *J. Chem. Crystallogr.*, 1995, **25**, 157 (*Chem. Abstr.*, 1995, **123**, 286 411q).

67   J. Kopf, G. Sheldrick, H. Reinke, D. Rentsch, and R. Mietchen, *Acta Crystallogr.*, 1995, **C51**, 1198.

68   L.K. Prikhodchencko, T.E. Kolosova, V.M. Adrianov, R.G. Dokuchaev and V.F. Sopin, *Kristallografiya*, 1994, **39**, 868 (*Chem. Abstr.*, 1995, **122**, 81 782).

69   A. Linden and C.K. Lee, *Acta Crystallogr.*, 1995, **C51**, 2320.

70   K. Panneerselvam and M. Soriano-Garca, *Acta Crystallogr.*, 1995, **C51**, 1646.

71   S. Tomic, B.P. van Eijck, B. Kojic-Prodic, J. Kroon, V. Magnus, B. Nigovic, G. Lacan, N. Ilic, H. Duddeck and M. Hiegemann, *Carbohydr. Res.*, 1995, **270**, 11.

72   S. Tomic, B. Kojic-Prodic, V. Magnus, G. Lacan, H. Duddeck and M. Hiegemann, *Carbohydr. Res.*, 1995, **279**, 1.

73   H. Prawat, C. Mahidd, S. Ruchirawat, U. Prawat, P. Tuntiwachwuttikul, U. Tootptakong, W.C. Taylor.C. Pakawatchal, B.W. Skelton and A.H. White, *Phytochem.*, 1995, **40**, 1167.

74   A. Linden and C.K. Lee, *Acta Crystallogr.*, 1995, **C51**, 1012.

75   A. Linden and C.K. Lee, *Acta Crystallogr.*, 1995, **C51**, 1007.

76   C.K. Lee and A. Linden, *J. Carbohydr. Chem.*, 1995, **14**, 9.

77   M. Ikegami, T. Sato, K. Suzuki, K. Noguchi, K. Okuyama, S. Kitamura, K. Takeo and S. Ohno, *Carbohydr. Res.*, 1995, **271**, 137.

78   O. Kanie, T. Takeda and K. Hatano, *Carbohydr. Res.*, 1995, **276**, 409.

79   A. Linden and C.K. Lee, *Acta Crystallogr.*, 1995, **C51**, 747.

80   A. Linden and C.K. Lee, *Acta Crystallogr.*, 1995, **C51**, 751.

81   K. Harata and M. Muraki, *Acta Crystallogr.*, 1995, **D51**, 718.

82   S. Karisen and E. Hough, *Acta Crystallogr.*, 1995, **D51**, 962.

83   P.K. Qasba, P.V. Balaji and V.S.R. Rao, *Glycobiology*, 1994, **4**, 805.

84   S. Raymond, B. Henrissat, D. Tran Qui, A. Kvick and H. Chanzy, *Carbohydr. Res.*, 1995, **277**, 209.

85   K. Itazu, K. Noguchi, K. Okuyama, S. Kitamura, K. Takeo and S. Ohno, *Carbohydr. Res.*, 1995, **278**, 195.

86   R.P. Millane and T.L. Hendrixson, *Carbohydr. Polym.*, 1994, **25**, 245 (*Chem. Abstr.*, 1995, **123**, 9792).

87   R. Borsali, M. Rinaudo and L. Noirez, *Macromolecules*, 1995, **28**, 1085.

88   S. Hanessian, A. Benalil, M. Simara and F. Belanger-Gariepy, *Tetrahedron*, 1995, **37**, 10149.

89   T. Steiner and W. Saenger, *J. Chem. Soc., Chem. Commun.*, 1995, 2087.

90   K. Harata, *Supramol. Chem.*, 1995, **5**, 231 (*Chem. Abstr.*, 1995, **123**, 340 612).

91   T. Steiner and W. Saenger, *Carbohydr. Lett.*, 1995, **1**, 143.

92   T. Steiner and W. Saenger, *Carbohydr. Res.*, 1995, **275**, 73.
93   C. Volkmann, A. Zeeck, O. Potterat, H. Zähner, F.-M. Bohnen and R. Herbst-Irmer, *J. Antibiotics*, 1995, **48**, 431.
94   T. Granier and A. Vasella, *Helv. Chim. Acta*, 1995, **78**, 1738.
95   J. Alzeer and A. Vasella, *Helv. Chim. Acta*, 1995, **78**, 177.
96   A. Cousson, I. Dancy and J.-M. Beau, *Acta Crystallogr.*, 1995, **C51**, 718.
97   J.L. Del Valle, T. Torroba, S. Marcaccini, P. Paoli and D.J. Williams, *Tetrahedron*, 1995, **51**, 8259.
98   I.A. Kennedy and T. Hemscheidt, *Can.J. Chem.*, 1995, **73**, 1329.
99   H. Kuno, S. Niihata, T. Ebata, and H. Matsushita, *Heterocycles*, 1995, **41**, 523.
100 T. Kawai, M. Isobe, and S.C. Peters, *Aust.J. Chem.*, 1995, **48**, 115.
101 G.J. Gainsford, R.H. Furneaux, J.M. Mason and P.C. Tyler, *Acta Crystallogr.*, 1995, **C51**, 2418.
102 J. Hirsch, E. Petrakova, M.S. Feather and C.L. Barnes, *Carbohydr. Res.*, 1995, **267**, 17.
103 C.K. Lee, A. Linden and A. Vasella, *Acta Crystallogr.*, 1995, **C51**, 1906.
104 M.J. Dianez, M.D. Estrada, A. Lopez-Castro and S. Perez-Garrido, *Z. Kristallogr.*, 1994, **209**, 506 (*Chem. Abstr.*, 1995, **122**, 291 393).
105 Z. Urbanczyk-Lipowska, *Acta Crystallogr.*, 1995, **C51**, 1184.
106 P. Dokurno, J. Lubkowski, H. Lyszka and Z. Smiatacz, *Z. Kristallogr.*, 1994, **209**, 808 (*Chem. Abstr.*, 1995, **122**, 265 863).
107 J.-P. Praly, D. Senni, R. Faure and G. Descotes, *Tetrahedron*, 1995, **51**, 1697.
108 M.E. Tate, I.M. Delaere, G.P. Jones and E.R.J. Tiekink, *Aust.J. Chem.*, 1995, **48**, 1059.
109 R. Herbst, T. Steiner, B. Pfannemuller and W. Saenger, *Carbohydr. Res.*, 1995, **269**, 29.
110 C. André, P. Luger, R. Bach and J.-H. Fuhrtop, *Carbohydr. Res.*, 1995, **266**, 15.
111 C. André, P. Luger, T. Gutberlet, D. Vollhardt and J.-H. Fuhrhop, *Carbohydr. Res.*, 1995, **272**, 129.
112 X.-L. Sun, T. Kai, M. Tanaka, H. Takayanagi and K. Furuhata, *Chem. Pharm. Bull.*, 1995, **43**, 1654.
113 P. Lei, Y. Ogawa, J.L. Flippen-Anderson and P. Kovác, *Carbohydr. Res.*, 1995, **275**, 117.
114 B. Kojic-Prodic, V. Milinkovic, J. Kidric, P. Pristovsek, S. Horvat and A. Jakas, *Carbohydr. Res.*, 1995, **279**, 21.
115 V.V. Mossine, G.V. Glinsky, C.L. Barnes and M.S. Feather, *Carbohydr. Res.*, 1995, **266**, 5.
116 C.E. Sowa, J. Kopf and J. Thiem, *J. Chem. Soc., Chem. Commun.*, 1995, 211.
117 J.M.J. Tronchet, M. Balkadjian, G. Zosimo-Landolfo, F. Barbalat-Rey, P. Lichtle, A. Ricca, I. Komaromi, G. Bernardinelli and M. Geoffrey, *J. Carbohydr. Chem.*, 1995, **14**, 17.
118 P. Herczegh, I. Kovacs, L. Szilagyi, F. Sztaricskai, A. Berecibar, A. Riche, A. Chiaroni, A. Olesker and G. Lukacs, *Tetrahedron*, 1995, **51**, 2969.
119 I. Leban, L. Golic, S. Kotnik, A. Resman and J. Zmitek, *Acta Chim. Slov.*, 1994, **41**, 405 (*Chem. Abstr.*, 1995, **122**, 187 966).
120 J.Y. LeQuestel, N. Mouhous-Riou and S. Perez, *Carbohydr. Res.*, 1995, **268**, 127.
121 M.J. Potrzebowski, M. Michalska, J. Blaszczyk and M.W. Wieczorek, *J. Org. Chem.*, 1995, **60**, 3139.
122 K. Peseke, H. Feist, W. Hanefeld, J. Kopf and H. Schulz, *J. Carbohydr. Chem.*, 1995, **14**, 317.

123   P. Franchetti, L. Cappellacci, M. Grifantini, A. Barzi, G. Nocentini, H. Yang, A. O'Connor, H.N. Jayaram, C. Carrell and B.M. Goldstein, *J. Med. Chem.*, 1995, **38**, 3829.

124   G. Argay, R. Csuk, Z. Gyorgydeak, A. Kalman and G. Snatzke, *Tetrahedron*, 1995, **51**, 12911.

125   W.H. Ojala, K.E. Albers, W.B. Gleason and C.G. Choo, *Carbohydr. Res.*, 1995, **275**, 49.

126   E.A. Yates, W. Mackie and D. Lamba, *Int.J. Biol. Macromol.*, 1995, **17**, 219 (*Chem. Abstr.*, 1995, **123**, 340 677).

127   W. Mackie, E.A. Yates and D. Lamba, *Carbohydr. Res.*, 1995, **266**, 65.

128   G.Y.S.K. Swamy, K. Ravikumar, A.V. Rama Rao, M.K. Gurjar and T. Ramadevi, *Acta Crystallogr.*, 1995, **C51**, 426.

129   A.J. Wood, P.R. Jenkins, J. Fawcett and D.R. Russell, *J. Chem. Soc., Chem. Commun.*, 1995, 1567.

130   X.Y. Wang, P.A. Seib, J.V. Paukstelis, L.L. Seib and F. Takusagawa, *J. Carbohydr. Chem.*, 1995, **14**, 1257.

131   R. Alen and J. Valkonen, *Acta Chem. Scand.*, 1995, **49**, 536.

132   A.S. Batsanov, M.J. Begley, R.J. Fletcher, A.A. Murphy and M.S. Sherburn, *J. Chem. Soc., Perkin Trans. 1*, 1995, 1281.

133   C. Di Nardo., O. Varela, R.M. de Lederkremer, R.F. Baggio, D.R. Vega and M.T. Garland, *Carbohydr. Res.*, 1995, **269**, 99.

134   H.-K. Fun, K. Sivakumar, H.-B. Ang, T.-W. Sam and E.-K. Gan, *Acta Crystallogr.*, 1995, **C51**, 1330.

135   M. Chandler, M.J. Bamford, R. Conroy, B. Lamont, B. Patel, V.K. Patel, I.P. Steeples, R. Storer, N.G. Weir, M. Wright and C. Williamson, *J. Chem. Soc., Perkin Trans. 1*, 1995, 1173.

136   M. Chandler, R. Conroy, C.A.W. J., B. Lamont, J.J. Scicinski, J.E. Smart, R. Storer, N.G. Weir, R.D. Wilson and P.G. Wyatt, *J. Chem. Soc., Perkin Trans. 1*, 1995, 1189.

137   P.J. Cox, R.A. Howie, O.A. Melvin and J.L. Wardell, *J. Organomet. Chem.*, 1995, **489**, 161.

138   J. Burger, C. Gack and P. Klüfers, *Angew. Chem., Int. Ed. Engl.*, 1995, **34**, 2647.

139   P. Klüfers and J. Schumacher, *Angew. Chem., Int. Ed. Engl.*, 1995, **34**, 2119.

140   K. Hegetschweiler, T. Kradolfer, V. Gramlich and R.D. Hancock, *Eur.J. Chem.*, 1995, **1**, 74.

141   R. Kivekäs, M. Klinga, J. Ruiz and E. Colacio, *Acta Chem. Scand.*, 1995, **49**, 305.

142   K. Tsuchiya, S. Kobayashi, T. Kurokawa, T. Nakagawa, N. Shimada, H. Nakamura, Y. Iitaka, M. Kitagawa and K. Tatsuta, *J. Antibiotics*, 1995, **48**, 630.

143   M.A. Shalaby, F.R. Fronczek, Y. Lee and E.S. Younathan, *Carbohydr. Res.*, 1995, **269**, 191.

144   T. Uchiyama, V.P. Vasilev, T. Kajimoto, W. Wong, C.-C. Linand and C.-H. Wong, *J. Am. Chem. Soc.*, 1995, **117**, 5395.

145   T. Kiguchi, K. Tajiri, T. Ninomiya, T. Naito and H. Hiramatsu, *Tetrahedron Lett.*, 1995, **36**, 253.

146   J.E. Anderson, S.J. Angyal, and D.C. Craig, *Carbohydr. Res.*, 1995, **272**, 141.

147   C.-H. Wong, L. Provencher, J.A. Porco, S.-H. Jung, Y.-F. Wang, L. Chen, R. Wang and D.H. Steensma, *J. Org. Chem.*, 1995, **60**, 1492.

148   M. Kiso, K. Ando, H. Inagaki, H. Ishida and A. Hasegawa, *Carbohydr. Res.*, 1995, **272**, 159.

149   C.E. Bryant, J.A.G. Drake and D.W. Jones, *J. Chem. Crystallogr.*, 1995, **25**, 123 (*Chem. Abstr.*, 1995, **123**, 314 375).

150   A.N. Checklov, *Zh. Struckt. Chim.*, 1995, **36**, 178 (*Chem. Abstr.*, 1995, **122**, 315 041).

151   L. Chen and B.M. Craven, *Acta Crystallogr.*, 1995, **B51**, 1081.

152   M.D. Bratek-Wiewiorowska, M. Alejska, M. Popenda, E. Utzig and M. Wiewiorowska, *J. Mol. Struct.*, 1994, **327**, 327.

153   A.K. Das and V. Bertolasi, *Acta Crystallogr.*, 1995, **C51**, 2408.

154   B.L. Partridge and C.E. Pritchard, *Acta Crystallogr.*, 1995, **C51**, 1929.

155   S. Napper, A.L. Stuart, S.V.P. Kumar, V.S. Gupta and L.T.J. Delbaere, *Acta Crystallogr.*, 1995, **C51**, 96.

156   M.P. Wallis, I.D. Spiers, C.H. Schwalbe and W. Fraser, *Tetrahedron Lett.*, 1995, **36**, 3759.

157   L.V. Meervelt, P.K.T. Lin and D.M. Brown, *Acta Crystallogr.*, 1995, **C51**, 137.

158   M.P. Groziak, R. Lin and P.D. Robinson, *Acta Crystallogr.*, 1995, **C51**, 1204.

159   S.S. Li, X.L. Sun, H. Ogura, Y. Konda, T. Sasaki, Y. Toda and H. Takayanagi, *Chem. Pharm. Bull.*, 1995, **43**, 144.

160   J.M.J. Tronchet, M. Iznaden, F. Barbalat-Rey, I. Komaromi, N. Dolatshaki and G. Bernardinelli, *Nucleosides Nucleotides*, 1995, **14**, 1737.

161   B. Doboszewski, N. Blaton, Rozenski, A. De Bruyn and P. Herdewijn, *Tetrahedron*, 1995, **51**, 5381.

162   I. Verheggen, A. Van Aerschot, L. Van Meervelt, J. Rozenski, L. Wiebe, R. Snoeck, G. Andrei, J. Balzarini, P. Claes, E. De Clerq, and P. Herdewijn, *J. Med. Chem.*, 1995, **38**, 826.

163   A.V. Bochkarev, E.V. Efimtseva, S.N. Mikhailov and G.V. Gurskaya, *Kristallografiya*, 1994, **39**, 635 (*Chem. Abstr.*, 1995, **122**, 106 351).

164   T. Sakai, K. Shindo, A. Odagawa, A. Suzuki, H. Kawai, K. Kobayashi, Y. Hayakawa, H. Seto and N. Otaka, *J. Antibiotics*, 1995, **48**, 899.

165   J. Schachtner, H.-D. Stachel and K. Polborn, *Tetrahedron*, 1995, **51**, 9005.

166   Y. Itoh, K. Haraguchi, H. Tanaka, H. Gen and T. Miyasaka, *J. Org. Chem.*, 1995, **60**, 656.

167   B.M. Burkhart, A. Papchikhin, J. Chattopadhyaya and M. Sundaralingam, *Acta Crystallogr.*, 1995, **C51**, 1462.

168   A.K. Das and S.K. Mazumdar, *Acta Crystallogr.*, 1995, **C51**, 1652.

169   H.-G. Schaible, J.W. Bats and J.W. Engels, *Acta Crystallogr.*, 1995, **C51**, 2410.

170   S. Van Calenbergh, C.L.M.J. Verlinde, J. Soenens, A. De Bruyn and M. Callens, *J. Med. Chem.*, 1995, **38**, 3838.

171   M.A.E. Sallam, L.B. Townsend and W. Butler, *J. Chem. Res.*, 1995, (S)54; (M)0467.

172   C.E. Briant, D.W. Jones and G. Shaw, *Nucleosides Nucleotides*, 1995, **14**, 1251.

173   V.A. Ostrovskii, E.P. Studentsov, V.S. Poplavskii, N.V. Ivanova, G.V. Gurskaya, V.E. Zavodnik, M.V. Jasko, D.G. Semizarov and A.A. Krayevsky, *Nucleosides Nucleotides*, 1995, **14**, 1289.

174   M.A. Amin, H. Stoeckli-Evans and A. Gossauer, *Helv. Chim. Acta*, 1995, **78**, 1879.

175   M. Bolli, P. Lubini and C. Leumann, *Helv. Chim. Acta*, 1995, **78**, 2077.

176   J.M.J. Tronchet, M. Zsely and M. Geoffroy, *Carbohydr. Res.*, 1995, **275**, 245.

177   D. Zeroka, J.O. Jensen and J.L. Jensen, *Gov. Rep. Announce. Index (U. S.)*, 1995, **95**, Abstr. 502 019 (*Chem. Abstr.*, 1995, **123**, 340 558).

178   D. Zeroka, J.O. Jensen and J.L. Jensen, *Gov. Rep. Announce. Index (U. S.)*, 1995, **95**, Abstr. 502 020 (*Chem. Abstr.*, 1995, **123**, 340 559).

179   Y.V. Morozov, N.P. Bazhulina, V.A. Bokovoy, A.V. Savitskiy, V.O. Chekhov and

V.L. Florentiev, *Mol. Biol. (Moscow)*, 1994, **28**, 1330 (*Chem. Abstr.*, 1995, **122**, 240 305).

180   T.D. James, H. Kawabata, R. Ludwig, K. Murata and S. Shinkai, *Tetrahedron*, 1995, **51**, 555.

181   N. Elloumi, L. Aguiar and M.L. Capmau, *Bull. Chim. Soc. Belg.*, 1994, **103**, 647 (*Chem. Abstr.*, 1995, **122**, 240 242).

182   Y. Mori and H. Furukawa, *Tetrahedron*, 1995, **51**, 6725.

183   Brittain, H. G., *Chirality*, 1994, **6**, 665 (*Chem. Abstr.*, 1995, **122**, 106 279).

184   S. Zhao and J.H.T. Luong, *J. Chem. Soc., Chem. Commun.*, 1995, 663.

185   M. Mikami and S. Shinkai, *J. Chem. Soc., Chem. Commun.*, 1995, 153.

186   T.D. James and S. Shinkai, *J. Chem. Soc., Chem. Commun.*, 1995, 1483.

187   P. Linnane, T.D. James and S. Shinkai, *J. Chem. Soc., Chem. Commun.*, 1995, 1997.

188   S. Yutaka, K. Kaoru, S. Miwako, T. Harad, K. Tsukagoshi and S. Shinkai, *Supramol. Chem.*, 1993, **2**, 11 (*Chem. Abstr.*, 1995, **122**, 10 395).

189   T. Umehara, Y. Tominanga, A. Hikida and S. Mashimo, *J. Chem. Phys.*, 1995, **102**, 9474.

190   C. Fringart, I. Tvaroska, K. Mazeau, M. Rinaudo and J. Desbrieres, *Carbohydr. Res.*, 1995, **278**, 27.

191   M. Catte, C.-G. Dussap and J.-B. Gros, *Fluid Phase Equilin.*, 1995, **105**, 1 (*Chem. Abstr.*, 1995, **123**, 56 401).

192   Y. Shi, J. Li, C. Sun and Z. Li, *Prog. Nat. Sci.*, 1995, **5**, 497 (*Chem. Abstr.*, 1995, **123**, 199 231).

193   D. Vetter, E.M. Tate and M.A. Gallop, *Bioconjugate Chem.*, 1995, **6**, 319 (*Chem. Abstr.*, 1995, **123**, 9820).

23
Separatory and Analytical Methods

1 General

Methods for the quantitative chromatographic analysis of inositol phosphates and inositol phospholipids have been reviewed.[1]

2 Chromatographic Methods

2.1 Gas-Liquid Chromatography. – In connection with analytical work on seaweed polysaccharides in which D- and L-galactose derivatives both appear, the enantiomers of galactose and their 3-, 4-, and 6-methyl ethers were readily separated after reductive amination using enantiomerically pure 1-amino-2-propanol followed by acetylation. The 2-methyl ethers could be separated when reductive amination was performed with (S)-α-methylbenzylamine.[2] During work on the synthesis of cationic starches 2-hydroxy-3-(trimethylammonio)propyl ethers were made and the positions of substitution were determined following glycosidic bond cleavage by methanolysis, and methylation of the free hydroxyl groups. Subsequently the cationic substituents were converted to 2-methoxy-2-propenyl ethers or 2-oxopropyl ethers prior to identification by capillary GLC/MS[3] For the purposes of monitoring bacterial peptidoglycan in house dust, muramic acid was released by hydrolysis and GC/MS analyses were carried out on the pertrimethylsilyl derivatives.[4]

Capillary GC/MS of trifluoroacetylated plant glycosides was applied to fruit and wine products using a negative ion chemical ionization method.[5] Capillary GC analyses of bile acid 3-glucosides and 3-glucuronides as various silylated derivatives were conducted using a thermostable stainless-steel capillary column.[6] A report of the GC/MS determination of 1,4:3,6- dianhydroglucitol-5-nitrate in human plasma has appeared, and a related study focused on the metabolites of the 2- and 5-mononitrates in plasma.[8] Fatty acid esters of sorbitan (a mixture of 1,4-anhydro- and 1,4:3,6-dianhydro-glucitol) used as emulsifiers and stabilizers in the food and cosmetic industries, were analysed by GC of their pertrimethylsilyl ethers.[9]

2.2 Thin-Layer Chromatography. – The combination of TLC separations and liquid secondary ion MS or MS-MS analyses for various carbohydrates has been covered in a review.[10]

Analysis of *N*-acetylchitooligosaccharides by the Iatroscan procedure has been reported. This system combines the efficiency of plate TLC for separating components and the sensitivity of flame ionization detection.[11] In related work disaccharides derived from heparin and chondroitin sulfate have been analysed by reductive amination with an aminolipid followed by high performance TLC separation and liquid secondary ion MS[12]

Two-dimensional TLC on poly(ethyleneimine)-cellulose plates of radiolabelled adenosine and 2'-deoxyadenosine and their metabolites have been described.[13] TLC was also used to separate hygromycin B from gentamicin and other aminoglycoside antibiotics, detection being done with a fluorescamine dip.[14] In related work the relative amounts of neomycin sulfates B and C were determined on TLC plates with 4-chloro-7-nitrobenzo-2-oxa-1,3-diazole as fluorescent reagent. This was followed by densitometric determination and the method was applied to a variety of commercial antibiotic samples.[15]

2.3 High-Pressure Liquid Chromatography. – References are grouped according to the class of sugar being analysed; not all of the important parameters pertaining to the separations are provided.

2.3.1 Detection methods. – An indirect UV method for mono-, di- and tri-saccharides uses an unsaturated acid, for example muconic or sorbic acid, as UV-absorbing component in the mobile phase. HPLC was conducted on a porous graphitized carbon column with aqueous alkaline eluant.[16]

2.3.2 Neutral sugars, alditols and derivatives thereof. – Reversed-phase HPLC studies have revealed a linear relationship between monosaccharide retention times and their Walkinshaw hydrophilicity indices. The method holds considerable promise for studies of carbohydrate solvation and hydrophobic interactions.[17] A system used to determine sugars and organic acids in food and drink samples uses on-line dialysis prior to HPLC separation.[18] HP anion exchange chromatography with pulsed amperimetric detection has been applied to marine sample hydrolysates for the determination of ten monosaccharides.[19] Cation exchange resins, on the other hand, have been used at 85 °C for alditol separations. A novel detection method employed in this work involves following the decrease in absorbance of a molybdate 2,5-dichloro-3,6-dihydroxy-1,4-benzoquinone anion complex caused by bonding of molybdate to the alditol.[20] Partially methylated 1,5-anhydro-D-galactitol benzoates, prepared during reductive cleavage analyses of polysaccharides, were separated by a mixture of reverse phase and normal phase HPLC and their GC and MS characteristics were reported.[21]

2.3.3 Glycosides and glycosylamines. – When methacrylic polymers were used as HPLC packing they showed selective binding for the β-isomer of *p*-aminophenyl galactoside tetraacetate.

Other α-anomers were also bound less effectively, and the amino group was required to permit association with the acid functions of the polymer.[22] Several reports have appeared on the HPLC analysis of plant glycosides: hydroxy-stilbene glycosides in wine,[23] steroidal and flavonoid glycosides of the Chinese medicine wuu-ji-san,[24] mono- and more complex acylated di- and tri-glycosides of anthocyanins,[25] various cardiac glycosides, for example digitoxin, digoxin and their metabolites,[26] and cardenolides and saponins in crude plant extracts.[27] In the last case both thermospray and continuous flow HPLC-MS techniques were employed and gave complementary information.

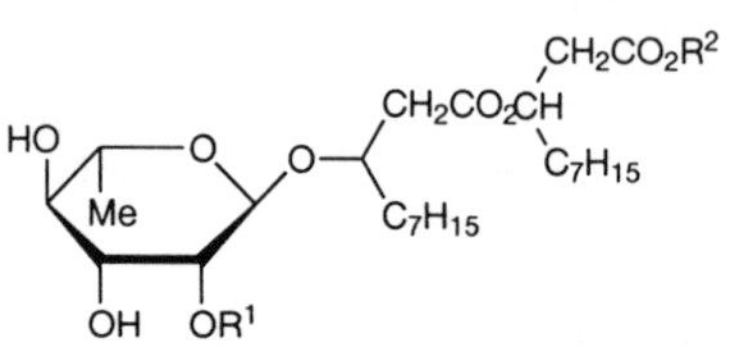

R^1 = H or α-L-Rhap
R^2 = H$_2$CCOC$_6$H$_4$Br-*p*

1

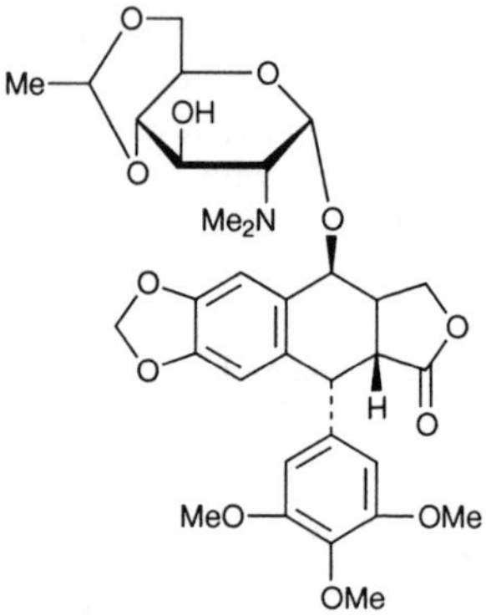

2

In the field of microbial products, the rhamnolipids **1** have been analysed by reverse phase methods,[28] and 2-*O*-α-D-glucopyranosylglycerol, a cyanobacterial 'osmolyte', has been analysed in the presence of several disaccharides, sugars and alditols. In the latter case reverse phase and calcium ion cation exchange resin columns were used.[29]

Reverse phase methods were also used to assay dimethylaminoetoposide (**2**) together with a stereoisomer and their *N*-demethylated metabolites in the urine of cancer patients.[30] A related paper discusses an analytical method for etoposide and isomers in plasma.[31] Other work describes the isolation of glucuronides of metronidazole (**3**) and its hydroxy metabolite (**4**) by preparative HPLC methods,[32] and the antipyrene metabolite derivatives **5-7** were identified by thermospray LC-MS methods.[33] In the area of the metabolites of narcotics, two papers have described the reverse phase-HPLC analysis of morphine and its 3- and 6-glucuronides in biological samples.[34,35] In the first case, fluorescence detection was used, in the latter electrospray MS procedures. Codeine and seven glycosidic metabolites have also been analysed by reverse phase-HPLC procedures with UV or electrochemical detection.[36]

The glycosylamine **8**, which is a new anti-cancer agent NB-506, can be assayed by a reverse phase-HPLC procedure.[37]

2.3.4 Di- and oligosaccharides and their derivatives. – Two HPLC-based methods have been used for the examination of unsaturated disaccharides released by enzymic digestion of heparin or heparan sulfate. In the first case various esters

	R^1	R^2	X
3 R^1 = R^2 = H	5	O-β-D-GlcAp	Me	H
4 R^1 = β-D-GlcAp	6	H	O-β-D-GlcAp	H
R^2 = O-β-D-GlcAp	7	H	Me	O-β-D-GlcAp

were examined,[38] while in the latter the sulfated disaccharides were examined on cation exchangers without derivitization.[39]

Reverse phase procedures were used for the separation of the major flavonoid disaccharide glycosides in orange and grapefruit concentrates,[40] and anion-exchange columns with pulsed amperometric detection allowed the separations of the oligosaccharides of honeys.[41] From this, fingerprints were developed for the types of honeys; none of the constituent oligomers was identified.[41]

The cellotriose and cellotetraose released from barleys with lichenase have been determined by a reverse phase procedure as a means of assaying for β-glucan content,[42] and HPLC-MS allowed the analysis of the maltosaccharides (DP 4-10) following reductive amination with 4-hexadecylaniline. The MS was conducted in the negative mode.[43]

2.3.5 Esters. – HPLC methods have been used to separate a wide range of sugar phosphates, bis-phosphates and nucleoside mono- and bis-phosphates. The method was based on the use of anion exchange resins and permits the detection of nanomole amounts of the esters.[44] A procedure for ion-pair reverse phase-HPLC analysis in plasma of gemfibrozil ester **9** avoids an intramolecular rearrangement or hydrolysis observed under physiological conditions.[45]

2.3.6 Sugar acids. – An extensive range of sugar acids and lactones has been separated on HPLC cation exchange columns,[46] and anion exchange resins have been used for determining ascorbic acid and galacturonic acid in fruit juices.[47] Ascorbic acid and dehydroascorbic acid have been assayed using ion exclusion columns,[48] and the magnesium salt of ascorbic acid 2-phosphate in cosmetic emulsions has been determined by a reverse phase procedure, it recently having been introduced as a skin whitener and radical scavenger.[49]

The neuraminidase inhibitor GC167 (**10**) has been assayed in human serum using reverse phase-HPLC and derivatization with benzoin to give **11**.[50]

2.3.7 Inositols. – Strong cation exchanger resins in the protonated form have been used for the determination of inositol in pharmaceutical formulations.[51]

2.3.8 Antibiotics. – The use of HPLC methods for the determination of the macrolide antibiotics roxithromycin and clarithromycin,[52] and also

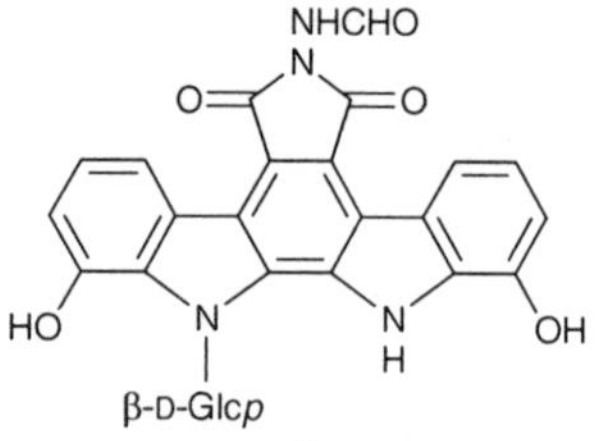

8

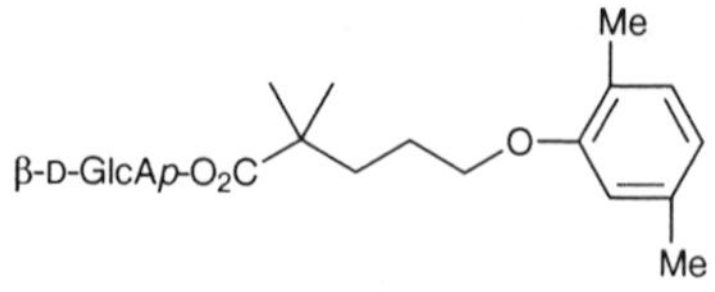

9

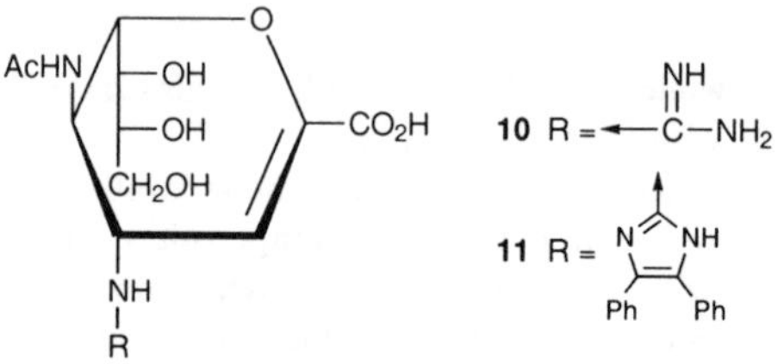

10 R = —C—NH₂ (with NH)

11 R = (imidazole with Ph, Ph)

spectinomycin in plasma,[53] has been reported, and an HPLC-thermospray MS assay for pirlimycin I (12) in bovine milk and liver samples has been developed.[54]

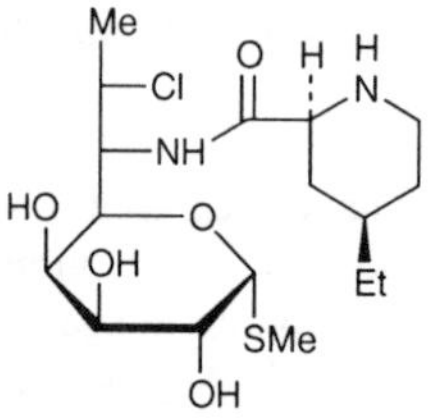

12

2.3.9 Nucleosides. – Adenosine was one of the compounds used to demonstrate a capillary HPLC-MS technique that employs a new 'in source' thermospray interface suitable for use with low molecular weight polar compounds.[55] Succinyladenosine and pseudouridine were amongst the purine and pyrimidine nucleosides detectable in an automated column switching HPLC method developed for examining nucleoside metabolic disorders.[56] Reverse phase procedures were used in the determination of diaminodideoxy analogues of 2′- and 3′-deoxyadenosine,[57] and the antiprotozoal drug sinefungin 13.[58] Similar procedures were used to assay thiopurine and methylthiopurine nucleosides and the corresponding nucleotides in cells. For low levels of thioguanine nucleotides in red blood cells the compounds were first oxidised with permanganate and detected fluorimetrically.[59]

In the field of pyrimidine nucleosides HPLC methods have been used to assay 1-β-D-arabinofuranosylcytosine in biological samples,[60] its 5′-stearyl phosphate derivative, which is a pro-drug,[61] and N^4-hexadecyl- and N^4-octadecyl derivatives which were prepared as pro-drugs for arabinosylcytosine.[62] Likewise an HPLC assay of the antitumour agent gemcitabine (2'-deoxy-2',2'-difluorocytidine) and the corresponding uridine derivative, to which it is metabolised, has been developed for use with human urine and plasma samples.[63]

A reverse phase HPLC thermospray MS method for detection of impurities in AZT found some of the 5'-trityl derivative and 3'-chloro-3'-deoxythymidine in tiny amounts; these are process impurities.[64] The AZT metabolite 3'-amino-3'-deoxythymidine can be assayed in plasma samples following derivatization with 9-fluorenylmethylchloroformate followed by fluorimetric detection. This metabolite is five times more toxic to bone marrow cells than is AZT.[65] 1-β-D-Arabinofuranosyl-*E*-5-(2-bromovinyl)uracil has been determined in urine using reverse phase columns and both manual and automated methods.[66]

2.4 Column Chromatography. – Enzyme-based and antibody-based post column detection systems for mono-, di- and oligosaccharides have been covered in a review of 'biospecific detection in liquid chromatography'.[67] In a fundamental study leading to methods for separation of macrolide antibiotics on silica gel Langmuir adsorption isotherms and rate parameters were determined.[68] Silica gel loaded with 8-hydroxyquinoline binding iron(III) has been used to selectively retain inositol 1,2,6-triphosphate and its *O*-phenylacetyl derivative for analysis in plasma.[69] Mixtures of various nucleosides were separated on copper(II)-complexed carbohydrate polymers such as Sephadex and modified cellulose.[70] The acidic products formed on treatment of saccharides under oxidizing conditions with alkali have been separated using cation exchange columns and the analyses led to a further assessment of the mechanism of the alkaline degradation of sugars. The well known Isbell mechanism was confirmed.[71] An improved method for determining the constituent monosaccharides of glycoproteins, following hydrolysis, used anion-exchange chromatography with pulsed amperimetric detection.[72] The proportions of glucose, maltose and maltotriose bound on a sodium-form polystyrene cation exchange resin at 60 °C increases with salt concentration and also with the size of the molecules.[73]

A porous graphite column has shown promise for separations of polar metabolites, such as glucuronides of AZT.[74] Micellar electrokinetic capillary chromatographic analysis of forty desulfoglucosinolates produced enzymolitically has been reported,[75] and a further study examined the liminoid glucosides in citrus seeds.[76]

13   **14**

3   Electrophoretic Methods

Almost all reports refer to the use of capillary electrophoresis which has now become an important analytical tool. In a comprehensive analysis of the aldohexoses the sixteen stereoisomers were separated by the method following reductive amination with (S)-1-phenylethylamine. In a separation of a more extensive range of compounds, nineteen out of twenty aldohexoses and tetroses were separated in a single run.[77] A combination of capillary electrophoresis in strongly alkaline media, coupled with off-column amperometric detection was effective for the separation and detection of monosaccharides with a detection limit of 1-2 µM.[78] Indirect UV detection of unsubstituted sugars under highly alkaline conditions was used in conjunction with their separation by capillary electrophoresis.[79] A further method operating at the nanomolar level involved derivatization of monosaccharides with the fluorogenic reagent 3-(p-carboxybenzoyl)quinoline-2-carboxaldehyde following reductive amination to 1-amino-1-deoxy alditols.[80]

High pressure capillary electrophoresis was used in the determination of mono-, di- and tri-sulfated unsaturated disaccharides released enzymically from glycosaminoglycans.[81] In related work, capillary electrophoresis and polyacrylamide gel electrophoresis were applied to the analysis of low molecular weight heparins and heparin oligosaccharides,[82] and the former method was used to monitor selective pivaloylation of a trisulfated unsaturated disaccharide derived from heparin.[83]

Capillary electrophoresis, coupled with laser-induced fluorescence detection has been used as an extraordinarily sensitive analytical procedure for oligosaccharides derived by enzymic degradation of polymers. The carbohydrates were linked to tetramethylrhodamine by way of a spacer group and the fluorescent products could be detected at the limit of 50 molecules, which must be taking analytical chemistry in this field to a new limit.[84] A similar procedure using 8-aminonaphthalene-1,3,6-trisulfonic acid was applied to neutral and basic oligosaccharides and operated with a detection limit of 5×10^{-8}M, derivatization being conducted in volumes as small as 2µl.[85] A further use of this approach has been with sialogangliosides labelled with 7-aminonaphthalene-1,3-disulfonic acid.[86]

Two reports have appeared on the use of capillary electrophoresis in the determination of ascorbic acid. One was for use with vitamin formulations and required L-cysteine be added as anti-oxidant,[87] while the other was for determination of the acid and also gluconic acid in foods and beverages.[88] Another application was to the separation of mono- and un-chlorinated analogues of compound **14** from this compound which is a novel phosphonate analogue of adenosine 5'-triphosphate.[89] A further study used the method coupled with on-line electrospray MS in the negative ionization mode to determine inositol mono- to hexa-phosphates and related compounds.[90]

A review has appeared on 'Fluorophore-assisted carbohydrate electrophoresis (FACE)' by which fluorescently-labelled carbohydrates are separated on polyacrylamide gels and a charged-coupled device camera is used to detect and quantify the products.[91]

4   Other Analytical Methods

A quantitative assay for D-galactose depends on an enzyme sensor based on D-galactose oxidase and a tris(2,2'-bipyridine) complex of osmium as a redox mediator developed on a carbon electrode.[92]

A range of physical methods including FTIR and ESR spectroscopy, magnetic susceptibility measurement and cyclic voltametry have been applied to transition metal saccharide complexes. A range of simple sugars were used in the study.[93]

Compound **15** has been used as a photo-induced electron transfer sensor for sugars which will detect a range of compounds at neutral pH values in aqueous media.[94]

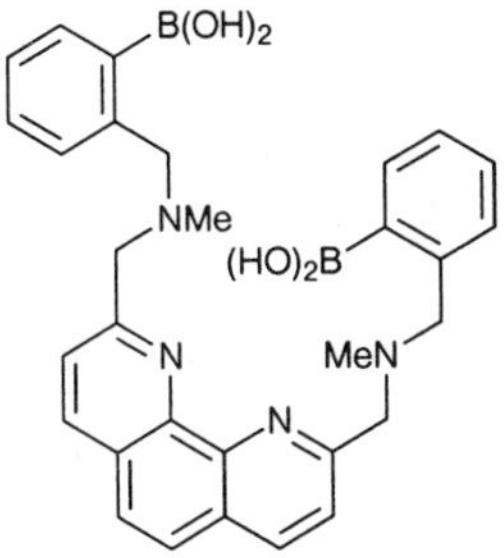

15

References

1   A.K. Singh and Y. Jiang, *J. Chromatogr. B*, 1995, **671**, 255.

2   M.R. Cases, A.S. Cerezo and C.A. Stortz, *Carbohydr. Res.*, 1995, **269**, 333.

3   O. Wilke, P. Mischnick, *Carbohydr. Res.*, 1995, **275**, 309.

4   Z. Mielniczuk, E. Mielniczuk and L. Larsson, *J. Chromatogr. B*, 1995, **670**, 167.

5   D. Chassagne, J. Crouzet, R.L. Baumes, J.-P. Lepoutre and C.L. Bayonove, *J. Chromatogr. A*, 1995, **694**, 441.

6   T. Iida, S. Tazawa, T. Tamura, J. Goto and T. Nambara, *J. Chromatogr. A*, 1995, **689**, 77.

7   C. Lauro-Marty, C. Lartigue-Mattei, J.L. Chabard, E. Beyssac, J.M. Aiache and M. Madesclaire, *J. Chromatogr. B*, 1995, **663**, 153.

8   I. Gremeau, V. Sautou, V. Pinon, F. Rivault and J. Chopineau, *J. Chromatogr. B*, 1995, **665**, 399.

9   J. Giacometti, C. Milin and N. Wolf, *J. Chromatogr. A*, 1995, **704**, 535.

10   G.W. Somsen, W. Morden and I.D. Wilson, *J. Chromatogr. A*, 1995, **703**, 613.

11   M. Esaiassen, K. Øvrbø and R.L. Olsen, *Carbohydr. Res.*, 1995, **273**, 77.

12   W. Chai, J.R. Rosankiewicz and A.M. Lawson, *Carbohydr. Res.*, 1995, **269**, 111.

13   E. Szabados and R.I. Christopherson, *J. Chromatogr. B*, 1995, **674**, 132.

14   M.B. Medina and J.J. Unruh, *J. Chromatogr. B*, 1995, **663**, 127.

15   E. Roets, E. Adams, I.G. Muriithi and J. Hoogmartens, *J. Chromatogr. A*, 1995, **696**, 131.

16   B. Lu, M. Stefansson and D. Westerlund, *J. Chromatogr. A*, 1995, **697**, 317.

17   N.W.H. Cheetham and K. Lam, *Carbohydr. Lett.*, 1995, **1**, 69.

18   E. Verette, F. Qian and F. Mangani, *J. Chromatogr. A*, 1995, **705**, 195.

19   P. Kerherve, B. Charriere and F. Gadel, *J. Chromatogr. A*, 1995, **718**, 283.

20   A.-M. Dona and J.-F. Verchère, *J. Chromatogr. A*, 1995, **689**, 13.

21   L.E. Elvebak, C. Abbott, S. Wall and G.R. Gray, *Carbohydr. Res.*, 1995, **269**, 1.

22   K.G.I. Nilsson, K. Sakaguchi, P.Gemeiner and K. Mosbach, *J. Chromatogr. A*, 1995, **707**, 199.

23   D.M. Goldberg, E. Ng, A. Karumanchiri, J. Yan, E.P. Diamandis and G.J. Soleas, *J. Chromatogr. A*, 1995, **708**, 89.

24   Y.-C. Lee, C.-Y. Huang, K.-C. Wen and T.-T. Suen, *J. Chromatogr. A*, 1995, **692**, 137.

25   M. Fiorini, *J. Chromatogr. A*, 1995, **692**, 213.

26   K.L. Kelly, B.A. Kimball and J.J. Johnston, *J. Chromatogr. A*, 1995, **711**, 289.

27   J.-L. Wolfender, K. Hostettmann, F. Abe, T. Nagao, H. Okabe and T. Yamauchi, *J. Chromatogr. A*, 1995, **712**, 155.

28   T. Schenk, I. Schuphan and B. Schmidt, *J. Chromatogr. A*, 1995, **693**, 7.

29   A. Schoor, N. Erdmann, U. Effmert and S. Mikkat, *J. Chromatogr. A*, 1995, **704**, 89.

30   M. de Fusco, M. D'Incalci, D. Gentili, S. Reichert and M. Zuchetti, *J. Chromatogr. B*, 1995, **664**, 409.

31   E. Liliemark, B. Pettersson, C. Peterson and J. Liliemark, *J. Chromatogr. B*, 1995, **669**, 311.

32   U.G. Thomsen, C. Cornett, J. Tjornelund and S.H. Hansen, *J. Chromatogr. A*, 1995, **697**, 175.

33   I. Velic, M. Metzler, H.G. Hege and J. Weymann, *J. Chromatogr. B*, 1995, **666**, 139.

34   J. Huwyler, S. Rufer, E. Kusters and J. Drewe, *J. Chromatogr. B*, 1995, **674**, 57.

35   R. Pacifici, S. Pichini, I. Altieri, A. Caronna, A.R. Passa and P. Zuccaro, *J. Chromatogr. B*, 1995, **664**, 329.

36   J.O. Svensson, Q.Y. Yue and J. Säwe, *J. Chromatogr. B*, 1995, **674**, 49.

37   N. Takenaga, Y. Ishii, S. Monden, Y. Sasaki and S. Hata, *J. Chromatogr. B*, 1995, **674**, 111.

38   R.J. Kerns and R.J. Linhardt, *J. Chromatogr. A*, 1995, **705**, 369.

39   K. Murata, A. Murata and K. Yoshida, *J. Chromatogr. B*, 1995, **670**, 3.

40   W.E. Bronner and G.R. Beecher, *J. Chromatogr. A*, 1995, **705**, 247.

41   I. Goodall, M.J. Dennis, I. Parker and M. Sharman, *J. Chromatogr. A*, 1995, **706**, 353.

42   A.M. Perez-Vendrell, J. Guasch, M. Francesch, J.L. Molina-Cano and J. Brufau, *J. Chromatogr. A*, 1995, **718**, 291.

43   L. Johansson, H. Karlsson and K.-A. Karlsson, *J. Chromatogr. A*, 1995, **712**, 149.

44   R. Swezey, *J. Chromatogr. B*, 1995, **669**, 171.

45   B.C. Sallustio and B.A. Fairchild, *J. Chromatogr. B*, 1995, **665**, 345.

46   K. Fischer, H.-P. Bipp, D. Bieniek and A. Kettrup, *J. Chromatogr. A*, 1995, **706**, 361.

47   G. Saccani, S. Gherardi, A. Trifiro, C.S. Bordini, M. Calza and C. Freddi, *J. Chromatogr. A*, 1995, **706**, 395.

48   T. Ito, H. Murata, Y. Yasui, M. Matsui, T. Sakai and K. Yamauchi, *J. Chromatogr. A*, 1995, **667**, 355.

49   A. Semenzato, R. Austria, C. Dall'Aglio and A. Bettero, *J. Chromatogr. A*, 1995, **705**, 385.

50   R.J. Stubbs and A.J. Harker, *J. Chromatogr. B*, 1995, **670**, 279.

51   R.W. Chong and B.J. Moore, *J. Chromatogr. A*, 1995, **692**, 203.

52   A. Hedenmo and B.-M. Eriksson, *J. Chromatogr. A*, 1995, **692**, 161.
53   N. Haagsma, P. Scherpenisse, R.J. Simmonds, S.A. Wood and S.A. Rees, *J. Chromatogr. B*, 1995, **672**, 165.
54   R.E. Hornish, A.R. Cazers, S.T. Chester and R.D. Roof, *J. Chromatogr. B*, 1995, **674**, 219.
55   A. Carrier, L. Varfalvy and M.J. Bertrand, *J. Chromatogr. A*, 1995, **705**, 205.
56   S. Sumi, K. Kidouchi, S. Ohba and Y. Wada, *J. Chromatogr. B*, 1995, **672**, 233.
57   G. Thoiti, A. Van Schepdael, P. Herdewijn, E. Roets and J. Hoogmartens, *J. Chromatogr. A*, 1995, **689**, 247.
58   C. Tharasse-Bloch, P. Brasseur, L. Farannec and J. Marchand, *J. Chromatogr. B*, 1995, **674**, 247.
59   C.W. Keuzenkamp-Jansen, R.A. de Abreau, J.P.M. Bökkerink and J.M.F. Trijbels, *J. Chromatogr. B*, 1995, **672**, 53.
60   M.L. Stout and Y. Ravindranath, *J. Chromatogr. A*, 1995, **692**, 59.
61   B. Ramsauer, J. Braess, M. Unterhalt, C.C. Kaufmann, W. Hiddemann and E. Schleyer, *J. Chromatogr. B*, 1995, **665**, 183.
62   R.M. Rentsch, R.A. Schwendener, H. Schott and E. Hanseler, *J. Chromatogr. B*, 1995, **673**, 259.
63   K.B. Freeman, S. Anliker, M. Hamilton, D. Osborne, P.H. Dhahir, R. Nelson and S.R.B. Allerheiligen, *J. Chromatogr. B*, 1995, **665**, 171.
64   A. Almudaris, D.S. Ashton, A. Ray and K. Valko, *J. Chromatogr. A*, 1995, **689**, 31.
65   H. Nattouf, C. Davrinche, M. Sinet and R. Farinotti, *J. Chromatogr. B*, 1995, **644**, 365.
66   D.B. Whigan and A.E. Schuster, *J. Chromatogr. B*, 1995, **664**, 357.
67   J. Emneus and G. Marko-Varga, *J. Chromatogr. A*, 1995, **703**, 191.
68   T. Kawai, R. Egashira, H. Itsuki and J. Kawasaki, *Kagaku Kogaku Ronbunshu*, 1995, **21**, 158 (*Chem. Abstr.*, 1995, **122**, 161 092).
69   B.A.P. Buscher, U.R. Tjaden, H. Irth, E.M. Andersson and J. van der Greef, *J. Chromatogr. A*, 1995, **718**, 413.
70   H. Sasaki and M. Ohara, *Carbohydr. Res.*, 1995, **271**, 235.
71   M.A. Shalaby, H.S. Isbell and H.S. El Khadem, *J. Carbohydr. Chem.*, 1995, **14**, 429.
72   S. Kishino, A. Nomura, M. Sugawara, K. Iseki, S. Kakinoki, A. Kitabatake and K. Miyazaki, *J. Chromatogr. B*, 1995, **672**, 199.
73   S. Adachi, T. Mizuno and R. Matsuno, *J. Chromatogr. A*, 1995, **708**, 177.
74   J. Ayrton, M.B. Evans, A.J. Harris and R.S. Plumb, *J. Chromatogr. B*, 1995, **667**, 173.
75   C. Bjergegaard, S. Michaelsen, P. Moller and H. Sorensen, *J. Chromatogr. A*, 1995, **717**, 325.
76   V.E. Moodley, D.A. Mulholland and M.W. Raynor, *J. Chromatogr. A*, 1995, **718**, 187.
77   C.R. Noe and J. Freissmuth, *J. Chromatogr. A*, 1995, **704**, 503.
78   X. Huang and W.T. Kok, *J. Chromatogr. A*, 1995, **707**, 335.
79   X. Xu, W.T. Kok and H. Poppe, *J. Chromatogr. A*, 1995, **716**, 231.
80   Y. Zhang, E. Arriaga, P. Diedrich, O. Hindsgaul and N.J. Dovichi, *J. Chromatogr. A*, 1995, **716**, 221.
81   N.K. Karamanos, S. Axelsson, P. Vanky, G.N. Tzanakakis and A. Hjerpe, *J. Chromatogr. A*, 1995, **696**, 295.
82   R. Malsch, J. Harenberg and D.L. Heene, *J. Chromatogr. A*, 1995, **716**, 259.
83   R.J. Kerns, I.R. Vlahov and R.J. Linhardt, *Carbohydr. Res.*, 1995, **267**, 143.
84   X. Le, C. Scaman, Y. Zhang, J. Zhang, N.J. Dovichi, O. Hindsgaul and M.M. Palcic, *J. Chromatogr. A*, 1995, **716**, 215.

356   *Carbohydrate Chemistry*

85   A. Klockow, R. Amado, H.M. Widmer and A. Paulus, *J. Chromatogr. A*, 1995, **716**, 241.
86   Y. Mechref, G.K. Ostrander and Z. El Rassi, *J. Chromatogr. A*, 1995, **695**, 83.
87   J. Schiewe, Y. Mrestani and R. Neubert, *J. Chromatogr. A*, 1995, **717**, 255.
88   C.H. Wu, Y.S. Lo, Y.-H. Lee and T.-I. Lin, *J. Chromatogr. A*, 1995, **716**, 219.
89   J.R. Dawson, S.C. Nichols and G.E. Taylor, *J. Chromatogr. A*, 1995, **700**, 163.
90   B.A.P. Buscher, R.A.M. van der Hoeven, U.R. Tjaden, E. Andersson and J. van der Greef, *J. Chromatogr. A*, 1995, **712**, 235.
91   G.-F. Hu, *J. Chromatogr. A*, 1995, **705**, 89.
92   K. Miyata, M. Fujiwara, J. Motonaka, T. Moriga and I. Nakabayashi, *Bull. Chem. Soc. Jpn.*, 1995, **68**, 1921.
93   K. Geetha, M.S.S. Raghavan, S.K. Kulshreshtha, R. Sasikala and C.P. Rao, *Carbohydr. Res.*, 1995, **271**, 163.
94   P. Linnane, T.D. James, S. Imazu and S. Shinkai, *Tetrahedron Lett.*, 1995, **36**, 8833.

24
Synthesis of Enantiomerically Pure Non-carbohydrate Compounds

1   Carbocyclic Compounds

An extensive review on the ring contraction of furanosides and pyranosides to cyclobutanols and cyclopentanols respectively, has appeared.[1]

The furanosides 1 and 4 react with magnesium methoxide to give the annelated cyclopropanes 2 and 5, and sodium borohydride reduction followed by detritylation affords 3 or 6 and 7 (Scheme 1). The reaction of 1 proceeds *via* an elimination occurring within intermediate 8, to generate magnesium enolate 9, which is internally alkylated by displacement of the C5 tosylate.[2]

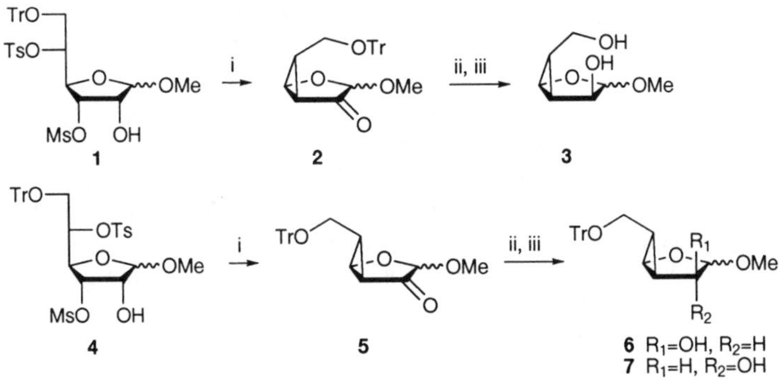

Reagents: i, Mg(OMe)$_2$, MeOH, PhH, 50-60 °C; ii, NaBH$_4$; iii, 80% AcOH.

Scheme 1

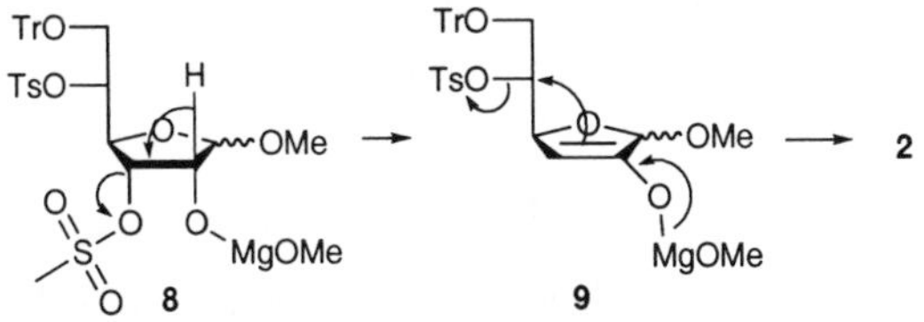

Intramolecular photochemical [2+2] ene-allene cycloaddition of 10 or homologue 11 generates the tricyclic methylene cyclobutanes 12 and 13 in 60-99%

yields. Using substrates with R=Me, Lewis acid catalysed introduction of an anomeric acetoxy group gives the bicyclic product **14**, further elaborated to *C*-allylic glycoside **15**. The related substrates with a tethered allylic alcohol **16** similarly undergo [2+2] addition to form the tricyclic cyclobutanes **17** in 62-95% yield.[3]

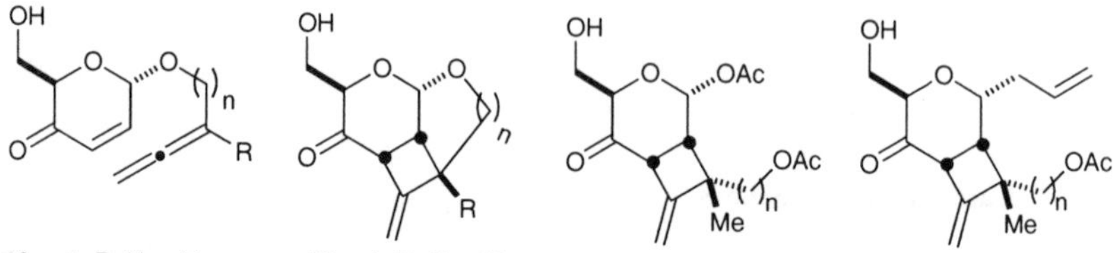

10 n=1, R=H or Me
11 n=2, R=H or Me

12 n=1, R=H or Me
13 n=2, R=H or Me

14

15

n =1,2

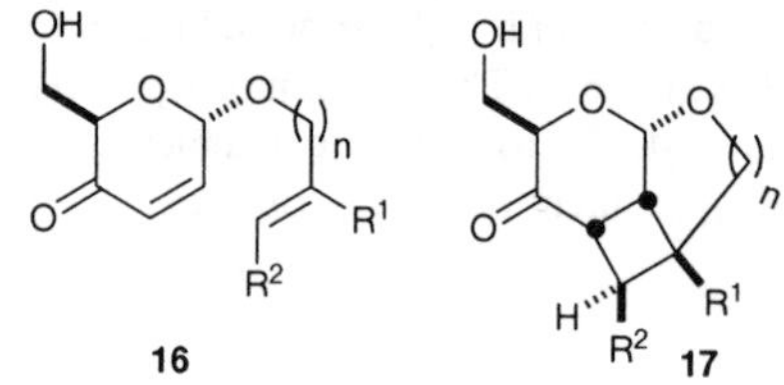

16

17

$R^1=R^2=H$
$R^1=Me, R^2=H$
$R^1-R^2= (CH_2)_3$
$R^1-R^2= (CH_2)_4$

Unsaturated aldonolactones **18** and **20** are converted in good yields to the functionalized homochiral cyclopentanoids **19** and **21** respectively, by intramolecular radical conjugate additions (Scheme 2).[4] Intramolecular radical conjugate addition has also been employed on the acyclic Wittig-derived **22** with concomitant lactonization to give the cyclopentanoid **23** in 64% yield (Scheme 3). The reaction was not diastereospecific, with C1 and C5 epimers of **23** also being formed.[5]

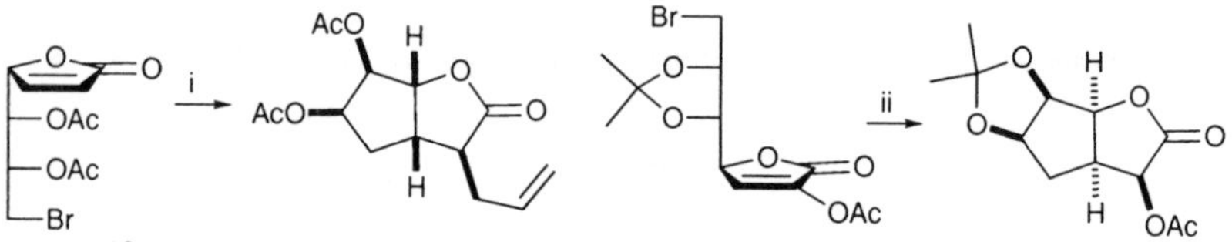

18

19

20

21

Reagents: i, ⟋⟍SnBu₃ , AIBN, EtOAc; ii, Bu₃SnH, AIBN, EtOAc.

Scheme 2

The fused cyclopentane **25** has been prepared by radical cyclization *via* deoxygenation of **24** (Scheme 4).[6] This ring system is a carbocyclic analogue related to the A/B ring system of the cytotoxic sesquiterpene, (+)-eremantholide.

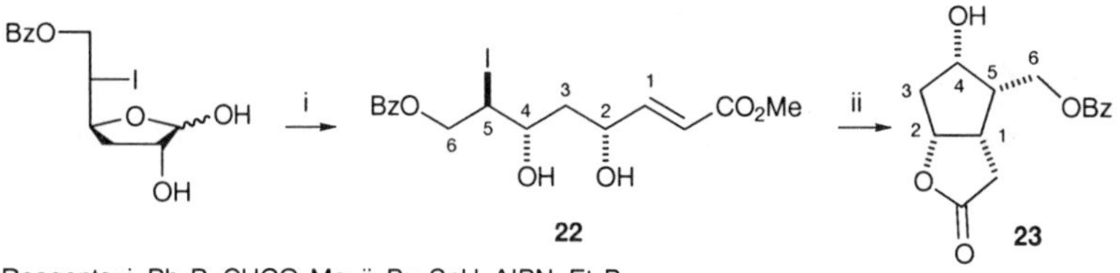

Reagents: i, Ph$_3$P=CHCO$_2$Me; ii, Bu$_3$SnH, AIBN, Et$_3$B.

Scheme 3

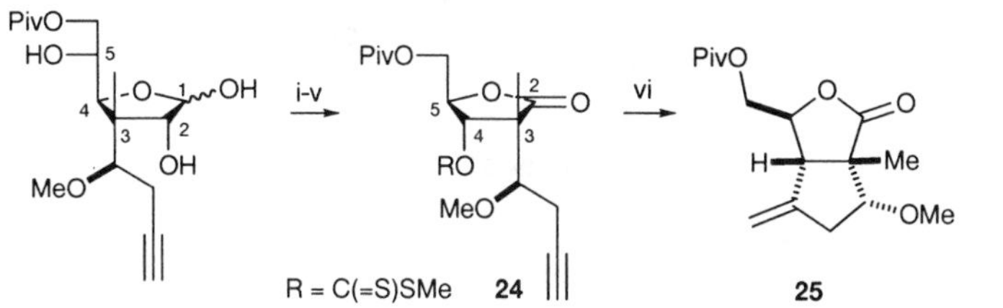

Reagents: i, IO$_4^-$; ii, PCC; iii, Et$_3$N, MeOH; iv, NaH, imidazole, CS$_2$; v, MeI; vi, Bu$_3$SnH, AIBN.

Scheme 4

The first example of an intramolecular aldol cyclopenta-annulation on a carbohydrate system has been described by Jenkins and co-workers.[7] Epoxide **26** was converted to **27** *via* axial epoxide opening with allyl magnesium chloride and subsequent diastereoselective enolate methylation, Wacker oxidation of **27** then providing diketone **28**. Treatment with potassium *t*-butoxide led to cyclopenta-annulation and dehydration of the intermediate aldol product gave **29**, confirmed by X-ray structure analysis.

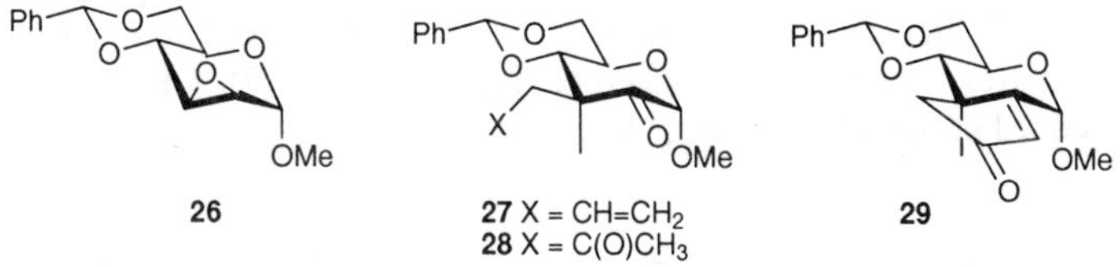

26

27 X = CH=CH$_2$
28 X = C(O)CH$_3$

29

The spiro-fused cyclopentenone **31** has been prepared by 3-aza-Cope rearrangement, intramolecular Mannich reaction, β-elimination reaction sequence from **30**. A mixture of enamine stereoisomers in the initial condensation leads to a mixture of spirocyclic epimers in **31** (Scheme 5).[8]

Chiara and co-workers have reported a new samarium diodide-catalysed synthesis of highly functionalized cyclopentanes, for example **32**, in good yields from hexosyl *O*-benzylhydroxylamines (Scheme 6).[9] Epimers were also formed in most cases. Terpenyl alditol derivative **34** was synthesized by Wittig reaction of D-ribose derivative **33** and further elaborations.[10]

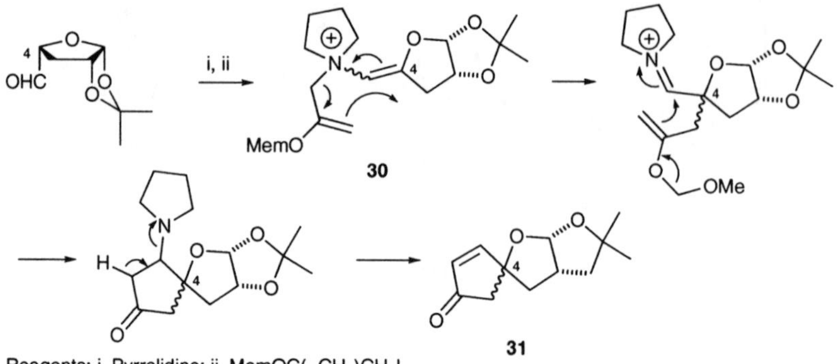

Reagents: i, Pyrrolidine; ii, MemOC(=CH₂)CH₂I.

Scheme 5

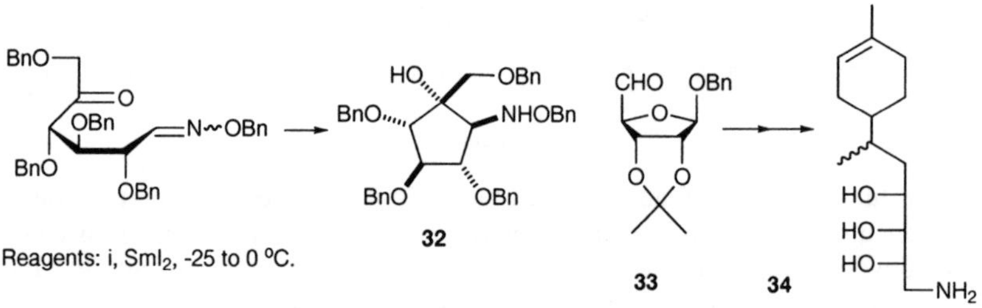

Reagents: i, SmI₂, -25 to 0 °C.

Scheme 6

Jung and Choe[11] have reported an efficient synthesis of cyclophellitol **36**, along with the 1(*R*),1(*S*) analogue **37**, from D-glucose derived **35**. The key transformations were Ferrier rearrangement, diastereoselective epoxidation and finally selective hydrogenation of the ring carbonyl group (Scheme 7). A similar strategy was described previously (Vol.28, p.349).

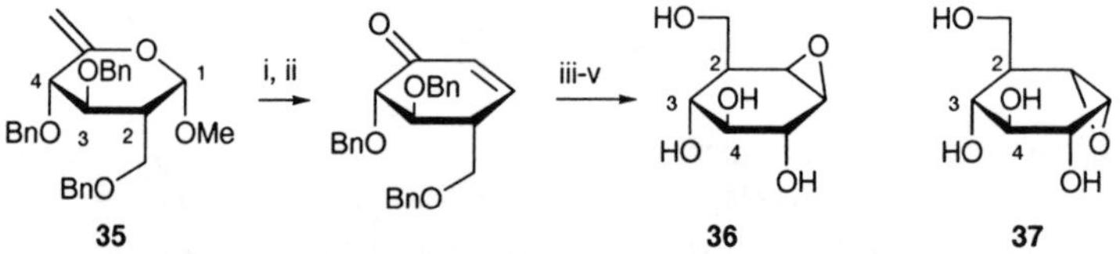

Reagents: i, HgCl₂, H₂O; ii, MsCl; iii, MCPBA; iv, K₂CO₃, MeOH; v, H₂, Pd.

Scheme 7

Ferrier rearrangement was also the key transformation in a synthesis of the 2-aminohexahydrobenzoxazole analogue **40** of trehazolin. Thus, reaction with

Hg²⁺, H₂O, oxidation then reduction of the derived ketone, followed by epoxidation and debenzylation converted **38** to **39** which was further elaborated to **40** (Scheme 8).[12]

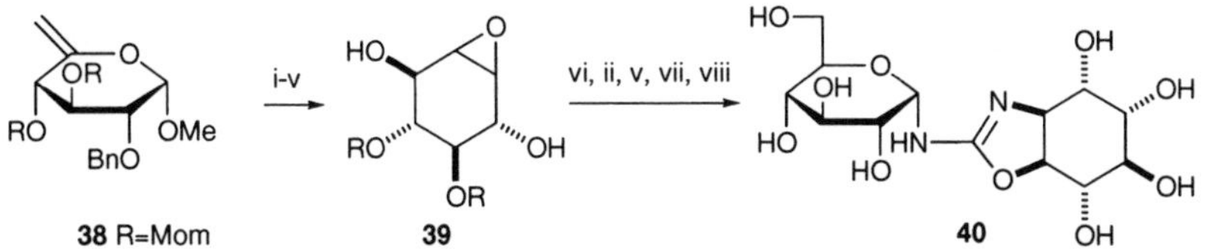

38 R=Mom   **39**   **40**

Reagents: i, Hg(O₂CCF₃)₂, Me₂CO, H₂O; ii, Ac₂O, Py; iii, NaBH₄, CeCl₃ ; iv, MCPBA; v, H₂, Pd(OH)₂/C; vi, NaN₃, NH₄Cl; vii, HCl, MeOH; viii, BnO—

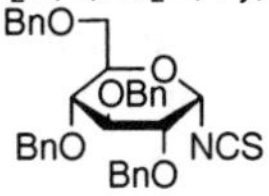

Scheme 8

Levoglucosenone was the starting material for studies towards the synthesis of the potent Na⁺-channel toxin (+)-tetrodotoxin. In model work, levoglucosenone was converted in 14 steps to **41** and then elaborated to **42** with new stereogenic centres introduced by allylic oxidation and epoxidation reactions. Lactone ring opening and elimination followed by transannular lactone formation through epoxide ring opening then gave the model compound **46**. Further work converted levoglucosenone to **43**. Epoxidation of the endocyclic alkene, followed by ozonolysis, addition of vinyl Grignard and benzoylation gave **44**. Oxidative cleavage of this new alkene then installed the carboxylate group. Lactonization analogous to that employed for the model compound gave **45** (Scheme 9).[13]

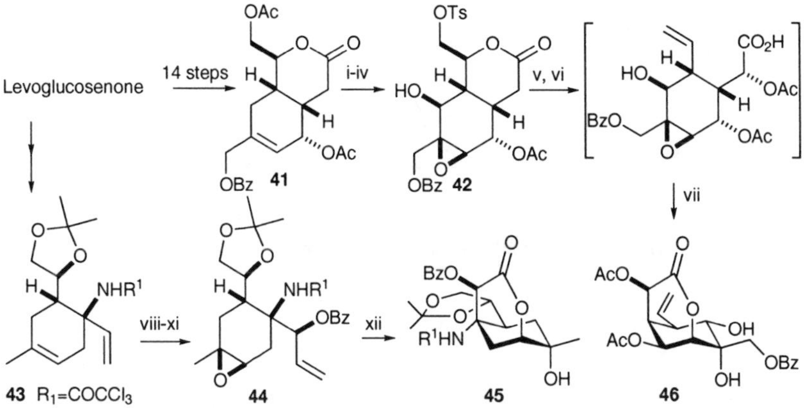

43 R₁=COCCl₃   **44**   **45**   **46**

Reagents: i, HCl, MeOH; ii, TsCl, Py; iii, SeO₂; iv, MCPBA; v, NaI; vi, Zn/Cu; vii, SiO₂; viii, MCPBA, NaHPO₄; ix, O₃, Me₂S; x, CH₂=CHMgBr; xi, BzCl; xii, cat. RuCl₃, NaIO₄.

Scheme 9

Shing and Wan have employed quinic acid as starting material for a synthesis of valiolamine **47** and several analogues, **48-50**, of this aminocyclitol analogue.[14]

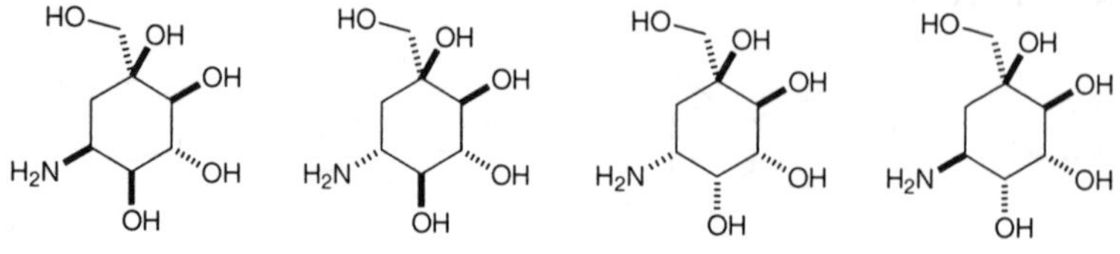

47   48   49   50

The antitumour alkaloids of the pancratistatin and narciclasine classes have been the subject of several synthetic efforts this year. The L-mannonolactone derivative **51** is the starting material for a synthesis of narciclasine analogue **55** and two related compounds. The key transformations are elaboration of the carbohydrate adduct **53**, derived by base-catalysed coupling of **51** with the anion derived from benzamide **52** to the cyclohexanone derivative **54**. This involves selective removal of the C5,C6 isopropylidene group, oxidative cleavage of the C5,C6 bond and an intramolecular aldol reaction involving the C5-derived aldehyde and the benzylic ketone enolate (Scheme 10).[15]

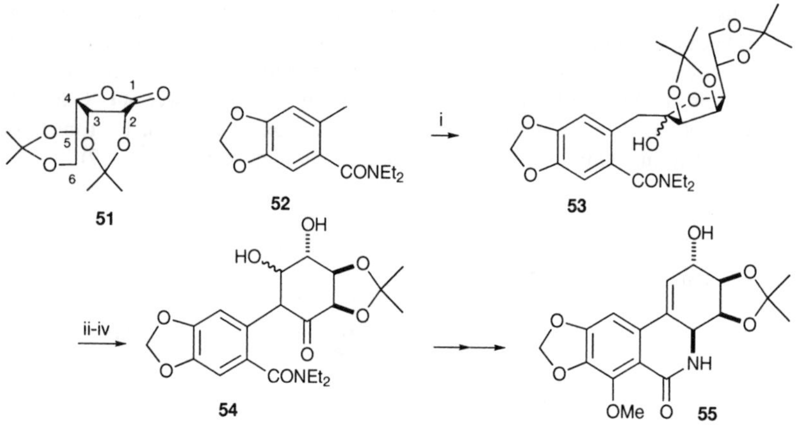

Reagents: i, s-BuLi; ii, AcOH; iii, NaIO₄; iv, DBU, THF, Na₂CO₃.

Scheme 10

D-Gulonolactone is the starting material for a synthesis of (+)-7-deoxypancra-tistatin (**60**). Thus, D-gulonolactone-derived **56** is elaborated to benzyloxyimine **57**, and thence to **58**. Intramolecular radical cyclization onto the benzyloxyimine function establishes the cyclohexylamine ring of the natural product in **59**, then elaborated to **60** (Scheme 11).[16] Hudlicky's group have reported the total synthesis of (+)-pancratistatin itself by elaboration of the diol obtained by *Pseudomonas aerogenosa* dihydroxylation of bromobenzene.[17]

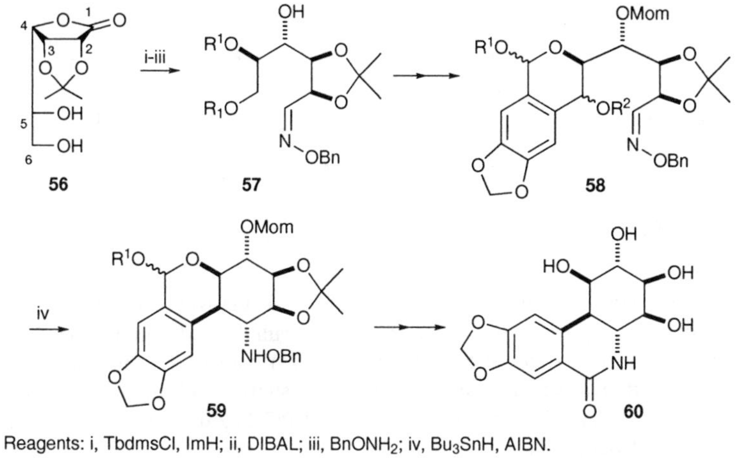

i-iii
iv
OH
OMom
R¹O
R₁O
OR²
N
OBn
NHOBn
NH
HO
56
57
58
59
60
Reagents: i, TbdmsCl, ImH; ii, DIBAL; iii, BnONH₂; iv, Bu₃SnH, AIBN.

Scheme 11

This year has also seen approaches towards components of the anticancer natural product taxol from carbohydrate starting materials. Paquette and Bailey have used the 5-deoxy-5-iodo-ribofuranoside **61** as starting material for synthesis of **63**, a precursor to the C ring of taxol. The carbohydrate-derived vinyl iodide **62** is transmetallated and the resulting vinyllithium reacts with terpenone **64** to afford **63** (Scheme 12).[18]

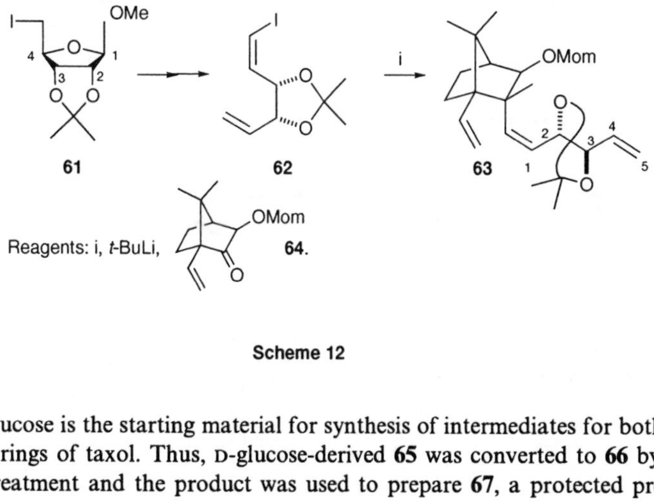

OMe
OMom
i
61
62
63
Reagents: i, t-BuLi,
64.

Scheme 12

D-Glucose is the starting material for synthesis of intermediates for both the A and C rings of taxol. Thus, D-glucose-derived **65** was converted to **66** by Hg^{2+}, H_2O treatment and the product was used to prepare **67**, a protected precursor directed towards the C ring of taxol.[19,20] This cyclohexanone served as an intermediate elaborated to **68**, a precursor directed towards synthesis of the A ring of taxol.

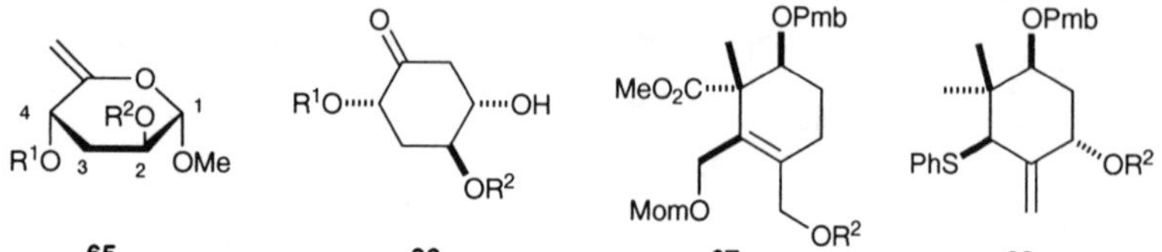

65
R[1] = Pmb
R[2] = Tbdms

66

67

68

The silyl-tethered compound **69** undergoes a tandem radical cyclization followed by desilylation and oxidation of the sulfide and elimination to give bicyclic **70**; further elaborations (by way of **71**) convert this intermediate to cyclohexane derivative **72**, transformed in five steps to **73**, an intermediate in Woodward's reserpine synthesis (Scheme 13).[21] PCC oxidation of the enol ether function of **72** yielded **74**.

D-Arabinose is the starting material in a synthesis of $1\alpha,2\beta$-dihydroxyvitamin D_3 (**78**). The derivative **75** is converted to the protected eneyne **76** in seven steps. This undergoes a palladium-catalysed cascade process on reaction with vinyl bromide to give **77**, yielding, after deprotection, the vitamin D_3 analogue (Scheme 14).[22] (See Vol 27, p.317 for synthesis of a similar chiron from D-mannitol).

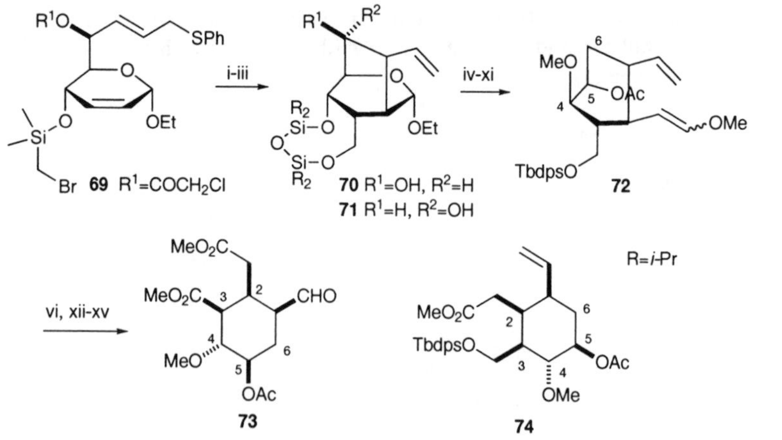

69 R[1]=COCH$_2$Cl

70 R[1]=OH, R[2]=H
71 R[1]=H, R[2]=OH

72

73

74   R=*i*-Pr

Reagents: i, Bu$_3$SnCl, NaCNBH$_3$, AIBN; ii, H$_2$O$_2$, KHCO$_3$, KF; iii, TipsCl; iv, PhOCOSCl, Py; v, Bu$_3$SnH, AIBN; vi, Bu$_4$NF; vii, Bu$_2$SnO, NBS; viii, NaBH(OAc)$_3$; ix, TbdpsCl, Et$_3$N; x, MeI, Ag$_2$O; xi, Ac$_2$O, Py; xii, AcOH; xiii, PDC; xiv, TmsCHN$_2$; xv, O$_3$.

Scheme 13

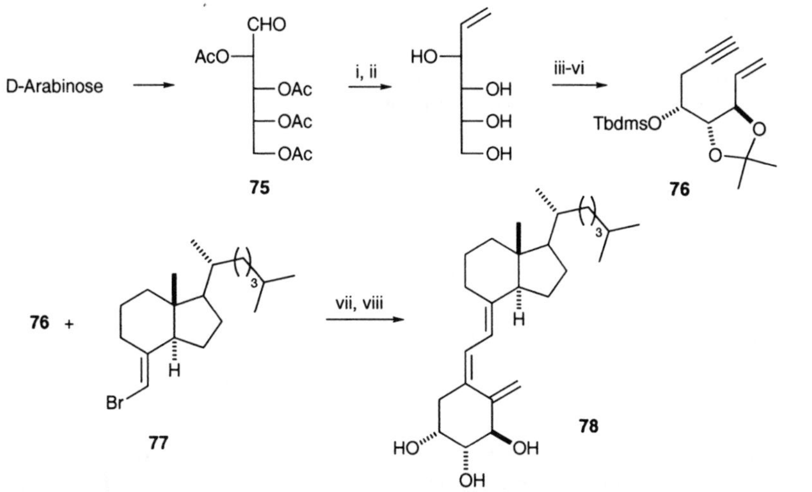

Reagents: i, Zn, CH$_2$I$_2$, AlMe$_3$; ii, HCl, MeOH; iii, 2,4,6-trimethylphenylsulfonyl chloride, Pyr; iv, Me$_2$CO, CuSO$_4$, TsOH; v, LiCCH.(CH$_2$NH$_2$)$_2$; vi, TbdmsCl; vii, Pd$_2$dba$_3$.CHCl$_3$; viii. HCl aq., THF.

Scheme 14

Tri-*O*-acetyl-D-glucal has been used for the synthesis of a number of functionalized *cis*-decalins, *via* pyranone **79**. Intramolecular Diels-Alder reaction of **79** gives the tricyclic intermediate **80**. Lithium aluminium hydride reduction affords a 1:1.7 mixture of alcohols which are separated and elaborated to **81** and **82** (Scheme 15). Ferrier rearrangement converts these to *cis*-decalins **83** the structures of which were proven by single crystal X-ray analysis.[23]

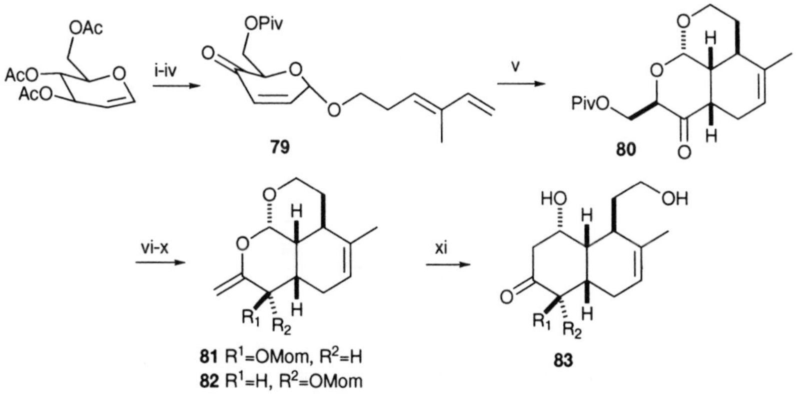

Reagents: i, BF$_3$.OEt$_2$, CH$_2$=C(Me)=CH(CH$_2$)$_2$OH; ii, NaOMe, MeOH; iii, *t*-BuCOCl; iv, MnO$_2$, CCl$_4$; v, PhH, hydroquinone, 155 °C; vi, LiAlH$_4$; vii, MsCl; viii, NaI, butanone; ix, MomCl, *i*-Pr$_2$NEt; x, AgF, Py; xi, Hg^{2+}, H$_2$O.

Scheme 15

2   Lactones

Sporothriolide **85** has been synthesized from carbohydrate derivative **84**, utilizing intramolecular radical cyclization onto a tethered acetylene, and 4-epi-ethisolide **87** has been prepared similarly from **86** (Scheme 16).[24]

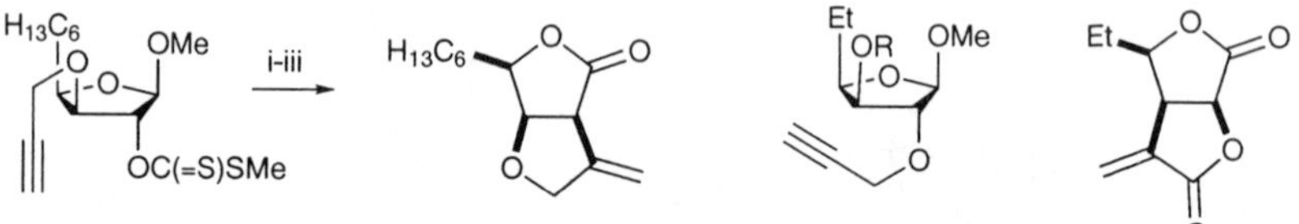

Reagents: i, Bu$_3$SnH, AIBN; ii, HCl; iii, PDC, NaOAc.   R = C(=S)SMe

(−)-Muricatacin **89** has been prepared from the D-isoascorbic acid-derived bis-epoxide equivalent **88** [SPR Vol.26, p.190, ref. 22] by two routes, one of which is shown in Scheme 17.[25] Similarly, L-ascorbic acid is the starting material in a synthesis of 6-hydroxy-δ-valerolactones, *via* the formal bis-epoxide equivalent **90**, epimeric with **88**, converted to epoxide **91** (with epoxide transposition), and thence to lactone **92**, the major mosquito oviposition attractant pheremone (Scheme 18).[26]

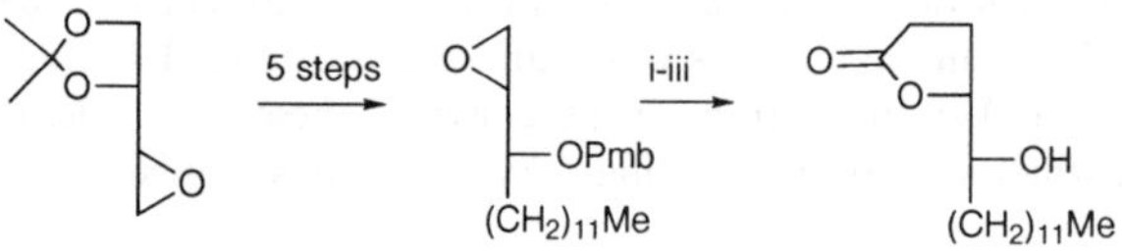

Reagents: i, CH$_2$(CO$_2$Et)$_2$, NaOEt; ii, MgCl$_2$.6H$_2$O, MeCONMe$_2$; iii, DDQ, CH$_2$Cl$_2$, H$_2$O.

Scheme 17

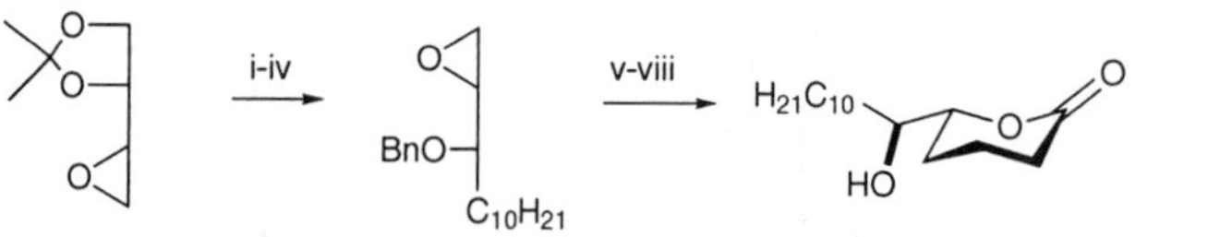

Reagents: i, C$_9$H$_{19}$MgBr, Li$_2$CuCl$_4$; ii, BnBr, NaH, Bu$_4$NI; iii, HOAc, H$_2$O; iv, Ph$_3$P, DIAD; v, LiCCCO$_2$Et, BF$_3$.OEt$_2$; vi, H$_2$, Pd/C; vii, K$_2$CO$_3$, MeOH then HCl; viii, 150 °C.

Scheme 18

Buszek and Jeong[27] have reported a synthesis of the 8-membered lactone **96**, a precursor of octalactin A **97** and B (which has an alkene in place of the side chain epoxide), from ascorbic acid. The known ascorbic acid-derived **93** undergoes conjugate addition and reductive elimination, followed by hydroboration and borane reduction to give the intermediate triol **94**. This was then elaborated to **95** and lactonized to the known compound **96** (Scheme 19).

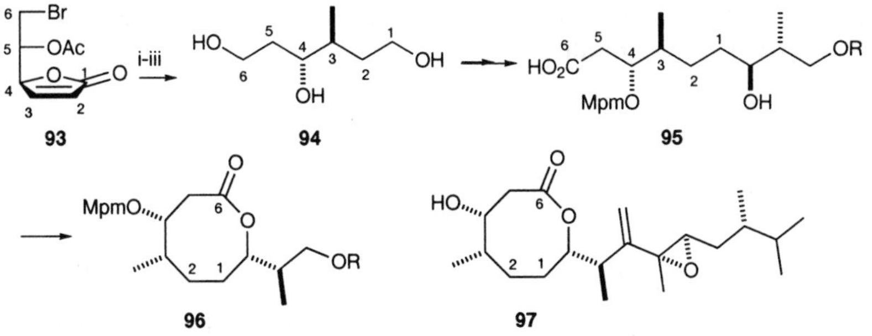

Reagents: i, Me$_2$CuLi; ii, Zn, HOAc; iii, 9-BBN, then BH$_3$ then H$_2$O$_2$.

Scheme 19

The two epimeric 3-hydroxy-4-hydroxymethyl-4-butanolides **98** and **99** have been prepared from levoglucosenone (Scheme 20).[28]

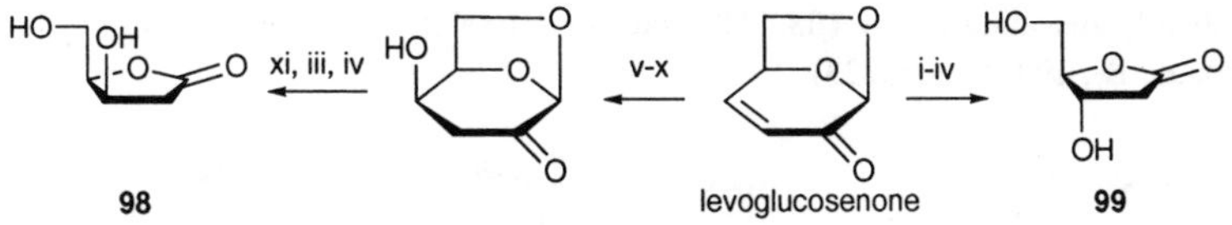

Reagents: i, H$_2$O, Et$_3$N; AcOOH, AcOH; iii, Me$_2$S; iv, HCl, MeOH; v, LAH; vi, AcOH, AcOAg.I$_2$; vii, NH$_3$.MeOH; viii, Swern; ix, (PhSe)$_2$, NaBH$_4$; x, AcOH, EtOH; xi, AcOOH, AcOH

Scheme 20

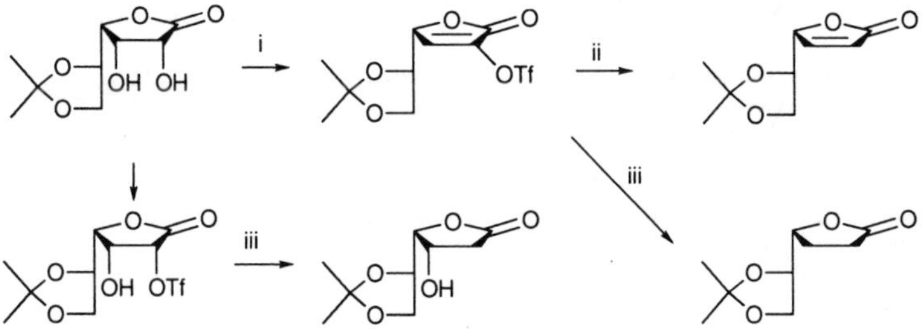

Reagents: i, Tf$_2$O, Py; ii, Me$_3$SnH, LiCl, Pd(PPh$_3$)$_4$; iii, H$_2$, Pd/C, i-Pr$_2$NEt.

Scheme 21

A number of useful chiral lactones have been prepared from L-mannonolactone (Scheme 21). D-Gulonolactone has also been similarly employed, and the enantiomers of these two groups of lactones have been obtained commencing with D-mannonolactone and L-gulonolactone, respectively.[29]

The lactone **103**, a key intermediate towards pseurotin A, is obtained in 10 steps from **100** (Scheme 22) through diastereoselective vinylation of **100**, and dihydroxylation of intermediate **101**.[30]

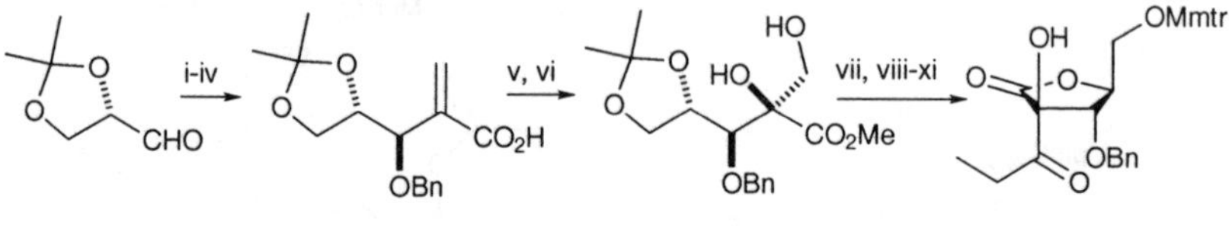

100 **101** **102** **103**

Reagents: i, BuLi, Br–C(=CH$_2$)–CH(OEt)$_2$; ii, KOH, BnBr, DMSO; iii HCl, THF; iv, $NaOCl_2$; v, 4-tolNNNHMe; vi, OsO_4, NMNO; vii, Swern; viii, EtMgCl; ix, HCl, THF; x, MmtrCl; xi, NMNO, $n\text{-}Pr_4NRuO_4$.

Scheme 22

Tri-*O*-acetyl-D-glucal is the starting material for a synthesis of lactone **108** *via* **104**. Palladium catalysed allylic substitution at C4 to give **105**, formed as a 72:28 mixture of epimers at the sulfonyl substituted centre, and desulfonation are the key steps. Intermediate **106** then undergoes iodoetherification giving **107** which is radically deiodinated to **108**. This lactone is related to a precursor of thromboxane B_2, **109** (Scheme 23).[31]

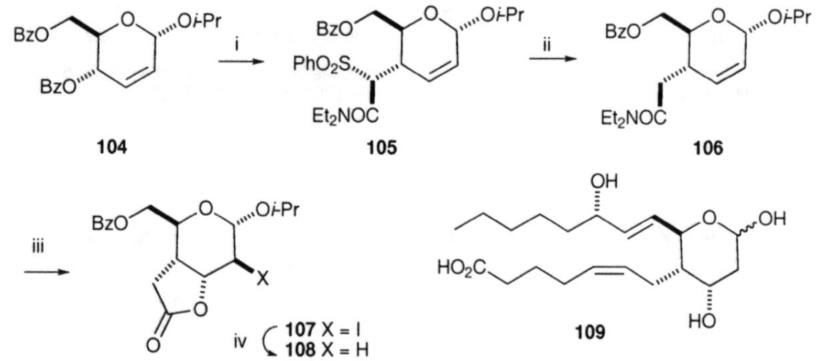

104 **105** **106**

107 X = I
108 X = H
109

Reagents: i, $NaPhSO_2CHCONEt_2$, $Pd(PPh_3)_4$; ii, 6% Na(Hg), MeOH, NaH_2PO_4; iii, I_2, $Ag(OCOCF_3)$, aq. THF; iv, Bu_3SnH, AIBN.

Scheme 23

D-Glucose is the starting material for a new synthesis of the immunosuppressive agent (–)-PA-48153C **110**, *via* the methyl glycoside **111**.[32]

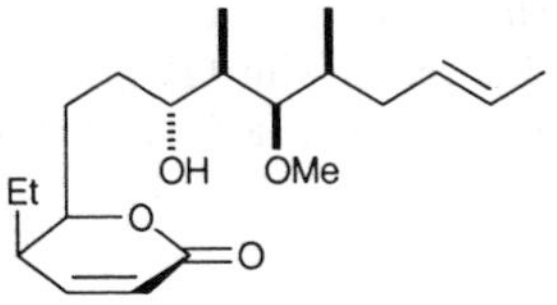

110 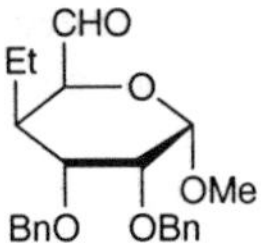   111

3   Macrolides, Macrocyclic Lactams and their Constituent Segments

The dispiroacetal component **122** (Scheme 25), a segment of tautomycin, has been synthesized from tri-*O*-acetyl-D-glucal and levoglucosenone in a convergent synthesis. The glucal is converted to the 2,3,4-trideoxysugar **112** and then to the protected *C*-glycoside **113**. Deprotection of the acetylene and hydrosilylation, followed by several steps involving methylation and sulfide oxidation gives **114** which is elaborated to epoxide **116**.

Levoglucosenone is converted to **117**, isomeric with **112**, in 5 steps and is then

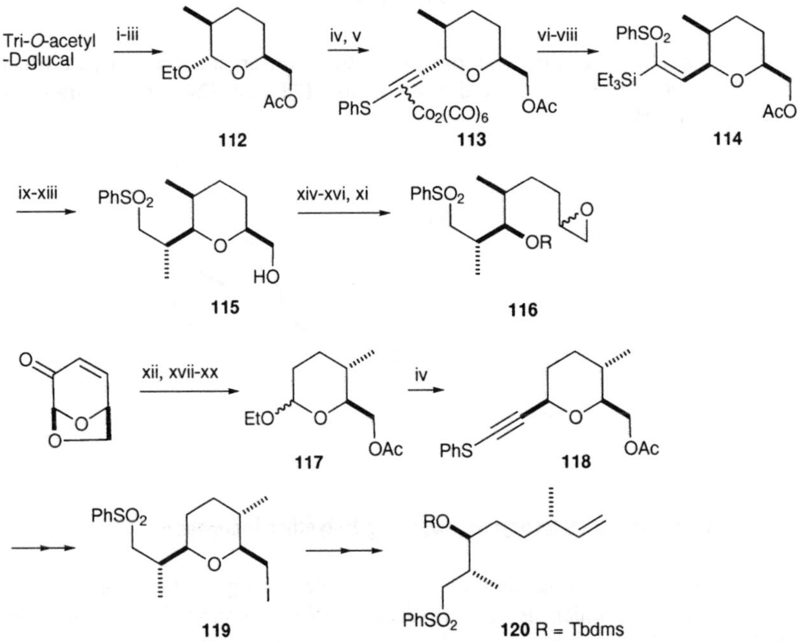

Reagents: i, BF$_3$.OEt$_2$, EtOH; ii, CuCN, MeLi; iii, H$_2$, 5% Pt/C; iv, TMS-CC-SPh, BF$_3$.OEt$_2$; v, Co$_2$(CO)$_8$; vi, TfOH; vii, I$_2$; viii, Et$_3$SiH, Na$_2$PtCl$_6$.6H$_2$O; ix, NaOMe, MeOH; x, TbdmsCl, ImH; xi, MCPBA; xii, MeLi.LiBr; xiii, TBAF; xiv, (PhO)$_3$PMeI; xv, Zn; xvi, TbdmsOTf, 2,6-lutidine; xvii, H$_2$NNH$_2$.H$_2$O, EtOH; xviii, NaH, DMSO; xix, Ac$_2$O, Py; xx, EtOH, TsOH.

Scheme 24

transformed to *C*-glycoside **118**, which is then elaborated to **119** and **120** (Scheme 24). The carbanion generated from **120** reacts with **116** giving **121**, which, after oxidation, deprotection and Wacker reaction affords the dispiroketal **122** (Scheme 25).[33]

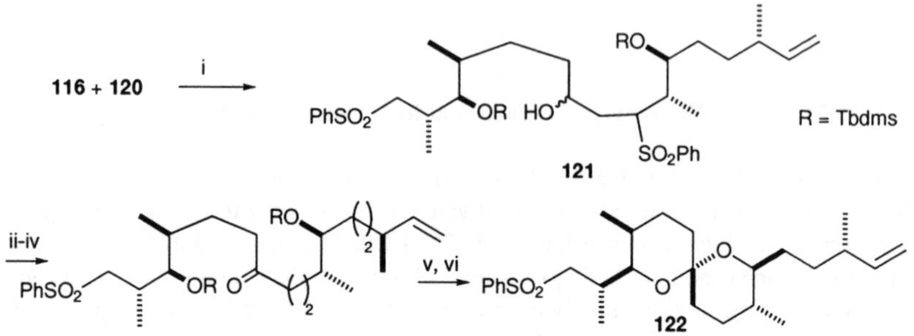
116 + 120
i
RO
PhSO2
OR
HO
SO2Ph
R = Tbdms
121
ii-iv
v, vi
122

Reagents: i, *n*-BuLi, cat. $BF_3.OEt_2$; ii, PCC; iii, DBU; iv, $[(Ph_3P)CuH]_6$; v, TBAF, TsOH; vi, $PdCl_2$, CuCl, O_2, DMF.

Scheme 25

D-Xylose is the starting material in another convergent synthesis, in which the two acyclic D-xylose derived components **123** and **124** are precursors to (−)-colletol **125**.[34]

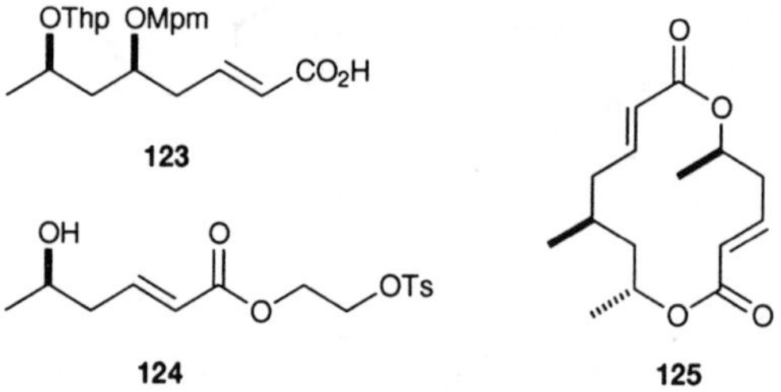
OThp OMpm
CO2H
123
OH
O
OTs
124
125

4 Other Oxygen Heterocycles, including Polyether Ionophores

A short review [8 refs] has appeared this year describing syntheses of zaragozic acid a number of which use carbohydrate starting materials.[35] D-Xylose is the starting material for synthesis of **126**, a model of the tricyclic core of the zaragozic acids. The key steps are thioacetal acetate exchange, and acetate displacement with a silyl enol ether, followed by acetal exchange to form the ketal ring of the final product (Scheme 26).[36]

A close analogue **129** of the bicyclic core of the zaragozic acids has been prepared from 2,3:4,6-di-isopropylidine-L-*xylo*-hexulosonic acid. Key carbon

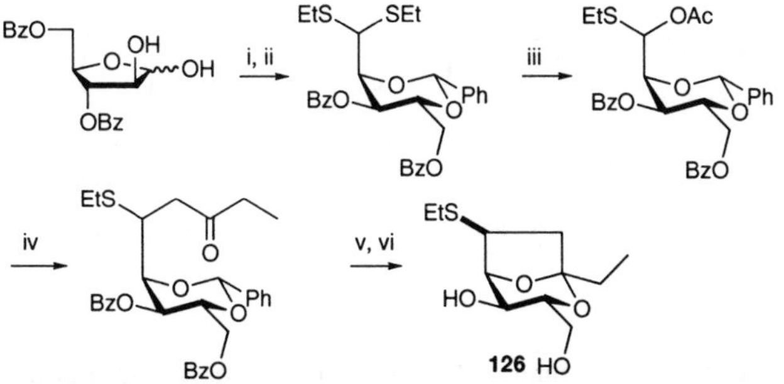

126

Reagents: i, EtSH, HCl; ii, $PhCH(OMe)_2$, H^+; iii, $Hg(OAc)_2$, HOAc; iv, $EtC(=CH_2)OTms$, TmsOTf, O °C; v, MeOH, H^+; vi, LAH.

Scheme 26

homologations are achieved through enolate hydroxymethylation of **127** and Wittig methylenation of the aldehyde from Swern oxidation of **128** (Scheme 27).[37] A modification of this approach has also been used to prepare **130**, which has two further carbon substituents as required for the zaragozic acids themselves.[38]

Carreira and Du Bois have provided full details of last year's report of a synthesis of (+)-zaragozic acid C from D-erythronic lactone **131**, *via* amide **132** and alkyne **133** (Scheme 26).[39]

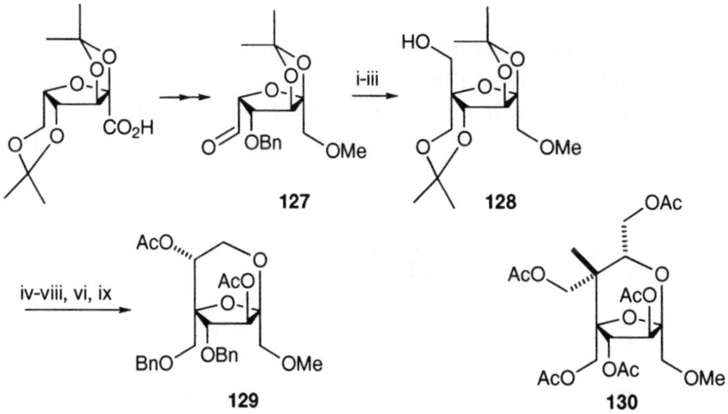

127 128 129 130

Reagents: i, HCHO, NaOH; ii, Li, NH_3; iii, $Me_2C(OMe)_2$; iv, Swern; v, $Ph_3P{=}CH_2$; vi, H^+; vii, NaH, BnBr; viii, OsO_4, NMNO; ix, Ac_2O, Py.

Scheme 27

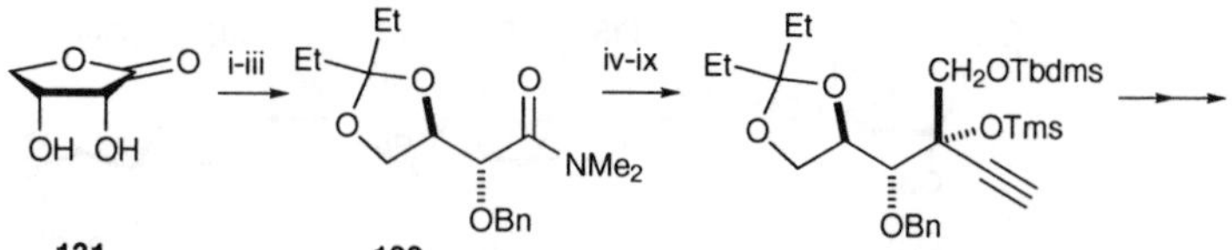

131 132 133

Reagents: i, Me_2NH; ii, $(MeO)_2CEt_2$; iii, NaH, BnBr; iv, $LiC(OEt){=}CH_2$; v, TMS-C≡C-MgBr; vi, O_3; vii, $NaBH_4$; viii, TbdmsCl; ix, TmsCl.

Scheme 28

The bicyclic compound **138** [R=allyl, phenyl or $CH_2OTbdps$], related to the core of the squalestatins and zaragozic acids, has been prepared from anhydro sugar **134** *via* **137** (Scheme 29). Notably, with models where R=Me or CO_2Et, **134** is converted to anhydro sugar **135** predominantly when R=Me, but **136** predominantes when R=CO_2Et. This is rationalized by the destabilizing effect of electron withdrawing carboethoxy group on the oxonium intermediate involved in the formation of **136** (carboethoxy α to O^+), relative to the 6-membered oxonium leading to **135** (carboethoxy β to O^+).[40]

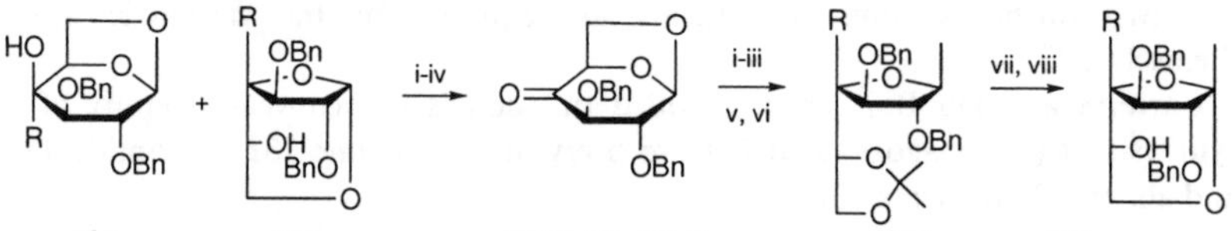

135 136 134 R=Me, CO_2Et 137 138

Reagents: i, RMgX; ii, CF_3CO_2H, Ac_2O; iii, NaOMe; iv, *p*-TsOH; v, Me_2CO, H^+; vi, PDC; vii, MeCeCl; viii, HCl.

Scheme 29

The dihydrofuran derivative **143** has been prepared ultimately from glucitol *via* the dianhydride ditosylate **139**. Reaction with sodium iodide in acetone gave a mixture of mono- and diiodides **140**, **141**, which were collectively converted to tosylate **142** and hence **143** on reaction with methyllithium (Scheme 30).[41]

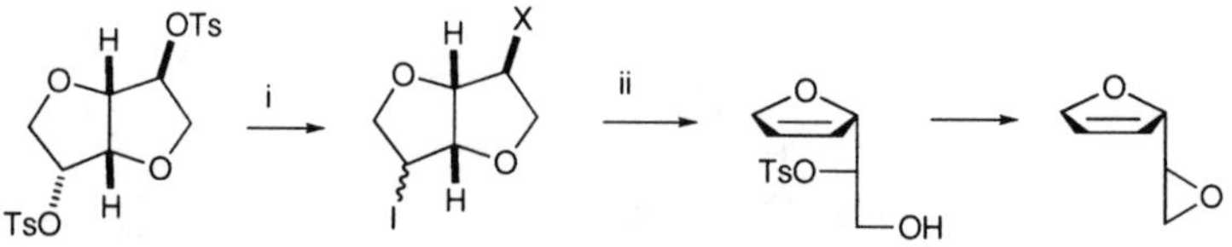

139 140 X=OTs 141 X=I 142 143

Reagents: i, NaI, acetone; ii, MeLi.

Scheme 30

Intramolecular 1,3-dipolar cycloaddition between the allyl ether and nitrone units in **144** provides the fused tetrahydrofuran **145**.[42] Interestingly, the analogous compound, containing *O*-allyl groups (in place of *O*-benzyl) potentially allowing formation of other fused ring systems, also gives the [3.3.0] bicyclic product. Methyl-4,6-*O*-benzylidine-α-D-glucopyranoside is the starting material for **146**, a model for the bicyclic unit of the miharamycins (discussed in chapter 19).[43]

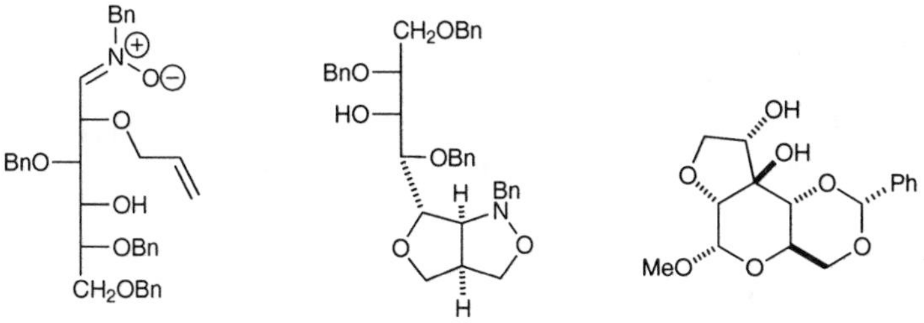

144   **145**   **146**

D-*Glycero*-D-*gulo*-heptono-γ-lactone from which **147** is derived, is the starting material for a synthesis of (+)-goniofufurone **148** (Scheme 31).[44]

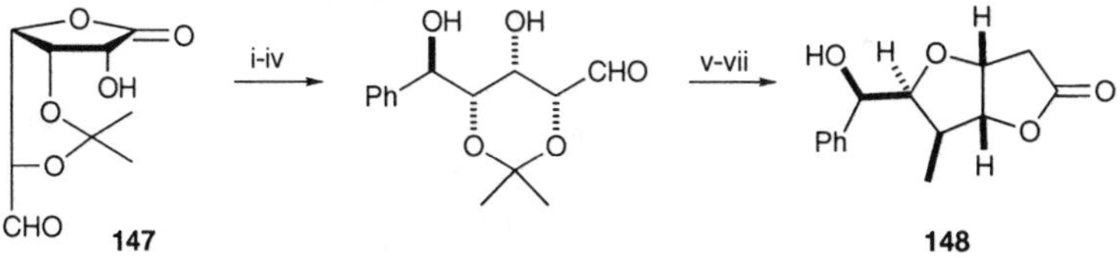

Reagents: i, PhMgBr; ii, Ac₂O, Py; iii, NaOH; iv, NaIO₄; v, Ph₃P=CHCO₂Me; vi, AcOH, H₂O; vii, DBU.

Scheme 31

The fused methylene tetrahydrofuran **150** is obtained in variable yield (20-72%) by palladium-catalysed cyclization of the vinyl bromides **149**, in the presence of NaBPh₄ or Bu₃SnH (in an attempt to trap the σ-palladium species with phenyl). The non-cyclized product **151** arising from phenyl coupling to the vinyl bromide function is also obtained, along with the endocyclic dihydrofuran **152** as a minor product (7-38% yield) when R=H.[45]

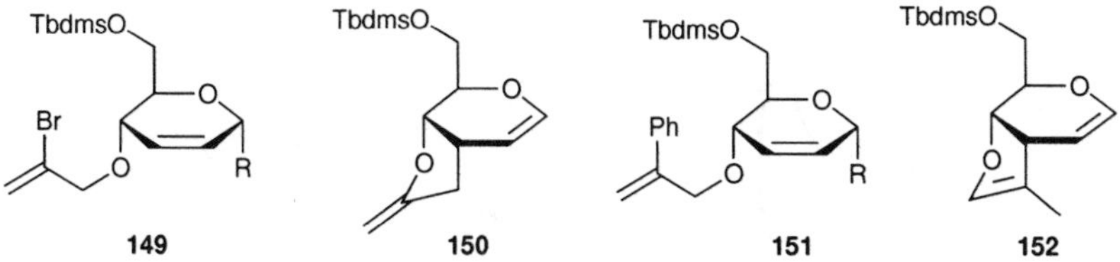

149   **150**   **151**   **152**

R = OEt, OBuᵗ, OC₆H₄Buᵗ

The isomeric epoxy triflates **153** and **157** undergo triflate displacement-epoxide opening with the dianion of methyl propanoyl acetate. Reaction of **153** generates the epimeric bicyclic tetrahydrofurans **154** and **155**, subsequent treatment with triflic acid leading to isomerization about the alkene bond to a mixture of **154-156**. Isomeric **157** under similar conditions gives **158** and **159**, with triflic acid catalysis leading to some of the alkene isomers **160** along with **158** (Scheme 32).[46]

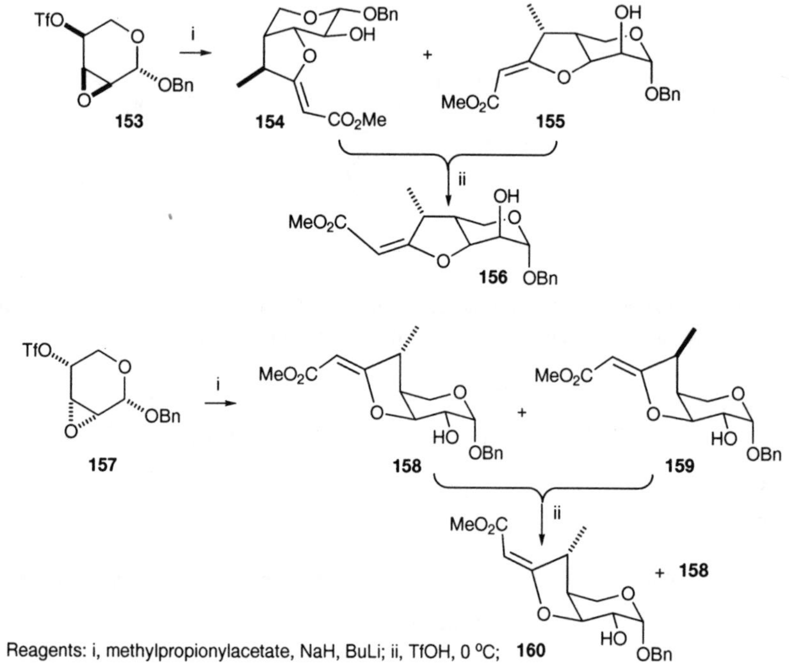

Reagents: i, methylpropionylacetate, NaH, BuLi; ii, TfOH, 0 °C;   **160**

Scheme 32

Epoxide **153** was also converted into the tricyclic ring systems **161** and **162**, using 2-carboxymethylcyclopentanone and 2-carboxymethylcyclohexanone, with NaH, BuLi, respectively.

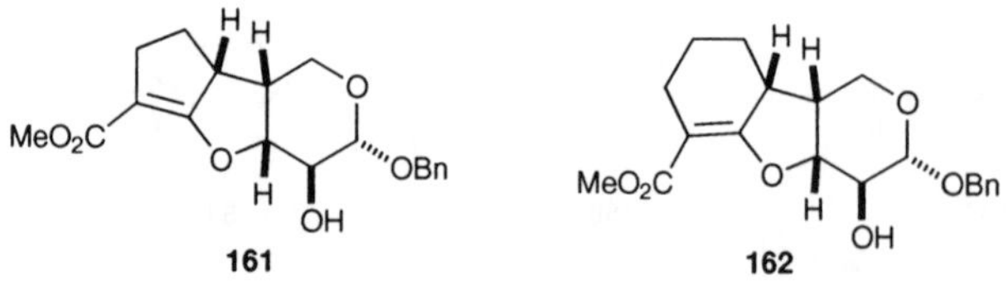

161   **162**

Tethered intramolecular [3+2] nitrile oxide and nitrone cycloadditions have been investigated starting from **163**.[47] In general, NO bond cleavage proved problematical, but the reaction was successful in some cases, compound **164** converted to the tetrahydrofurans **165** and **167** (Scheme 33). Analogous cyclo-addition chemistry was used to prepare the pyrrolidine analogue **278** [section 5].

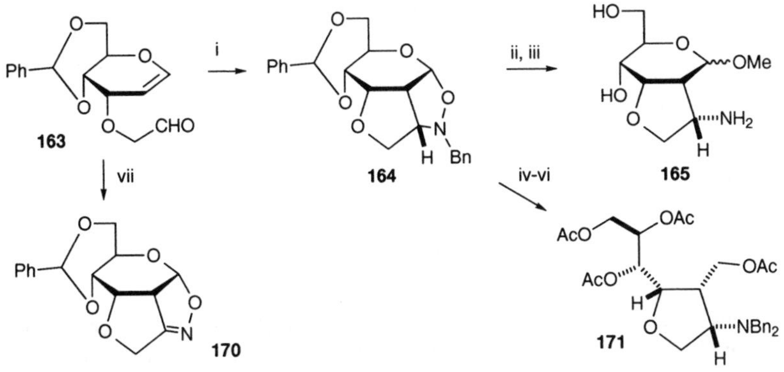

Reagents: i, BnNHOH; ii, H₂, Pearlman's catalyst; iii, MeOH, HCl, Ultrasound; iv, BnBr; v, LAH; vi, Ac₂O; vii, NH₂OH; viii, NCS.

Scheme 33

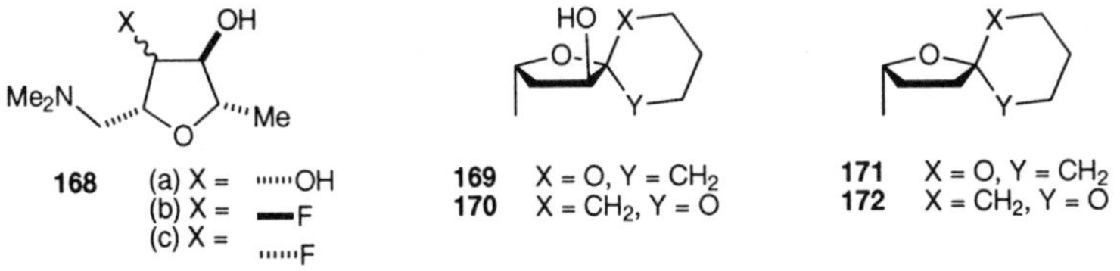

The synthesis and structure-activity relationships of a series of 3-substituted muscarines **168** have been described. The *cis*-3-fluoromuscarine **168(b)** showed selectivity and binding affinity similar to the muscarines, while the *trans* epimer **168(c)**, and either epimer of the 3-hydroxy, 3-azido or the 3,4-epoxymuscarine all showed substantially lower binding affinity and activity.[48] L-Sorbose is the starting material for synthesis of the tetrahydrofurans **169** and **170**, which were deoxygenated *via* their *O*-xanthates to give **171** and **172**. Epimerization was observed so that either **169** or **170** generated mixtures of epimers **171** and **172**.[49]

Levoglucosenone is the starting material for a new, shorter route to (−)-δ-multistriatin **175**.[50] The key steps involve firstly conjugate addition of nitro-methane anion to levoglucosenone, followed by addition of another equivalent of this anion to the ring ketone. Radical deoxygenation and denitration of the derived thiocarbonate **173**, and a two carbon homologation at C2 *via* alkylation of the intermediate dithiane **174**, is then followed by final anhydro ketose formation (Scheme 34).

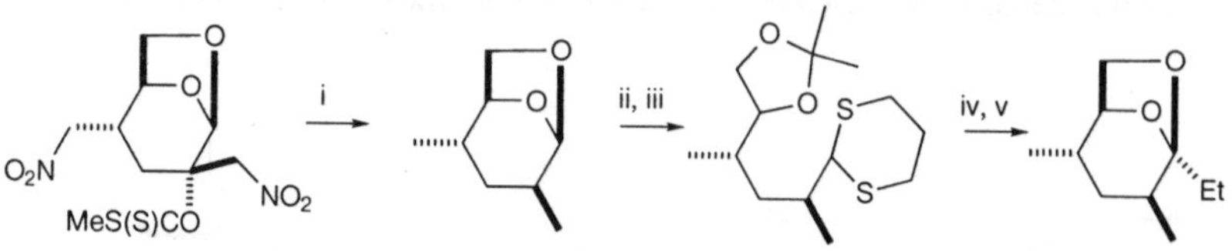

173   **174**   **175**

Reagents: i, Bu$_3$SnH, initiator; ii, BF$_3$, HS(CH$_2$)$_3$SH; iii, Me$_2$C(OMe)$_2$, H$^+$; iv, BuLi, EtI, v. Cd(CO$_3$)$_2$, HgCl$_2$, MeCN.

Scheme 34

Diels-Alder addition of dimethyl acetylenedicarboxylate to ascorbic acid-derived furan **176** (through (+)-Eu(hfc)$_3$ catalysis) provides a route to the bicyclic system **177** (and its diastereoisomer).[51] Chain-extended 2,3-dideoxypyranosides **178** (both D-gluco- and D-galacto- series) serve as precursors to 2,5-disubstituted tetrahydrofurans in an IDCP-mediated reaction. *Cis* isomers predominate when using *E*-alkenes **178** (E=Pr, Z=H) **179**, and in the case of the Z-alkene isomer of **178**, *cis* products again generally predominated, though a 1:1 *cis:trans* mixture was obtained in the D-galacto case with R=*t*-Bu.[52] Remote stereocontrol directs introduction of the new chiral centre. When the alkene bears only terminal hydrogens, a 1.5:1 mixture of C2 epimers is obtained.

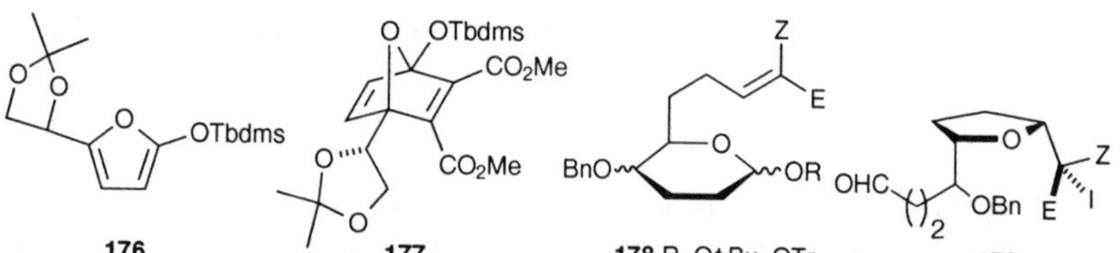

176   **177**   **178** R=O*t*-Bu, OTr   **179**

D-Glucose is the starting material for a synthesis of an intermediate towards zincophorin **186**, a natural product with high activity against Gram-positive bacteria. Thus, palladium-catalysed allylic substitution of **180**, followed by reductive desulfurisation and 6-OH protection affords **181**. Iron(III) chloride *C*-glycosylation followed by reduction of the ring double bond generates a near equal mixture of the diastereoisomers **182** and **183**, which are converted to **184** and **185** respectively (Scheme 35). Product **185** has the appropriate stereochemistry for the tetrahydropyran ring of zincophorin.[53]

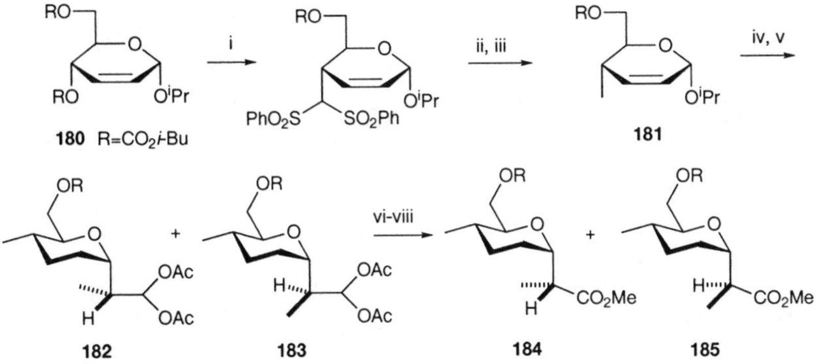

Reagents: i, NaCH(SO$_2$Ph)$_2$, Pd(PPh$_3$)$_4$; ii, Mg, MeOH; iii, ClCO$_2$^{i}Bu, Py; iv, FeCl$_3$, Ac$_2$O, MeCH=CHOAc; v, H$_2$, Pd/C; vi, NaOMe, MeOH; vii, Jones; viii, CH$_2$N$_2$.

Scheme 35

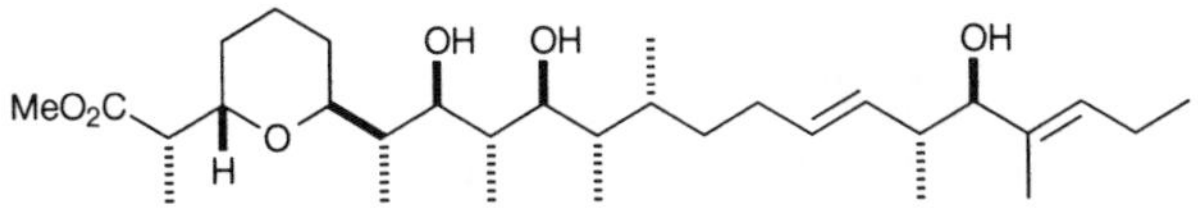

186, zincophorin

A synthesis of (+)-hongocinin **188** from L-rhamnal derivative **187**, has established that the natural product is the enantiomer of **188** (Scheme 36).[54]

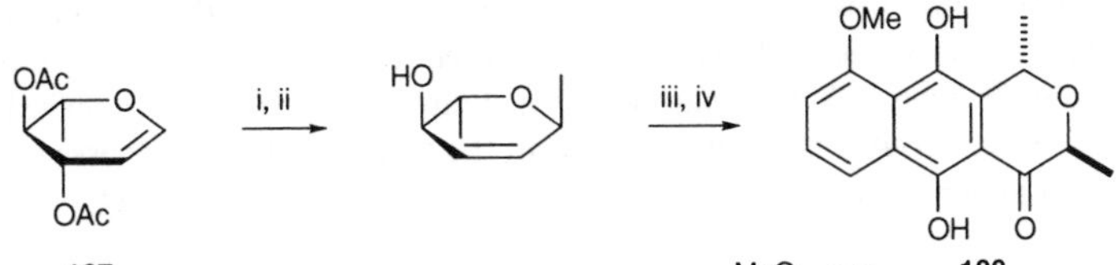

Reagents: i, TiCl$_4$, AlMe$_3$, -78 °C; ii, NH$_3$, MeOH; iii, PDC, iv,

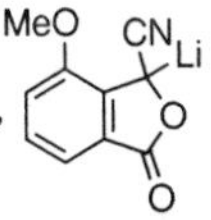

Scheme 36

Great interest in enediyne targets has continued. A [7.3.1]oxabicyclo analogue **193** of the esperamicin/calicheamicin aglycon has been prepared from the thioglycoside **189**. The diyne component is introduced by using a cerium acetylide addition to the C4 keto derivative of the sugar, giving **190**. Elaboration of the sugar to the unsaturated sulfone **191**, hydroxymethylation at C2, conversion of the second acetylene to an iodoacetylene, desulfonation and hydroxymethyl oxidation affords **192**, which undergoes a Nozaki-type ring closure to the cyclic diyne **193** (Scheme 37).[55]

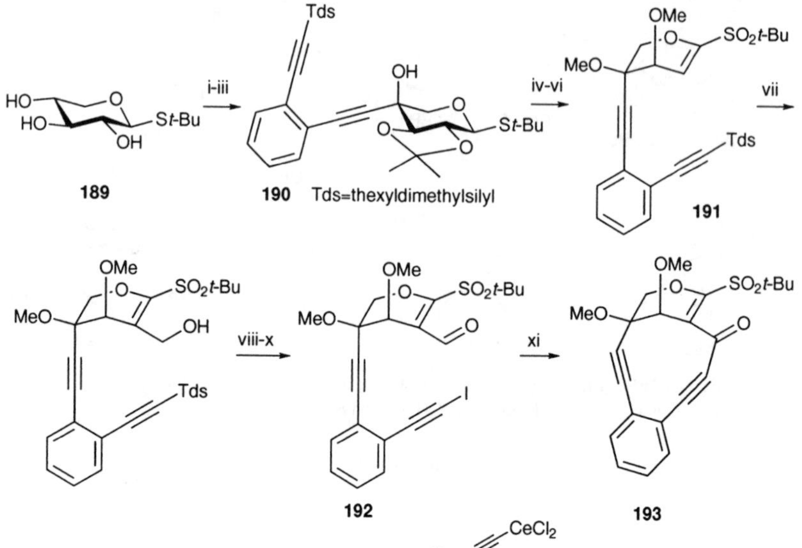

189   **190**   Tds=thexyldimethylsilyl   **191**

192   **193**

Reagents: i, CH₂=CMe(OMe); ii, Dess-Martin; iii, 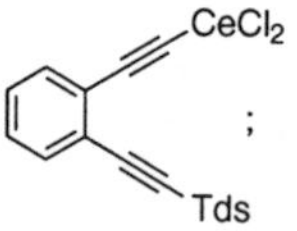
iv, MCPBA, NaHCO₃; v, BuLi; vi, MeI, NaH;
vii, LDA then HCHO; viii, TBAF; ix, I₂, morpholine;
x, PCC; xi, CrCl₂, NiCl₂.

Scheme 37

Staurosporine **196** has been prepared from 6-*O*-triisoprylsilyl-L-glucal **194** by Danishefsky and co-workers,[56] *via* the oxazolidinone epoxide **195**. The key epoxide stereochemistry was established by diastereoselective epoxidation using dimethyl-dioxirane (Scheme 38). The enantiomer of staurosporine was also prepared by the same synthetic route but commencing with the D-sugar. These syntheses corrected the prior assignment of absolute configuration of the natural product.

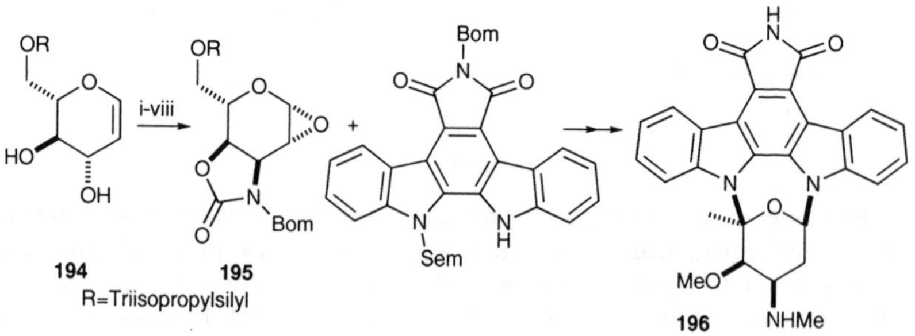

194   **195**   **196**   NHMe
R=Triisopropylsilyl

Reagents: i, NaH, CCl₃CN; ii, BF₃.OEt₂; iii, TsOH, Py, H₂O; iv, NaH, 0 °C to r.t.; v, NaH, BomCl;
vi,TBAF; vii, NaH, PmbCl; viii, DMDO.

Scheme 38

A number of approaches to components of polyether natural products have been reported. Several diastereoisomers of a part of the 32-ring polyether antibiotic maitotoxin have been prepared to establish configurations in the C63-C68 alicyclic portion of the natural product (by NMR comparisons).[57] The D-glucose derived **197** [*Tetrahedron Lett.* 1994, **35**, 5023] was converted to **198** and **199**, which were then coupled to afford **200** and 3 diastereoisomers. NMR indicated that **200** has the relative stereochemistry of the natural product.

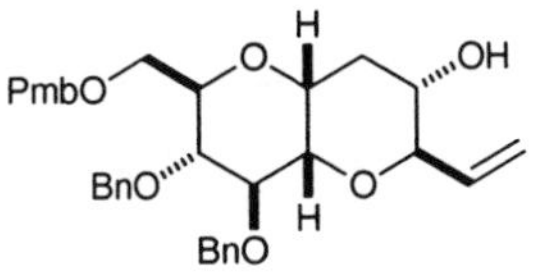

197 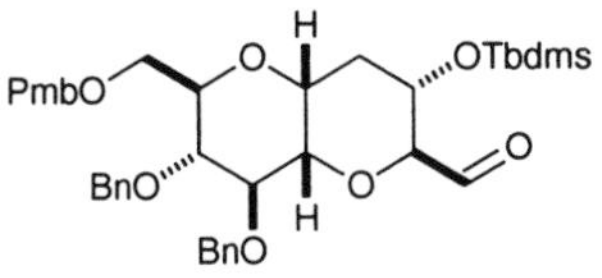   **198**

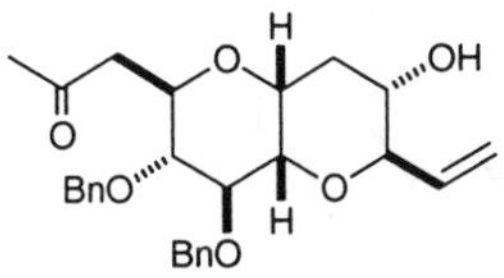

199 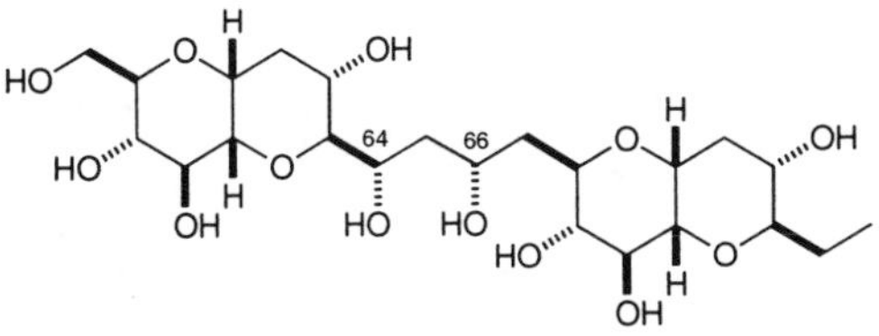   **200**

The same group have used a similar tactic to determine stereochemistry at C35-C39 by preparing **201** and **202** from D-glucose (*via* known intermediate derivatives) and elaborating these to **203**.[58]

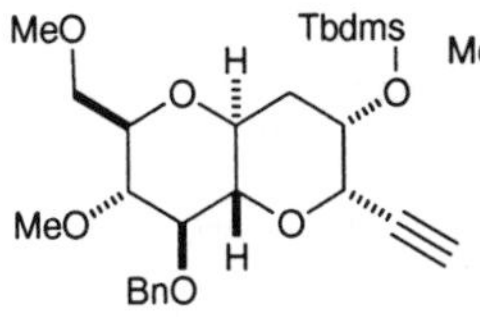

201 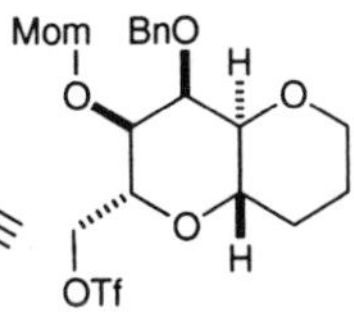   **202** 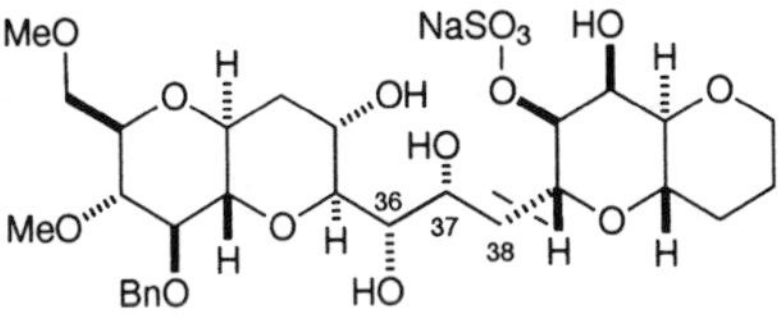   **203**

The A/B ring system of the polyether marine toxin ciguatoxin has been synthesized from methyl D-glucopyranoside-derived *C*-allyl glycoside **204**. This was elaborated by hydroboration-oxidation and protection group reorganization to **205**. Palladium-catalysed *O*-allylation, reduction and stannylation, changing the O3 benzyl to *p*-bromobenzyl, desilation and oxidation then affords **206**. Boron trifluoride-catalysed intramolecular allylation provided **207**, elaborated then to the ciguatoxin A/B ring component **208** (Scheme 39).[59]

Another synthesis of the ciguatoxin A/B ring system has been prepared from tri-*O*-acetyl-D-glucal from which **209** was made. Lithium Tms acetylide substitution, replacement of the O4 ethoxyethyl with acetate and anomeric deoxygenation gives **210**. Titanium dichloride-catalysed reaction with 3*R*,4*S*-

dipivaloyloxy-4,5-dihydropyran and acetylene protection as its cobalt hexacarbonyl complex then gives **211**.[60] Ring opening is directed by the cation-stabilizing effect of the cobalt-complexed acetylene; pivaloylation and deacylation affords **212**, and boron trifluoride-catalysed cyclization is then effected onto the cationic intermediate stablised as its cobalt complex. Removal of the alkyne protecting group and selective hydrogenation affords the A/B ring system analogue **213** (Scheme 40). The C2 epimer was also prepared by employing 3*R*,4*R*-dipivaloyloxy-4,5-dihydropyran.

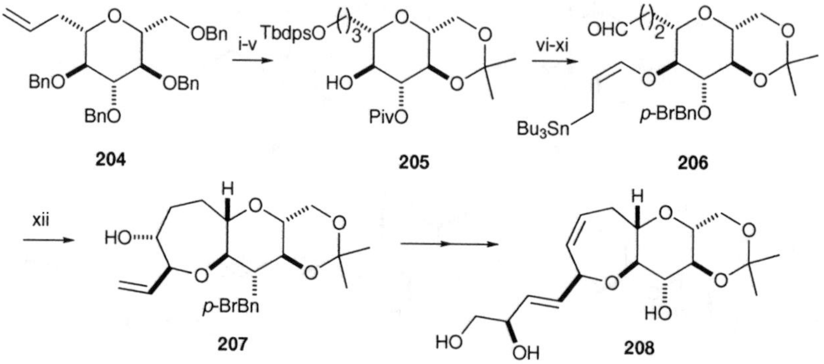

Reagents: i, 9-BBN, sonication then H_2O_2, NaOH; ii, TbdpsCl, ImH; iii, H_2, Pd(OH)$_2$; iv, Me$_2$C(OMe)$_2$, H$^+$; v, PivCl, py, DMAP; vi, allylcarbonate, Pd(dibenzylidene acetone)$_3$.CHCl$_3$, Ph$_2$P(CH$_2$)$_4$PPh$_2$; vii, DIBAL; viii, Bu$_3$SnCl, *n*-BuLi, HMPA; ix, *p*-BrBzCl, Et$_3$N, DMAP; x, TBAF; xi, SO$_3$.Py, Et$_3$N, DMSO; xii, BF$_3$.OEt$_2$.

Scheme 39

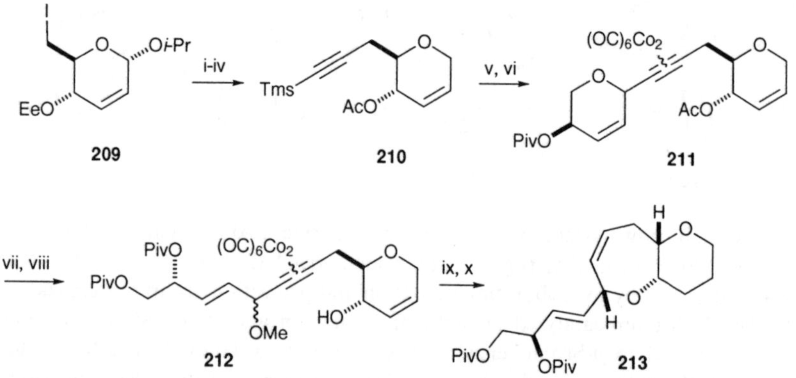

Reagents: i, Li-C≡C-TMS; ii, cat. PPTS; iii, Ac$_2$O, Py; iv, BF$_3$.OEt$_2$, HSiEt$_3$; v, TiCl$_4$, 3,4-dipivaloyloxy-4,5-dihydropyran, -20 °C; vi, Co$_2$(CO)$_8$; vii, Piv$_2$O, TfOH then MeOH; viii, K$_2$CO$_3$, MeOH; ix, BF$_3$.OEt$_2$; x, H_2, RhCl(PPh$_3$)$_3$

Scheme 40

The herbicidins continue to attract interest, and this year Vogel and Emery have utilized an aldol reaction of bicyclic lactone **217** and D-xylose derivative **214** as a key carbon-carbon bond forming step to give **215**, as an epimeric mixture. Further elaboration then provides a route to the herbicidin analogue **216**.[61]

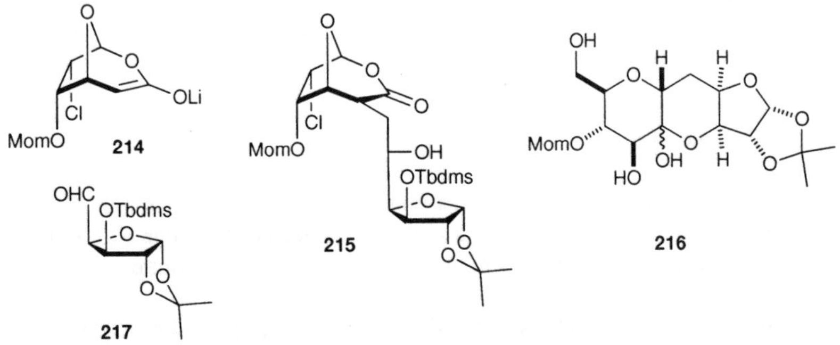

Two reports detailing the synthesis of carbohydrate-containing macrocycles have been reported this year. Thus, *O*-allyl glycoside **218**, derived from D-glucose, undergoes oxidative cleavage, reduction and tosylation to afford **219**. Substitution of the tosylate with the precursor alcohol **220** followed by reductive benzylidene opening with sodium cyanoborohydride provides **221**. Macrocyclization is then achieved by reaction with *bis*-tosyloxy di(ethylene glycol) to give *bis-gluco*[22]crown-8 **222**.[62]

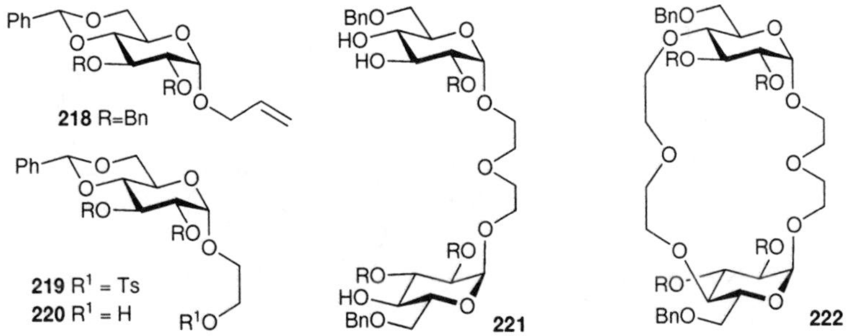

D-Threitol is the starting material for a cryptand. Thus, carbohydrate-derived **223** reacts with diacyl chlorides **224** and borane reduction affords macrocycles **225** (n=1,2). Similarly, **223** reacts sequentially with two equivalents of dimesylate **227** to afford cryptand **226**.[63]

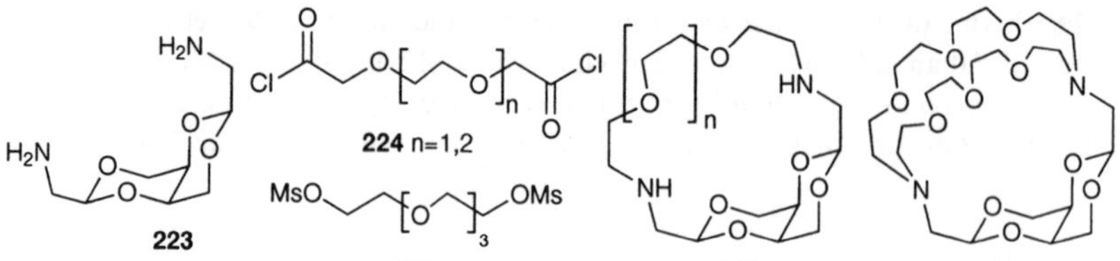

223 **224** n=1,2 **227** **225** **226**

5   N-Heterocycles

A survey of the utility of 1-thiazoles as masked aldehyde equivalents in amino-carbohydrate synthesis has appeared and describes routes to a range of 2-amino-sugars, azasugars and higher amino-sugars (including desmotic acid and lincosamine).[64]

A number of reports have described syntheses of quinolizidines related to castanospermine, swainsonine and analogues. Hetero-Diels-Alder reaction of carbohydrate-derived imines with Danishefsky's diene has provided a route to several castanospermine analogues.[65] D-Glucose-derived imine **228** provides heterocycle **229** with 9:1 diastereoselectivity. Borohydride reduction, selective O5,6 deprotection and C5,C6 cleavage with lead tetraacetate, followed by intramolecular reductive amination affords castanopsermine analogue **230**. In a similar manner, L-arabinose and D-mannose serve as starting materials for the synthesis of **231** and **232** respectively. This same hetero-Diels-Alder strategy has previously been used by the same group for synthesis of swainsonine and isomers [SPR, Vol.28, p.365-6, ref. 74].

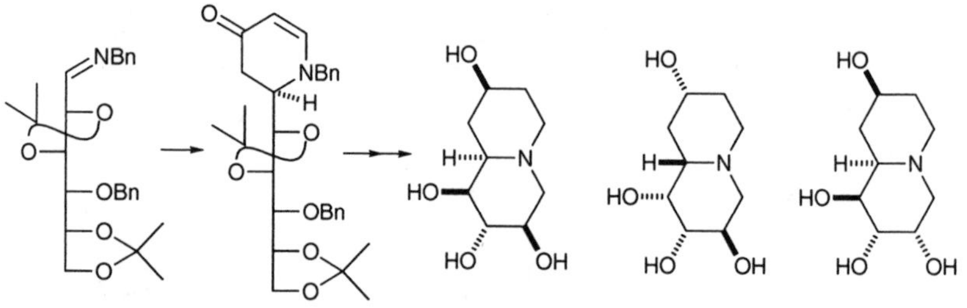

228   **229**   **230**   **231**   **232**

Martin and co-workers[66] have utilized 1-lithiofuran addition as a key carbon-carbon bond-forming step for construction of 8,8a-di-*epi*-castanospermine. Intermediate furan adduct **233** is elaborated to **234** *via* furan oxidation and thence to the castanospermine analogue **235** (Scheme 41).

A number of C8 modified castanospermine analogues **237** have been prepared by displacement of 8-sulfonate derivative **236**.[67] Retention of stereochemistry is

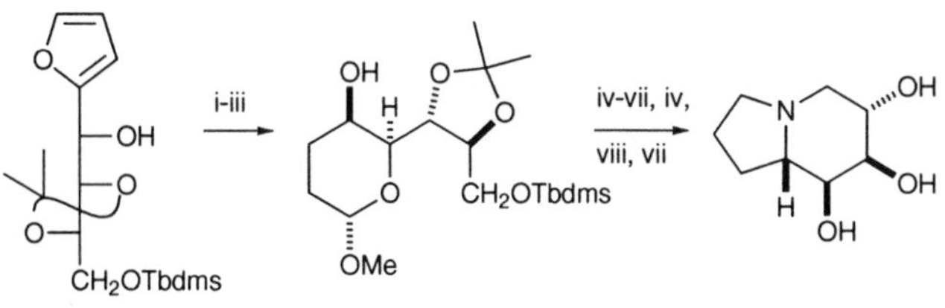

Reagents: i, *t*-BuOOH, VO(acac)$_2$; ii, MeI, Ag$_2$O; iii, K-selectride; iv, MsCl, Et$_3$N; v, NaN$_3$, DMF; vi, TBAF; vii, H$_2$, Pd/C; viii, CF$_3$CO$_2$H.

Scheme 41

observed due to intramolecular participation by the tertiary amine nitrogen which accounts also for the further rearrangement products **238**. The glycosidase inhibitory efficacy of a range of bicyclic and monocyclic azasugars have been reported. Thus, **241** and **242** (which are stereochemically related) are less potent β-glucosidase inhibitors than castanospermine **240**, but better β-galactosidase inhibitors than swainsonine **239**. Compound **243** and swainsonine show similar inhibition of α-mannosidases.[68]

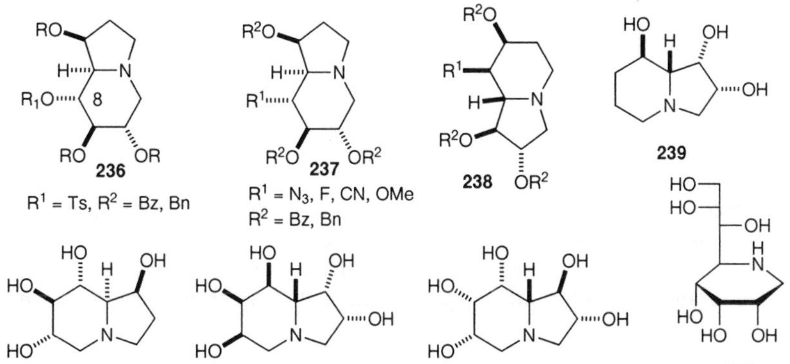

L-Ribose derivative **244** undergoes addition by 1-lithio-2-*N*-tritylimidazole to give **245**, which is then converted to fused imidazole **246** and thence to amino sugar analogue **247** (R=NHAc, OH), which shows glycosidase inhibitory properties (Scheme 42).[69]

The fucose-related imino alditol **249** has been prepared from **248**,[70] a series of 1,5-dideoxy-1,5-iminoheptitols have been prepared from 2-bromoaldonolactones derived from D-glucose, D-mannose and D-galactose,[71] D-*Arabino*- and D-*gluco*-

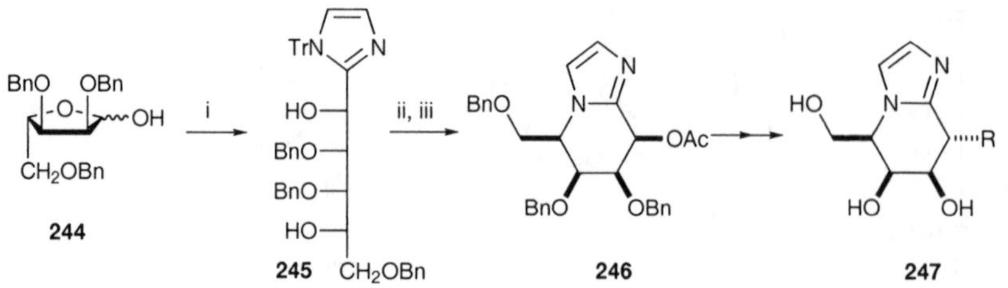

Reagents: i, 1-lithio-2-*N*-tritylimidazole; ii, $BnSO_2Cl$, Py; iii, Ac_2O.

Scheme 42

glycosylamines have been elaborated to 5- and 6-membered ring products **250** and **251**,[72] while D-Mannitol is the ulitmate starting material for pyrrolidine and piperidine derivatives **252** and **253** [see Chapter 18 for full synthetic discussion].[73]

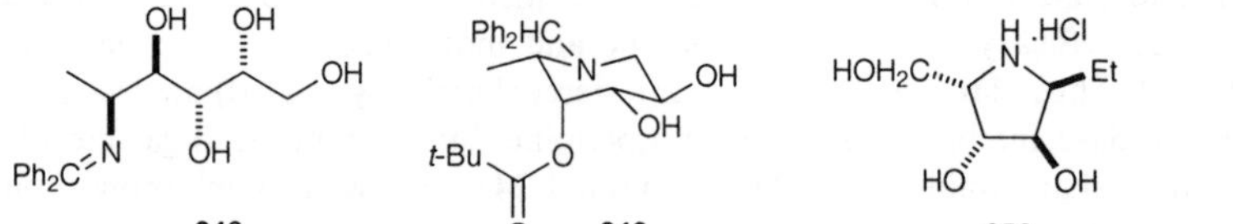

248 **249** **250**

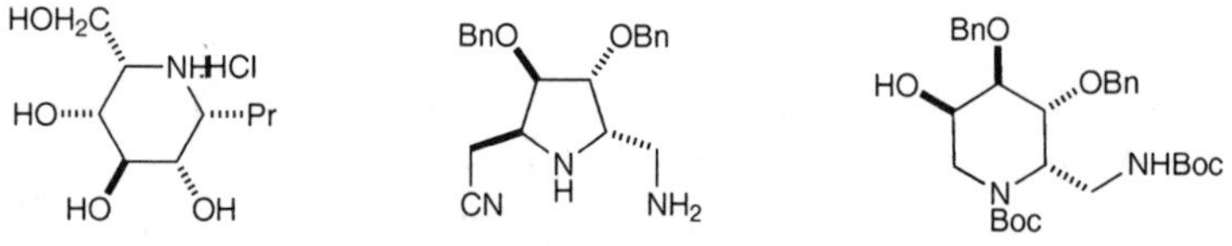

251 **252** **253**

Synthesis of the morpholine-derivative renin inhibitor BW-175 (**254**) and the spiroammonium derivative **255** are described in chapter 18.[74] The antibiotic (−)-anisomycin **258** has been prepared from glycosylamine **256**. Grignard addition, oxidation and cyclization affords pyrrolidinone **257** which was elaborated by known means to **258**.[75]

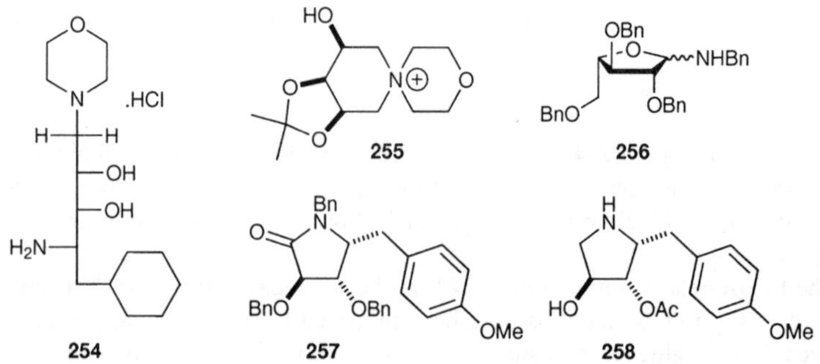

254 **255** **256** **257** **258**

D-Glucose is the starting material for a total synthesis of (+)-lactacystin **264**, the first non-protein neurotrophic factor. Selective protection and secondary alchohol oxidation of **259**, derived from D-glucose, provides **260**, elaborated to the trichloro-acetimidate **261**. Overman rearrangement to **262** is a key step in the conversion to **263**, which is converted in eight further steps to (+)-lactacystin (Scheme 43).[76]

Quayle and co-workers have examined the hetero-Diels-Alder reactions of **266** and related dienes (derived by Stille coupling of the precursor stannyl glycals) with **265** to provide tricycles **267-270**.[77]

A number of chiral piperazines e.g. **273** and **271** have been prepared from amino acid amide derivatives of D-glucosamine **272** (Scheme 44).[78]

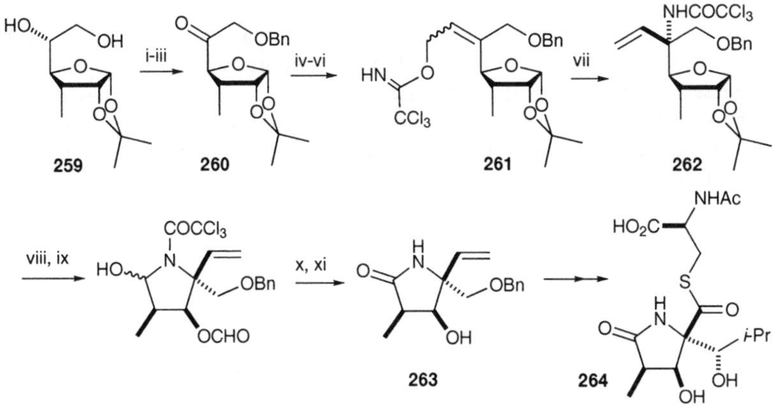

Reagents: i, Bu₂SnO, D; ii, BnBr, CsF; iii, CrO₃, dil. H₂SO₄; iv, Ph₃P=CHCO₂Et; v, DIBAL; vi, Cl₃CCN, NaH; vii, 150 °C; viii, TFA, H₂O; ix, NaIO₄, MeOH-H₂O; x, Jones; xi, NaBH₄, MeOH.

Scheme 43

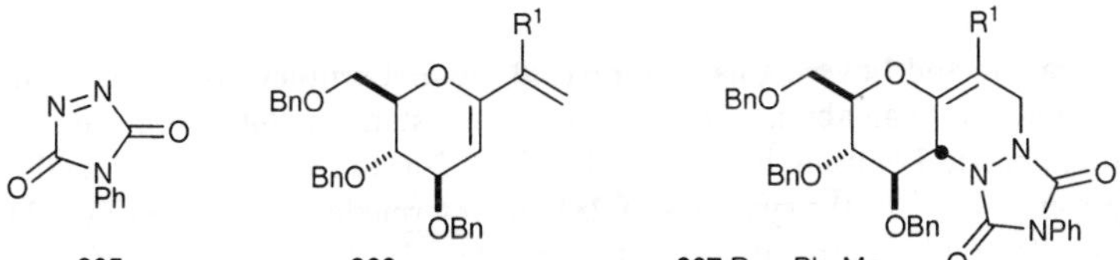

265

266

267 R₁ = Ph, Me

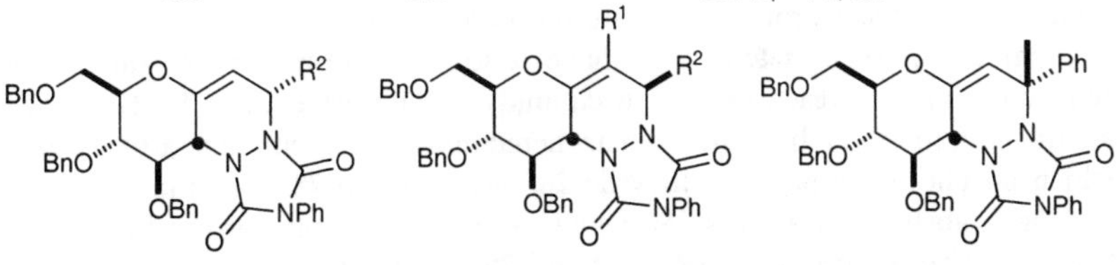

268 R² = Ph, n-Bu

269 R¹ = H, Me; R² = CO₂Me

270

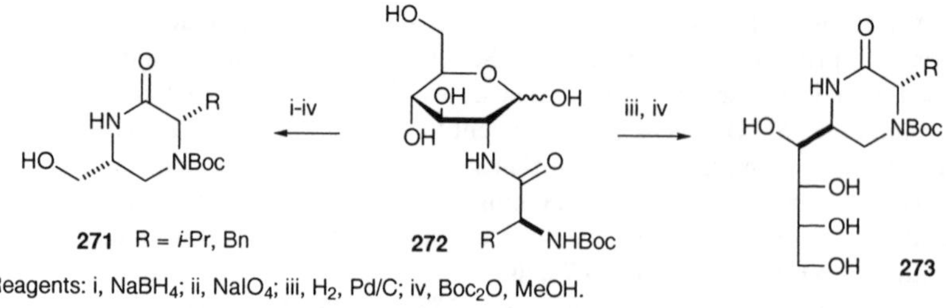

271 R = *i*-Pr, Bn   **272**   **273**

Reagents: i, NaBH₄; ii, NaIO₄; iii, H₂, Pd/C; iv, Boc₂O, MeOH.

Scheme 44

1-Oxacephem **275** has been prepared from the L-arabinal adduct **274**.[79] Fleet's group have provided full details [preliminary report: Vol.23, p.273, ref. 63] of syntheses of quinuclidine diol **276** and of octahydro-2-furo[2,3-*c*]-pyridinol **277** from D-arabinose.[80] The fused pyrrolidine **278** has been prepared by an intramolecular cycloaddition route the oxa-analogue chemistry of which is described above in section 4.[47]

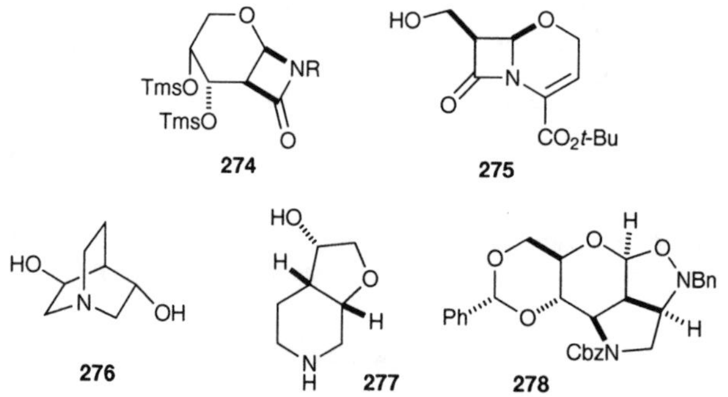

274   **275**

276   **277**   **278**

Pearson and Lovering have completed the first synthesis of the amaryllidaceae alkaloids (−)-amabiline **281** and (−)-augustamine **280**, from the common intermediate **279** using intramolecular [3+2] 2-azaallyl anion based methodology (Scheme 45).[81] In the synthesis of **281** the intermediate is trapped by a Mannich type electrophilic aromatic substitution, while for **280**, trimethyl orthoformate is used for electrophilic aromatic substitution-based cyclization.

Complex tricyclic heterocycles have also been prepared by interesting sequential radical reactions of succinimidyl *N*-glycoside **284**. 1,6-Hydrogen abstraction followed by transannular hemiaminal fragmentation affords **282**, which on photolysis provides tricyclic **285** *via* biradical **283**. Alternatively, 1,7-hydrogen abstraction affords **286** which does not undergo further fragmentation.[82] The structures of **285** (R=Me) and **286** were proved by X-ray structural analysis.

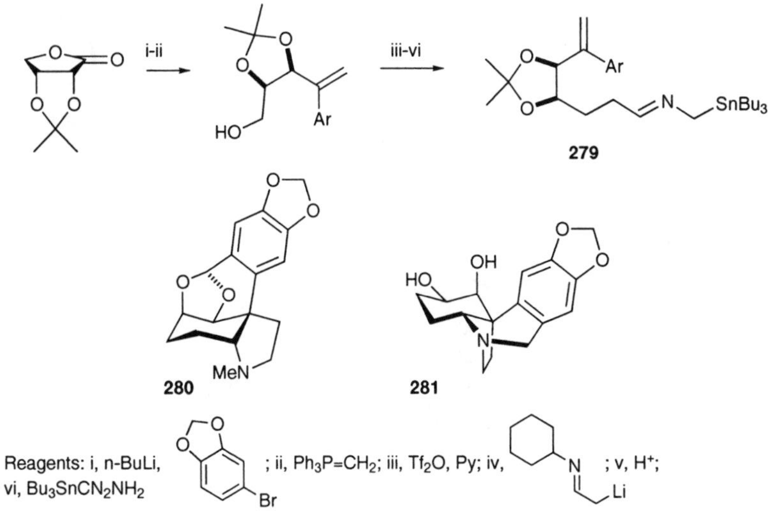

Scheme 45

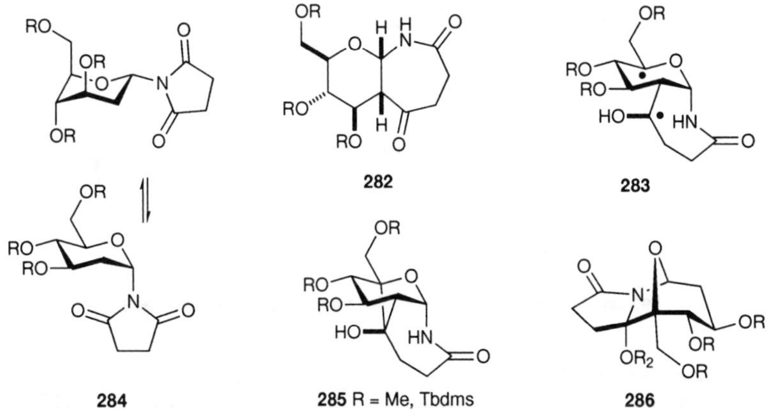

L-Ascorbic acid has been converted into isomers of 1,4-dimethyl-1,4-diazepin-2-one, the key heterocyclic component of the liposidomycin nucleoside antibiotics. The common intermediate **288** is converted to epimeric **289** and **290**, which are elaborated by parallel means (i.e. conversion of **290** to **291** involves the same reactions as conversion of epimeric **289** to **292**) to heterocycles **292** and **291** (Scheme 46).[83]

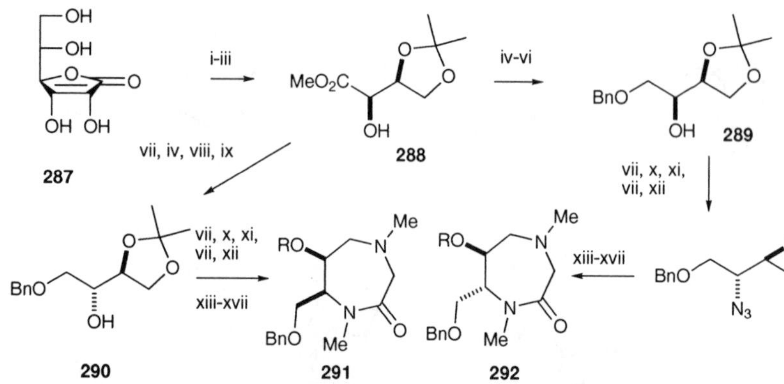

Reagents: i, AcCl, MeCOMe; ii, H_2O, $CaCO_3$; iii, MeI, $AcNMe_2$; iv, $NaBH_4$; v, Bu_2SnO; vi, BnBr; vii, TsCl, Py; viii, K_2CO_3, MeOH; ix, NaOBn; x, NaN_3, DMF; xi, HCl; xii, K_2CO_3, DMF; xiii, $MeNHCH_2CO_2H$; xiv, H_2, Pd/C; xv, DCC; xvi, TbdpsCl; xvii, MeI, DMF.

Scheme 46

6 Acyclic Compounds

A review on the reactivity of sugar-derived acyclic sulfones has appeared, and describes conjugate additions, desulfonations, vinyl deprotonation, allylic transposition and dipolar cycloaddition reactions[84]

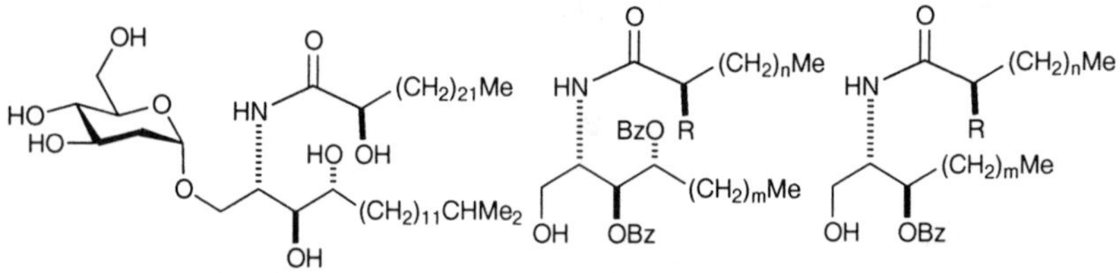

293 agelasphin-9b **294** R = H, OAc **295**

As in previous years, the synthesis of ceramides, sphingosines and related compounds has seen continued attention. Analogues of agelasphin-9b **293**, an antitumour agent from a marine sponge [Vol.27, p.21] have been described. Thus, 3,4,6-tri-*O*-benzyl-D-galactose and 3,5-di-*O*-benzyl-D-xylose have been elaborated into the aminopolyol derivatives **294** and **295**, which are coupled with tetrabenzyl-D-galactosyl fluoride to provide the ceramide analogues.[85] Structure-activity data against B-16 tumour cells in mice were reported.

The sphingosine **296** has been prepared from D-mannose in 10 steps.[86] Levoglucosenone is the starting material for a synthesis of (2*S*,3S,4*R*)-2-amino-1,3,4-octadecanetriol **300** (see Scheme 47), by carbonyl reduction, diastereoselective *cis*-oxyamination and reoxidation to **297**. Regioselective Baeyer-Villiger reaction and reduction of the resulting lactone **298** and Wittig homologation of the terminal hydroxyl provides the key steps to **300**.[87]

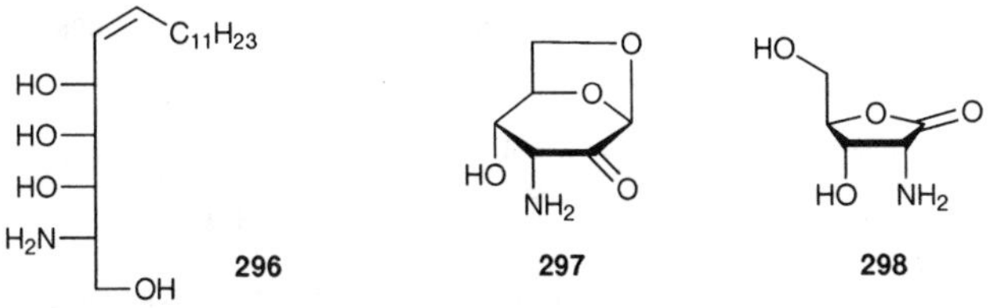

The same D-*ribo* phytosphingosines **300** and also the L-*lyxo*- analogue **303** have been prepared starting from D-galactose and D-xylose, respectively. The known D-galactose derived **299** (R=Me) underwent stereoselective [*J. Org. Chem.* 1966, **31**, 220] propargylation and further homologation, while xylose was elaborated *via* the known dithioacetal **301** to **302**, and then using recently reported [*J. Org. Chem.* 1993, **58**, 5576] terminal acetal opening methodology, followed by mesylation, azidation, acetylene alkylation, deprotection and acetylene reduction, **303** was produced (Scheme 47).[88]

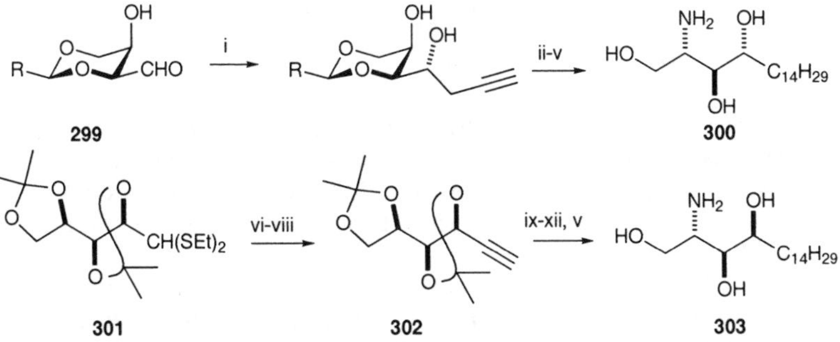

Reagents: i, Zn, propargyl bromide; ii, $C_{11}H_{23}Br$, BuLi; iii, Tf_2O, Py; iv, NaN_3; v, TFA then H_2, Pd/C; vi, HgO, $BF_3.OEt_2$; vii, CBr_4, PPh_3, Zn; viii, BuLi; ix, MeMgI; x, MsCl, Py; xi, NaN_3; xii, $C_{12}H_{25}Br$, LDA.

Scheme 47

D-Galactose is also the starting material for another sphingosine synthesis *via* **299** (R=Ph), proceeding analogously but using aldehyde vinylation (rather than propargylation). The allylic alcohol products **304** and **305**, on reaction with the appropriate orthoformate, then undergo a Claisen rearrangement to **306**. This can then be converted to sphingosine **308** or deoxy analogue **307** (Scheme 48).[89]

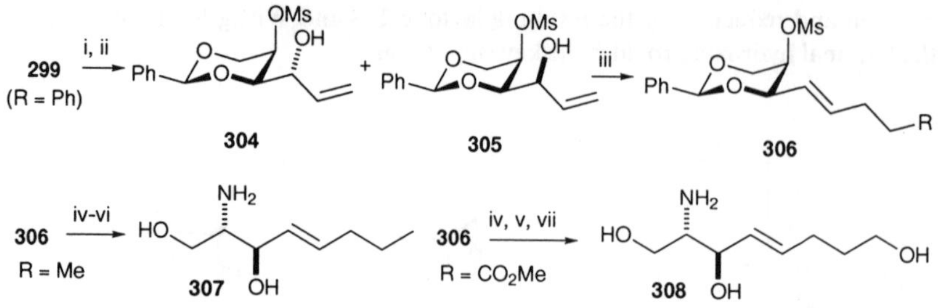

Reagents: i, CH_2=CHMgBr; ii, MsCl, Py; iii, $RCH_2C(OMe)_3$, D; iv, tetramethylguanidinium azide, DMF; v, TsOH, MeOH; vi, H_2S, Py; vii, LAH.

Scheme 48

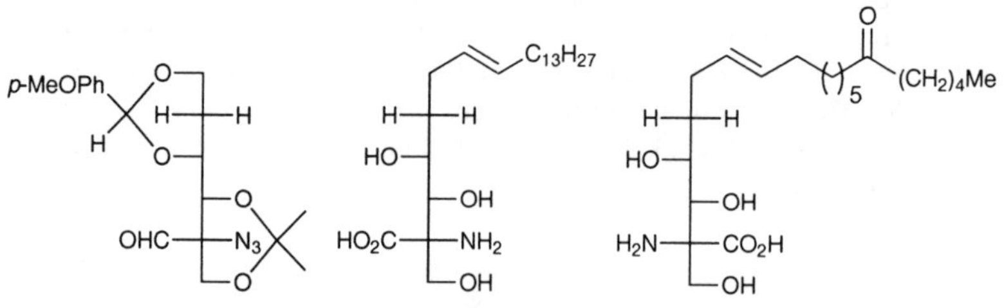

309 **310** **311**

Synthesis of nine analogues of the potent immunosuppressive agent myriocin have been made, starting from 2-deoxyglucose *via* **309** as a common intermediate. Amongst the analogues were 2-*epi*-myriocin **310** and 14-deoxymyriocin **311**. The latter proved to be 30 times more active than the natural product in a mouse model.[90,91]

(2*S*,3*R*)-3-Hydroxyleucine **313** has been prepared from D-glucose *via* known ketone **312**.[92] Dipeptide isosteres **315** (e.g. Y=CH_2Ph, X=OMe, R^2=Boc, R^1=Bn) are available from D-glucose-derived **314**.[93]

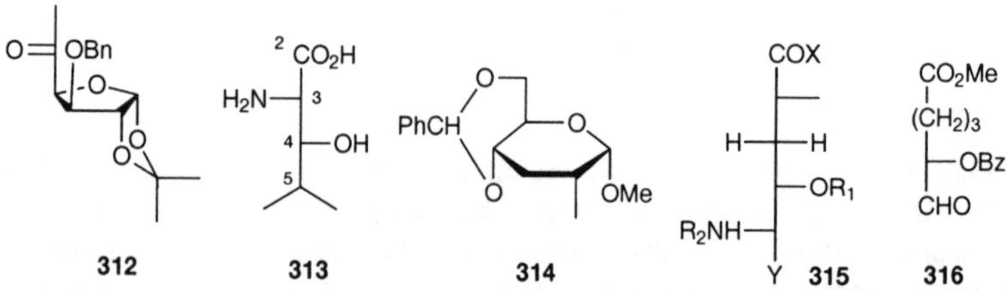

312 **313** **314** **315** **316**

D-Arabinose has been converted to the aldehyde **316**, used in synthesis of the eicosatetraenoic acid (leukotriene) 12-oxo-LTB_4[94] The D-arabinose-derived dithioacetal **317** has been used to construct **319**, which is related to the C12-C18 portion of the macrolide antitumour antibiotic laukacidin C (Scheme 49). Asymmetric crotylation (using Brown's *E*-crotyl diisopinocamphenyl borane reagent) of the aldehyde obtained from *p*-methoxybenzyl ether protection and dithioacetal deprotection of **317** provides intermediate **318**.[95] The same dithioacetal has also been converted to **321** by White and Jensen (Scheme 50). This vinyl iodide is an intermediate for a synthesis of cyclopropane containing marine eicosanoids.[96]

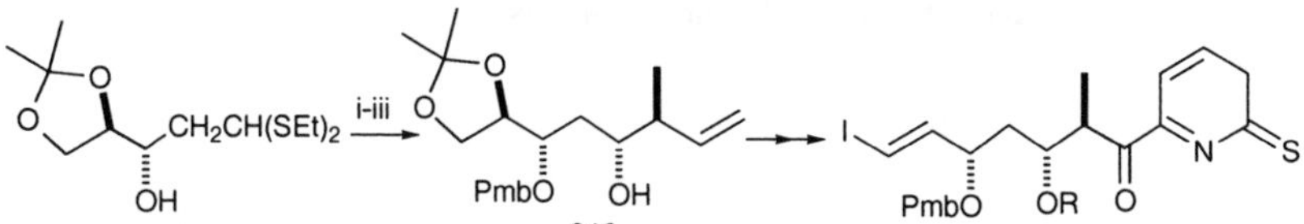

317 **318** **319**
R = Tbdps

Reagents: i, NaH, PmbCl; ii, $HgCl_2$, HgO; iii, 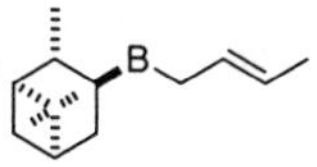

Scheme 49

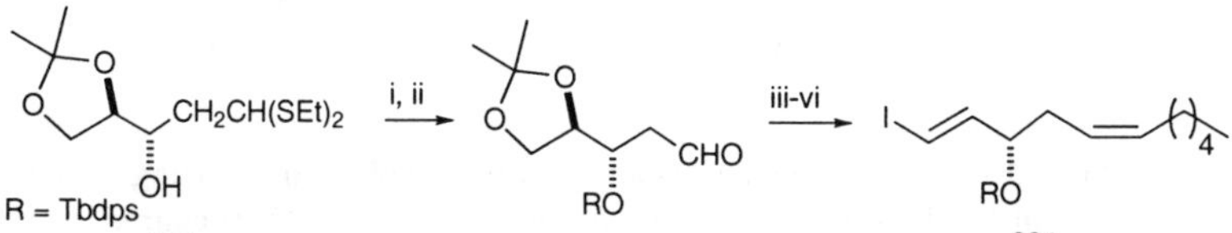

R = Tbdps
317 **320** **321**

Reagents: i, NaHMDS, TbdpsCl; ii, NCS, $AgNO_3$; iii, $Ph_3P{=}C_6H_{12}$; iv, TFA; v, $Pb(OAc)_4$; vi, CHI_3, $CrCl_2$.

Scheme 50

D-Mannitol has been converted into 1-stearoyl-2[(*Z,Z,Z*)-9,12,15-linolenoyl]-sn-glycerophosphocoline **323**, *via* **322**.[97] 1,6-Anhydro-D-glucose has been converted into aldehyde **324**.[98]

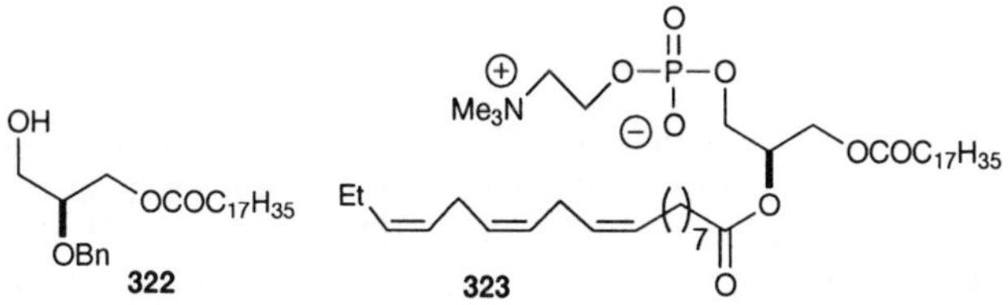

322 **323**

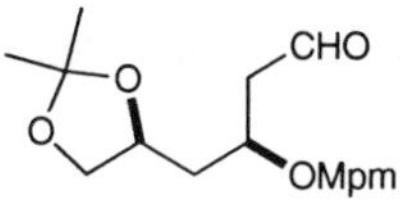

324

Acyclic eneynes have been prepared by Nicholas ring opening of cobalt hexacarbonyl complexed *C*-glycosylacetylene **325**. Silver tetrafluoroborate-catalysed reaction with pivaloyl chloride followed by nucleophilic trapping of the intermediate α-acetylenic cation leads to isomer **326** using alcohols or allylic nucleophiles, but to a mixture of **326** with isomeric **327** (R=SPh) when thiophenol is used as the nucleophile. Use of water as the nucleophile leads to competing acyl migration and thus products **327** [R=OH] and **328**. This suggests participation of the acetoxy group in cation stabilization as in **329**.[99]

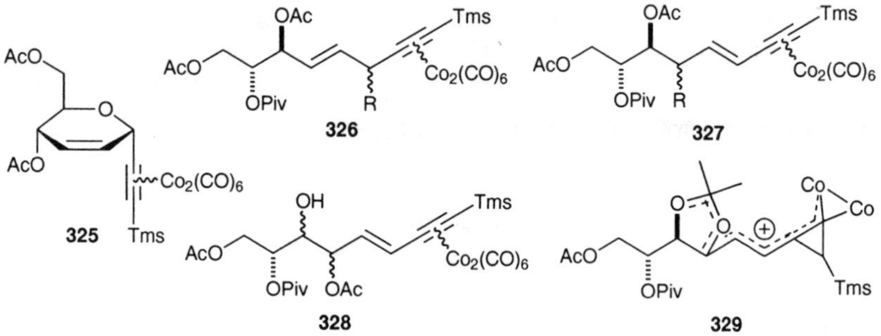

D-Mannitol *bis-p*-methoxybenzylacetal derivative **330** is the starting material for synthesis of the lipid-like ether stereoisomers **331** and **332**, prepared as ^{125}I labelled materials for radiodiagnostic tumour imaging.[100]

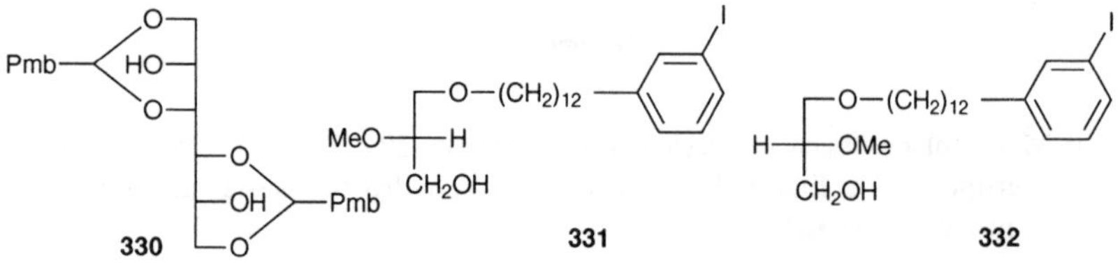

All possible stereoisomers of *E*-4,5-dihydroxydec-2-enal **333** have been prepared from D- and L-arabinose, D-ribose and L-lyxose. Key reactions involve C1 Wittig homologation, reduction and C4,C5 oxidative cleavage.[101] The eneyne **336**, an intermediate for a synthesis of 1α,25-dihydroxyvitamin D₃, has been prepared from D-xylose *via* lactone **334** and **335**. Lithioacetylene used for alkyne introduction and further elaboration provides **336** (Scheme 51).[102]

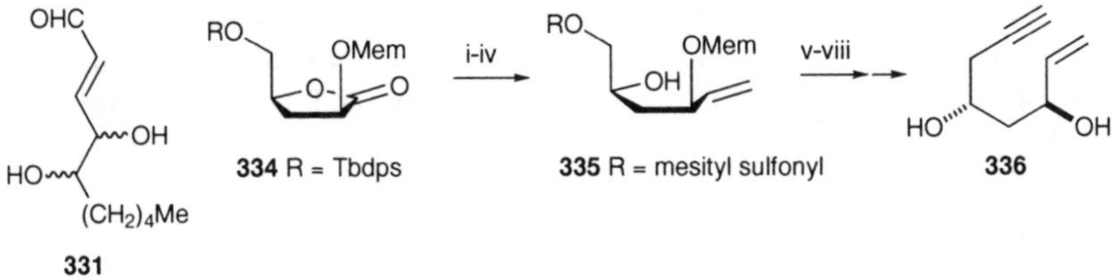

Reagents: i, DIBAL; ii, $Ph_3P^+CH_3Br^-$, *t*-BuOK; iii, TBAF; iv, 2-mesitylenesulfonyl chloride, Py; v,K_2CO_3; vi, HC≡CLi; vii, MemCl.

Scheme 51

1,6-Anhydro-D-glucose has been converted to **339**, a prostaglandin precursor, by way of **337**, which was epoxide ring-opened by an acetylide anion. Standard manipulation affords **338** and hence the target compound (Scheme 52).[103]

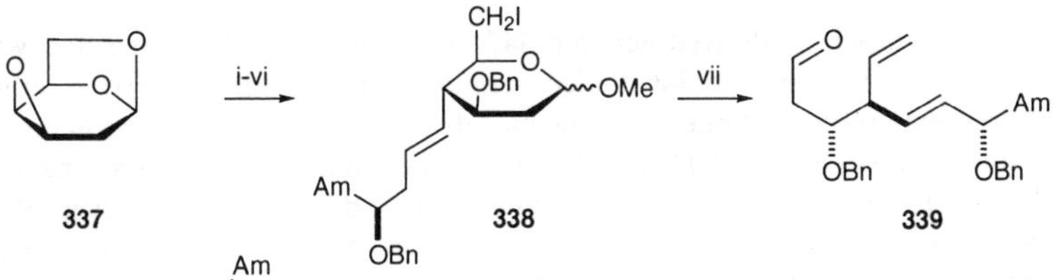

Reagents: i, ⟨Am, OThp⟩ ; ii, LAH; iii, NaH, BnBr; iv, HCl, MeOH; v, MsCl, Py; vi, NaI, Me_2CO; vii, Zn/Cu, $BF_3.OEt_2$.

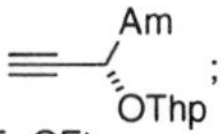

Scheme 52

The protected glyceric acid derivatives **340** and **341** are obtained from D-isoascorbic acid by ruthenium catalysed oxidative degradation.[104]

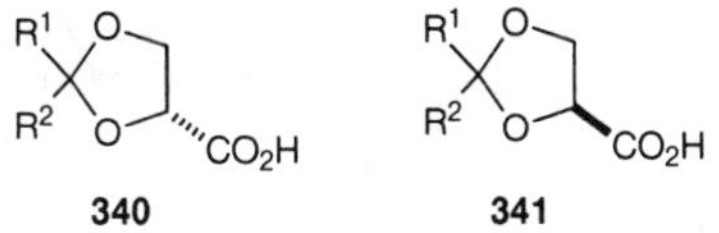

$R^1 = R^2 = Me$
$R^1 = R^2 = $ cyclohex
$R^1 = H, R^2 = Ph$

340   **341**

7   Carbohydrates as Chiral Auxiliaries, Reagents and Catalysts

A further report on uses of anomerically-linked D-glucose systems as auxiliaries in hetero-Diels-Alder reactions has appeared this year. In this case, the sugar served as the dieneophile component **342** (R^*=tetra-O-acetyl-β-D-glucopyranosyl) which reacted with Danishefsky's diene to afford enone **344** as the major epimer (9:1 ratio) after TFA-catalysed elimination. Oxidative removal of the auxiliary provided the chiral hydroxy diacid **345**. The reaction was shown to proceed *via* the initial Mukaiyama aldol product **343**, with TFA catalysing the subsequent intramolecular conjugate addition and elimination reactions (Scheme 53).[105]

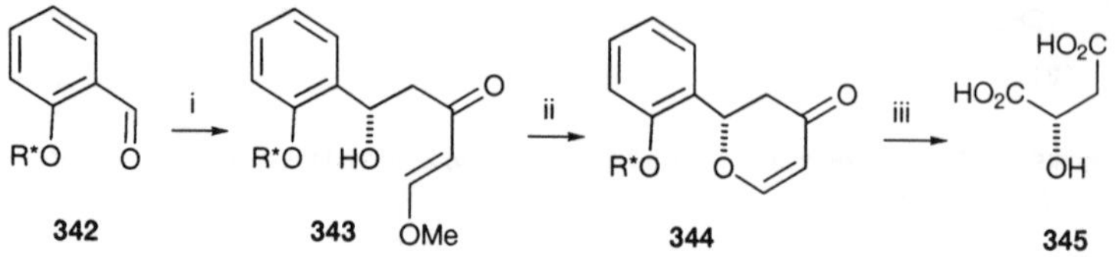

Reagents: i, BF₃, [structure] ; ii, TFA; iii, O₃ then H₂O₂.

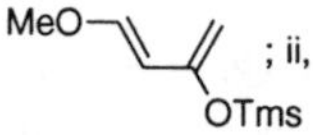

Scheme 53

The L-quebrachitol-derived acrylate **347** undergoes Diels-Alder reaction with cyclopentadiene, affording **346** under titanium tetrachloride catalysis in ether, and predominantly **348** under tin tetrachloride catalysis in toluene. The nature of the solvent was shown to play a major role in influencing the stereochemical outcome. The facial selectivity under tin tetrachloride catalysis was reversed on using ether as solvent, providing mainly **346**, however, using aluminium trichloride or diethylaluminium chloride leads to mainly **346** in either toluene or ether. The results were rationalized by suggesting that coordinating solvents lead to *Re*-face addition on conformer **349**, while non-coordinating solvents allow *Si*-face addition *via* chelate **350**. Similar results were obtained using the dicyclohexylidene acetal analogue of **347** (Scheme 54).[106]

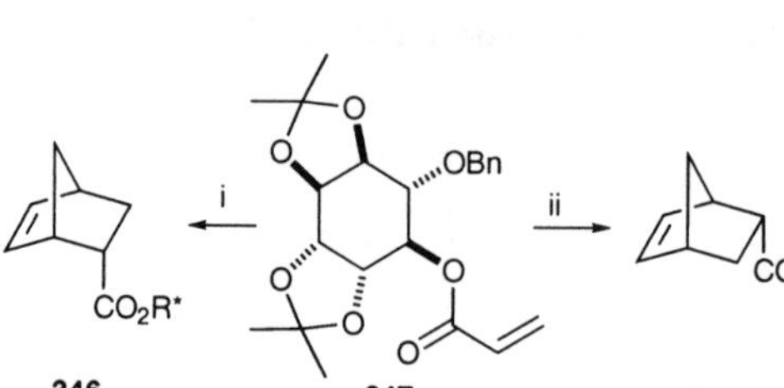

Reagents: i, TiCl₄, Et₂O, -78 °C; ii, SnCl₄, PhH, -78 °C.

R* = L-quebrachitol

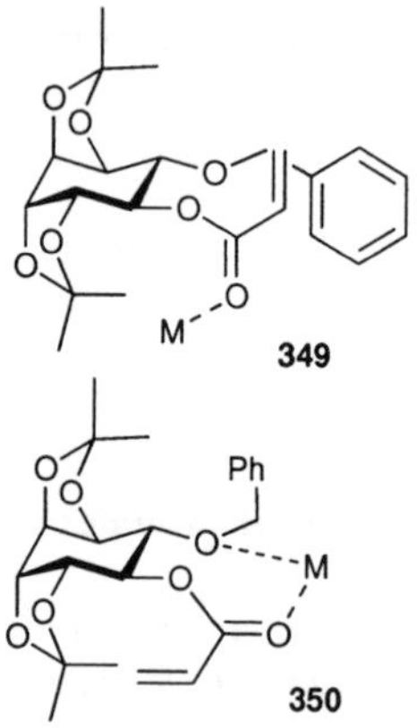

Scheme 54

The carbohydrate derived nitroalkene **351** undergoes Diels-Alder reaction with 1,3-butadiene to afford **352**, while **354** leads to a reversal of the diastereoselectivity giving **355** as the major products. However, the diastereoselectivity was modest with ratios of 78:22 and 67:33 for these two reactions. Denitration, acidic deacylation, sodium periodate cleavage and final reduction provided the (*S*) and (*R*) alcohols **353** and **356** respectively (Scheme 55).[107]

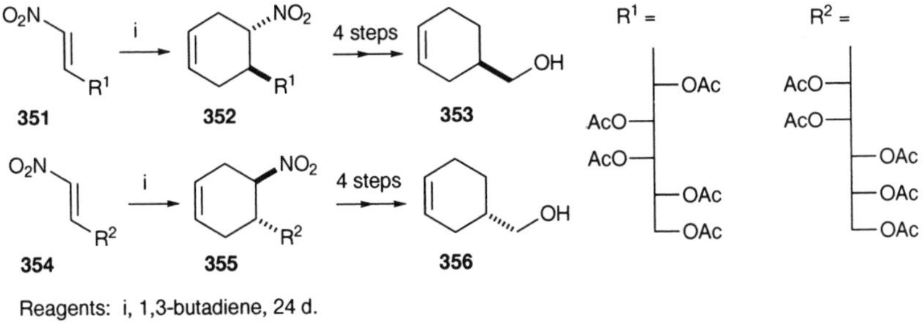

Reagents: i, 1,3-butadiene, 24 d.

Scheme 55

Vinyl ether **357** undergoes [2+2] cycloaddition with chlorosulfonyl isocyanate to give the β-lactams **358** and **359**, with selectivities of 54:46 to >97:3. The best results were obtained with small R^1 and R^2 groups and the stereoselectivity was rationalized by invoking π-π stacking of the electrophilic aryl ring and the vinyl ether olefin, blocking the *Si* face. Notably, when the sulfonyloxy group was replaced by a hydrogen, thereby producing a C6 methyl group, the facial selectivity was reversed.[108]

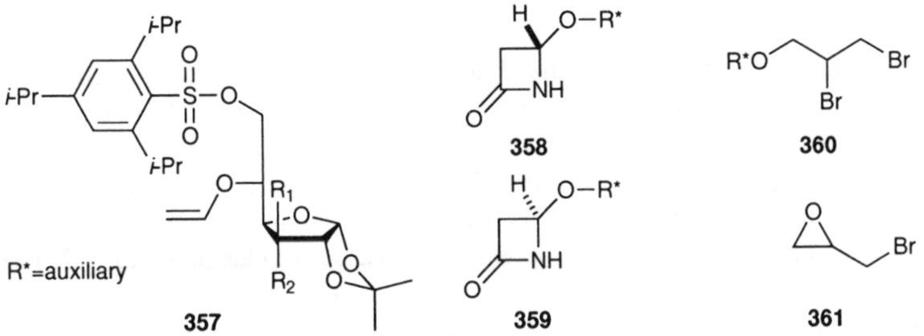

Diastereoselective bromination of a series of galactose and glucose-derived allyl glycosides, several of which are new compounds, yields dibromide **360**. Glycosidic cleavage affords chiral bromohydrins which can be cyclized to the bromomethyl epoxide **361** in up to 60% e.e.[109]

This year two groups have described the use of sugar auxiliaries to control diethylzinc-catalysed cyclopropanations. Fructose-derived **362** undergoes cyclopropanation, and after auxiliary removal and aldehyde reduction, gives the chiral cyclopropane **363** in up to 90% e.e. (Scheme 56).[110]

Charette and co-workers have described the synthesis of the glucose-derived allylic ether stereoisomers **364** and **365** from the precursor trichloracetimidate. Cyclopropanation and auxiliary removal gave the 1,1,2-trisubstituted cyclopropanes **366** and **367** respectively with high selectivities (Scheme 62). These were elaborated to provide all four stereoisomers of the cyclopropyl α-amino acid coronamic acid, **368**, a component of vivotoxin and coronatine.[111,112]

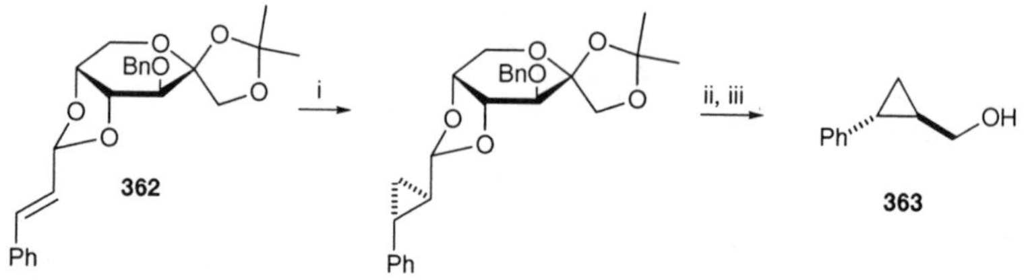

362

363

Reagents: i, Et$_2$Zn, CH$_2$I$_2$; ii, H$_3$O$^+$; iii, NaBH$_4$.

Scheme 56

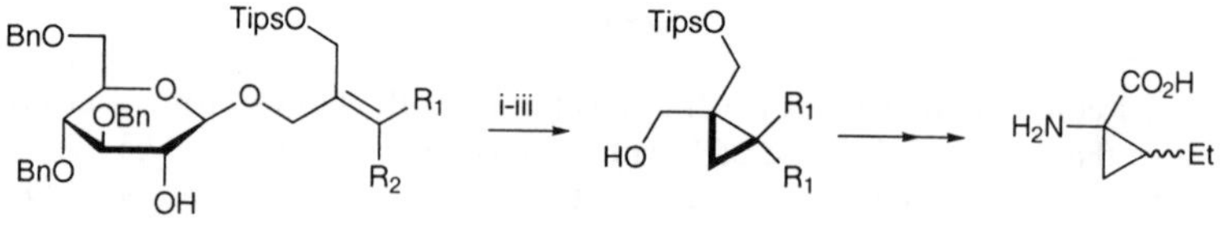

364 R$_1$=H, R$_2$=Et
365 R$_1$=Et, R$_2$=H

366 R$_1$=H, R$_2$=Et
367 R$_1$=Et, R$_2$=H

368

Reagents: i, Et$_2$Zn, CH$_2$I$_2$; ii, Tf$_2$O, py; iii, DMF, py, H$_2$O, 120 °C.

Scheme 57

A further example of Grignard additions to diacetone-D-glucose derived sulfinates for the synthesis of chiral sulfoxides has appeared this year, extending previous work [see Vol 26, p.321 and Vol 28, p.375] to include *p*-tolyl sulfoxides and improving large scale synthesis.[113] Thus, reaction of sulfinate **370** with methylmagnesium iodide, or of **371** with *p*-tolyl magnesium bromide, afforded, after cleavage of the auxiliary, the enantiomeric sulfoxides **369** and **372**, respectively (Scheme 58). The stereochemical outcome agrees with the revised analysis reported previously [Vol 28, Ch. 24, p.375].

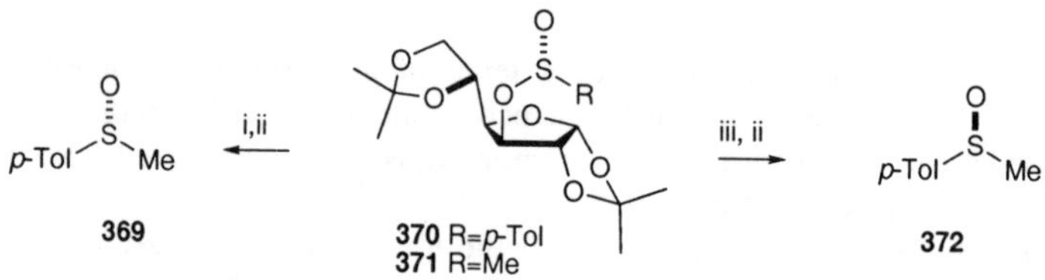

369

370 R=*p*-Tol
371 R=Me

372

Reagents: i, MeMgI (R=*p*-Tol); ii, TFA, MeCN-H$_2$O, iii, *p*-TolMgBr (R=Me).

Scheme 58

Diacetone-D-glucose has also served as auxiliary for synthesis of (*S*)-2-haloalk-anoic acids, **374**. Silyl enol ethers **373** were brominated or chlorinated (NBS or NCS) with 75-96 % d.e., oxidative cleavage of the auxiliary then yielding the (*S*)-2-chloro- and (*S*)-2-bromoalkanoic acids (Scheme 59).[114]

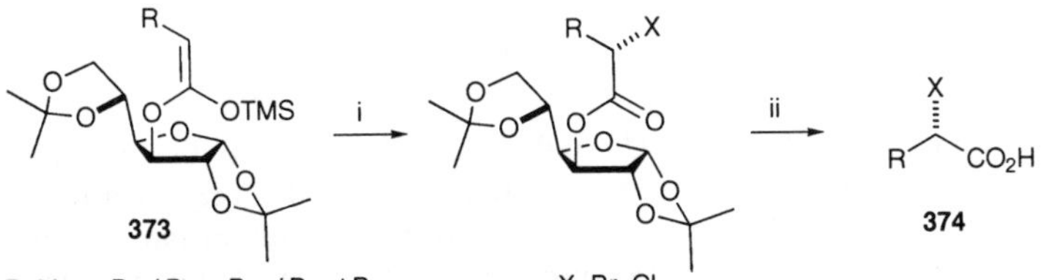

R=Me, *n*-Pr, *i*-Pr, *n*-Bu, *i*-Bu, *t*-Bu   X=Br, Cl
Reagents: i, NXS, THF; ii, LiO$_2$H, THF-H$_2$O.

Scheme 59

An elegant use of D-glucose as an auxiliary for controlling [2,3]-Wittig rearrangements has been reported. Treatment of **375** with *n*-butyllithium effected [2,3]-Wittig rearrangement to give **376** in >99% d.e., auxiliary cleavage providing the homochiral propargylic alcohol **377**. This methodology was extended to the homologue **378**, which underwent rearrangement to afford a 9:1 mixture of **379** and **380**, then elaborated to acetals **381** and **382**.[115]

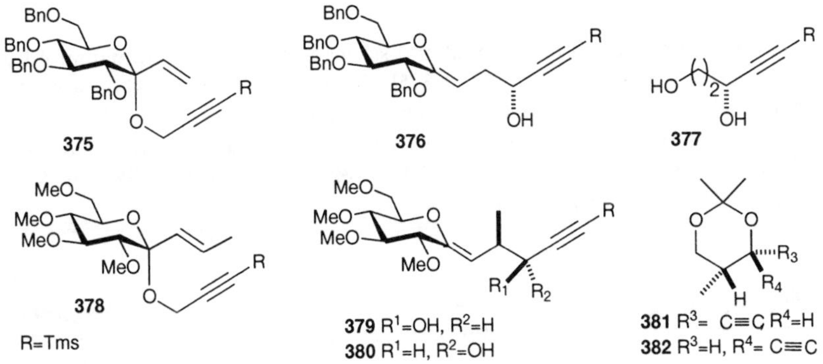

A four component Ugi-type reaction was employed for the synthesis of a range of dipeptides **383** with diastereomeric excesses of 92 to >99%, although the compounds described all retained the sugar auxiliary (Scheme 60).[116]

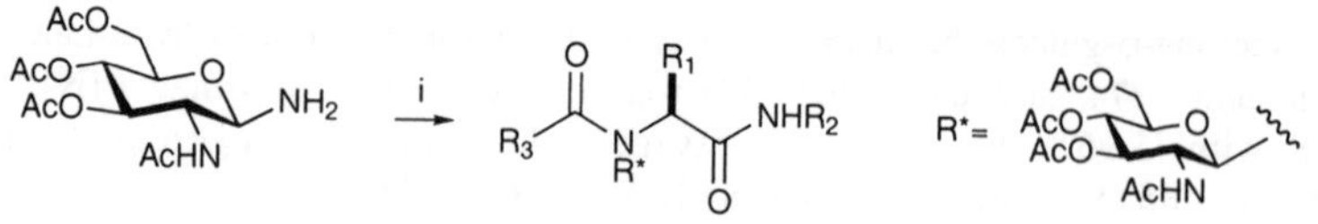

383

Reagents: i, R_1CHO, R_2NC, R_3CO_2H, $ZnCl_2.OEt_2$.

Scheme 60

Photolysis of tri-*O*-benzyl glucal-derived **384** generated the acetaloxyalkyl radical **385**, which undergoes conjugate addition to methyl acrylate affording the product of an overall atom transfer process **386**. Desulfuration and auxiliary cleavage with concomitant lactonization affords the (known) lactone **387** (Scheme 61).[117] The diastereoisomeric excess of **386** was not reported.

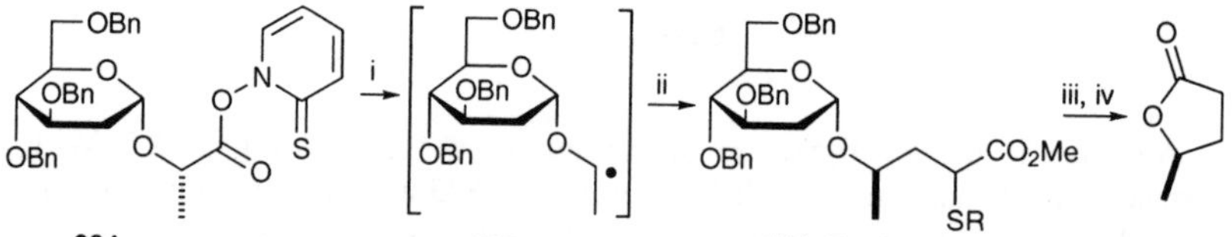

384 **385** **386** R = 2-pyridyl **387**

Reagents: i, hu; ii, CH_2=$CHCO_2Me$; iii, Raney Ni; iv, PPTS, MeOH.

Scheme 61

A further application of C_2-chiral aryl selenides derived from D-mannitol has appeared this year [see Vol 28, p.375]. Reagent **389** (from **388**) promotes enantioselective selenoetherifications and selenolactonizations.[118] Homoallylic alcohols **390** and **391** are converted to cyclic ethers **392** and **393** with >98% and 94% d.e., while β,γ-unsaturated acids **394** and **395** formed lactones **396** and **397** in >98% and 92% d.e., respectively.

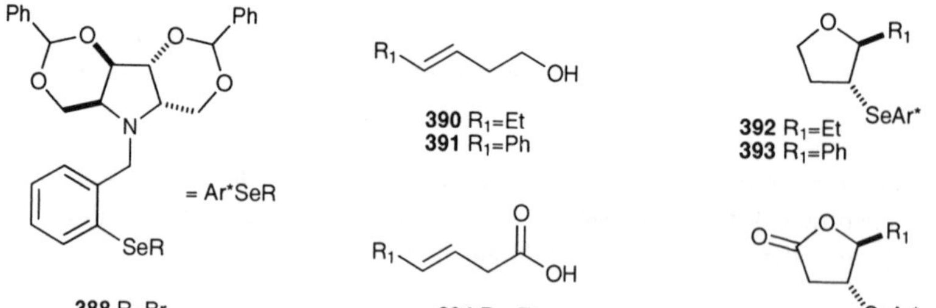

388 R=Br
389 R=PF_6

390 R_1=Et
391 R_1=Ph

392 R_1=Et
393 R_1=Ph

394 R_1=Et
395 R_1=Ph

396 R_1=Et
397 R_1=Ph

New chiral oxazolidin-2-ones **399** and **400** have been prepared from the D-xylose derivative **398**, and evaluated as chiral derivatizing agents for sulfonic and carboxylic acids.[119]

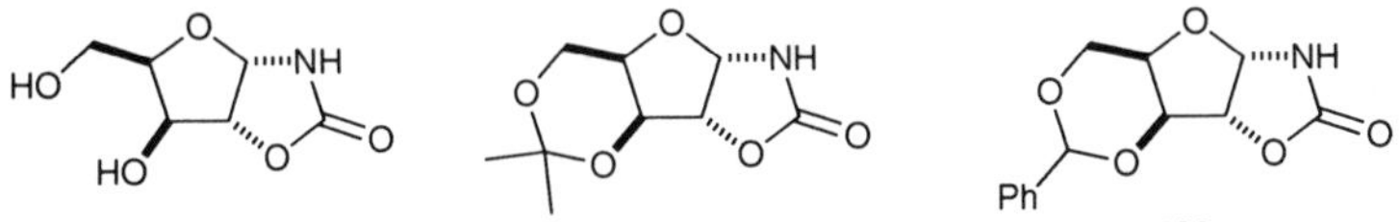

398   399   400

Two reports have apeared in which enantioselective recognition within the environment of cyclodextrins accelerates the rate of a reaction. The L,L-dipeptide **401** (R=Me) cyclizes to form the piperazine **402** in the presence of β-cyclodextrin, while the D,D- and L,D- isomers do not cyclize. This is rationalized as being due to a combination of favourable π-π stacking and chiral recognition by the cyclodextrin. In the case of the L,D- isomer, cyclization requires bond rotations forcing the aromatic systems apart.[120]

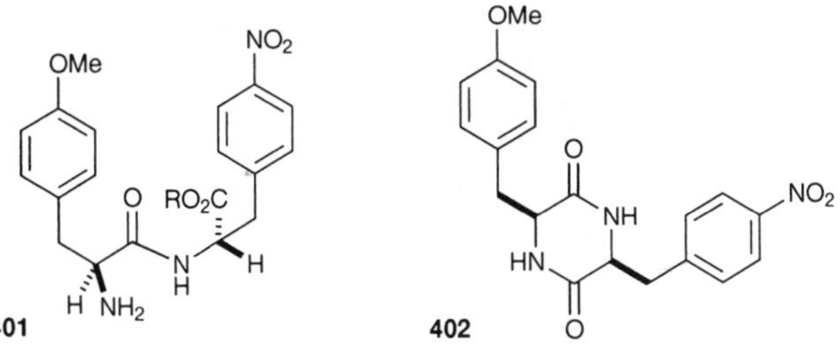

401   402

The rhodium chelate of 2,3-*bis*-(*O*-diphenylphosphino)-β-D-glucopyranoside **403** catalysed hydrogenation of the enantiomeric substrates **404** and **405** with >96% d.e, with the same sense of stereoinduction. This is of note since one case will involve matching, and one mismatching, double diastereoselection, but catalyst stereocontrol is clearly dominant. The α-anomer of the catalyst [see Vol 28, p.377] also gave the same diastereoisomeric outcome but with low selectivity. The opposite sense of diastereoselectivity was obtained using **406**.[121]

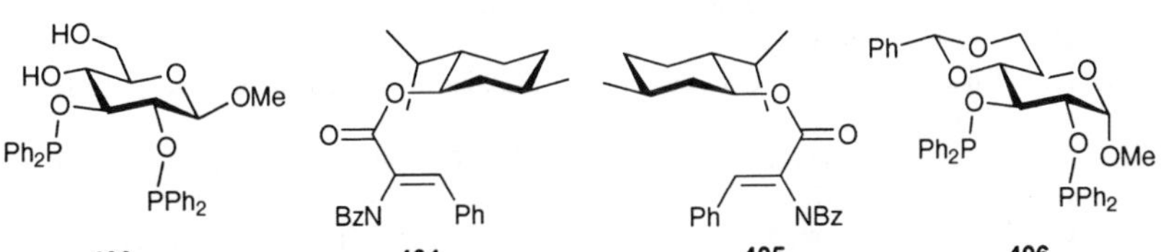

403   404   405   406

A range of 5-dialkylamino-5-deoxy-1,2-isopropylidene-α-D-xylofuranoses **407** have been evaluated for catalysis of diethyl zinc additions. Addition to benzaldehyde gave *R* products with all catalysts in 30-90% e.e. with all but one reaction proceeding in ≥64% e.e.[122] A range of monosaccharide derivatives (with free OH groups) were investigated as chiral additives in sodium borohydride reductions of prochiral ketones, the best result being the reduction of benzylacetone in 61.9% e.e. using diacetone-D-glucose as additive.[123]

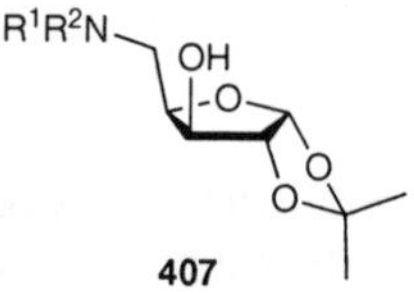

R¹=R²=Me, Et, *n*-Bu, Bn
R¹=Me, R²=CH₂CH₂NMe₂
R¹,R₂= -CH₂CH₂CH₂CH₂-

407

D-Arabinose has been elaborated into 2-alkoxytetrahydropyranyl auxiliaries for reductions, Grignard additions and allylations. All reactions utilize magnesium bromide etherate as catalyst. Thus, **408** [R¹=H, Me] undergoes diastereoselective reaction with certain Grignard reagents to give **409** with ≥30:1 selectivity. In addition, **408** [R¹=Me, R²=Bn] undergoes allylation with allyltributyltin with 25:1 selectivity (allyl Grignard gave poor selectivity) to give **409** [R¹=Me, R²=Bn, R³=CH₂CH=CH₂]. Addition of phenyl Grignard reagent to **408** [R¹=Me, R²=Tips] gives **410** showing opposite facial selectivity to Grignard or allyltin additions to **408** [R¹=Me, R²=Bn].[124,125]

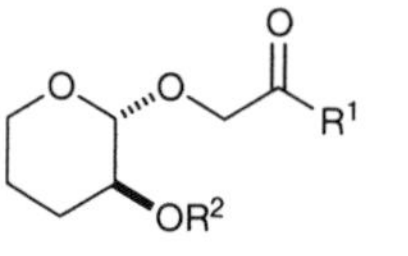

408 R₂ = Bn, Si(*i*-Pr)₃ 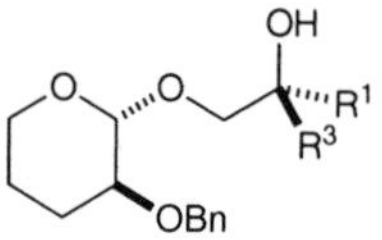   **409** 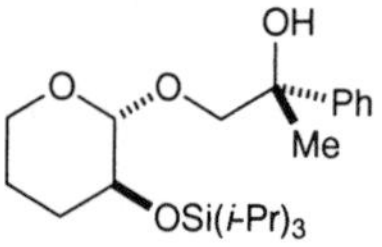   **410**

References

1   Y. Hanzawa, H. Ito and T. Taguchi, *Synlett.*, 1995, 299.

2   M. Kawana and H. Kuzuhara, *Synthesis*, 1995, 544.

3   A. Tenaglia and D. Barillé, *Synlett*, 1995, 776.

4   A.M. Horneman, I. Lundt and I. Søtofte, *Synlett*, 1995, 918.

5   B. Rondot, T. Durand, J.-P. Vidal, J.-P. Girard and J.-C. Rossi, *J. Chem. Soc., Perkin Trans. 2*, 1995, 1589.

6   K.-i. Takao, H. Ochai, K.-i. Yoshida, T. Hashizuka, H. Koshimura, K.-i. Tadano and S. Ogawa, *J. Org. Chem.*, 1995, **60**, 8179.

7   A.J. Wood, P.R. Jenkins, J. Fawcett and D.R. Russel, *J. Chem. Soc., Chem. Commun.*, 1995, 1567.

8   C. Kuhn, G. Le Gouadec, A.L. Skaltsounis and J.-C. Florent, *Tetrahedron Lett.*, 1995, **36**, 3137.

9   J.-L. Chiara, J. Marco-Contelles, N. Khiar, P. Gallego, C. Destabel and M. Bernabé, *J. Org. Chem.*, 1995, **60**, 6101.

10   T. Duvold, G.W. Francis, and D. Papaioannou, *Tetrahedron Lett.*, 1995, **36**, 3153.

11   M.E. Jung and S.W.T. Choe, *J. Org. Chem.*, 1995, **60**, 3280.

12   H. Miyazaki, Y. Kobayashi, M. Shiozaki, O. Ando, M. Nakajima, H. Hanzawa and H. Haruyama, *J. Org. Chem.*, 1995, **60**, 6103.

13   N. Yamamoto, T. Nishikawa and M. Isobe, *Synlett*, 1995, 505.

14   T.K.M. Shing and L.H. Wan, *Angew. Chem., Int. Ed. Engl.*, 1995, **34**, 1643.

15   M. Khaldi, F. Chrétien and Y. Chapleur, *Tetrahedron Lett.*, 1995, **36**, 3003.

16   G.E. Keck, S.F. McHardy and J.A. Murry, *J. Am. Chem. Soc.*, 1995, **117**, 7289.

17   X. Tian, T. Hudlicky and K. Königsberger, *J. Am. Chem. Soc.*, 1995, **117**, 3643.

18   L.A. Paquette and S. Bailey, *J. Org. Chem.*, 1995, **60**, 7849.

19   M.S. Ermolenko, T. Shekharam, G. Likavs and P. Potier, *Tetrahedron Lett.*, 1995, **36**, 2465.

20   M.S. Ermolenko, T. Shekharam, G. Likavs and P. Potier, *Tetrahedron Lett.*, 1995, **36**, 2461.

21   A.M. Gómez, J.C. López and B. Fraser-Reid, *J. Org. Chem.*, 1995, **60**, 3859.

22   R.M. Moriarty and H. Brumer, III, *Tetrahedron Lett.*, 1995, **36**, 9265.

23   C. Taillefumier, Y. Chapleur, D. Bayeul and A. Aubry, *J. Chem. Soc., Chem. Commun.*, 1995, 937.

24   G.V.M. Sharma and K. Krishnudu, *Tetrahedron Lett.*, 1995, **36**, 2661.

25   M. Saniere, I. Charvet, Y. Le Merrer and J.-C. Depezay, *Tetrahedron*, 1995, **51**, 1653.

26   C. Gravier-Pelletier, Y. Le Merrer and J.-C. Depezay, *Tetrahedron*, 1995, **51**, 1663.

27   K.R. Buszek and Y. Jeong, *Tetrahedron Lett.*, 1995, **36**, 7189.

28   K. Matsumoto, T. Ebata, K. Koseki, K. Okano, H. Kawakami and H. Matsushita, *Bull. Chem. Soc. Jpn.*, 1995, **68**, 670.

29   I. Kalwinsh, K.-H. Metten and R. Brückner, *Heterocycles*, 1995, **40**, 939.

30   Z. Su and C. Tamm, *Helv. Chim. Acta*, 1995, **78**, 1278.

31   J.F. Booysen and C.W. Holzapfel, *Synth. Commun.*, 1995, **25**, 1461.

32   K. Yasui, Y. Tamura, T. Nakatani, K. Kawada and M. Ohtani, *J. Org. Chem.*, 1995, **60**, 7567.

33   Y. Jiang, Y. Ichikawa and M. Isobe, *Synlett*, 1995, 285.

34   G.V.M. Sharma, A.V.S. Raja Rao and V. S. Murthy, *Tetrahedron Lett.*, 1995, **36**, 4117.

35   U. Koert, *Angew. Chem., Int. Ed. Engl.*, 1995, **34**, 733.

36   G.A. Kraus and H. Maeda, *J. Org. Chem.*, 1995, **60**, 2.

37   M.K. Gurjar, S.K. Das and K.S. Sadalapure, *Tetrahedron Lett.*, 1995, **36**, 1933.

38   M.K. Gurjar, S.K. Das and A.C. Kunwar, *Tetrahedron Lett.*, 1995, **36**, 1937.

39   E.M. Carreira and J. Du Bois, *J. Am. Chem. Soc.*, 1995, **117**, 8106.

40   S. Caron, A.I. McDonald and C.H. Heathcock, *J. Org. Chem.*, 1995, **60**, 2780.

41   C. Paolucci, C. Mazzini and A. Fava, *J. Org. Chem.*, 1995, **60**, 169.

42   R. Mukhopadhyay, A.P. Kundu and A. Bhattacharjya, *Tetrahdeon Lett.*, 1995, **36**, 7729.

43   A.J. Fairbanks and P. Sinaÿ, *Synlett*, 1995, 277.

44   T.K.M. Shing, H.-C. Tsui and Z.-H. Zhou, *J. Org. Chem.*, 1995, **60**, 3121.

45   J.-F. Nguefack, V. Bolitt and D. Sinou, *J. Chem. Soc., Chem. Commun.*, 1995, 1893.

46   T.H. Al-Tel and W. Voelter, *J. Chem. Soc., Chem. Commun.*, 1995, 239.

47   A.T. Hewson, J. Jeffery and N. Szczur, *Tetrahedron Lett.*, 1995, **36**, 7731.

48   D. Brown, D. Liston, S.J. Mantell, S. Howard and G.W.J. Fleet, *Carbohydr. Lett.*, 1994, **1**, 31.

49   I.I. Cubero, M.T.P. Lopez-Espinosa and N. Kami, *Carbohydr. Res.*, 1995, **268**, 187.

50   Z.J. Witczak and Y. Ki, *Tetrahedron Lett.*, 1995, **36**, 2595.

51   C.-K. Sha, C.-Y. Shen, R.-S. Lee, S.-R. Lee and S.-L. Wang, *Tetrahedron Lett.*, 1995, **36**, 1283.

52   H. Zhang, P. Wilson, W. Shan, Z. Ruan and D.R. Mootoo, *Tetrahedron Lett.*, 1995, **36**, 649.

53   J.F. Booysen and C.W. Holzapfel, *Synth. Commun.*, 1995, **25**, 1473.

54   P.P. Deshpande, K.N. Price and D.C. Baker, *Bioorg. Med. Chem. Lett.*, 1995, **5**, 1059.

55   I. Danay, T. Skrydstrup, C. Crevisy and J.-M. Beau, *J. Chem. Soc., Chem. Commun.*, 1995, 799.

56   J.T. Link, S. Raghanvan and S.J. Danishefsky, *J. Am. Chem. Soc.*, 1995, **117**, 552.

57   M. Sasaki, T. Nonomura, M. Murata and K. Tachibana, *Tetrahedron Lett.*, 1995, **36**, 9007.

58   M. Sasaki, N. Matsumori, M. Murata, K. Tachibana and T. Yasamoto, *Tetrahedron Lett.*, 1995, **36**, 9011.

59   H. Oguri, S. Hishiyama, T. Oishi and M. Hirama, *Synlett.*, 1995, 1252.

60   S. Hosokawa and M. Isobe, *Synlett*, 1995, 1179.

61   F. Emery and P. Vogel, *J. Org. Chem.*, 1995, **60**, 5843.

62   P.P. Kanakamma, N.S. Mani and V. Nair, *Synth. Commun.*, 1995, **25**, 377.

63   K. Frische, M. Greenwald, E. Ashkenasi and N.G. Lemcoff, *Tetrahedron Lett.*, 1995, **36**, 9193.

64   A. Dondoni, S. Franco, F. Junquera, F.L. Merchán, P. Merino, T. Tejera and V. Bertolasi, *Chem. Eur. J.*, 1995, **1**, 505.

65   P. Herczegh, I. Kovacs, L. Szilagyi, F. Sztaricskai, A. Berecibar and C. Riche, *Tetrahedron*, 1995, **51**, 2969.

66   S.F. Martin, H.-J. Chen and V.M. Lynch, *J. Org. Chem.*, 1995, **60**, 276.

67   R.H. Furneaux, G.J. Gainsford, J.M. Mason, P.C. Tyler, O. Hartley and B.G. Winchester, *Tetrahedron*, 1995, **51**, 12611.

68   S. Picasso, Y. Chen and P. Vogel, *Carbohydr. Lett.*, 1995, **1**, 1.

69   K. Tatsuta, S. Miuro, S. Ohta and H. Gunji, *Tetrahedron Lett.*, 1995, **36**, 1085.

70   D. Sames and R. Polt, *Synlett.*, 1995, 552.

71   I. Lundt and R. Madsen, *Synthesis*, 1995, 787.

72   L. Cipola, L. Lay, F. Nicotra, C. Pangrazio and L. Panza, *Tetrahedron*, 1995, **51**, 4679.

73   J. Fitremann, A. Dureault and J.-C. Depezay, *Synlett.*, 1995, 235.

74   Z. Shi and G. Lin, *Tetrahedron*, 1995, **51**, 2427.

75   H. Yoda, T. Nakajima, H. Yamazaki and K. Takabe, *Heterocycles*, 1995, **41**, 2423.

76   N. Chida, J. Takeoka, N. Tsutsumi and S. Ogawa, *J. Chem. Soc., Chem. Commun.*, 1995, 793.

77   A. Abas, R. L. Beddoes, J.C. Conway, P. Quayle and C.J. Urch, *Synlett*, 1995, 1264.

78   T. Kolter, C. Dahl and A. Giannis, *Liebigs Ann. Chem.*, 1995, 625.

79   J. Grodner and M. Chmielewski, *Tetrahedron*, 1995, **51**, 829.

80   M.P. Varquez-Tato, J.A. Seijas, G.W.J. Fleet, C.J. Mathews, P.R. Hemmings and D. Brown, *Tetrahedron*, 1995, **51**, 959.

81   W H. Pearson and F.E. Lovering, *J. Am. Chem. Soc.*, 1995, **117**, 12336.

82   C.E. Sowa, J. Kopf and J. Thiem, *J. Chem. Soc., Chem. Commun.*, 1995, 211.

83   K.S. Kim, I.H. Cho, Y.H. Ahnand and J.I. Park, *J. Chem. Soc., Perkin Trans. 1*, 1995, 1783.

84   C. Marot and P. Rollin, *Phosphorus, Sulfur, Silicon Relat. Elem.*, 1995, **95 & 96**, 503.

85   M. Morita, K. Motoki, K. Akimoto, T. Natori, T. Sakai, E. Sawa, K. Yamaji, Y. Koezuka, E. Kobayashi and H. Fukushima, *J. Med. Chem.*, 1995, **38**, 2176.
86   Y.-L. Li and Y.-L. Wu, *Tetrahedron Lett.*, 1995, **36**, 3875.
87   K. Matsumoto, T. Ebata and T. Matsushita, *Carbohydr. Res.*, 1995, **279.**, 93.
88   Y.-L. Li, X.-H. Mao and Y.-L. Wu, *J. Chem. Soc., Perkin Trans. 1*, 1995, 1559.
89   R.R. Schmidt, T. Bär and R. Wild, *Synthesis*, 1995, 868.
90   M. Yoshikawa, Y. Yokokawa, Y. Okuno, N. Yagi and N. Murakami, *Chem. Pharm. Bull.*, 1995, **43**, 1647.
91   M. Yoshikawa, Y. Yokokawa, Y. Okuno and N. Murakami, *Tetrahedron*, 1995, **51**, 6209.
92   J.S. Yadav, S. Chandrasekhar, Y.R. Reddy and A.V.R. Rama Rao, *Tetrahedron*, 1995, **51**, 2749.
93   T.K. Chakraborty, K.A. Hussain and D. Thippeswamy, *Tetrahedron*, 1995, **51**, 3873.
94   S.P. Khanapure, S. Manna, J. Rokach, R.C. Murphy, P. Wheelan and W.S. Powell, *J. Org. Chem.*, 1995, **60**, 1806.
95   A.S. Kende, K. Liu, I. Kaldor, G. Dorey and K. Koch, *J. Am. Chem. Soc.*, 1995, **117**, 8258.
96   J. D. White and M.S. Jensen, *J. Am. Chem. Soc.*, 1995, **117**, 6224.
97   J. Xia and Y.Z. Hui, *Bioorg. Med. Chem. Lett.*, 1995, **5**, 1919.
98   J.D. White, G.L. Bolton, A.P. Dantanarayana, C.M.J. Fox, R.N. Hiner, R.W. Jackson, K. Saksuma and U.S. Warrier, *J. Am. Chem. Soc.*, 1995, **117**, 1908.
99   S. Tanaka and M. Isobe, *Synthesis*, 1995, 859.
100 M.A. Rampy, A.N. Rinchuk, J.P. Weichert, R.W.S. Skinner, S.J. Fisher, R.L. Wahl, M.D. Gross and R.E. Counsell, *J. Med. Chem.*, 1995, **38**, 3156.
101 P. Alleui, P. Ciuffreda, G. Tarocco and M. Anastasia, *Tetrahedron: Asymmetry*, 1995, **6**, 2357.
102 R.M. Moriarty, J. Kim and H. Brumer, *Tetrahedron Lett.*, 1995, **36**, 3651.
103 K. Takatori and M. Kajiwara, *Synlett*, 1995, 280.
104 P.H.J. Carlsen, K. Misund and J. Røe, *Acta Chem. Scand.*, 1995, **49**, 297.
105 R.P.C. Cousins, A.D.M. Curtis, W.C. Ding and R.J. Stoodley, *Tetrahedron Lett.*, 1995, **36**, 8689.
106 T. Akiyama, N. Horiguchi, T. Ida and S. Ozaki, *Chem. Lett.*, 1995, 975.
107 E. Roman, M. Banos, J.I. Gutierrez and J.A. Serrano, *J. Carbohydr. Chem.*, 1995, **14**, 703.
108 Z. Kaluza, B. Furman and M. Chmielewski, *Tetrahedron: Asymmetry*, 1995, **6**, 1719.
109 G. Belluci, C. Chiappa and F. D'Andrea, *Tetrahedron: Asymmetry*, 1995, **6**, 221.
110 J. Kang, G.J. Lim, S.K. Yoon and M.Y. Kim, *J. Org. Chem.*, 1995, **60**, 564.
111 A.B. Charette and B. Côté, *J. Am. Chem. Soc.*, 1995, **117**, 12721.
112 A.B. Charette, B. Côté, S. Monroc and S. Prescott, *J. Org. Chem.*, 1995, **60**, 6888.
113 V. Guerrero-de la Rosa, M. Ordoñez, J.M. llera and F. Alcudia, *Synthesis*, 1995, 761.
114 P. Angibaud, J.L. Chaumette, J.R. Desmurs, L. Duhamel, G. Ple, J.Y. Valnot and P. Duhamel, *Tetrahedron: Asymmetry*, 1995, **6**, 1919.
115 K. Tomooka, Y. Nakamura and T. Nakai, *Synlett*, 1995, 321.
116 S. Lehnhoff, M. Goebel, R.M. Karl, R. Klösel and I. Ugi, *Angew. Chem., Int. Ed. Engl.*, 1995, **34**, 1104.
117 P.P. Garner, P.B. Cox and S.J. Klippenstein, *J. Am. Chem. Soc.*, 1995, **117**, 4183.
118 K. Fujita, K. Murata, M. Iwaoka and S. Tomoda, *J. Chem. Soc., Chem. Commun.*, 1995, 1641.
119 P. Koll and A. Lutzen, *Tetrahedron:Asymmetry*, 1995, **6**, 43.

120   T.-L. Ho, P.-Y. Liao and K.-T. Wang, *J. Chem. Soc., Chem. Commun.*, 1995, 2437.
121   U. Behrens, C. Fischer and R. Selke, *Tetrahedron: Asymmetry*, 1995, **6**, 1105.
122   B.T. Cho and N. Kim, *Synth. Commun.*, 1995, **25**, 167.
123   L. Sharma and S. Singh, *Indian J. Chem., Sect. B: Org. Chem. Incl. Med. Chem.*, 1995, **33B**, 1183.
124   A.B. Charette, A.F. Benslimane and C. Mellon, *Tetrahedron Lett.*, 1995, **36**, 8557.
125   A.B. Charette, C. Mellon and M. Motamedi, *Tetrahedron Lett.*, 1995, **36**, 8561.

Author Index

Rondot, B. (24) 5
Rongere, P. (18) 5
Ronnow, T.E.C.L. (3) 117
Roof, R.D. (23) 54
Rosankiewicz, J.R. (23) 12
Rose, L. (13) 22
Rosen, S.D. (4) 45
Rosenberg, I. (22) 14
Rosenski, J. (20) 98, 285, 287, 288
Rossello, A. (20) 172
Rossi, J.-C. (24) 5
Rossi, M. (3) 36
Roth, S. (4) 6
Rotteveel, F.T.M. (18) 61
Roush, W.R. (3) 53; (9) 43
Rousseau, R.W. (4) 216
Roussi, G. (9) 77
Roussin-Bouchard, C. (3) 292
Routledge, A. (3) 166; (20) 183
Roy, N. (4) 58, 98, 106, 107, 118; (18) 17
Roy, R. (3) 140, 143; (10) 19, 21, 66
Rozanas, C.R. (7) 5; (18) 34
Rozenski, J. (7) 49; (13) 13, 14; (20) 125, 171; (22) 161, 162
Rozners, E. (20) 190
Ruan, Z. (24) 52
Ruch, T. (4) 191, 199
Ruchirawat, S. (22) 73
Rudd, B.A.M. (19) 54
Rudge, A.J. (18) 44
Rudolf, M.T. (18) 155
Rudyk, H. (10) 84
Rufer, S. (23) 34
Ruigt, G.S.F. (16) 45
Ruijtenbeek, R. (7) 87; (19) 24
Ruiz, J. (22) 141
Ruohtula, T. (4) 165
Rupitz, K. (11) 16
Rush, C.P. (19) 54
Russ, P. (3) 74
Russell, A.L. (21) 73
Russell, D.R. (14) 47; (22) 129; (24) 7
Russo, G. (3) 246, 273; (4) 94, 95; (9) 37; (10) 44; (21) 76
Rutherford, T.J. (4) 108; (18) 163; (21) 69
Rycroft, A.D. (20) 144

Saady, M. (20) 173
Saarikangas, A. (4) 51
Saarma, M. (20) 78
Saavedra, O.M. (18) 73
Sabesan, S. (4) 144; (9) 92

Saccani, G. (23) 47
Sadalapure, K.S. (24) 37
Sadana, K.L. (20) 215
Sadeghi, M.M. (20) 213
Sadler, I.H. (19) 31; (21) 97
Sadler, N.P. (16) 66
Saegusa, Y. (8) 16
Saenger, W. (22) 89, 91, 92, 109
Saewe, J. (23) 36
Saez, V. (4) 83
Sagaki, S. (3) 291
Sah, M.P. (2) 59, 60
Saha, A.K. (20) 50, 249
Saha, S. (3) 286; (19) 38
Saha, U.K. (3) 140; (10) 19
Sahu, N.P. (21) 87
Saiakhov, R. (20) 208
Saimoto, H. (2) 15
Sainsbury, M. (3) 93
Saint-Marcoux, G. (4) 173
Saito, I. (20) 319
Saitoh, M. (3) 105
Sakaguchi, K. (17) 19; (23) 22
Sakai, T. (3) 77; (19) 74-76; (22) 164; (23) 48; (24) 85
Sakairi, H. (20) 91
Sakairi, N. (3) 87; (4) 182, 183; (18) 141
Sakaki, T. (4) 214
Sakakibara, J. (7) 33
Sakakibara, T. (5) 36; (10) 54; (14) 45
Sakamoto, H. (3) 129
Sakamoto, K. (20) 178
Sakamoto, M. (8) 15
Sakano, Y. (4) 90
Sakharov, A.M. (2) 50
Saki, S. (4) 80
Saksuma, K. (24) 98
Sakthivel, K. (20) 106
Sakuda, S. (4) 87
Sakuno, T. (4) 167
Sala, L.F. (2) 55-57; (16) 6
Salamonczyk, G.M. (18) 107
Salazar, J.A. (2) 33
Saleh, M.A. (3) 224; (10) 76, 77
Salimath, P.V. (4) 93
Sallam, M.A.E. (10) 89; (22) 171
Sallustio, B.C. (23) 45
Salomon, H. (20) 301
Salvador, L.A. (3) 66
Salvadori, P. (4) 185
Salyan, M.E.K. (4) 164; (10) 74
Salzner, U. (21) 7
Sam, T.-W. (22) 134
Samadi, M. (19) 57
Samain, E. (9) 15; (11) 22; (15) 1

Sambandam, A. (7) 41
Sames, D. (18) 89; (24) 70
Samreth, S. (3) 236
Samuel, C.J. (4) 38
Samuelsson, B. (14) 16; (20) 132, 133, 166
Sanai, Y. (3) 139
Sanceau, J.-Y. (10) 67; (19) 71
Sancho, M.-R. (9) 83
Sandanayake, K.R.A.S. (17) 31, 32
Sandermann, H., Jr. (3) 113
Sanders, J.K.M. (3) 215
Sandstrom, A. (21) 21
Sancyoshi, M. (20) 52, 91
Sanghvi, Y.S. (20) 281
Saniere, M. (24) 25
Sano, A. (3) 13, 128, 129
Sano, H. (10) 30, 31; (19) 69, 70, 72, 73
Santaella, C. (3) 67
Santoyo-Gonzales, F. (11) 5
Sarabia-Garcia, F. (2) 36; (14) 29
Sarazin, H. (9) 53; (18) 86
Sarbajna, S. (18) 17
Sarfati, R.S. (20) 269
Sarkar, A.K. (3) 152
Sarries, N. (3) 72
Sasahara, M. (6) 15
Sasai, H. (3) 20, 132
Sasaki, H. (23) 70
Sasaki, M. (24) 57, 58
Sasaki, S. (18) 136
Sasaki, T. (3) 198; (20) 116; (22) 159
Sasaki, Y. (19) 66; (23) 37
Sashida, Y. (3) 109; (4) 104; (21) 86
Sashiwa, H. (2) 15
Sasikala, R. (17) 20; (23) 93
Sata, N.U. (20) 2
Sato, A. (20) 143
Sato, K. (3) 46; (9) 13; (12) 19; (14) 8
Sato, S. (3) 258
Sato, T. (2) 22, 23; (4) 174; (22) 77
Sato, Y. (19) 41
Satomi, Y. (4) 104; (21) 86
Sautou, V. (23) 8
Savitskiy, A.V. (22) 179
Sawa, E. (3) 77; (24) 85
Sawa, R. (19) 36
Sawa, T. (19) 36, 39
Sawada, M. (22) 20
Sawada, T. (18) 152
Sawamoto, T. (19) 53
Sayer, B.G. (12) 3